Towards Energy Smart Homes

Stephane Ploix • Manar Amayri • Nizar Bouguila
Editors

Towards Energy Smart Homes

Algorithms, Technologies, and Applications

Springer

Editors
Stephane Ploix
G-SCOP lab
Grenoble Institute of Technology
University of Grenoble Alps
Grenoble, France

Manar Amayri
CNRS, Grenoble, France

Nizar Bouguila
CIISE
Concordia University
Montreal, QC, Canada

ISBN 978-3-030-76476-0 ISBN 978-3-030-76477-7 (eBook)
https://doi.org/10.1007/978-3-030-76477-7

This Springer imprint is published by the registered company Springer Nature Switzerland AG
The registered company address is: Gewerbestrasse 11, 6330 Cham, Switzerland

Foreword to the Book: Towards Energy Smart Homes

The Global Issues for Smart Buildings (SBs)

The first reason for this book is that buildings stand as a key pillar of environmental transition, not only their envelopes and HVAC systems but also their inhabitants. As written by K. B. Janda (2011), "Buildings Don't Use Energy: People Do" in *Architectural Science Review*, *54*(1). Buildings are unique inhabited systems where occupants spend about 90% of their time: they impact the physics of the buildings with their complex activities driven by beliefs, desires, and intentions, and at the same time, occupants have to decide on retrofitting and everyday management, i.e., occupants have to take a decision that depends on their behaviors. Like all the book's authors, I will start with some key figures and analyze the relationship between the potential of consumption, production, and flexibility at a different scale: starting from the building to the eco-district scale. It can be summarized with a couple of words: buildings with their inhabitants are, on the one side, the biggest consumers of energy (all energies but electricity in particular in a grid context) and could be, at the same time, among the most significant producers of renewable energy, mostly thanks to the concept of positive energy buildings. Additionally, at the same time, they offer an essential degree of flexibility in energy demand, which can compensate for the intermittence due to the increasing part of renewables in the production. For instance, households can improve the matching between their local production and their power consumption.

SBs: The Biggest Source of Consumption

The main portion of the final consumed energy in France and over the world is indeed consumed in buildings. An ADEME report indicates that buildings are among the most significant emitters of CO_2, with a 20% ratio, and represent 45% of the total final energy consumption. France is far from being an exception; all over the world, buildings are among the main energy consumers with globally an increase in the needs, and this tendency is independent of the place in the world and of the kind of climate. This point is recalled by most of the authors in the introductions to the majority of the chapters of this book.

SBs: One of the Possible Biggest Producer of Renewable Energy

The study of the ADEME agency proposes an estimation of the potential of renewable energy in France for producing electricity. It gives more especially an idea of the PV Panel potential on the top of building roofs. In this study, which aimed at estimating whether it is possible to go toward a pure electric renewable energy mix in France for 2050, it comes out that:

- Firstly, the global potential of renewable energy in France is estimated at 1268 TWh per year, knowing that the report estimates the energy consumption for 2050 in France at 422 TWh. In other words, the potential is estimated to be three times higher than the anticipated electrical energy needs.
- Secondly, the report tries to anticipate different scenarios with different renewable energy mixes like solar, wind, and hydraulic. An interesting scenario envisaging an energy produced 100% from renewables, could use the existing non-cultivated surfaces, i.e. mainly the roofs of the buildings. In this case, PV could be one of the primary energy production capabilities by representing up to 34.8% of the network's global capacity with 68.3 GW for a global estimated power of 196 GW.

This importance of the production capacity of buildings is not the focus of the book. Still, it is fundamental to understand why buildings and their inhabitants are probably the most critical key for the energy transition.

SBs: The Key Issue Regarding Flexibility in Energy Demand and Consumption

Flexibility and demand response are, respectively, the capability of adapting and decreasing the energy demand. This is a beneficial property that can be offered by smart buildings (SBs) in smart grids (SGs). Here again, SBs are the primary source of flexibility and demand response. This is confirmed by different studies and experience feedbacks:

- Modulating the heating energy use of new and existing buildings could provide 10–20 GW of flexible load in France according to the literature and demonstrator projects.
- A French study by ADEME estimates the available flexibility at a national level to 18 GW for households equipped with 4 GW of electric hot water tanks, 14 GW of electric heaters and an air conditioning system, but also 0.695 GW supplied by an oven and a washing machine.
- A more recent report even evaluates the flexibility of buildings associated with the coming electric vehicles to an intraday potential of up to 47 GW at horizon 2050.

This potential must be compared, at the French level, to the global need for power. If we take the historical pic demand of 102 GW of 8 February 2012, the potential flexibility of SBs represents then from 10 to 17%, and even near from 50%. This brief evaluation confirms the real potential of SBs as a principal and significant source of flexibility and demand response.

This importance of the production capacity of buildings is not the focus of the book. Still, it is fundamental to understand why buildings and their inhabitants are probably the most critical key for the energy transition.

But There Are No Smart Buildings Without Smart Users: The Need of a "Human-in-the-Loop" Approach

Besides the energy and technological entry, this book must be read by keeping in mind that there will be no smart buildings without smart users as established by Social and Human Science. This means a complex multidisciplinary research in which the inhabitants must be involved as "prosumers," i.e., active and implicated designers and users. Some elements explaining why such an approach is mandatory are going to be given. It explains why the book essentially focused on algorithms, applied mathematics, applied physics, or even deep learning and Artificial Intelligence with a necessary "Human-in-the-loop" approach.

About the Desire of the Final Users for Being Involved from the Individual Level to the Collective Level

About the Individual Level

First, there is no smart building without smart users because it seems to exist, at an individual level, a clear expectation of the final energy consumer to become a prosumer, in other words, a producer of renewable energy. It has been demonstrated by opinion polls that take place every year to measure the French citizens' evolution regarding their relationship with the emergence of renewable energies and the perspective of being involved as an individual and a collective energy producer. One of the main results is that "*a very large majority of French people (88%, stable from one year to another) would prefer to consume their electricity rather than selling it if they had solar panels.*"

About the Emergence at Collective Level

The same survey reveals that "energy pooling improves social and solidarity relationships between inhabitants: 23% of French people would choose to sell or exchange electricity surplus (+2% from 2018 to 2019) and 11% would offer it to an association or to people in fuel poverty (+1% from 2018 to 2019)."

About the Necessity to Involve the Final User from the Individual Level to the Collective Level

About the Individual Level

Users are the primary sources of unpredictability and uncertainty. Completely identical houses can have heating consumptions varying with a two to three times factor depending on user practices, and comparing households living in similar houses, electricity consumptions can vary with a factor of 5. It points out that human practices are determinant for consumption. Because automation cannot cope with all occupant services, SBs imply occupants' involvement thanks to interactions and cooperative search for good solutions. Putting humans "in the loop" is therefore becoming a necessity, and it is probably the core feature of a SB.

To the Collective Level

"Putting humans in the loop", both individually and collectively, is being pushed at a European level, where the concept of local energy market opened for energy communities through two recent directives enacted by the European Council. These directives must now be adopted by each country of the European Community so that those CECs and RECs can have access to the energy market and share and exchange energy by using the public networks. Typically, CECs aim to be a legal entity based on voluntary and open participation, effectively controlled by members and stakeholders, and have the primary purpose of providing environmental, economic, or social community benefits. REC is defined similarly and is more focused on valorization of renewable energy, locally produced by means (Photovoltaïc Panels, windmills, ...) typically implemented near or on buildings' roofs.

That Is Why It Is a Timely Book for Exploring Innovative and High-Level Solutions Putting Humans in the Loop Approach

About a Formal and Linear Description of the Content of the Book

The book contains 17 chapters.

In chapter "Energy Sobriety: A Behavior Measurement Indicator for Fuel Poverty Using Aggregated Load Readings from Smart Meters", Fergus and Chalmers discuss current fuel poverty strategies across the European Union. The authors propose a new and foundational behavior measurement indicator to assess and monitor fuel poverty risks in households via smart meters, customer access device data, etc., deploying machine learning. In particular, they show that it is possible to spot early signs of financial difficulty by detecting daily living activities. ***Here, we can link with the opinion pool showing that 11% of French people are ready to offer their excess of locally produced energy to an association or people in fuel poverty.***

In chapter "Energy Sobriety: A Behaviour Measurement Indicator for Fuel Poverty Using Aggregated Load Readings from Smart Meters", Santos Silva and Costa describe urban energy modeling tools' contribution to the development and exploitation of smart urban contexts such as smart energy districts and smart energy cities. An overview of the main urban modeling frameworks is made. A review of applications of the use of these tools is performed.

Chapter "Standards and Technologies from Building Sector, IoT, and Open-Source Trends", written by Delinchant and Ferrari, is devoted to standards and technologies from the building sector, IoT, and open-source trends, and several essential notions are explained and discussed. ***Here, the questions of numerical standards and technologies, especially around IoT, are addressed.***

Chapter "Formalization of the Energy Management Problem and Related Issues", by Ploix and Alyafi, is an introductory chapter that states the main issues that are appearing in building energy management based on practical examples. ***The limitations of theory-driven approaches are especially underlined.***

In chapter "Dynamic Models for Energy Control of Smart Homes", Ghiaus discusses dynamic models for smart homes' energy control and presents two algorithms: assembling thermal circuits and extracting the state-space representation from the thermal circuit. ***This is a contribution for smartness seen as mathematical and algorithmic development in a typical theory-driven approach.***

A detailed survey of several machine learning techniques applied for activity recognition in smart buildings is presented in chapter "Machine Learning for Activity Recognition in Smart Buildings: A Survey" by Amayri, Ali, Bouguila, and Ploix. ***This is again a contribution for smartness seen as mathematical and algorithmic development in a typical data-driven approach, which is compliant with this new trend and significant movement linked to Artificial Intelligence***.

In chapter "Characterization of Energy Demand and Energy Services Using Model Based and Data-Driven Approaches", Santos Silva, Amayri, and Basu describe state-of-the-art methods to forecast energy consumption and energy services in buildings. ***Here again, several data-driven and model-based approaches are discussed. A proposal of an interactive learning approach is given, which is an interesting way for putting the humans in the loop.***

Pal and Bandyopadhyay propose, in chapter "Occupant Actions Selection Strategies Based on Pareto-optimal Schedules and Daily Schedule for Energy Management in Buildings", a framework for occupant action selection strategies based on Pareto-optimal schedules and daily schedules to manage energy in buildings***. It is addressed to the users of SBs, as future "Humans in the Loop" to which efficient and ergonomic decision tools must be provided.***

Different approaches to generate an optimal energy management system strategy in terms of energy consumption reduction have been proposed and discussed in chapter "Generation of Optimal Strategies for Complex Living Places" by Ngo, Joumaa, and Jacomino. ***It is addressed to the designers of the numerical and technical systems of SBs***.

In chapter "Distributed and Self-Learning Approaches for Energy Management", Joumaa, Jneid, and Jacomino discuss distributed and self-learning approaches for energy management. In particular, they present a reinforcement learning solving-based approach. ***It is also addressed to the designers, and it is a contribution to what can be the algorithmic, Artificial Intelligence, and data-driven approaches for SBs.***

In chapter "Modelling, Forecasting and Control for Smart Buildings", Thilker, Junker, Bacher, Bagterp, and Madsen introduce the grey-box modeling principle and illustrate how it can be used for multilevel control systems of smart buildings and neighborhoods. ***This contribution is addressed to designers. It is also a contribution to the debate for the necessity of theory-driven approaches besides the emerging data-driven strategies in this movement of adoption of techniques coming from Artificial Intelligence. A novel stochastic dynamical model is introduced to take advantage of local weather disturbances to assist occupants in better understanding their energy management systems.***

Alyafi, Reignier, and Ploix present an approach to generate explanation with knowledge models in chapter "Explanations Generation with Knowledge Models". ***This contribution deals with the involvement of the inhabitants of SBs developing humans in the loop approaches. Indeed, energy management is conceived as a symetric learning between humans and artificial system, which provides explanations to empower inhabitants about better energy management strategies, humans keeping the control of the final decision.***

In chapter "The Mondrian User Interface Pattern: Inspiring Eco-responsibility in Homes", Laurillau, Coutaz, Calvary, and Van Bao present the Mondrian user interface pattern intended to help system developers structure and populate the interactive components of systems that cooperate with residents combined with digital behavior changes intervention to inspire eco-responsibility. ***Again, it is addressed to the users of SBs, as future "Humans in the Loop" to which efficient and ergonomic decision tools must be provided.***

A challenging problem to tackle in building systems is fault diagnosis and maintenance. This problem is discussed in chapter "Faults and Failures in Smart Buildings: A New Tools for Diagnosis" by Najeh, Singh, and Ploix, who propose several new diagnosis tools. ***It is addressed to the humans involved as designers or the actors in charge of the maintenance of SBs.***

The problem of analyzing load profiles in commercial buildings using smart meter data is tackled in chapter "Analyzing Load Profiles in Commercial Buildings Using Smart Meter Data" by Basu, Mishra, and Maulik via the automatic segmentation and symbolic representation of time series.

Kashif, Ploix, and Dugdale develop in chapter "A Nouvelle Approach to Validate Representative Behavior Models in Energy Simulations" a novel approach to validate representative behavior models in energy simulation whose primary goal is to analyze the impact of inhabitants' behavior on energy consumption in domestic situations. ***This contribution proposes modeling of the behavior of the inhabitants of SBs.***

Shalbart, Vorger, and Peuportier propose in chapter "Stochastic Prediction of Residents' Activities and Related Energy Management" macroscopic and statistical models of users. ***It is a set of tools for the designers in order to give a macroscopical and statistical view of the final users as humans in the system.***

To a Transversal Analysis of the Main Messages of the Book

With a global transversal perspective of the content of the book, the book covers:

- Tools and approaches from the urban scale to the building scale
- Question of the interaction with the building actors, considering the necessity of putting "Humans in the Loop."
- Question of the involvement of inhabitants of SBs.

With Some of the Main Scientific and Fundamental Debate Linked on SBs put on the Table by the Book

It must also be noticed that the book is addressing some essential questions related to the so-called SBs. Among the most interesting, we will remember the following ones.

Compromise Between Data-Driven Approach and Theory-Driven Approach?

This question is typically addressed for instance in chapters "Formalization of the Energy Management Problem and Related Issues", "Dynamic Models for Energy Control of Smart Homes", "Machine Learning for Activity Recognition in Smart Buildings: A Survey", "Distributed and Self-Learning Approaches for Energy Management", "Modelling, Forecasting and Control for Smart Buildings", and "Explanations Generation with Knowledge Models" showing alternatively the power and the limits of the data-driven approach, as well as of the theory-driven model. The perspectives are then probably to explore a mix of both paradigms, and this can be well organized by designers of the system, and well understood and well used by the final users of the systems.

Compromise Between Complete Automatic Delegation to Numerical System, or Buildings and Dwellings Entirely (or only) Controlled by Inhabitants?

This question is put on the table by all the chapters talking about data-driven approaches. Behind all those approaches is the question of the complement of human and artificial smartnesses. The chapters dedicated to the generation of explanations, to the involvements, and to the interactions with the SB users and the designers, suggest that there are no "smart buildings" without "smart users" and "smart designers."

The Need to Introduce an Interdisciplinary Approach: The Book Is a Main Contribution of the Cross-Disciplinary Program Eco-SESA

In conclusion, this book is a clear demonstration of the need to develop an interdisciplinary approach. There is no way for developing SBs only with an energy/technical/numerical approach. This book demonstrates this with high-level contributions in mathematical and computer science approaches for designing systems in interaction with designers, and final users are becoming prosumers. This approach is compliant with the CDP eco-SESA projects in which many contributors of the book have been associated, and that is studying with an interdisciplinary approach what can be the new trends and solutions for valorization of locally produced energy, from the building level, to the eco-district level, with an efficient interaction between social, economic, territorial, energy, numerical, and technical networks.

That is why, as co-leader of the Cross-Disciplinary Program eco-SESA of the Université Grenoble Alpes Institut D'Excellence, I warmly congratulate the contributors of this book, and I am proud to propose this foreword of this very important book for our interdisciplinary community.

Contents

Urban Modeling and Analytics in a Smart Context

Francisco P. Costa and Carlos A. Santos Silva

1 Introduction

The twenty-first century is characterized by a strong urbanization trend. In the report from [26] it has been estimated that in 2018 55% of the world population already lived in cities and it is expected that this number grows to 68% by 2050. This urbanization trend is driven by the potential access to better jobs and salaries, education, health and cultural services. However, the urban lifestyle requires a high level of resources demand—food, water, energy—and generates a significant amount of waste flows—material waste, wasted water, emissions. Cities are today responsible for 70% of the resources, 67–76% of total energy consumption, and 71–76% of CO_2 emissions [2]. Therefore, the challenge of addressing climate change by developing more sustainable energy systems is entangled with the urbanization trend, which means that the energy transition needs to be addressed at the urban level.

Building occupants, especially in urban contexts, spend in average 87% of the time inside buildings [11]. Several papers emphasize the importance for the society to design better the living spaces, in order to make them more comfortable to occupants, while respecting the environmental constraints in an economical way, in a symbiotic balance between all these dimensions [16]. Hence, it is important to understand and model this particular sector of consumption in order to design better urban spaces.

For decades, the energy analysis of the built environment has been done at the individual building level, but many solutions are more economically and environmentally viable if implemented at a larger scale (neighborhood, district, or

F. P. Costa · C. A. S. Silva (✉)
IN+ Centre for Innovation, Technology and Policy Research, LARSYS, Instituto Superior Técnico, Universidade de Lisboa, Lisboa, Portugal
e-mail: carlos.santos.silva@tecnico.pt

S. Ploix et al. (eds.), *Towards Energy Smart Homes*,
https://doi.org/10.1007/978-3-030-76477-7_1

even city scale). Therefore, over the last years, a new set of tools, generally called Urban Building Energy Models (UBEM), have been developed as tools that model the energy of built environment at larger scales [18].

The information generated by these tools is especially useful for different stakeholders, from architects or engineers that develop the technical solutions, to investors and promoters that finance the projects, to policy makers and governmental agencies that license the urban developments. This type of tool—UBEMs—helps all these stakeholders to make better informed decisions, in construction, design, and planning. The advantage is that these tools enable the modeling of large areas—from neighborhoods, districts, or the whole city—and therefore estimate and analyze the impact of interaction of multiple buildings, which does not correspond to the simple extrapolation of the impact of a single building. As examples, consider the analysis of the impact in PV generation of a building due to the shadow effects of neighbor buildings, or the impact in the energy supply network of the combined energy demand of buildings with different heating needs due to their orientation.

Despite its importance, the evolution and impact of UBEM is not yet sufficiently covered in literature. Not only is a very recent field—its roots come from the first attempts at modeling buildings energy in an urban environment, but there have been very few attempts to organize the evolution of this field—actually one of the first articles explicitly using this expression was released only in 2016 [18].

This chapter discusses the contribution of urban energy modeling tools to the development and exploitation of smart urban contexts, like smart energy districts or smart energy cities. We start by the definition of what is a smart urban context and then we introduce the concept of urban energy modeling tools and how these tools can be used to design, implement, and evaluate smart contexts. An overview of the main urban modeling frameworks is made and a review of applications of the use of these tools is done. It can be concluded that all the urban energy modeling tools share a common integration framework, even if the tools that are used are different. Further, the available analytics are still centered only in general energy demand and generation indicators but complementary analytics like life cycle assessment or indoor comfort are becoming available. Finally, the number of applications and case studies has been exponentially increasing, which demonstrates the advantage of using these tools. At the end, we discuss and summarize a potential evolution roadmap for these tools in the coming years.

2 Definitions

In this section we introduce the core concepts of smart urban contexts and urban energy models.

2.1 Smart Cities and Smart Urban Contexts

Bibri and Krogstie [2] present a detailed overview of the origin of the use of the term "Smart Cities" in the literature and demonstrate that there is yet an inconsistent use of it. Still, most of the literature converges in the use of ICT across urban different urban infrastructures and its application to manage operations or services provision and design to citizens across multiple domains, such as traffic management, street lighting, air quality monitoring, or energy networks monitoring.

In parallel, it is possible to find in the literature the term "sustainable city." The term is still mostly used in the context of urban planning [2, 15, 16], as an urban environment that contributes to environmental quality and protection, and also and to social equity and well-being by adopting sustainable development strategies to foster advancement and innovation in built environment, infrastructure, operational functioning, planning, and ecosystem and human service provisioning. However, Taipale et al. [24] had already highlighted that to develop sustainable environments, the use of ICT is key factor for success. This interlink between smart cities and sustainable cities has been captured by the "smart sustainable cities" concept [2] which basically emphasizes that ICT are enabling technologies that will induce transformational effects in urban environment to increase its sustainability.

In this work, in line with [2], we sustain also that to achieve sustainable urban environments, it is necessary to develop a smart urban context, i.e. a context where ICT is present in different spatial scales in urban infrastructure—buildings, vehicles, water and energy networks, waste bins, traffic and street lights—to collect real time information to be used by applications that provide services to citizens, manage operations, support decision making systems in multiple domains, from mobility, to energy or healthcare.

2.2 Urban Energy Modeling Tools

The first use of "Urban Energy Modeling" (UBEM) tools as such was in [18], to coin a nascent field related to the development of tools that are able to model energy consumption in urban environments. Still, this definition is not established, and in the literature we can find many works that fall under this category but are classified simply as urban energy simulation tools [8], framework for the analysis and optimization of energy systems in neighborhoods and city districts [6], or Urban Scale Energy Models [22]. Here we will use the UBEM term.

Originally, UBEM was defined as a tool that applies physical models of heat and mass flows in and around buildings to predict operational energy use, as well as indoor and outdoor environmental conditions for groups of buildings [18]. In this work, we broaden the definition to frameworks that can estimate the energy consumption of more than one building taking into consideration the existing interaction between them. This relaxation allows to include in the definition data driven models [8] and to include other potential outcomes such as retrofit measures [8], or Life Cycle analysis [6].

We can trace the goal of developing urban modeling tools to the late 1990s, such as the work of [9] and since then, some systematic reviews can be found in [10, 18] or more recently [22]. Here, we take a different approach from those reviews to characterize the evolution and impact of the field, by performing a systematic review of the literature and map the research front in this field by performing a bibliometric analysis coupled with a content analysis, using a set of analytical rules of documents [25]. In this case, some themes were retrieved and automated analysis of titles and abstracts of the articles was performed to characterize its contents. The analysis uses the database of Web of Science (WoS) [1], which is a daily updated dataset that contains more than one billion searchable cited references in more than 12,000 of journals, along with conference proceedings, book contents and other materials.

The search method was fully updated in 10-02-2020, and consisted in using the keywords "urban," "building," "energy" and "model," and their variations in title, abstract, or keywords of the documents. The search returned 3024 results that were screened to remove incomplete records, and the sample narrowed down to 2960 articles. All idioms or languages were selected for the results, but only peer-reviewed documents were selected: journal articles, reviews, proceedings papers, data papers, hardware reviews, and software reviews. The results are presented in Figs. 1 and 2.

Figure 1 shows that 1995 was the first year where a publication fitting all the search criteria was identified [3], and in 1998, the number of publications per year reached more than ten. The UBEM field has grown steadily surpassing 100 publications per year in 2006.The number of different authors per publication yearly

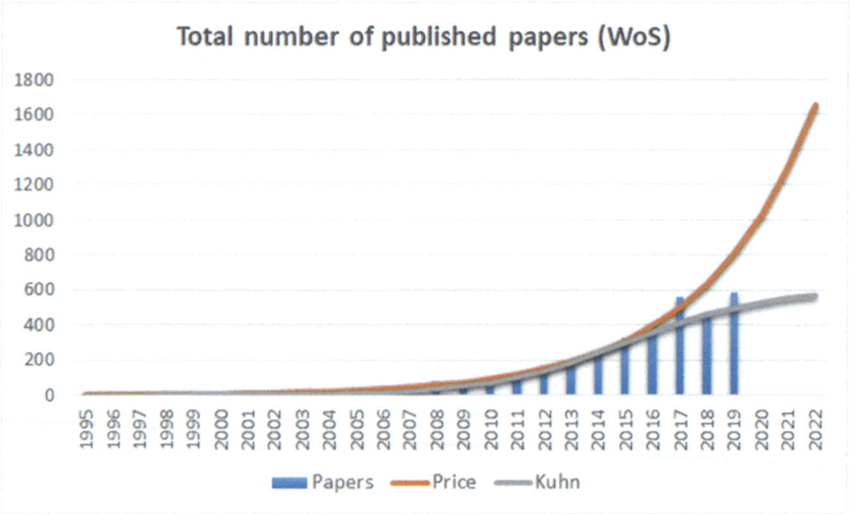

Fig. 1 Total number of published papers in the period 1995–2019

has varied between 2 and 3, but lately it has stabilized to 3 in this field. The number of citations has grown significantly, with peaks in 2012 and 2015.

The variation in annual publications can be explained with high coefficient of determinations (R^2) both by:

- the exponential logistic growth function that describes the productivity phase introduced in [23] (97.3% considering the curve $Y(t) = 0.278 \times e^{0.2385(t-1995)}$;
- the logistic curve following Kuhn's paradigm shift theory to explain scientific revolutions [12] (98.8% considering the curve $Y(t) = \frac{600}{1+e^{(-0.4\times(t-2015))}}$.

However, we can see that for the last 2 years, 2018 and 2019, the exponential curve is no longer well adjusted, while the logistic curve fits if we consider a ceiling of 600 publications per year and that the inflection point occurred in 2015. This means that the field is mature.

A word frequency analysis was also performed to identify the more frequently used words by the authors in the keywords, abstract, and title. The results are displayed in Fig. 2. In the picture we cluster the results by methodology (represented in lines), by application (represented in dotted lines) and by scale (represented by columns). Regarding the scale, we can see that most papers focus on mesoscale, which in the urban context usually means large areas of the cities, but more recently the number of papers restricted to smaller scales (campus, small districts or neighborhoods) has been increasing. Regarding the application, most of the papers refer to the study of urban heat island effect closely followed building shading, while the application in the smart grid context is very recent and still not very significant. Regarding the methodological approach, GIS is often referred as the

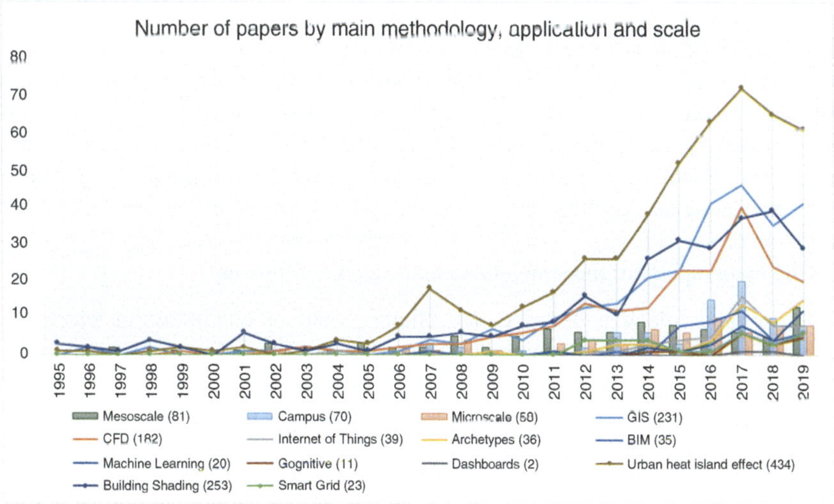

Fig. 2 Total number of published papers by methodology (continuous), application (dashed) and scale (columns)

main technology, followed by CFD, but over the last years, the use of archetypes or BIM has been increasing, together with IoT and Machine Learning or Cognitive technologies. This shows that also in this field of study the importance of data is being acknowledge. A result that is very interesting is the fact that dashboard is rarely used as a keyword, demonstrating that the representation advantage of these tools is not particularly highlighted in the scientific community yet.

From this analysis we can conclude that the development and application of urban energy modeling tools is already a mature field, with a significant number of publications per year. The field has evolved from the foundational methodologies of GIS and CFD to model urban heat island effect and building shading effects to focus more on the use of data collected by IoT devices, the use of machine learning techniques applied to the models, which demonstrates that the researchers are more or more using these tools in the framework of the smart urban contexts.

3 Urban Energy Modeling Tools

Based on the literature survey, in this section we define the general framework of an Urban Energy Modeling tool and we analyze in detail the current tools that fit this framework, discussing the similarities and differences between them.

3.1 Framework

Figure 3 presents the basic UBEM framework.

As the UBEM primary objective is to simulate the energy consumption of buildings at an urban scale, i.e. more than one building, the core module is a simulation engine for the building energy consumption. There are many tools for building energy simulation, as thoroughly described in [7]. Most of these tools focus on thermal modeling, as in general, more than half of the buildings' energy use is spent in heating and cooling. The most used tool is Energyplus [5], but many other tools can be used [7, 22].

The main inputs for the simulation engine are the following:

- the weather file that describes the climatic outdoor conditions in which the building is located, like temperature, humidity, wind speed, and radiation;
- the geometric information of the building, like the orientation of the walls—for instance, to include the impact of solar radiation—or the existence of different thermal zones—for instance, to distinguish areas that need to be heated like rooms and areas that do not need heating, like attics or basements;
- building physical characteristics from the constructive solutions—like the thermal resistances, widths of the materials, and areas of application;

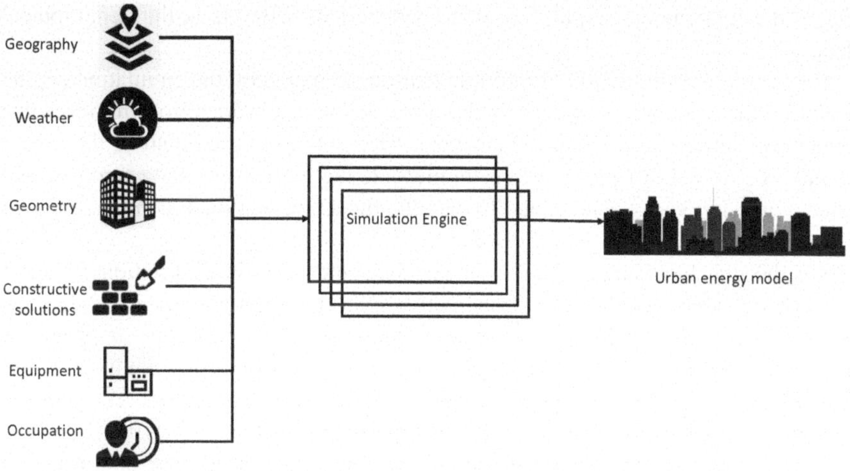

Fig. 3 Urban Energy Modeling tools framework

- the existing equipment, with the description of nominal powers, efficiency, and thermal losses;
- the schedule of occupation is also very important, to infer about the use of the different appliances and systems, the set-points definition, and the calculation of the internal gains.

There are other energy uses in buildings which are not necessarily tackled by thermal simulation tools, like lighting, cooking, hot water, or electrical appliances. In general, most of the tools used for thermal simulation do not consider other energy uses directly, unless they influence thermal behavior like lighting or internal gains from electrical appliances. An example of this is the use of RC models, which use an electric system as an analogue model of a thermal system. However, for a complete modeling of energy consumption, all uses should be considered, thus it is also necessary to provide information on the existing equipment and its use (also through the schedule of occupation), in order to guarantee that the building energy simulation tool accounts for all the energy uses.

This information is in general enough to grasp the basic thermal behavior of a single building, but for the simulation at the urban scale, the fundamental difference from an Urban Building Energy Model (UBEM) and a Building Energy Model (BEM) lays on two dimensions: the fact that it is necessary to take into consideration the effects of other buildings in the envelope of each building and then apply this model to more than one building.

Regarding the first aspect, the use of Geographic Information Systems (GIS) allows that the simulation tool knows the exact location of each building (and use, for example, the closest available weather file) and provides the thermal simulation tool automatically with the information regarding the relative distance and orientation of the different buildings under analysis. Therefore, as indicated in

Fig. 3, the fundamental input data that characterizes UBEMs is the geographical data.

Regarding the simulation of multiple buildings, in general the simulation engine performs the simulation of each building taking into consideration the influence of the remaining buildings in the envelope. Depending on the detail of the model (simple RC model or Eplus model with multiple thermal zones), the computational time might be very significant. To handle the computational complexity, two different approaches may be followed.

On the one hand, instead of using detailed and specific descriptions of each building information (geometry, constructive solutions, equipment, and schedules), we can archetypes, which are building definitions that represent a group of buildings with similar properties (geometry, construction solutions, equipment, and schedules). In this case, only one archetype can be simulated and the results of the archetype are assigned to all the buildings represented by that archetype.

Another approach is to simulate all the buildings, but considering simplification assumptions in terms of geometry (simple rectangular polyhedron geometry, with singe thermal zone and simplified equipment and schedules).

At the end, the results may be presented by building, usually using a GIS tools, 2D, or 3D. The model outputs, the urban energy model, may be the total consumption of each building, disaggregated or not by energy vector (electricity, heat) and/or the energy use (heating, cooling, appliances), but other outputs are possible like radiation in the different surfaces (which can be used to estimate solar PV generation potential), indoor temperature, interior light levels, etc.

In the following subsection, we analyze the existing UBEM tools.

3.2 UBEM Tools

There are currently several tools that follow the baseline framework described in the previous section, which represents the minimum set of features of an UBEM. The most mature tools, by chronological order of seminal publications, are: City Energy Analyst (CEA) [6], Urban Modeling Interface (UMI) [18], City Building Energy Simulation (CityBES) [8]. Many other projects are being developed, such as Tool for Energy Analysis and Simulation for Efficient Retrofit (TEASER) [19], Urban planning decision support tool (URBio) [20], or URBANopt [14], but the level of maturity is lower or the framework does not include all the previous features. Therefore, in this work, we analyze in detail the tools UMI, CIYBES, and CEA, since these tools that have already some track-records of application to different case studies, in different areas. The analysis is detailed in Tables 1, 2, and 3 according to the framework described in Sect. 3.1.

In the next subsections, further detail is provided for each tool.

Table 1 UBEM tools comparison—CEA

	CEA
URL	cityenergyanalyst.com
Overview paper	Fonseca et al. [6]
User Interface	Desktop IDE
Simulation Engine	RC model
Geography	Open Street Maps
Weather	Eplus weather file
Database tool	QGIS
Outputs	Energy demand, Energy Supply, LCA, Comfort
Applications	Zurich, Singapore and Amsterdam

Table 2 UBEM tools comparison—UMI

	UMI
URL	web.mit.edu/sustainabledesignlab/projects/umi/index.html
Overview paper	Reinhart and Cerezo Davila [18]
User Interface	Rhinoceros and Web Browser
Simulation Engine	Eplus
Geography	Rhinoceros
Weather	Eplus weather file
Database tool	SQLite
Outputs	
Applications	Chicago, Lisbon, Kuwait, Boston

Table 3 UBEM tools comparison—CityBES

	CityBES
URL	citybes.lbl.gov
Overview paper	Hong et al. [8]
User Interface	Web Browser
Simulation Engine	Eplus
Geography	CityGML
Weather	Eplus weather file
Database tool	QGIS
Outputs	
Applications	San Francisco, Chicago, Boston and 7 other US cities

3.2.1 CEA

The CEA framework can be freely downloaded. The user interface is a Desktop IDE built on top of Python and uses Open Street map as the main source for GIS data. The weather data is based on Eplus weather files and the simulation engine runs a RC model of the buildings. The Desktop IDE allows the users to parametrize the archetypes of buildings (constructive solutions, equipment for multiple uses, schedules), but QGIS can also be used to change the databases. It has been successfully used for different locations such as Singapore or Zurich.

3.2.2 UMI

The UMI framework requires Rhinoceros 3D, a commercial tool, and includes an Application Programming Interface (API) for developers. The weather data is based on Eplus weather files and the simulation engine runs also Eplus for archetypes of the buildings. The main limitation at the moment is the need to use Rhinoceros 3D.

3.2.3 CityBES

The CityBES framework supports online based simulations in the cloud. The weather data is based on Eplus weather files and the simulation engine runs also Eplus for all the buildings. The databases have to be build using CityGML and at the moment, provided to the team managing the software. In the web browser, it is possible to use the software upon urban environments previously uploaded. The main limitation is the need to develop the urban database in CityGML and upload to the system.

4 Case Studies

In this section, we show the application of the different tools to urban case studies. The case studies are not directly comparable (since due to limitations of the different tools, it is not easy to apply them to the same exact case study). In this sense, we evaluate the tools using available case studies, to clearly highlight the possible applications of each tool, their advantages, and disadvantages.

4.1 CEA

The CEA tool, from the three analyzed tools, is the one which is more mature, with a broader spectrum of applications, and the one which is easier to use for none experts. The tool is subjected to constant update releases and the current version has an autonomous, user-friendly dashboard that enables to develop case studies easily in any city in the world.

Figure 4 presents a picture of the dashboard, for an application in Zurich.

The tool starts by importing the shape file from open street maps. In case the file is not available, it is possible to go to open street maps and manually define the shape files, for example, in the case of buildings which are yet to be built. This can also be done in QGIS. It also allows the user to easily define the individual characteristics of the buildings (number of floors, height, construction age, schedule, type of use, equipment). There are archetypes previously defined for two geographic areas (Switzerland and Singapore), but from those it is easy to define archetypes for different areas (this is an area that in terms of user friendliness could definitively improve). The user can also easily choose the weather file and the flow of sequential steps required to perform an analysis is straightforward to follow.

The tool has a broad spread of features that can be evaluated. Apart from the energy demand, which relies on RC models and not on Eplus, it evaluates the fuels use, and the disaggregated energy uses. In terms of energy supply, it allows to test not only the PV panels, as in UMI and CityBES, but also solar thermal and even hybrid PVT systems. In Figs. 5 and 6, it is possible to observe the results for a particular house, namely the energy demand and the potential for PV generation in each surface. Like UMI and CItyBES, it also provides a Life Cycle Assessment tool.

The tool includes many additional features compared to UMI and CItyBEs, in particular, the possibility to model district heating and cooling networks, electrical networks and also to evaluate the shallow geothermal potential. The design of the network is done using a multiobjective algorithm, based on genetic Algorithms, and also performs a sensitivity analysis.

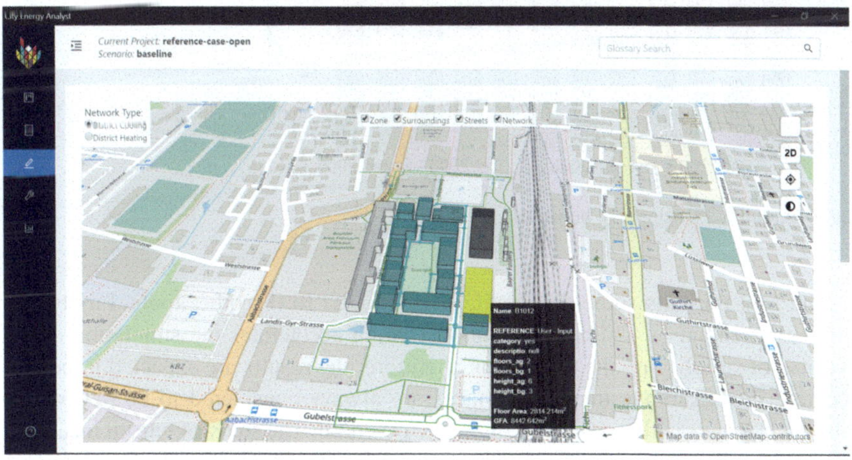

Fig. 4 CEA results for a case study in Zurich, Switzerland

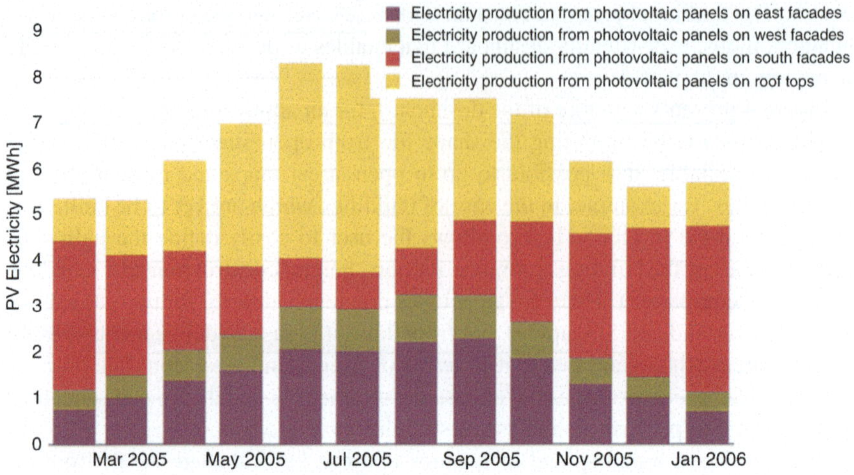

Fig. 5 Monthly energy demand of household "B108" in CEA

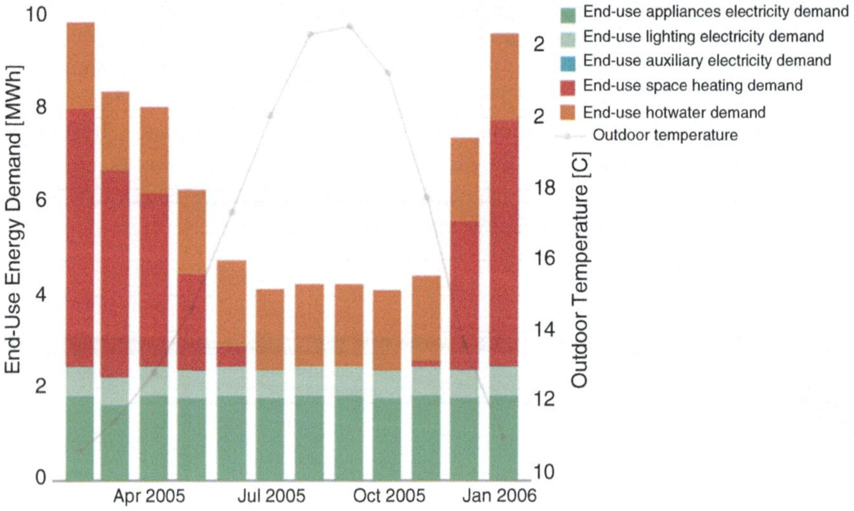

Fig. 6 Monthly PV supply of household "B108"

The tool presents some disadvantage, compared to the other tools. Firstly, the simulation engine is a simple RC model, which accelerates the simulation at the expense of simplifying the conditions, therefore, the results are expected to be less accurate. For example, the roof surfaces are always considered to be flat, which introduces errors in the evaluation of the PV potential. The study of retrofit measures can only be done indirectly by changing the archetypes definitions.

Overall, it is definitely the easiest tool to use from the three, especially for a non-specialist, and the one that provides currently a broader spectrum of applications specifically related to energy. However, the simulation engine is the least accurate. This easiness of use and the fact that the simulation engine is less accurate makes this tool the perfect tool to perform preliminary analysis, both for existing areas or new urban areas, to estimate the potential impact in energy networks, with or without local energy generation. Then, for detailed retrofit studies, moving from CEA to CityBEs would be advisable, or to UMI, especially when other studies, like lighting or air quality are required. Thus, although from a general overview the tools are currently competitive, they cannot yet be used exactly for the same applications and the choice of the tool should be driven by the final application purpose.

4.2 *UMI*

As previously explained, UMI is a Rhinoceros plug-in, so the location, geometry, weather, constructive solutions, schedules, and equipment are defined in the tool. In Fig. 7, we can see a figure of the web interface of UMI for an application in Boston (Seafront area), with the results of new buildings for that area, in particular, regarding energy use intensity. It is also possible to check other results, like the life cycle assessment of the buildings, the daylight, walkability and lately there are other tools, like a tool to access the potential for urban agriculture.

The main advantage of UMI is that, as a plug-in of Rhinoceros 3D—a software for 3D modeling—it enables users to design detailed geometric information and constructive solutions, and in that way, perform thorough analysis of the thermal behavior, but also lighting, ventilation, and other physical parameters that directly influence the energy consumption in building. From the three analyzed tools, UMI is the one that allows for more rigorous analysis of the buildings. Further, it allows the users to develop analysis in other fields related to urban environment, like transportation (walkability or bikeability) or sustainability on a broader scope. The main disadvantage is the need to have the commercial tool Rhinoceros, which is not free and which requires some experience to work width and up to a certain extend

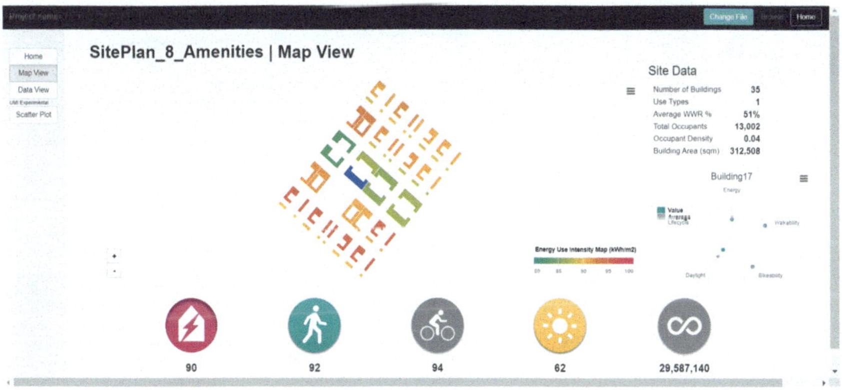

Fig. 7 City Energy Analyst dashboard for Boston, USA

is only accessible to practitioners, like architects. Further, although it can be used to model the potential for energy supply, especially solar technologies, currently it is limited to explore the design of the supply energy system of a district, when compared to CEA.

In conclusion, UMI is great tool for architects to work in the urban design of the cities and evaluate the different impacts, not only in energy but also in other areas related to the quality and sustainability of urban environments, based on rigorous simulations. It is a tool that can be used for preliminary studies, but due to the current complexity, definitely should be used to develop the final projects of urban designs.

4.3 CItyBES

In [13] several examples can be found of the use of CityBES for different cities in the USA, namely San Francsico, Chicago, Boston in a total of 10 cities. Within each city, we can find examples of particular districts and in each district, we can choose the buildings to be analyzed, by type of use (e.g. office or residential), by year of construction or area. An example can be found for San Francisco (downtown district) for small office buildings in Fig. 8.

In the dashboard, we can see results from energy consumption (e.g. energy intensity like in UMI or CEA) using baseline information, but also considering different types of retrofit measures or even a package of measures, centered in the envelope or systems, including PV generation. The tool also shows the urban climate and allows the simulation also of Life Cycle Assessment.

The main advantage of CityBES is the capacity to test different retrofit measures of the buildings, by allowing the parametrization of the simulation in Eplus, like the definition of the thermal zones. The dashboard is easy to use and intuitive.

The main disadvantage has to do with the fact that it requires the definition of the areas in cityGML, an open data model and XML-based format for the storage

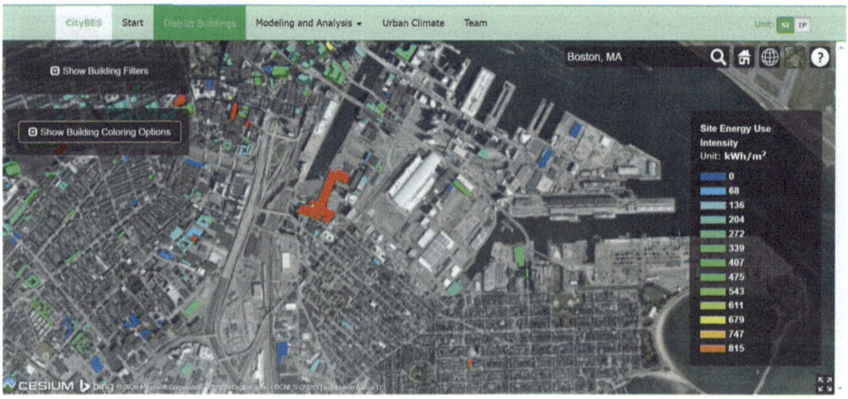

Fig. 8 CityBES results for San Francisco, USA

and exchange of virtual 3D city models, which presents a significant barrier even for practitioners, as is not yet a wide adopted standard. And even if the model is available, it is not yet possible to upload it directly in the platform. Further, as the CityBES approach is to simulate all buildings individually, the computational effort associated with the simulation is quite high. In fact, due to this, not all case studies can be simulated in the platform. Finally, like UMI, it can be used to model the potential for solar PV, but the tool is limited to explore the design of the supply energy system of a district, when compared to CEA.

In conclusion, CityBES is good platform to test retrofit packages in buildings environments, based on rigorous simulations using Eplus. However, from the three analyzed tools is the one which has more limited features, and the one which presents more barriers to be used.

5 The Evolution of UBEM Systems for Smart Urban Contexts: The Cognitive UBEMs

As it discussed in the previous sections, UBEM tools rely currently on static data (data that does not change after being recorded), like shape files, constructive solutions data, or even weather data files based on historical time-series. However, in the framework of a smart urban context, data is updated periodically so the UBEM tools must evolve to include dynamic, real time datasets.

As examples, the weather data can use real time data from local weather stations, or the constructive solutions can be based on updated information on municipality records or energy performance certificates after retrofit. The access to real time updated data can then be used to give UBEM the capability to adapt the models to new changing conditions, i.e. adding a cognitive layer. In this sense, we advocate that the UBEM systems need to evolve to provide some of features that are discussed in the following subsections. To demonstrate these capabilities, we will resort to the COgnitive UBEM tool being currently develop at IST [4].

5.1 The Use of 3D and Time

The tools analyzed in Sect. 3 do not fully explore yet the 3D representation capabilities, as now they mostly provide a representation of the results—e.g. energy demand—in the urban shape for one period (typically 1 year). However, the fact that we have access to representation in 3D allow us to differentiate the different units inside the same building and see specific information about that. In Fig. 9, we provide an example of the energy performance certificate (EPC) information for a particular household of a building with two households in the neighborhood of Encarnação in Lisbon, Portugal.

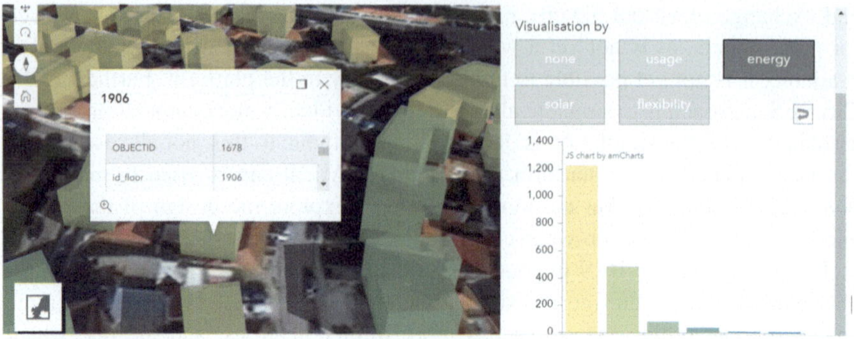

Fig. 9 3D visualization of the energy performance certificate (EPC) information for different households in a building

Fig. 10 Real Time visualization of environmental conditions in real time at IST campus amphitheater GA1

The tools can also be used to visually display the evolution of the different results in time, e.g. as an animated sequence of frames, to represent the time-based information.

5.2 *Use of Real Time Data*

Another obvious development of UBEM tools is the integration of real time data. This data can be collected from energy or environment conditions monitoring devices. In Fig. 10, we provide an example of the real time environmental conditions for amphitheater GA1 of IST Alameda Campus, in Lisbon, Portugal, which combines the use of 3D information and use of real time data.

5.3 Integration of Machine Learning Models with UBEM

As shown previously, the number of articles using machine learning algorithms in the field of UBEM has been growing. An example on how these tools can be used is the one in [21], where Bayesian models are used to calibrate unknown or uncertain parameters in archetype descriptions as probability distributions, using measured energy data. Other examples can be the energy demand forecast using the real data [17] and the models or energy generation forecast [27].

5.4 UBEM for Operational Real Time Energy Management

At the end, these new features will enhance UBEM tools in a way that will enable its use beyond planning purposes, in particular, for operational real time energy management. As described in Sect. 3, these tools present in general a dashboard to present the results and enable the users to test different parameters. Thus, the combination of the previous features—spatial 3D resolution, real time information and the integration of machine learning models—in a dashboard-like tool will enable UBEM to be used to compare real time consumption data with forecast or baseline data, to test in real time different operation strategies and therefore contribute to a more efficient real time energy management of campus, micro-grids, or neighborhoods operating as energy communities.

6 Conclusions

The current trend of increasingly using information and communication technologies in urban environments is creating a context that enables the development and provision of more sustainable solutions to citizens, like aiding in the decision of the best energy efficiency measures or the best local renewable generation solutions.

Urban Energy Modeling (UBEM) tools have been developed over the last decade to provide these types of answers at the urban scale. Currently there are at least three tools that have reached the maturity—CEA, UMI, and CityBES and have proven to be able to simulate energy consumption of neighborhoods and therefore can be used to aid decision makers regarding the design ore retrofit of new neighborhoods in different parts of the world.

The tools work currently with static, historical data, but the smart urban context that is being developed can induce the development of new features. In particular, we propose that the next evolution of the UBEMs should focus on the use of 3D and time dimensions to represent information, the update of the input data with real time data in the models and the integration of machine learning modules in the analysis.

In this way, we envisage that in a near future, these tools can be used to optimize in real time the operation of energy systems like campus, micro-grids, or energy communities in a smarter, more sustainable way.

Acknowledgments This project is supported by the National Foundation of Science and Technology through MIT Programme under grant MIT-EXPL/SUS/0015/2017 and through the Pessoa2019 grant Comportamento Interactivo para Flexibilidade Urbana.

References

1. C. Analytics, Web of science (2020). http://wokinfo.com/. Accessed 03 March 2020
2. S.E. Bibri, J. Krogstie, Smart sustainable cities of the future: an extensive interdisciplinary literature review. Sustain. Cities Soc. **31**, 183–212 (2017). https://doi.org/10.1016/j.scs.2017.02.016
3. V. Ca, T. Asaeda, M. Ito, S.W. Armfi, Characteristics of wind field in a street canyon. J. Wind Eng. Ind. Aerodyn. **1**(57) (1995). https://doi.org/10.1016/0167-6105(94)00117-V
4. F. Costa, CogUBEM - Cognitive Urban Energy Modeling (2020). http://luxmagna.org/. Accessed 03 March 2020
5. Energy of USD, EnergyPlus (1996). https://energyplus.net/. Accessed 03 March 2020
6. J.A. Fonseca, T.A. Nguyen, A. Schlueter, F. Marechal, City Energy Analyst (CEA): integrated framework for analysis and optimization of building energy systems in neighborhoods and city districts. Energy Build. **113**, 202–226 (2016). https://doi.org/10.1016/j.enbuild.2015.11.055
7. V.S. Harish, A. Kumar, A review on modeling and simulation of building energy systems. Renew. Sust. Energ. Rev. **56**, 1272–1292 (2016)
8. T.L. Hong, Y.L. Chen, S.H.L. Lee, M.P.L. Piette, Y.L. Chen, M.P.L. Piette, CityBES: a web-based platform to support city-scale building energy efficiency, in *5th International Urban Computing Workshop, at San Francisco*, July 2016, p. 10. https://doi.org/10.1016/j.jclepro.2015.05.105
9. P.J. Jones, J. Williams, S. Lannon, An energy and environmental prediction tool for planning sustainability in cities, in *Green Building Challenge '98 Conference: An International Conference on the Performance Assessment of Buildings* (1998)
10. J. Keirstead, M. Jennings, A. Sivakumar, A review of urban energy system models: approaches, challenges and opportunities. Renew. Sust. Energ. Rev. **16**(6), 3847–3866 (2012). https://doi.org/10.1016/j.rser.2012.02.047
11. N. Klepeis, W. Nelson, W. Ott, J. Robinson, A.M. Tsang, P. Switzer, The National Human Activity Pattern Survey (NHAPS): a resource for assessing exposure to environmental pollutants. J. Expo. Sci. Environ. Epidemiol. **11**, 231–252 (2001). https://doi.org/doi.org/10.1038/sj.jea.7500165
12. T.S. Kuhn, *The Structure of Scientific Revolutions* (University of Chicago Press, Chicago, 1996)
13. Lab LBN, CityBES (2016). https://citybes.lbl.gov/. Accessed 03 March 2020
14. National Renewable Energy Laboratory, URBANopt advanced analytics platform (2020). https://www.nrel.gov/buildings/urbanopt.html. Accessed 03 March 2020
15. J. Pedro, C. Silva, M.D. Pinheiro, Scaling up LEED-ND sustainability assessment from the neighborhood towards the city scale with the support of GIS modeling: Lisbon case study. Sustain. Cities Soc. **41**(November 2017), 929–939 (2018). https://doi.org/10.1016/j.scs.2017.09.015

16. J. Pedro, C. Silva, M.D. Pinheiro, Integrating GIS spatial dimension into BREEAM communities sustainability assessment to support urban planning policies, Lisbon case study. Land Use Policy **83**(January), 424–434 (2019). https://doi.org/10.1016/j.landusepol.2019.02.003
17. H. Pombeiro, R. Santos, P. Carreira, C. Silva, J.M. Sousa, Comparative assessment of low-complexity models to predict electricity consumption in an institutional building: linear regression vs. fuzzy modeling vs. neural networks. Energy Build. **146**, 141–151 (2017). https://doi.org/10.1016/j.enbuild.2017.04.032
18. C.F. Reinhart, C. Cerezo Davila, Urban building energy modeling - a review of a nascent field. Build. Environ. **97**, 196–202 (2016). https://doi.org/10.1016/j.buildenv.2015.12.001
19. P. Remmen, M. Lauster, M. Mans, M. Fuchs, T. Osterhage, D. Müller, Teaser: an open tool for urban energy modelling of building stocks. J. Build. Perform. Simul. **11**(1), 84–98 (2018). https://doi.org/10.1080/19401493.2017.1283539
20. N. Schüler, S. Cajot, 11 - Application of the planning support system URBio, in *Urban Energy Systems for Low-Carbon Cities*, ed. by U. Eicker (Academic Press, London, 2019), pp. 415 – 445. https://doi.org/10.1016/B978-0-12-811553-4.00011-1
21. J. Sokol, C. Cerezo Davila, C.F. Reinhart, Validation of a Bayesian-based method for defining residential archetypes in urban building energy models. Energy Build. **134**, 11–24 (2017). https://doi.org/10.1016/j.enbuild.2016.10.050
22. A. Sola, C. Corchero, J. Salom, M. Sanmarti, Simulation tools to build urban-scale energy models: a review. Energies **11**(12) (2018). https://doi.org/10.3390/en11123269
23. D.J. de Solla Price, *Little Science, Big Science... and Beyond* (Columbia University Press, New York, 1986)
24. K. Taipale, C. Fellini, D. Le Blanc, Challenges and way forward in the urban sector. Tech. rep., United Nations, 2012
25. R. Todeschini, A. Baccini, *Handbook of Bibliometric Indicators Quantitative Tools for Studying and Evaluating Research* (Wiley, New York, 2016)
26. United Nations, Department of Economic and Social Affairs, World Urbanization Prospects: The 2018 Revision. Tech. rep., United Nations, 2018. https://doi.org/10.4054/demres.2005.12.9
27. C. Voyant, G. Notton, S. Kalogirou, M.L. Nivet, C. Paoli, F. Motte, A. Fouilloy, Machine learning methods for solar radiation forecasting: a review. Renew. Energy **105**, 569–582 (2017). https://doi.org/10.1016/j.renene.2016.12.095

Energy Sobriety: A Behaviour Measurement Indicator for Fuel Poverty Using Aggregated Load Readings from Smart Meters

Paul Fergus and Carl Chalmers

1 Introduction

Fuel poverty describes members of a household that cannot afford to adequately warm their home or run the necessary energy services needed for lighting, cooking, hot water, and electrical appliances [1]. It is estimated that between 50 and 125 million households are affected in Europe (EPEE, 2009). In the UK, approximately four million households are classified as being fuel poor (15% of all households)—613,000 in Scotland (24.9% of the total); 291,000 in Wales (23% of the total); 160,000 in Northern Ireland (22% of the total); and 2.55 million in England (11% of the total) [2]. The problem is complex but is typically caused by three factors: low income, high energy costs, and energy-inefficient homes [1, 3–5].

In the UK, financial support is provided for low-income households through the Warm Home Discount Scheme, Cold Weather Payments, and Winter Fuel Payments (similar support is provided in other EU member states) [6]. According to a UK report written in 2018, the government provided £1.8 billion in funding annually for Winter Fuel Payments, £320 million for the Warm Homes Discount Scheme, and £600 million for the Energy Company Obligation Scheme [7]. Schemes like this provide temporary relief, but do not tackle the underlying causes of fuel poverty [8, 9].

Currently, fuel bills in the UK cost on average £1813 a year, a 40% increase from £1289 in 2015 [10]. The Office of Gas and Electricity Markets (Ofgem) caps the maximum price that consumers can pay for electricity and gas; however, the recent lifting of price caps has seen a £1.7bn increase in consumer bills [11]. Subsequently,

P. Fergus (✉) · C. Chalmers
Department of Computer Science, Liverpool John Moores University, Liverpool, UK
e-mail: p.fergus@ljmu.ac.uk; c.chalmers@ljmu.ac.uk

S. Ploix et al. (eds.), *Towards Energy Smart Homes*,
https://doi.org/10.1007/978-3-030-76477-7_2

rising energy prices force more people to live in fuel poverty rather than easing the financial pressures fuel-poor households already have [12].

Alongside low income and rising fuel costs, a substantial share of the residential housing stock in Europe is older than 50 years with many in use reportedly hundreds of years old [13]. More than 40% were constructed before the 1960s when energy regulations were limited [14]. The performance of buildings depends on the installed heating system and building envelope, climatic conditions, indoor temperatures and fuel poverty [15]. This means that largest energy savings often come from improving older buildings, particularly poorly insulated properties built before the 1960s.

In the UK, the energy efficiency of homes is measured using the Standard Assessment Procedure (SAP) rating [16]. During the winter months colder weather lowers the energy efficiency of the property and increases domestic energy demand. The performance of the heating system, appliances, and the number of people living in the property (and how long they say in the home throughout the day) determine the household fuel bill. In low-income and energy-inefficient homes the winter months are particularly problematic and a source of constant worry for occupants about debt, affordability, and thermal discomfort [17]. The impact this has on health is significant given that fuel-poor households spend increased amounts of time in the cold. Hence, poor health among this social group is prevalent [18]. In fact, evidence shows us that fuel-poor occupants are more likely to experience poor health, miss school [19–24], and report absences from work [17, 25].

According to the E3G, the UK has the sixth-highest rate of Excessive Winter Deaths (EWD) of the 28 EU member states—a large number have been directly linked to cold homes [19, 26]. EWD is the surplus number of deaths that occur during the winter season (in the UK this is between the 22nd of December and 20th of March) compared with the average number of deaths in non-winter seasons [19]. The main causes of EWD are circulatory and respiratory diseases [27]. It is estimated that about 40% of EWD are attributable to cardiovascular diseases, and 33% to respiratory diseases [22]. According to the Office of National Statistics (ONS), there were 50,100 EWDs in England and Wales in the 2017–2018 winter period, the highest recorded since the winter of 1975–1976 [28]. Cold homes have also been linked with high blood pressure [29], heart attacks, and pneumonia, particularly among vulnerable groups such as children and older people [22, 23, 30–33]. This often leads to inhabitants experiencing loss of sleep, increased stress, and mental illness [17].

Alongside serious health outcomes, cold homes are uninviting leaving inhabitants stigmatized, isolated, and embarrassed because they are often forced to put on additional clothing, wrap up in duvets or blankets, and use hot water bottles to stay warm [34]. This undoubtably increases the likelihood of depressions and other mental illness. Epidemiological studies show that occupants in damp homes are more likely to have poorer physical and mental health [35]. According to the Building Research Establishment (BRE) poor housing costs the National Health Service (NHS) £1.4 billion each year [36]. The World Health Organization (WHO) commissioned a comprehensive analysis of epidemiological studies and concluded that a relationship exists between humidity and mould in homes and health-related problems [37].

Fuel poverty is a focal point for the EU; however, as the figures show, current policy has had/is having little effect on reducing the number of fuel-poor households. This is hardly surprising given the EU does not provide a common definition of fuel poverty or a set of indicators to measure it [38]. This means that fuel poverty numbers vary depending on what measurement indicator is implemented.

2 Measuring Fuel Poverty

Measurement indicators are used to identify which households are considered to be in fuel poverty—in the UK, this is the responsibility of the Department for Business, Energy & Industrial Strategy (BEIS) [39]. A detailed report, commissioned by the EU in 2014, found that 178 indicators exist: of which 58 relate to income or expenditure and 51 to physical infrastructure [40]. Indicators related to energy demand and demographics amount to 10 and 15, respectively. 139 are single metric indicators and 39 combinatory or constructed indicators, representing 22% of the total and mostly falling under the category of income/expenditure. Among the identified energy poverty metrics, 10 are consensual-based, 42 expenditure-based, and 11 outcome-based, while another 14 indicators comprise a combination of metrics. The two main approaches used today are expenditure-/consensual-based. Only the most common indicators within both approaches will be considered in this chapter. For a more detailed discussion the reader is referred to [40].

2.1 Expenditure-Based Indicators

Expenditure-based indicators focus primarily on the proportion of the household budget used to pay for domestic fuel [41]. The best-known indicator is the 10% rule proposed by Boardman in the early 1990s [1] which was adopted in the UK in 2001. A household is classed as being fuel poor if more than 10% of its income is spent on fuel to maintain an acceptable heating regime [42]. The indicator uses a ratio of modelled fuel costs and a Before Housing Costs (BHC) measure of income [43]. Modelled fuel costs are derived from energy prices and a modelled consumption figure that includes data about property size, the number of people in the property, the household's energy efficiency rating, and the types of fuel used. Fuel-poor households are those with a ratio greater than 1:10 (10%).

The Hills report in 2011, commissioned by the Department of Energy and Climate Change (DECC) (now BEIS), triggered a replacement of the 10% indicator with the Low Income High Cost (LIHC) indicator [44]. LIHC is now used in the UK to measure fuel poverty and has attracted considerable attention within different national contexts [43, 45–47]. The LIHC indicator is calculated using a national income threshold and a fuel cost threshold [42, 44]. A household is classified as fuel poor if it exceeds both thresholds. The fuel cost threshold is a weighted median of the fuel costs for all households, weighted according to the number of people in a property. This average fuel cost value is the assumed cost of achieving an adequate level of comfort. The threshold is the same for all households of equivalent size. The income threshold is calculated as 60% of the weighted national median for income After Housing Costs (AHC) are accounted for. The income figure for each household is also weighted to account for the number of people living in the property. This figure is combined with the weighted fuel costs of the household. The income threshold is therefore higher for those that require a greater level of income to meet larger fuel bills.

2.2 *Consensual-Based Indicators*

Consensual-based indicators on the other hand assess whether a person is in fuel poverty by asking them. The approach was initially based on Townsend's early relative poverty metric [48] and later on the consensual poverty indicator proposed in [49, 50]. The fundamental principle is centred on a person's inability '*to afford items that the majority of the general public considered to be basic necessities of life*' [50].

Using surveys, household occupants are asked to make subjective assessments about their ability to maintain and adequately warm their home and pay their utility bills on time. The EU has adopted the core principles of the consensual model and implemented the Survey on Income and Living Conditions (EU-SILC) [51]. EU-SILC includes a set of questions that asks whether the household (a) is able to keep their home warm during winter days, (b) has been in arrears with utility bills, and (c) has leakages or damp walls [52]. The recommendation was launched in 2003 and was the first micro-level data set to provide data on income and other social and economic aspects of people living in the EU [51].

EU-SILC has a rotating panel that lasts 4 years; a quarter of the sample is replaced each year by new subsample members [53]. During the 4 years, households are contacted up to four times. The consensual approach has been acclaimed for being easy to implement and less complex, in terms of collecting data, than expenditure-based indicators. A key feature of the EU-SILC dataset is that it provides an important basis for identifying and understanding fuel poverty and the differences that exist across all EU member states [54].

2.3 *Limitations*

Fuel poverty measures have several limitations, primarily because of the multidimensional nature of the phenomenon, which makes it difficult to adequately capture or measure it using a single indicator [40]. Additionally, most indicators have been disparaged for focusing solely on fuel expenditure without consideration for under-consumption which has led to governments underestimating the real extent of fuel poverty [44, 55]. In the case of expenditure-based approaches, the main issue is the lack of available data, particularly on the contributing factors needed to assess the extent of fuel poverty. This is alleviated with consensus-based approaches given that micro-level data is collected. However, the approach has also been criticized for being too subjective and exclusive [56].

In the case of the 10% rule, it does not respond to variations in income, fuel prices, or energy efficiency improvements [57] and this has led to skewed results [58]. Hills suggested that '*flaws in the 10% indicator have distorted policy choices, and misrepresented the problem*'. Therefore, relatively well-off households in energy-inefficient properties have been identified as being fuel poor [57, 59].

The LIHC indicator on the other hand excludes low-income, single-person households [59, 60]. Moore argues that this indicator obscures increases in energy prices, as its introduction has led to a fall in fuel-poor households, in spite of significant increases in energy costs during the same period [58]. This has been described by some as an attempt to move the goalposts in order to justify missing targets for the eradication of fuel poverty, which was a target for all households by 2016 [61]. Middlemiss adds that the LIHC prioritizes energy efficiency as a solution to fuel poverty distracting from other drivers, such as the wider failure of the energy market to provide an affordable and appropriate energy supply to homes [62].

Finally, the EU-SILC consensus-based approach has been criticized for (a) only including specific household types, (b) containing anomalies in the data collected (i.e. missing data), (c) being subjective due to self-reporting, and (d) providing a limited understanding of the intensity of the issue due to the binary character of the metrics [56]. Participants do not view judgements like 'adequacy of warmth' in the same way while some households may not even identify themselves as being fuel poor due to pride even though they have been characterized as being fuel poor under other measures [56]. It is not unusual for fuel-poor residents to deny the reality of their situation and report that they are warm enough when they are in fact not.

3 Smart Meters

Residential homes consume 23% of the total energy delivered worldwide (29% in the UK) [63]. Industries consume 37%, and this is closely followed by transportation which is 28% [64]. Household energy consumption is considered a multidimensional phenomenon rooted within a socio-cultural and infrastructure context, and as such occupant behaviour is complex. Existing measurement indicators, as we have seen, fail to capture the behavioural traits associated with individual households. Yet, with the current smart meter rollout well underway in many developed countries which facilitates the automatic reporting of energy usage, it is now possible to capture the behavioural aspects of energy consumption through data provided by

CADs paired with smart meters [65]. CADs provide data every 10 s for all energy consumed within the home at the aggregated level [66]. This data combined with advanced data analytics allows us to determine whether a house is occupied, what electrical appliances are operated, and when they are being used [67, 68]. Such insights provide the based for routine formation which we will return to later in the chapter.

3.1 Smart Meter Infrastructure

Smart meters measure gas and electricity consumption and send usage information to energy suppliers and other interested parties. This (a) removes the need for home visits and manual meter readings and (b) allows consumption data to be used by the smart grid, to balance energy load and improve efficiency [69]. According to the International Energy Agency (IEA), smart grids are essential to meet future energy requirements [70], given that worldwide energy demand is expected to increase annually by 2.2%, eventually doubling by 2040 [71].

Energy consumption data in the smart grid is received directly from smart meters and stored, managed, and analysed in the Meter Data Management System (MDMS) [66]. The MDMS is implemented in the data and communications layer of the Advanced Metering Infrastructure (AMI) and is a scalable software platform that provides data analytic services for AMI applications, i.e. data and outage management, demand and response, remote connect/disconnect, smart meter events, and billing [66]. Data contained in the MDMS is shared with consumers, market operators, and regulators.

Smart meters in the UK collect and transmit energy usage data to the MDMS every 30 min [72]. Higher sample rates are possible, but this increases the costs for data storage and processing. Data transmitted through a smart meter consists of (a) aggregated energy data in watts (W), (b) a Unix date/time stamp, and (c) the meters personal identification number (PID). The energy distribution and automation system collects data from sensors dispersed in the smart grid. Each sensor generates up to 30 readings per second and includes (a) voltage and equipment health monitoring and (b) outage voltage and reactive power management information. External data sets by third-party providers are also used to facilitate demand and response subsystems. OS/firmware software provides a communication link between the MDMS and smart technologies and this allows geographically aggregated load readings to be analysed to ensure-efficient grid management. The OS/firmware system also manages OS/firmware version patching and updating. Figure 1 shows a typical MDMS system and its common components.

Information stored in the MDMS is a significant data challenge that requires data science tools to maintain optimal operational function [73, 74] and derive insights from the information collected [75, 76]. This allows decision-making and service provisioning to be implemented directly atop the smart meter infrastructure [77–

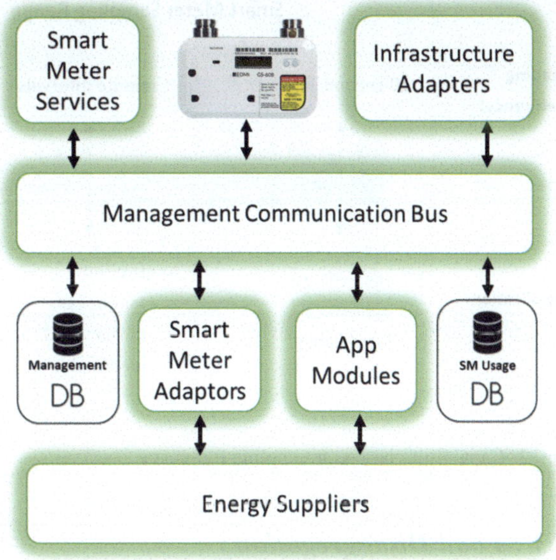

Fig. 1 Meter data management system for processing home energy usage and automated billing

81]. Services exploit the smart grid infrastructure to provide application support in different domains, i.e. health, climate change, and energy optimization [82].

3.2 Smart Meter Sampling Frequencies

Most studies do not use actual smart meter data for monitoring. Smart meter readings are provided every 30-min in the UK (other countries have different sample frequencies) [83]. With 30-min data it is possible to detect occupancy; however no reliable appliance information can be noticed at this frequency [84]. Therefore, electricity monitors are either paired with the smart meter using a consumer access device (CAD), CT Clip, or sensor plugs attached to the actual appliance when higher sample frequencies are required as shown in Fig. 2.

CADs are an inexpensive way to obtain whole-house measurements at higher sampling rates (i.e. readings every 10 s in the UK). With a CAD you can detect when high-energy appliances, such as an oven, kettle, and microwave, are being operated. CT Clips are used when either a smart meter has not yet been installed in a household or when sample frequencies higher than every 10 s are required. CT Clips, clamped around the power cable (live or natural), can sample the aggregated energy feed thousands of times every second. However, the approach is more costly than a CAD as additional hardware and software need to be installed. With a CT Clip, it is possible to detect faulty appliances and overlapping use, including low-energy appliances, such as lights and audio equipment. Device types will be discussed in more detail later in the chapter.

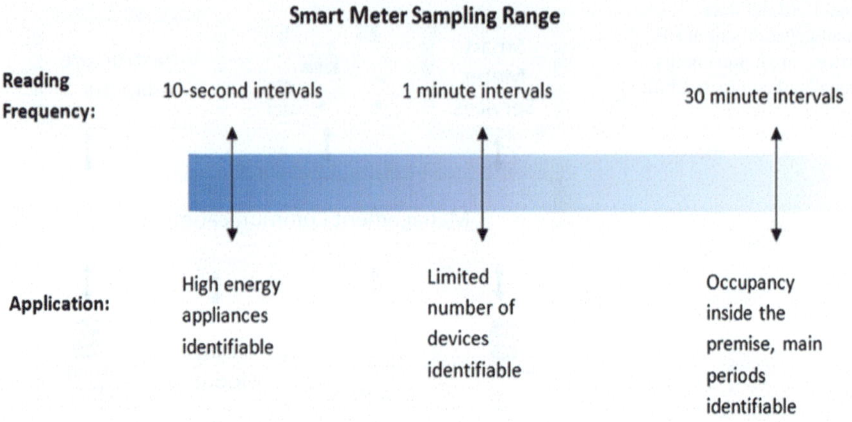

Fig. 2 Capabilities based on sampling frequency

3.3 *Load Disaggregation*

Load disaggregation is a broad term used to describe a range of techniques for splitting a household's energy supply into individual electrical appliance signatures, for example, a kettle, microwave or oven [68]. There are a number of reasons why load disaggregation is important. In the context of fuel poverty, appliance detections provide the basis for habitual appliance usage patterns, which manifest as routine household behaviours [68, 83]. Through an understanding of normal routine behaviour it is possible to identify anomalies and assess whether they are linked to fuel poverty indictors—more on this later [83].

Disaggregating electrical device usage is called Appliance Load Monitoring (ALM) [85]. ALM is divided into two types: Non-Intrusive Load Monitoring (NILM) [86] and Intrusive Load Monitoring (ILM) [87]. NILM is a single point sensor, such as a smart meter or CT clip. In contrast, ILM is a distributed sensing method that uses multiple sensors—one for each electrical device being monitored [87]. ILM is more accurate than NILM as energy usage is read directly from sensors attached to each electrical appliance being measured. The practical disadvantages however include high costs, multi-sensor configuration, and complex installation [88]. More importantly, ILM sensors can be moved between different devices and this can skew identification and classification results.

NILM on the other hand is less accurate than ILM and more challenging as appliances are identified from aggregated household energy readings [89]. NILM was first developed in the mid-1980s [90]. Since then academic interest in the field has increased rapidly [91]. More recently there has been significant commercial interest [92]. This has been primarily driven by an increased focus on energy demand combined with significant reductions in the cost of sensing technology, and equally, improvements in machine learning algorithms. Commercial interest is directly linked with the huge commercial potential of services that exploit the smart metering infrastructure, for example, in health, energy management, and climate change.

3.4 *Electrical Device Types*

Electrical appliances, alongside their normal on-off states, run in multiple modes. Many devices have low power requirements or standby modes, while appliances like ovens operate using several control functions. Understanding different device categories is important in NILM, as they define different electrical usage characteristics. Device categories include Type 1, Type 2, Type 3, and Type 4. The associated signals for each are illustrated in Fig. 3.

The characteristics for each appliance type are described as:

- Type 1 devices are either on or off. Examples include kettles, toasters, and lighting. Figure 4 illustrates a power reading for a kettle—(a) shows a series of devices being used in conjunction or in close succession; while (b) presents evenly distributed single device interactions.
- Type 2 devices, known as Multi-State Devices (MSD) or finite state appliances, operate in multiple states and have more complex behaviours than Type 1 devices. Devices include washing machines, dryers, and dishwashers.
- Type 3 devices, known as Continuously Variable Devices (CVD), have no fixed state. There is no repeatability in their characteristics and as such they are problematic in NILM. Example devices include power tools such as a drill or electric saw.

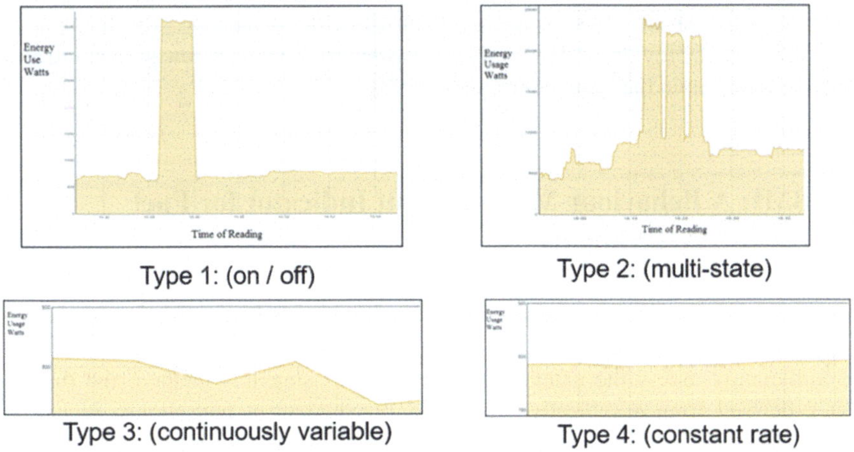

Fig. 3 Appliance type energy readings

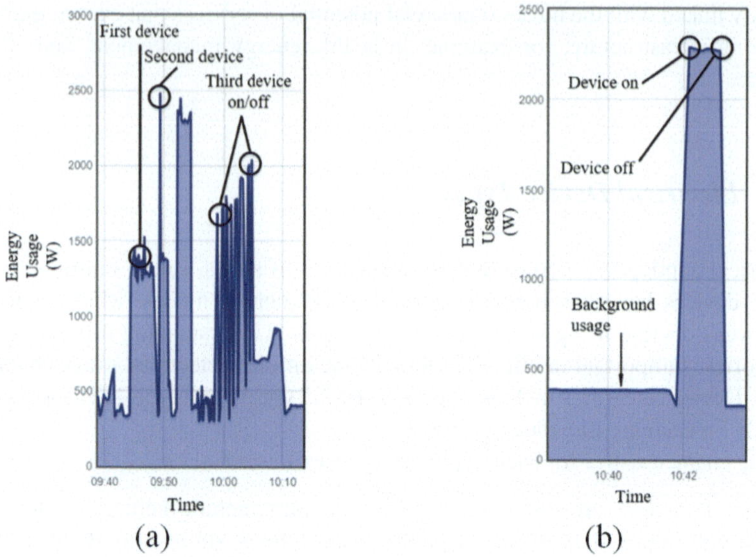

Fig. 4 Aggregated load readings highlighting unique device signatures

- Type 4 are fairly new in terms of device category. These devices are active for long periods and consume electricity at a constant rate—they are always on. Hence, there is no major events to detect other than small fluctuations. Such devices include smoke detectors and intruder alarms.

Understanding device types is important in any load disaggregation system, as electrical appliances are often used in combination, typically when preparing meals. This can affect the performance in classification tasks due to the boundaries that exist between device classes, making them difficult to identify. The boundaries between classes provide guidance on what classifiers to use (i.e. linear, quadratic, or polynomial) within the same feature space [93].

4 BMI: A Behaviour Measurement Indicator for Fuel Poverty Assessments

Measuring and monitoring household fuel poverty is challenging as we have seen [40]. Expenditure-based approaches lack data on all the contributing factors needed to sufficiently assess the extent of fuel poverty. Using this method, the data is often derived from a subjective and generalized view of households, including their occupants and how energy is consumed. In fact, data is often skewed or contaminated given that households may not even identify themselves as being in fuel poverty due to pride [94]. The remainder of this chapter proposes a different point of view that incorporates personalized household behaviour monitoring using activities of daily living. By doing this it is possible to understand the unique characteristics of each household in terms of what, when, and how often electrical appliances are used. The hope is to derive some useful insights and provide a more objective measure of fuel poverty from a socio-behavioural view point to better support the occupants and their energy needs.

4.1 BMI Framework

The Behaviour Measurement Indicator (BMI) proposed was initially developed and evaluated in partnership with Mersey Care NHS Foundation Trust to measure appliance usage in dementia patients and derive routine behaviours for social care support [83, 95]. Here we consider an extension to the existing framework and build on the behavioural monitoring aspects of the system to provide a household BMI indicator for fuel poverty assessment.

The BMI builds on the existing smart meter infrastructure. Smart meters in households, paired with a CAD using the ZigBee Smart Energy Profile (SEP) [96], provide access to aggregated power usage readings every 10 s. This sample frequency allows high-powered appliances associated with ADLs to be detected and used to establish household behavioural routines. Appliances such as a kettle, microwave, washing machine, and oven are regarded as necessary appliances used by occupants to live a normal life (ADLs). Therefore, appliances such as TVs, mobile chargers, computers, and lighting are of limited interest as they do not contribute to ADL assessment, for example, TVs are often left on for background noise and provide no information about what an occupant in a household is doing [83].

The BMI operates in three specific modes in order to achieve this: device training mode, behavioural training mode, and prediction model.

- **In device training mode** power readings are obtained from the CAD and recorded to a data store. Readings alongside device usage annotations are used to train the machine learning algorithms to classify appliances from aggregated load readings. Features automatically extracted using a one-dimensional convolutional neural network (discussed in more detail later in the chapter) act as input vectors to a fully connected multi-layer perceptron (MLP) for device classification.
- **In behavioural training mode** features from device classifications are extracted to identify normal and abnormal patterns in behaviour. The features allow the system to recognize the daily routines performed by occupants in a household, including their particular habits and behavioural trends.
- **In prediction mode** both normal and abnormal household behaviours are detected and remediated.

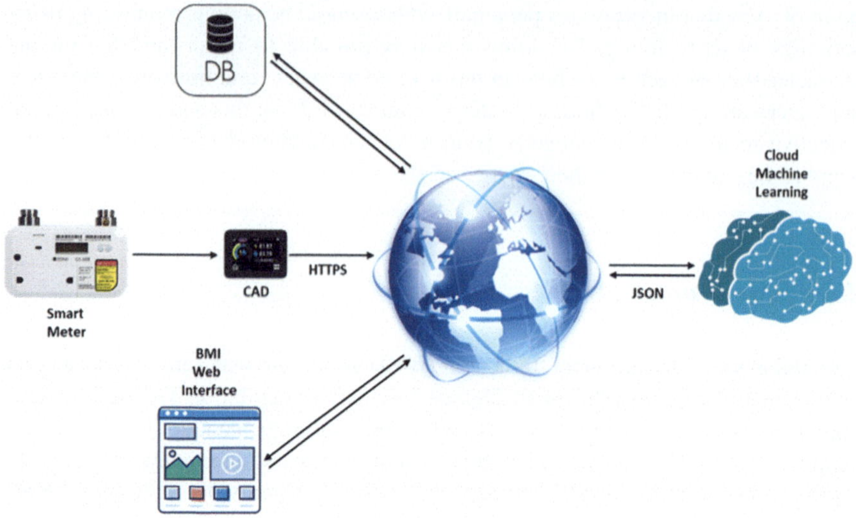

Fig. 5 System framework showing the end-to-end components

The framework implements web services for machine-to-machine communications using enterprise-ready protocols, Application Programming Interfaces (API's) and standards. The monitoring application interfaces with web services to receive real-time monitoring alerts about the household's status (i.e. green for normal behaviour, amber for unusual behaviour, and red when drastic changes occur). The complete end-to-end system is shown in Fig. 5.

4.2 Data Collection

The training dataset for device classification is constructed using energy monitors (i.e. a CAD paired with a household smart meter). CAD payload data contains the aggregated energy readings generated every 10 s. To detect ADLs, a kettle, microwave, washing machine, oven, and toaster are used, although others could be included if required, such as an electric shower depending on the relapse indicators of interest in fuel poverty.

Generating device signatures is achieved using a mobile app to record when each appliance is operated (annotation). Time-stamped recordings are compared with mobile app recordings to extract specific appliance signatures. Each signature is labelled and added to the training data and subsequently used to train the machine learning algorithms for appliance classification.

4.3 Data Pre-processing

CAD energy readings are filtered and transformed before they are used to train machine learning algorithms. A high-pass filter is implemented to remove background noise below 300 watts (although this value needs to be personalized based on individual household energy usage as each home will be different)—signals below this threshold typically represent Type 4 electrical appliances which cannot be detected using CAD data.

Device signatures are obtained by switching appliances on and off individually and filtering normal background noise. Individual appliance signatures are combined to generate new appliance usage patterns that represent composite appliance usage. For example, Fig. 6 shows that when the individual energy readings for three appliances (kettle, microwave, and toaster) are combined (i.e. they are operated in parallel) a 'Total Load' signature is produced.

The aggregated signature (total load) describes the three appliances being used in parallel. Repeating this process for all device combinations yields different aggregate signatures that describe which devices are on and which are not. Hence, a dataset is built containing individual and combined appliance usage signatures and used to train and detect which of the ADL appliances are in use.

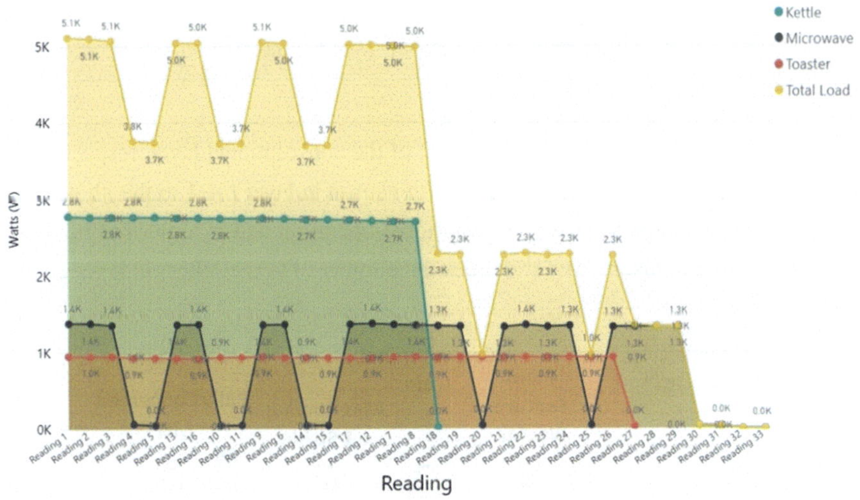

Fig. 6 Whole household aggregated power consumption and individual device power consumption

4.4 CAD NILM Machine Learning Model for Appliance Disaggregation

In contrast to manually extracted features based on input from domain knowledge experts (i.e. peak frequency and sample entropy), features can automatically learn from appliance energy signatures using a one-dimensional convolutional neural network (1DCNN) [97]. Appliance signatures are input directly to a convolutional layer in the 1DCNN. The convolutional layer detects local features along the time-series signal and maps them to feature maps using learnable kernel filters (features). Local connectivity and weight sharing are adopted to minimize network parameters and overfitting [98]. Pooling layers are implemented to reduce computational complexity and enable hierarchical data representations [98]. A single convolutional and pooling layer pair along with a fully connected MLP comprising two dense layers and softmax classifier output (an output for each appliance being classified) completes the 1DCNN network as the time-signals are not overly complex. The proposed architecture is represented in Fig. 7.

The network model is trained by minimizing the cost function using feedforward and backpropagation passes. The feedforward pass constructs a feature map from the previous layer to the next through the current layer until an output is obtained. The input and kernel filters of the previous layer are computed as follows:

$$z_j^l \sum_{l-1}^{M^{l-1}} 1dconv\left(x_i^{l-1}, k_{ij}^{l-1}\right) + b_j^l$$

where x_j^{l-1} and Z_j^l are the input and output of the convolutional layer, respectively, and k_{ij}^{l-1} the weight kernel filter from the *i*th neuron in layer $l-1$ to the *j*th neuron in layer l; $1dconv$ represents the convolutional operation and b_j^l describes the bias of the *j*th neuron in layer l. M^{l-1} defines the number of kernel filters in layer $l-1$.

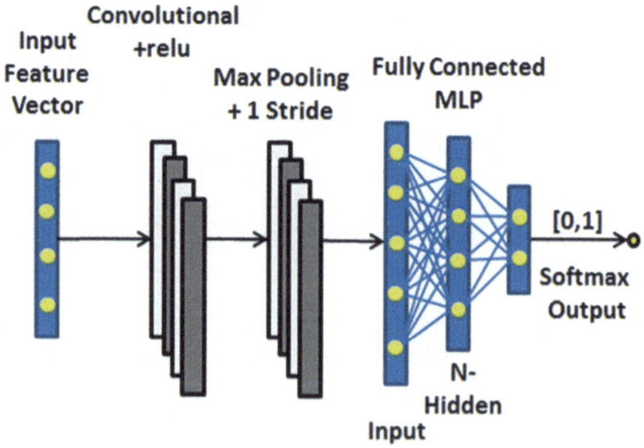

Fig. 7 One-dimensional convolutional neural network

A ReLU activation function is used for transforming the summed weights and is defined as:

$$x_j^l = ReLU\left(z_j^l\right)$$

where x_j^l is the intermediate output at current layer l before downsampling occurs. The output from current layer l is defined as:

$$y_j^l = downsampling\left(x_j^l\right)\ x_j^{l+1} = y_j^l$$

where *downsampling()* represents a max pooling function that reduces the number of parameters, and y_j^l is the output from layer l and the input to the next layer $l + 1$. The output from the last pooling layer is flattened and used as the input to a fully connected MLP. Figure 8 shows the overall process.

The error coefficient E is calculated using the predicted output y:

$$E = -\sum_n \sum_i (Y_{ni} \log (y_{ni}))$$

where Y_{ni} and y_{ni} are the target labels and the predicted outputs, and i the number of classes in the nth training set. The learning process optimizes the network's free parameters and minimizes E. The derivatives of the free parameters are obtained and the weights and biases are updated using the learning rate (η). To prompt rapid convergence, *Adam* is implemented as an optimization algorithm and *He* for weight

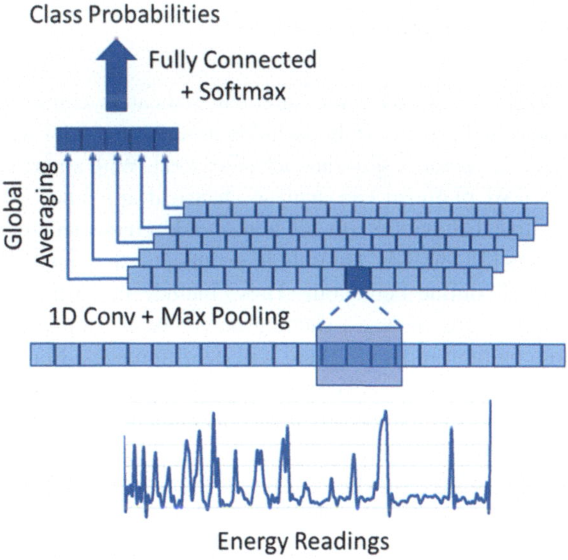

Fig. 8 Convolution and max pooling process

initialization. The weights and bias in the convolutional layer and fully connected MLP layers are updated using:

$$k_{ij}^{l} = k_{ij}^{l} - \eta \frac{\partial E}{\partial k_{ij}^{l}} b_{j}^{l} = b_{j}^{l} - \eta \frac{\partial E}{\partial b_{j}^{l}}$$

Small learning rates reduce the number of oscillations and allow lower error rates to be generated. Rate annealing and rate decay are implemented to address the local minima problem and control the learning rate change across all layers.

Momentum start and ramp coefficients are used to control momentum when training starts and the amount of learning for which momentum increases—momentum stable controls the final momentum value reached after momentum ramp training examples. Complexity is controlled with an optimized weight decay parameter, which ensures that a local optimum is found.

The number of neurons and hidden layers required to minimize E, including activation functions and optimizers, can be determined empirically. Input and hidden layers are also determined empirically depending on data and the number of softmax outputs required for classification. The network's free parameters can be obtained using the training and validation sets over a set number of epochs and evaluated with a separate test set comprising unseen data.

The 1DCNN approach allows the unique features from single appliance and composite appliance energy signatures to be automatically extracted and used in subsequent machine learning modelling for classification tasks. This removes the need for manual feature engineering and simplifies the data analysis pipeline.

4.5 Measuring Behaviour

Current fuel poverty measurement indicators cannot directly collect, monitor, or assess fuel poverty in households in real time. ADL is a term used in healthcare to assess a person's self-care activities [99]. With smart meters, CADs and 1DCNNs, the BMI platform can analyse electrical appliance interactions and detect ADLs (routine behaviours) in all households connected to the smart grid using smart meters [78–80, 84]. Household occupants carry out ADLs each day as part of their normal routine behaviour. These include preparing breakfast, lunch, and dinner, making cups of tea, switching on lights, and having a shower. While such tasks are common to us all, there will be differences. For example, one household may use the toaster to make toast for breakfast, while another might use the cooker to make porridge. Some might boil the kettle to make tea in the evening after finishing work, while others might prefer to have a glass of wine. Some households might use the shower (likely at different times of the day and frequency, i.e. one or two showers a day), while others might prefer to have a bath.

These activities can be easily detected through ongoing interactions with home appliances. This is useful for deriving normal routine behaviours within households, but more importantly to detect anomalies, for the purpose of safeguarding vulnerable homes against fuel poverty risks. How we interact and use energy in our home will likely be affected by our circumstances, i.e. having a baby, children moving out of the family home, gaining employment (or losing a job) as well as caring for an elderly family member who has moved in.

Such circumstantial changes directly alter our routine use of electrical appliances. For example, in the case of having a baby, the microwave, kettle, or oven hob may be operated throughout the night for a period of time to heat the milk required to bottle-feed babies. In the unfortunate situation where a person has lost their job, household occupants may have to substitute fresh food cooked using the oven and hob for more cheaper food options, such as microwave meals. These are clues that household circumstances have changed. Families experiencing financial difficulties may have to cut heating-based appliance usage and ration hot water—this will lead to an overall dip in energy consumed by that household.

Significant changes in behaviour will act as key indicators and facilitate decision-making strategies to support struggling households. For example, appliances operated during abnormal times of the day (when this is not normal behaviour for that household) may indicate that occupants are experiencing difficulties (i.e. making tea in the early hours of the morning could be due to sleep disturbances possibly caused through financial worry; conversely occupants staying in bed for longer periods of time or not cooking meals may indicate severe financial difficulty or energy disconnection issues). The BMI system can detect significant changes in behaviour like these as we see in the next section.

4.5.1 Vectors for Behavioural Analysis

Individual device detections classified by the CAD NILM machine learning model are combined as feature vectors for behaviour analysis. Predicted classes are given a unique device ID and assigned to an observation window depending on the time of day the appliance is used, i.e. during breakfast or evening meal times.

Observation windows can be defined and adjusted to meet the unique behaviours of each household. This is performed automatically following a baseline learning period for each household connected to the smart grid. Observation windows capture routine behaviour and act as placeholders for the fuel poverty relapse indicators being measured and monitored (these will need to be defined by fuel poverty experts). This allows the system to construct a personalized representation of each household and assign device usage to specific observation windows common to that household. Continually repeating this process allows routine behaviours to be identified and anomalies in behaviour to be detected. Figure 9 describes seven possible observation windows in a 24-h period. Each observation window is configurable to meet the unique needs of the application or service.

The order of device interactions is not necessarily important unless there is a clear deviation from normal behaviour. From the behaviour vectors it is possible to see the degree of correlation between appliance usage and the hour-of-day (strong routine behaviour). Figure 10 shows the correlations for different home appliances used over a 6-month period [100].

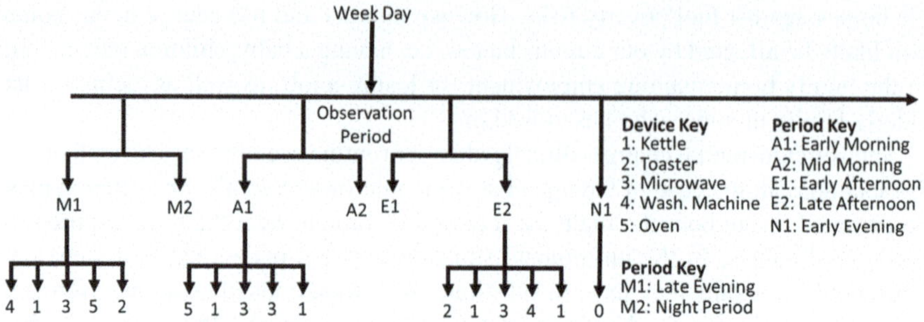

Fig. 9 Device assignment for identifying key activities within significant observation periods

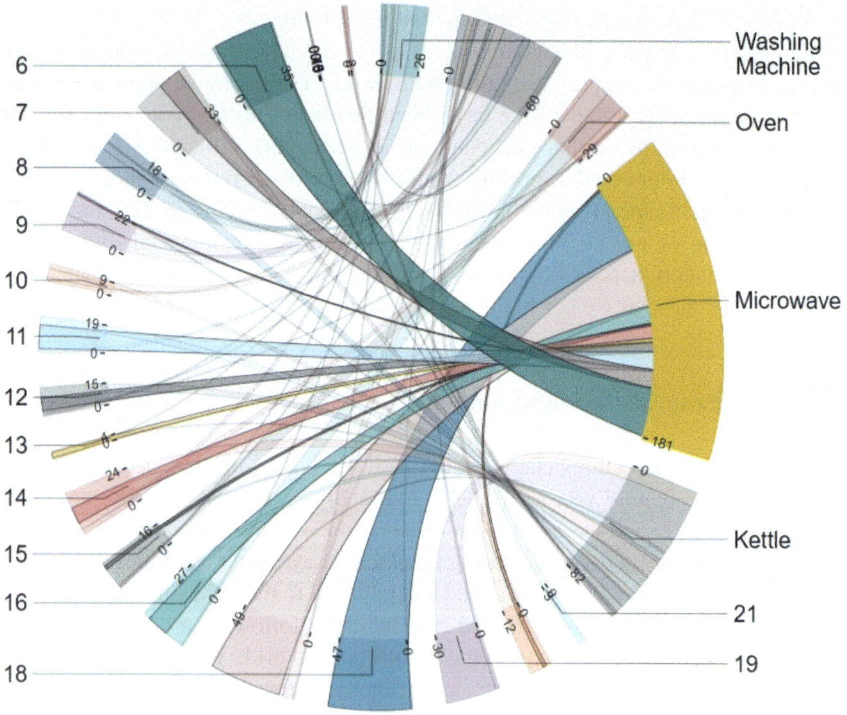

Fig. 10 Degree of correlation between device usage and hour

The figure shows quantitative information relating to flows, including relationships and transformations. The lines between appliances and time-of-day, like ant pheromone trails, show the established routine behaviour for a particular home. For example, it is possible to see that the microwave is mostly used at 06:00 h and 18:00 h. Alternations in either link proportionality or association may indicate the early signs of circumstantial change which could be linked to fuel poverty risk factors. Anomalies are progressed through a traffic light system—red would

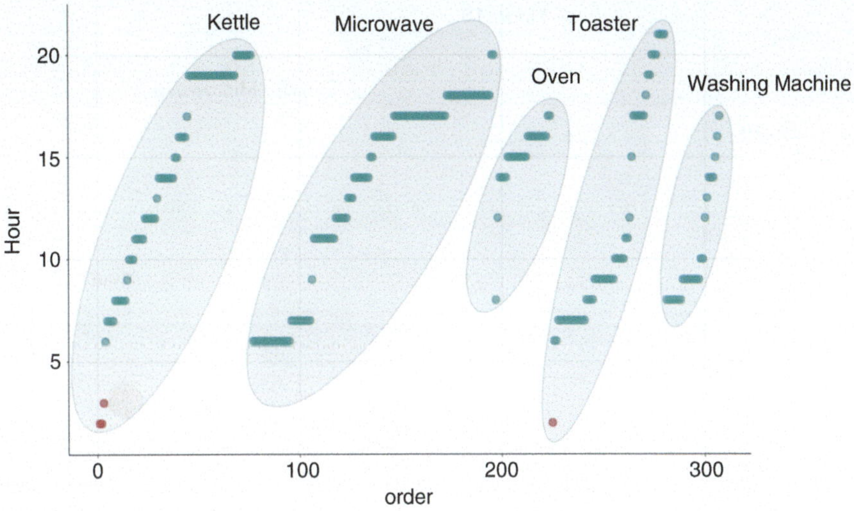

Fig. 11 Sleep disturbances for an occupant using Z-score anomaly detection

suggest a sustained change in routine behaviour over a period of time (time period would be set by expert in fuel poverty) and may or may not indicate that the house is experiencing financial difficulties. Conversely, green would show that normal routine behaviour has been observed and that no support or intervention is required. Amber would flag the house as worrisome (this does not necessarily mean the house is transitioning into a fuel poverty state, simply that a change in behaviour has been detected). This could be caused by circumstantial changes, i.e. people coming to stay or household occupants going on holiday. Viewing Fig. 6 periodically we would expect to see changes between correlations and their associated strengths for those households experiencing significant changes in normal routine behaviours.

Anomalies in device usage can be seen with the Z-score technique to describe data points in terms of their relationship to the mean and the standard deviation of a group of points. Figure 7 shows the inliers in green which represent normal appliance interactions for that household. Each cluster represents a specific appliance class. The outliers are depicted in red where both the kettle and toaster classes in this case reside outside the household's normal routine behaviour. Figure 11 shows that in total three kettles were used on three separate occasions between the hours of 00:00 and 05:00 and a single interaction with a toaster was detected during the same observation period. In the context of fuel poverty such results may provide interesting insights when managing fuel poverty households. As the household continues to struggle financially, we would expect routine behaviour to become more erratic (or even disappear for long periods) leading to an increase in the number of anomalies detected.

The BMI framework presents the first platform of its kind that capitalizes on the smart meter infrastructure to describe a behaviour measurement indicator for

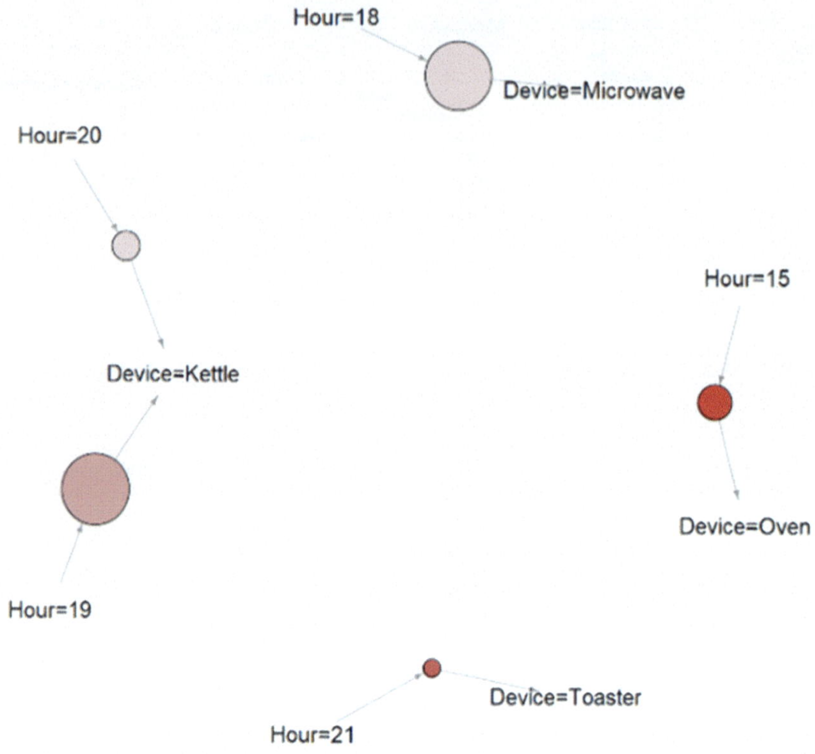

Fig. 12 Association rule mining for the identification of behavioural patterns

use in fuel poverty assessments. It has been designed to exploit the smart metering infrastructure and provide foundational services to more accurately assess fuel poverty in real time within individual households [77]. Obviously, future trials are required to test the applicability of the BMI system and evaluate whether it has any real potential in tackling fuel poverty. Based on our previous use of the system in dementia, the technology is a powerful tool for assessing routing behaviour and detection anomalies. We therefore think the solution will lend itself to household behaviour analysis (in terms of electricity consumption) in fuel poverty assessment [83].

The use of association rule mining within load disaggregation is also an interesting technique that can uncover relationships and their associated strengths using transactional data. Identifying device relationships (what devices are commonly used together or in sequence) and their relationship with the time of day can expose strong behavioural traits within the dwelling. Reoccurring deviation from identified routine patterns or the weakening of common relationships could be used to trigger an intervention where fuel poverty is suspected. Figure 12 highlights the use of association rule mining to determine the relationship strength between an appliance and time of day.

Association rule mining can be used to provide a more abstracted view above and beyond the aggregated load level of a dwelling. Instead, the collective behaviour of entire regions could be monitored to assess the impact of shifting financial and social economic changes, for example, raising fuel prices or the closure of large employers (retail/manufactures) and reduction in the associated foot flow to a region. By using association rule mining the impact can be objectively measured and the effectiveness of any intervention/recovery passively monitored.

5 Discussion

As this chapter has highlighted, fuel poverty affects a significant number of households in Europe and indeed globally. The problem is primarily caused by a combination of low income, high energy costs, and energy-inefficient homes. In the UK, four million households are currently in fuel poverty, which, among other things, contributes to poor health and premature winter deaths. Poor-quality housing has also been linked with fuel poverty which is hardly surprising given that a substantial share of the residential stock in Europe is older than 50 years.

The problem is recognized by governments; however, the EU has not yet adopted a common definition of fuel poverty, nor a set of common indicators to measure it, making a standardized approach difficult to implement. Many households move in and out of fuel poverty but there are households that find themselves persistently trapped in fuel poverty [101]. Measuring and monitoring fuel poverty is challenging as we have seen [40], and while Expenditure-based approaches have been proposed, they lack data on all the contributing factors needed to sufficiently assess fuel poverty. Consensus-based approaches on the other hand have data, but this is only from snap shots in time, meaning data is often outdated, subjective, and exclusive in nature.

Of the 178 measurement indicators reported in the literature, many do not respond to variations in income, circumstantial changes, fuel prices, or energy efficiency improvements. They exclude low-income and single-person households [59, 60] and this has distorted policy choices, and misrepresented the problem. Against this negative backdrop and an overall distrust of government bodies and energy providers, fuel-poor customers feel that the intensity of the issue is not fully understood by those developing policies to combat it [56].

We proposed the BMI system to monitor a household's activities of daily living and understand routine behaviour in order to gain insights into how energy is consumed [78–80, 84]. Households behave in different ways. While there may be common tasks, such as meal preparation, there will be differences. By detecting ADLs using appliance interactions, it is possible to derive routine behaviour for each household. This makes BMI highly personalized and sensitive to the unique characteristics of each household connected to the smart grid.

Changes in behaviour can be identified and investigated and support services provided if and when they are needed. Modelling ADLs in households will allow the onset of fuel poverty issues to be identified much earlier. When households are identified, appropriate packages can be put in place to help mitigate the adverse effects fuel poverty has on fuel-poor occupants. Detecting self-disconnect in households, particularly among the most vulnerable in society, such as young children and the elderly, would allow appropriate support services to be put in place to ensure homes are appropriately warm.

The identification of expected behaviour and relapse indicators aids in the selection of appropriate analytical techniques. Establishing routines facilitates the detection of abnormal behaviour. Combining this with unique energy signatures within each household a new and foundational fuel poverty indicator is possible that is adaptable and reflective of household circumstances. We believe that the BMI system could contribute significantly to the fuel poverty domain. To the best of our knowledge BMI is the first of its kind as currently there is no fuel poverty measurement indicator that can measure household energy usage interactions and derive routine behaviour in every home fitted with a smart meter. The approach is highly personalized and closely aligned with the different routines households exhibit despite the size of the house or the number of occupants. Once routine behaviour has been established, BMI is highly sensitive to change; using a traffic light system it is therefore possible to target and support households classified as being fuel poor.

6 Conclusions

This chapter discussed the many aspects of fuel poverty and the government policies put in place to combat it. The key message is that cold homes waste energy and harm their occupants. Most fuel-poor indicators are derived from generalized estimates disconnected from the unique characteristics of individual households. Houses and occupants do not behave the same—they have their own socio-behavioural characteristics that affect how and when energy is consumed. Therefore, coupled with the household envelope and the many other factors that influence household behaviours, there is a disparity between existing measurement indicators and fuel poverty prevalence.

The only way to fully understand fuel poverty is to measure high-risk households and the unique characteristics and behaviours they exhibit in terms of energy consumption and ADLs. We believe that the BMI system can do this with minimal installation requirements as the solution exploits the existing smart meter infrastructure to provide appropriate services. System operation requires no input from household occupants as BMI is based on the assessment of ADLs (the everyday things that people do in their home in order to survive) captured through normal appliance interactions.

The BMI has been previously evaluated in a clinical trial with Mersey Care NHS Foundation Trust to model the ADLs of dementia patients [83]. However, it has been possible to extend the system to include fuel poverty risk factors following minor changes to observation periods and fuel poverty related relapse indicators. Future work will focus on a trial to evaluate the BMI system in fuel and non-fuel poverty homes. Cases will include households that find themselves in and out of fuel poverty. Controls will be those households that have not previously experienced fuel poverty or had difficulties with paying bills and keeping their home warm. The measurable outputs will be to evaluate whether the BMI system can detect which houses are in or likely to be in fuel poverty and those that are not.

To the best of our knowledge this is the first fuel poverty measurement indicator that builds on the existing smart meter infrastructure and associated CAD technology to carry out NILM and personalized ADL monitoring in every household connected to the smart grid that is designed to safeguard households and occupants against fuel poverty.

Acknowledgements This work was inspired by an event run in Liverpool where the authors were invited to present at the 'Better at Home' workshop run by the National Energy Action (NEA) organization. We would like to particularly thank Matt Copeland and Dr. Jamie-Leigh Rosenburgh at NEA who asked whether we could extend our dementia smart meter framework to include a behaviour measurement indicator for fuel poverty.

References

1. B. Boardman, Fuel poverty is different. Policy Stud. **12**(4), 30–41 (1991)
2. NEA, NEA Action for Warmer Homes, Warm and Safe Homes Advice (2020), https://www.nea.org.uk/. Accessed 3 Jan 2020
3. S. Bouzarovski, Energy poverty in the European Union: Landscapes of vulnerability. Wiley Interdiscip. Rev. Energy Environ. **3**(3), 276–289 (2014)
4. EPEE, *European Seminar of 8 October 2009: Common Initiatives to Combat Fuel Poverty* (Brussels, Belgium, 2009)
5. C. Waddams Price, K. Brazier, W. Wang, Objective and subjective measures of fuel poverty. Energy Policy **49**, 33–39 (2012)
6. C. Jensen, *Energy Demand Challenges in Europe* (Palgrave Macmillan, 2019)
7. Committee on Fuel Poverty, Committee on Fuel Poverty: Third Annual Report, Nov 2018
8. S. Pye, A. Dobbins, C. Baffert, J. Brajković, P. Deane, R. De Miglio, Energy poverty across the EU: Analysis of policies and measures. Analysis of policies and measures. Eur. Energy Transit. Insights Policy Mak. **1**, 261–280 (2017)
9. J. Rosenow, R. Platt, B. Flanagan, Fuel poverty and energy efficiency obligations—A critical assessment of the supplier obligation in the UK. Energy Policy **62**, 1194–1203 (2013)
10. C. Skopeliti, Britons paying 40% more for energy than in 2015, analysis reveals, The Guardian (2019), https://www.theguardian.com/money/2019/dec/30/britons-paying-40-more-for-energy-than-in-2015-analysis-reveals. Accessed 5 Jan 2020
11. M. Ioannidou, D. Mantzari, The UK domestic gas electricity (Tariff cap) act: Re-regulating the retail energy market. Mod. Law Rev. **82**(3), 488–507 (2019)
12. FPA, Fuel Poverty Action (2018), https://www.fuelpovertyaction.org.uk/home-alternative/media/. Accessed 5 Jan 2020
13. S. Nicol, M. Roys, D. Ormandy, V. Ezratty, *The Cost of Poor Housing in the European Union* (2014), p. 76

14. J. Sousa, L. Bragança, M. Almeida, P. Silva, Research on the Portuguese building stock and its impacts on energy consumption—An average U-value approach. Arch. Civ. Eng. **59**(4), 523–546 (2013)
15. BPIE, *Europe's buildings under the microscope, Buildings Performance Institute Europe* (2011)
16. G.B. Murphy, M. Kummert, B.R. Anderson, J. Counsell, A comparison of the UK Standard Assessment Procedure and detailed simulation of solar energy systems for dwellings. J. Build. Perform. Simul. **4**(1), 75–90 (2011)
17. C. Liddell, C. Guiney, Living in a cold and damp home: Frameworks for understanding impacts on mental well-being. Public Health **129**(3), 191–199 (2015)
18. J. Bosch, L. Palència, D. Malmusi, M. Marí-Dell'Olmo, C. Borrell, The impact of fuel poverty upon self-reported health status among the low-income population in Europe. Hous. Stud. **34**(9), 1377–1403 (2019)
19. J.D. Healy, J.P. Clinch, 04/02182 Quantifying the severity of fuel poverty, its relationship with poor housing and reasons for non-investment in energy-saving measures in Ireland. Fuel Energy Abstr. **45**(4), 299 (2004)
20. P. Howden-Chapman et al., Effects of improved home heating on asthma in community dwelling children: Randomised controlled trial. BMJ **337**(7674), 852–855 (2008)
21. C.D. Maidment, C.R. Jones, T.L. Webb, E.A. Hathway, J.M. Gilbertson, The impact of household energy efficiency measures on health: A meta-analysis. Energy Policy **65**, 583–593 (2014)
22. M. Marmot, Marmot Review Team. The Health Impacts of Cold Homes and Fuel Poverty (2011), https://www.foe.co.uk/sites/default/files/downloads/cold_homes_health.pdf
23. Z. Kmietowicz, GPs should identify and visit people at risk from cold homes, says NICE. BMJ **350**, 1–2 (2015)
24. G. O'Meara, A review of the literature on fuel poverty with a focus on Ireland. Soc. Indic. Res. **128**(1), 285–303 (2016)
25. P. Howden-Chapman, H. Viggers, R. Chapman, K. O'Sullivan, L. Telfar Barnard, B. Lloyd, Tackling cold housing and fuel poverty in New Zealand: A review of policies, research, and health impacts. Energy Policy **49**, 134–142 (2012)
26. E3G, UK has sixth-highest rate of excess winter deaths in Europe (E3G, 2020), https://www.e3g.org/news/media-room/uk-has-sixth-highest-rate-of-excess-winter-deaths-in-europe. Accessed 4 Jan 2020
27. S. Hajat, R.S. Kovats, K. Lachowycz, Heat-related and cold-related deaths in England and Wales: Who is at risk? Occup. Environ. Med. **64**(2), 93–100 (2007)
28. ONS, Excess winter mortality in England and Wales: 2016 to 2017 (provisional) and 2015 to 2016 (final). Stat. Bull. Off. Natl. Stat. **2018**, 1–23 (2018)
29. R. Jevons, C. Carmichael, A. Crossley, A. Bone, Minimum indoor temperature threshold recommendations for English homes in winter—A systematic review. Public Health **136**, 4–12 (2016)
30. J. Teller-Elsberg, B. Sovacool, T. Smith, E. Laine, Fuel poverty, excess winter deaths, and energy costs in Vermont: Burdensome for whom? Energy Policy **90**, 81–91 (2016)
31. X. Bonnefoy, Inadequate housing and health: An overview. Int. J. Environ. Pollut. **30**(3–4), 411–429 (2007)
32. C. Liddell, C. Morris, Fuel poverty and human health: A review of recent evidence. Energy Policy **38**(6), 2987–2997 (2010)
33. IHE, *Fuel Poverty and Cold Home-related Health Problems* (2014), pp. 10–15
34. C. Hughes, S. Natarajan, 'The Older I Get, the Colder I Get'—Older People's Perspectives on Coping in Cold Homes. J. Hous. Elderly **33**(4), 337–357 (2019)
35. L. S. of H. and T. M. PHE, UCL, Excess winter deaths Review 1 (2014)
36. D. Buck, S. Gregory, Housing and Health Opportunities for sustainability and transformation partnerships. Kings Fund, 1–38 (2018)
37. WHO, *WHO Guidelines for Indoor Air Quality: Dampness and Mould* (Denmark, 2009)

38. J. Rosenow, R. Cowart, E. Bayer, M. Fabbri, Assessing the European Union's energy efficiency policy: Will the winter package deliver on 'Efficiency First'? Energy Res. Soc. Sci. **26**, 72–79 (2017)
39. BEIS, Fuel Poverty Strategy for England, Government, https://www.gov.uk/government/consultations/fuel-poverty-strategy-for-england. Accessed 5 Jan 2020
40. K. Rademaekers et al., *Selecting Indicators to Measure Energy Poverty* (2016)
41. P. Heindl, Measuring fuel poverty: General considerations and application to German Household Data. SSRN Electron. J. **13** (2013)
42. Department of Energy & Climate Change, *Fuel Poverty Statistics* (Fuel Poverty Stat, 2015)
43. C. Robinson, S. Bouzarovski, S. Lindley, 'Getting the measure of fuel poverty': The geography of fuel poverty indicators in England. Energy Res. Soc. Sci. **36**, 79–93 (2018)
44. J. Hills, Getting the measure of fuel poverty. Final Report of the Fuel Poverty Review. Centre for the Analysis of Social Exclusion (2012), p. 19
45. M. Lis, A. Miazga, K. Sałach, Location, location, location: What accounts for the regional variation of energy poverty in Poland? Energy Poverty Vulnerability A Glob. Perspect., 119–140 (2017)
46. I. Imbert, P. Nogues, M. Sevenet, Same but different: On the applicability of fuel poverty indicators across countries—Insights from France. Energy Res. Soc. Sci. **15**, 75–85 (2016)
47. J.I. Levy, Y. Nishioka, J.D. Spengler, The public health benefits of insulation retrofits in existing housing in the United States. Environ. Heal. A Glob. Access Sci. Source **2**, 1–16 (2003)
48. P. Townsend, *Poverty in United Kingdom* (1970)
49. J. Mack, S. Lansley, Poor Britain **305**(6851) (1985)
50. D. Gordon et al., *Poverty and Social Exclusion in Britain* (Joseph Rowntree Foundation, York)
51. M. Iacovou, H. Levy, Using EU-SILC data for cross-national analysis: Strengths, problems and recommendations. Inst. Soc. Econ. Res. **2012**, 1–21 (2012)
52. H. Thomson, C. Snell, C. Liddell, Fuel poverty in the European Union: A concept in need of definition? People Place and Policy Online **10**(1), 5–24 (2016)
53. K. Krell, J.R. Frick, M.M. Grabka, Measuring the consistency of cross-sectional and longitudinal income information in EU-SILC. Rev. Income Wealth **63**(1), 30–52 (2017)
54. T. Goedemé, How much confidence can we have in EU-SILC? Complex sample designs and the standard error of the Europe 2020 Poverty Indicators. Soc. Indic. Res. **110**(1), 89–110 (2013)
55. E. Fahmy, D. Gordon, D. Patsios, Predicting fuel poverty at a small-area level in England. Energy Policy **39**(7), 4370–4377 (2011)
56. H. Thomson, S. Bouzarovski, C. Snell, Rethinking the measurement of energy poverty in Europe: A critical analysis of indicators and data. Indoor Built Environ. **26**(7), 879–901 (2017)
57. B. Boardman, Fuel poverty synthesis: Lessons learnt, actions needed. Energy Policy **49**, 143–148 (2012)
58. R. Moore, Definitions of fuel poverty: Implications for policy. Energy Policy **49**, 19–26 (2012)
59. L. Middlemiss, R. Gillard, *How Can You Live Like That?: Energy Vulnerability and The Dynamic Experience of Fuel Poverty in the UK* (2013)
60. R. Walker, P. McKenzie, C. Liddell, C. Morris, Estimating fuel poverty at household level: An integrated approach. Energ. Buildings **80**, 469–479 (2014)
61. S.M. Hall, Energy justice and ethical consumption: Comparison, synthesis and lesson drawing. Local Environ. **18**(4), 422–437 (2013)
62. L. Middlemiss, A critical analysis of the new politics of fuel poverty in England. Crit. Soc. Policy **37**(3), 425–443 (2017)
63. BEIS, *UK Energy Statistics, 2016 & Q4 2016* (2017)
64. OECD/IEA, *Key World Energy Statistics 2016, International Energy Agency, Paris* (International Energy Agency, Paris, 2016), pp. 1–77

65. N. Mogles et al., How smart do smart meters need to be? Build. Environ. **125**, 439–450 (2017)
66. D. Niyato, P. Wang, Cooperative transmission for meter data collection in smart grid. IEEE Commun. Mag. **50**(4), 90–97 (2012)
67. R. Razavi, A. Gharipour, M. Fleury, I.J. Akpan, Occupancy detection of residential buildings using smart meter data: A large-scale study. Energ. Buildings **183**, 195–208 (2019)
68. C. Chalmers, W. Hurst, M. Mackay, P. Fergus, Smart meter profiling for health applications. Proc. Int. Jt. Conf. Neural Networks **2015**, 1–7 (2015)
69. J.A. Momoh, in *Smart Grid Design for Efficient and Flexible Power Networks Operation and Control*, 2009 IEEE/PES Power Systems Conference and Exposition, PSCE 2009 (2009), pp. 1–8
70. IEA, Will a Smarter Grid Lead to Smarter End Users—or Vice versa: Smart Grid Applications at End-User Points (2015), pp. 1–9
71. International Energy Agency, *2018 World Energy Outlook: Executive Summary* (OECD/IEA, 2018), p. 11
72. B. Stephen, S.J. Galloway, Domestic load characterization through smart meter advance stratification. IEEE Trans. Smart Grid **3**(3), 1571–1572 (2012)
73. J.S. Chou, N.T. Ngo, Smart grid data analytics framework for increasing energy savings in residential buildings. Autom. Constr. **72**, 247–257 (2016)
74. P.D. Diamantoulakis, V.M. Kapinas, G.K. Karagiannidis, Big data analytics for dynamic energy management in smart grids. Big Data Res. **2**(3), 94–101 (2015)
75. Y. Zhang, T. Huang, E.F. Bompard, Big data analytics in smart grids: A review. Energy Informatics **1**(1), 1–24 (2018)
76. T. Wilcox, N. Jin, P. Flach, J. Thumim, A big data platform for smart meter data analytics. Comput. Ind. **105**, 250–259 (2019)
77. W. Hurst, C. Chalmers, M. Mackay, P. Fergus, Analysing Energy/Utility Usage, US 2019/0180389 A1, 2019
78. BEIS, *Unlocking the Future* (2018)
79. BEIS, *Smart Meter Rollout—Cost Benefit Analysis 2019* (2019)
80. Smart Energy GB, *Smart Meters. Paving the Way to a Smarter Future* (2020)
81. Smart Energy GB, *The Smart E-Home of the Future*
82. V. Giordano, G. Fulli, A business case for smart grid technologies: A systemic perspective. Energy Policy **40**(1), 252–259 (2012)
83. C. Chalmers, P. Fergus, C. A. C. Montanez, S. Sikdar, F. Ball, B. Kendall, *Detecting Activities of Daily Living and Routine Behaviours in Dementia Patients Living Alone Using Smart Meter Load Disaggregation* (2019)
84. M. Fell, H. Kennard, G. Huebner, M. Nicolson, S. Elam, and D. Shipworth, "Energising health: A review of the health and care applications of smart meter data," 2017
85. L. Yu, H. Li, X. Feng, J. Duan, Nonintrusive appliance load monitoring for smart homes: Recent advances and future issues. IEEE Instrum. Meas. Mag. **19**(3), 56–62 (2016)
86. J. Zhang, X. Chen, W.W.Y. Ng, C.S. Lai, L.L. Lai, New appliance detection for nonintrusive load monitoring. IEEE Trans. Ind. Informatics **15**(8), 4819–4829 (2019)
87. T. Petrovic, H. Morikawa, in *Active Sensing Approach to Electrical Load Classification by Smart Plug*, 2017 IEEE Power Energy Soc. Innov. Smart Grid Technol. Conf. ISGT 2017 (2017)
88. O. Hamid, M. Barbarosou, P. Papageorgas, K. Prekas, C.T. Salame, Automatic recognition of electric loads analyzing the characteristic parameters of the consumed electric power through a Non-Intrusive Monitoring methodology. Energy Procedia **119**, 742–751 (2017)
89. G.W. Hart, Nonintrusive appliance load monitoring system. Proc. IEEE **80**(12), 1870–1891 (1992)
90. G.W. Hart, *Prototype Nonintrusive Appliance Load Monitor* (1985), pp. 1–170
91. D. G.-U. Daniel Precioso, E. Alba, P.J. Zufiria, *Non-Intrusive Load Monitoring Poster* (2019)
92. Sustainability First, *Energy for All—Innovate for All* (2017)
93. C.M. Bishop, *Pattern Recognition and Machine Learning*, 1st edn. (Springer Berlin Heidelberg, New York, NY, 2008)

94. R. Galvin, Letting the Gini out of the fuel poverty bottle? Correlating cold homes and income inequality in European Union countries. Energy Res. Soc. Sci. **58**, 101255 (2019)
95. Mersey Care NHS Foundation Trust, *Unlocking the Social Benefits of Smart Meters for Uses in Assistive Living and Healthcare* (Article, 2019)
96. C. Gezer, C. Buratti, A ZigBee smart energy implementation for energy efficient buildings. IEEE Veh. Technol. Conf., 1–5 (2011)
97. J. Oh, J. Wang, and J. Wiens, in *Learning to Exploit Invariances in Clinical Time-Series Data using Sequence Transformer Networks*, Proceedings of Machine Learning (2018), pp. 1–15
98. T. Brosch, Y. Yoo, L.Y.W. Tang, R. Tam, *Deep Learning of Brain Images and Its Application to Multiple Sclerosis*, vol 1 (Elsevier, 2016)
99. A. Fleury, M. Vacher, N. Noury, SVM-based multimodal classification of activities of daily living in health smart homes: Sensors, algorithms, and first experimental results. IEEE Trans. Inf. Technol. Biomed. **14**(2), 274–283 (2010)
100. A. Ng, Sparse autoencoder. CS294A Lect. Notes **72**, 1–19 (2011)
101. D. Roberts, E. Vera-Toscano, E. Phimister, Fuel poverty in the UK: Is there a difference between rural and urban areas? Energy Policy **87**, 216–223 (2015)

Standards and Technologies from Building Sector, IoT, and Open-Source Trends

Benoit Delinchant and Jérôme Ferrari

1 From Home Automation to Smart Home

In the 2010s, everything had to be smart. This qualifier can be applied to systems capable of capturing, communicating, and acting. In particular, in 2009, with the explosion of the theme of smart grids (Fig. 1), linked to the decentralization of the electricity network, a direct consequence of the need to improve the penetration of renewable energies, such as solar photovoltaics, distributed on the territory. The same goes for Smart Home, the use of the term listed by Google [1] takes off again during this period and is followed by the smart city which integrates energy, transport but also security issues with especially facial recognition systems.

It all began in the twentieth century with the introduction into the house of motorized equipment to relieve tedious tasks within the house such as washing clothes (1904), cooking (oven, refrigerator, robot ...), washing dishes, clean (vacuum cleaner, etc.), etc. This equipment introduces the concept of home automation or "home automation" which appears as the contraction of the Latin word domus (home) and the word robotics. But beyond this individual equipment, automating previously manual activities, home automation introduces the concept of automated and centralized control. The very first protocol for controlling home appliance was the X10 developed in 1975 by Pico Electronics of Glenrothes, Scotland. Its objective was to remotely operate the power supply to light fixtures or electrical equipment plugged into an outlet. Its principle is based on the power line carrier (Powerline Carrier Systems (PCS)). X10 is a technology still existing, which is also able to use other media such as radio frequency, but which retains characteristics that can be called archaic, sending for instance the signal three times to increase the probability of being received.

B. Delinchant · J. Ferrari (✉)
G2Elab, CNRS, Grenoble INP, University of Grenoble Alpes, Grenoble, France

S. Ploix et al. (eds.), *Towards Energy Smart Homes*,
https://doi.org/10.1007/978-3-030-76477-7_3

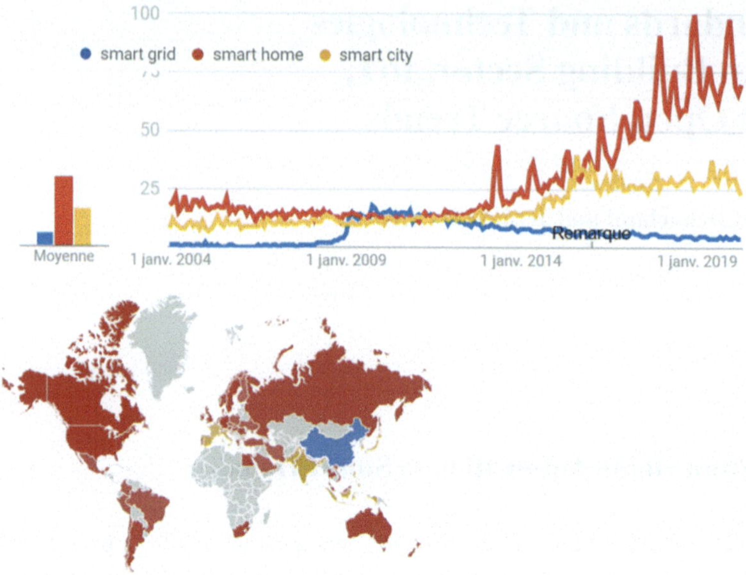

Fig. 1 Google trends for smart grid, smart home, and smart city (https://trends.google.com/trends/explore?date=all&q=smart%20grid,smart%20home,smart%20city)

In the early 1990s, home automation had to overcome a large number of challenges such as high costs at all levels, from development to maintenance, including manufacturing and installation [2]. In addition, their use was reserved for interested people mastering the technologies and agreeing to spend time on GUIs that are not very ergonomic.

In a more recent study [3], the authors always indicate a high overall cost, a complexity linked to incompatible equipment, a lack of reliability, and always a great difficulty of handling by an average user.

Associated with Building Automation, the term Intelligent Building appeared in the 1980s. The vision, and therefore the definition, that we attribute to intelligent building is largely dependent on the authors and their point of view, but we can also consider that they evolve over time. In particular, Wigginton and Harris have listed 30 different definitions [4]. The first were purely technological, such as in 1983 when Cardin refers to Intelligent Buildings as "Buildings which have fully automated building service control systems." Others have a broader vision which is part of a compromise between the well-being of the occupant and the optimization of resources. "A building that creates an environment that maximizes the efficiency of the occupants of the building while at the same time allowing effective management of resources with minimum life-time costs" [Intelligent Buildings International (IBC) cited by Wigginton and Harris [4]]. The latter definition is also used in numerous scientific publications on the intelligent building.

According to Ghaffarianhoseini [5] "Current definitions of IBs have gradually considered the users' interactions and even the social values of users [. . .] raise the idea that intelligent living environments must be aware of and responsive to their occupants' demands and activities".

Wilson in [6] is dividing research topics on smart homes into three categories:

1. **Functional**: considers the smart home as a way of better managing the daily living demands.
2. **Instrumental**: focuses on managing and reducing energy demand in households towards a low-carbon future.
3. **Socio-technical**: considers smart homes as the next wave of development regarding daily life digitization.

Responding to the well-being of the occupant is therefore a first challenge, and this requires an environment capable of capturing and acting according to the needs of the occupants and their well-being. Generally speaking, it is the interaction with the occupants which makes it possible to achieve the objectives of the smart building, which can go beyond the well-being of the occupant. This is particularly the case for the current objectives linked to the energy transition, to which this work is addressed.

2 Standards and Performance Indicators from Building Sector

2.1 Standards to Deploy Energy-Efficient Technologies

Directives and associated standards have been in place for several years aimed at massively deploying energy performance technologies in buildings. In 2018, the European Parliament updated directives 2010/31/EU on the energy performance of buildings and 2012/27/EU on energy efficiency. In particular, emissions from European buildings will have to be reduced by 80–95% by 2050, compared to 1990. This challenge can partly be met by the residential sector and monitoring, interaction, and piloting technologies.

The new 2018 directives on the energy performance of buildings (2010/31/EU) and energy efficiency (2012/27/EU) are translated into standards (see Fig. 2). Throughout this chapter, we will have the opportunity to cite a number of these standards and see how smart home technologies are adapted to them or not.

We can already evoke with Fig. 2 that a certified building (ISO 50001), that is to say having implemented an energy management system, is exempt from the obligation of energy audit (EN 16247). In France, for example, a company building with more than 250 people or a condominium with more than 50 dwellings must undergo an energy audit every 4 years. This is an exhaustive examination and analysis of the energy use and consumption of a site or a building, with the aim of identifying energy flows and potential for improving the energy efficiency and then report on it. Individual houses are subject to a much less restrictive Energy Performance Certificate (in French "DPE") at the time of sale. We could imagine that an equivalent system could be developed in smart home in order to lead to a continuous improvement in the energy performance of all housing.

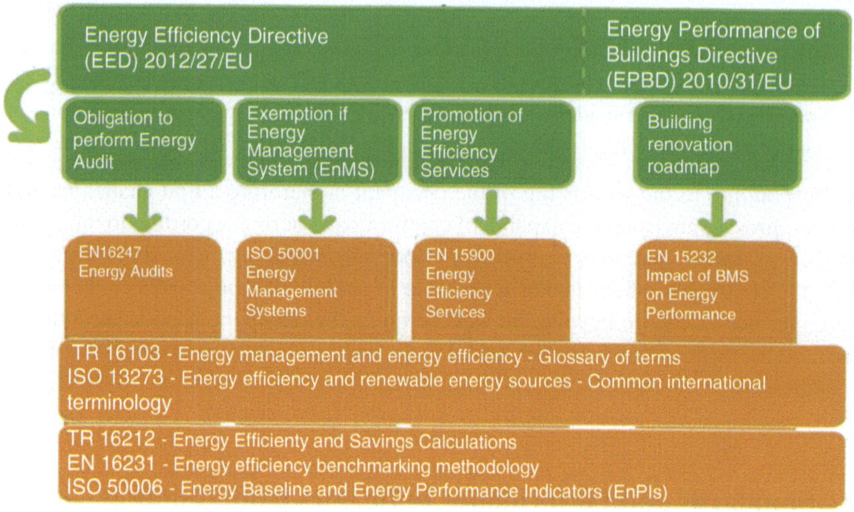

Fig. 2 Consistency between regulations and standardization—European Level. (Source: Obara [7])

In Sect. 3.2.3, we will discuss the energy management methodology defined in the ISO **50001** standard as well as the associated services that can be offered by Energy Service Company (ESCO) in the framework of EN 15900: 2010—Energy efficiency services.

Finally, we will more particularly develop energy management systems (EMS) and their performance defined in standard EN 15232.

2.2 *The Benefits of Energy Monitoring*

2.2.1 Measure and Verification (M&V), from Design to Real Performances

A delivered building does not operate efficiently immediately. The building and its equipment must be regulated and its occupants informed and accompanied during the first year of operation according to the uses and seasons. This year of learning the building is very important to avoid consumption drifts and/or user discomfort. The commissioning and monitoring of technical installations must not be left to chance. Only energy and technical An effective building monitoring will be able to identify deviations, anomalies and give indications to reduce or even eliminate them in the best of cases. This monitoring is all the more essential since it still happens that, despite the drifts observed and reported, certain problems remain unsolved 2 years after their commissioning. Good practice requires that M&V is well integrated into the process of identifying, developing, procuring, installing, and operating energy conservation measures (ECM).

There is often a distortion between the consumption assumptions, studied in design, and the actual consumption of a building in operation. The causes of these distortions can be numerous: deviations in particular related to the choice of materials and their implementation, the climate, the conditions of occupation or management of equipment, the varied behavior of users, etc.

Post Occupancy Evaluations (POE) based on questionnaires and on-site physical measurements is used to improve the ways that buildings are used to support productivity and well-being. Specifically it is used to account for building construction quality. According to Leitner [8], the most used evaluation methods are questionnaires and on-site physical measurements. The most frequently evaluated criteria were lighting, internal temperature and thermal comfort, and acoustic comfort. In Di Giuda [9] authors are exploring the potential application of IoT sensors and Machine Learning techniques to POE.

An effective building monitoring will reduce the majority of overconsumption, dysfunctions, and therefore the satisfaction of the occupants, but also improve knowledge of the systems in use and their daily operation. The monthly invoice control is the first monitoring tool to be implemented. However, this control is not sufficient to identify precisely the causes of possible overconsumption. To maintain the level of performance and the loads linked to the building, only real follow-up after reception makes it possible to identify and understand the origin of these distortions in order to reduce them quickly.

2.2.2 Energy Performance Contracting

In order to ensure that the performance defined in the design for a new construction or renovation will be achieved, an energy performance contract (EPC) can be established between a project owner and an operator in order to set an energy efficiency objective.

An international standard call International Performance Measurement and Verification Protocol (IPMVP®) defines best practice for quantifying the results of energy efficiency investments and increase investment in energy, demand management and renewable energy projects.

Indeed, energy savings cannot be directly measured, because savings represent the absence of energy consumption. Instead, savings are determined by comparing measured consumption before and after implementation of a program, making suitable adjustments for changes in conditions. The comparison of before and after energy consumption should be made on a consistent basis, using the following general equation:

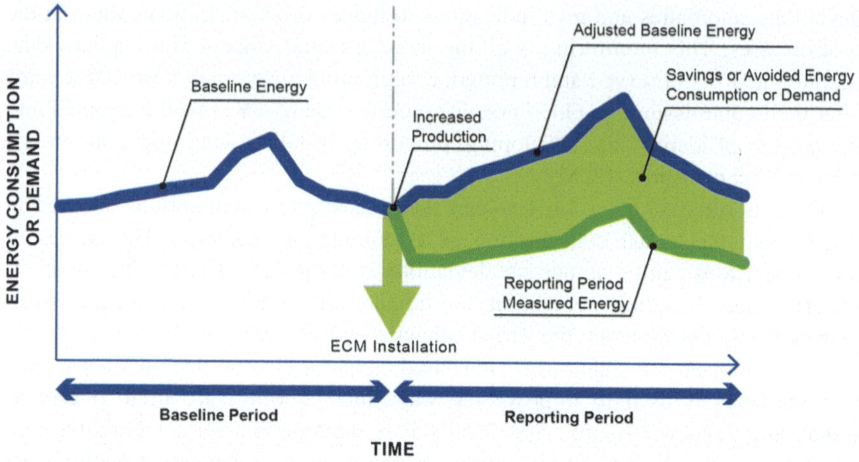

Fig. 3 International performance measurement and verification protocol (IPMVP®). Source: Efficiency Valuation Organization (https://evo-world.org/en/products-services-mainmenu-en/protocols/ipmvp)

$$\text{Savings} = (\text{Baseline Period Energy} - \text{Reporting Period Energy}) \pm \text{Adjustments}$$

IPMVP's framework requires certain activities to occur at key points in this process and describes other important activities that must be included as part of good M&V practice (Fig. 3).

Some researchers are looking for key factors like competence, integrity, communication, or reciprocity in generating trust and cooperation in energy performance contracting (EPC) [10]. Other are looking for Blockchain and smart contracts to provide a trading platform that enables the execution and enforcement of agreements between untrusted parties without involving a trusted third party [11]. This kind of new smart contract can be easier to extend to smart home.

2.2.3 Impact of End-User Energy Consumption Feedbacks

Numerous studies have analyzed the effects of providing feedback to the occupant on energy consumption. Ehrhardt-Martinez et al. [12] in particular identified 36 studies carried out between 1995 and 2010 and categorized the impact of the technology used on energy saving. The category called indirect feedback is characterized by global information such as the monthly invoice, provided with a consequent delay of several months. The energy savings linked to this feedback are relatively low, up to 8% with daily information. The other category corresponds to real-time feedback, that is to say directly related to the occupants' action. This instantaneousness is much more beneficial in improving occupant behavior, which can reach average savings of more than 10% (Fig. 4).

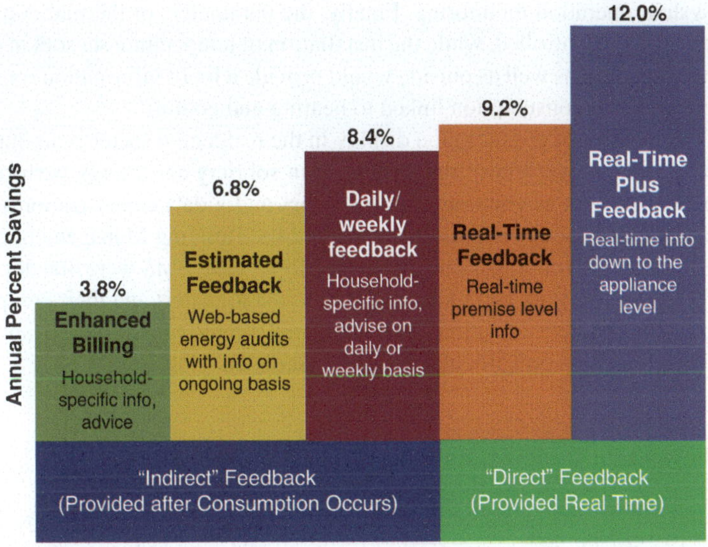

Fig. 4 Average Household electricity savings (4–12%) by feedback type. (Source: Ehrhardt-Martinez et al. [12])

Table 1 French RT2012 monitored usages

Heating (except for individual wood systems)	For the individual systems, the consumptions of the auxiliaries can be taken into account in this item or in the "other" item. For collective systems, auxiliaries are not taken into account
Cooling	
Dwelling Hot Water	
Electrical outlets	These are devices connected to electrical outlets. The specialized circuits intended for the cooking department are counted in the "other" item
Other	These are real estate lighting, specialized circuits for the hob and nonelectric oven (gas) of the individual air conditioning or individual automations

New French buildings are subject to thermal regulation (e.g., RT2012), which require them to have an energy consumption monitoring system that informs occupants, at least monthly, of their energy use, by energy type. However, sub-metering by usage or by dwelling is not required if a mathematical desegregation method is defined and indicated to the occupant. The usages considered are heating, cooling, and dwelling hot water (DWH), as well as electrical outlets and others (see Table 1).

The regulations do not impose a technical solution and give full freedom to go beyond. In particular, certain items are not considered (lighting for accommodation, thermal auxiliaries, parking lighting, exterior lighting, inverter, etc.) or for collective buildings, consumption linked to common areas (elevators, ventilation boxes, lighting of common areas, etc.). In addition, there is no obligation to install sensors for the system operation monitoring. Finally, the parameters of thermal comfort in all seasons are not controlled, while the installation of temperature sensors in several representative zones as well as outside would provide a lot of information for a more detailed analysis of consumption linked to heating and cooling.

This requirement of consumption display in the residential sector is an important advance for the involvement of the occupants in sobriety and energy performance. Supervision, via a set of communicating sensors and a data centralization system, is the first tool of a more complex system called the Building Management System (BMS) or Technical Building Management (TMB). These BMS were first developed in large buildings (tertiary and industrial) due to the high cost and their complexity. Although in a different situation, the residential sector must take advantage of existing developments, in particular existing standards.

2.2.4 Beyond the Building, Interaction with the Power Grid

The classical steps for building design from bioclimatic design to energy system design and renewable energy production on site can be extended to the district level. Step 4 in Fig. 5 includes Smart Building into the urban district/urban system. The goal is to address the overall system of district/city/region, including the interaction of the building with the infrastructure (building to grid). Then several optimizations can be done regarding load management, building as energy storage, building as energy consumer and producer (prosumer), and control engineering to facilitate interaction with the occupant and (energy) infrastructure (smart grids).

This is also the concept of *transactive energy networks* which could turn homes from passive energy consumers into intelligent, active energy storage and service providers for the future grid [14] as described in Fig. 6.

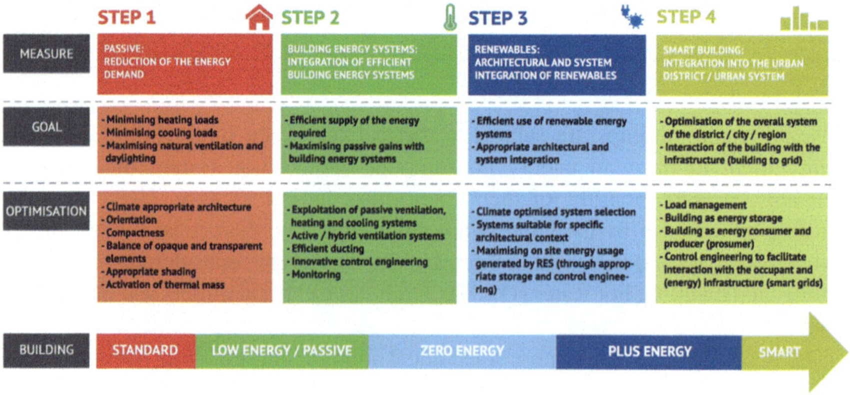

Fig. 5 Methodology for efficient and sustainable building design including the integration into the urban district/urban system. (Source: Märzinger and Österreicher [13])

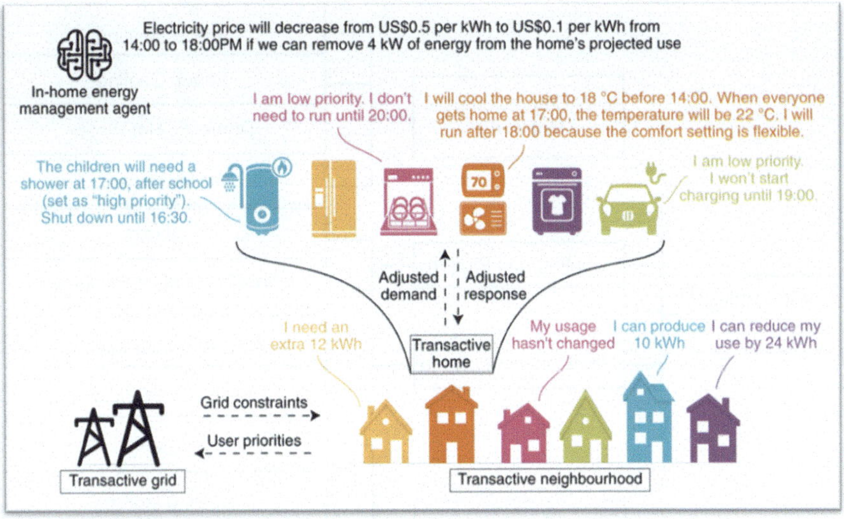

Fig. 6 Illustration of transactive energy networks. (Source: Wang [14])

2.3 *Standardized Performance Indicators*

Higher-level measures and indicators must be defined in order to quantify how far the building is to the targets and then to the objectives. An indicator used to set targets corresponds, for example, to the consumption in kWh per unit of use. The nature of the energy consumed, the duration considered, and the unit representative of the use are indicated. For example, this may be the electrical energy consumption for air conditioning in kWh/m^2/year of occupied room. These indicators can be corrected by climatic conditions, they can be corrected by the actual durations of use of the premises. Other indicators are used to monitor performance drifts in energy systems. These are, for example, the efficiency of fuel generators or the performance coefficients of thermodynamic equipment (COP, EER), or the consumption of ventilation which may indicate overconsumption probably linked to the fouling of filters. In the following we will draw a review on available indicators from smart home.

2.3.1 Disaggregation of Overall Consumption and Categorization

We saw previously that a real-time display by usage could be important. A building's energy use can be divided into heating, cooling, domestic hot water, air movement, lighting, household/office equipment, indoor transportation, auxiliary devices, and cooking, as shown in Fig. 7.

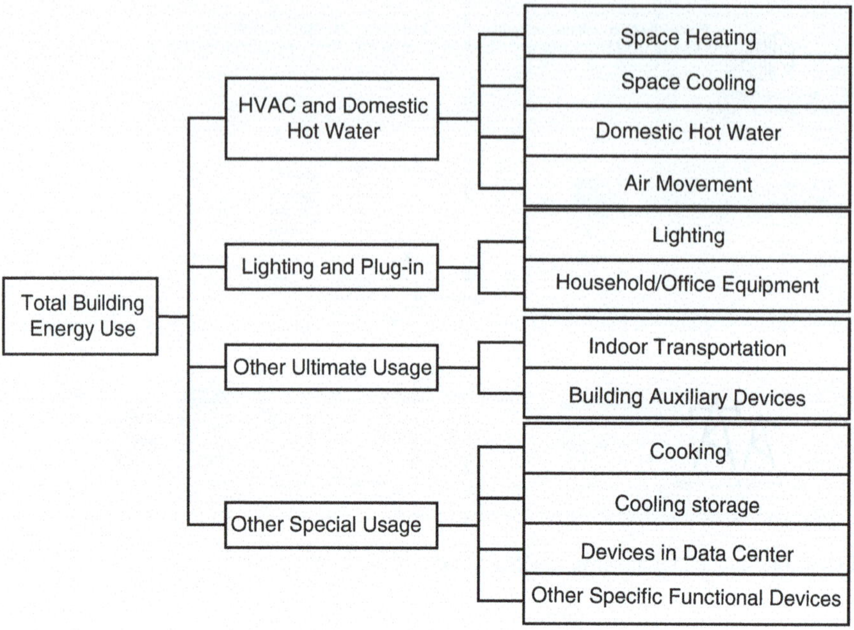

Fig. 7 Classification of end-use energy consumption. (Source: Jin [15])

Fig. 8 Energy consumption of a building by boundaries in ISO 12655:2013

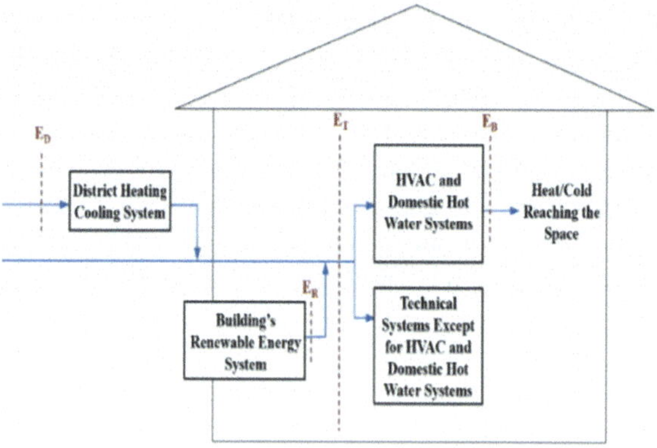

International Standard ISO 12655:2013, about presentation of measured energy use of buildings, is considering boundaries described in Fig. 8, which is used to decompose the energy flows. This standard was last reviewed and confirmed in 2018 and remains current. But with the arrival of Zero Energy Buildings, other boundaries can be considered, especially to calculate indicators such as Load Matching indicators and Grid interaction Indicators [16] (see Fig. 8).

2.3.2 Standards on Performance Indicators

The European Committee for Standardization (CEN, French: Comité Européen de Normalisation) published in 2007 two standards on performance indicators. EN 15603 provides methods for measuring and calculating the energy use of buildings, and EN 15217 tells how to represent it on a scale or a label. These two standards led to two ISO standards:

- ISO 16346: Energy performance of buildings—Assessment of the overall energy performance (an improvement and generalization of EN 15603).
- ISO 16343: Energy performance of buildings—Methods for expressing energy performance and for energy certification of buildings, which succeeds to EN 15217.

They have been replaced recently by (EN) ISO 52003-1 and (CEN) ISO/TR 52003-2. ISO 52003 defines overall energy performance feature, such as total primary energy use, nonrenewable primary energy use, renewable primary energy use, renewable energy ratio, greenhouse gas emissions, and annual energy costs. It also defines numeric indicator such as total primary energy use per useful floor area [kWh/m^2], total primary energy use [kWh], and nonrenewable primary energy use per useful floor area [kWh/m^2].

In order to generalize indicators definition, IEA presents in IEA [17] a conceptual framework for the development of building energy performance metrics with examples for metric parameters (Fig. 9).

Inputs	Final energy (total, electricity, gas, etc.) Primary energy (total, electricity, gas, etc.) Energy cost (total, electricity, gas, etc.)
	Per
Outputs	Persons served (total population, occupants, employees, etc.) Floor area served (total, occupied, heated, cooled, enclosed) Buildings served (total, grid-connected, etc.) Service level provided (amount of heating, cooling, lighting, etc.) Economic value (GDP, property value, etc.)
	For
Scope	Sector (all buildings, residential sub-sector, services sub-sector [commercial and public]) Building types (single-family, multi-family, office, healthcare, etc.) End uses (heating, cooling, water heating, lighting, appliances, cooking, etc.) Region (country, state, city, etc.)
	Normalised by
Normalisation factors	Climate (ground temperature, heating degree days, cooling degree days) Economic indicators (purchasing power parity, currency, etc.) Time (percent change from baseline date, lifecycle)

Fig. 9 Conceptual framework for the development of building energy performance metrics. (Source: IEA [17])

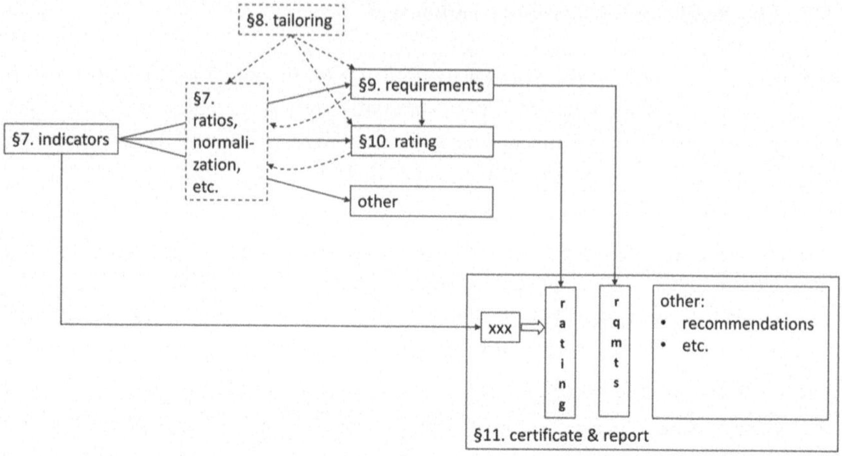

Fig. 10 Schematic description of indicators usage from ISO 52003 [18]

The framework uses four basic metric parameters: input, output, scope, and normalization factors. The input is the amount or cost of the energy by fuel source; this could be expressed as final (also known as delivered or site) energy, or as primary energy, or as the cost of energy. It is also possible to express these input into environmental impact such as CO_2 emissions. The output reflects the service provided by the energy, and can include the building space (floor area) served, the number of people or number of buildings served, or the amount of cooling and heating provided. The scope is a classification of the metric, such as the portion of the buildings sector under consideration (e.g., the entire buildings sector, or certain building types, or energy end uses). Finally, the normalization factors are used to modify the basic input-per-output metric values, such as economic purchasing power differences among regions, climate differences that impact heating and cooling energy use, and change in time relative to a reference or base year.

And finally, Van Orshoven and van Dijk [18] provides a description of EN ISO 52003 to make intelligent use the Energy Performance Building (EPB) assessment outputs based on indicators (Fig. 10).

2.3.3 Smart Readiness Indicator (SRI)

The 2018 revision of the European Energy Performance of Buildings Directive (EPBD) aims to further promote smart building technologies, in particular through the establishment of a Smart Readiness Indicator[1] (SRI) for buildings. As part of indictors, SRI aims at providing simple and understandable criteria to rank smart

[1] https://smartreadinessindicator.eu

buildings regarding several criteria in order to reach a better energy performance. As described in Fig. 11, the expected advantages of smart technologies in buildings are:

- Optimized energy use as a function of (local) production
- Optimized local (green) energy storage
- Automatic diagnosis and maintenance prediction
- Improved comfort for residents via automation

The three functionalities of smart readiness in buildings are:

- Readiness to adapt in response to the needs of the occupant
- Readiness to facilitate maintenance and efficient operation
- Readiness to adapt in response to the situation of the energy grid. As described in Fig. 12, Verbeke et al. [19] outlines a quantitative approach based on the load shifting potential using BMS and energy storages, and then a subsequent

Fig. 11 Expected advantages of smart technologies in buildings. (Source: Verbeke et al. [19])

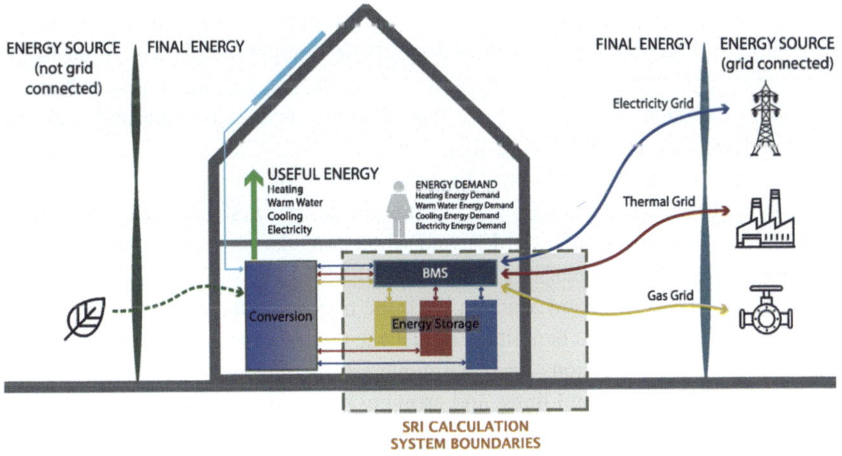

Fig. 12 System boundaries of SRI calculation. (Source: Verbeke et al. [19])

active interaction of the building (energy use and production) with energy grids, including electrical, thermal, and gas.

Most of smart services are related to the automation of the control of technical building systems, as defined in the technical standard EN 15232 already introduced in part *Standards to deploy energy-efficient technologies*.

According to Hogeling [20] SRI will characterize the ability of a building:

- To manage itself
- To interact with its occupants
- To take part in demand response and contribute to smooth, safe, and optimal operation of connected energy assets

Other initiatives, such as the one of Smart Building Alliance[2] (SBA), a French association, is also addressing SRI questions since several years, defining certification schemes for smart buildings such as *Ready2Services* for commercial buildings and *Ready2Grids*.

3 Building Automation and Control System

3.1 Introduction to BACS

The global and integrated vision of Building Automation and Control Systems (BACS) is essential and goes beyond the simple regulation implemented on isolated equipment. Indeed, multiple interactions take place in a building and the centralized vision of the BACS makes it possible to obtain control aimed at optimizing all the criteria at the same time. For example, if you consider HVAC (Heating Ventilation and Air Conditioning) systems that deal with both heating/cooling and ventilation, you could say that opening a window disrupts its operation. Likewise lighting or obscuring solar gain also disturbs thermal regulation. The lighting is also subject to the windows shutter. We therefore see that all these devices interact and a global centralized vision is necessary (Fig. 13).

Building Automation and Control Systems (BACS) or simply Building Automation System (BAS) aim at providing smart functionalities. BACS is generalized with Building Management System (BMS), and when considering energy explicitly, it is called Building Energy Management System (BEMS) [21].

According to Research and Markets [22], the major factors that drive the market for BMS are significant cost benefits to industrial, commercial, and residential users, simplified building operations and maintenance, increasing demand for energy-efficient and eco-friendly buildings, and growing integration of IoT. It places energy and IoT as a central key words of building management systems.

Regarding standards, the ISO 16484-1:2010 defines the operational implementation of BACS during the different phases of a project:

[2] https://www.smartbuildingsalliance.org/

Fig. 13 Integrated global control is required in order to cope with physical interactions

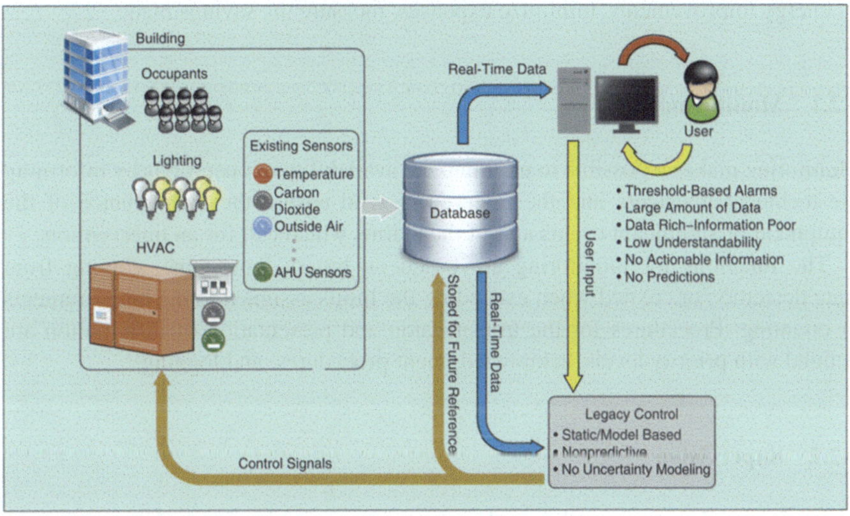

Fig. 14 Typical building automation system. (Source: Manic [23])

- Design (determination of project requirements and production of design documents including technical specifications)
- Engineering (detailed function and hardware design)
- Installation (installing and commissioning of the BACS)
- Completion (handover, acceptance, and project finalization)

It is addressing the various aspects relating to project specification and implementation, hardware, functions, data communication protocol, and data communication conformance testing.

The following figure (Fig. 14) illustrates the operation of a classic BACS offering standard functionalities that will be detailed in the next section.

3.2 Building Management System Functions

The functions of the BMS can be categorized as follows:

- **Monitoring**: Functions which facilitate the tasks of technicians who maintain the functionality of the equipment, with the aim of minimizing downtime.
- **Supervision**: Functions that allow technical managers to know, record and manage operations. The objective is to control the installations as close as possible to occupations and uses, to know the operations, the consumption of the equipment, the interventions to be carried out and those that have been carried out.
- **Energy efficiency continuous improvement**: Functions that measure efficiency and thus provide elements to maintain and improve it. The objective is to establish consumption indicators, adapt supplies as closely as possible, implement energy improvements, minimize expenses, measure the savings made.

3.2.1 Monitoring

Monitoring makes it possible to maintain the availability of operations by informing the technical managers and the interveners who ensure the maintenance of the equipment of significant events and of the alarms which call for an intervention.

The functions for monitoring are based on binary input state, coming from state detectors, or created when exceeding the limits assigned to the measurements or counting. Procedures for the transmission and presentation of information are defined with priority levels, acknowledgment procedures, and logging.

3.2.2 Supervision

Supervision allows you to know the status of the equipment and to control its operation:

- Inform in real time the stakeholders who perform technical management and operational tasks. For this, the measurements, counts, operating states, and events are centralized, transmitted remotely, and presented in dashboards.
- Adapt the management of the equipment to the uses by easy means of action to: control, remotely adjust, configure the operating conditions, and derogate from the automated systems.
- Record technical data for technical management, operating tasks and allow monitoring for energy efficiency to be carried out.

The supervised equipment can be, among other things, the power supply (electrical board); emergency power (generators, batteries), lighting, heating, ventilation and air conditioning (HVAC), plumbing (lift pumps, tanks, etc.), access control, fire devices (alarms, extinction).

We generally categorize physical data according to their discrete or continuous type:

- States (operation modes of equipment, position, command return, etc.);
- Measurements (temperature, operating time, number of failures, etc.).

We can extract from these physical points other information such as alarms (failure, abnormal stop, measurement exceeding a threshold, etc.) or additional quantities such as the thermal power calculated from flow rate and supply/return temperatures.

A dashboard is presenting the dynamic states of points on images corresponding to their location: plans of the premises, photos of the equipment, or block diagrams of the installations. The displayed points: operating states, commands, settings, measurements, or counts are selected according to their interest for the supervisors. Meters can be grouped according to their nature (electrical energy, thermal energy, water) and their location (general and on each zone) (Fig. 15).

The means of action can be manual commands such as starting, stopping, and/or operating an appliance at partial load. It can also be a programming of the intermittences of the equipment. Several pieces of appliances are controlled by recorded periodic programs (day, week, or year) such as thermal equipment, lighting, domestic hot water, fans, elevators, load shedding of power stations ...[3]

It can also involve the offloading of electricity consuming stations to reduce costs by adapting their operations to tariff signals, or by a forecast algorithm (chilled water generators with hourly or daily storage, kitchen equipment ...), while retaining the possibility of temporarily derogating from this automation.

Advanced management strategies can also be implemented, such as night ventilation for cooling the structures; natural cooling by introducing outside air; free cooling by outdoor air-water exchange; integrated control of solar protection,

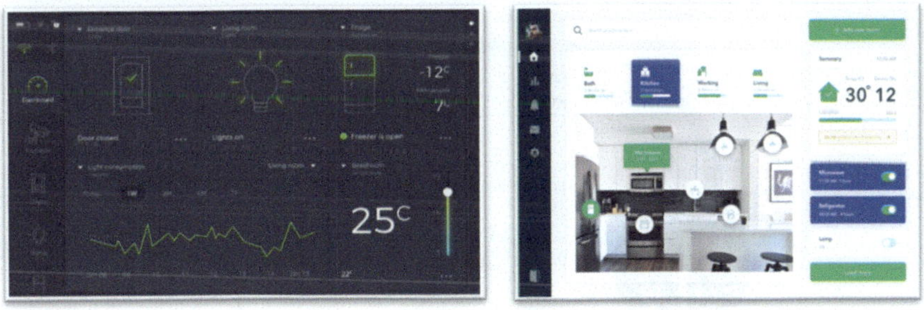

Fig. 15 Examples of smart home dashboards. (Source: Manic [23]) (https://www.pinterest.com/pin/302726406202699573/)

[3]https://dribbble.com/shots/4612989-Smart-home-dashboard-concept

lighting and air conditioning terminals. The parameters of these automations are adjusted to adapt them to weather conditions or to the particularities of use.

Finally, regulation algorithms are implemented to maintain set points such as ambient temperatures. Rising and lowering of temperatures can be optimized, that is to say, anticipated or delayed, taking into account ambient and/or outdoor temperature measurements.

All of the time-stamped data can be recorded in a trend log. The conservation period is to be configured according to the needs and capacities of the available equipment. These time series can be consulted a posteriori to analyze and optimize monitoring.

3.2.3 Energy Efficiency Continuous Improvement

A continuous improvment process allows a more efficient use a continuous improvement process for more efficient use of energy through an energy monitoring plan as well as energy analyses based on the monitoring and supervision data described above. This approach directly linked to the ISO 50001 energy management standard, is based on the PDCA (Plan-Do-Check-Act) methodologies. PDCA is an iterative four-step management method used in business for the control and continuous improvement of processes and products (Fig. 16).

Plan: An energy plan is the determination of the initial energy baseline, the energy performance indicators, the strategic and operative energy objectives, and the action plans. These data and evaluations form the basis of the following improvement processes. They also make it possible to identify potential for improvement of energy efficiency.

Do: In this phase, planning and action takes place, improvements are aimed for and implemented. Indicators and objectives for energy performance are defined on

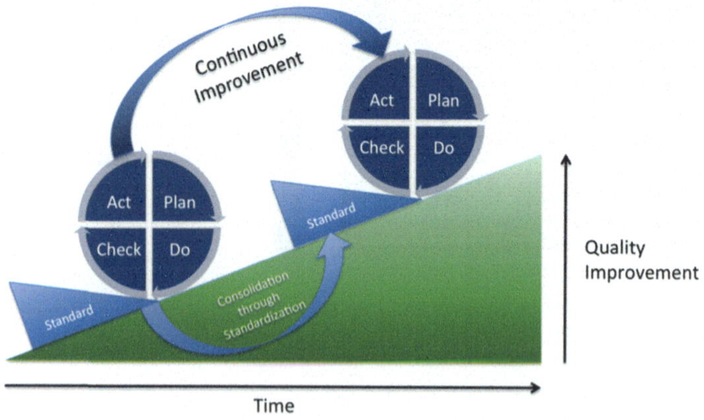

Fig. 16 Continuous quality improvement with PDCA. (Source: Johannes Vietze, CC BY-SA 3.0)

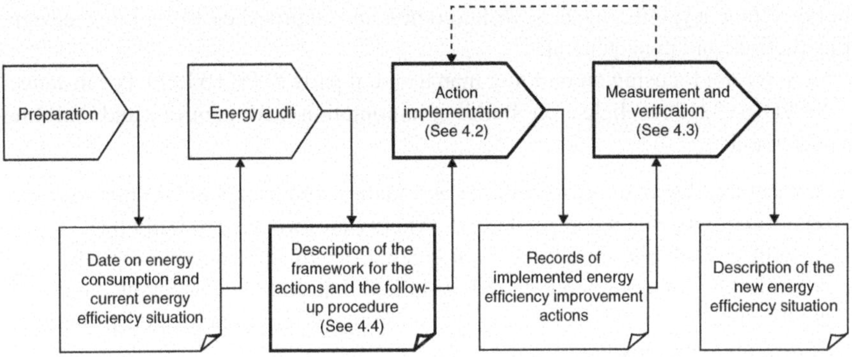

Fig. 17 Typical process of providing energy efficiency services (EN 15900:2010)

the basis of the results of the energy assessment. In doing so, action plans are also created, with which the objectives for the improvement of energy performance can be achieved.

Check: The plans executed in the "Do" phase must continually be checked to ensure that they are effective. To do this, core processes that are significant to the energy-related performance are monitored and measured in this phase. The results are compared to the previously established objectives.

Act: The constant measurements are broken down in reports. These form the basis for further studies, in order to improve the energy-related performance and the BEMS.

The data must be kept in order to conduct comparative analyses over different periods (annual in general). It is important to regularly update the use of the building, in order to adapt the targets depending on some changes like space allocation, energy contracting prices, etc. Finally, it is important to communicate these analyses, in particular with occupants. Whether it is information concerning the energy consumption observed but also the actions that make it possible to reduce it.

Figure 17 presents a typical process of providing energy efficiency services (EN 15900:2010).

3.3 Energy Management Algorithms

3.3.1 Overall Energy Performance Assessment

Since 2017, the ISO 52000 series aims to reorganize all the standards relating to the energy performance of buildings. ISO 52000 contains a comprehensive method of assessing energy performance as the total primary energy used for heating, cooling, lighting, ventilation, and domestic hot water of buildings. According to Elizabeth Gasiorowski-Denis [24], it will help accelerate progress in building energy efficiency utilizing new technologies and approaches to building design, construction, and management.

Here we are focusing on building management parts of ISO 52000. For instance:

ISO/DIS 52120-1 helps on building automation and controls and building management:

- A structured list of control, building automation and technical building management functions which contribute to the energy performance of buildings.
- A method to define minimum requirements or any specification regarding the control, building automation and technical building management functions, contributing to energy efficiency of a building, to be implemented in building of different complexities.
- A factor-based method to get a first estimation of the effect of these functions on typical buildings types and use profiles.
- Detailed methods to assess the effect of these functions on a given building.

ISO/DIS 52127-1 "Building management system" is relative to operational activities, overall alarming, fault detection and diagnostics, reporting, monitoring, energy management functions, functional interlocks, and optimizations to set and maintain energy performance of buildings (equivalent to standard EN 16947).

In Fig. 18, Hogeling [20] presents the set of standards supporting the implementation of the Energy Performance of Building Directive in Europe and specifically

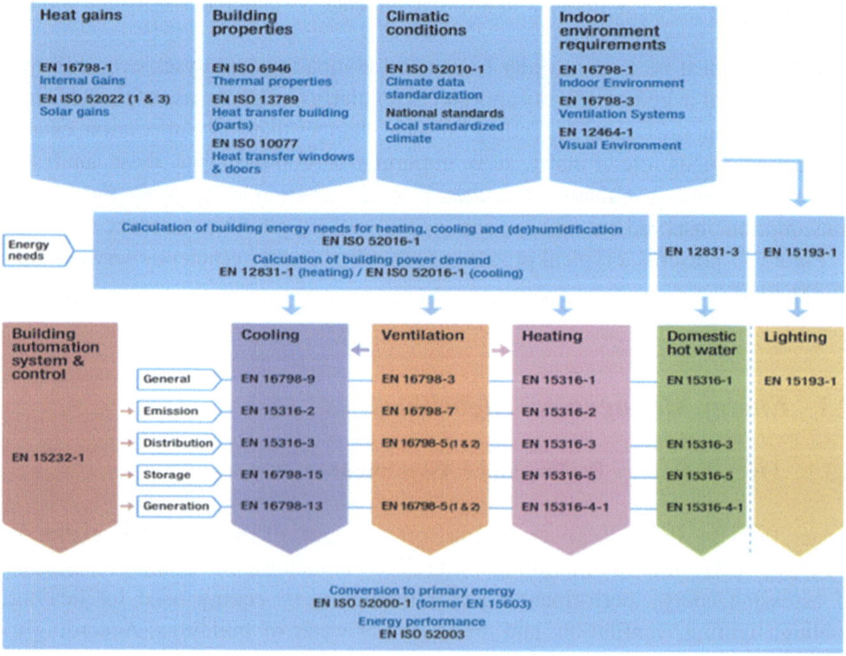

Fig. 18 The set of standards supporting the implementation of the Energy Performance of Building Directive in Europe

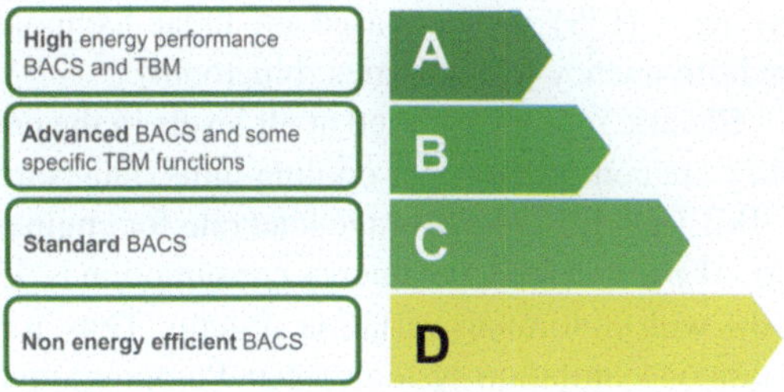

Fig. 19 Four energy-efficient BACS and TBM according to EN-15232

addressing the relation between ISO52000 and other standards. One can notice that standard EN-15232 on Building Automation & Control Systems is contributing transversally to most of energy domains. This standard is detailed in the next section.

3.3.2 Impact of Building Automation

EN 15232-1: 2017 is on Impact of Building Automation, Controls and Building Management. It lists the regulation functions, with a residential/nonresidential distinction, and realizes a distribution in efficiency classes (A, B, C, and D) (see Fig. 19). It also defines methods (detailed or simplified estimation) for calculating the impact of BACS and Technical Building Management (TBM) functions on energy performance.

3.3.3 Control of Energy Systems

Among the systems that have an impact on consumption, HVAC is certainly the most important. The current control systems must make it possible to modulate ventilation, heating, and cooling as required. Conventionally based on a calendar, the control of HVAC systems is increasingly becoming a function of sensors which allow real needs to be achieved. This has, for example, been the case with thermostats for heating systems, with temperature set point avoiding to control directly the power, which must adapt automatically by a regulation loop including a temperature sensor which activates or modulates the power.

Similarly, we are now looking to be as close as possible to the needs of the occupants, with radiant heating with presence detector which is activated only in occupied rooms. It is the same idea with the lighting, first we modulate according to the needs (close to the bay windows or rather in the dark corners), then we activate only if there is occupation, more or less fine (at room level, or on each lamp).

Occupancy detection capacity is therefore increasingly important for integrating energy systems into a control loop. This function can be performed by various devices such as infrared sensors (PIR motion sensor) or even measurements of the CO_2 concentration in the air. Even more sophisticated systems of Indoor Positioning

Systems (IPS) appear based on radar technologies or indoor geolocation using radiofrequency technologies (bluetooth).

Electric motors are used at all levels in the energy systems of buildings, whether they are compressors for cooling units, fans, circulators, or pumping. According to GIMELEC [25], the average load rate for engines under 500 kW is around 55–60%. At 80% of the load, the energy consumption is 95% of the nominal on/off operation, 50% with continuous variation systems. Thus, it is possible to optimize the operating point on ventilation systems (resp. Pumping) up to 50% of energy savings generated (resp. 30%), and a payback time of less than 2 years (resp. 3 years).

However, these conventional methods of regulation are not optimal because of the systems' dynamics, in particular those related to heating due to thermal inertia. Indeed, this can range from a few minutes for radiant systems to several hours for a heated floor for example. It is therefore necessary to anticipate to better control.

3.3.4 Adaptive Behavior, Predictive Automation, Control, and Maintenance

The regulatory functions presented above offer significant efficiency gains and can be implemented without the need to deploy complex and smart technologies. Intelligence in buildings are features such as anomaly detection, predictive modeling, optimization, and perhaps one of the most important premises of artificial intelligence, learning on their own. With the increasing amounts of diverse and dynamically changing data, extracting relevant and actionable information through legacy BEMS is difficult. This leads to a flood of data and decreased situational awareness, which may result in suboptimal building behavior.

Furthermore, the control strategies employed are often static and non-predictive; hence, they fail to adapt to changing environments and deteriorating building states. According to Manic [23], BEMS needs to be adaptive in order to cope with constant changes inside and outside the building such as occupancy patterns, aging of materials and equipment, floor plan changes, etc. As described in Fig. 20, BEMS functionalities are going further than simple regulation, they may use adaptability, predictive modeling, multisensor fusion, dynamic optimization, state-awareness, providing actionable information, etc.

Thus, techniques based on Model Predictive Control (MPC) (Artiges, [26]) are developing more and more. Commercial connected thermostat solutions (Netatmo, Qivivo, Nest, etc.) anticipate, for example, the heating restart in order to reach the set point taking into account the building's thermal inertia. The models being, for these thermostats, obtained by machine learning methods on the basis of the history of the data collected.

Finally, it should be noted that the challenges linked to the introduction of renewable energies go far beyond the regulation and control functions which we

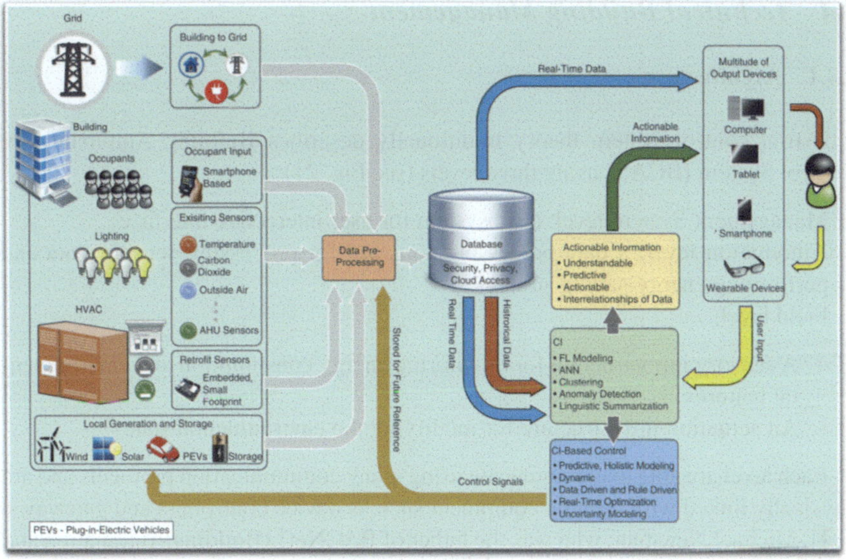

Fig. 20 Smart BEMS architecture. (Source: Manic [23])

have just discussed. Measures to reduce consumption can also be combined with intelligent management of costs depending on the energy sources available. In particular, the photovoltaic production during the day to operate, for example, a washing machine.

Energy systems such as HVAC are intended to be regularly maintained and controlled. In a house with dual-flow ventilation, it is, for example, recommended to change the filters every 6 months. Malfunction or degradation of HVAC system components causes reduced comfort on the one hand, and approximately 15 30% waste of energy on the other hand (Katipamula, 2005).

In the same way that a gas boiler is annually checked by a technical agent, it becomes necessary to generalize these scheduled preventive maintenance operations to all equipment. On the other hand, these visits can be costly; moreover, it is possible that a defect appears between two visits, or even that a defect is not detected by a visual examination.

Based on real-time monitoring of energy systems, it is possible to perform Fault Detection and Diagnostics (FDD) but also predictive maintenance. It is indeed possible by artificial intelligence algorithms to detect drifts and intervene before it is too late. Numerous research studies still make it possible to develop these techniques to make them truly operational, in particular to adapt easily to changes in piloting mode and to different configurations of the building (Verbert, [27]).

3.4 Technical Building Management

3.4.1 Architecture

Industrial control system theory traditionally describes Building Automation & Control System (BCAS) using three levels (see Fig. 21):

- Management System level: a subsystem for user interactive interface
- Automation level: a data processing software for processing sensory data and performing energy-saving strategies
- Field level:
  - A sensory infrastructure for monitoring energy consumption and environmental features
  - An actuation infrastructure for modifying the environmental state

Each level are able to communicate using many communication protocols and are physically linked with specific equipment such as device controllers and gateway.

H. Michael Newman, who was the father of BACNet[4] (Building Automation and Control Networks), defines these three levels:

The management level is where the majority of operator interface functions reside. Additional functions include communication with controllers, monitoring,

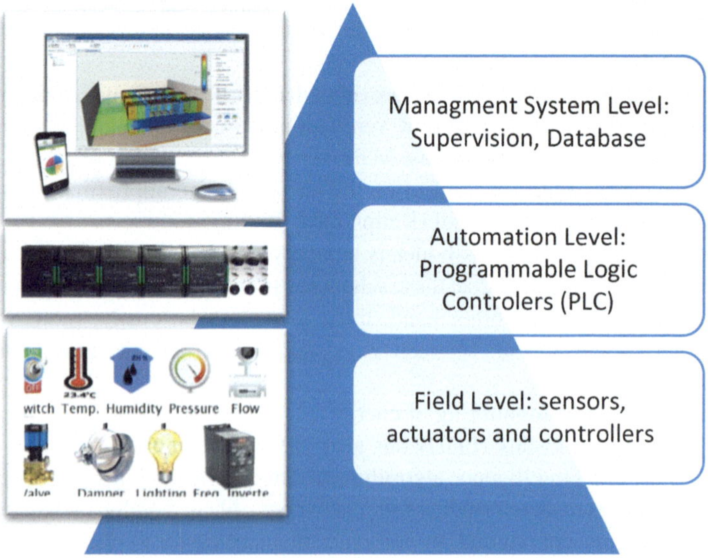

Fig. 21 Architecture of a building automation and control system

4https://www.big-eu.org/en/news/news-press-releases/news/mike-newman-the-father-of-bacnet-has-passed-away alarm, trend logging and statistical analysis, centralized energy management functions, and communication with, or coordination of, dedicated non-HVAC systems such as fire alarm and security control. As a practical matter, most of the devices at this level are personal computer workstations.

The automation level is where the majority of real-time control functions are carried out. The devices tend to be general-purpose, programmable controllers.

The field-level contains the devices that connect to sensors and actuators. We would tend to think of field-level devices as unitary or application-specific controllers.

Through years, supervisory control and data acquisition (SCADA) has been more and more connected[5]:

- First generation: "monolithic/standalone"
- Second generation: "distributed" across multiple stations, which were connected through a LAN
- Third generation: "networked" with standardized communication protocols
- Fourth generation: "web-based" through Internet and without dedicated software.

As illustrated in Fig. 22, in such a fourth generation, energy systems such as chiller, Air Handling Unit (AHU), Uninterruptible Power Supply (UPS), and Diesel Generators (DG sets) are still connected to BMS but also to Cloud platform in order to access data and alert from everywhere. It allows also nonfunctional services such as data safety and security that do not have to be addressed by the building itself.

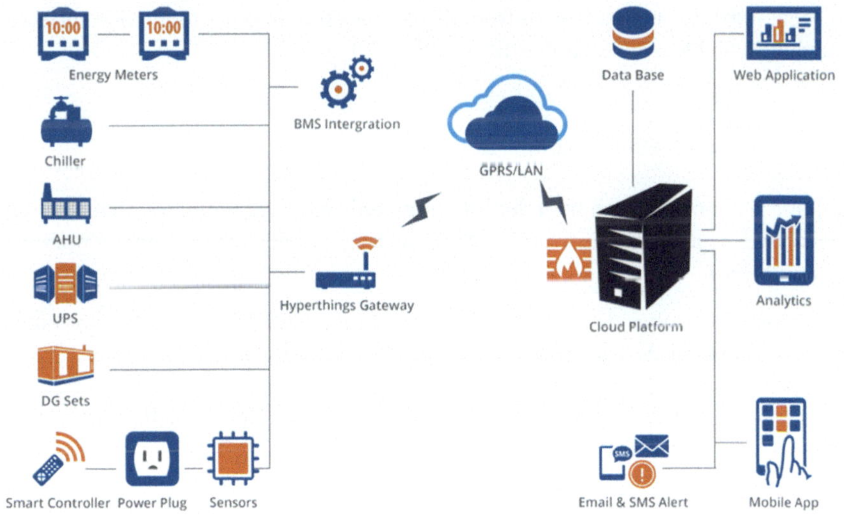

Fig. 22 Web-based SCADA. (Source: Calvert Controls [28])

[5]https://en.wikipedia.org/wiki/SCADA

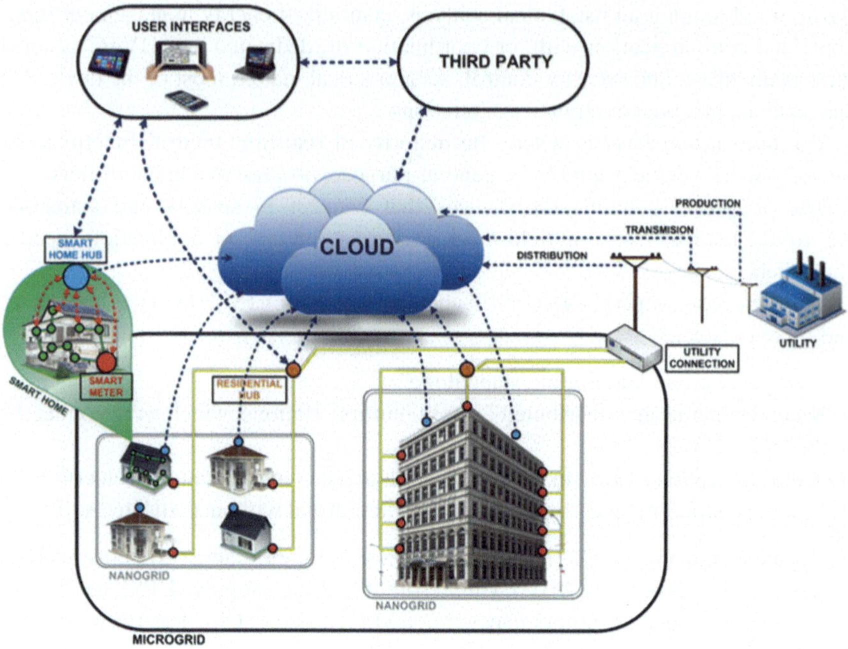

Fig. 23 Third-party services on top of web based. (Source Totonchi [29])

Then, third parties can benefit from this architecture in order to provide services in a more scalable manner as described in Fig. 23.

3.4.2 Communication Technologies

Building Automation count a lot of communication technologies that can be categorized depending on several criteria such as the openness, centralization, or versatility [30]:

- *Openness* describes dependency of a system on a manufacturer. Open protocols are BACnet, KNX, LonWorks, DALI, OpenTherm, EnOcean... Nowadays most of protocols tends to be open. Proprietary protocols are developed by one or a consortium but does not open specifications such as Universal Powerline Bus (UPB).
- *Centralization* describes the degree of independency of each component of the communication network. Centralized systems are based on PLC using protocols such as Modbus, while decentralized or distributed systems allow peer-to-peer (P2P) communication such as KNX.
- *Versatility* represents the ability of a system to cover one or more control tasks in building and home automation. KNX or LON are able to control HVAC, lights,

shutters/blinds... While DALI bus is specialized on light control or OpenTherm focused on heating control.

Home networks can use either wired or wireless technologies to connect endpoints. But classical building automation is using mainly wired technologies:

- Twisted pair of copper cables is mostly common wired physical support. This medium provides for instance the physical connectivity between the Ethernet interfaces present on a large number of building IP-aware devices, but also RS845 widely used for field buses.
- Fiber optics offer much higher bandwidth and/or lower latency characteristics associated with end-to-end optical signaling.
- Power lines are also used to communicate over existing power wiring, using devices also known as HomePlug.

Protocols can address field, automation, or management levels as shown in Fig. 24 for some protocols. There can be high-level versatile protocols such as BACNet (Building Automation and Control Networks) or WEB-Service oBIX (open Building Information Exchange) which are designed for applications such as HVAC, lighting control, access control, and fire detection and provides mechanisms for building automation devices to exchange information, regardless of the particular building service they perform.

There can be field-level protocols as shown in Fig. 25 that can also be versatile protocols such as Modbus or specific ones like M-Bus for Automatic meter reading or DALI for light control.

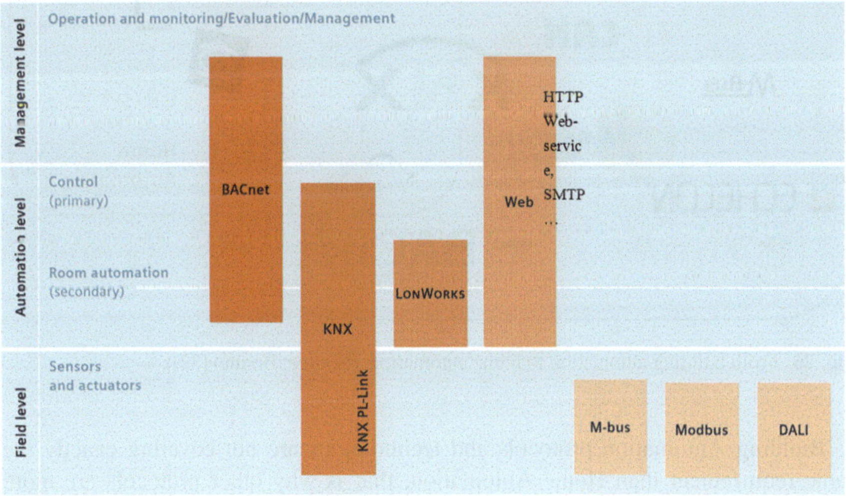

Fig. 24 Communication protocols can cover different level of BMS. (Source Siemens Building Technologies [31])

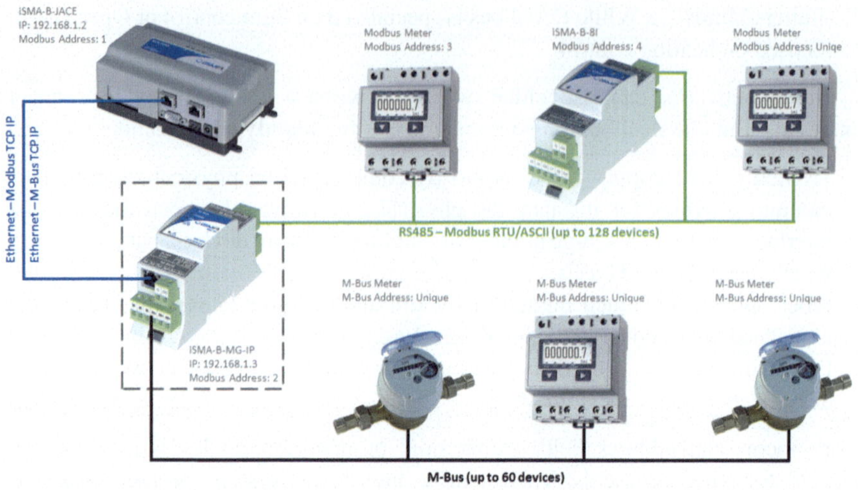

Fig. 25 Automatic meter reading using Filed bus (M-Bus & Modbus) connected to the automation level using Ethernet physical support and TCP/IP protocols. (Source: http://energycare.dk/portfolio-item/meter-gateway/)

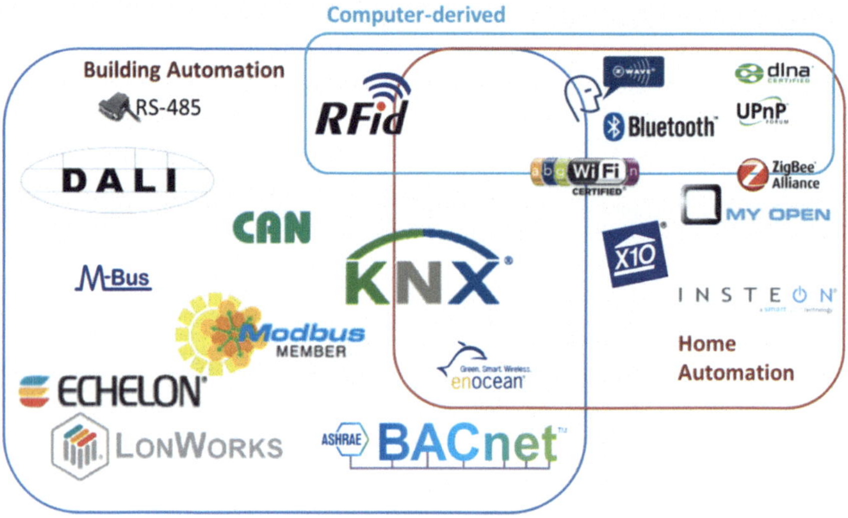

Fig. 26 From building automation to home automation. (Source: Bonino [32])

Building Automation protocols and technologies are not covering exactly the same requirement than Home Automation, that is why other protocols are more dedicated to Home Automation, and some of them are coming from Internet technologies or Computer technologies. Wireless technology are especially more developed for Home Automation than in Building Automation as described in Fig. 26.

4 Home Automation Technologies

4.1 Home Automation Market

4.1.1 Home-Specific Constraints

The principles of energy efficiency in buildings are based on technological solutions that we have just reviewed, but these are expensive and complex to deploy and therefore mainly reserved for the tertiary sector, hotels or high residential standing. With lower resources, the installation costs of smart home technologies must be drastically reduced, hence the emergence of free/open-source solutions, Do It Yourself (DIY) solutions. The market is less structured and the standards presented before are not really used. Smart home solutions are built as occupants need them, exacerbating the need for interoperability and adaptable wireless technologies. The generalization to all housing and the advent of the smart-home can only be done through a technological revolution, a paradigm shift such as that of the wireless Internet of things (IoT), Cloud services, and open-source/open-hardware as it can be represented in the following figure (Fig. 27).

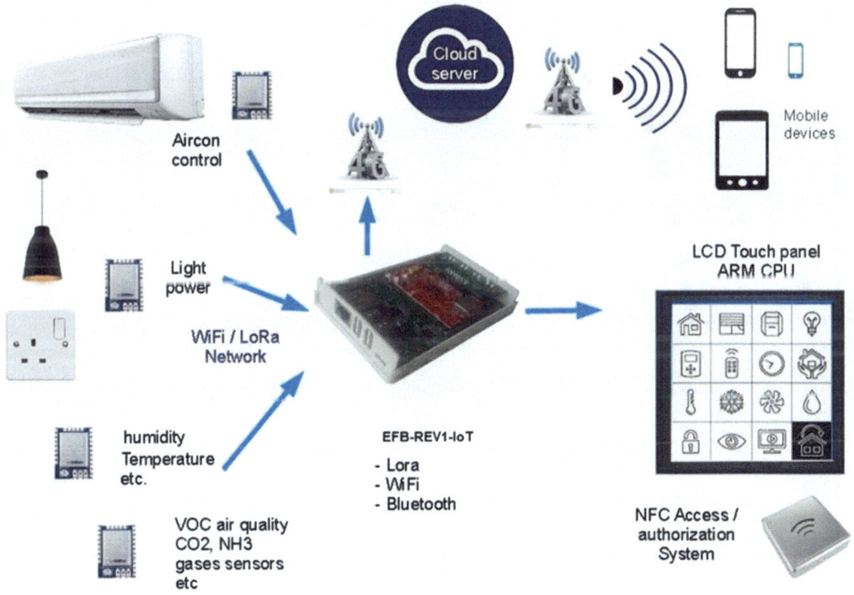

Fig. 27 Example of home automation architecture. (Source: Hackaday, https://hackaday.io/project/166414-home-automation-architecture-based-on-lora)

4.1.2 Smart Home Key Technologies and Market

The Hype cycle for emerging technologies (Fig. 28) shows that "Connected Home" and "IoT Platform" have already reached the Peak of Inflated Expectations and are now going to the stable position (Plateau) between 2023 and 2028. "Virtual assistants" also reached the peak but is going faster to mature products (before 2023). This technology can advise end-user, and interact with natural language processing (NLP) such as chat-based solutions that uses text or audio commands, to monitor and control the home appliances [34]. But the next revolution with artificial intelligence will be in the capacity of virtual assistant to really advice end-users and not only to control home appliance.

According to the Gartner report [33], the increasing number of IoT and related services is leading to a growth of 700 million smart homes in 2020. According to Research and Markets [35], the smart home technology market is expected to reach an estimated $112.8 billion by 2024 with a consumer need for simplicity and personalized experience, and the growing adoption of cloud-based technologies. According to another source [36], using Statista[6] data, the size of the world smart home market is about 33.4 billion US dollars in 2017 and is expected to rapidly increase to 78.2 billion US dollars by 2022.

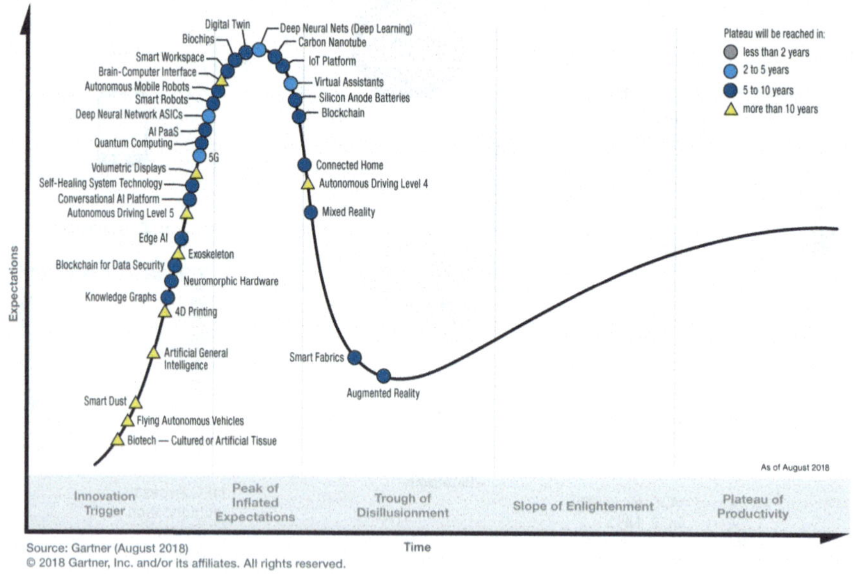

Fig. 28 Hype cycle for emerging technologies—Gartner [33] (https://blogs.gartner.com/smarterwithgartner/files/2018/08/PR_490866_5_Trends_in_the_Emerging_Tech_Hype_Cycle_2018_Hype_Cycle.png)

[6]https://www.statista.com/

Emerging trends, which have a direct impact on the dynamics of the smart home industry, include development of voice assistant technology for high-end automated households or emergence of air quality sensor devices that measure volatile organic compounds (VOCs).

According to Qolomany [37], 33% of IoT smart building market will be supplied by artificial intelligent technologies by 2023, and automation systems of smart building will grow up to 48.3% CAGR (Compound Annual Growth Rate) from 2018 to 2023. By 2025, the growth of connected home living will reach 3.7 billion smartphones, 700 million tablets, 520 million wearable health-related devices, and 410 million smart appliances in the connected person world.

According to Lobaccaro [38], IoTs are becoming increasingly embedded in the society by allowing faster and more efficient interaction between users and both public and private environments. IoT development has been recognized as having significant potential to create an interactive energy management system for homes.

4.2 *Internet of Things (IoT) Technology*

4.2.1 IoT Definition and Characteristics

The Smart Sustainable Cities group (SSC) of the International Telecommunication Union (ITU) defines Internet of things in ITU-T SSCIoT 2 [39] as a global infrastructure for the information society, enabling advanced services by interconnecting (physical and virtual) things based on existing and evolving interoperable information and communication technologies. Through the exploitation of identification, data capture, processing, and communication capabilities, the IoT makes full use of things to offer services to all kinds of applications, while ensuring that security and privacy requirements are fulfilled. From a broader perspective, the IoT can be perceived as a vision with technological and societal implications.

According to the Cluster of European Research Projects on the Internet of Things [CERP-IoT] [40], an autonomous home network has to be intelligent and capable of sensing and adapting to environment changes while performing self-capabilities (e.g., configuration, healing, optimization, protection). Autonomy will make home network architecture highly dynamic and distributed enabling the interworking of several devices and systems. Interworking of home networking systems and devices with other systems and devices external to the intranet will be achieved via Personal Virtual Private Networks (VPN). Any device or thing that has human input controls can be used to securely interface with the building's services to monitor status and change its settings. Using home automation devices with wireless communication technologies, all of building's "things" can have two-way communication with each other. For example, the thermostat can be controlled from the console of the refrigerator, or the detection of a mobile phone entering a room allows to configure automatically the light atmosphere. The washing machine can stop if the oven is switched on in order to respect the electric subscription limit, etc.

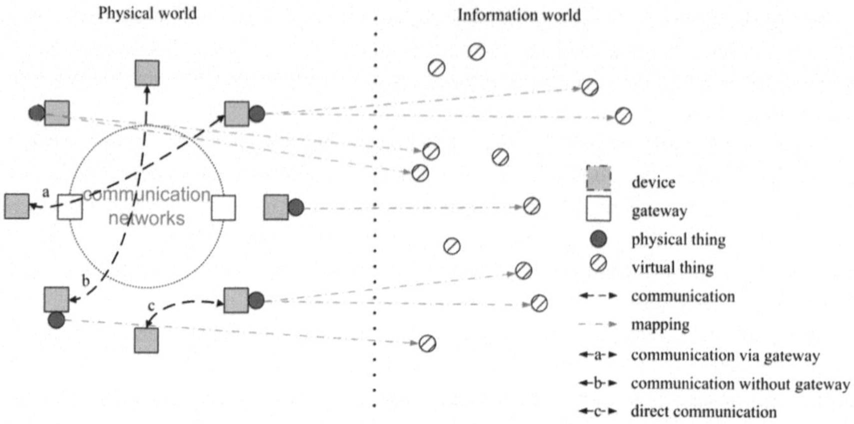

Fig. 29 Technical overview of the IoT and interactions between physical devices [39]

The IoT is expected to greatly integrate leading technologies, such as technologies related to advanced machine-to-machine communication, autonomic networking, data mining and decision-making, security and privacy protection and cloud computing, with technologies for advanced sensing and actuation.

As represented in Fig. 29, a physical thing may have a mapping with one or more virtual things in the information world. Virtual thing can also exist without any associated physical thing. A device has the mandatory capabilities of communication and optional capabilities of sensing, actuation, data capture, data storage, and data processing. Devices communicate with other devices through the communication network, with (case a) or without (case b) a gateway, or directly (case c).

The **fundamental characteristics** of the IoT are defined by [39] as follows:

- **Interconnectivity**: With regard to the IoT, anything can be interconnected with the global information and communication infrastructure.
- **Things-related services**: Such as privacy protection and semantic consistency between physical things and their associated virtual things.
- **Heterogeneity**: The devices in the IoT can be heterogeneous, based on different hardware platforms and networks.
- **Dynamic changes**: The number of devices, their own state and their context change dynamically (e.g., sleeping/waking up, connected/disconnected, location, speed, etc.)
- **Enormous scale**: number of connected devices, communication triggered, data generated and processed can be huge.

Associated to the previous fundamental characteristics, IoT may provide **high-level requirements such as context-awareness, interoperability, security, or privacy** that can be detailed here.

4.2.2 Context-Aware IoT

Ubiquitous or context-aware computing has proven to be successful in understanding sensor data. Collection, modeling, reasoning, and distribution of context in relation to sensor data plays critical role in adding value to raw data and to help understanding it.

According to Perera [41], a system is **context-aware** if it uses context to provide relevant information and/or services to the user, where relevancy depends on the user's task. Where **context** is any information that can be used to characterize the situation of an entity. Where an **entity** is a person, place, or object that is considered relevant to the interaction between a user and an application, including the user and applications themselves.

Figure 30 presents contextual information **who** (identity), **where** (location), **when** (time), **what** (activity), in two different columns, depending on how they are obtained. First column is about information retrieved directly from sensors (e.g., GPS sensor readings as location information), while second column corresponds to virtual sensors, where information are computed using primary context, using sensor data fusion operations or data retrieval operations such as web service calls.

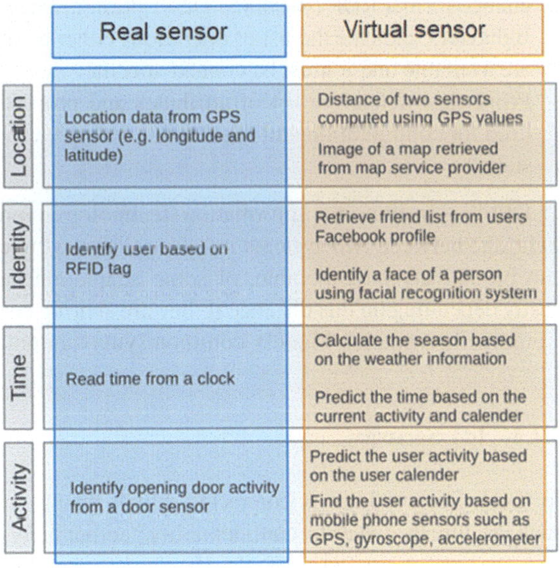

Fig. 30 Context information collected by real sensor or calculated by virtual sensor. (Adapted from Perera [41])

4.2.3 IoT Interoperability

Interoperability needs to be ensured among heterogeneous and distributed systems for provision and consumption of a variety of information and services. Interoperability can be first achieved using standards defined by organizations such as IP Smart Objects (IPSO) alliance, European Telecommunications Standard Institute (ETSI), AllSeen alliance with AllJoyn open-source software framework, Open Interconnect Consortium (OIC) with IoTvity open-source software framework or IoT-A forum with OpenIoT open-source platform for connecting physical and virtual sensors to the Cloud. Interoperability has also been attempted to be solved by using Semantic Ontologies which provide deeper understanding of the raw sensor data enabling machines to take decisions based on simple rules [42]. Semantics provide a different dimension to the data interoperability at a higher level than what just raw data gives; Sensor Semantic Network (SSN) provides a comprehensive set of ontologies for interpreting the sensor data. SenML (Sensor Markup Language) has proposed to provide a common media-type for sensor data exchange.

Tayur and Suchithra [42] defines the application layer interoperability requirements:

- Protocol: defines preferred language of communication such as HTTP, CoAP, or MQTT
- Message: specifies the encoding and structure of the data based on JSON, XML, or Binary
- Semantics: allow interpretation of the meaning and context via established ontologies which work on the raw sensor data converted into web data interchange format RDF (Resource Description Framework)
- Behaviors: indicate the list of operations either configuration or management that are available and it must be context specific.
- Properties: define the list of attributes and properties of the device that can be used for configuration and operations (example can be vendor name, light on/off status).

OFFIS—Institute for Information Technology represent in Fig. 31 the distance to bridge between two systems interface that need to communicate. These systems can be natively interoperable, of some adaptation have to be done more or less easily depending on this distance if they are sharing common semantic model (CIM: Common Information Model), common syntax, or nothing [43].

4.2.4 IoT Security

In the IoT, everything is connected which results in significant security threats, such as threats towards confidentiality, authenticity, and integrity of both data and services. A critical example of security requirements is the need to integrate different security policies and techniques related to the variety of devices and user networks in the IoT.

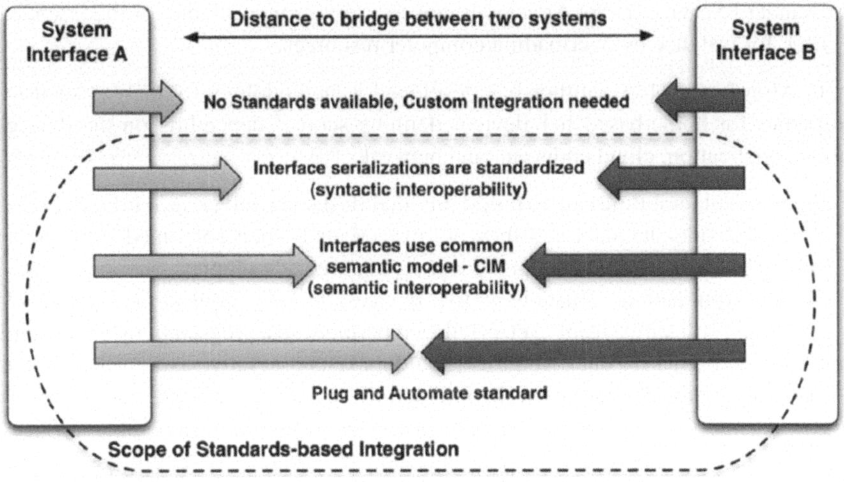

Fig. 31 Semantic integration distance for interoperability. (Source: OFFIS)

The 2017 attack by the Triton malware, which targeted critical systems of petrochemical plant in Saudi Arabia through the Triconex safety controller (Schneider Electric), showed the potential destruction that these types of threats can bring. With the convergence of operations technology (OT) and internet technology (IT), as well as the robust adoption of the Industrial Internet of Things (IIoT) by ICS operators, risks have grown. IoT that provides external access of home network are a gateway for criminals who can attack personal data such as bank access codes, or demand ransom for intimate data.

Based on the literature survey [44], a few threats to the SCADA systems in IoT-cloud environments are defined below:

- Advanced Persistent Threats: An unauthorized person attempts to gain access to the system using zero-day attacks (unknown attack) with the intention of stealing data rather than causing damage to it.
- Lack of Data Integrity: Data integrity is lost when the original data are destroyed, and this could happen through any means such as physical tampering or interception.
- Man-in-the-Middle (MITM) Attacks (spoofing and sniffing attacks): In a spoofing attack, a program or person masquerades as another program or person to gain illegitimate access to the system or the network. In a sniffing attack, the intruder monitors all the messages being passed and all the activities performed by the system.
- Replay Attacks: A valid message containing some valid data is repeated again and again; in some cases, the message may repeat itself. These attacks affect the performance of SCADA systems and can be serious threats when a replay attack delays messages sent to physical devices.

- Denial of Service (DoS) Attacks: It makes a service unavailable for the intended user, for instance by overloading computer resources.

In Alrawi et al. [45], author has developed a methodology to analyze security properties for home-based IoT devices defining scores[7] depending on the device, mobile application, cloud endpoint, and network.

- Device: Vulnerabilities in IoT systems manifest themselves in hardware, software, and side-channels and they are exacerbated when combined. Mitigating vulnerabilities relies on vendors, adopting mature technologies.
- Mobile Application: Trusted by IoT devices, mobile applications are attack points and still suffer from over-privileged permissions, programming errors, and hard-coded sensitive data. Vendors should make conservative assumptions about the trust relationship and limit the interactions with core services.
- Cloud endpoint: Are suffering from misconfiguration and vulnerable services that can be properly secured using industry standards. Third-party cloud providers play an important role by offering securely managed IoT platforms, which vendors are adapting.
- Communication: IoT devices may still rely on insecure protocols that do not offer confidentiality or integrity but mitigate them by using TLS/SSL protocols. Many devices lack encryption on the LAN, which leave them susceptible to MITM attacks.

For instance, to assess and analyze the old but still used Modbus communication protocol's vulnerability and risks, Byres [46] used an attack tree model, revealing that the Modbus protocol is weak and lacks basic security requirements such as integrity, confidentiality, and authentication. They recommended using firewalls, Intrusion Detection Systems (IDSs), and encryption techniques for secure communications.

More recently, Fouladi and Ghanoun [47] performed MITM Attacks on a Z-Wave door lock, causing a lot of turmoil in the security and home automation world. They were able to intercept the unencrypted packets being sent between devices and the controller, and easily retrieve the home and node IDs. They could easily dissect packets for timestamps, home IDs, sources, and targets, as none of this information is encrypted. Using this information, the team was able to spoof the controller, sending raw packets to devices that appeared to come from the real controller. This attack relies greatly on the lack of encryption in the first generation of Z-Wave. Therefore, in the later generations, Z-Wave radio chips support encryption to increase security.

Protocols are more and more secured, but some attacks are not related to technology issues but user lack of awareness. For instance weak password or manufacturer default password can be used to easily access home network.

[7] https://yourthings.info

4.2.5 IoT Privacy

Identification and access control technologies provide link between data and user's devices. Sensed data of things may contain private information concerning their owners or users. Profiling methods based on linked records can reveal unexpected details about users' identity and private life. The IoT needs to support privacy protection during data transmission, aggregation, storage, mining, and processing.

Privacy is more than security; for instance, it has been proved in Caputo [48] that cloud traffic analysis allows to detect the presence of a person in a house equipped with a Google Home device, even if the same person does not interact with the smart device.

Moreover, vendor or third parties have grant access to your data; that is why the EU General Data Protection Regulation (GDPR) provides an essential guidance to achieve a fair balance between the interests of IoT providers and users. But many challenges are still to be met, such as those detailed in Wachter [49] about:

- Profiling, inference, and discrimination
- Control and context-sensitive sharing of identity
- Consent and uncertainty
- Honesty, trust, and transparency

4.3 *IoT Architecture*

4.3.1 Four Layers Architecture

It is usual to represent communication system in the classical seven level OSI model. Here IoT is simplified to four layers levels (Fig. 32):

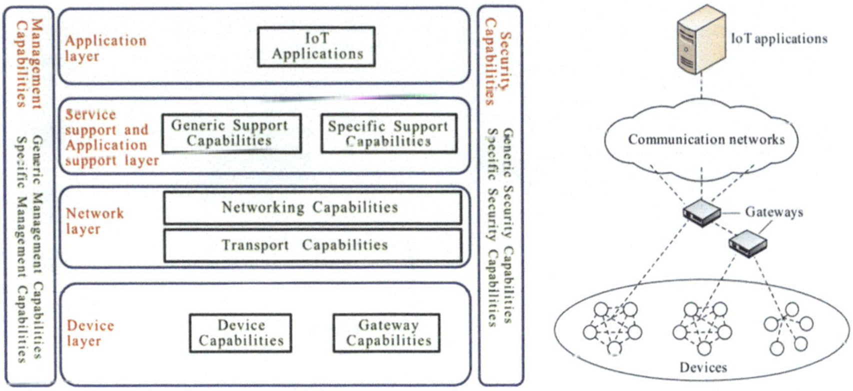

Fig. 32 Four level layers of IoT reference model [39]

- Application layer contains IoT application.
- Service support and application support layer contains generic support capabilities such as data processing or data storage and specific ones.
- Network layer contains networking and transport capabilities, providing relevant control functions of network connectivity, such as access and transport resource control functions, mobility management or authentication, authorization and accounting (AAA), and transport capability for data information, IoT-related control and management information.
- The device layer contains device capabilities such as direct and indirect interaction with the communication network to gather and upload information with or without gateway, Ad-hoc networking construction (to increased scalability and quick deployment), Sleeping and waking-up... The device layer contains also gateway capabilities such as multiple interfaces support (e.g., USB, ZigBee, Bluetooth, or Wi-Fi), and protocol conversion at device level (e.g., ZigBee and Bluetooth) or at both device and network layer (e.g., ZigBee and 4G)

In a similar way to traditional communication networks, IoT **management capabilities** cover the traditional fault, configuration, accounting, performance, and security (FCAPS) classes, i.e., fault management, configuration management, accounting management, performance management, and security management.

Security capabilities are present at each layers. It can be authorization, authentication, data confidentiality and integrity protection, privacy protection, security audit and anti-virus, and integrity protection and validation, access control...

4.3.2 Cloud-Based Architecture

Smart Home solutions architecture (Fig. 33) is mostly based on a gateway with an intelligent part to manage local automation and a cloud part to manage user interaction and database.

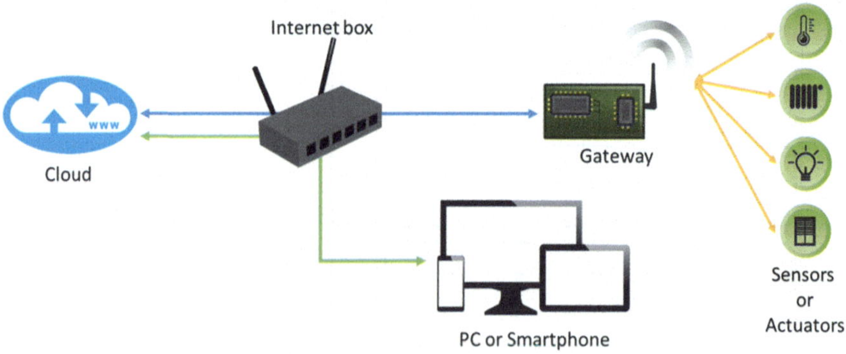

Fig. 33 Classical architecture of Smart Home solutions

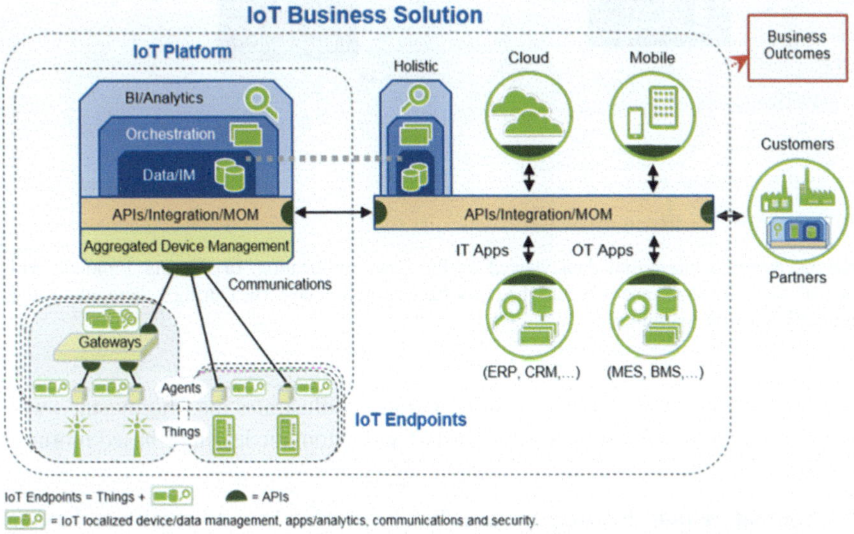

Fig. 34 IoT business solution. (Source: Gartner [50])

Fig. 35 IoT deployment in a cloud-based architecture. (Source: Alrawi et al. [45])

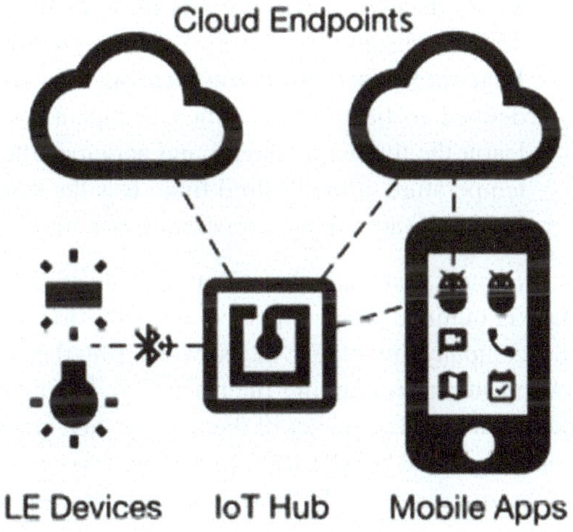

Moreover, third parties' access grant is part of the new architectural model of IoT (Fig. 34).

Installation of such a device follows a classical scheme (Fig. 35), where low-energy (LE) devices are connected (e.g., Bluetooth) to an IoT hub or directly to the vendor mobile application in order to configure network connection, then able to access cloud platform.

Let us consider Netatmo, a French company that offers a connected thermostat, a weather station, and a face recognition security system. The data can be consulted by

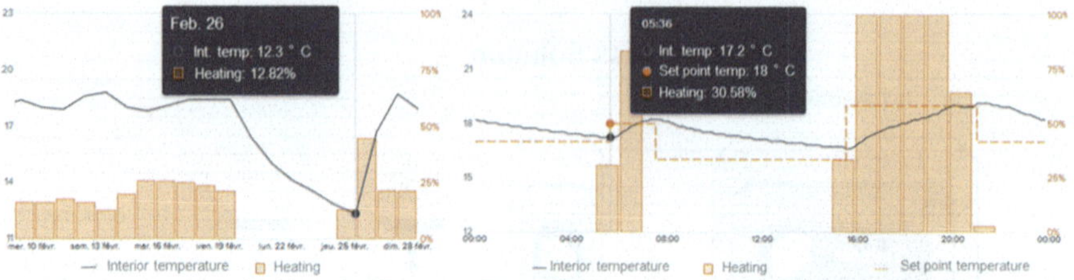

(a) (b)

Fig. 36 Netatmo thermostat data viewed on the cloud application. (**a**) During 1 month, with manual remote heating restart before the come back from holidays. (**b**) During 1 day with heating restart by machine learning

smartphone application, and by website. Figure 36 illustrates the Thermostat's web interface with manual remote heating restart and automatic learning-based morning restart:

(a) **Manual remote heating restart**: has been used here to reactivate heating system before coming back to home at the end of holidays. The graph shows a temperature decrease from 19 °C to 12.3 °C in one week, and a rise back to 19 °C during 2 days. It leads to no heating consumption during a week, and double consumption during 1 day to retrieve the initial temperature.
(b) **Learning-based morning heating restart**: While user set point has been defined to have 18 °C when occupants wake-up (7 am), the controller has learnt the thermal behavior and anticipate the heating at 05:36 in order to reach temperature at the desired time. It is the same for the evening, starting heating from 3:30 am taking temperature rise into account.

We have just seen that Netatmo data is accessible from a remote server. The data is automatically uploaded from the IoT device and is accessible as long as the company maintains the service. But the data is also available through API (Application Programing Interface[8]) so that anyone can develop applications using these sensors. The owner of the IoT can then grant access to his data for a third-party application and benefit from new services.

As discussed in IoT privacy part above, it is important to know about the data access possibilities of connected systems that appear on the market. The first precaution concerns the property of the data which must remain with the owner rights, including the data remove. Then, the security of the data, if those data are available on a server, they must be accessible by secured manner, and if accesses are authorized, to know the treatments and objectives of these treatments.

In Table 2 below are summarized some manufacturers that can be currently met on the home automation market in terms of monitoring and control.

[8]https://dev.netatmo.com/

Table 2 Some example of home-automation device manufacturers and prices

Providers	Protocols	Approximated prices
Xiaomi Home	Zigbee Wifi	Gateway: 20 €. Device price is varying from 10 € the temperature/humidity sensor up to 300 € the robot vacuum cleaner
Fibaro	Zwave	Gateway: 40 €. Device price between 50€ and 100€
IKEA	Zigbee	Gateway: 30 €. Device price from 15 € for a socket plug, up to 150 € for connected blinds
Somfy	433mhz	Very varied price
DeltaDore	X3D	Gateway: 100€. Device 50€
Smappee	Wifi, bluetooth, Modbus, MQTT	Gateway: 300€. Device 50€

4.3.3 Typical IoT for Home Energy Monitoring

The following figures illustrate typical monitoring for three specific flux, namely electrical, gas, and water fluxes.

Figure 37 is relative to typical IoT for home electrical consumption monitoring. They can be placed on the electrical distribution board or directly on appliances. It can also use existing measurements such as the main electricity meter such as Linky smart meter or pulse-based electrical counter (Table 3).

Figure 38 is focusing on electricity monitoring for PV system generation. A storage can be added and energy flux can be monitored with different control strategies such as self-consumption (Table 4).

Figure 39 is relative to typical IoT for gas-based heating and hot water consumption monitoring (Table 5).

Figure 40 is relative to typical IoT for water consumption monitoring (Table 6).

4.4 *Wireless Communication Energy Consumption*

4.4.1 Wireless Characteristics

As described in Fig. 41, wireless protocols have to be chosen depending on data rate and range required by application domain. For instance, home automation range is from 10 to 100 m, then technologies such as Z-Wave or EnOcean are widely used, associated with a mesh network which uses nearby devices to piggyback communication to devices in all buildings rooms. Depending on the range, the network is called a Wireless Local Area Network (WLAN, e.g., Wifi) or a Wireless Personal Area Network (WPAN, e.g., bluetooth). Multimedia communication requires high-rate protocol such as wifi, but for home automation, low-rate (about 100 kb/s) is still enough. The target is then to reach as close as possible the targeted application.

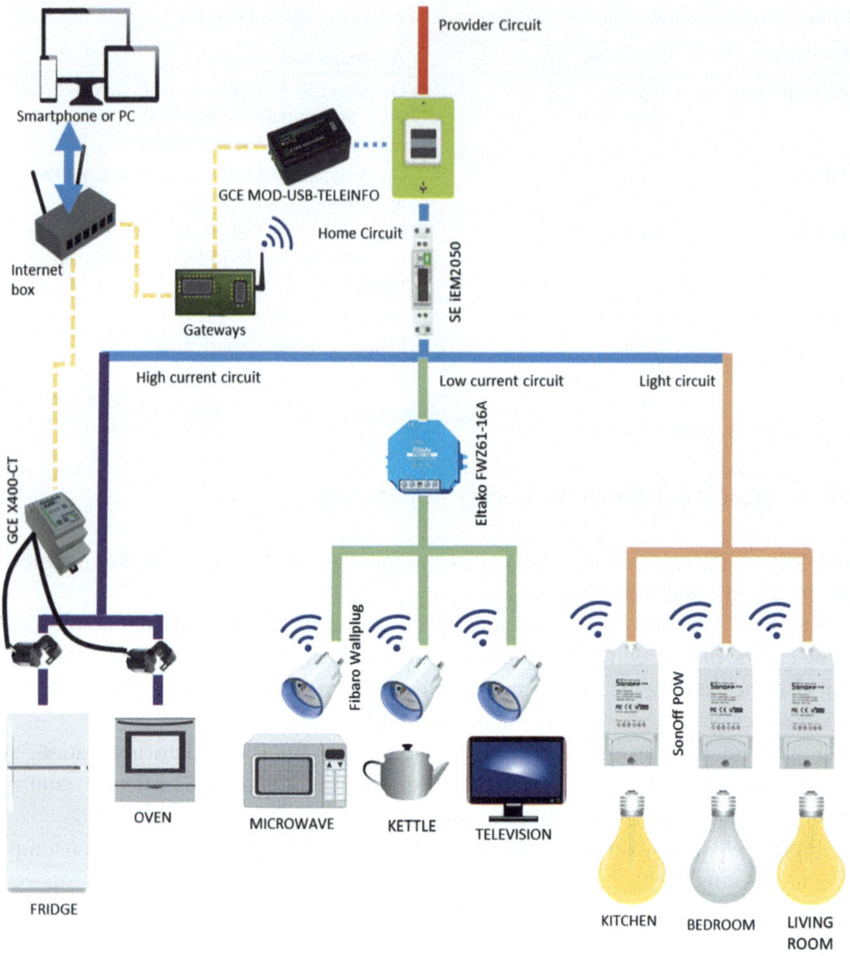

Fig. 37 Typical IoT for home electrical consumption monitoring

Indeed, if one increases the range or data rate, the consumption is increasing too. That is why we are focusing on Low-Rate Wireless Personal Area Network (LR-WPAN). Other protocols are available for LR-WPAN such as 6LoWPAN, BLE (Bluetooth low energy), Thread, UWB (Ultra WideBand), ZigBee, and ANT+.

IEEE 802.15.4 is a technical standard which defines the operation of LR-WPAN. It specifies the physical layer and media access control for LR-WPAN, and is maintained by the IEEE 802.15 working group, which defined the standard in 2003. It is the basis for the Zigbee, ISA100.11a, WirelessHART, MiWi, 6LoWPAN, Thread, and SNAP specifications.

Table 3 References and properties of electrical meters

Meter	Reference	Measurements	communication
Main electricity meter	GCE electronics TELEINFO	– All data provided by main counter (Linky) such as index, power, high/low period . . .	USB
Main electrical circuit meter (distribution board)	Schneider electric iEM2050	– Active and reactive energy – Active and reactive power – Power factor – Current – Voltage – Frequency	Modbus
Sub circuit meter (distribution board)	GCE electronics X400-CT	– Current (hot plug using tore)	Ethernet
sub circuit meter	Eltako FWZ61-16A	– Active energy – Active power	Enocean
sub circuit meter & switch	SonOff POW R2	– Active energy – Active and reactive power – Power factor – Current – Voltage	Wifi
socket meter & switch	Fibaro FGWPE-102-ZW5	– Active energy and power – Remote switch	Zwave

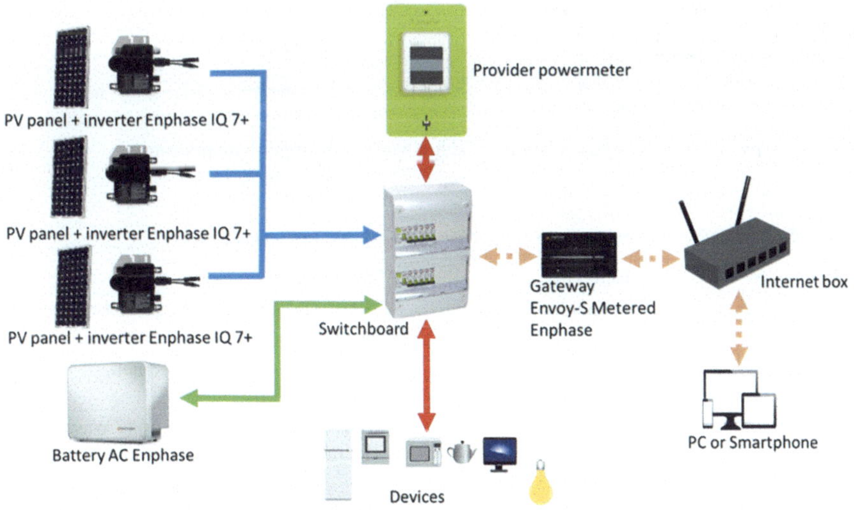

Fig. 38 Autoconsumption system monitoring composed of PV production based on micro-inverters and energy storage system

Table 4 References and properties of photovoltaic production and storage system

Meter	Reference	Measurements	Communication
Electrical production and storage	Enphase Envoy-S Metered	– Transmit PV production for each micro-inverter and Battery storage	Wifi
PV panel + Inverter	Enphase IQ7	– Monitoring of production	Power-line communication (PLC)
Battery	Enphase AC Battery	– Control strategies: • Self-consumption • Dynamic tariff adaptation • Limitation of energy injection	Power-line communication (PLC) and TCP/IP

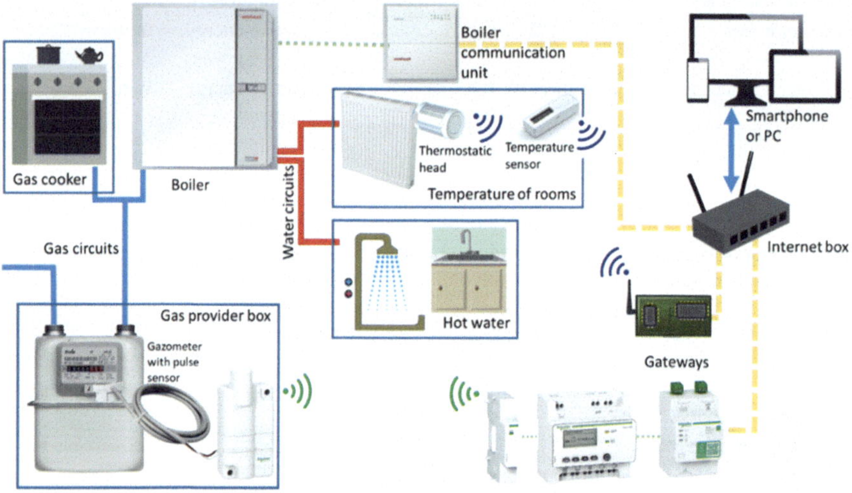

Fig. 39 Typical IoT for gas-based heating and hot water consumption monitoring

Table 5 References and properties of gas-based heating and hot water meters

Meter	Reference	Measurements	communication
Main gas meter	Schneider electric Wiser Link	– Pulse-based gas consumption (m^3)	Radio + Ethernet
Boiler proprietary meter	Weishaupt WCM-COM	– All data provided by boiler	Ethernet
Heater regulator	Danfoss Living connect Z LC13 POPP	– Heater temperature	Radio (ZWave)
Ambiance sensor	NODON NOD_STP-2-1-05	– Temperature – Humidity	Radio (EnOcean)

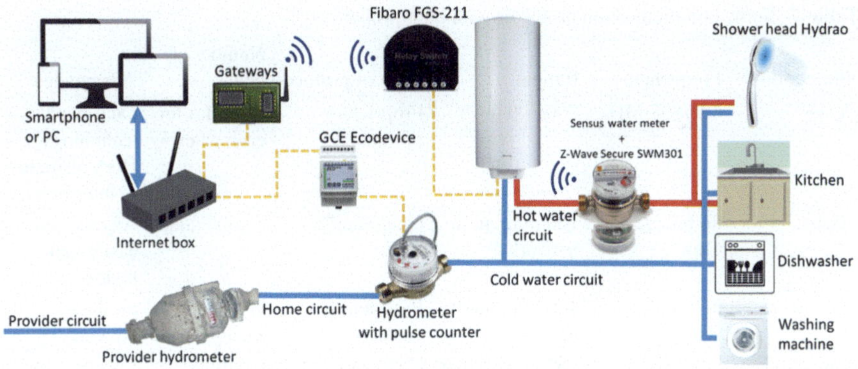

Fig. 40 Typical IoT for water consumption monitoring

Table 6 References and properties of water meters

Name	Provider	Measurements	Communication
Main water meter	GCE electronics ecodevice	– Pulse meter	Ethernet
Domestic hot water meter	Sensus SWM301	– Pulse meter	Z-Wave
Shower meter	Hydrao shower head	– Consumption	Bluetooth
Hot water switch	Fibaro FGS-211	– Smart switch	Z-Wave

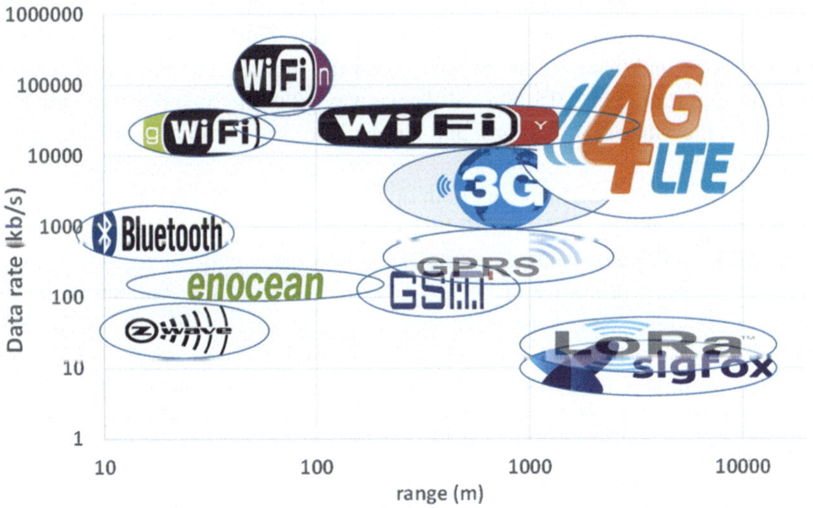

Fig. 41 Range/data rate of the main wireless protocols

4.4.2 Will There Be Only One Standard?

Manufacturer are using different standards for wireless communication protocols. A non-exhaustive list with some properties is given in Table 7.

Table 7 Some communication protocols with their properties

Protocols	Frequencies	Range	Type of protocol	Number of nodes	Advantages
X3D	434/868 MHz	200/300 m	Proprietary protocol	16 nodes for each gateway	Dual-band technology. High resilience to interference
Zigbee	868mhz/**2.4 Ghz**	100 m in mesh network	OpenSource	65,000 nodes	Anyone can easily make nodes
Zwave	868 MHz in Europe	100 m in mesh network	Proprietary protocol	232 nodes for each gateway	meshed network
Enocean	868 MHz in Europe	200 m with max 2 jumps	Open and interoperable	unlimited	Energy-efficient technology
Bluetooth Low Energy	2.4 GHz	100 m	Open	unlimited	low latency
Wifi	2.4Ghz/5 GHz	250 m	Open	256	Technology found in almost all buildings

In the past few years, there has been a battle for the short-range, low-power protocol for smart home IoT applications between ZigBee and Thread. ZigBee started in 2005 and has millions of devices on the market. Thread is from Google Nest Labs and started in 2015. Thread/Weave was a Google/Nest play but now joins forces with the other two: Amazon and Apple. The new standard will be managed under Zigbee[9].

ZigBee (3.0/pro) and Thread are both open standard builds on the same physical and link layer protocol stacks (IEEE 802.15.4). Whereas their biggest competition in this space, Z-Wave, is using a proprietary Z-Wave standard.

Zigbee operates primarily in the 2.4 GHz radio band; however, some devices operate in the higher end of the MHz range (e.g., 868 MHz in EU, 915 MHz in the US).

Z-Wave is another mesh network; however, it operates at a lower frequency band of 918/860 MHz. This allows for a better device-to-device signal range at the cost of reduced data rates.

In contrast to the above technologies, Thread is a much younger and less established mesh networking standard. It is also built on IEEE 802.15.4 using 2.4 GHz radiofrequency. It is defined up until the Application Layer, which means that other application layer protocols such as MQTT can be used [51].

[9]Hui Fu, The IoT Smarthome Battlefield: A Jointly Endorsed IoT Standard for the Home Area Network (HAN), IoT for all, February 12, 2020. Ref: https://www.iotforall.com/connected-home-over-ip/

The question of one unique protocol is still open, but the history shows that it is better to invest on open and interoperable standards instead of waiting for the Holy Grail.

4.4.3 Energy Harvesting

The EnOcean solution has particularly benefited from the fact that it integrates energy recovery solutions to power its sensors making them autonomous and without intervention. This is, for example, the EnOcean PTM210 switch with an ECO 200 harvester (Fig. 42) which by mechanical-magnetic conversion will generate a pulse of electrical energy sufficient to supply the transmission and reception of a radio signal frequency.

For the EnOcean heater thermostat, the Seebeck effect (inverse of the Peltier effect) is used to transform a temperature gradient into electrical voltage. A very low voltage conversion module (ECT 310) is then necessary to exploit this energy to transmit the RF signal.

A more common energy recovery is that coming from light radiation by the use of photovoltaic cells integrated into the sensor. These include the EnOcean temperature and humidity measurements (Fig. 43a) or the Z-Wave anemometer (Fig. 43b).

In this area, the French company Enerbee[10] has developed an innovative solution for generating energy from all types and speeds of movement, for comfort and air quality applications based on HVAC control. Energy peaks are converted to useable energy delivering energy in the 100 μW to 10 mW range, which can be stored in a supercapacitor and managed using ultra-low leakage power management (Fig. 44).

These battery-less power solutions still pose some difficulties. Regarding the temperature sensor, it is necessary for the O2Line model (Fig. 43a) to have an average brightness greater than 100Lux (i.e., > 300 Lux, 8 h per day). However, certain areas of the buildings are particularly dark like some corridors. Thus, data

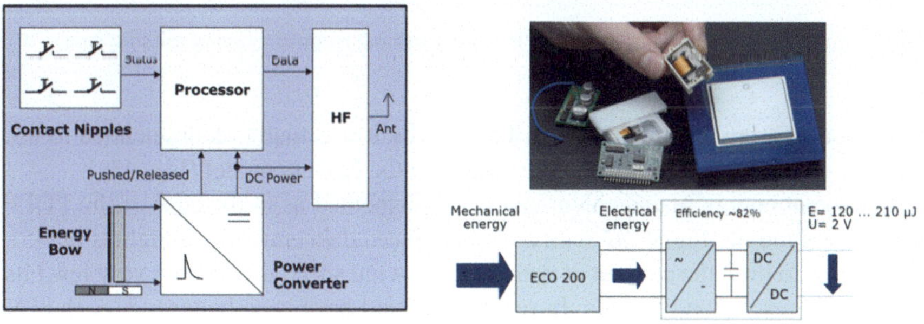

Fig. 42 EnOcean, Energy harvesting. (Source: EnOcean)

[10] www.enerbee.fr

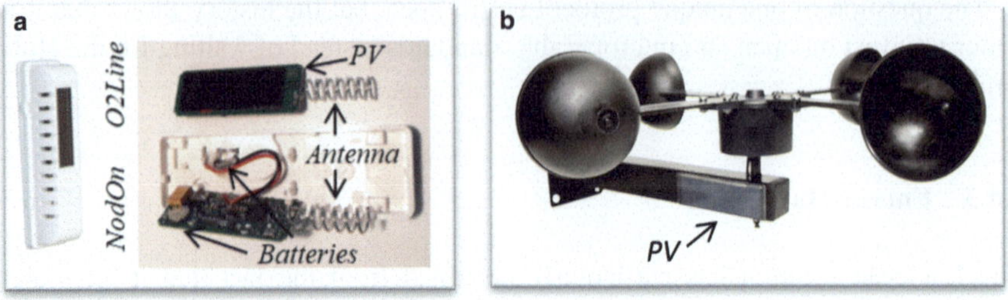

Fig. 43 Photovoltaic (PV) energy harvesting. (**a**) EnOcean temperature and humidity sensor (NodOn and O2Line). (**b**) Z-Wave anemometer (POPP Z-Weather)

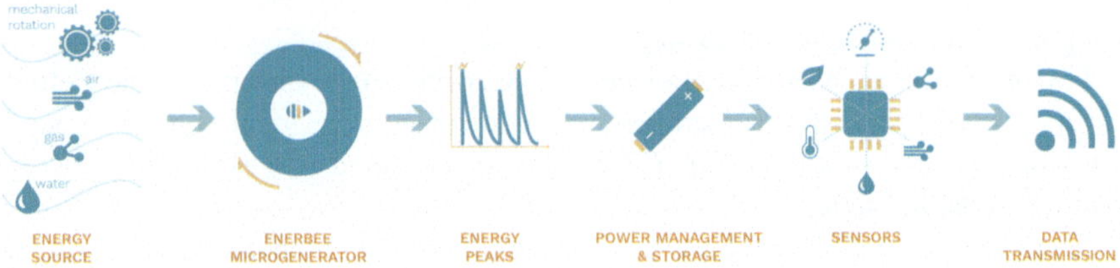

Fig. 44 Mechanical rotation energy recovery. (Source: Enerbee)

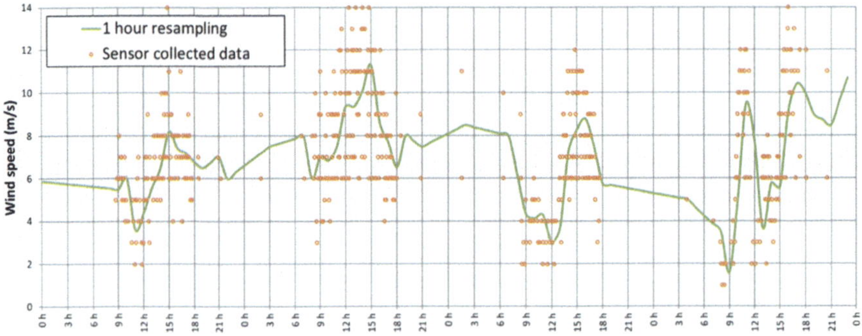

Fig. 45 Anemometer measurements (POPP Z-Weather) and reconstruction of missing data

gaps are appearing during the night. This is why it is actually recommended to use versions including batteries, as is the case with the NodOn model (Fig. 43a).

Intelligent energy management can be implemented as in the case of the POPP Z-Weather module (Fig. 43b), with a much reduced data emission at night (Fig. 45). In addition, to limit the amount of data sent, the wind speed is coded on very few bits with a poor resolution of 1 m/s. In reality, it would have been better to size the PV cell and a storage allowing the transmission of the measurements more frequently because the reconstruction of the missing data is of very poor quality (Fig. 45).

5 Open-Source Home Automation

5.1 Open-Source Projects

5.1.1 Home Automation Software

We have seen in the previous section that commercial home automation proprietary solutions are available (Table 2), based on IoT with wireless communication and a cloud architecture. But free and Open-Source Software (OSS) home automation platform are numerous today as it is possible to find many[11] like: **openHAB** which can integrate with over 1500 devices and which has one of the biggest community with 33,000 members; **Home Assistant**, similar to openHAB, very flexible from the developer side; **Jeedom**, well known mostly for French community; **Domoticz**, with many step-by-step guidance on their web site.

These software have become more and more popular thanks to cheap nano-computers such as the well-known Raspberry Pi where these solution can be installed easily by end-users. The classical architecture presented in Fig. 33 is becoming the following one (Fig. 46) with an internal structure based on a main Core System using different plugins or add-on and a local database.

Compared to a fully cloud-based platform, these solutions are keeping data locally with the possibility or not to expose them on the Internet. Some solutions are easy to administer, others require more skills in network configuration and time on community forums. The IoT can then be connected to this home automation server,

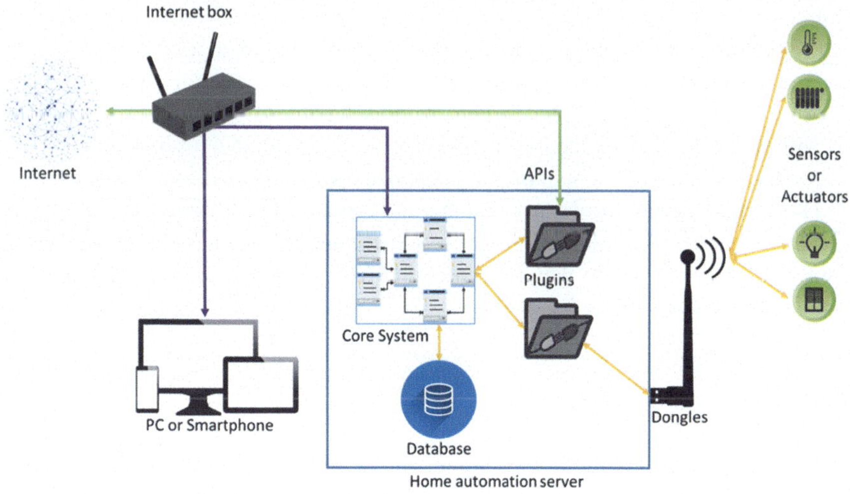

Fig. 46 Smart Home solution with local server implement of low-cost nano-computer

[11] https://ubidots.com/blog/open-source-home-automation/

Table 8 Smart Home open-source project programming standards and technologies

	Programming language	Database	Configuration
Openhab	Java	RRD4J	Textual and graphic
Home Assistant	Python	SQLite	Yaml files
Jeedom	PHP	Mariadb	Web interface
Domoticz	LUA	SQLite	Web interface

which must therefore communicate in all the protocols involved, and have physical equipment supporting the communication. Conventionally, a module (expansion card or dongle) is required per protocol, hence the need to limit the number of protocols in a single installation to facilitate interoperability.

The software is then able to code/decode the communication frames transmitted to/by the IoT. This layer can be provided with the device, available as OSS (OpenZWave), or reimplemented in specific environments. This is the case, for example, with openHAB and Jeedom where the EnOcean drivers have been written from specifications. The main feature of these environments is to integrate the different technologies in an agnostic way (Abras, [52]) in order to treat them in the upper layers independently of the wireless communication layer.

They generally allow to interact directly with the system in read/write (sensor/actuator), to archive the data in databases, and to visualize them. OSS generic solutions are also available for time series database (e.g., InfluxDB) and for visualization (e.g., Grafana).

Most of home automation project use similar architecture but with different standards and technologies for programming language, database, and configuration (Table 8).

Programming technologies are becoming major characteristics for OSS since users are interested in understanding the code and may contribute to the software. If a user wants a plug and play system without having to do much programming, he will probably choose Jeedom or Domoticz, while another user wants to fully customize its interface, he will probably choose openHAB or Home Assistant.

This can also bring third-party commercial services. For instance, the simplicity of Jeedom is counterbalanced through some plugins which are not free (e.g., EnOcean protocol), while it is free in openHAB, but require more time and tips to be implemented.

5.1.2 Low-Cost Hardware

As already illustrated with nano-computer and the success of Raspberry Pi for home automation servers, recent advances in wireless technologies and embedded systems, based on Open-Hardware (Arduino), ANT wireless technology (nRF24L01+ module), low-cost wireless sensors and actuators network (WSAN) for building energy services is available [53]. Moreover, unlike commercial products, this WSAN is customizable and easy to be extended for adapting different research

Fig. 47 Arduino Pro mini and nRF24L01+ module

Fig. 48 nRF52 chip and its communication antenna which can also be used for energy harvesting

situations. RF24Network is a network layer for Nordic nRF24L01+ radios running on Arduino-compatible hardware (Fig. 47).

On the other hand, connected things are not necessarily intended to fit into such a wireless network, but can simply be connected to the Internet through the mobile phone using bluetooth then 4G, or through the home automation box using Wifi then ADSL or fiber. Solutions integrating this type of communication become very affordable with the rise of the IoT and announce very low consumption. For example, Nordic Semiconductor's NRF52 (Fig. 48) is based on an ARM Cortex-M4 processor incorporating a 2.4 GHz transmitter for Bluetooth Low Energy (BLE) communication, as well as the Thread protocol. For its part, the Chinese company Espressif offers the ESP32 (about 5 €), a more powerful solution (double heart clocked at 240 MHz backed by 4 MB flash memory) integrating in addition to BLE, WiFi and a cryptographic chip supporting the latest data security standards.

Both software bricks and hardware components are available on the shelf in order to create innovative open-source project, with active community, in order to promote energy efficiency, sobriety and make our home smarter and ready for the energy transition.

5.1.3 Definition and History of Open-Source Projects

First of all, it is important to understand the difference between OSS and free software. OSS can be the base of commercial software (e.g., sold with hardware and/or services), and free software can be proprietary software. The key difference between OSS and proprietary software is that the OSS publishes the source code whereas the proprietary software retains the source code.

Motivations to create OSS are various among a desire for improved transparency, or new business models based on service rather than in software, but it is probably mainly for involving many people within a dedicated home energy management community. In the current context of struggle against climate change, it is a great challenge to develop new collaborative and open organizations, involving citizen through energy communities. A good home automation needs an equally strong community that is willing to back it up and improve upon its initial state.

The general idea of Open Source (OS) dates back to the 1970s through projects driven by electronic enthusiasts such as the Homebrew Computer Club[12], which aimed to exchange ideas and components in order to create computers. In the 1980s, the movement weakened as most activists joined Silicon Valley businesses. Although Open-Source Software (OSS) was well established, it was not until the early 1990s that Open-Source Hardware (OSH) regained a second youth thanks to the advent of FPGA (Field-Programmable Gate Array).

Since the concept of OS, it has widened and has touched more and more areas. In recent years and with the advent of the Arduino project [54], we have seen an explosion of Do It Yourself (DIY) projects where the spirit of open source is a real driver. This OSS and OSH card project opened access to the "smart" part of the projects because it was designed like a real electronic Swiss Army knife and can be made by anyone with a minimum of hardware and programming knowledge.

It is important to distinguish hardware from software because OSH only covers the creation of material products[13]. However, if software is necessary to operate OSH, it may be required by the various existing licenses that the interfaces be sufficiently detailed to allow writing OSS to ensure its essential functions.

In recent years, a debate has been animating the community through to find out the difference between the term free software and OSS. Fortunately, the nuance is clearer with regard to the hardware because an OSH is supplied with plans and diagrams allowing everyone to be able to reproduce it, while an open hardware is supplied with complete specifications allowing a user to interact with it without necessarily knowing what is going on inside. Most of the time, OSHW depends on open hardware.

As an example, 3D printers use stepper motors which one can have all the specifications to operate them without needing the information necessary for their production.

[12]https://www.computerhistory.org/revolution/personal-computers/17/312

[13]https://www.oshwa.org/definition/

5.1.4 Efficient Open-Source Projects

The community structure that is generally found in Open-Source project is called a contributor funnel[14]. It is like the four basic sales funnel stages (Awareness, Interest, Decision, Action) but applied on members roles (Users, Contributors, Maintainers) as described in Fig. 49. It means that the project has to bring awareness and interest to users, which can decide to become a contributor and make actions until becoming a maintainer, with higher rights on the project.

In order to bring user from the top of the funnel to the way down, it is important to ensure that users have easy victories as a contributor to encourage them to do more. Efforts are also made in terms of documentation because the majority of open-source contributors are "occasional contributors," because they do not necessarily have time to get to know the whole project.

In 2017, GitHub conducted a survey on open source[15] and demonstrated that incomplete or confused documentation is an obstacle for most open-source users. This is why the projects that work have good documentation, invites people to interact with the project and to contribute to it. Gathering points are also created through the establishment of forums and are also an important part of this kind of

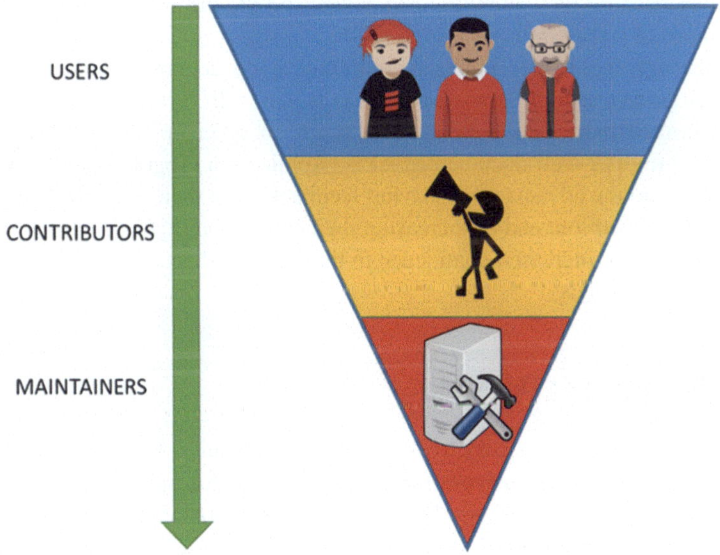

Fig. 49 Open-source community, contributor funnel. (Source: McQuaid [55])

[14] https://mikemcquaid.com/2018/08/14/the-open-source-contributor-funnel-why-people-dont-contribute-to-your-open-source-project/

[15] https://opensourcesurvey.org/2017/

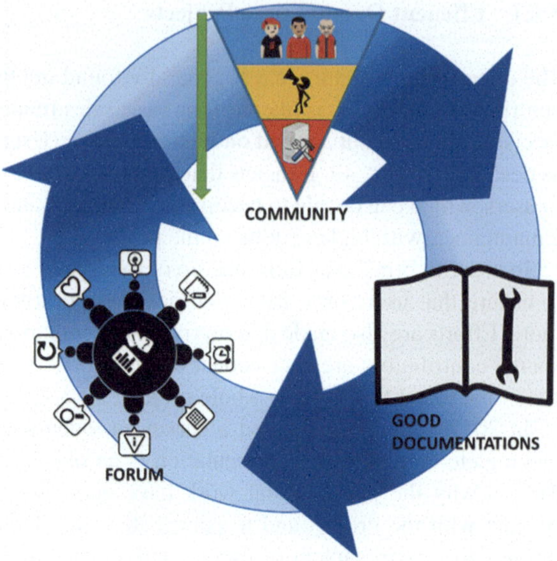

Fig. 50 Ingredients for an effective open-source project

community because they allow to open debates on the project and to submit ideas on the future of it (Fig. 50).

It is through these three pillars (community, good documentation, and forum) that the virtuous circle of open-source projects exists. In recent years, this scheme has encouraged more and more manufacturers to open up the sources of their product and to offer well-documented APIs (Application Programming Interfaces). Companies find the advantages of having feedback from their users, of being more reactive to competition, and of increasing their brand image. At the user level, this also translates into increased confidence in the purchased products and a feeling of listening to their needs.

5.1.5 How to Protect This Model?

Most of the time when a creative work is done (code, plans, ...) the law indicates that only the author has an exclusive copyright by default. This implies that no one else can use, copy, distribute, or modify this work without risking litigation and legal consequences. Regarding open source, the author expects just the opposite because he wants others to use, modify, and share his work. It is because of this legal flaw that the author of an open-source project needs a license that explicitly states these permissions.

There are several types of licenses more or less open rights to the use, modification, and distribution of the work that has been made by a contributor. There are

Table 9 Creative commons licensing

	(BY) Attribution	(ND) No Derivatives	(NC) Non-Commercial	(SA) Share Alike
CC0				
CC-BY	X			
CC-BY-ND	X	X		
CC-BY-NC-ND	X	X	X	
CC-BY-NC	X		X	
CC-BY-NC-SA	X		X	X
CC-BY-SA	X			X

different criteria in order to choose the appropriate open-source license[16]; it will depend on the strategy, commercial and/or community aim of the project.

In the software domain, there are two main licenses:

- The MIT License which is the simplest and most permissive because it allows people to do almost anything they want with your project, like making and distributing closed source versions.
- The GNU GPLv3 license which will protect and ensure that improvements and modifications made by someone are always distributed open.

In the non-software domain, there are also creative commons licenses[17] which, depending on the options chosen, will more or less authorize certain rights of use, modification, or distribution. It is thanks to these licenses that Open-Source projects have a legal framework and guarantee freedom to share the ideas emerging in them. The six main ones are listed in Table 9 based on the following criteria:

- Attribution (BY): Licensees may copy, distribute, display, and perform the work and make derivative works and remixes based on it only if they give the author or licensor the credits (attribution) in the manner specified by these.
- Share-alike (SA): Licensees may distribute derivative works only under a license identical ("not more restrictive") to the license that governs the original work (copyleft).

[16]https://choosealicense.com/

[17]https://creativecommons.org/

- Non-commercial (NC): Licensees may copy, distribute, display, and perform the work and make derivative works and remixes based on it only for non-commercial purposes.
- No Derivative Works (ND): Licensees may copy, distribute, display, and perform only verbatim copies of the work, not derivative works and remixes based on it.

5.2 Review of Some Smart Home Projects

In this last part, we would like to detail some existing open-source projects relative to smart home.

5.2.1 Smart Citizen Kit

The main objective of the Smart Citizen project[18] is to offer citizens easy-to-access measurement tools so that they can get involved in local environmental pollutions. This project gives free access to an Open-Source kit capable of capturing and analyzing various environmental data in real time. This project offers the possibility of purchasing the components and/or of assembling them itself as well as of making modifications in the code provided. Measures are about weather conditions (Temperature, Humidity, Air pressure), Light pollution, Air Quality (Indoor/Outdoor), Noise Pollution.

It is a project resulting from the collaboration between the Institute of Advanced Architecture of Catalonia and the "Fab Labs" of Barcelona. Currently, there are 230 active stations around the world that can be localized in a map[19] (Fig. 51).

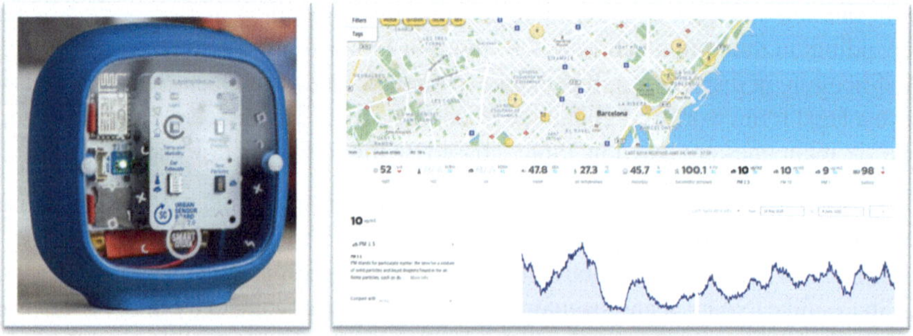

Fig. 51 Smart Citizen kit and Internet world map for online monitoring

[18]https://www.seeedstudio.com/smartcitizen

[19]https://smartcitizen.me/kits/1352

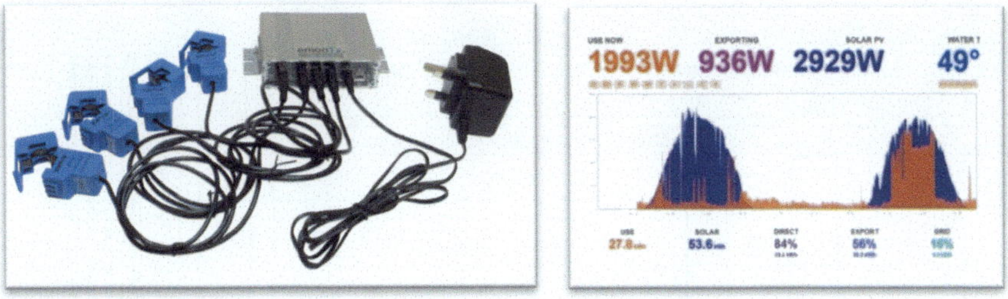

Fig. 52 Open Energy Monitor device with four current measurement, and monitoring

5.2.2 Open Energy Monitor

This project[20] aims to develop tools to help people who want to understand energy systems and their use. It is aimed at all types of profiles, from the novice who can buy the material already made to the expert who can adapt the source codes as well as the hardware in order to match his needs.

It is more a project aimed at monitoring and understanding energy in the building than a system dedicated to home automation.

The measurement systems currently available are:

- Electricity, gas, and water consumptions monitoring
- PV production
- Electric vehicle charging monitoring
- Monitoring for heat pump
- Monitoring of climatic conditions

This project is supported by fifteen participants from all backgrounds and draws on feedback from its user community (Fig. 52).

5.2.3 A4H Smart Home

The smart home of Amiqual4Home[21] is a space of about 90 m^2 simulating a home. It serves as a tool for usage experiments actors who work in the field of research and innovation on smart housing. The 87 m^2 was renovated and equipped with home automation systems, multimedia, water and electricity meters, and means for observing human activity. It is also equipped with all the actuators capable of controlling all the devices present such as the kettle, the lights, or the roller shutters. All measurements and equipment are linked to a central home automation system which allows an operator to act as the wizard of OZ during the experiments.

[20] https://openenergymonitor.org/

[21] https://amiqual4home.inria.fr/tools/smart-home/

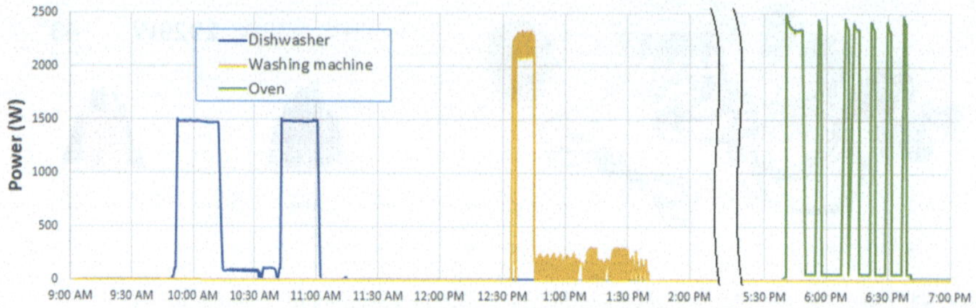

Fig. 53 Three main power consumption profiles of home electric appliances

This smart home was for instance used to publish open dataset[22] (CC-BY) containing real-life sensor data of a person living in a smart home. It is a high-quality dataset with a dense but nonintrusive sensor infrastructure [56]. A deep sensing approach was used with over 200 variables measured. These include all doors (rooms, cabinets, fridge) state, light states, temperature, CO_2 levels, noise levels, weather, appliances (oven, stove, TV, coffee maker...) state. Nine daily living activities are self-annotated (taking shower, using toilet, sleeping, cooking, going outside, washing dishes, eating, and working).

Figure 53 plots three main home appliances consumption, namely the dishwasher, the washing machine, and the oven, extracted for this open data base.

5.2.4 G2Elab Smart Home and Open-Source Tutorials

G2Elab Smart Home project provides data from a 120 m^2 household where a five people family is living in. This project grant an access to about 340 measuring points for scientists, accessible in real time through a Grafana portal with Influxdb database. There are measures of:

- Electricity, gas, and water consumption of each device
- Temperature, humidity, brightness of each common room
- Opening position of each door and window
- Motion sensors
- The state of each light
- Air analysis of each room
- Outdoor weather conditions

Expe-Smarthouse is developed based on Open-Source Hardware and Software. Many tutorials have been made based on this house and from students projects from G2Elab (Grenoble Electrical Engineering lab), and posted in miniprojets.net website

[22]https://data.mendeley.com/datasets/fcj2hmz5kb

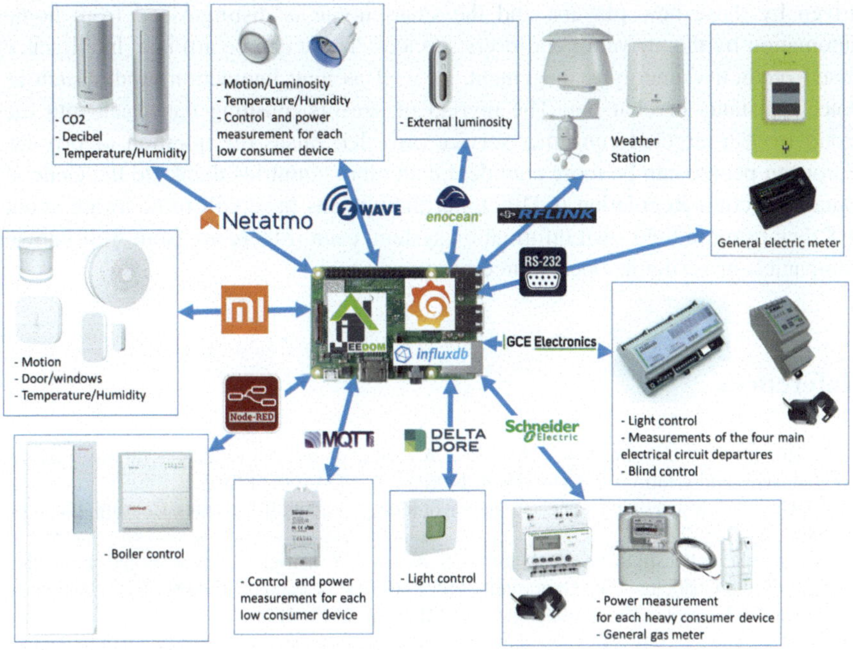

Fig. 54 G2Elab Smarthouse devices and multi-protocol gateway

(in French). It is then accessible for citizens who want to set up their own smart home with commercial or homemade hardware and software (Fig. 54).

6 Conclusions

Is open source the solution of the future for the democratization of the smart home? We wanted to finish with this last part on open source to show that movements complementary to commercial solutions could find their place in this highly technological field. The massive arrival on the market of low-cost IoT, connecting directly to the Internet, or through home automation gateways greatly modifies the previous monolithic paradigm of a single solution provider.

Interoperability and the openness of standards has for several years shown its interest, and now open-source continues this advance to offer ever more accessible services to citizens. Standards from the world of building automation, necessary to structure and optimize an industrial organization, have given way to agile solutions exploiting plug-and-play and cloud infrastructure. These solutions do not require the intervention of experts in situ, and thus optimize the benefits.

IT players have entered this market historically occupied by manufacturers from building management systems. Today, developments and innovations are largely driven by these new players, and the smart home is distinguished from home automation by the arrival of end-users services. These can be artificial intelligence based predictive energy management, as well as new interaction modes such as voice assistants or chat bot. The myriad of sensors, pushing data back into the cloud from a targeted consumer service provider, raises the question of privacy. European people can be more confident than other countries thanks to the General Data Protection Regulation (GDPR), but the citizens have still to be aware about IoT they bring in their own information system, even if there are from well-known companies, or verifiable open-source solutions.

References

1. L. Biljana, R. Stojkoska, K.V. Trivodaliev, A review of Internet of Things for smart home: challenges and solutions. J. Clean. Prod. **140**(Part 3), 1454–1464 (2017)
2. J.J. Greichen, Value based home automation or today's market. IEEE Trans. Consum. Electron. **38**(3), 34–38 (1992)
3. A.J. Bernheim Brush, B. Lee, R. Mahajan, S. Agarwal, S. Saroiu, C. Dixon, Home automation in the Wild: Challenges and Opportunities, in CHI'11 Proceedings of the SIGCHI Conference on Human Factors in Computing Systems (2011), pp. 2115–2124
4. M. Wigginton, J. Harris, *Intelligent Skins* (Routledge, London, 2002). https://doi.org/10.4324/9780080495446
5. A. Ghaffarianhoseini, U. Berardi, H. Alwaer, S. Chang, E. Halawa, A. Ghaffarianhoseini, D. Clements-Croome, What is an intelligent building? Analysis of recent interpretations from an international perspective. Archit. Sci. Rev. **59**, 338–357 (2015). https://doi.org/10.1080/00038628.2015.1079164
6. C. Wilson, T. Hargreaves, R. Hauxwell-Baldwin, Smart homes and their users: a systematic analysis and key challenges. Personal Ubiquitous Comput. **19**(2), 463–476 (2015). https://doi.org/10.1007/s00779-014-0813-0
7. H. Obara, Schneider Electric – "Standardization in EE in buildings, in Sustainable Places 2015 Conference, Savona, Italy, 16–18 September 2015
8. D.S. Leitner, N.C. Sotsek, A. de Paula Lacerda Santos, Postoccupancy evaluation in buildings: systematic literature review. J. Perform. Constr. Facil. **34**(1) (2020)
9. G.M. Di Giuda, L. Pellegrini, M. Schievano, M. Locatelli, F. Paleari, BIM and post-occupancy evaluations for building management system: weaknesses and opportunities, in *Digital Transformation of the Design, Construction and Management Processes of the Built Environment*, ed. by B. Daniotti, M. Gianinetto, T. S. Della, (Research for Development. Springer, Cham, 2020)
10. W. Jiang, X. Zhao, Trust and the intent to cooperate in energy performance contracting for public buildings in China. Eng. Constr. Archit. Manag. **28**, 372–396 (2019). https://doi.org/10.1108/ECAM-07-2019-0385e
11. A.G. Aoun, Blockchain application in energy performance contracting. Int. J. Strat. Energy Environ. Plann. **2**(2) (2020)
12. K. Ehrhardt-Martinez, K.A. Donnelly, J. Laitne, *Advanced Metering Initiatives and Residential Feedback Programs: A Meta-Review for Household Electricity-Saving Opportunities* (American Council for an Energy-Efficient Economy, Washington, DC, 2010)
13. T. Märzinger, D. Österreicher, Supporting the smart readiness indicator—a methodology to integrate a quantitative assessment of the load shifting potential of smart buildings. Energies **12**, 1955 (2019)

14. N. Wang, Transactive control for connected homes and neighbourhoods. Nat. Energy **3**, 907–909 (2018). https://doi.org/10.1038/s41560-018-0257-2
15. H.-S. Jin, B.-H. Choi, J.-K. Kang, S.-I. Kim, J.-H. Lim, S.-Y. Song, Measurement and Normalization Methods to Provide Detailed Information on Energy Consumption by Usage in Apartment Buildings. Energy Procedia **96**, 881–894 (2016). https://doi.org/10.1016/j.egypro.2016.09.161
16. J. Salom, A.J. Marszal, J. Widén, J. Candanedo, K.B. Lindberg, Analysis of load match and grid interaction indicators in net zero energy buildings with simulated and monitored data. Appl. Energy **136**, 119–131 (2014). https://doi.org/10.1016/j.apenergy.2014.09.018
17. IEA, *Building Energy Performance Metrics* (IEA, Paris, 2015)., https://www.buildingrating.org/sites/default/files/1448011796IEA_IPEEC_BEET4_Final_Report.pdf
18. D. Van Orshoven, D. van Dijk, EPB standard EN ISO 52003: How to put the EPB assessment outputs to intelligent use, REHVA J. (2016)
19. S. Verbeke, Y. Ma, P. Van Tichelen, S. Bogaert, V. Gómez Oñate, P. Waide, K. Bettgenhäuser, J. Ashok, A. Hermelink, M. Offermann, et al., *Support for Setting Up a Smart Readiness Indicator for Buildings and Related Impact Assessment, Final Report; Study Accomplished under the Authority of the European Commission DG Energy 2017/SEB/R/1610684* (VITO NV, Mol, Belgium, 2018)., https://smartreadinessindicator.eu
20. J. Hogeling, The set of EN and EN ISO EPB standards: supporting the implementation of the EPB Directive in Europe, in EBC-Annex 71 Symposium The Building as Cornerstone of our Future Energy Infrastructure, Bilbao, Spain, 10–11 April 2019. https://dynastee.info/wp-content/uploads/2019/06/6_Hogeling-Presentatie2019-04-Bilbao-EBC-Annex71.pdf
21. H.A. Gabbar, *Building Energy Management Systems (BEMS), Part1, Chapter2 in Energy Conservation in Residential, Commercial, and Industrial Facilities* (Wiley, 2018). https://doi.org/10.1002/9781119422099.ch2
22. Research and Markets, Global Building Management System Market 2016–2023 (2017). https://www.prnewswire.com/news-releases/global-building-management-system-market-2016-2023-300492250.html
23. M. Manic, D. Wijayasekara, K. Amarasinghe, J.J. Rodriguez-Andina, Building energy management systems: the age of intelligent and adaptive buildings. IEEE Ind. Electron. Mag. **10**, 25–39 (2016). https://doi.org/10.1109/MIE.2015.2513749
24. Elizabeth Gasiorowski-Denis, ISO 52000 leads the way on clean energy building solutions, ISO News, 28 June 2017. https://www.iso.org/news/ref2196.html
25. GIMELEC, Mener à bien un projet d'efficacité énergétique Bâtiments et collectivités, Guide De Bonnes Pratiques, Efficacité Energétique. https://www.ac-paris.fr/portail/jcms/p1_529163/guide-ee-batiment-avril2008
26. N. Artiges, A. Nassiopoulos, F. Vial, B. Delinchant, Calibrating models for MPC of energy systems in buildings using an adjoint-based sensitivity method, Energy and Buildings, 109647 (2020). ISSN 0378-7788. https://doi.org/10.1016/j.enbuild.2019.109647
27. K. Verbert, R. Babuška, B. De Schutter, Combining knowledge and historical data for system-level fault diagnosis of HVAC systems, Engineering Applications of Artificial Intelligence **59**, 260–273 (2017). ISSN 0952-1976. https://doi.org/10.1016/j.engappai.2016.12.021
28. Calvert Controls, Internet of Things – "IoT" – the next evolution in smart building management (2020). http://calvertcontrols.com/index.php/2019/05/21/internet-of-things-iot-the-next-evolution-in-smart-building-management/
29. A. Totonchi, *Internet of Things for Smart Home: State-of-the-Art Literature Review* (2018)
30. O. Nývlt, Buses, Protocols and Systems for Home and Building Automation, Department of Control Engineering Faculty of Electrical Engineering Czech Technical University in Prague, 2009–2011. http://www.tecnolab.ws/pdf/Buses,%20Protocols%20and%20Systems%20for%20Home%20and%20Building%20Automation.pdf
31. Siemens Building Technologies. Communication in building automation, Answers for infrastructure and cities. https://www.downloads.siemens.com/download-center/Download.aspx?pos=download&fct=getasset&id1=A6V10209534

32. D. Bonino, F. Corno, D. Russis, Luigi, A semantics-rich information technology architecture for smart buildings. Buildings **4**, 880–910 (2014). https://doi.org/10.3390/buildings4040880
33. Gartner, Inc., Market Insight: The Move From the Connected Home to the Intelligent Home, Technical Report G00355564, Jul. 2018. [Online]. https://www.gartner.com/doc/3876868/market-insight-connected-home-intelligent
34. O. Hamdan, H. Shanableh, I. Zaki, A.R. Al-Ali and T. Shanableh, IoT-based interactive dual mode smart home automation, in 2019 IEEE International Conference on Consumer Electronics (ICCE), Las Vegas, NV, USA, 2019, pp. 1–2, https://doi.org/10.1109/ICCE.2019.8661935
35. Smart Homes Market Report: Trends, Forecast and Competitive Analysis, Research and Markets, Global Report n°4846240, September 2019. https://www.researchandmarkets.com/reports/4846240/smart-homes-market-report-trends-forecast-and
36. Z. Yang, J.H. Cho, Application and development trend of smart home in residential interior design Journal of Physics: Conference Series, Volume 1487, in 2020 4th International Conference on Control Engineering and Artificial Intelligence (CCEAI 2020) 17–19 January 2020, Singapore (2020)
37. B. Qolomany et al., Leveraging machine learning and big data for smart buildings: a comprehensive survey. IEEE Access **7**, 90316–90356 (2019)
38. G. Lobaccaro, S. Carlucci, E. Lfstrm, G. Lobaccaro, S. Carlucci, and E. Lfstrm, "A review of systems and technologies for smart homes and smart grids," Energies, vol. 9, no. 5, p. 348, 2016. https://www.mdpi.com/1996-1073/9/5/348
39. ITU-T SSCIoT 2, *Unleashing the Potential of the Internet of Things* (Implementing ITU-T International Standards to Shape Smart Sustainable Cities, International Telecommunication Union (ITU), 2016). https://www.itu.int/pub/T-TUT-SSCIOT-2016-2/fr
40. Vision and Challenges for Realizing the Internet of Things, CERP-IoT (Cluster of European Research Projects on the Internet of things), Publication Office of The European Union, March 2010. http://bookshop.europa.eu/en/vision-and-challenges-for-realising-the-internet-of-things-pbKK3110323/
41. C. Perera et al., Context aware computing for the internet of things: a survey. IEEE Commun. Surveys Tutorials **16**(1), 414–454 (2014)
42. V.M. Tayur, R. Suchithra, Review of interoperability approaches in application layer of internet of things, in International Conference on Innovative Mechanisms for Industry Applications (ICIMIA 2017)
43. M. Uslar, S. Rohjans, M. Specht, Technical requirements for DER integration architectures. Energy Procedia **20**, 281–290 (2012). https://doi.org/10.1016/j.egypro.2012.03.028
44. A. Sajid, H. Abbas, K. Saleem, Cloud-assisted IoT-based SCADA systems security: a review of the state of the art and future challenges. IEEE Access **4**, 1375–1384 (2016). https://doi.org/10.1109/ACCESS.2016.2549047
45. O. Alrawi, C. Lever, M. Antonakakis, F. Monrose, SoK: security evaluation of home-based IoT deployments, in 2019 IEEE Symposium on Security and Privacy (SP), San Francisco, CA, 2019, pp. 1362–1380. https://doi.org/10.1109/SP.2019.00013
46. E. Byres, M. Franz, D. Miller, The use of attack trees in assessing vulnerabilities in SCADA systems, in Proc. IEEE Int. Infrastruct. Survivability Workshop (IISW), Lisbon, Portugal, Dec. 2004
47. B. Fouladi, S. Ghanoun, Security evaluation of the Z-wave wireless protocol", Black Hat USA, vol. 1, pp. 1–6, Aug. 2013 "Black Hat 2013 – Honey, I'm Home!! – Hacking Z-Wave Home Automation Systems." Online video clip. YouTube. YouTube, 19 Nov 2013. Web 1 Nov. 2016. https://www.youtube.com/watch?v=KYaEQhvodc8
48. D. Caputo, L. Verderame, A. Ranieri, A. Merlo, L. Caviglione, Fine-hearing Google Home: why silence will not protect your privacy. J. Wireless Mobile Netw. Ubiquit. Comput. Depend. Appl. **11**, 35–53 (2020). https://doi.org/10.22667/JOWUA.2020.03.31.035
49. S. Wachter, Normative challenges of identification in the Internet of Things: Privacy, profiling, discrimination, and the GDPR. Comp. Law Secur. Rev. **34**(3), 436–449 (2018). https://doi.org/10.1016/j.clsr.2018.02.002
50. B.I. Gartner, Summit Summary: Internet of Things (IoT)—Buttigieg.org (2015)

51. G.M. Toschi, L.B. Campos, C.E. Cugnasca, Home automation networks: a survey. Comput Stand. Interf. **50**, 42–54 (2017). https://doi.org/10.1016/j.csi.2016.08.008
52. A. Shadi, C. Thomas, P. Stephane, D. Benoit, W. Frederic, M.P. Singh, Power management of laptops batteries in dynamic heterogeneous environments using iPOPO, IBPSA, 20–21 mai 2014, Arras (2014)
53. B. Delinchant, H.A. Dang, H.T.T Vu, D.Q. Nguyen, Massive arrival of low-cost and low-consuming sensors in buildings: towards new building energy services, in 2019 IOP Conf. Ser.: Earth Environ. Sci., Vol. 307, Conf. 1, 2019. https://doi.org/10.1088/1755-1315/307/1/012006
54. D. Kushner, The Making of Arduino, in IEEE Spectrum, Oct. 26th, 2011. https://spectrum.ieee.org/geek-life/hands-on/the-making-of-arduino
55. M. McQuaid, The Open Source Contributor Funnel: Turning Users Into Maintainers, in CodeConf 2016
56. P. Lago, F. Lang, C. Roncancio, C. Jiménez-Guarín, R. Mateescu, N. Bonnefond, The ContextAct@A4H Real-Life Dataset of Daily-Living Activities, in *Modeling and Using Context. CONTEXT 2017. Lecture Notes in Computer Science*, ed. by P. Brézillon, R. Turner, C. Penco, vol. 10257, (Springer, Cham, 2017)

Formalization of the Energy Management Problem and Related Issues

Stephane Ploix and Amr Alzouhri Alyafi

1 Introduction

Coping with climate challenges is usually dealing with reduction of energy needs, i.e. with better energy efficiency and sobriety. Nevertheless, reduction of environmental impacts also leads to the massive development of distributed energy production means to collect renewables all over a large territory. Except for hydropower plants, most renewable energy production means are fatal: they cannot be consumed when needed but they have to be used when energy is available. Consequently, until solutions for massive energy storage within a day, but also between seasons, become feasible and economically viable, energy has to be consumed when it is produced. Therefore, for now, tackling climate changes are not only a matter of reduction; it is also a matter of balancing consumption with production at any time. As seen in Chap. "Energy Sobriety: A Behavior Measurement Indicator for Fuel Poverty Using Aggregated Load Readings from Smart Meters", adjustments of consumption are named *demand response* mechanism. It aims at increasing the flexibility of energy use, which is very critical for electricity. Indeed, in France, for instance, 68% of the electricity is feeding residential and office buildings: it's obvious that they have an important role to play and that it will impact the way of life in buildings.

Various automatic "demand response" solutions have been considered: deactivation of electric heating and cooling systems for domestic hot water production [1],

S. Ploix (✉)
G-SCOP lab, Grenoble Institute of Technology, University of Grenoble Alps, CNRS UMR5272, Grenoble, France
e-mail: stephane.ploix@grenoble-inp.fr

A. A. Alyafi
LIGlab, University of Grenoble Alpes, CNRS UMR, Grenoble, France
e-mail: amr.alzouhri-alyafi@imag.fr

S. Ploix et al. (eds.), *Towards Energy Smart Homes*,
https://doi.org/10.1007/978-3-030-76477-7_4

for heating or cooling buildings, etc... [2]. However, these solutions remain difficult to implement today because of the diversity of appliances and their usage (domestic boilers are not always present, and if so, they are not always electric and, not always used when consumption reduction is needed, and not always controllable). Interruption of such appliances leads to a rebound effect to compensate the temperature decrease or increase [4]. It increases the risk of discomfort for users and the competition between energy distributors is discouraging these kinds of incentives because resellers prefer to set up seducing offers to attract new customers. As an alternative to these injunctive approaches, other approaches aim to involve and empower electricity users: for example, sending messages (nudges) indicating the green hours (it is recommended to consume more) and the red hours (try to reduce consumption) [6, 11] or analyses of practices, explanations, and suggestions for action. The latter approaches aim at involving energy consumers in the energy process rather than doing without their participation. This potentially opens up new horizons.

Encouraging self-consumption is an additional way to make people more concerned by energy mechanisms. Increasing the part of the energy produced locally in the consumption leads to a better involvement of actors in the everyday energy because the money they expend in energy is dependent of their own behavior. For instance, launching a washing machine is more interesting during sunny periods. Therefore, not only the quantity of energy matter but also the time it is used does also. Europe is promoting the massive use of energy from renewables with the 2018/2001 directive voted in December 11th, 2001 asking the European countries to allow collective self-consumption, i.e. the pooling of means of production on a micro-grid connected to the distribution network, in association with energy users, who must coordinate each other to consume preferably when renewable is available. For energy users, the consequence is that not only the quantity of energy matter, the time it is consumed but also what the neighbor are doing also comes up into the decisions of actions. However, this new context, when it concerns buildings and residential areas, cannot be easily implemented because the investment capacities of these types of actors are limited and prohibit expensive approaches for having the best use of residential micro-grids. A building embedding an artificial system, supporting the inhabitants of a building in the everyday energy management, is an *energy Smart-Building*, *energy Smart-Home*, or *energy Smart-Office*. The upcoming increasing complexity justifies such solutions: INSEE, a French survey institute, stated that people spend about 90% of daytime located in buildings, while smart-phone applications are mainly supporting outdoor activities like GPS, shopping, leisure time,... almost nothing is available for the remaining 90% of people time. Because of sobriety, flexibility, and coordination within communities, the complexity is becoming higher and user-friendly decision aiding solutions become more important.

2 Sobriety and Flexibility Issues at Dwelling Scale

A building is designed to provide services to inhabitants like protecting from the rains, snow, solar radiations,..., providing thermal comfort and good air quality at an affordable price. A building contains also specific appliances providing specialized services: washing and drying clothes, keeping and cooking the food, lighting, multimedia,... Therefore, a building can be seen as a place where a set of services are provided to people. Residential and office buildings are usually distinguished for several reasons that lead to different solutions for aiding people. In office building, CAPEX is usually higher and the diversity of inhabitant activities more limited: automation based on PLC, SCADA, Building Energy Management Systems more or less advanced are quite common. Indeed, because of the limited diversity of activities, it is relatively easy to guess occupants' intentions. Additionally, contrary to residential buildings, inhabitants are not building office managers. If automation seems relevant for office building, it is more questionable for residential buildings for exactly the opposite reasons: little CAPEX, high diversity of activities and people more demanding of comfort because they're generally the home building managers but with little expertise. In homes, automations are frequently limited to HVAC systems and possibly to triggering of boilers for Domestic Hot Water if there are storage tanks. Home solutions for aiding people in everyday energy are therefore more complex to design: they have to guess occupants' intentions; they have to be cheap and user-friendly. Whereas main services like heating, cooling, lighting can be automatized in office buildings, many services related to multimedia, gaming, cooking,... cannot and flexibility requirements should involve occupants in decisions.

Taking good decisions might seem easy for occupants but actually, it's not. Most phenomena are not visible (air quality, energy fluxes going through walls, doors and windows) and complex. For instance, opening a window can be justified during winter for improving air quality but for a given time, heating or cooling a room, which is rarely occupied, is an energy waste, the door position might highly impact the air quality, the temperature, and the power consumption in a room. Occupying a room in a low consumption building might also have a strong impact. [20] did a Morris sensitivity analysis in an efficient building and showed that the model parameters mainly impacting the energy performances are dealing with human practices. Phenomena are also complex because they are highly dependent on human practices, which are most of the time, not conscious: sociologists pointed out [3] that human practices are mainly routines not involving cognitive mechanisms. This is complex because residents are often unaware of their own practices; this is especially true if an area is shared by several people. Additionally, because occupants lack of information and possibly knowledge, they might develop fake beliefs like the windows have to be kept opened for 1 h every day to remove microbes... Phenomena will grow in complexity when the timing of an action will become important due to the requirements of flexibility and coordination.

Building optimized energy management points out that without occupants' comfort expectations, building doesn't require energy [7].

To face the increasing complexity, new decision aiding services are needed. Two complementary approaches can be followed: (1) improvement of existing *automations* and (2) better *involvement* of inhabitants. Approach (1) requires better recognition of people expectations and intentions in order to apply the most relevant strategy to each context in order to maximize the occupant satisfaction while minimizing different kinds of costs like environmental, economic, cognitive costs,... It requires estimation from sensor data, learning of user profiles and adaptation of the building configuration, with possible anticipation (see Model Predictive Control approaches in related chapters), to each context. The difficulty in this approach is due to the specificities of building systems: they are composed of envelope, appliances but also of occupants which are both highly affecting the overall behavior with their actions and metabolisms, and are also decision maker assessing the quality of the services provided by the energy management system. Therefore, building systems are very special because there are highly human-centric. The second approach aims at involving occupants in the everyday building energy management. Indeed, occupants' assets are complementary to the ones of an information system and the final decisions are co-determined thanks to human-machine interactions:

- occupants:
  - have a more global qualitative perception of the current building system because sensors are limited in number and cannot perceive all what human can easily do (bad smell, broken device, people visit, ...)
  - can easily change a configuration, like a door position, ...
  - have an updated representation of the building system global behavior because of their perception capabilities
  - know occupants' expectations and intentions
- information systems:
  - have a partial precise perception of the building system thanks to sensors
  - can precisely adjust set-points and possibly change configurations providing that actuators are available
  - embed a more or less precise, but costly to update, numerical representation of the behavior of building systems, at least a model implicit to control rules, and sometimes an explicit causal model

The second approach aims at developing cooperation between occupants and information system to get the best of each for emerging solutions yielding best quality of services for occupants and costs compromises. It can be simply a GUI displaying mirroring indicators for the occupants to figure out what they are doing and what are their impacts of the building system. It can also be a system sending nudges to influence the energy consumption of the occupants. Because occupant behavior is determined mostly by routines, i.e. unconscious habits, a set of pinned indicators for improvement can be focused to follow a step-by-step way to better sobriety, flexibility, and coordination. More advanced system could also propose building energy management strategies to inhabitants either in the past (what I could have done) or in the present and future (what I could do). Additionally, generation of explanation to justify a strategy can also be done (see related chapter).

3 Illustrative Examples

The first illustrative example that will be used all along this book is an office named "H358" containing 4 desks. It is particularly interesting because both it is simple, in that such, it can be used as a first illustrative example, and it is equipped with 18 sensors, appearing in the square boxes of Fig. 1. Data are available from the beginning of 2015 till now but with gaps (see Chap. "The Mondrian User Interface Pattern: Inspiring Eco-Responsibility in Homes" for handling). It is representative of real life because ENOCEAN sensors are leading to time series:

- with irregular time samples: to avoid useless repetitive data, low consumption radio sensors send data when a significant change with the previous measurement data is detected according to sensor resolution threshold. Collected data should be resampled at regular sample times or time quantum.

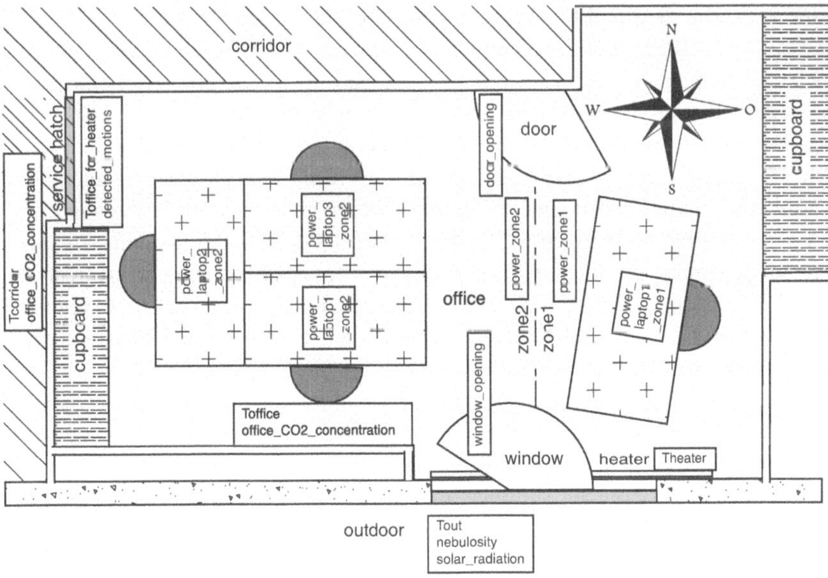

Fig. 1 H358 office map with available sensors

- with a regular sample time after resampling which might vary depending on the problem to solve. Therefore, resampling should be a dynamic process.
- with gaps: sequences of data without gaps have to be detected. After more than 4 years of recording, the length of the longest contiguous set of data series is lasting 40 days!
- with outliers: they have to be detected and weird values replaced by realistic ones. Outlier detection can be done by detection of a sudden significant increase and then a sharp decrease of consecutive data values but also when data values are out of the meaningful data value range.

In the office, the heating level is considered as proportional to the difference between the temperature close to surface of the radiator (Theater) and the room temperature of the office (Toffice_reference): it is named “dT_heat = Theater - Toffice_for_heater”. The proportional gain is estimated using a physical model together with a global parameter estimation approach. The occupancy, i.e. the average number of occupants during a sample time is deduced from computer laptop consumptions and possibly from motion detections, CO_2 concentration, average acoustic pressure during a time quantum (sample time) and much more (see Chap. “Occupant Actions Selection Strategies based on Pareto-Optimal Schedules and Daily Schedule for Energy Management in Buildings”). Nebulosity, outdoor temperature, humidity,... are collected through the Internet from weather forecast webservices, and used in particular to calculate the solar power coming through the windows to the office by using a model of the sun w.r.t. earth position [5]. During winter, the door and the window (one big window composed of all the 5 windows) positions can be modified by occupants providing there are present ($\zeta_{door} \in [0, 1]$ and $\zeta_{window} \in [0, 1]$, corresponding to the fraction of the sample period during which the door or one of the windows remained open). A remotely controllable thermostat is available for the heater, therefore automatic control can be considered during winter. Heating system is off during summer time, of course. The office is relatively closed to home context: occupants must be in the energy management loop because at least, they are the only ones to be able to open the door or the windows.

A second example is given to illustrate more modern buildings with advanced air handling units, dual duct system, and heat recovery from stale air to fresh air. Figure 2a shows the global HVAC system connected to the MHI classroom platform dedicated to experiments. It is equipped both by legacy sensors and by Z-wave and Enocean sensors like for “H358” office. The air handling unit (AHU) is located at the rooftop. In and out filters and blowers can be seen as well as the rotating honeycombs wheels exchanging from extracted stale air to injected fresh air in the middle of the AHU. There is a coil fed by warm or cold water, 25–30°C during winter and 19–23°C during summer, to pre-heat or pre-cool the air in the duct feeding the rooms with a set-point equal to 22°C during daytime and weekdays. At the AHU level, there is also a pressure control for each local room air renewal not to influence others. In the inlet duct, the pressure is set to 120hPa during daytime and working days and to 110hPa in the outlet duct. Therefore, the blowers of the AHU adapt their speed in order to keep a pressure difference of 10hPa between

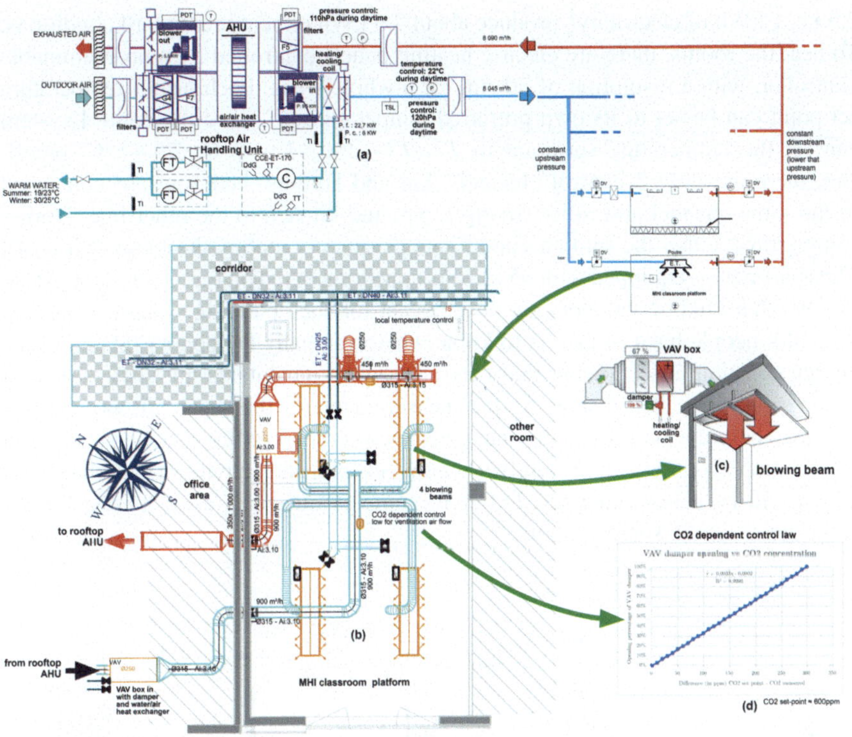

Fig. 2 MHI classroom platform and its HVAC system

the air injection and extraction of each connected room. Among the connected rooms, there is the so-called *Monitoring and Habitat Intelligent* (MHI) classroom platform in which the fresh air arrived through passive blowing beams located at the ceiling Fig. 2c and stale air is removed thanks to air outlets in red in Fig. 2b. The blue air inlet duct passes through a variable air volume (VAV) box, which is composed of a damper to control the air flow and a coil fed by the same water than the one use in the coil of the main AHU. It is used to control the local temperature of the room by adding or removing heat, respectively, to the pre-heat or pre-cool air flow coming from the AHU. Figure 2d shows the control law that is used to provide good air quality to room occupants: the ventilation air flow is dependent of the CO_2 concentration in the room. The system decouples the room controls from the global AHU systems: each connected room receives the same pressures and the air temperature. For such systems, automation with presence anticipation is suitable but in home context, a better involvement of occupants is generally more relevant. Let's study a third example.

Figure 3 deals with a typical 4 rooms plus kitchen apartment. To keep the temperature at an acceptable level, there are 2 systems for heating. An air-to-air heat pump split system in the living room given with a coefficient of performance equal to

2.5 i.e. 1 kWh of electricity[1] produce about 2.5 kWh of heat in standard conditions. To heat the rooms, there are electric heating panels controlled by a programmable controller, with a resolution of 30 min for a whole week. Each heater temperature set-point can be set to its own preferred temperature (T^*) and the control system can set the temperature set-point to T^*, T^*—1°C, T^*—3°C, T^*—5°C, or off. But, there are only 2 control channels: kid and Parents' bedroom are connected to the same channel and office, living room, and kitchen to the other one. During summertime, either the split air conditioning system is cooling the room or it is off. The apartment is equipped with 65 sensors appearing in red on Fig. 3. PLC, SCADA, or any other kind of actuators are not relevant for such building systems. The first difficulty people have to face is to know what is actually happening related to the energy consumption knowing that they are several inhabitants not always present at the same time. Just displaying the measurements of the sensors is not enough because there are 65 curves that have to be cross-analyzed to extract meaningful information about impact of family's actions (events) and activities (with duration), which will be called practices. Different complex questions arise for the family, some are seasonal, some are not:

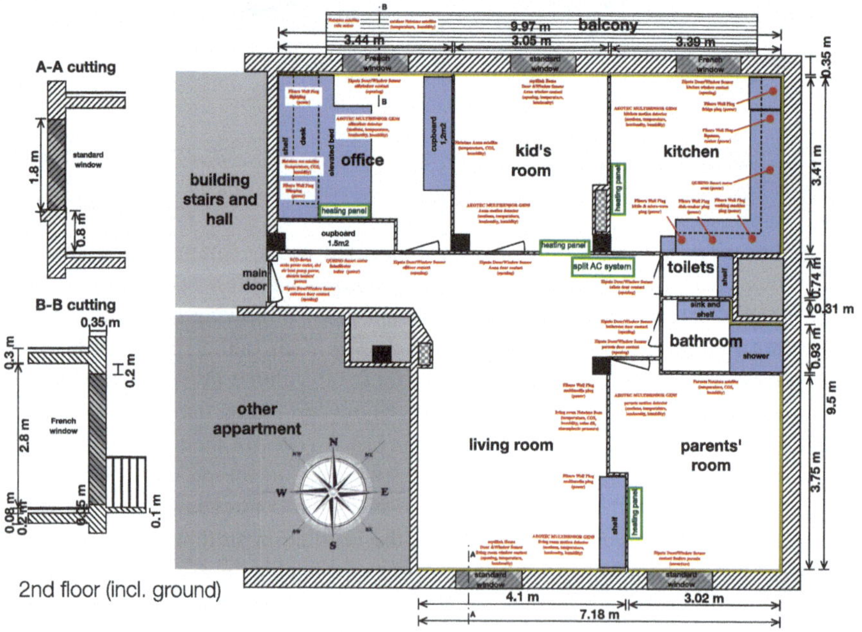

Fig. 3 Standard flat with sensors

[1]Depending on the outdoor temperature: the closer to the indoor temperature, the best.

- the electricity tariff is time-varying: a cheap tariff, about 6 cents/kWh, from 11pm to 7am and from 12:30 to 2pm, and a normal tariff, about 12 cents/kWh, for the rest. The question is: does the family practices adapted enough to the tariff to meet the expected compromise between comfort (services achieved on time,...) and cost. This problem can even be more complex in case of photovoltaic panels with a self-consumption schema that encourages people to consume their own energy. Mirroring occupant practices at a relevant level making it easy to check whether practices are suitable or not.
- during winter, there are 2 heating systems: an air/air heat pump located in the living room and electric radiative heating panels in the other rooms. How to determine set-points of the 2 groups of heating panels but also the set-points of the heat pump? what are the best temperature profiles taking into account the maximum power of the heating devices, the outdoor temperature, the solar gains through the windows, the positions of the doors, windows, and shutters, the room occupancy and the occupants' intentions and preferences?
- during summer, the split air conditioning system can be switched on or off, and the set-point can be adjusted but how to adjust these degrees of freedom taking into account the outdoor temperature, the solar gains, the positions of the doors, windows, and shutters, the room occupancy and the occupants' intentions and preferences?
- additionally, metabolisms and breaths generate CO_2, whose concentration indicates a level of confinement. It's also representative of the level of dusts with mites, and the smells. Because the air is extracted from the toilets, bathroom, and kitchen in the apartment, if the room doors and windows are kept closed, the concentration of CO_2 in the bedrooms can reach very high values, till 5000 ppm at the end of the night. But keeping the windows and doors opened has a strong influence on the heating system consumption. What is the best strategy depending on occupancy and on thermal phenomena?

There are much more complex questions that people sometimes do not figure out regarding the usage of domestic hot water vs heating cold water, the usage of the fridge and freezer,[2] the best room to stay,... They are problems not easy to solve because most of the phenomena are not perceivable and because there are all inter-dependent. Most of all, people use to follow routines, i.e. they don't frequently question about the environmental impacts of their practices. The decision-aided services coming with the smart-homes are going to support the solving of all these problems; everyone has to solve on an everyday basis.

[2]Which use to cause the main consumption in many home settings if the HVAC system is set apart.

4 Problem Statement of Energy Management in Smart Buildings

To state the smart building energy management problem in a general way, let's distinguish different kinds of phenomena that can be modeled by variables: there are the causes and the consequences but also the performances (see Fig. 4):

- causes:
  - *context variables* model phenomena that cannot be modified and impact effect variables
  - *action variables* model phenomena that can be controlled and impact effect variables
- consequences:
  - *intermediate variables* model phenomena that are consequences of causes but that are not appearing directly in the performance evaluation
  - *effect variables* model phenomena that are consequences of causes and that are appearing directly in the performance evaluation
- *performance variables* model both the quality of a service delivered to occupants and the economic, environmental, and/or cognitive costs. It depends on effect variables but it might also depend on context and action variables.

For instance, Fig. 4 represents the different types of phenomena related to H358 office. Contextual variables are related to the temperature and CO_2 concentration of the juxtaposed corridor, outdoor temperature, occupancy, and solar energy coming through the windows. Action variables can be context-dependent like door and window openings as a ratio of chosen time quantum[3] considered as time samples. Indeed, actions cannot be carried out in case of absence, i.e. it is dependent on the occupancy. The heating level, as a ratio of the maximum heating power, is not context-dependent because it is operated by an automatic control system. The average temperature of the envelope cannot be measured because it is an equivalent temperature but it can be estimated. It is an internal state variable which is set to intermediate variable. Temperature and CO_2 concentration in the office are the effect variables because they are consequences of the actions that directly determined the performances of the living area.

In Fig. 4 are shown the main variables influencing the thermal phenomena and the air quality in the office H358 (see Fig. 1). The context variables are: the temperature of the nearby corridor, the CO_2 concentration in the corridor, the outdoor temperature, the occupancy (that globalizes body metabolisms and personal electricity consumption of laptops), and the solar radiations coming through the

[3]Typ. 1 h but it could also be 30 min or even 10 min even if weather forecasts are usually given per hour.

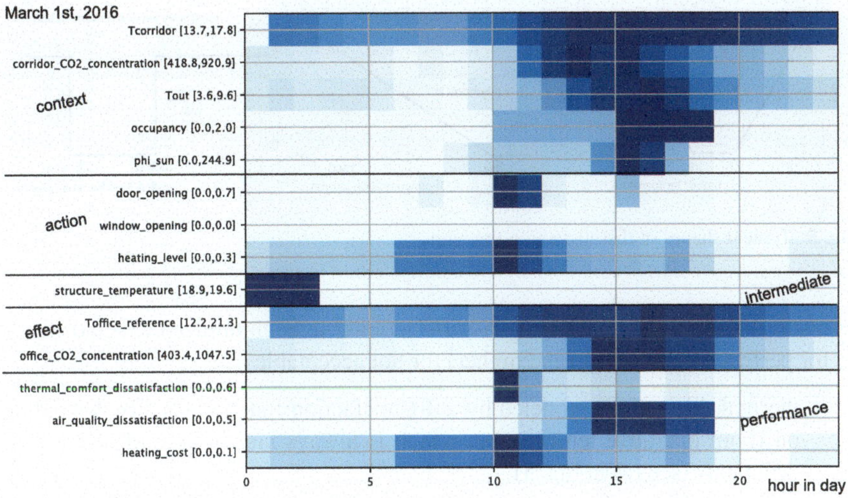

Fig. 4 Different kinds of phenomena modeled by variables

window. The possible actions are limited: opening or not the office door and opening the window (0 means always closed during the considered sample time and 1 always opened, with intermediate values possible), but it can be opened only in case of someone is present in the office. Moreover the level of heating (0 for no heating and 1 for full power) can be changed during winter, when the central boiler is on. The structure temperature is here the equivalent temperature of the heavy parts of the office, like slab and concrete walls: it is an intermediate phenomenon, which provides inertia for indoor temperature. The resulting effect phenomena are the office temperature and the resulting concentration of CO_2. Performances are deduced from the context (occupancy) and the effect phenomena. It is dealing with thermal comfort and air quality dissatisfactions together with the energy cost. It will be detailed in the next.

Notice that the decomposition into types of variables is not ontological: it's just a way of modeling. For instance, if room occupancy can be changed, asking people to move to another place, then occupancy will be modeled by an action variable instead of a context variable. Causality is also not ontological but a part of the modeling. For example, the indoor temperature is an effect variable when there is no HVAC system. When an HVAC system is operating, the indoor temperature can be equated to the setpoint temperature and, assuming perfect control, the temperature becomes an action while the energy consumed by the HVAC system becomes an effect variable.

Typical performance variables are:

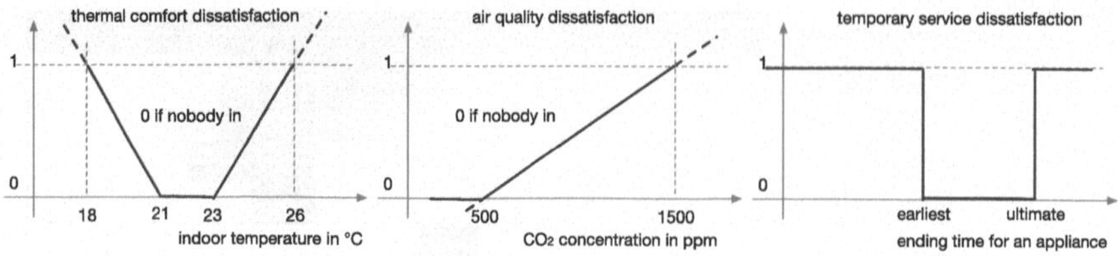

Fig. 5 Typical shapes of dissatisfaction functions

- quality of services, modeled by dissatisfactions[4] to be minimized where 0 stands for perfect quality and 1 (and higher) for not acceptable
  - thermal discomfort, modeled by a dissatisfaction function usually depending on room operative temperature, which is always null in case of absence of occupants (see (a) in Fig. 5)
  - air quality discomfort, modeled by a dissatisfaction function usually depending on CO_2 concentration, which is always null in case of absence of occupants (see (b) in Fig. 5)
  - possibly temporary service discomfort, to help deciding when to start an appliance like a washing machine, modeled by a dissatisfaction function usually depending on the expected ending time of the related appliance (see (c) in Fig. 5)
- costs, which should be normalized to get a cost scale comparable to dissatisfactions, belonging to [0, 1]
  - energy cost, corresponding to the quantity of the needed energy
  - economical cost, corresponding to money expended for the needed energy
  - environmental cost, corresponding, for instance, to the equivalent quantity of CO_2 rejection in the atmosphere
  - cognitive cost, corresponding, for instance, to the number of actions occupants have to do to adapt the configuration of his living place.

The concept of dissatisfactions is interesting because it normalizes comfort performances, but is it possible to determine an overall dissatisfaction D from specific dissatisfactions D_i like those related to air quality or indoor temperature? Should we sum up, multiply? Let's state the problem:

- $D \in [0, 1]$ and $D_i \in [0, 1]; i \in \{0, n-1\}$, where 0 stands for totally satisfied and 1 for totally dissatisfied.
- the global dissatisfaction is a function of specific ones: $D = D(D_0, \dots, D_{n-1})$
- an increase of a specific dissatisfaction cannot decrease the global dissatisfaction: $D(D_0, \dots, D_i, \dots, D_{n-1}) \geq D(D_0, \dots, D_i + \epsilon, \dots, D_{n-1})$ with $\epsilon > 0$, i.e.

[4]Dissatisfaction can be minimized like cost, whereas satisfaction, equal to 1 minus dissatisfaction, has to be maximized.

$$\forall i, \frac{d}{dD_i} D(D_0, \ldots, D_i, \ldots, D_{n-1}) \geq 0$$

- a priori, specific dissatisfactions should impact differently the global dissatisfaction: $i \neq j \rightarrow \frac{dD}{dD_i} \neq \frac{dD}{dD_j}$
- if an occupant is totally dissatisfied with all the specific dissatisfactions, he will be globally totally dissatisfied and same with total satisfaction. It yields: $D(0, \ldots, 0) = 0$ and $D(1, \ldots, 1) = 1$.

Linear functions for merging are good candidates although more complex merging function can be imagined. A linear function for $D(D_0, \ldots, D_i, \ldots, D_{n-1})$ that satisfies the statements are:

$$D(D_0, \ldots, D_i, \ldots, D_{n-1}) = \sum_i \alpha_i D_i \text{with} \sum_i \alpha_i = 1$$

It is therefore a multi-objective problem where comfort criteria are modeled by dissatisfactions and costs modeled by normalized functions. The best compromise is reached when the following objectives are minimum:

$$\forall i, \min_{\text{actions}} \sum_k D_i(k) \tag{1}$$

$$\forall j, \min_{\text{actions}} \sum_k C_j(k) \tag{2}$$

where i stands for a type of comfort, j a type of cost, and k a sample time belonging to a considered time horizon. There is an implicit contextual[5] Pareto front (Fig. 6) that represents all the contextual optimal cost-comfort compromise, i.e. there is no one optimal compromise for a given context but one best compromise for each activities and presences. Let's normalize the costs by solving the following problems:

$$\mathrm{C}_j(\text{actions}_j^*) = min_{\text{actions}} (C_j) ; j \in J \tag{3}$$

and define:

$$\bar{C}_j = max_{\forall l \neq j} C_j(\text{actions}_l^*) \tag{4}$$

Consequently, all the criteria can be normalized. The normalized problem to be solved is

[5]For a given set of values of contextual variables over a given time period.

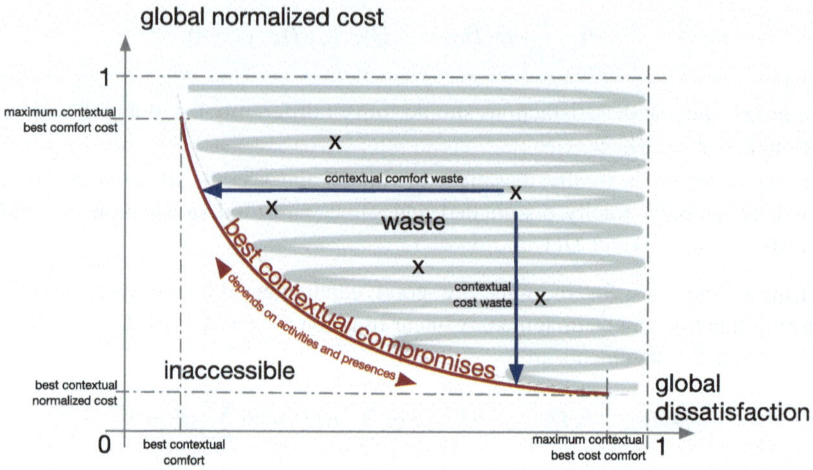

Fig. 6 Avoiding waste can be modeled as a distance to a theoretical contextual Pareto front

$$\Pi_{\gamma_j; j \in J}(\text{actions}^*) = min_{\text{actions}} \left(\lambda \sum_k \sum_i \alpha_i D_i(k) + (1 - \lambda) \sum_{j \in J} \gamma_j \frac{C_j - \underline{C}_j}{\bar{C}_j - \underline{C}_j} \right) \tag{5}$$

$$\sum_i \alpha_i = 1; \forall i, \sum_i \gamma_j = 1; \forall j, \lambda \in [0, 1] \tag{6}$$

Of course, $\underline{C}_j$ and $\bar{C}_j$ can be estimated without solving an optimization problem for each cost criteria. α_i represent the relative occupant preferences between comfort dissatisfactions, γ_j the relative occupant preferences between normalized costs and λ, the relative importance of comfort versus cost.

The optimization problem cannot be decomposed in time because of inertia in thermal phenomena but also in CO_2 concentration: the current state depends on the previous state. Moreover, decomposing in space can also be problematic because phenomena might not be independent depending on the situation. For instance, the temperature or the CO_2 concentration in a nearby room considered as a limit condition can highly be influenced by the studied room for computation of best energy strategies.

Solving an optimization problem might not provide a helpful service to inhabitants because:

- models might not exist and if so, they might be imprecise or far from reality
- sensors might not be enough or could be faulty with generally gaps in the data
- occupant preferences and intentions might not be known
- automatized control systems might not be able to solve optimization problems and apply a best energy strategy

- in home settings, actuators are usually occupants: they act on their home. Therefore interactive problem solving approaches have to be designed.

Therefore, energy smart-home services cannot be reduced to computing actions yielding best compromises between comforts and costs according to occupants' preferences. It is just a theoretical formalization of the best way to manage energy in a dwelling. Perfect sobriety doesn't mean no consuming any energy but it means avoiding energy waste in being as close as possible to the Pareto front of a best comfort/cost compromise consistent with occupants' preferences and intentions. As mentioned earlier, there are other energy services that reflect occupant behaviors and their impacts. Conversely to common perceptions, an energy smart-home or smart-office, which will call indistinctly energy smart-home for the sake of readability in the next, is not a place with an intelligence guessing inhabitant intentions and assisting them in the everyday life, but rather a place equipped with perception means, using sensors or interactions with inhabitants, that mirrors users' practices, suggests and explains what are the consequences of recorded actions. The management of some services can be partially delegated to an automation system but a full delegation for all services rather looks like a nightmare for most people because automation has to assume inhabitant intentions to take the right decision. Many people experimented situations with motion detector assuming absence or automated shutters opening when intention is to display slides. Guessing intentions is a lot more tough in residential settings where diversity of human practices is much higher.

5 The Model Issue

In order to be able to determine the consequences of actions, a model is needed but what kind of models? and most importantly, how easy it is to get it? This is a critical point for the spreading of the so-called energy smart-home. Indeed, without cause-effect contextualized representation of the impacts of actions, how is it possible to aid occupants in their everyday decisions? If setting up a model is too costly, it is going to slow down the deployment of such technologies. Only dashboard with mirroring functionalitics would sprcad. Figurc 7 points out that thcrc are different natures of models composing a global behavioral model for a home (or office) system. Firstly, there is the inhabitant model that relates perceptions yielding dissatisfactions and appreciations of costs to decisions of actions. Because it is usually several inhabitants sharing a place, such a model should represent routines and homeostatic unconscious behaviors, beliefs, desires, intentions (committed plan of actions) and also deliberative behaviors representing social interactions. Kashif et al. [9] proposed a multi-agent model for the inhabitant model. Modeling building behavior yields to the so-called *performance gap* whose one of the most important cause is the difficulty to model human behavior. Research is currently developing tools for representing human behavior (IEA Annex 66). Different approaches have

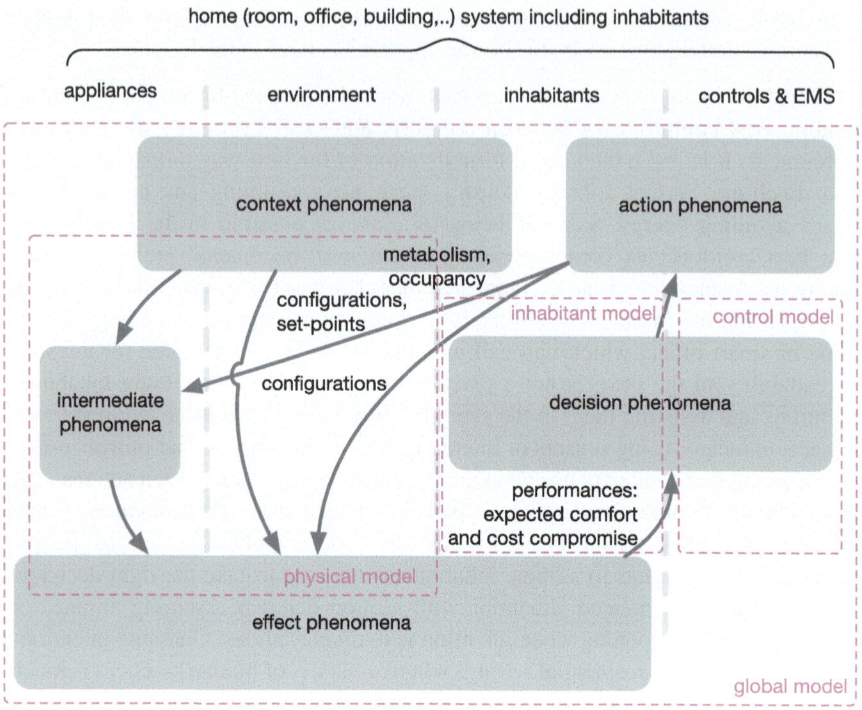

Fig. 7 Different kinds of models for a home system

been proposed, which can be split into 2 categories: the average general hybrid models that combine time of use surveys, and specific observation models. The average general hybrid models can be average profiles or stochastic models using, for instance, Hidden Markov models [16], whereas specific observation models rely on similar buildings with similar household. Belief-Desire-Intention Multi-agent systems with cognitive and deliberative capabilities [8] have been used but the number of parameters to tune is huge: it requires questionnaires and parameter estimations but also a long observation period. Alternatively, simplified observation models have been also used based on Bayesian networks learnt from sensor data and inhabitant feed-back about their activities [19]. Even if these models are mostly used for the design of buildings, it is interesting to note that the behavior of inhabitants is highly influencing the energy consumption of a low consumption building.

The control model is required when occupants delegate part of the energy management to controllers and more generally to energy management systems (EMS). These automations are generally dedicated to HVAC systems and to the management of domestic hot water (DHW), batteries and energy generators. The control model establishes a relationship between the context variables, the automated actions managed by the EMS, and the resulting effects.

The third model, called *physical model*, is the most useful for decision support systems because it links user actions to effects, taking into account contextual phenomena. It includes the control model with possible simplifications because the time for control systems is much faster than psychological time for inhabitants. It allows us to determine the values of the intermediate phenomena and the resulting effects. The modeling step is more subtle than for the design because the building system is existing and the simulation results have to match on-site phenomena, whereas for design results have to be realistic enough with sufficient details to determine in which way each design option will modify the overall performances. There are different papers REF that compare the simulation results obtained by common building simulators for the same building: it comes out that the overall performances can vary up to 300%... with the same assumptions on usage profiles! Nevertheless, even if such discrepancies are known, the most important is to obtain the trend of an option on the overall results. Let's summarize the differences in the model requirements for design and for runtime management.

Typical requirements	Design	Runtime
Time horizon	1 year	1 day
Time step	1 h	10 min and possibly less for automated control systems, to 30 min and 1 h in case of human interactions
Number of parameters	No limit provided values can be assigned (from experiments or knowledge)	Limited because values have to be adjusted in order to match measurements: too many values might yield identifiability issues
Accuracy	Expected in trends	Should comply with actual phenomena, according to inhabitant perceptions
Representativity	Models represent internal phenomena that can be reduced in complexity for the sake of simulation speed: there are context independent	Models contain virtual parameters that can represent each several phenomena: there are more or less dependent on the current context, i.e. parameter values don't match all the contexts
Self-learning	Not meaningful	Strong requirement
Usage	All-year simulation, sensitivity analysis, design parameter optimization	Parameter estimation, day simulation, action optimization, state estimation, generation of explanations

Basically, there are two main kinds of physical models, with respective advantages:

- knowledge model, whose structure is inspired from physical knowledge. The structure of these models is dependent of the underlying physics and it is generally well-representing the relationships between phenomena but conversely it is not adapted to the way the model is going to be used. Nguyen Hong et al. [15] showed that parameter estimation is not an easy task when using such models: minimizing the output simulation error cannot be managed with descent approaches and using meta-optimization technics to avoid getting stuck in local minima is not always successful. Moreover, these models require knowledge to be designed and also updated: it implies manpower that can be relatively costly particularly in multi-zone home systems.
- learnt model, whose structure is depending of the type of regressor used, like, for instance, linear or artificial neural network based regressors. The structure is regular and generally suitable for learning of parameter values from collected dataset. Because their structure can be fairly representative of the physical phenomena, these models might fail to represent accurately the physics whatever the context is.

Another type of model is under development (see Sect. 8): the case-based data model, based on historical data. This is a promising approach because it requires little design work, but a large amount of sensor data together with a causal modeling of phenomena. Other alternative can also be considered like context-dependent linear regressor... but let's focus on existing solutions. To illustrate the capability of both approaches for representing runtime behavior, let's analyze the real H358 office platform with recorded measurements: detailed measurements over a long time period with real occupancy are not that common in the scientific literature.

5.1 *Modeling from Knowledge*

Let's discuss the right type of model for energy smart-homes and use the H358 office (Fig. 1) to illustrate the differences. It's an interesting example because more than 4 years of recorded sensor data are available from January 1st, 2015 till the end of 2019 but with regular gaps due to missing data. Before comparison, let's start by establishing a simple knowledge model representing the relationships between phenomena. The model complexity is limited to one differential equation for representing the thermal inertia and another one for the evolution of CO_2 concentration, although some authors [12] recommend more, but with less parameters, identifiability issues are less because the number of parameters to learn is less, but depending on the data, the discussion is still open. Although there is no mechanical ventilation system in the office, the knowledge model for the office system is nonlinear but it is modeled by a time-varying state space model. Figure 4 points

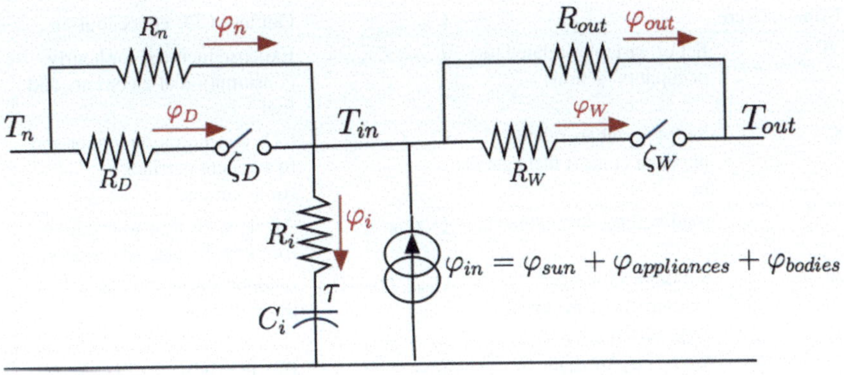

Fig. 8 Lumped element model for the H358 office

out the phenomena used by the model. Thermal phenomena are represented by a lumped element model in Fig. 8:

- outdoor temperature is written T_{out}, τ is the equivalent structure temperature, which cannot be measured.
- Tcorridor (in Celsius degrees), written T_n: it is the air temperature in the corridor, which is assumed to be not influenced by the office room phenomena.
- Q_w and Q_D stand for the air flows passing, respectively, through the *window* to outdoor and through the *door* to the corridor. CO_2 concentration in the corridor is considered as not influenced by the office CO_2 concentration. Q_W and Q_D are assumed to be in an affine relation with the ratios of opening time z_W and z_D, which are deduced from contact sensors: $Q_W = Q_W^0 + \zeta_W Q_W^1$ and $Q_D = Q_D^0 + \zeta_D Q_D^1$. $\zeta_W = 0$ stands for window closed and $\zeta_D = 0$ stands for door closed, and other values are constant parameters to be estimated.
- φ_i, φ_D, φ_W, φ_{out}, and φ_{in} stand for heat flows in Watt. φ_i is the power exchange with the office structure. φ_D and φ_W are heat flows going through the door and windows that depend on air flows as shown in Eqs. (7a), (7c), (7h), (7b), (7c), and (7g). φ_{out} is the power exchanged through the wall with outdoor.
- φ_{in} is the sum of φ_{sun} (phi_sun), the solar power passing through the window calculated by a solar model using cloud cover and outdoor temperature provided by weather forecasts, $\varphi_{body} = n p_{\text{metabolism}}$ where n is the number of occupants (see Chapter 6 for estimation) and $p_{\text{metabolism}}$ the average metabolic power generated by a body, $\varphi_{\text{appliances}}$ the power generated by electrical appliances and $\varphi_{heat} = K_{heat}(T_{heater} - T_{in})$ is the power of the heater where K_{heat} is a parameter to be estimated.

The following equations correspond to the lumped element model given in Fig. 8:

$$Q_D = Q_D^0 + \zeta_D Q_D^1 \tag{7a}$$

Nomenclature		C_{out}	Outdoor CO_2 concentration
$\mathcal{A}$	Set of variables related to occupants' actions	E_{elec}, E_{fuel}	Expense incurred for hourly consumption of electricity and fuel
C	Set of variables related to physical context that cannot be	C_{in}, C_n	CO_2 concentration indoor and in adjacent corridor (time-variant)
	modified by the occupants	$Q, Q_{in}, Q_n,$	Air flow, air flow to inside of room, with adjacent corridor,
$\mathcal{S}$	Set of variables for assessment of occupants' satisfaction (effects)	Q_{out}, Q_W, Q_D	outdoor, through window, through door
$\mathcal{I}$	Set of intermediate variables having two subsets I_1 and I_2	S_{CO_2}	Breath production in CO_2 (8.73×10^{-6} mol.m^{3}s^{-1} per person
I_1	Set of variables estimated through physical models like φ, Q, etc.		per mol of air)
I_2	Set of some variables measured through sensors like T_{in}, C_{in}, etc.	σ_{temp}	Thermal dissatisfaction (time-variant)
ζ_D	Opening of door (time-variant)	σ_{air}	CO_2 based air quality dissatisfaction (time-variant)
ζ_W	Opening of window (time-variant)	σ_{cost}	Overall energy consumption related expense (time-variant)
ζ_H	Switching on/off of heater (time-variant)	$\Pi(.)$	Quantitative to qualitative transformation function, i.e. amount of
ζ_{pair}	Opening of door/window, i.e. $\zeta_{pair} = (\zeta_W, \zeta_D)$ (time-variant)		change in a variable is quantized into seven levels using thresholds
δ_{WD}	Binary variable for changes in recommended actions		at $\upsilon_{-3}, \upsilon_{-2}, \upsilon_{-1}, \upsilon_1, \upsilon_2, \upsilon_3$
	(time-variant)	Δx	Amount of change in a variable
n	Number of occupants at office room (time-variant)	$x^\star$	Optimal value of a variable
τ	Average temperature of building envelope	$\tilde{x}$	Usual value of a variable (historical value)
$R_n, R_{out},$	Thermal resistances with neighboring zones, outdoor	$\mathcal{H}$	Historical database
R_W, R_D	windows and doors	V	Volume of the room
R_i, C_i	Equivalent resistance and capacitance due to inertia	t, k	Continuous and discrete time variables
R_{eq}	Overall equivalent resistance of thermal model	S_{avg}	Array of occupants' satisfaction values averaged over a day
$\rho_{air}, c_{p,air}$	Air density and specific heat of air at room temperature	$\mathcal{A}^{PS}$	Pareto-optimal set of routines (occupants' actions)

T_{in}, T_n, T_{out}	Temperatures inside, in adjacent corridor and outside	S^{PF}_{avg}	Pareto front (quantified effects of optimal actions)
	(time-variant)	S^{ref}_{avg}	Reference value for decision-making from a set of trade-offs
$\varphi, \varphi_{bodies},$	Heat flow, human body metabolism, total indoor energy gains,	w_1, w_2, w_3	Weights representing preference of each kind of
$\varphi_{in}, \varphi_n, \varphi_{out}$	heat flow with adjacent corridor, outdoor	and w_4	satisfaction (optimization objectives)
P_{elec} or	Electric power consumption from work-associated routine	$\mathcal{T}$	Differential explanations table
$\varphi_{appliances}$	appliance heat power (time-variant)	t_{start}, t_{end}	Start and end time for differential explanations table
P_{fuel}	Energy drawn to heat water for heater (time-variant)	$\hat{\mathcal{A}}^j$	A routine (occupants' actions) identical to Pareto-optimal routine
P^{max}_{heater}	Energy consumption for hourly heater usage		except the action at the jth time is replaced by the historical action

$$Q_W = Q_W^0 + \zeta_W Q_W^1 \tag{7b}$$

$$R_D(\zeta_D) = \frac{1}{\rho_{air} c_{p,air} Q_D} \tag{7c}$$

$$R_W(\zeta_W) = \frac{1}{\rho_{air} c_{p,air} Q_W} \tag{7d}$$

$$T_n - T_{in} = R_n \varphi_n \tag{7e}$$

$$T_{in} - T_{out} = R_{out} \varphi_{out} \tag{7f}$$

$$T_{in} \quad T_{out} = R_W(\zeta_W)\varphi_W \tag{7g}$$

$$T_n - T_{in} = R_D(\zeta_D)\varphi_D \tag{7h}$$

$$T_{in} - \tau = R_i \varphi_i \tag{7i}$$

$$C_i \frac{d\tau}{dt} = \varphi_i \tag{7j}$$

$$\varphi_n + \varphi_D + \varphi_{in} = \varphi_i + \varphi_{out} + \varphi_W \tag{7k}$$

$$\tau(0) = . \tag{7l}$$

with $c_{p,air} = 1004$ and $\rho_{air} = 1.204$.

Some authors suggest that a first order model is not enough and recommend higher order. Actually, it depends on the context, on the sampling time, on the time horizon, but most of all on the identifiability of the parameters and on the performance of the parameter estimation method because conversely to building design where assumptions can be easily done, in energy management, the building is existing and simulated data have to match recorded data. Too many parameters might lead to poor results because of bad parameter estimation. Indeed, spectral analysis of measurement signals reveals poor dataset with a dominant frequency corresponding to 24h, then less energy at 12h and eventually a little energy at 6h period. Spectral energy at other frequencies is almost not visible: it points out how tough is the parameter estimation problem also self-tuning of physical models is a major issue.

The previous equations can be combined in order to get the following state space model where some parameters are time-varying:

$$\frac{d\tau}{dt} = \frac{R(\zeta_D, \zeta_W) - R_i}{R_i^2 C_i}\tau + \frac{R(\zeta_D, \zeta_W)}{R_i C_i}\left(\frac{1}{R_{out}} + \frac{1}{R_W(\zeta_W)}\right) T_{out} + \dots$$

$$\dots + \frac{R(\zeta_D, \zeta_W)}{R_i C_i}\left(\frac{1}{R_n} + \frac{1}{R_D(\zeta_D)}\right) T_n + \frac{R(\zeta_D, \zeta_W)}{R_i C_i}\varphi_{in}$$

$$T_{in} = \frac{R(\zeta_D, \zeta_W)}{R_i}\tau + R\left(\frac{1}{R_{out}} + \frac{1}{R_W(\zeta_W)}\right) T_{out} + \dots$$

$$\dots + R(\zeta_D, \zeta_W)\left(\frac{1}{R_n} + \frac{1}{R_D(\zeta_D)}\right) T_n + R(\zeta_D, \zeta_W)\varphi_{in}$$

with:

$$\frac{1}{R(\zeta_W, \zeta_D)} = \frac{1}{R_i} + \frac{1}{R_{out}} + \frac{1}{R_n} + \frac{\zeta_W}{R_W} + \frac{\zeta_D}{R_D}$$

$$\varphi_{in} = P_{sun} + P_{elec} + P_{body} \times \sum_j (P_{laptop,j} > 15W)$$

In energy management problems, the differential equations are not used directly but there are integrated over a constant time quantum, named sample time, to get recurrent equation able to model the behavioral evolution from one ample time to another. There are different points of views regarding the most relevant sample time for home systems. Some people recommend 10 min and, at the other extreme, there is 1 h. Let's discuss these 2 options:

- finest weather forecast is given with a 1-h time step
- changing the configuration of a home system can be tiring for people if changes are perceivable
- 1 h time step is a long time period considering adjustments to unpredicted changes in occupancy and in occupant activities
- predicting occupant behavior with a 10 min time resolution is much more challenging than with a 60 min time step.

According to these facts, it comes out that:

- 10 min or even less is in a better option for automatized controlled systems dealing with not perceivable changes for the occupants like in HVAC, DHW, battery, and power generator management. A sliding time window with a time horizon of a couple of hours is generally used to anticipate scheduled and more certainly ongoing events. A permanent readjustment is then used to cope with unpredicted events [10].
- 1 h is a better option for involving occupants in the energy management and interacting with them. Recommendations can be computed once a day and occupants will adjust it themself depending on the actual context.

In the next, a 1-h time resolution is used but it can easily change to 10 min.

Using Euler approximation of the derivative[6] with a sampling time Δ, it yields

$$\tau_{k+1} = \frac{\Delta R_k + (1-\Delta)R_i}{R_i}\tau_k + \Delta R_k \varphi_{in,k} + \Delta R_k \left(\frac{1}{R_{out}} + \frac{\zeta_{W,k}}{R_W}\right) T_{out,k} + \ldots$$
$$\ldots + \Delta R_k \left(\frac{1}{R_n} + \frac{\zeta_{D,k}}{R_D}\right) T_{n,k} \tag{8a}$$

$$T_{in,k} = \frac{R_k}{R_i}\tau_k + R_k \varphi_{in,k} + R_k \left(\frac{1}{R_{out}} + \frac{\zeta_{W,k}}{R_W}\right) T_{out,k} + R_k \left(\frac{1}{R_n} + \frac{\zeta_{D,k}}{R_D}\right) T_{n,k} \tag{8b}$$

Air quality has also to be modeled. Even if there are many pollutants like carbon monoxide, fine particles, microbes, VOC, radon, ozone,..., the CO_2 concentration is usually used to estimate a level of confinement like in the ICONE indicator [14]. The level of confinement is meaningful for regularly occupied houses but for others, VOC and radon, for instance, can accumulate during absence and require a deep air renewal before occupying the place. Because the office is regularly occupied, a CO_2 concentration model is established:

$$\begin{aligned}
\Delta n_{CO_2,room} &= V\frac{dC_{in}}{dt} \\
\Delta n_{CO_2,out} &= Q_W\left(C_{out} - C_{in}\right) \\
\Delta n_{CO_2,cor} &= Q_D\left(C_{cor} - C_{in}\right) \\
\Delta n_{CO_2,room} &= \Delta n_{CO_22,out} + \Delta n_{CO_2,cor} + S_{CO_2} n(t)
\end{aligned}$$

where S_{CO_2} stands for average breath production of CO_2 per person, $n(t)$ for the average number of occupants during a sampling period, Δn for variations in the number of molecules of CO_2 and C_x for CO_2 concentrations.

It yields the following differential equation:

$$V\frac{dC_{in}}{dt} = -\left(Q_W + Q_D\right) C_{in} + Q_W C_{out} + Q_D C_{cor} + S_{CO_2} n(t) \tag{9}$$

[6] Exact discretization can also be done numerically.

It can be discretized in time using Euler approximation as well or exact integration over a sample time. With Euler approximation, it comes

$$C_{in,k+1}=\frac{1-\Delta(Q_{W,k}+Q_{D,k})}{V}C_{in,k}+\frac{\Delta Q_{W,k}}{V}C_{out}+\frac{\Delta Q_{D,k}}{V}C_{cor,k}+\frac{\Delta S_{CO_2}}{V}n_k \tag{10}$$

with $C_{out} \approx 400\,\text{ppm}$, $Q_{D,k} = Q_D^0 + \zeta_{D,k} Q_D^1$, and $Q_{W,k} = Q_W^0 + \zeta_{W,k} Q_W^1$

In the next, the difference between knowledge model and regressive observation model is going to be discussed.

5.2 Learning Parameters of Knowledge Models

Parameter values can be assigned from physical knowledge but because they represent several phenomena by only few parameters, like the resistances or the capacitance, their value can just be roughly estimated. Therefore, since the model is highly nonlinear in the parameters, descent approaches such as sequential quadratic programming do not yield reproducible results. A metaoptimization approach has been developed, but repeatability of the estimated parameter values is still not achieved and the best parameter values are often at the limits of the possible value domains for the parameters, even if the value domains are extended. Some authors like [18] used a Bayesian Inference approach of parameter estimation in order to take into account an a priori knowledge of the parameter values to master the convergence but such estimations are not that easy to obtain because most parameters represent multiple physical phenomena. An exploratory approach is going to be used because it led to the most repetitive meaningful results for knowledge model: it's time consuming (about 10 min of computations), but the convergence is repeatable and parameter values don't slide to the bounds. In Fig. 9, a differential evolution optimization has been used. It belongs to the family of genetic algorithms. For the thermal part, a state observer has been used to replace the state once a day, the equivalent temperature for structure, by a measurement corrected better estimation at 0am. Indeed, anticipation is re-performed everyday taking into account weather forecasts, occupancy previsions, etc...

Figure 9 shows 2 situations of learning stage and validation stage for the parameter adjustment of models defined by (8) and (10). In the left hand side, both datasets are consecutive, and in the middle and right side, the datasets are separated by 3 months. Although a knowledge model is supposed to be context independent and applies independently of the time, the second case reveals a poor estimation quality during validation for the indoor temperature. Some would say that the model is not complex enough but complexity makes convergence of nonlinear optimizations more difficult. Let's remind that establishing simplified

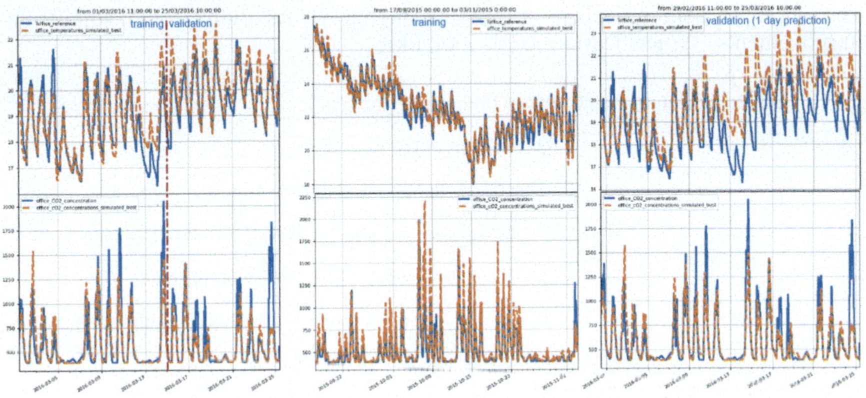

(a) validation in the aftermath (b) validation later

Fig. 9 Parameter estimation of physics inspired model of H358 office is context-dependent

runtime knowledge models and calculated meaningful parameter values is costly, even for monozone offices like H358.

It would be useful if regression could lead to good predictions but their structure is not adapted to the physics. Context-dependent models can be expected: let's investigate this approach.

5.3 *Learning Parameters of Regressive Models*

The structures of linear (ARX, ARMAX, ...) and nonlinear (artificial neural networks) regressive models are such that parameter estimation is easier than knowledge models, which are usually highly nonlinear in the parameters. For the H358 model of the office, because temperature and CO_2 concentration are relatively independent phenomena (with common causes like door and window openings), 2 linear regressions are going to be used: one for anticipating indoor temperature and another one for indoor CO_2 concentration. Let's analyze the results to point out the specificities of regressive models. Firstly, let's notice that regressive models embed a little physical knowledge in the choice of inputs and outputs but also in their structure. Both linear regressions have inputs and outputs inspired by physical knowledge:

- thermal regressive model has inputs: "Tout", "Tcorridor", "window_opening", "door_opening", "total_electric_power", "phi_sun", "dT_heat", "occupancy" and output: "Toffice_reference"
- air quality regressive model has inputs: "corridor_CO2_concentration", "window_opening", "door_opening", "occupancy" and output: "office_CO2_concentration"

The order of both models is 1 because higher values didn't yield significant improvements. Typical learnt models look like:

$$
\begin{aligned}
\text{Toffice\_reference}_k &= +0.896665\text{Toffice\_reference}_{k-1} \\
&\ldots + 0.037271\text{Tout}_k - 0.030218\text{Tout}_{k-1} \\
&\ldots + 0.628310\text{Tcorridor}_k - 0.534000\text{Tcorridor}_{k-1} \\
&\ldots - 0.236362\text{window\_opening}_k + 0.056120\text{window\_opening}_{k-1} \\
&\ldots - 0.250519\text{door\_opening}_k + 0.135190\text{door\_opening}_{k-1} \\
&\ldots + 0.003684\text{total\_electric\_power}_k - 0.000847\text{total\_electric\_power}_{k-1} \\
&\ldots + 0.000377\text{phi\_sun}_k - 0.000250\text{phi\_sun}_{k-1} \\
&\ldots + 0.024356\text{dT\_heat}_k - 0.003150\text{dT\_heat}_{k-1} \\
&\ldots - 0.065434\text{occupancy}_k + 0.055590\text{occupancy}_{k-1} \\
&- 0.795588 \\
\text{office\_CO2\_concentration}_k &= +0.671016\text{office\_CO2\_concentration}_{k-1} \\
&\ldots + 0.451932\text{corridor\_CO2\_concentration}_k \\
&\ldots - 0.209482\text{corridor\_CO2\_concentration}_{k-1} \\
&\ldots - 154.093078\text{window\_opening}_k + 45.894794\text{window\_opening}_{k-1} \\
&\ldots + 47.667825\text{door\_opening}_k - 27.245011\text{door\_opening}_{k-1} \\
&\ldots + 97.997477\text{occupancy}_k + 8.604379\text{occupancy}_{k-1} \\
&+ 28.190620
\end{aligned}
$$

The left hand side and the middle curves of Fig. 10 show that for the same non-consecutive periods that before, excellent results are obtained for the training dataset, comparable to those obtained with knowledge model but with validation dataset. Results for temperature anticipation are very poor. Nevertheless, it's not directly comparable because the model is not adjusted once a day to match the measurements. To better match the objective of a multi-24 h prediction, jumping (sliding day by day) models have been run: they anticipate one day using the 8 previous days for training with a maximum memory of 12 days. The right hand side of Fig. 10 shows excellent results for both previsions. The regressive models are permanently adapting to the context. For instance, the resulting pole for temperature model is belonging to [0.79, 0.83], which leads to mode attenuation between 13 and 17 h.

Jumping linear regressors operate very well for the H358 office: it's a good news because it makes self-learning approaches possible. The approach can be easily extended to multi-zone homes like the one presented at the beginning of the chapter. Regarding more complex systems like the MHI classroom platform with adaptative

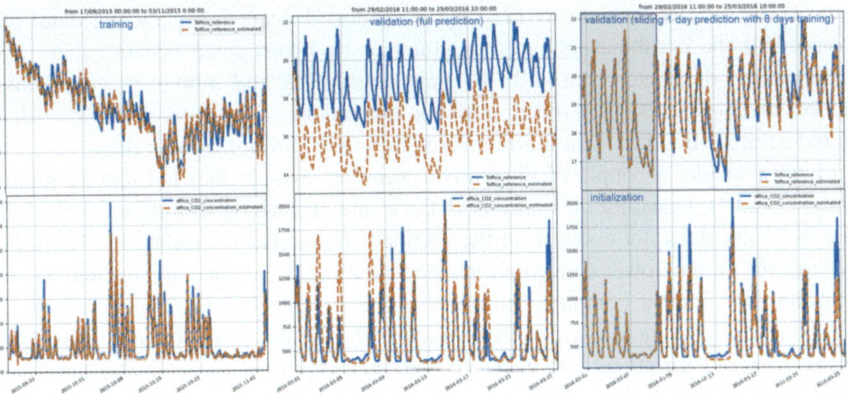

Fig. 10 Linear regressions are also context-dependent but with a day "jumping" window, results are excellent for H358 office

ventilation, recursive multi-layer artificial neural networks can be used to represent nonlinear phenomena but datasets have to be much more important (8 days were enough for H358 office) because the number of parameters to be tuned is huge comparing to the available data in building.

6 Mirroring Inhabitant Service

Avoiding energy and comfort wastes has been formalized as a contextual multi-objective problem were best compromises have to be met as shown in Fig. 6. It's not easy to meet because of unconscious phenomena, fake beliefs, and routines. A first approach, which does not require complex technologies except sensors and relevant Graphic User Interfaces (see Chap. "Explanations Generation with Knowledge Models"), is to mirror occupant activities together with their relative impacts. It is a way to support occupants to become more conscious of what's going on in their home. Waste can be detected using general indicators, whose values depend on inhabitant activities and presences. According to the illustrative examples at the beginning of the chapters, 20–80 IoT sensors are common for an energy smart-home to get a fine analysis of people practices. Nevertheless, such a big number of sensor data are quite long to analyze. Let's consider that 10 s are required to analyze the data of a sensor, and that the meaning of curves results from correlating curves, like presence, openings, temperatures together, which requires also 10 s for each data correlation. If n is the number of sensors, the number of combinations of curves to watch is

$$\sum_{p=1}^{n} \binom{n}{p} = 2^n - 1$$

For 20–80 sensors, it leads to 121 days to almost 4 years for analyzing data! Of course, all the combinations are not going to be investigated because there are independent data, etc…but still, it's time consuming and moreover, values and curves are not necessarily meaningful for inhabitants. Therefore, what is needed is indicators extracting meaningful information from multiple sensor data.

Let's consider a couple of illustrative examples, which does not require input-output models. Consider the analysis of the usage and the performance of a fridge: the difficulty does not come from the number of signals to correlate but from the complexity of the power consumption signal (see Fig. 11). The daily average power consumption can be computed but does a greater consumption means a lack of efficiency of a fridge, or an expensive usage? The indicator is computed on a daily basis, to avoid variations due to intermittent consumption, by removing the basic consumption, which dependents on the room temperature. Some questionable data have been removed for household 2 using a 3-clustering approach, keeping the cluster with the most days. Then, the basic power consumption is obtained with a linear interpolation of the Pareto front (green points on the left side of Fig. 11), considering the minimum consumption for each temperature. The slope R_{fridge} of the red line $T_{\text{room}}(d) = R_{\text{fridge}} P_{\text{base}}(d) + T_0$ stands for the fridge sensitivity around average room temperature. Right hand side curves show the usage power: $P_{\text{usage}}(d) = P_{\text{fridge}(d)} - P_{\text{base}}(T_{\text{room}}(d)$. The following indicators facilitate the comparison between the 2 usages of fridges:

Indicator	Household 1	Household 2	Comment
Average base power consumption (W)	84	51	Average for records
Average usage power consumption (W)	10	8	Average for records
Thermal sensitivity R_{fridge} (W/°C)	6.9	1.3	

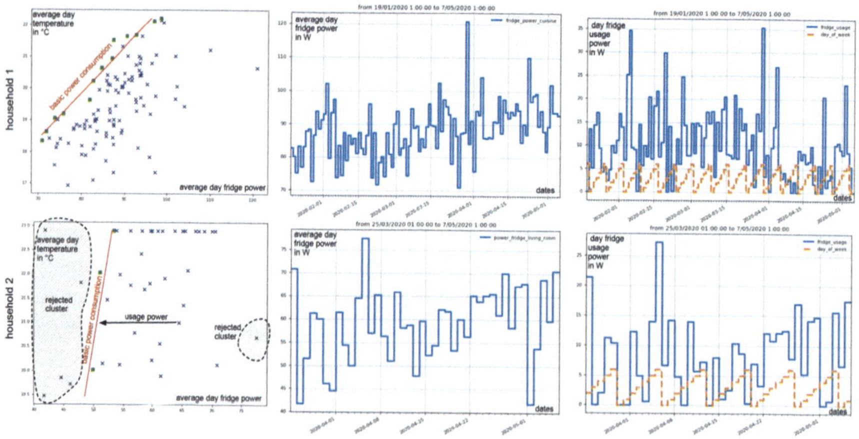

Fig. 11 Example of fridge usage indicator applied to 2 households

These types of indicator are also used for washing and drying machine, dishwashers,... by distinguishing the average consumption per cycle, from the average number of cycles a week. Many others can be imagined to aid inhabitants to get a better understanding of their practices.

There is another kind of indicators that extract information from different measurement signals. Let's take another example dealing with energy waste through a window. When a window is opened, the power loss is given by $P_{loss} = \zeta_w \dot{m} c_p (T_{out} - T_{in})$ where ζ_w is the opening ratio of the window and $\dot{m}$ the mass flow through the window. Because $\dot{m}$ is not easy to measure, the loss can be expressed in °C.h/day: $\sum_{h \in \text{day}} (T_{out}(h) - T_{in}(h))$ considering $\dot{m}$ as a constant value. This loss becomes a waste if a heating (cooling) system is operating ($O_{\text{HVAC}} = 1$) AND air quality is poor, which can be expressed by a dissatisfaction function $D(\text{CO}_2)$ like in Fig. 5. The day waste can be expressed in °C.h/day, as:

$$W_w(d) = \sum_{h \in \text{day}} O_{\text{HVAC}} D(\text{CO}_2(h)) \left(T_{out}(h) - T_{in}(h)\right)$$

O_{HVAC} can be obtained either from a power meter or, more easily, from a temperature sensor properly installed. Figure 12 shows this indicator assessing the usage of a main entrance door leading to a living room. The focus was on May 3, 2020 as this day was detected as particularly important. It points out the hours with questionable behaviors. It can also be applied for monitoring the window usages but because the air mass flow is not known, comparison is not meaningful.

Other indicators related to the estimation of people activities (Chap. "Occupant Actions Selection Strategies Based on Pareto-Optimal Schedules and Daily Sched-

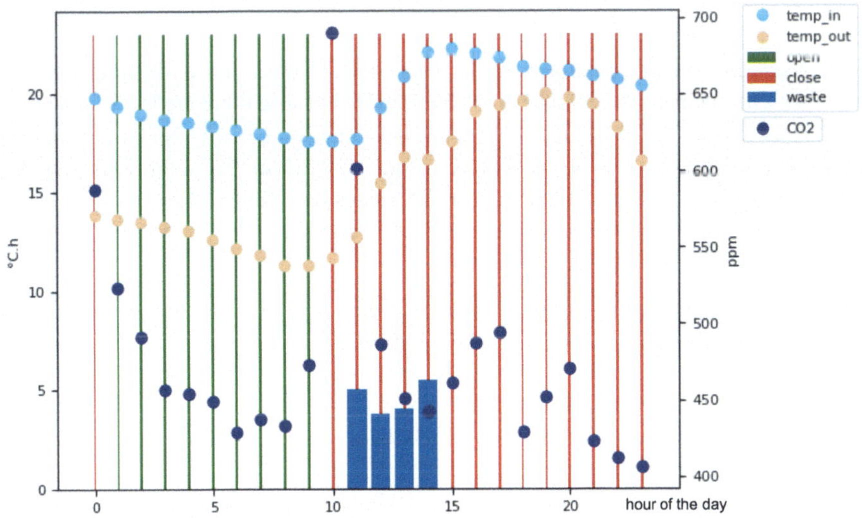

Fig. 12 Assessment of wastes through an entrance door leading to a living room on May, 3rd 2020

ule for Energy Management in Buildings") can be designed for better adaptation to practices, like the preferred temperature during meal, reading,...

Remembering Fig. 6 raises a question: how to determine how far are the practices from the Pareto front of best compromises without input-output models? Historical data can be used assuming best compromises have been met sometimes in the past. Best compromises are context-dependent: therefore, with a day basis, context-similar days have to be gathered but variable values should be normalized to be comparable. Because the value domains are usually open-scaled, a sensitive index is introduced for the sake of normalization of each variable. Let v_i be a variable value from the n-dimensional variable space $\mathcal{S} = (v_i)_{i \in \{1,\ldots,n\}}$ and Δ_i be a minimum sensitive difference for the i-th value of $v \in \mathcal{S}$. A sensitive distance between 2 vectors $v_1, v_2 \in \mathcal{S}^2$ is given by

$$d_\Delta(v_1, v_2) = \frac{1}{n} \sum_{i=1}^{n} \frac{|v_{1,i} - v_{2,i}|}{\Delta_i} \tag{11}$$

Thanks to this sensitive distance, context-similar days can be found by clustering. Concerning the redline H358 office example, the following value sensitive resolutions for context variables have been used:

Context variable V	Granularity Δ_V
Tout	1 °C
Tcorridor	0.5 °C
Occupancy	0.2 person
phi_sun, total_electric_power	50 W
Corridor_CO2_concentration	200 ppm

Considering the periods from 17/09/2015 to 3/11/2016 with a 1h sample time, together with the period 29/02/2016 to 25/03/2016, because there are lots of missing data in between, a hierarchical clustering has been used to gather similar days according to the contextual sensitive distance defined in Eq. (11). Each day is modeled by a data vector of 144 elements: 24 values for "Tout", "Tcorridor", "occupancy", "phi_sun", "total_electric_power", and "corridor_CO2_concentration". 15 clusters of similar days have been found:

- cluster 0 (2 elements): 12/03/2016, 13/03/2016
- cluster 1 (2 elements): 05/03/2016,06/03/2016
- cluster 2 (7 elements): 01/03/2016, 02/03/2016, 03/03/2016, 04/03/2016, 08/03/2016, 09/03/2016, 11/03/2016
- cluster 3 (2 elements): 10/03/2016, 14/03/2016
- cluster 4 (3 elements): 15/03/2016, 17/03/2016, 18/03/2016
- cluster 5 (7 elements): 16/03/2016, 19/03/2016, 20/03/2016, 21/03/2016, 22/03/2016, 23/03/2016, 24/03/2016
- cluster 6 (1 elements): 07/03/2016

- cluster 7 (9 elements): 14/10/2015, 15/10/2015, 16/10/2015, 17/10/2015, 18/10/2015, 19/10/2015, 20/10/2015, 21/10/2015, 22/10/2015
- cluster 8 (18 elements): 03/10/2015, 04/10/2015, 07/10/2015, 08/10/2015, 09/10/2015, 10/10/2015, 11/10/2015, 13/10/2015, 23/10/2015, 24/10/2015, 25/10/2015, 26/10/2015, 27/10/2015, 28/10/2015, 29/10/2015, 30/10/2015, 31/10/2015, 01/11/2015
- cluster 9 (1 elements): 12/10/2015
- cluster 10 (2 elements): 05/10/2015, 06/10/2015
- cluster 11 (3 elements): 23/09/2015, 28/09/2015, 02/10/2015
- cluster 12 (5 elements): 24/09/2015, 25/09/2015, 29/09/2015, 30/09/2015, 01/10/2015
- cluster 13 (6 elements): 19/09/2015, 20/09/2015, 21/09/2015, 22/09/2015, 26/09/2015, 27/09/2015
- cluster 14 (1 elements): 18/09/2015

It can be noticed that although beginning and ending winter days in France appears to be closed, actually the days from these 2 periods are not mixed into clusters: they are not that similar.

For the sake of clarity, let's focus on 2 small clusters: number 0 with 2 similar days and number 8 with 3 similar days. Figure 13 shows the contextual variables for cluster 0: it gathers cold days without occupancy. The left side of Fig. 16 shows

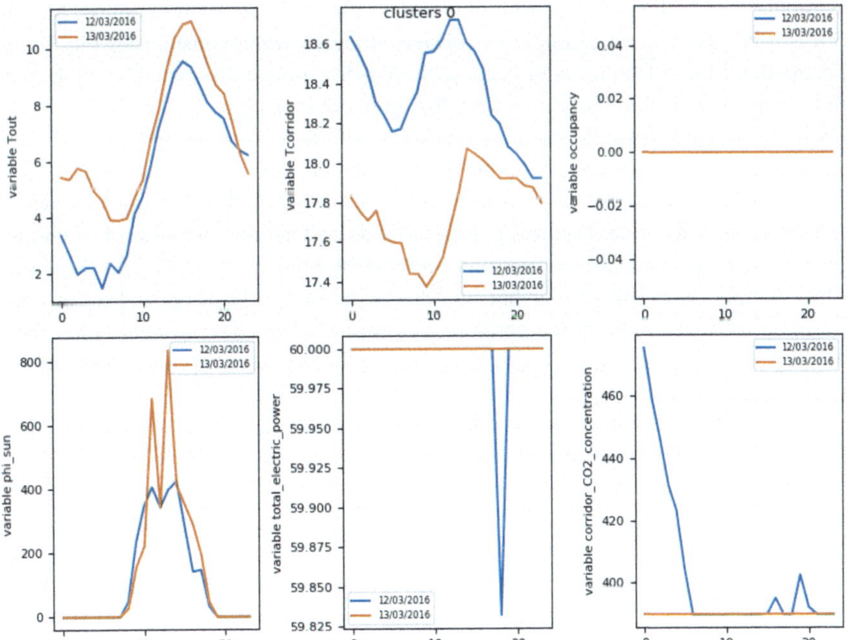

Fig. 13 Context variables for cluster 0 of similar contextual sensitive days

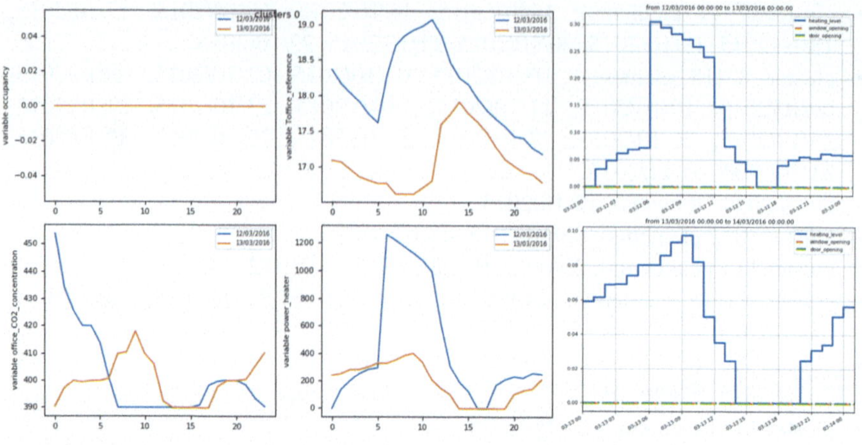

Fig. 14 Effects and actions during days in cluster 0

the corresponding performances. Comfort dissatisfaction is calculated with the 2 functions on the left and on the middle of Fig. 5 provided there is a presence detected, otherwise comfort dissatisfaction is set to 0. Weight has been set to 0.5 for each kind of dissatisfaction. Cost corresponds to energy consumed by the heater. March 12th is clearly dominated by March 13th because comfort is perfect in both cases but on 13th, the heater consumption is less than half. So, in this context, March 13th can be given as an example of sobriety. Indeed, when considering Fig. 14, it appears that March 13th is better because there is less power dissipated by the heater while there is no one in the office. Conclusion is obvious here.

Cluster 4 stands for mild sunny winter days with late office occupancy as shown in Fig. 15. Right side of Fig. 16 points out that March 17th is dominated but March 15th and 18th, which are both dominating compromises. It's not obvious when looking at recorded data. Figure 17 shows effects and actions in cluster 4: it comes out that on 17th, the heater is operating almost like during the 18th and therefore indoor temperatures are similar but on 18th, window has been opened during the afternoon while the outdoor temperature was around 15 °C. The air quality has then been significantly improved. During the 15th, the heater consumption is less but the air quality is poorer.

These results show that it's possible to advice inhabitants about better practices without tuning any input-output model: it's a case-based reasoning approach. Nevertheless, it must be verified that same causes, including context and action variables, lead to the same effects. A checking methodology is presented in Sect. 8. For instance, if a non-measured electric heater is operating, conclusions won't be valid. Figure 18 summarizes the principle of the case-based model validity checking and prevision. Each color for the context variables (c), the action variables (a), or for effect variables stand for similar value sets (there is no connection between colors of different types of variables). The validity checking consists in checking whether the same colored (c) + (a) yields the same colored (e). It's represented by red, green,

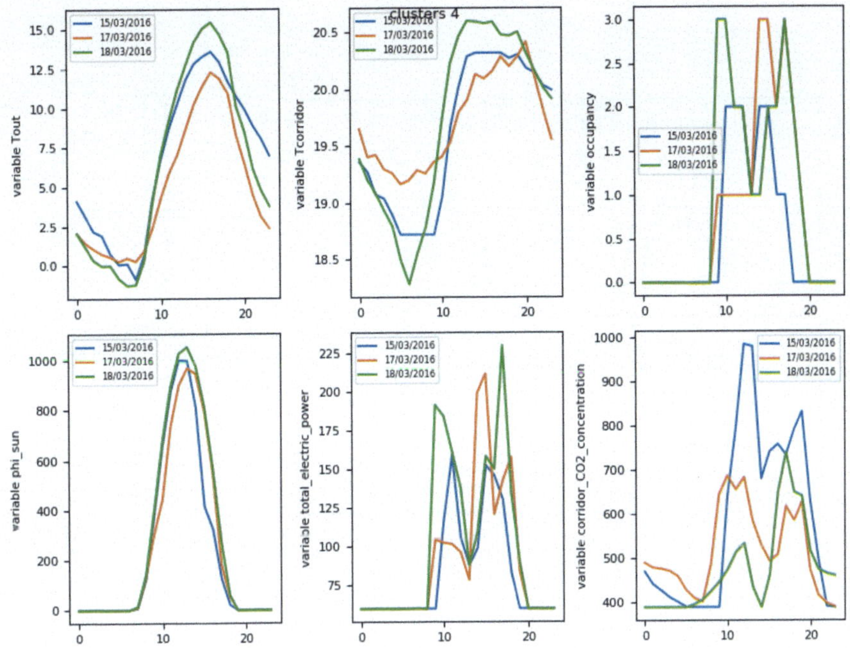

Fig. 15 Variables for cluster 4 of similar contextual sensitive days

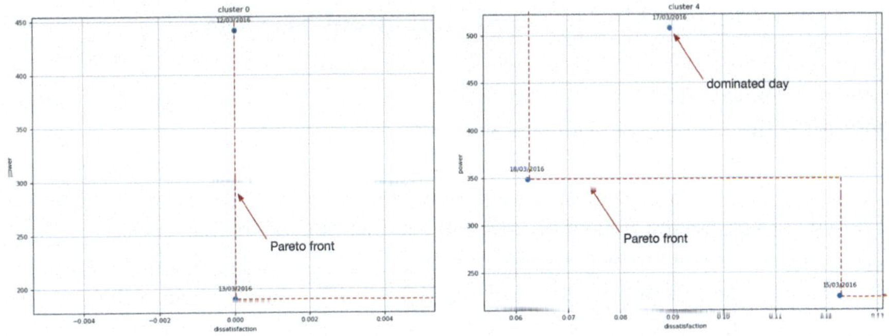

Fig. 16 Performances of days from clusters 0 and 4 with Pareto front

and grey circles in Fig. 18. Case-based advice in a given context (c) consists in searching among the same contexts (c) of historical data, which actions (a) led to the best effects (e).

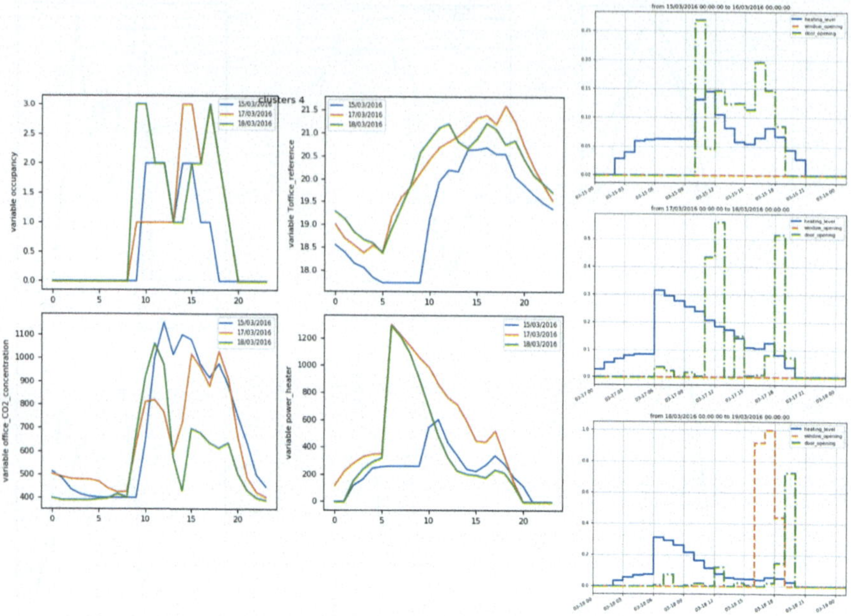

Fig. 17 Effects and actions during days in cluster 4

7 Input-Output Model Based Inhabitant Services

Figure 16 showed that March 17th, 2016 was a day whose energy strategy can be improved. Similar days like 15th or 18th can be followed but Fig. 15 shows that days are not that similar and additionally, there are 3 days in the cluster, i.e. experiences for these kinds of days are limited. Adjusted models given in Sects. 5.2 or 5.3 make it possible to assess the impact of different actions: this inhabitant service is called *replay service*. It is a fast way to experiment an energy strategy without waiting for a similar day. Let's exemplify this with the 1-day sliding regressive model shown in Fig. 10.

Figure 19 represents 3 different strategies for the upper plots and the bottom left corner one:

- in blue, what has been measured
- in orange, what has been simulated with same actions than the ones recorded
- plots ending with "_door_closed" stands for same actions than before but door is kept closed
- plots ending with "_door_closed_window_opened" stands for same actions than before but window is opened 1 h during 6 to 7pm.

Keeping the door closed doesn't change a lot the comfort. The bottom right corner plot is a simulation with door closed and window opened 3 h from 4 to 6pm.

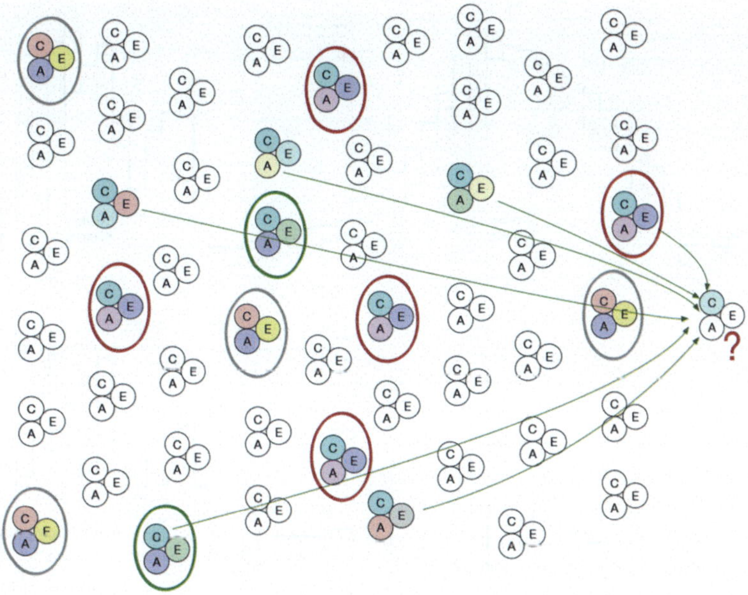

Fig. 18 Principle of case-based model validity checking and prevision

The CO_2 concentration decreases significant, even more than it could be in reality because values below 400 ppm can't be reached. It points out a drawback of the regressive models: their structure does guarantee the consistency with physics.

Let's now illustrate another inhabitant service named *suggest* that solves the so-called inverse problem instead of the direct one. It can be used for a past day like March 17th, 2016, or for the upcoming day provided context variables can be predicted (weather forecast services are available and SARIMA approaches, for instance, can be used for other signals). When the approach is automatized with a sliding time window anticipating the future, it is called Model Predictive Control in the literature (see Chap. "Faults and Failures in Smart Buildings: A New Tools for Diagnosis"). Here, we focus on inhabitant services with a 1 h-time step and on 24 h time horizon, but, from an algorithm perspective, the discussion can easily be extended to other problems. This time, we are going to use a knowledge based model but input-output regressive models can be used alternatively. The inverse problem consists in finding a best cost-comfort weighted compromise for a day by determining the optimal values for action variables while respecting the model, seen as a set of constraints, and according to the context variables. Many optimization algorithms can be used. Here is a couple of them with their features:

Most optimization algorithms are widely used. Let's focus on dynamic programming, which is a general approach for problem solving adapted to optimization. It relies on graph problem formulation and amount to search for the shortest path composed of valued edges in a graph. The algorithm applies to a problem if the

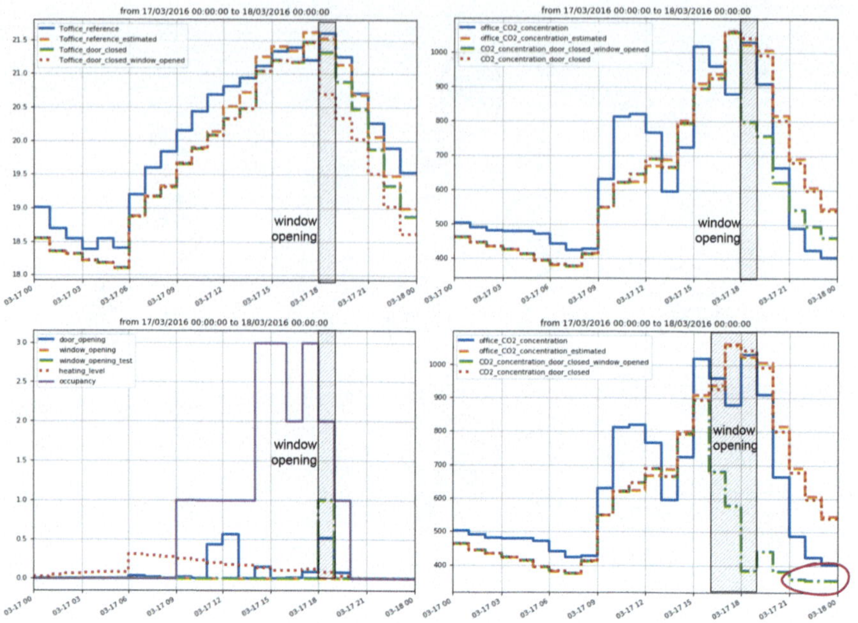

Fig. 19 Different strategies tested to discover how to improve days like March 17th, 2016 in H358 office

Bellmann-Ford lemma is satisfied. Let N_i and N_j be 2 nodes of a graph. The lemma stating that the shortest path corresponds to the global optimum assumes that:

- each edge joining 2 nodes has a positive value, named distance
- each node is discrete, represented by discrete values
- the distance from N_i to N_j is equal to a sum of distances along a path joining N_i to N_j
- let p be a path from N_i to N_j, then the minimum distance from N_i to N_j is lower or equal to the distance along path p.

For energy strategy calculation, each node is related to a stage corresponding to a time step in the solving as shown on Fig. 20. A node corresponds to a state belonging to a discrete grid with pre-defined resolution. The values of context variables are known in between 2 time stages. All possible discrete actions can be applied to each state of stage k to lead to a continuous state belonging to a cell of stage $k + 1$. The state can be composed of state variables for state space models, or to estimated outputs for input-output models. Each edge is weighted by the cost-comfort criterion (stage cost and dissatisfaction) resulting from the actions leading from an original state at stage k to the state corresponding to the consecutive state at stage $k + 1$. All the paths leading to cells of stage $k+1$ are shrunk: the sum of cost-comfort criterion along the path from the initial state are compared within a same grid cell and only the path leading to the smallest value is kept per cell. The solving is going one

Algorithm	Cont. variables	Disc. variables	Optimum	Speed	Complexity	Comment	References
Simulated annealing	Yes	Yes	Local	Medium	Limited	Local search with diversification	
Sequential Quadratic Programming	Yes	No	Local for non convex problem	Fast	Limited	Descent algorithm	
Differential evolution	Yes	No	Global	Slow	Limited	Genetic algorithm	
Dynamic programing	No	Yes	Approx. global	Fast	Limited	Specific, explained in the next	
Mixed Integer Linear Programing	Yes	Yes	Global	Medium/slow	Efficient	Requires a complex reformulation of the model	

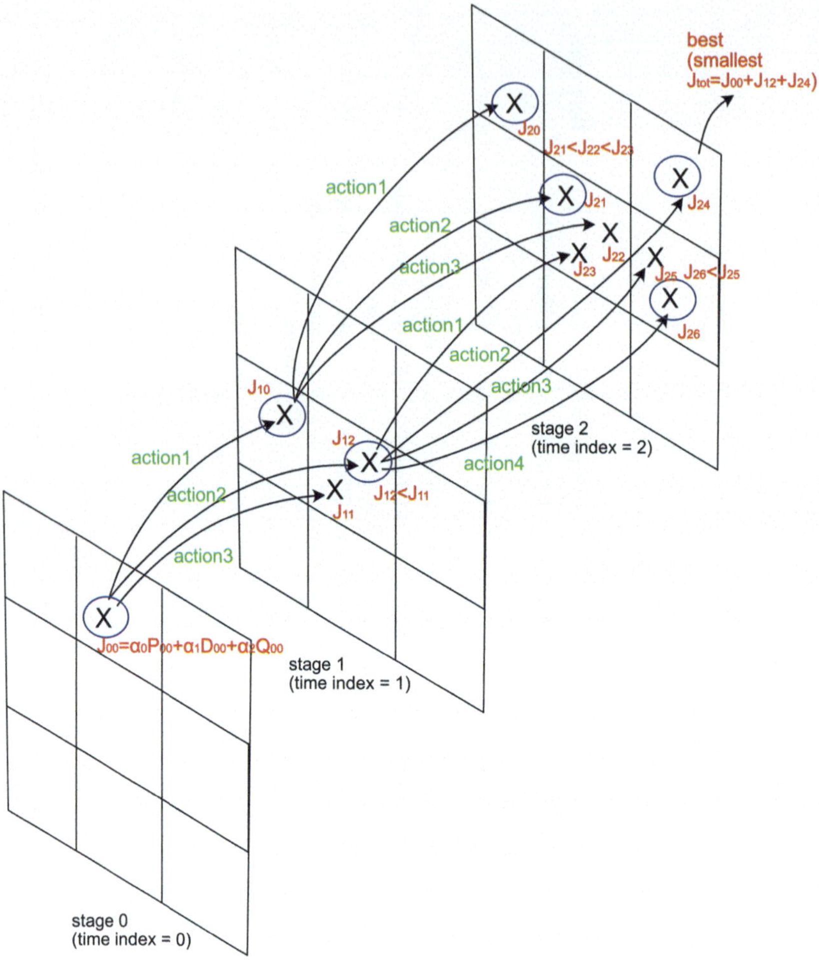

Fig. 20 Adaptation of dynamic programming to computation of best energy strategies

step forward and is recursively solved. A backward solving is not possible because without shrinking, the number of possible states is diverging.

For small to medium size problems, dynamic programming proved to be very efficient. For large size problem, MILP solving is the recommended approach although it is complex to setup. For human interactions, energy strategies usually cover an 24 h horizon with a 30 min/1 h time step. If a state space knowledge model is used, the initial state has to be estimated from input but also output recorded measurement values: setting up a state observer [13] is useful.

The next table, different performance indicators of the best energy strategies obtained with different optimization algorithms are appearing for March 17th, 2016 in H358 office with a 1-h sampling time. Possible actions are adjustment of temperature set-point of the heater, window, and door openings (percentage of the time it is opened). It has been obtained with an input-output knowledge model whose parameters have been retrained with the preceding 2 weeks data. The blue curves stand for actual recorded data and the orange curves (init) for the re-simulated data. There is no difference for the cause (context and action) variables but differences can be seen in the CO_2 concentration and in the indoor temperature. This shows that there can be discrepancies between what is predicted by the model and reality: this can be explained by an inappropriate model structure (this is why the model is re-trained just before the strategy calculation with a period from February 29, 2016 to March 16, 2016 included with a time step of 3600 s), but also by unmeasured events such as undetected visitors, apertures detected as open but in reality closed, a wrong estimation of the initial temperature of the structure,. . . . . . It can be stated that because of these gaps, the computed energy strategies should not be considered as truth but as suggestions that could be tested. Actual accuracy of the suggested plan should be evaluated for occupants to trust or not the energy management decision-aided system. The computation of energy strategy can be seen as a complementary mechanism to case-based reasoning (see Sect. 8) because it can only cope with the analysis of similar contexts, whereas strategy computation can drive occupants in exploring new strategies. On March 17, 2016, from a numbers perspective, the best strategy is the one computed by the sequential quadratic programming algorithm, followed by differential evolution, which takes much longer but leads to the same results, although it is less sensitive to local minima. Simulated annealing is not operating well here. From an occupant perspective, dynamic programming might seem more interesting because the action is discretized: close or open door and window, which is more easy to apply than something like "open the window 35% of the hour. Temperature set-points are also discretized into 5 levels, just like the thermostatic valve is graduated. Dynamic programming is all the more interesting here that it guarantees the global optimum and it is very fast in terms of computation times. Nevertheless, problem complexity might reduce the interest of dynamic programming, in this case, more complex approaches requiring problem automatic reformulation into mixed integer linear programming can be used.

In Fig. 21, it can be noticed how the energy strategy is adapting to the occupancy of the H358 office.

There is another paradigm than the "doing with occupants", which puts humans into the loop, the "doing instead" but it's not relevant in this case because doors and windows cannot be moved automatically, but it can be applied in some specific contexts dealing with the adjustment of temperature set-points. It is usually solved using shorter time horizon, smaller sampling time and sliding optimization time window with possibly a short-term reactive mechanism that is adjusting the control to cope with unexpected discrepancies, while recomputing a long-term energy strategy taking into account the current context [17].

Indicator	Initial	SQP	SA	DIFEVO	DYNPROG
Global objective	0.049	0.026	0.043	0.027	0.032
Average thermal dissatisfaction	0.022	0.0005	0.022	4.635	0.008
Average heating level	0.123	0.053	0.096	0.060	0.069
Average temperature during presence (Celsius)	21.3	21.1	21.3	21	21.3
Average 10% lowest temperature during presence	20.4	21	20.4	21	20.8
Average 10% highest temperature during presence	22.1	21.5	22.2	21.3	22
Average CO2 concentration during presence (ppm)	582	582	582	582	581
Average 10% lowest CO2 concentration during presence	489	489	489	489	489
Average 10% highest CO2 concentration during presence	668	668	668	669	666
Computation time in seconds	NA	13	18	203	0.02

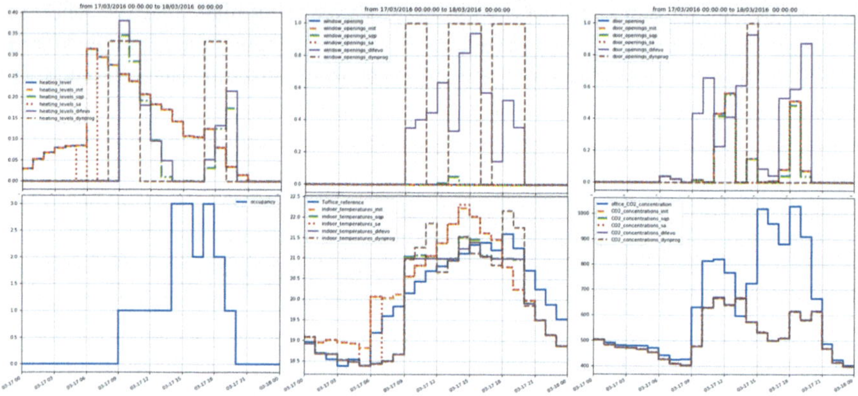

Fig. 21 Energy strategies computed with different optimization algorithms

8 Case-Based Inhabitant Services

Let's investigate the computation of energy strategies without designing input-output model (neither from knowledge nor from observations), which is often a big issue for the deployment of smart-home technologies.

8.1 Proposed Approach

Let's push the clustering of Sect. 6 one step further ahead in order to implement the case-based approach represented in Fig. 18. In order to cluster vectors of heterogeneous variables dealing with a specific day, a sensitive distance has been introduced in Eq. (11). It is a way to normalize different types of variables like temperature, power, CO_2 concentration,...

The principle of the *case-based reasoning* is to do clusters of *context-similar days*, i.e. days for which context variable values are closed. Then, the search for a good energy strategy among days similar in context to the day studied, allows us to determine whether the actions recorded led to better performance on the effect variables. It's a self-experimental approach, where occupants experiment different strategies and the energy management decision-aided system determine what was the best strategy ever experimented.

Nevertheless, the number of measurements has to be big enough to ensure that same recorded context variable values and same recorded action variable values lead to the same recorded variable effects. To establish this completeness property, the distance is highly determinant but the sensitive distances are arbitrary defined by an expert: it's not satisfactory. This distance is expanded to a weighted sensitive distance.

Definition: Weighted Sensitive Distance

$$d_{\Delta,W}(v_1, v_2) = \frac{1}{n}\sum_{i=1}^{n} w_i \frac{|v_{1,i} - v_{2,i}|}{\Delta_i} \tag{12}$$

$$W \in \mathbb{R}^n / \sum_{i=1}^{n} W_i = 1$$

The vector of weights W is going to be adjusted so that the same context and the same actions lead to the same effects, as much as possible. In order to determine whether 2 vectors of effects are similar, a threshold is needed. The expert-defined sensitive value Δ_i related to each effect variable is going to be used: it stands for the human perceivable sensitivity to a variation of Δ_i in the value of the variable v_i.

Definition: Effect-Similar Vectors Two vectors of effect variables V and V' are similar if

$$\forall i \in \{1, \ldots, \dim(V)\}, |V_i - V_i'| < \Delta_i \tag{13}$$

Establishing the completeness of a case-based reasoning problem consists in computing whether there exists a weight vector W such as when the W-weighted sensible distance between 2 cause vectors (context and action variables) is small, it implies that the resulting effects are similar. An optimization approach is proposed:

1. set $i = 0$
2. choose an initial weight vector W_i for causes (for instance, n times $1/n$), i.e. context and action variables

3. gather the causes, which are closed, using a k-mean clustering approach with the weighted sensitive distance $d_{\Delta,W_i}(v_1, v_2)$ and initially $k = n$.
4. adjust W_i to W_{i+1} so that the clusters' magnitude, i.e. the average sensitive distance within the effects in each cluster of similar causes is as small as possible, and the number of clusters is as small as possible
5. if W_{i+1} is almost same than W_i, stop and check among each cluster whether all the couples of effect vectors v_1 and v_2 are similar, i.e. $\forall i, |V_i - V_i'| < \Delta_i$; otherwise, increment i and go to point 3.

Once the weighted sensitive distance defined and completeness verified, similar context days of a current day context $v \in \mathcal{V}_c$ can be selected. Let's consider the maximum weighted sensitive distance T_c delimiting a neighborhood around the current day context v: $\mathcal{N}_c(v, \eta) = \{v_i; d_{\Delta,W}(v, v_i) \leq \eta; v \in \mathcal{V}_c, v_i \in \mathcal{V}_c, \forall i\}$. Then among this neighborhood, the best effects will be selected (let's say day i^*) and it will determine the best actions ever seen.

8.2 Results for H358 Office

This section presents a demonstration for the proposed approach applied to the H358 office. The sensors data and weather conditions have been recorded for a period of time starting from April 2015 and going till October 2016 with a 1-h time step. Table 1 describes the different variables, their units and in which group they were allocated (Context, actions, effects) depending on their roles.

During the summer (from April to September 2015), the heater does not work, and some of the data are missing specially during some holidays such as 29/05/2015–01/06/2015 and 20/06/2016–22/06/2016. Only working days were chosen (no weekends or closed days, as there are no occupants in these days) for periods: 01/05/2015–28/05/2015, 01/06/2015–23/07/2015, 11/05/2016–31/05/2016 and 02/06/2016–19/06/2016; after filtering, 100 days have been found for training the model.

Table 1 Table of features in the building

Feature	Unit	Description	Type
Toffice_reference	°C	Indoor temperature	Effect
office_CO2_concentration	ppm	Indoor CO2 concentration	Effect
Tcorridor	°C	Corridor temperature	Context
Illuminance (new,old)		Luminosite in office	Context
solar radiation	W/m2	Solar radiation in office	Context
wind speed	m/s	Outside wind's speed	Context
corridor_CO2_concentration	ppm	Corridor CO2 concentration	Context
power (block east, west, total)		electricity consumption	Context
occupacy		Number of occupacy in office	Context
nebulosity	%	Outside nebulosity	Context
Tout	°C	Outside temperature	Context
window_opening	minute	The duration of opening window	Action
door_opening	minute	The duration of opening door	Action

When applying the proposed approach, using a Mac laptop with Core-i5 (2,7 Ghz) processor and 8 Giga Bytes of memory, the genetic algorithm needed 6900 s about 2 h with 11 iterations. The algorithm leads to the following W vector:

$$W = [0.031813, 0.962732, 0.450447, 0.501815, 0.008006, \ldots$$
$$\ldots 0.001168, 0.006802, 0.039470, 0.7701]$$

It represents the relative weights for the different variables, respectively: Tout, Tcorridor, window opening, door opening, total electric power, phi sun, occupancy, nebulosity, Corridor CO_2 concentration.

With weighted sensitive distance, it is possible to get the similar days and extract from the recorded history the day with the best satisfaction and recommend it to the occupants as in Figs. 22 and 23.

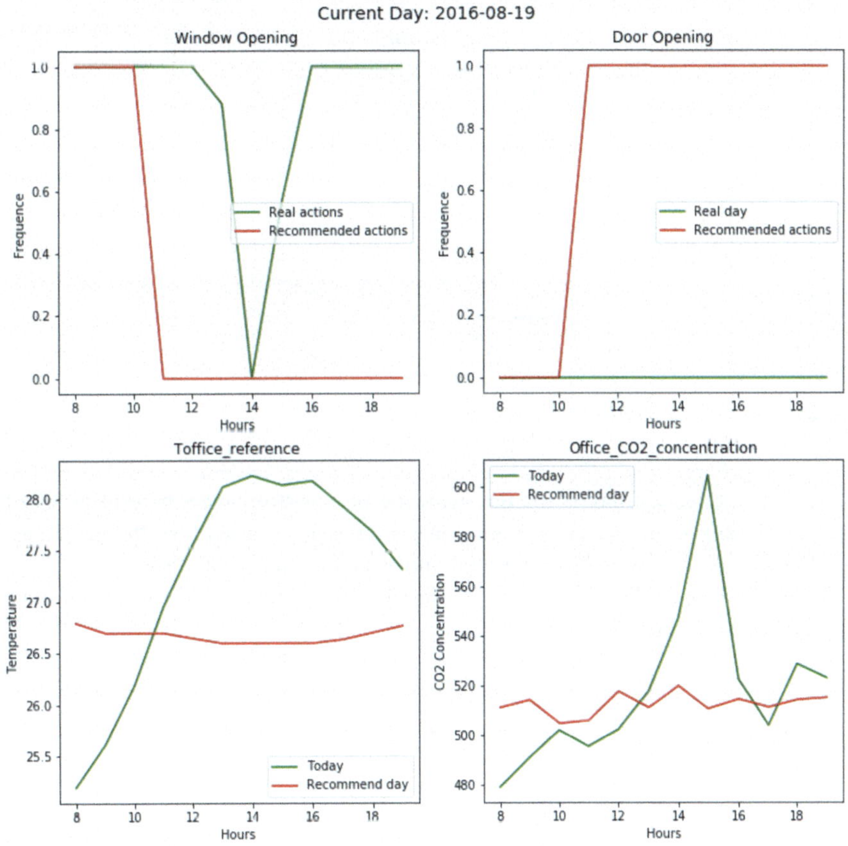

Fig. 22 Applying the case-based approach on a day 19/8/2016

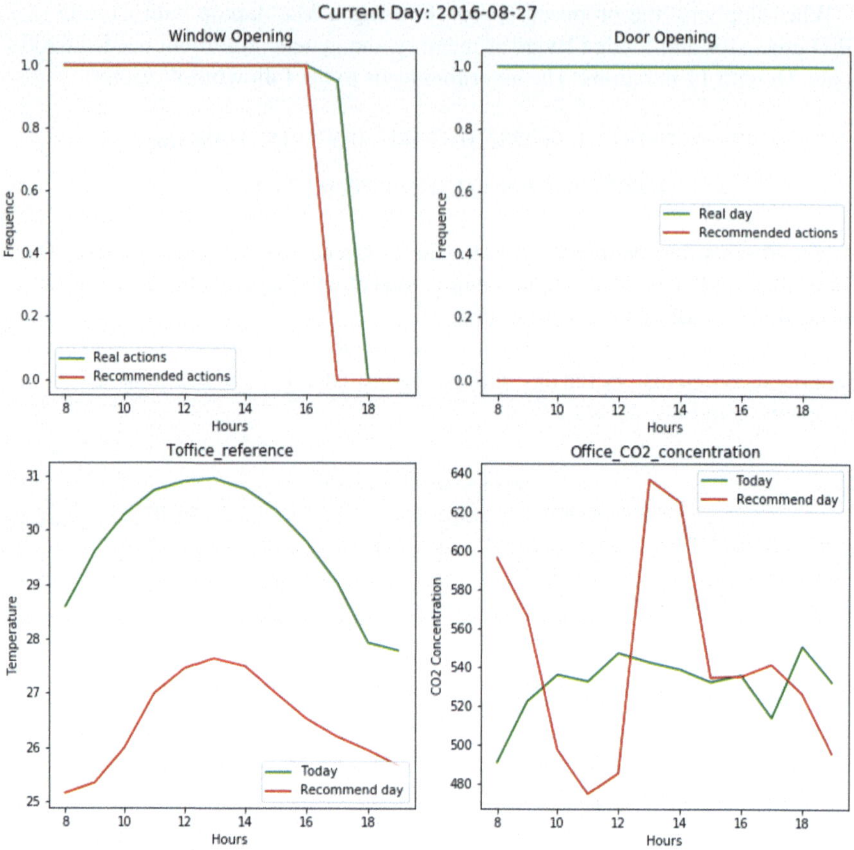

Fig. 23 Applying the case-based approach on a day 27/8/2016

The recommended day is the day with the best performance, here in terms of user satisfactions, among the similar context days. It presents how the occupants can change his actions and the expected improvement on his satisfaction.

In the next, the solving approach is validated and it reveals how it can improve the user satisfaction.

8.3 Validation for H358 Office

Validating the proposed case-based approach for H358 office is carried out in 2 steps: validation of the completeness and validation of the effect improvement. Nevertheless, to avoid wrong conclusions due to modeling errors, the recorded effect variables have been replaced by re-simulated effect variables using the knowledge model given by Eqs. (8) and (10).

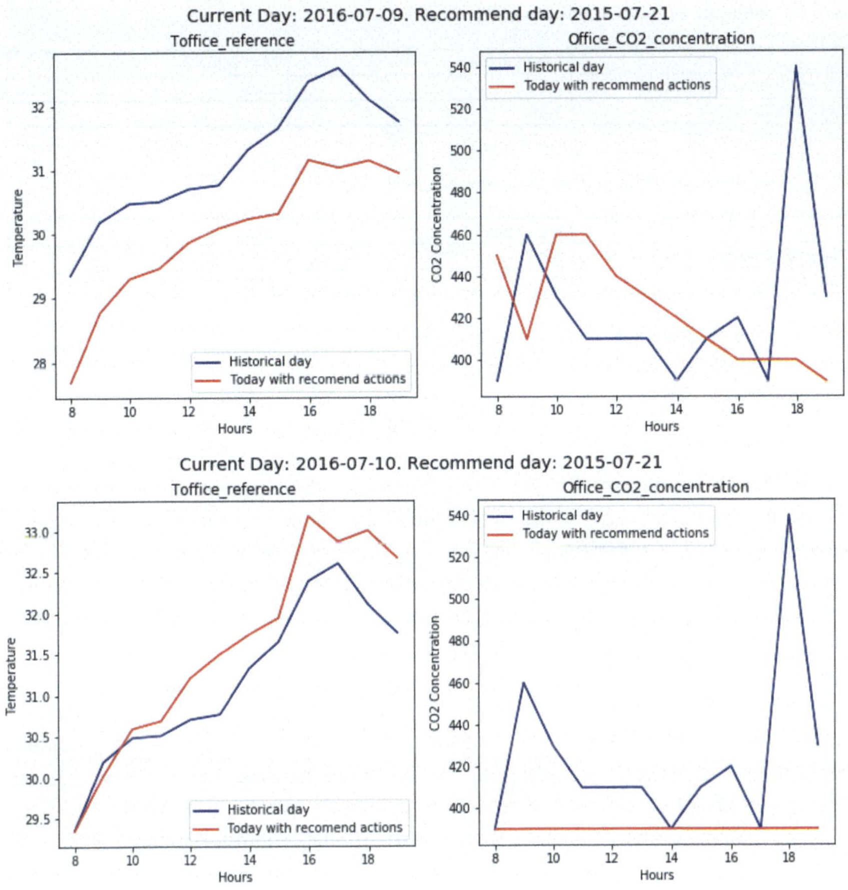

Fig. 24 Results of applying recommended actions

Let's validate the completeness of the available data for H358 for the following set of days: From 22/06/2016 to 30/07/2016, …according to the weighted distance W resulting from GA optimization. Figure 24 shows that when applying the same set of actions to two similar context days, their effects are similar. For example, with 09/07/2016, the proposed approach finds 21/07/2015 as similar day. When applying the set of actions from 21/07/2015 to 09/07/2016, the obtained indoor temperature and CO_2 concentration are similar to those of 21/07/2015, the indoor temperature difference between the two days is about 1.5 °C and this metric for the CO_2 level is less than 150 ppm.

For each day i of the 34 testing days, suppose that $\Delta E = |E^* - \bar{E}|$ and ΔT, ΔCO_2 are the difference of indoor temperature and CO_2 concentration of ΔE, respectively. ΔT and ΔCO_2 are computed by the maximum difference between each hour of two days, respectively. It has been tested with 34 days, and the result is showed in Fig. 25. The mean difference of temperature—ΔT is about 1.56 °C and

ΔT - The difference of indoor temperature (°C)

	Mean value	Max value
Mean difference	0.59	1.26
Max difference	1.56	2

$\Delta CO2$ - The difference of indoor CO2 concentration (ppm)

	Mean value	Max value
Mean difference	55	139
Max difference	199	400

Fig. 25 Table of results for validating the hypothesis

the mean difference of CO_2—ΔCO_2 is about 199 ppm with all pairs of days. The maximum value of ΔT is 2 °C and the maximum value of ΔCO_2 is 400 ppm.

The second validation step consists in checking whether there are improvements in resulting best effects but using re-simulated effects to avoid impacts of modeling errors. For each day, it consists in determining the best actions from recorded similar context days and to simulate the resulting effects from action and context variables in order to compare the resulting performances between recorded data and best effects. How much is the improvement?

For a specific day i with context $v_{c,i} \in \mathcal{V}_c$, let's denote $| = \{j; i \neq j, v_{c,j} \in \mathcal{N}_c(v_{c,i}, \eta)\}$, the context-similar days for day i. Let's denote $i^* \in \mathcal{J}$, the best context-similar effect days from historical recorded data. $v_{a,i^*} \in \mathcal{V}_a$ denotes the best set of actions according to the recorded history. Let's verify that the recommended set of actions enhance the occupant's satisfaction. To do that, let's compare the best effect v_{e,i^*} of the day i^* to re-simulated similar context days to the re-simulated effect $v_{e,i}$ with the real actions $v_{a,i}$ of occupants during the day. From that, it could estimate the enhancement of recommended actions to the occupant's satisfaction defined in Fig. 5. The concept of this validation method is shown in Fig. 26.

For this validation test, the effect vector contains the resulting occupant satisfaction $S(v_{e,i})$, which is composed of thermal and air quality satisfactions. The global satisfaction is given by

$$S(v_{e,i}) = \alpha * TS(v_{e,i}) + (1 - \alpha) * CS(v_{e,i}) \tag{14}$$

with $\alpha \in (0, 1)$. It represents the relative importance of thermal satisfaction (TS) with the air quality satisfaction (CS) (as shown in Fig. 5).

Suppose that $S(v_{e,i^*})$ is the occupant's satisfaction obtained for effect v_{e,i^*}. The enhancement obtained when applying the recommended set of actions is computed as follows:

$$H(S(v_{e,i}), S(v_{e,i^*})) = \frac{S(v_{e,i^*}) - S(v_{e,i})}{| S(v_{e,i}) |} \tag{15}$$

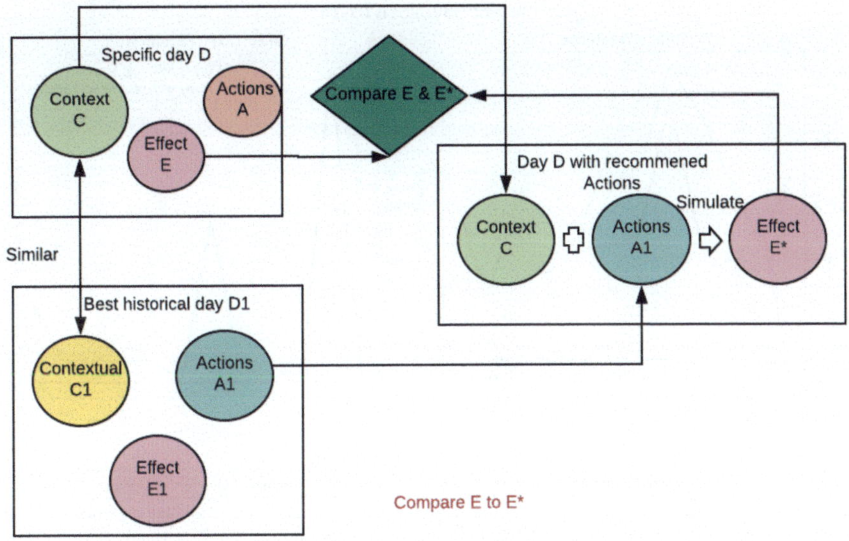

Fig. 26 The concept of recommended actions validation

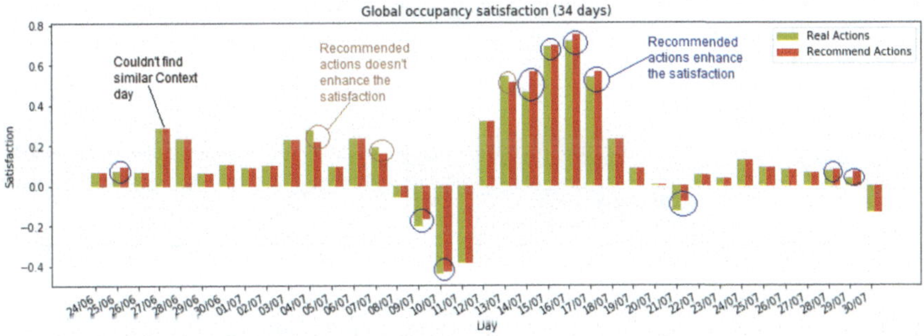

Fig. 27 Results of recommended actions to enhance global satisfaction (When no similar context day is not found the same actions of the test day are used for simulation)

If $H(S, S^*) > 0$, it means that the recommended actions enhanced the satisfaction of occupants compared to the real actions. However, due to the limited number of data (34 testing days), we could recommend a set of actions to only 13 days, so there are 21 days we could not find similar days from the past days. This is the result of comparing recommended satisfaction with real satisfaction in this case.

From Fig. 27, it can be seen that, with 13 days, the recommended set of actions could enhance the satisfaction for 10 days (77%). There are 21 days where the approach could not find similar Context days and recommend actions due to the limitation of data.

In different days, the recommended actions could improve from 8 to 18% occupants' global satisfaction. Regarding the example shown in Fig. 28, when

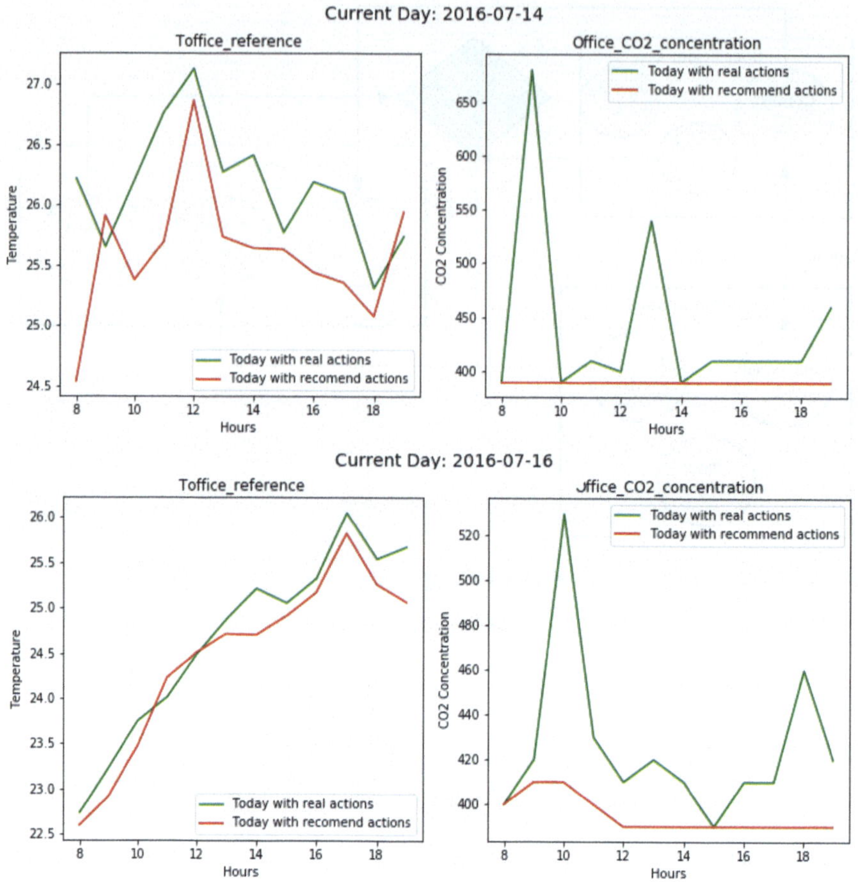

Fig. 28 Results of applying recommended actions between 14/07/2016 and 16/07/2016

applying recommended actions between 14/07/2016 and 16/07/2016, the global satisfaction was better by 18 and 11%, respectively, than real satisfaction; the detailed results of these days are shown in Fig. 28.

It can be seen in Fig. 28, for 14/07/2016, when applying recommended actions, the recommended actions could reduce the indoor temperature by about 1 ^{circ}C and with 16/07/2016, the indoor temperature could decrease by 0.5 ^{circ}C.

Limitation of data is the challenge of this approach. For some days, the work could not find any day having similar context features because of the restriction of data. It is necessary to have enough significant data to find similar days and choose the best set of actions.

9 Conclusion

In this introductory chapter, the different issues to be solved for decision-aided systems dedicated to energy smart-homes have been pointed out and the different kinds of problems to be solved stated. It has been shown that variables belong to different categories: context and action variables among the causes, and effect and performance variables among the consequences. It has been discussed that a living area (home/office) is providing services to inhabitants, some are energy consuming and the energy management problem consists in supporting people in taking relevant decisions taking into account that most phenomena are invisible, preferences are personal, sobriety and flexibility are important issues. The modeling problem was presented as a major barrier to the diffusion of energy smart building technologies. Knowledge models are difficult to establish because their structure is generally not suited for parameter estimation but they are by nature yielding physically consistent simulated data. Conversely, regressive models are easy to adapt to recorded data but they are highly dependent of the content of the dataset used for parameter adjustment: physical inconsistencies might appear, even for sliding models with adaptative learning. Case-model based only on data, without formalized relations between data, is a promising approach. Mirroring services have been exemplified as well as *replay* and *suggest* services where different optimization algorithms can be used with discussed specificities.

All the topics of this chapter are further investigated in the book. One important issue in building sector is to cope with gaps into datasets, which are quite common, and faulty component. It's discussed in chapter 13.

Acknowledgments This work benefits from the support of the INVOLVED ANR-14-CE22-0020-01 project of the French National Research Agency. It is also supported by the French National Research Agency in the framework of the EquipEx program AmiQual4Home ANR -11-EQPX-00 and the "Investissements d'avenir" program (ANR-15-IDEX-02) as the cross disciplinary program Eco-SESA. This work is also partially supported by the project (DST-INRIA/2015-02/BiDEE/0978) of Indo-French Centre for the Promotion of Advanced Research (CEFIPRA|IFCPAR).

References

1. B. Béjannin, T. Berthou, B. Duplessis, D. Marchio, Evaluation of water heaters control strategies for electricity storage and load shedding at national scale, in *4th Building Simulation and Optimization Conference, Cambridge, UK* (2018). https://hal.archives-ouvertes.fr/hal-01981401
2. S. Carloganu, Evaluation de produits d'effacement sur un ensemble de consommateurs par modélisation bottom-up d'un parc de logements. Ph. D. Thesis (2016). http://www.theses.fr/2016PSLEM064/document. 2016PSLEM064
3. A. Crabtree, T. Rodden, Domestic routines and design for the home. Comput. Supported Coop. Work **13**(2), 191–220 (2004)

4. D. Da Silva, Analytical study of household electricity demand flexibility. Theses, École Nationale Supérieure des Mines de Paris (2011). https://pastel.archives-ouvertes.fr/pastel-00678316
5. J.A. Duffie, W.A. Beckman, *Solar Engineering of Thermal Processes*, 3rd edn. (Wiley, Hoboken, 2006)
6. J. Guibourdenche, P. Salembier, G. Poizat, Y. Haradji, M. Galbat, A contextual approach to home energy management systems automation in daily practices, in *Proceedings of the European Conference on Cognitive Ergonomics 2015*, pp. 1–4 (2015)
7. L.D. Ha, S. Ploix, E. Zamaï, M. Jacomino, Realtimes dynamic optimization for demand-side load management. Int. J. Manag. Sci. Eng. Manag. **3**(4), 243–252 (2008)
8. A. Kashif, J. Dugdale, S. Ploix, Simulating occupants' behaviour for energy waste reduction in dwellings: a multi agent methodology. Adv. Complex Syst. **16**(04–05), 37pp. (2013)
9. A. Kashif, S. Ploix, J. Dugdale, X. Binh Le, Simulating the dynamics of occupant behaviour for power management in residential buildings. Energy Build. **56**, 85–93 (2013)
10. M.Y. Lamoudi, M. Alamir, P. Béguery, Distributed constrained model predictive control based on bundle method for building energy management, in *2011 50th IEEE Conference on Decision and Control and European Control Conference* (IEEE, Piscataway, 2011), pp. 8118–8124
11. S. Lannez, G. Foggia, M. Muscholl, J.C. Passelergue, C. Lebosse, K. Mercier, Nice grid: smart grid pilot demonstrating innovative distribution network operation, in *2013 IEEE Grenoble Conference* (IEEE, Piscataway, 2013), pp. 1–5
12. S. Lisa, P. Stephane, B. Pierre, W. Etienne, Model tuning approach for energy management of office and apartment settings, in *Building Simulation (BS2017), San Francisco, August 7–9* (2017)
13. M. Lundh, M. Molander, State space models in model predictive control. ABB Automation Products AB, ABB library (2002). https://library.e.abb.com/public/c8687aec93abed2e85256f9b0054e8fb/TP_Lundh_Molander.pdf
14. N. Michelot, C. Marchand, O. Ramalho, V. Delmas, M. Carrega, Monitoring indoor air quality in French schools and day-care centers. HvAC R Res. **19**(8), 1083–1089 (2013)
15. Q. Nguyen Hong, A. LeMounier, V.-B. Dinh, B. Delinchant, S. Ploix, F. Wurtz, Meta-optimization and scattering parameters analysis for improving on site building model identification for optimal operation, in *Building Simulation* (2017)
16. J. Page, D. Robinson, N. Morel, J.-L. Scartezzini, A generalised stochastic model for the simulation of occupant presence. Energy Build. **40**(2), 83–98 (2008)
17. M. Pal, A.A. Alyafi, S. Ploix, P. Reignier, S. Bandyopadhyay, Unmasking the causal relationships latent in the interplay between occupant's actions and indoor ambience: a building energy management outlook. Appl. Energy **238**, 1452–1470 (2019)
18. S. Rouchier, T. Busser, M. Pailha, A. Piot, M. Woloszyn, Hygric characterization of wood fiber insulation under uncertainty with dynamic measurements and Markov chain Monte-Carlo algorithm. Build. Environ. **114**, 129–139 (2017)
19. K. Tijani, Q.D. Ngo, S. Ploix, B. Haas, J. Dugdale, Dynamic Bayesian networks to simulate occupant behaviours in office buildings related to indoor air quality, in *14th International Conference of the International Building Performance Simulation Association, Hyderabad* (2015)
20. E. Vorger, P. Schalbart, B. Peuportier, Integration of a comprehensive stochastic model of occupancy in building simulation to study how inhabitants influence energy performance, in *30th International Plea 2014 Conference* (2014)

Dynamic Models for Energy Control of Smart Homes

Christian Ghiaus

1 Introduction

Dynamic models are widely used for optimization of energy consumption in buildings, both in design and operation phase. A vast literature exists on modelling and software implementation for thermal simulation of buildings based of physical laws of energy and mass transfer [1–3].

The design of control systems requires thermal models of the building. Linear Time Invariant (LTI) models, such as state-space representations, transfer functions or zero-pole-gain models, are widely used for this aim. However, obtaining LTI models for the thermal behaviour of the buildings is difficult for two reasons. First, buildings are complex objects modelled by a set of partial differential equations and non-linear functions that describe the heat transfer by conduction, convection and radiation and energy transport by advection. For the range of variables involved in heat transfer occurring in buildings, these equations can be considered linear or are locally linearizable. The problem is to obtain the models for a whole building by using the models of the components. Second, the models used in heat transfer are thermal networks, while the models used in control theory are LTI models. The problem is to transform models with hundreds of parameters from thermal networks to LTI representation.

This chapter deals with these two problems. It presents two algorithms, one for assembling the thermal circuits and the other for extracting the state-space representation from the thermal circuit. Although the examples are given for thermal models of buildings, the two algorithms are of general interest.

C. Ghiaus (✉)
INSA-Lyon, CETHIL UMR5008, Villeurbanne, France
e-mail: christian.ghiaus@insa-lyon.fr

S. Ploix et al. (eds.), *Towards Energy Smart Homes*,
https://doi.org/10.1007/978-3-030-76477-7_5

2 Thermal Networks

There are three forms of heat transfer: conduction, convection and radiation. Heat is also carried by mass flow or advection. Heat transfer can be modelled by thermal circuits or networks composed of nodes and branches. The nodes, which may have capacities, are connected by conductances. There are two types of sources in the thermal networks: temperature (on the branches) and heat flow rate sources (connected to nodes).

2.1 *Heat Sources*

2.1.1 Temperature Sources

A temperature source (Fig. 1a) represents a difference of temperature on a branch which does not change with the flow rate crossing the branch. In the case of buildings, typically they model:

- Outdoor air which does not change its temperature with the temperature of the surfaces of the buildings.
- Temperature of the ground at a depth at which it is not influenced by the building.
- Adjacent spaces that have their temperature controlled by thermostats.

2.1.2 Heat Flow Rate Sources

A heat flow rate source (Fig. 1b) represents a thermal energy rate that does not change with the temperature node in which it is entering. In the case of buildings, typically they model:

- Solar (or short wave) radiation on the surfaces of the building
- Electrical power from household appliances which is transferred to the indoor air through convection

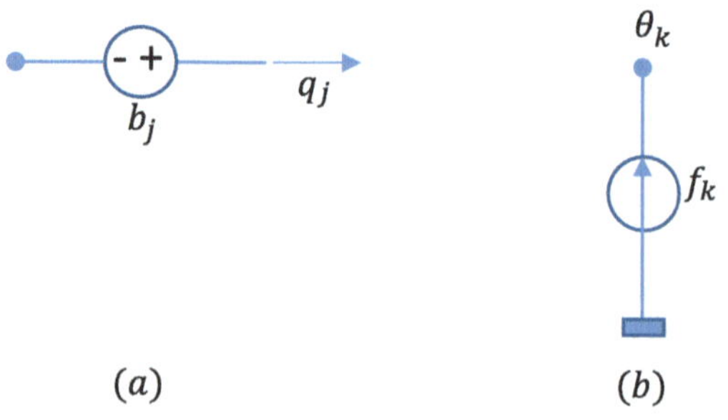

Fig. 1 Sources of: (**a**) temperature; (**b**) heat flow rate

– Heat transferred through convection from occupants to the indoor air.

The solar (or short wave) radiation in a building is entering through the windows, touches some surface of the indoor walls and then it is multi-reflected. This heat flow rate is modelled by heat flow rate sources. For finding the values of these sources, let us consider radiation exchange between opaque, diffuse, grey surfaces in an enclosure formed by $j = 1 \ldots n$ surfaces S_j with reflection coefficient ρ_j. The view factors between the surfaces are F_{ij}. The direct irradiance of each surface is E_j^0. The total thermal flux received by the surface S_i directly and after reflection is

$$S_i E_i - S_i E_i^0 + \sum_{j=1}^{n} F_{ji} S_j \rho_j E_j \tag{1}$$

By using the reciprocity relation for view factors, $F_{ij}S_i = F_{ji}S_j$, Eq. (1) becomes:

$$E_i - \sum_{j=1}^{n} F_{ij} \rho_j E_j = E_i^0 \tag{2}$$

The set of Eq. (2) can be written in matrix form

$$\begin{bmatrix} 1-\rho_1 F_{11} & -\rho_2 F_{12} & \ldots & -\rho_n F_{1n} \\ -\rho_1 F_{21} & 1-\rho_2 F_{21} & \ldots & -\rho_n F_{2n} \\ \ldots & \ldots & \ldots & \ldots \\ -\rho_1 F_{1n} & -\rho_2 F_{2n} & \ldots & 1-\rho_n F_{nn} \end{bmatrix} \begin{bmatrix} E_1 \\ E_2 \\ \ldots \\ E_n \end{bmatrix} = \begin{bmatrix} E_1^0 \\ E_2^0 \\ \ldots \\ E_n^0 \end{bmatrix} \tag{3}$$

or

$$(\mathbf{I} - \mathbf{F}\boldsymbol{\rho})\,\mathbf{E} = \mathbf{E^0} \tag{4}$$

where

$$\mathbf{I} = \begin{bmatrix} 1 & 0 & \ldots & 0 \\ 0 & 1 & \ldots & 0 \\ \ldots & \ldots & \ldots & \ldots \\ 0 & 0 & \ldots & 1 \end{bmatrix}; \mathbf{F} = \begin{bmatrix} F_{11} & F_{12} & \ldots & F_{1n} \\ F_{21} & F_{21} & \ldots & F_{2n} \\ \ldots & \ldots & \ldots & \ldots \\ F_{1n} & F_{2n} & \ldots & F_{nn} \end{bmatrix}; \boldsymbol{\rho} = \begin{bmatrix} \rho_1 & 0 & \ldots & 0 \\ 0 & \rho_2 & \ldots & 0 \\ \ldots & \ldots & \ldots & \ldots \\ 0 & 0 & \ldots & \rho_3 \end{bmatrix} \tag{5}$$

and the vectors

$$\mathbf{E} = \begin{bmatrix} E_1 \\ E_2 \\ \ldots \\ E_n \end{bmatrix}; \mathbf{E^0} = \begin{bmatrix} E_1^0 \\ E_2^0 \\ \ldots \\ E_n^0 \end{bmatrix} \tag{6}$$

The vector of irradiance of the surfaces is then

$$\mathbf{E} = (\mathbf{I} - \mathbf{F}\boldsymbol{\rho})^{-1}\,\mathbf{E^0} \tag{7}$$

Then, the heat flow rate source for each surface S_i is $q_i = E_i S_i$.

Calculating the view factors F_{ij} may be complicated [4]. A simple but rough estimation for rooms with planar surfaces is

$$F_{ij} \cong \frac{S_j}{\sum_{k \neq j} S_k};\ F_{ii} = 0 \tag{8}$$

which complies with the summation rule of view factors, $\sum_{j=1}^{n} F_{ij} = 1$, but not with the reciprocity relation, $F_{ij}S_i = F_{ji}S_j$. Another simplified estimation,

$$F_{ij} \cong \frac{S_j}{\sum_k S_k} \tag{9}$$

complies with the summation rule and with reciprocity relation but results in $F_{ii} \neq 0$.

2.2 *Heat Resistances (or Conductances)*

The three modes of heat transfer (conduction, convection and radiation) and the heat advection can be modelled by thermal resistances or conductances (Fig. 2).

2.2.1 Conduction

Thermal conduction is the heat diffusion in solids in the direction of the temperature gradient (Fig. 2a). Fourier law, the equation relating the thermal heat flow rate in a

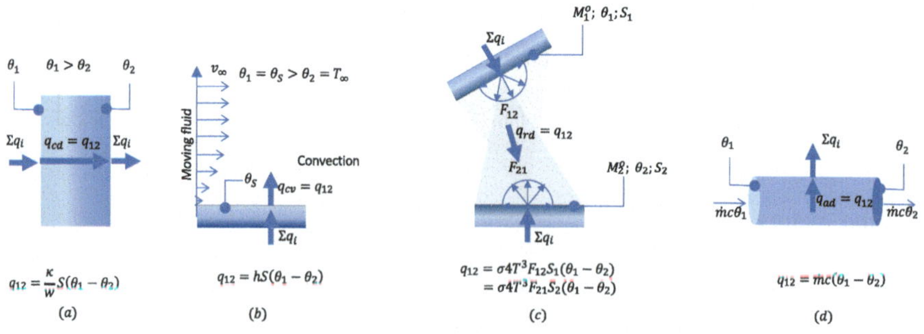

Fig. 2 Conductances in heat transfer and energy advection: (**a**) conduction; (**b**) convection; (**c**) radiation; (**d**) advection

direction , q_x, to the temperature gradient, $d\theta/dx$, in the direction x is:

$$q_x = -\kappa S \frac{d\theta}{dx} \tag{10}$$

where S is the area of the surface perpendicular to the heat flow rate q_x. The minus sign shows that heat transfer is from high to low temperature.

Let us consider stationary conduction in a stream tube in a homogenous and isotropic material without internal heat sources (Fig. 3a). Since the heat flow rate q is conserved, Fourier law in section s of the streamline is

$$q = -\kappa S \frac{d\theta}{ds} \tag{11}$$

where the conductivity $\kappa = \kappa(s)$ and the area surface $S = S(s)$ depend on the curvilinear coordinate s. By separating the variables, Eq. (11) becomes:

$$q \frac{ds}{\kappa S} = -d\theta \tag{12}$$

By integrating Eq. (12) from s_0 to s, the temperature variation with the distance,

$$q \int_{s_0}^{s_1} \frac{ds}{\kappa S} = -\int_{\theta_0}^{\theta_1} d\theta \tag{13}$$

can be written as

$$qR = \theta_0 - \theta \tag{14}$$

where

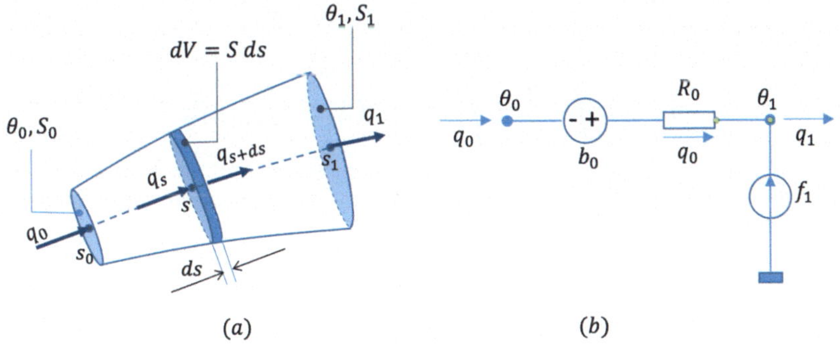

Fig. 3 Steady-state thermal conduction: (**a**) stream tube; (**b**) thermal network model

$$R = \int_{s_0}^{s} \frac{\mathrm{d}s}{\kappa S} \tag{15}$$

is the thermal resistance of the stream tube between s_0 and s.

If there are internal sources, the variation of the heat flow rate along the curvilinear coordinate ds is

$$\mathrm{d}q = p\,\mathrm{d}V \tag{16}$$

where dV is the infinitesimal volume. If d$s \to 0$, then d$V = S$ ds + dS ds. By integrating Eq. (16) between s_0 and s, $\int_{q_0}^{q} \mathrm{d}q = \int_{s_0}^{s} p\ S\ \mathrm{d}s$, it becomes:

$$q(s) = \int_{s_0}^{s} p\ S\ \mathrm{d}s + q_0 \tag{17}$$

The flow rate getting out through the surface S_1 is:

$$q_1 = \int_{s_0}^{s_1} p\ S\ \mathrm{d}s + q_0 \tag{18}$$

where q_0 is the heat flow rate entering through the surface S_0. Substituting (18) in (11), we obtain after integration

$$\theta_1 = \int_{s_0}^{s_1} \frac{1}{\kappa S}\left(\int_{s_0}^{s} - p\ S\ \mathrm{d}s'\right)\mathrm{d}s - q_0 \int_{s_0}^{s_1} \frac{1}{\kappa S}\mathrm{d}s + \theta_0 \tag{19}$$

By substituting in Eq. (19) the expression of thermal resistances given by Eq. (15),

$$R_0 = \int_{s_0}^{s_1} \frac{\mathrm{d}s}{\kappa S} \tag{20}$$

we obtain

$$\theta_1 = -\int_{s_0}^{s_1} \frac{1}{\kappa S}\left(\int_{s_0}^{s} p\ S\ \mathrm{d}s'\right)\mathrm{d}s - R_0 q_0 + \theta_0 \tag{21}$$

Equations (18) and (21) can be represented by the thermal circuit presented in Fig. 3b where the heat rate low source is

$$f_1 = \int_{s_0}^{s} p\ S\ \mathrm{d}s \tag{22}$$

and the temperature source is

$$b_0 = \int_{s_0}^{s_1} \frac{1}{\kappa S} \left(\int_{s_0}^{s} p\ S\ \mathrm{d}s' \right) \mathrm{d}s \tag{23}$$

Equation (23) can be integrated by parts. By noting $u \equiv \int_{s_0}^{s_1} pS\mathrm{d}s$ and $v' \equiv \frac{1}{\kappa S}$, the integration by parts $\int uv' ds = uv - \int u' v\ \mathrm{d}s$ of Eq. (23) becomes

$$b_0 = -\left(\int_{s_0}^{s_1} pS\ \mathrm{d}s \right) \left(\int_{s_0}^{s_1} \frac{1}{\kappa S} \mathrm{d}s \right) + \int_{s_0}^{s_1} pS \left(\int_{s_0}^{s} \frac{1}{\kappa S} \mathrm{d}s' \right) \mathrm{d}s \tag{24}$$

By substituting in Eq. (24) the expressions of R given by Eq. (15) and R_0 given by Eq. (20), we obtain:

$$b_0 = -R_0 \int_{s_0}^{s_1} p\ S\ \mathrm{d}s + \int_{s_0}^{s_1} R\ p\ S\ \mathrm{d}s \tag{25}$$

With these notations, Eq. (18) becomes

$$q_1 = q_0 + f_1 \tag{26}$$

and Eq. (21) becomes

$$\theta_0 - \theta_1 + b_0 = R_0 q_0 \tag{27}$$

where

$$e_0 = \theta_0 - \theta_1 + b_0 \tag{28}$$

is the difference of temperature across the thermal resistance R_0.

2.2.2 Convection

Convection implies heat transfer in fluids (Fig. 1b). If steady state is considered, Newton's law of convection is used, which is a phenomenological simplification having an expression similar to Fourier law:

$$q = hS\,(\theta_0 - \theta_1) \tag{29}$$

where h is the convective coefficient determined experimentally [4]. Typical values of the heat convection coefficient are $h_i = 8$ W/m^{2}K inside and $h_o = 25$ W/m^{2}K outside the building [1].

2.2.3 Long-Wave Radiation

Long-wave radiation exchange is between two surfaces that are facing each other and that have different temperatures (Fig. 2c). The radiative heat flow rate between two black body surfaces i and j is [4]

$$q_{ij} = S_i F_{ij} \left(M_i^o - M_j^o \right) \tag{30}$$

where

S_i is the area of the surface;
F_{ij}—view factor between surface i and surface j;

M_i^o and M_j^o—the black body radiant emittance of the surfaces i and j, respectively.

By using Stefan–Boltzmann law,

$$M = \sigma T^4 \tag{31}$$

where σ is Stefan–Boltzmann constant and T is the temperature of the surface expressed in kelvin, Eq. (30) becomes

$$q_{ij} = S_i F_{ij} \sigma \left(T_i^4 - T_j^4 \right) \tag{32}$$

The two temperatures T_i^4 and T_j^4 may be linearized around a mean value $\overline{T}$,

$$T_i^4 = \overline{T}^4 + 4\overline{T}^3 \left(T_i - \overline{T} \right) \tag{33}$$

and

$$T_j^4 = \overline{T}^4 + 4\overline{T}^3 \left(T_j - \overline{T} \right) \tag{34}$$

By subtracting Eq. (34) from (33), we obtain

$$T_i^4 - T_j^4 = 4\overline{T}^3 \left(T_i - T_j \right) \tag{35}$$

The exact value of the mean value $\overline{T}$ can be obtained from the equivalence of Eq. (35) with

$$T_i^4 - T_j^4 = \left(T_i^2 + T_j^2 \right) \left(T_i + T_j \right) \left(T_i - T_j \right) \tag{36}$$

as

$$\overline{T} = \sqrt[3]{\frac{1}{4} \left(T_i^2 + T_j^2 \right) \left(T_i + T_j \right)} \tag{37}$$

Substituting (37) in (32), we obtain the linear expression of the heat flow rate:

$$q_{ij} = S_i F_{ij} \sigma 4 \overline{T}^3 \left(T_i - T_j\right) \tag{38}$$

where $S_i F_{ij} \sigma 4 \overline{T}^3$ may be considered as a thermal conductance. For $15^\circ \mathrm{C} < \theta < 30^\circ \mathrm{C}$, i.e. 288.15 K < $\overline{T}$ < 303.15 K, the value of $\sigma 4 \overline{T}^3$ is about 6 W/m^{2}K, more exactly $5.41 < \sigma 4 \overline{T}^3 < 6.31$ W/m^{2}K.

The radiosity of a surface represents the radiative fluxes leaving an opaque, diffuse, grey body surface

$$J_i = \varepsilon M_i^o + \rho_i E_i \tag{39}$$

where

$\varepsilon_i M_i^o$ is the emitted radiant flux, with ε_i the emissivity and M_i^o the black body radiant emittance of the surface i;

$\rho_i E_i$—the reflected radiant flux, with ρ_i the reflectivity and E_i the incident radiant flux on the surface.

The radiative exchange between two opaque, diffuse, grey surfaces may be expressed by an equation similar to (30):

$$q_{ij} = S_i F_{ij} \left(J_i - J_j\right) \tag{40}$$

Following the same reasoning as before, the linear expression of (40) is

$$q_{ij} = S_i F_{ij} 4 \sigma \overline{T}^3 \left(J_i - J_j\right) \tag{41}$$

where $\overline{T}$ is given by (37).

An example of transforming a radiative network into a thermal network is given in Fig. 4. The conductances for the radiation network are:

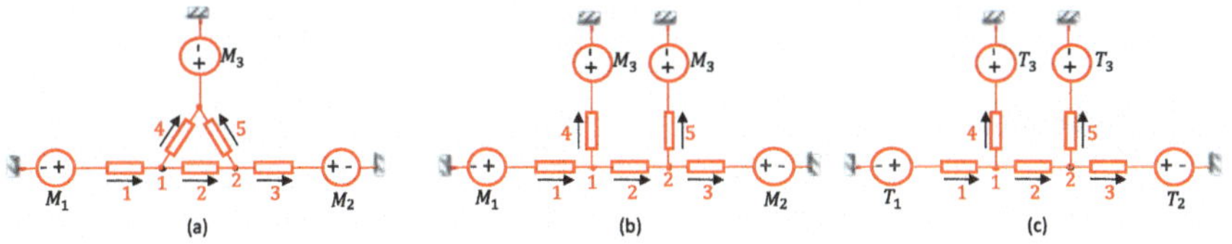

Fig. 4 Radiation networks and their transformation to thermal networks: (**a**) an example of radiation network; (**b**) same network as in (**a**) but arranged to have a source of emittance on branch; (**c**) thermal network of the radiation network from (**b**)—the sources and the conductances are changed

$$G_1 = \frac{\varepsilon_1}{1-\varepsilon_1} S_1; \; G_2 = F_{12} S_1; \; G_3 = \frac{\varepsilon_2}{1-\varepsilon_2} S_2; \; G_4 = F_{13} S_1; \; G_5 = F_{23} S_2 \tag{42}$$

The oriented incidence matrix **A**, the conductance matrix **G** and the vector of inputs **b** are:

$$\mathbf{A} = \begin{bmatrix} 1 & 0 \\ -1 & 1 \\ 0 & 1 \\ -1 & 0 \\ 0 & -1 \end{bmatrix}; \mathbf{G} = \begin{bmatrix} G_1 & 0 & 0 & 0 & 0 \\ 0 & G_2 & 0 & 0 & 0 \\ 0 & 0 & G_3 & 0 & 0 \\ 0 & 0 & 0 & G_4 & 0 \\ 0 & 0 & 0 & 0 & G_4 \end{bmatrix}; \mathbf{b} = \begin{bmatrix} M_1^o \\ 0 \\ M_2^o \\ -M_3^o \\ -M_3^o \end{bmatrix} \tag{43}$$

respectively (see Sect. 3.2 for their definition). The radiosities $\mathbf{j} \equiv [J_1 \;\; J_2]^T$ are given by:

$$\mathbf{j} = \left(\mathbf{A}^T \mathbf{G} \mathbf{A}\right)^{-1} \mathbf{A}^T \mathbf{G} \mathbf{b} \tag{44}$$

(see Sect. 4.1 for details).

The radiative network, in which the unknowns are radiosities, can be transformed into a thermal network, in which the unknowns in the nodes are temperatures, by using the transformations for emittances and radiosities.

$$M^o = \sigma T^4 \text{ and } J = \sigma \theta^4 \tag{45}$$

where T is a temperature source and θ is an unknown temperature. The conductances for the temperature network become:

$$G_1 = \frac{\varepsilon_1}{1-\varepsilon_1} S_1 \sigma 4 \overline{T}^3; \; G_2 = F_{12} S_1 \sigma 4 \overline{T}^3; \; G_3 = \frac{\varepsilon_2}{1-\varepsilon_2} S_2 \sigma 4 \overline{T}^3; \; G_4 = F_{13} S_1 \sigma 4 \overline{T}^3;$$
$$G_5 = F_{23} S_2 \sigma 4 \overline{T}^3 \tag{46}$$

2.2.4 Advection

Energy advection is the transport of energy by a mass flow rate entering and leaving a control volume (Fig. 1d). For example, this is the case of energy transported by air in ventilation. The heat flow rate transported is:

$$q_{12} = \dot{m} c \left(\theta_1 - \theta_2\right) \tag{47}$$

where $\dot{m}$ is the mass flow rate and c is the heat capacity of the fluid.

Table 1 Values of airflow rates in air changes per hour as a function of the position of the window [5] Table 1.12.1-4

Position of the window	Air changes per hour
Closed windows, closed doors	0–0.5
Tilt window, closed jealousy	0.3–1.5
Tilt window, no jealousy	0.8–4.0
Half-opened window	5–10
Full-opened window	9–15
Windows and French-windows fully opened	About 40

Table 2 Typical values of airflow rate per person [5] §3.5.1.1.2

Typical situations	Volumetric airflow m^3/h per person
Theatre, concert, cinema, exhibition halls, supermarkets, museums, gyms	20
Restaurants, rest area, conference rooms, classrooms, auditorium	30
Office rooms	40
Open office rooms	60

The mass flow rate is calculated from the volumetric flow rate $\dot{V}$:

$$\dot{m} = \rho \dot{V} \tag{48}$$

where ρ is the fluid density. The volumetric flow rate $\dot{V}$ is obtained from hydraulic or aeraulic calculations. the infiltration (or the airtightness) of the building is measured by a blower door. It may be expressed in "air changes per hour" which represents the number of volumes of air contained by the building which are vehiculated in an hour. Some indicative values are given in Table 1. Typical values for airflow per person are given in Table 2.

2.3 *Heat Capacities*

The heat capacity C of a control volume is the amount of heat ΔQ that needs to be added in order to increase the temperature by $\Delta\theta$:

$$C = \lim_{\Delta\theta \to 0} \frac{\Delta Q}{\Delta\theta} \tag{49}$$

For a homogeneous object of mass m having the specific heat c,

$$C = mc \tag{50}$$

From Eq. (49), the flow rate entering the body is

$$q = C\dot{\theta} \tag{51}$$

3 Assembling of Thermal Networks

Buildings are systems composed of elements such as walls, windows and doors connected through heat and mass transfer. Therefore, the models of whole buildings are obtained by combining the models of individual components. Two important methods for obtaining large models are coupling and assembling. In coupling, the system of equations, which is obtained from the models of each element, is solved iteratively [6–8]. For example, building energy software tools use iterative methods to solve the coupled equations: EnergyPlus uses Gauss-Seidel successive substitution or Newton-Raphson method [2, 9], TRNSYS uses successive substitutions and Powell's method [3], ESP-r solves independently the domain equations and then the coupling [6], IDA ICE uses a modular approach [10].

The assembling of models is very different of coupling. While in coupling the models of the elements are separate, in assembling the complex model is a system of linear equations, at least at a certain moment during an iterative solving procedure. Assembling is an important reason for the use of models such as transfer functions in thermal modelling of buildings. More generally, input-output linear time invariant (LTI) models, such as state-space, transfer function, zero-pole-gain models [11] or two-port networks [12–15] may be connected to obtain a new, more complex, model. The model obtained by assembling has the advantage that can be analysed (e.g. find the eigenvalue and the time constants, the static gain, stability, controllability, observability, identifiability). However, these techniques are not applicable to networks or circuits that model transport phenomena in which the connections are done by conservation laws (such as conservation of mass, energy, momentum and electrical charge). The usual technique used for network models is coupling.

Circuits, networks or bond graphs are widely used for modelling transfer phenomena [16–17]. The method of thermal networks (or circuits) is present in almost any primer on heat transfer. The heat conduction equation, introduced by Fourier, has been used for about two centuries to describe diffusion phenomena in dynamical physical systems. Ohm work on electricity was inspired by Fourier's heat conduction model; he considered the flow of electricity as being exactly analogue to the flow of heat. Fick also used an analogy with Fourier equation for transient diffusion of solutes in liquids. Models influenced by the diffusion equation are used for diffusion of gases, Brownian motion, flow in porous materials, random walk, etc. [18]. Therefore, it is important to have a procedure for assembling the networks (or the circuits).

There are well-established algorithms for assembling models represented by finite elements [19–21]. However, for thermal networks the solution is obtained by solving iteratively the set of equations of the elementary models [22].

This section proposes a data structure for thermal networks and an assembling procedure. Then, aspects of software implementation are discussed. The example is given for a very simple, yet relevant, network on which the procedure can be checked by hand.

3.1 Defining the Problem of Circuit Assembling

Given a number of thermal circuits, TC_1, TC_2, ..., TC_n, and knowing that some of their nodes are in common, find the assembled circuit TC. A simple example is given in Fig. 5. There are four thermal circuits, TC_1, TC_2, TC_3, TC_4, having in common some of their nodes: the node 5 of TC_1 is common with the node 1 of TC_2, the node 2 of TC_2 is common with node 2 of TC_3 and the node 3 of TC_2 is in common with the node 2 of TC_4 (Fig. 5b). Find the model of the assembled circuit TC shown in Fig. 5a.

From conservation of energy, it results that if there is a flow source in the node of the assembled circuit TC, it needs to be the sum of the sources in the respective nodes of each circuit TC_k. For example, the flow source in the node 5 of the assembled circuit from Fig. 5a is the sum of flow sources present in node 5 of TC_1 and the flow source present in node 1 of TC_2. Since the thermal capacity is proportional to mass, from the conservation of mass, it results that if there is a capacity in a node of the assembled circuit, it needs to be the sum of the capacities in the respective nodes of each circuit TC_k. For example, the capacity in node 5 of the assembled circuit from Fig. 5a is the sum of capacities present in node 5 of TC_1 and in node 1 of TC_2 shown in Fig. 5b.

To exemplify the procedure, we will use a toy model representing a building formed by an insulated concrete wall and a glass wall. The room is ventilated and its air temperature is controlled by a P-controller. Auxiliary load is added to the room (Fig. 6). The toy model is used to show specific aspects of the assembling procedure, not for the correctness of the modelling.

We would like to construct separate models for concrete wall, glass wall, ventilation, and room air (Fig. 7) and to assemble them into one model (Fig. 6).

3.2 Algebraic Description of the Thermal Circuits

A circuit is a weighted oriented graph with node representing temperatures, branches representing heat flows and the weights representing the thermal conductances. Some nodes have thermal capacities, but not all of them. Some branches have temperature sources and some nodes have flow sources, but not all of them.

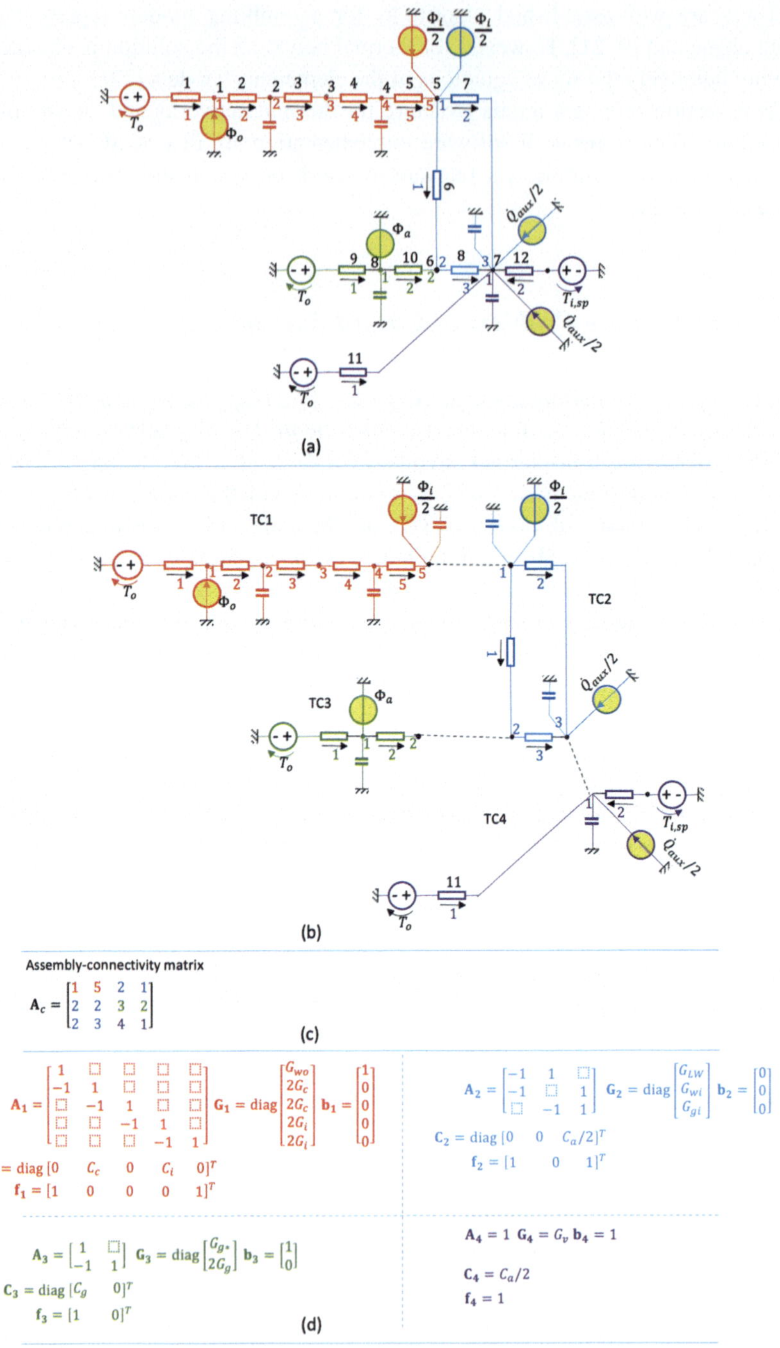

Fig. 5 Example of the problem of assembling thermal circuits: given four circuits, assemble them knowing the common nodes. (**a**) Assembled circuit. (**b**) Four disassembled circuits. (**c**) Assembling matrix. (**d**) Algebraic description of each disassembled circuit

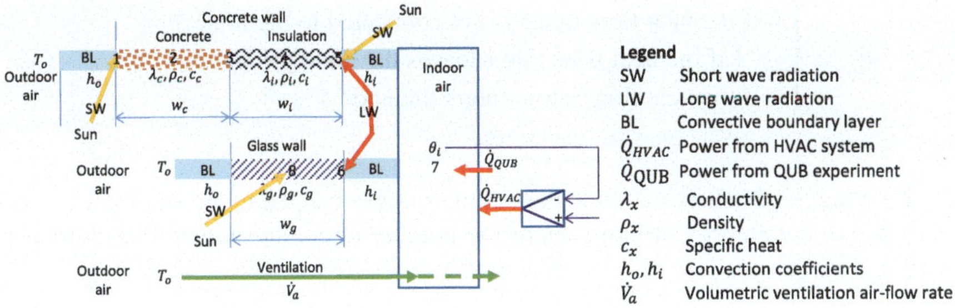

Fig. 6 Toy model used for example

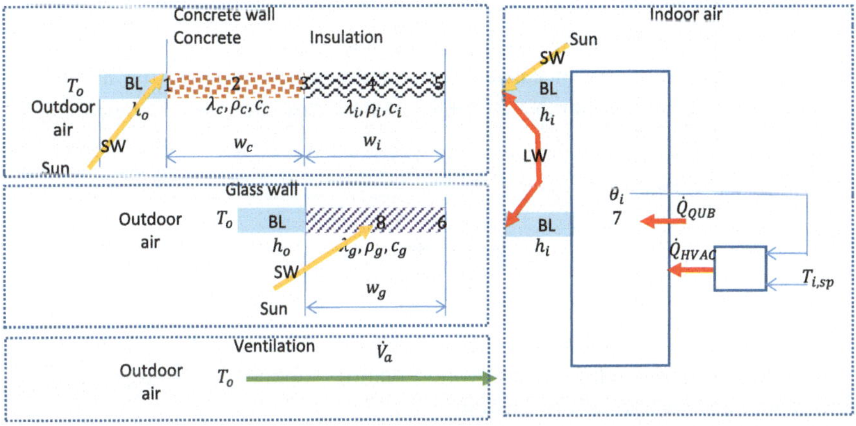

Fig. 7 Model for components to be assembled

The sources represent the input (i.e. the independent) variables of the model. The temperatures of the nodes and the flows in the branches represent the unknowns for which the problem is solved. Usually, only some of the temperatures of the nodes and/or flow in the branches are of practical interest and represent the output of the model. If the heat flow rate of a branch is considered as an output, then the temperatures in the nodes of the branch need to be found; the flow in the branch is calculated as the product between the conductance and the difference of temperatures.

A thermal circuit may be described by three matrices and three vectors. The matrices are:

1. **A** is an oriented incidence matrix with the number of rows equal to the number of branches and the number of columns equal to the number of nodes of the thermal circuit. The elements of matrix **A** are:

$$a_{ij} = \begin{cases} 0 \text{ if the heat flow rate } i \text{ is not connected to the node } j \\ -1 \text{ if the heat flow rate } i \text{ leaves the node } j \\ 1 \text{ if the heat flow rate } i \text{ enters the node } j \end{cases} \tag{52}$$

2. **G** is a diagonal matrix of conductances of dimension equal to the number of rows of **A**, i.e. the number of branches or the number of conductances. The elements of matrix **G** are:

$$g_{ij} = \begin{cases} R_i^{-1} \text{ for } i = j \\ 0 \text{ for } i \neq j \end{cases} \tag{53}$$

   Note that each branch needs to have a conductance.
3. **C** is a diagonal matrix of capacitances of dimension equal to the number of columns of **A,** i.e. the number of nodes of the thermal circuit. The elements of matrix **C** are:

$$c_{ij} = \begin{cases} C_i \text{ for } i = j \\ 0 \text{ for } i \neq j \end{cases} \tag{54}$$

Note that not all nodes have a thermal capacity. Since the thermal capacity is proportional to mass and the mass proportional to volume, a node representing a surface will always have a zero capacity.

The vectors are:

1. **b** is a vector indicating the branches which have temperature sources. Its size is equal to the number of rows of matrix **A**, i.e. the number of branches. Its elements are:

$$b_i = \begin{cases} 1 \text{ if there is a temperature source on branch } i \\ 0 \text{ otherwise} \end{cases} \tag{55}$$

2. **f** is a vector indicating the nodes which have a heat flow rate sources. Its size is equal to the number of columns of matrix **A,** i.e. the number of nodes. Its elements are:

$$f_i = \begin{cases} 1 \text{ for flow source in node } i \\ 0 \text{ otherwise} \end{cases} \tag{56}$$

3. $\mathbf{y}$ is a vector indicating the temperatures that are considered as outputs. Its size is equal to vector $\mathbf{f}$. Its elements are:

$$y_i = \begin{cases} 1 \text{ for temperature of node } i \text{ as output variable} \\ 0 \text{ otherwise} \end{cases} \tag{57}$$

Any thermal circuit TC can be described by the list of arrays: $\{\mathbf{A}, \mathbf{G}, \mathbf{C}, \mathbf{b}, \mathbf{f}, \mathbf{y}\}$ (Fig. 5d).

3.3 *Numbering the Thermal Circuits*

The construction of the list of arrays $\{\mathbf{A}, \mathbf{G}, \mathbf{C}, \mathbf{b}, \mathbf{f}, \mathbf{y}\}$ requires the numbering of circuits. In principle, the numbering of the nodes and branches can be done arbitrarily. Once the numbering of the elementary circuits is done, the numbering of the assembled circuit is automatic.

3.3.1 Numbering Elementary Circuits

The connections between nodes are indicated by the oriented incidence matrix $\mathbf{A}$. Since numbering becomes tedious for large circuits, the following rules may be adopted (Fig. 5b):

- Number the nodes in order (from left to right or from right to left).
- Number the branches in increasing order of nodes and orient them from the lower to the higher node. Note: reference temperature is node 0.

As an example, for the thermal circuit TC_1 (red in Fig. 5b), the branches are:

First node	Second node	Branch
0	1	1
1	2	2
2	3	3
3	4	4
4	5	5

For TC_2 (blue in Fig. 5b), the branches are:

First node	Second node	Branch
1	2	1
1	3	2
2	3	3

For TC_3 (green in Fig. 5b), the branches are:

First node	Second node	Branch
0	1	1
1	2	2

For TC_4 (violet in Fig. 5b), the nodes have the same numbers:

First node	Second node	Branch
0	1	1
0	1	2

3.3.2 Numbering the Assembled Circuit

When assembling the thermal circuits, some nodes are put in common. Therefore, the number of nodes in the assembled circuit will be smaller than the sum of the nodes of elementary circuits. The number of branches will not change. The nodes and the branches of the assembled circuit will be in the order of assembling (Fig. 5a, Table 3).

The assembling of the circuits is indicated by the assembling matrix. Each row of this matrix has four elements that indicate two nodes that will be put together:

- Number of circuit 1
- Node of circuit 1
- Number of circuit 2
- Node of circuit 2

For our example, the assembling matrix is:

$$\mathbf{Ass} = \begin{bmatrix} 1 & 5 & 2 & 1 \\ 2 & 2 & 3 & 2 \\ 2 & 3 & 4 & 1 \end{bmatrix} \tag{58}$$

The description of the assembled circuit, given by the list $TC = \{TC_1, \ldots, TC_i\}$ of list of arrays $TC_i = \{\mathbf{A}_i, \mathbf{G}_i, \mathbf{C}_i, \mathbf{b}_i, \mathbf{f}_i, \mathbf{y}_i\}$ (Fig. 5c), and the assembling matrix **Ass** contain all the necessary information for obtaining the assembled circuit.

Table 3 Local and global indexing of nodes

Thermal circuit	TC1	TC2	TC3	TC4
Local node index	1 2 3 4 5	1 2 3	1 2	1
Global node index	1 2 3 4 5	5 6 7	8 6	7

3.4 Assembling the Circuits

The analysis (or the direct problem) of a thermal circuit TC_i is to solve for $\mathbf{q}_i$ and $\boldsymbol{\theta}_i$ the equation:

$$\begin{bmatrix} \mathbf{G}_i^{-1} & \mathbf{A}_i \\ -\mathbf{A}_i^T & \mathbf{C}_i s \end{bmatrix} \begin{bmatrix} \mathbf{q}_i \\ \boldsymbol{\theta}_i \end{bmatrix} = \begin{bmatrix} \mathbf{b}_i \\ \mathbf{f}_i \end{bmatrix} \quad (59)$$

or find $\mathbf{u}_i$ from equation:

$$\mathbf{K}_i \mathbf{u}_i = \mathbf{a}_i \quad (60)$$

where

$$\mathbf{K}_i = \begin{bmatrix} \mathbf{G}_i^{-1} & \mathbf{A}_i \\ -\mathbf{A}_i^T & \mathbf{C}_i s \end{bmatrix}; \mathbf{u}_i = \begin{bmatrix} \mathbf{q}_i \\ \boldsymbol{\theta}_i \end{bmatrix}; \mathbf{a}_i = \begin{bmatrix} \mathbf{b}_i \\ \mathbf{f}_i \end{bmatrix}.$$

Let's note the dissembled block vectors $\mathbf{u}_d$, $\mathbf{a}_d$ and matrix $\mathbf{K}_d$:

$$\mathbf{u}_d = \begin{bmatrix} \mathbf{u}_1 \\ \dots \\ \mathbf{u}_n \end{bmatrix}; \mathbf{a}_d = \begin{bmatrix} \mathbf{a}_1 \\ \dots \\ \mathbf{a}_n \end{bmatrix}; \mathbf{K}_d = \begin{bmatrix} \mathbf{K}_1 & \mathbf{0} & \mathbf{0} \\ \mathbf{0} & \ddots & \mathbf{0} \\ \mathbf{0} & \mathbf{0} & \mathbf{K}_n. \end{bmatrix}$$

There is a disassembling matrix $\mathbf{A}_d$ which transforms the assembled vectors (i.e. the block vector of elementary circuits) into disassembled vectors:

$$\mathbf{u}_d = \mathbf{A}_d \mathbf{u}; \mathbf{a}_d = \mathbf{A}_d \mathbf{a}; \quad (61)$$

The assembled matrix and vectors are obtained by using disassembling matrix $\mathbf{A}_d$:

$$\mathbf{K} = \mathbf{A}_\mathbf{d}^\mathbf{T} \mathbf{K}_\mathbf{d} \mathbf{A}_\mathbf{d} \quad (62)$$

$$\mathbf{u} = \mathbf{A}_d^T \mathbf{u}_d \quad (63)$$

$$\mathbf{a} = \mathbf{A}_d^T \mathbf{a}_d \quad (64)$$

The elements of the assembled circuit, **A, G, C, b, f, y,** are then obtained from:

$$\mathbf{K} = \begin{bmatrix} \mathbf{G}^{-1} & \mathbf{A} \\ -\mathbf{A}^T & \mathbf{C}s \end{bmatrix}; \mathbf{u} = \begin{bmatrix} \mathbf{q} \\ \boldsymbol{\theta} \end{bmatrix}; \mathbf{a} = \begin{bmatrix} \mathbf{b} \\ \mathbf{f} \end{bmatrix} \tag{65}$$

3.5 Algorithm

3.5.1 Obtaining the Global Indexes of the Assembling Matrix

In order to indicate the common nodes of the circuits, it is convenient to give the assembling matrix **Ass**, of which an example is given in Eq. (58), with four elements on each line (Fig. 8):

1. Number of the first circuit
2. Local number of the node of the first circuit
3. Number of the second circuit
4. Local number of the node of the second circuit

We need to obtain an assembling matrix **Ass** of two columns of global disassembled nodes that are put in common. For our example (Fig. 5):

- The node 5 of TC_1 is put in common with the node 1 of TC_2, which has the global value $5 + 1 = 6$ ($5 =$ number of nodes of TC_1, $1 =$ local index in TC_2).
- The node 2 of TC_2 (global value $5 + 2$) is put in common with the node 2 of TC_3 (global value $5 + 3 + 2 = 10$, where $5 =$ number of nodes of TC_1, $3 =$ number of the nodes of TC_2, 2 local index in TC_3).
- The node 3 of TC_2 (global value $5 + 3 = 8$) is put in common with the node 1 of TC_4 (global value $5 + 3 + 2 + 1 = 11$, where $5 =$ number of nodes of TC_1, $3 =$ number of the nodes of TC_2, $2 =$ number of nodes in TC_3, $1 =$ local index in TC_4).

From

$$\mathbf{AssX} = \begin{bmatrix} 1 & 5 & 2 & 1 \\ 2 & 2 & 3 & 2 \\ 2 & 3 & 4 & 1 \end{bmatrix} \tag{66}$$

we obtain:

$$\mathbf{Ass} = \begin{bmatrix} 5 & 6 \\ 7 & 10 \\ 8 & 11 \end{bmatrix} \tag{67}$$

The information on the number of branches and nodes for each thermal circuit TC_k is taken from the length of the vector **b.**

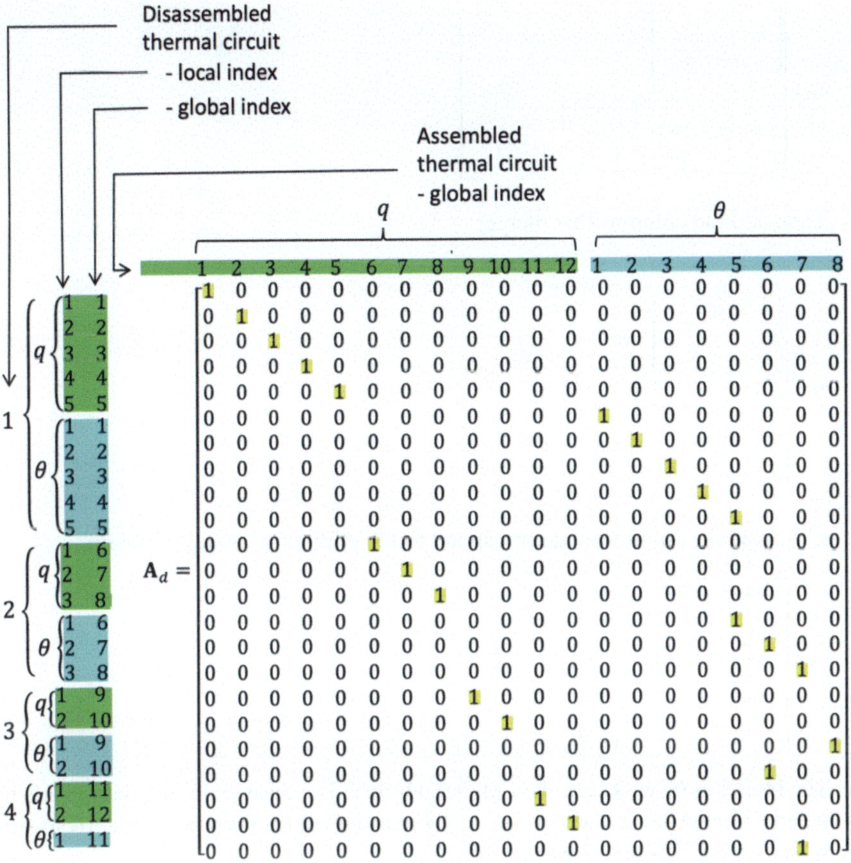

Fig. 8 The disassembling matrix: rows correspond to disassembled circuits; columns correspond to assembled circuits

3.5.2 Obtaining the Disassembling Matrix

The disassembling matrix $\mathbf{A}_d$ transforms the assembled vectors into dissembled vectors (i.e. the block vector of elementary circuits):

$$\mathbf{u}_d = \mathbf{A}_d \mathbf{u};\ \mathbf{a}_d = \mathbf{A}_d \mathbf{a}; \tag{68}$$

The assembling implies that some of the nodes are merged: their number decreases and their "global" index changes.

First, create a block matrix that keeps the indexes of the temperature nodes:

Then add the columns that merge:

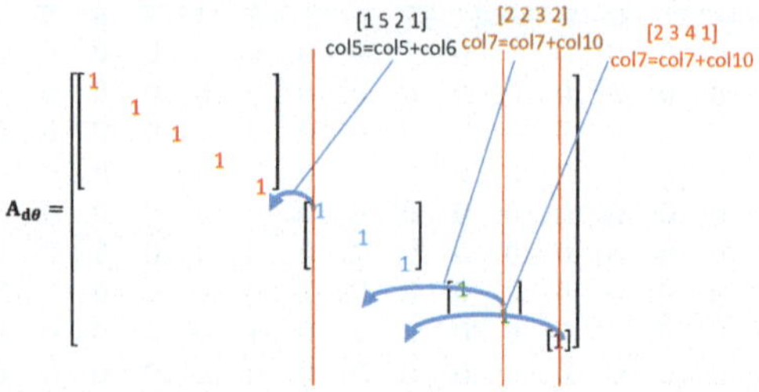

Eliminate the columns that correspond to the eliminated nodes to obtain

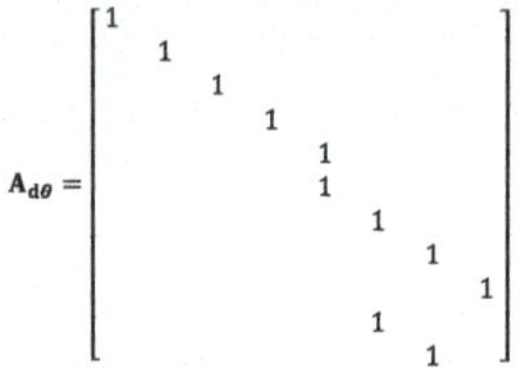

The branches (flows) keep their global number. The disassembling matrix is then obtained from the block matrix (Fig. 9a) by rearranging the rows in order correspond to the vector (Fig. 9b): $\left[\, \mathbf{q}_1^T \; \boldsymbol{\theta}_1^T \; \mathbf{q}_2^T \; \boldsymbol{\theta}_2^T \; \mathbf{q}_3^T \; \boldsymbol{\theta}_3^T \; \mathbf{q}_4^T \; \boldsymbol{\theta}_4^T \,\right]^T$.

3.5.3 Algorithm for the Disassembling Matrix

Having the disassembling matrix $\mathbf{A}_d$, the assembling

$$\mathbf{K} = \mathbf{A}_d^T \mathbf{K}_d \mathbf{A}_d \tag{69}$$

needs the matrix $\mathbf{K}_d$ which is a block matrix

$$\mathbf{K}_d = \begin{bmatrix} \mathbf{K}_1 & \mathbf{0} & \mathbf{0} \\ \mathbf{0} & \ddots & \mathbf{0} \\ \mathbf{0} & \mathbf{0} & \mathbf{K}_n \end{bmatrix}$$

of block matrices of each thermal circuit TC_i

$$K_i = \begin{bmatrix} G_i^{-1} & A_i \\ -A_i^T & C_i s \end{bmatrix}$$

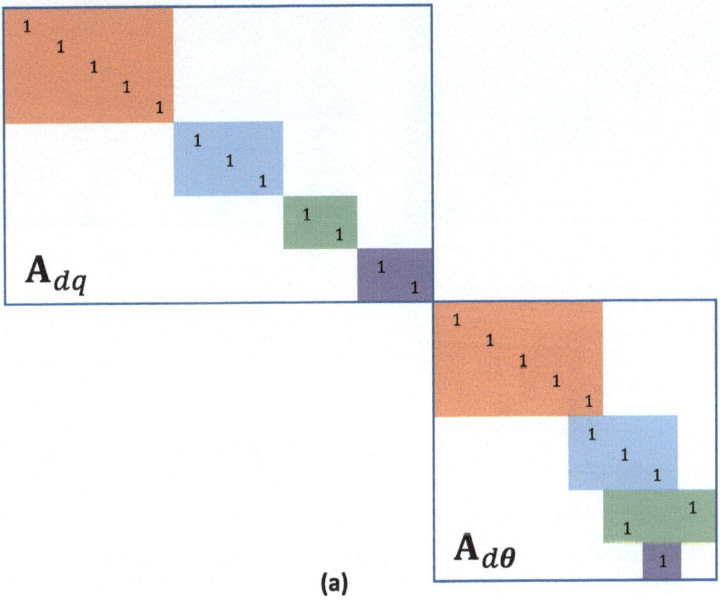

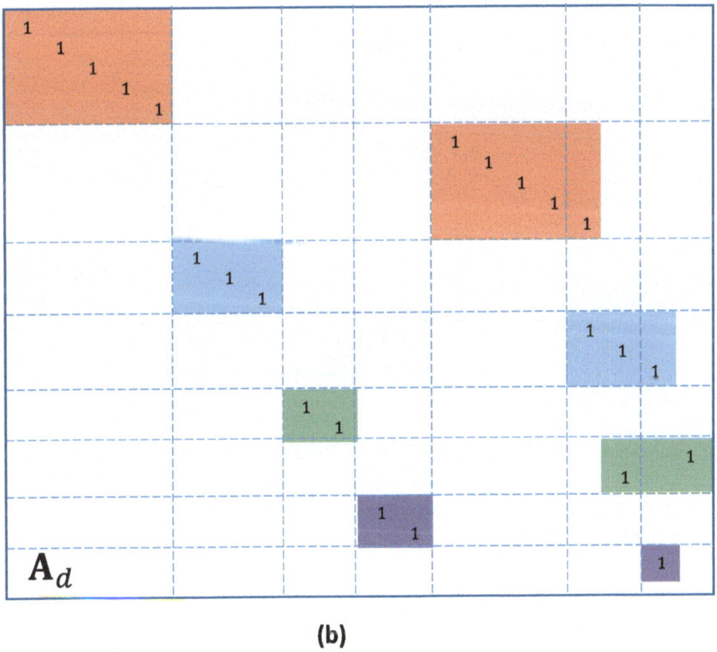

Fig. 9 Obtaining the disassembling matrix: (**a**) block matrix for $[q_1^T\ q_2^T\ q_3^T\ q_4^T\ \theta_1^T\ \theta_2^T\ \theta_3^T\ \theta_4^T]^T$. (**b**) Block matrix rearranged for $[q_1^T\quad \theta_1^T\quad q_2^T\quad \theta_2^T\quad q_3^T\quad \theta_3^T\quad q_4^T\quad \theta_4^T]^T$

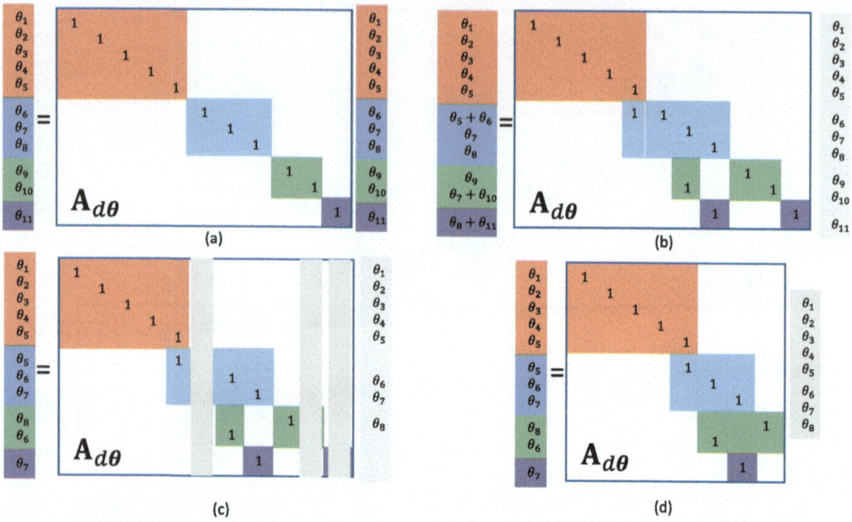

Fig. 10 Obtain matrix temperature nodes $A_{d\theta}$: (**a**) create diagonal matrix; (**b**) add the column of the second node to the column of the first node in merging: [1 5 2 1] ➔ col5 = col5 + col6; [2 2 3 2] ➔ col7 = col7 + col10; [2 3 4 1] ➔ col8 = col. 8 = col11; (**c**) delete the columns of the second node in merging; (**d**) obtain matrix $A_{d\theta}$

Fig. 11 Obtain matrix for flow branches A_{dq}

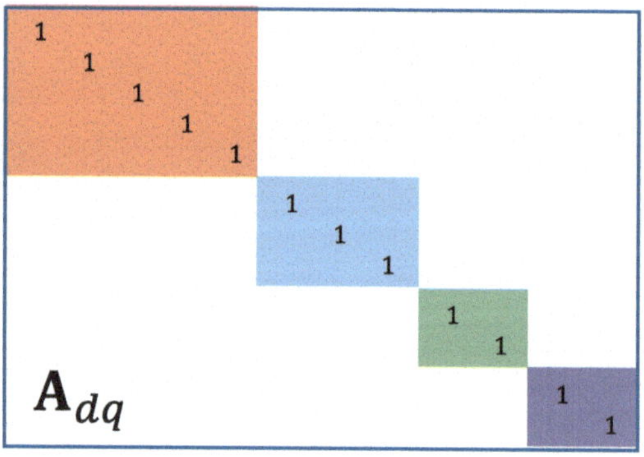

Algorithm for obtaining the dissembling matrix (Figs. 10, 11, and 12)

3.5.4 Assembled Circuits

The assembled matrix and vector are obtained by using the disassembling matrix $\mathbf{A}_d$:

$$\mathbf{K} = \mathbf{A}_d^T \mathbf{K}_d \mathbf{A}_d \tag{70}$$

The elements of the assembled circuit, **A, G, C, b, f, y,** are then obtained from:

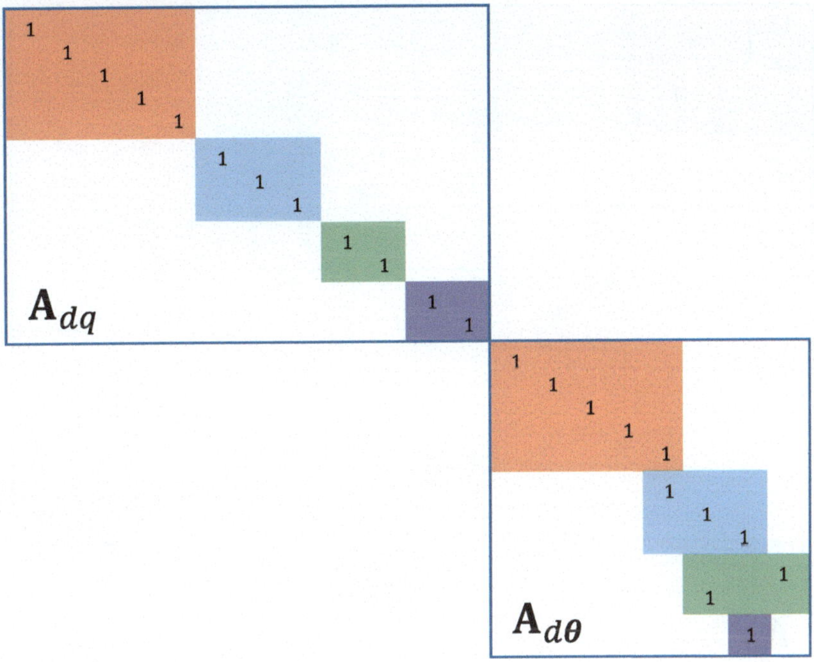

Fig. 12 Block matrix for $\begin{bmatrix} q_1^T & q_2^T & q_3^T & q_4^T & \theta_1^T & \theta_2^T & \theta_3^T & \theta_4^T \end{bmatrix}^T$ (first flow branches, then temperature nodes)

$$\mathbf{K} = \begin{bmatrix} \mathbf{G}^{-1} & \mathbf{A} \\ -\mathbf{A}^{\mathbf{T}} & \mathbf{C}_S \end{bmatrix}; \mathbf{u} = \begin{bmatrix} \mathbf{q} \\ \boldsymbol{\theta} \end{bmatrix}; \mathbf{a} = \begin{bmatrix} \mathbf{b} \\ \mathbf{f} \end{bmatrix} \tag{71}$$

3.5.5 Global Assembled Indexes

The global indexes of the assembled circuit (Fig. 13) result from the calculation of the dissembling matrix $\mathbf{A}_d$ (Fig. 14).

4 Transforming Thermal Circuits into State-Space Representation

Thermal circuits are linear models with constant coefficients. However, in control theory, the state-space representation is widely used. The aim of this chapter is to transform a thermal circuit, formed by resistors, capacities, temperature sources and heat flow sources, into its state-space representation. The problem in this transformation is that some capacities may be zero.

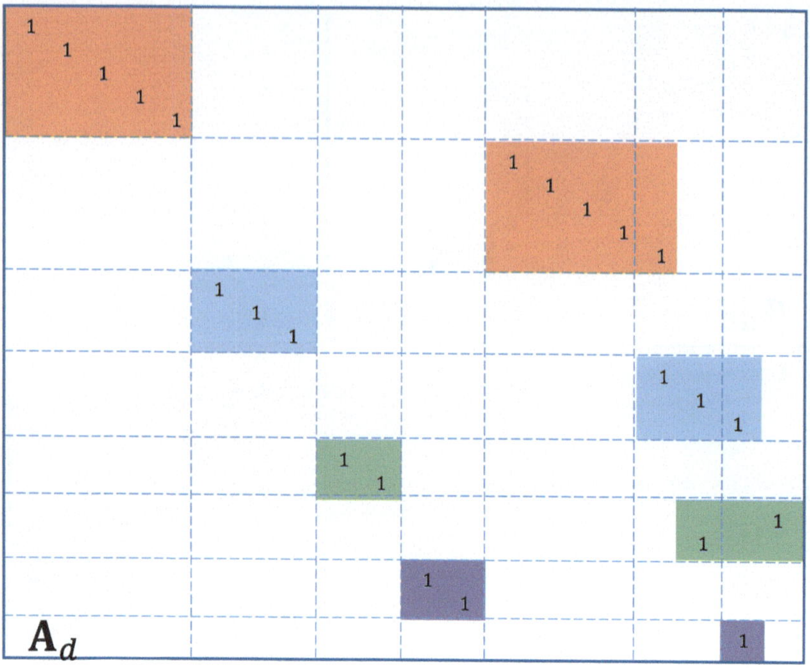

Fig. 13 Matrix A_d for $[q_1^T \quad \theta_1^T \quad q_2^T \quad \theta_2^T \quad q_3^T \quad \theta_3^T \quad q_4^T \quad \theta_4^T]^T$ after rearranging in order to obtain in the order of thermal circuits

Fig. 14 The relation between the local indexes of the blocks of circuits and the global indexes of the assembled circuit

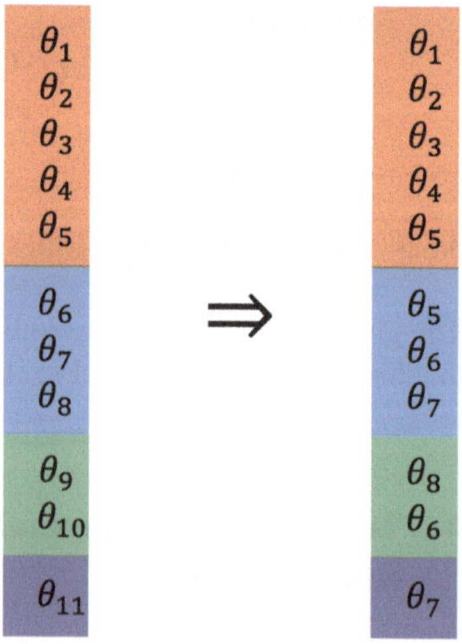

In this section we will use a very simple model of heat transfer through a wall (Fig. 15). The model is very simplified (e.g. the number of meshes in the wall is too small) in order to keep the presentation manageable by hand calculations (Figs. 16 and 17).

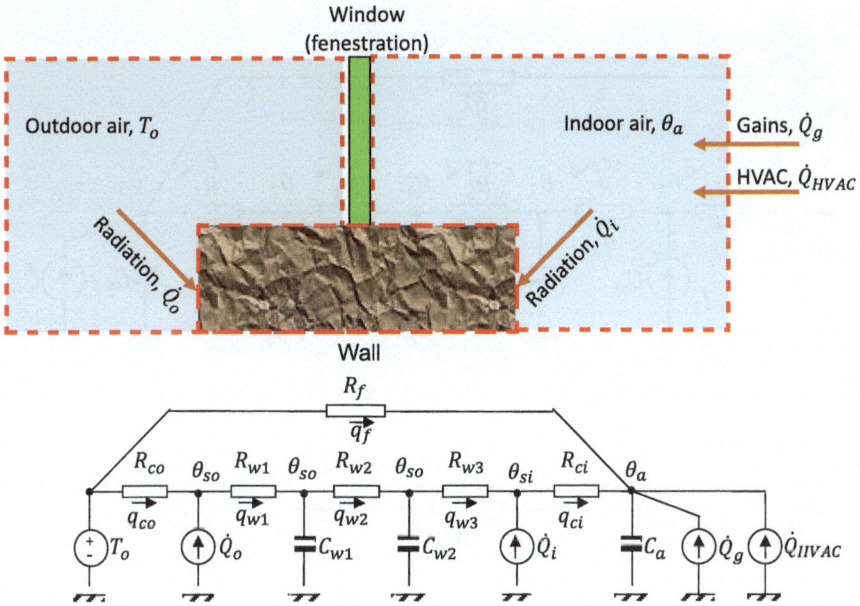

Legend

Temperature source

T_o outdoor temperature

Heat flow sources

$\dot{Q}_o$ absorbed incident solar and long wave radiation on the outdoor surface;

$\dot{Q}_i$ short wave radiation from lights, transmitted solar, long wave radiation exchange with other surfaces, and long wave radiation from internal sources on the indoor surface;

$\dot{Q}_g$ heat flow gained by convection from the internal sources;

$\dot{Q}_{HVAC}$ heat flow from the HVAC system, i.e. the thermal load.

Temperatures in the thermal network

θ_{os}, θ_{is} outdoor and indoor surface temperatures, respectively,

θ_{w1}, θ_{w1} temperatures in the wall.

Heat flows in the thermal network

q_v heat flow rate through the window (advection, conduction, convection);

q_{co}, q_{ci} convection from outside and inside air, respectively;

q_{w1}, q_{w2}, q_{w3} conduction through the wall.

Fig. 15 Typical thermal circuit for heat balance method: (**a**) usual representation; (**b**) representation in the form of typical branches

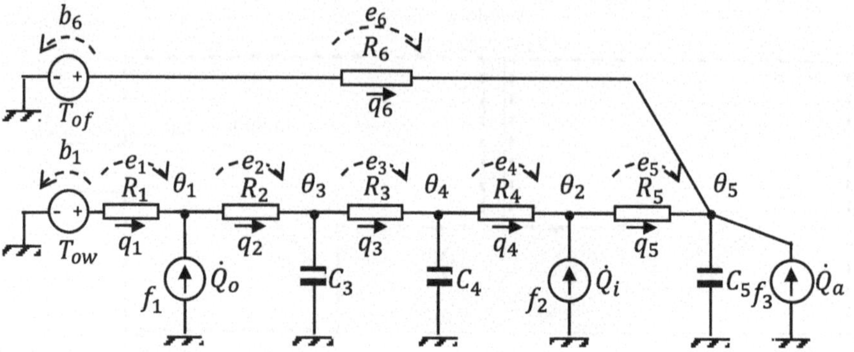

Fig. 16 Temperature nodes are ordered so that the capacity matrix contains zero-blocks

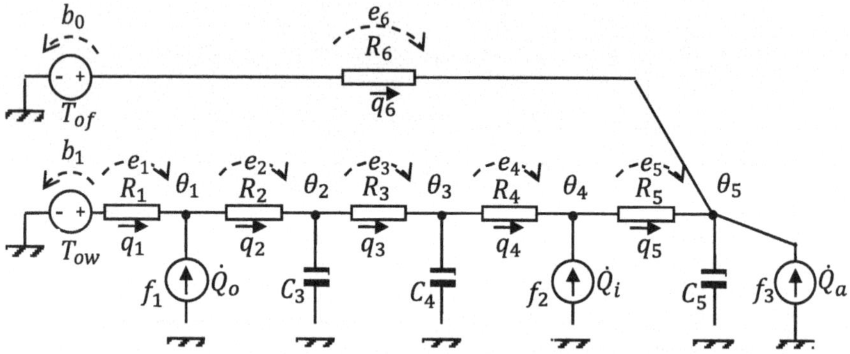

Fig. 17 Temperature nodes need to be re-ordered

4.1 Obtaining the System of Differential-Algebraic Equations

The system of equations corresponding to a thermal circuit may be obtained by using the Kirchhoff's laws and the constitutive laws for thermal transfer [17, 23]. The steps needed to obtain the differential-algebraic system of equations may be synthetized in the following algorithm:

1. Group the temperatures according to the type of node: without capacitance (θ_{so}, θ_{si}, θ_a), and with capacities (θ_{w1}, θ_{w2}):

$$\boldsymbol{\theta} = \left[\theta_{so} \; \theta_{si} \; \theta_a \; \theta_{w1} \; \theta_{w2} \right]^T \tag{72}$$

2. Write the matrices describing the circuit (see an example in Figs. 18 and 19 for the thermal circuit given in Fig. 15):
   (a) Transform the thermal circuit in an oriented graph by indicating the direction of the heat transfer rate for each branch. The directions are arbitrary (if the nodes are numbered, can be in increasing order of the node numbering).
   (b) Write the oriented incidence matrix **A,** Eq. (52), conductance matrix **G**, Eq. (53), and capacitance matrix **C**, Eq. (54).
   (c) Write the vectors of temperature sources **b** given by Eq. (55), of flow rate sources **f** given by Eq. (56) and of outputs **y** given by Eq. (57).
3. Apply Kirchhoff's laws and the constitutive laws to obtain the differential-algebraic system of equations:

$$\mathbf{C}\,\dot{\boldsymbol{\theta}} = -\mathbf{A}^T\,\mathbf{G}\,\mathbf{A}\,\boldsymbol{\theta} + \mathbf{A}^T\,\mathbf{G}\,\mathbf{b} + \mathbf{f}$$

By writing the differences of temperature according to Eq. (28) (equivalent to the Kirchhoff's voltage law),

$$\mathbf{e} = -\mathbf{A}\,\boldsymbol{\theta} + \mathbf{b} \qquad (73)$$

the balance of heat rates in nodes (equivalent to the Kirchhoff's current law),

$$\mathbf{C}\,\dot{\boldsymbol{\theta}} = \mathbf{A}^T\,\mathbf{q} + \mathbf{f} \qquad (74)$$

and the constitutive laws for heat transfer,

$$\theta_{so} \quad \theta_{si} \quad \theta_{w1} \quad \theta_{w2} \quad \theta_a$$

$$\begin{matrix} q_{co} \\ q_{w1} \\ q_{w2} \\ q_{w3} \\ q_{ci} \\ q_f \end{matrix} \quad \mathbf{A} = \begin{bmatrix} 1 & 0 & 0 & 0 & 0 \\ -1 & 0 & 1 & 0 & 0 \\ 0 & 0 & -1 & 1 & 0 \\ 0 & 1 & 0 & -1 & 0 \\ 0 & -1 & 0 & 0 & 1 \\ 0 & 0 & 0 & 0 & 1 \end{bmatrix};\ \mathbf{G} = \begin{bmatrix} R_{co}^{-1} & 0 & 0 & 0 & 0 & 0 \\ 0 & R_{w1}^{-1} & 0 & 0 & 0 & 0 \\ 0 & 0 & R_{w2}^{-1} & 0 & 0 & 0 \\ 0 & 0 & 0 & R_{W3}^{-1} & 0 & 0 \\ 0 & 0 & 0 & 0 & R_{ci}^{-1} & 0 \\ 0 & 0 & 0 & 0 & 0 & R_f^{-1} \end{bmatrix};\ \mathbf{b} = \begin{bmatrix} T_{ow} \\ 0 \\ 0 \\ 0 \\ 0 \\ T_{of} \end{bmatrix}$$

$$\mathbf{C} = \begin{bmatrix} 0 & 0 & 0 & 0 & 0 \\ 0 & 0 & 0 & 0 & 0 \\ 0 & 0 & C_{w1} & 0 & 0 \\ 0 & 0 & 0 & C_{w2} & 0 \\ 0 & 0 & 0 & 0 & C_a \end{bmatrix} \qquad \begin{cases} \mathbf{e} = -\mathbf{A}\,\boldsymbol{\theta} + \mathbf{b} \\ \mathbf{q} = \mathbf{G}\,\mathbf{e} \\ \mathbf{C}\,\dot{\boldsymbol{\theta}} = \mathbf{A}^T\,\mathbf{q} + \mathbf{f} \end{cases}$$

$$\mathbf{f} = [\ \dot{Q}_o \quad 0 \quad 0 \quad \dot{Q}_i \quad \dot{Q}_a\ ]^T \qquad \mathbf{C}\,\dot{\boldsymbol{\theta}} = -\mathbf{A}^T\,\mathbf{G}\,\mathbf{A}\,\boldsymbol{\theta} + \mathbf{A}^T\,\mathbf{G}\,\mathbf{b} + \mathbf{f}$$

$$\dot{Q}_a = \dot{Q}_g + \dot{Q}_{HVAC}$$

Fig. 18 Obtaining the differential-algebraic equations for the circuit from Fig. 17

$$\begin{array}{c} \\ q_{co} \\ q_{w1} \\ q_{w2} \\ q_{w3} \\ q_{ci} \\ q_f \end{array} \quad \mathbf{A} = \begin{array}{c} \theta_{so} \quad \theta_{si} \quad \theta_{w1} \quad \theta_{w2} \quad \theta_a \\ \begin{bmatrix} 1 & 0 & 0 & 0 & 0 \\ -1 & 0 & 1 & 0 & 0 \\ 0 & 0 & -1 & 1 & 0 \\ 0 & 1 & 0 & -1 & 0 \\ 0 & -1 & 0 & 0 & 1 \\ 0 & 0 & 0 & 0 & 1 \end{bmatrix} \end{array}; \quad \mathbf{G} = \begin{bmatrix} R_{co}^{-1} & 0 & 0 & 0 & 0 & 0 \\ 0 & R_{w1}^{-1} & 0 & 0 & 0 & 0 \\ 0 & 0 & R_{w2}^{-1} & 0 & 0 & 0 \\ 0 & 0 & 0 & R_{W3}^{-1} & 0 & 0 \\ 0 & 0 & 0 & 0 & R_{ci}^{-1} & 0 \\ 0 & 0 & 0 & 0 & 0 & R_f^{-1} \end{bmatrix}; \quad \mathbf{b} = \begin{bmatrix} T_{ow} \\ 0 \\ 0 \\ 0 \\ 0 \\ T_{of} \end{bmatrix}$$

$$\mathbf{C} = \begin{bmatrix} 0 & 0 & 0 & 0 & 0 \\ 0 & C_{w1} & 0 & 0 & 0 \\ 0 & 0 & C_{w2} & 0 & 0 \\ 0 & 0 & 0 & 0 & 0 \\ 0 & 0 & 0 & 0 & C_a \end{bmatrix} \qquad \begin{cases} \mathbf{e} = -\mathbf{A}\,\boldsymbol{\theta} + \mathbf{b} \\ \mathbf{q} = \mathbf{G}\,\mathbf{e} \\ \mathbf{C}\,\dot{\boldsymbol{\theta}} = \mathbf{A}^T\,\mathbf{q} + \mathbf{f} \end{cases}$$

$$\mathbf{f} = [\ \dot{Q}_o \quad 0 \quad 0 \quad \dot{Q}_i \quad \dot{Q}_a\ ]^T \qquad \mathbf{C}\,\dot{\boldsymbol{\theta}} = -\mathbf{A}^T\,\mathbf{G}\,\mathbf{A}\,\boldsymbol{\theta} + \mathbf{A}^T\,\mathbf{G}\,\mathbf{b} + \mathbf{f}$$

$$\dot{Q}_a = \dot{Q}_g + \dot{Q}_{HVAC}$$

Fig. 19 Obtaining the differential-algebraic equations for the circuit from Fig. 17

$$\mathbf{q} = \mathbf{G}\ \mathbf{e} \tag{75}$$

we obtain the differential algebraic equations describing the thermal circuit:

$$\mathbf{C}\ \dot{\boldsymbol{\theta}} = \mathbf{K}\ \boldsymbol{\theta} + \mathbf{K}_b\ \mathbf{b} + \mathbf{f}; \tag{76}$$

where

$$\mathbf{K} \equiv -\mathbf{A}^T\ \mathbf{G}\ \mathbf{A}\ \boldsymbol{\theta} \text{ and } \mathbf{K}_b \equiv \mathbf{A}^T\ \mathbf{G} \tag{77}$$

If the diagonal matrix **C** has elements on the diagonal which are zero, the system of Eq. (76) is a system of differential algebraic equations.

4.2 Obtaining the State-Space Representation from the Thermal Circuit

If the thermal circuit contains nodes without capacitance, the matrix **C** is singular. In order to obtain the state-space model, the equations corresponding to the nodes without capacitance need to be eliminated from the system of Eq. (76) [23]. By partitioning the matrix **C**,

$$\mathbf{C} = \begin{bmatrix} \mathbf{0} & \mathbf{0} \\ \mathbf{0} & \mathbf{C}_C \end{bmatrix} \tag{78}$$

where $\mathbf{C}_C$ corresponds to the nodes having capacities, the set of Eq. (76) may be written as:

$$\begin{bmatrix} \mathbf{0} & \mathbf{0} \\ \mathbf{0} & \mathbf{C}_C \end{bmatrix} \begin{bmatrix} \dot{\boldsymbol{\theta}}_0 \\ \dot{\boldsymbol{\theta}}_C \end{bmatrix} = \begin{bmatrix} \mathbf{K}_{11} & \mathbf{K}_{12} \\ \mathbf{K}_{21} & \mathbf{K}_{22} \end{bmatrix} \begin{bmatrix} \boldsymbol{\theta}_0 \\ \boldsymbol{\theta}_C \end{bmatrix} + \begin{bmatrix} \mathbf{K}_{b1} \\ \mathbf{K}_{b1} \end{bmatrix} \mathbf{b} + \begin{bmatrix} \mathbf{I}_{11} & \mathbf{0} \\ \mathbf{0} & \mathbf{I}_{22} \end{bmatrix} \begin{bmatrix} \mathbf{f}_0 \\ \mathbf{f}_C \end{bmatrix} \quad (79)$$

where

$\boldsymbol{\theta}_0$ and $\mathbf{f}_0$ correspond to the nodes without thermal capacity;
$\boldsymbol{\theta}_C$ and $\mathbf{f}_C$ correspond to the nodes with thermal capacity;
$\mathbf{C}_C$ is the bloc of the partitioned matrix $\mathbf{C}$ for which the elements on the diagonal are non-zero;
$\mathbf{K}_{11}$, $\mathbf{K}_{12}$, $\mathbf{K}_{21}$ and $\mathbf{K}_{22}$ are blocs of the partitioned matrix $\mathbf{K}$ obtained according to the partitioning of the matrix $\mathbf{C}$;
$\mathbf{K}_{b1}$ and $\mathbf{K}_{b2}$ are blocs of the partitioned matrix $\mathbf{K}_b$ obtained according to the partitioning of the matrix $\mathbf{C}$;
$\mathbf{I}_{11}$ and $\mathbf{I}_{22}$ are identity matrices.

The state equation of the state-space model is

$$\dot{\boldsymbol{\theta}}_C = \mathbf{A}_S \boldsymbol{\theta}_C + \mathbf{B}_S \mathbf{u} \quad (80)$$

where the state matrix is

$$\mathbf{A}_S = \mathbf{C}_C^{-1} \left(-\mathbf{K}_{21} \mathbf{K}_{11}^{-1} \mathbf{K}_{12} + \mathbf{K}_{22} \right) \quad (81)$$

and the input matrix is

$$\mathbf{B}_S = \mathbf{C}_C^{-1} \left[-\mathbf{K}_{21} \mathbf{K}_{11}^{-1} \mathbf{K}_{b1} + \mathbf{K}_{b2} \;\; -\mathbf{K}_{21} \mathbf{K}_{11}^{-1} \;\; \mathbf{I} \right] \quad (82)$$

For the numerical example of the thermal circuit shown in Fig. 15 with the values of its parameters given in Table 4, the state variables are

$$\boldsymbol{\theta} = \left[\theta_a \; \theta_{w1} \; \theta_{w2} \right]^T \quad (83)$$

and the bloc vector of inputs is:

$$\mathbf{u} = \left[\mathbf{b} \; \mathbf{f}_0 \; \mathbf{f}_C \right]^T \quad (84)$$

The numerical values of the matrices of the model are:

Table 4 Parameter of the thermal network shown in Fig. 1

Parameter	Value
Indoor air capacity, C_a	$82 \bullet 10^3$ J/K
Wall capacity, $C_{w1} = C_{w2}$	$2 \bullet 10^6$ J/K
Thermal conductance of the wall, R_w^{-1}	1.45 W/K
Thermal conductance of one third of the wall, $R_{w1}^{-1} = R_{w2}^{-1} = R_{w3}^{-1} = 3\ R_w^{-1}$	4.35 W/K
Thermal conductance of the window and due to losses by ventilation, R_v^{-1}	38.3 W/K
Outdoor convection conductance, R_{co}^{-1}	250.0 W/K
Indoor convection conductance, R_{ci}^{-1}	125.0 W/K

$$\mathbf{C} = \left[\begin{array}{cc|ccc} 0 & 0 & 0 & 0 & 0 \\ 0 & 0 & 0 & 0 & 0 \\ \hline 0 & 0 & 82\cdot 10^3 & 0 & 0 \\ 0 & 0 & 0 & 2\cdot 10^6 & 0 \\ 0 & 0 & 0 & 0 & 2\cdot 10^6 \end{array}\right] = \begin{bmatrix} \mathbf{0} & \mathbf{0} \\ \mathbf{0} & \mathbf{C_C} \end{bmatrix} \tag{85}$$

$$\mathbf{K} = \left[\begin{array}{cc|ccc} -254.35 & 0 & 0 & 4.35 & 0 \\ 0 & -129.35 & 125 & 0 & 4.35 \\ \hline 0 & 125 & -163.30 & 0 & 0 \\ 4.35 & 0 & 0 & -8.70 & 4.35 \\ 0 & 4.35 & 0 & 4.35 & -8.70 \end{array}\right] = \begin{bmatrix} \mathbf{K}_{11} & \mathbf{K}_{12} \\ \mathbf{K}_{21} & \mathbf{K}_{22} \end{bmatrix} \tag{86}$$

$$\mathbf{K}_b = \left[\begin{array}{cccccc} 250 & -4.35 & 0 & 0 & 0 & 0 \\ 0 & 0 & 0 & 4.35 & -125 & 0 \\ \hline 0 & 0 & 0 & 0 & 125 & 38.3 \\ 0 & 4.35 & -4.35 & 0 & 0 & 0 \\ 0 & 0 & 4.35 & -4.35 & 0 & 0 \end{array}\right] = \begin{bmatrix} \mathbf{K}_{b1} \\ \mathbf{K}_{b2} \end{bmatrix} \tag{87}$$

Substituting these matrices in Eq. (81), we obtain the state matrix:

$$\mathbf{A}_S = \begin{bmatrix} -5.18\cdot 10^{-4} & 0 & 5.13\cdot 10^{-5} \\ 0 & -4.31\cdot 10^{-6} & 2.17\cdot 10^{-5} \\ 2.10\cdot 10^{-6} & 2.17\cdot 10^{-5} & -4.28\cdot 10^{-5} \end{bmatrix} \tag{88}$$

The input matrix $\mathbf{B}_S$ has 11 columns, corresponding to the input vector

$$\mathbf{u} = \left[\mathbf{b}^T\ \mathbf{f}_0^T\ \mathbf{f}_C^T\right]^T = \left[T_{ow}\ 0\ 0\ 0\ 0\ T_{ov}\ \dot{Q}_o\ \dot{Q}_i\ \dot{Q}_{aux}\ 0\ 0\right] \tag{89}$$

The inputs corresponding to zeros are useless and can be eliminated. Therefore, from the 11 columns of the input matrix $\mathbf{B}_S$ it can be retained only the columns corresponding to inputs 1, 6, 7, 8 and 9:

$$\mathbf{B}_S = \begin{bmatrix} 0 & 4.67 \cdot 10^{-4} & 0 & 1.18 \cdot 10^{-5} & 1.22 \cdot 10^{-5} \\ 2.14 \cdot 10^{-6} & 0 & 8.55 \cdot 10^{-9} & 0 & 0 \\ 0 & 0 & 0 & 1.68 \cdot 10^{-8} & 0 \end{bmatrix} \quad (90)$$

The output matrix of the state-space representation extracts the output, i.e. the indoor air temperature θ_a which is the first element from the state vector:

$$\mathbf{C}_S = \begin{bmatrix} 1 \; 0 \; 0 \end{bmatrix} \quad (91)$$

For the thermal circuit given in Fig. 15, the feedthrough matrix is zero,

$$\mathbf{D}_S = \mathbf{0} \quad (92)$$

The complete state-space representation of the thermal circuit from Fig. 15, with the values of its parameters given in Table 4, is

$$\begin{cases} \dot{\boldsymbol{\theta}}_C = \mathbf{A}_S \boldsymbol{\theta}_C + \mathbf{B}_S \boldsymbol{\theta}_C \\ \theta_a = \mathbf{C}_S \boldsymbol{\theta}_C + \mathbf{D}_S \mathbf{u} \end{cases} \quad (93)$$

with the values of the matrices given by Eqs. (88), (90) and (91).

If the term $C_a = m_a c_a \dot{\theta}_a$ is zero, then the state vector is

$$\dot{\boldsymbol{\theta}} = \begin{bmatrix} \theta_{w1} \; \theta_{w2} \end{bmatrix}^T \quad (94)$$

which implies that the matrices in Eqs. (85)–(87) are partitioned correspondingly. In our numerical example, the first three equations need to be eliminated from the system of Eq. (76). The state matrix $\mathbf{A}_S$, obtained with the expression (81), is

$$\mathbf{A}_S = \begin{bmatrix} -4.31 \cdot 10^{-6} & 2.17 \cdot 10^{-6} \\ 2.17 \cdot 10^{-6} & -4.07 \cdot 10^{-6} \end{bmatrix} \quad (95)$$

The input matrix $\mathbf{B}_S$, obtained with the expression (82) and retaining only the columns corresponding to inputs 1, 6, 7, 8 and 9, is:

$$\mathbf{B}_S = \begin{bmatrix} 2.14 \cdot 10^{-6} & 0 & 8.55 \cdot 10^{-9} & 0 & 0 \\ 0 & 1.89 \cdot 10^{-6} & 0 & 6.46 \cdot 10^{-8} & 4.95 \cdot 10^{-8} \end{bmatrix} \quad (96)$$

The observation equation can be obtained from the first row of Eq. (79) [23]

$$\boldsymbol{\theta}_0 = -\mathbf{K}_{11}^{-1} (\mathbf{K}_{12} \boldsymbol{\theta}_C + \mathbf{K}_{b1} \mathbf{b} + \mathbf{I}_{11} \mathbf{f}_0) = -\mathbf{K}_{11}^{-1} \left(\mathbf{K}_{12} \boldsymbol{\theta}_C + \begin{bmatrix} \mathbf{K}_{b1} \; \mathbf{I}_{11} \; \mathbf{0} \end{bmatrix} \begin{bmatrix} \mathbf{b} \\ \mathbf{f}_0 \\ \mathbf{f}_C \end{bmatrix} \right) \quad (97)$$

Then, the output equation is

$$\mathbf{C}_S = -\mathbf{K}_{11}^{-1}\,\mathbf{K}_{12} \tag{98}$$

and the feedthrough matrix is

$$\mathbf{D}_S = -\mathbf{K}_{11}^{-1}\left[-\mathbf{K}_{b1}\;\mathbf{I}_{11}\;\mathbf{0}\right] \tag{99}$$

Keeping only the non-zero inputs $\left[\,T_{ow}\;T_{ov}\;\dot{Q}_o\;\dot{Q}_i\;\dot{Q}_{aux}\,\right]$ from the input vector $\mathbf{u}$ given by Eq. (89), the output and feedthrough matrices for our numerical example are:

$$\mathbf{C}_S = \left[\,0\;9.89\cdot 10^{-2}\,\right] \tag{100}$$

and

$$\mathbf{D}_S = \left[\,0\;9.01\cdot 10^{-1}\;0\;2.27\cdot 10^{-2}\;2.35\cdot 10^{-2}\,\right] \tag{101}$$

5 Conclusions

Thermal networks are widely used to model heat transfer. The phenomena of conduction, convection, radiation and advection can be linearized; as a consequence, the thermal networks, which are weighted oriented graphs, can be represented by matrices and vectors.

State-space representation is widely used in the analysis and synthesis of control systems. Linear time invariant models may be used as local linearized models of non-linear systems. Therefore, the linear algebra representation of state-space models has a large field of applications.

This chapter described succinctly the modelling of heat transfer by thermal networks and emphasized the fact that the temperature and heat-flow rate sources are inputs (or independent variables), the temperatures in nodes and the flow rates are outputs (or dependent variables), and the resistances and the capacities are parameters of a model structure. A data structure composed of matrices and vectors was proposed. The novelty of the data structure is the definition of an output vector which represents a subset of temperatures that are needed as observables.

A second novelty presented in this chapter is the assembling of thermal circuits. This technique allows us to construct large models from constitutive blocs. For example, the model of a complex building may be obtained by interconnecting typical blocs such as walls, floors, doors and windows. Complex systems can be obtained also by coupling the equations of the typical blocs and solving iteratively the system of equations. The advantage of assembling is that the model of the whole system is a single thermal network that can be analysed. The key point in assembling is obtaining the disassembling matrix. An algorithm for obtaining it is presented.

The third novelty is the transformation of thermal circuit in state-space representation. While examples for simple circuits are abundant and other methods (such as Kirchhoff laws and nodal analysis) are available, the method proposed is directly related to the matrix representation of the thermal circuits. The principal characteristic of the method is Gauss elimination of the block matrices and vectors related to node temperatures that do not have capacities (i.e. that are not state variables).

Assembling thermal circuits and obtaining state-space models from them can be used in at least two important fields. The first is Building Information Modelling (BIM): each component has its model and the model of the building can be obtained by assembling the models of components. The second is system theory in which the state space is a suitable form of the model for system analysis and synthesis: eigenvalue decomposition, model order reduction, model predictive control, observability, controllability and identifiability, etc.

Acknowledgement The author highly appreciates the discussions with Florent Alzetto, research engineer at Saint Gobain Research. His feed back and comments were very valuable.

References

1. J.A. Clarke, *Energy Simulation in Building Design*, 2nd edn. (Butterworth–Heinemann, Oxford, 2001)
2. EnergyPlus, *Engineering Reference* (DOE, University of Illinois & LBNL, 2015a)
3. TRNSYS 17, *Mathematical Reference* (Thermal Energy System Specialists, LLC, Madison, 2009)
4. T. Bergman, A. Lavine, F. Incropera, D. Dewitt, *Fundamentals of Heat and Mass Transfer*, 7th edn. (Wiley, 2011)
5. H. Recknagel, E. Sprenger, E.-R. Schramek, *Génie Clmatique*, 72nd edn. (Dnod, Paris, 2007)
6. J. Clarke, N.J. Kelly, D. Tang, A review of ESP-r flexible solution approach and its application to prospective technical domain developments. Adv. Build. Energy Res. **1**(1), 227–247 (2007)
7. K. El Khoury, G. Mouawad, G. El Hitti, M. Nemer, The component interaction network approach for modeling of complex thermal systems. Adv. Eng. Softw. **65**, 149–157 (2013)
8. J. Ma, M. Le, A new method for coupling of boundary element method and finite element method. Appl. Math. Model. **16**(1), 43–46 (1992)
9. D.E. Ficher et al., *A Modular Loop-Based Approach to HVAC Energy Simulation and Its Implementation in EnergyPlus* (Building Simulation, Kyoto, Japan, 1999), p. B-31
10. P. Sahlin, *Modelling and Simulation Methods for Modular Continuous Systems in Buildings* (Royal Institute of Technology, Stockholm, 1996)
11. MathWorks, Model Interconnection (2017), https://mathworks.com/help/control/model-interconnection.html
12. Y. Chen, A. Athienitis, K. Galal, Frequency domain and finite difference modeling of ventilated concrete slabs and comparison with field measurements: Part 1, modeling methodology. Int. J. Heat Mass Transf., 948–956 (2013)
13. S. Ghosh, *Network Theory: Analysis and Synthesis* (Prentice Hall of India, New Delhi, 2005)

14. D. Maillet et al., *Thermal Quadrupoles: Solving the Heat Equation Through Integral Transforms* (Wiley, 2000)
15. E. Piotrowska, A. Chochowski, Representation of transient heat transfer as the equivalent thermal network (ETN). Int. J. Heat Mass Transf. **63**, 113–119 (2013)
16. D.C. Karnopp, D.L. Margolis, R.C. Rosenberg, *System Dynamics: A Unified Approach* (Wiley, 1990)
17. G. Strang, *Computational Science and Engineering* (Wellesley-Cambridge Press, Wellesley, CA, 2007)
18. T.N. Narasimhan, Fourier's heat conduction equation: history, influence, and connections. Rev. Geophys. **37**, 151–172 (1999)
19. A. Cueva-Zepeda, J. Avalos-Garcia, The role of a connectivity matrix in the assemblage process of the finite element method. Adv. Eng. Softw. **37**(11), 721–727 (2006)
20. G. Nikishkov, *Programming Finite Elements in Java* (Springer, 2010)
21. A.A. Ramabathiran, S. Gopalakrishnan, Automatic finite element formulation and assembly of hyperelastic higher order structural models. Appl. Math. Model., 2867–2883 (2014)
22. F. Ruiz-Calvo et al., Coupling short-term (B2G model) and long-term (g-function) models for ground source heat exchanger simulation in TRNSYS. Application in a real installation. Appl. Thermal Eng. **102**, 720–732 (2016)
23. C. Ghiaus, Causality issue in the heat balance method for calculating the design heating and cooling load. Energy **50**, 292–301 (2013)

Machine Learning for Activity Recognition in Smart Buildings: A Survey

Manar Amayri, Samer Ali, Nizar Bouguila, and Stephane Ploix

1 Introduction

New buildings (e.g., commercial, residential, public) are now generally equipped with a variety of smart sensors and smart meters. This makes these buildings smart with improved capacities, opportunities, and applications related, for instance, to energy management systems which main goal is to decrease waste mainly due to irresponsible human behaviors [1]. Indeed, energy deficiency represents a global problem. Hence, energy generation increase and consumption efficiency are two active areas of research [2]. Nonetheless, energy usage is ever in demand particularly given the various technological advances that rely on electrical power for operating.

The automatic reduction of energy requirements in buildings has received a lot of attention recently [3] and early attempts included automatic regulation of light or heating in home automation. However, these approaches were deemed inappropriate due to improper reaction to the expectations of the users; i.e., the occupants [4]. The majority of recent studies has shown the importance of putting the users in the energy-saving loop while ensuring their comfort [5]. Occupants' behavior has a major influence on building energy consumption [6–9]. Hence, [10] introduced methods for modeling occupant behavior and quantifying its impact on building

M. Amayri (✉)
CNRS, Grenoble, France
e-mail: manar.amayri@grenoble-inp.fr

S. Ploix
G-SCOP lab, Grenoble Institute of Technology, University of Grenoble Alps, CNRS UMR5272, Grenoble, France
e-mail: stephane.ploix@grenoble inp.fr

S. Ali · N. Bouguila
CIISE, Concordia University, Montreal, QC, Canada
e-mail: nizar.bouguila@concordia.ca

S. Ploix et al. (eds.), *Towards Energy Smart Homes*,
https://doi.org/10.1007/978-3-030-76477-7_6

energy use. The major themes include advancements in analytic data collection techniques, modeling methods, and applications that provide insights into behavior-related energy saving's potential and impact [11]. There is a large gap between the predicted energy demand and the actual consumption, once the building is in use [12]. According to [13], occupants' behaviors account for significant uncertainty in building energy use. One cause could be that occupant behavior might not fit with the energy concept and thus cause counterproductive effects [14]. Occupants have influence due to their presence and activities in the building and due to their control actions, which aim to improve indoor environmental conditions (thermal, air quality, light, noise). Consequently, the weight of the user behavior on the energy balance of a building increases [15]. Indeed, several studies suggest huge energy savings in buildings just by detecting occupancy (presence/absence) as shown, for instance, in [10] where motion sensors and magnetic door switches are used to detect occupancy in offices and HVAC (Heating, Ventilation, and air-conditioning) control, thereby estimating potential energy savings from 10–15%. Similarly, [16] focus on how to estimate the number of occupants in a room by processing CO_2 concentration, temperature, and HVAC actuation levels in order to identify a dynamic model. Additionally, there is a lot of potential for energy savings and increasing occupants' comfort by detecting activities and this motivates to carry forward the activity recognition task [17]. Methods investigated for finding occupancy using common sensors vary from basic single feature classifiers that distinguish between two classes (presence and absence) [17, 18] to multi-sensor multi-feature models [16, 19–22]. A primary approach, which is prevalent in many commercial buildings, is to use passive infrared (PIR) sensors for occupancy estimation. However, motion detectors fail to detect presence when occupants remain relatively still. This is quite common during activities like regular deskwork. Furthermore, drifts of warm or cold air on objects can be interpreted as motion leading to false positive detection. This makes the use of PIR sensors alone, less attractive for occupancy counting purposes. Fusion of PIR sensor data with other sensors can be useful as discussed in [10]. As such, motion sensors are usually paired with magnetic reed switches for occupancy detection in order to increase the efficiency of HVAC systems in smart buildings in a simple and non-intrusive manner. Acoustic sensors may also be used [23]. However, environment audio signals may cause many false positives when no support from other sensors is available. The use of pressure, PIR, and acoustic sensors to detect occupancy in single desk offices has been discussed in [24]. Further tagging of activities is based on this knowledge, where a pressure sensor detects chairs occupancy with the offices filmed and then the footage is used to manually classify the activities of people over time.

Smart buildings related tasks in general and activity recognition in particular have been widely approached using classic optimization models (e.g., meta-heuristics, linear programming, dynamic programming, etc.). Unfortunately, these approaches do not take full advantage of the large-scale data generated by smart buildings settings. In order to extract and exploit the knowledge hidden in these data, recent trend and efforts in smart building applications have been based on data mining and machine learning techniques [25]. The main goal is to build specific models from the available data with respect to the task to tackle. A typical data mining and machine learning framework is generally based on the following steps. The first step is data collection from the available sensors. The second step concerns preprocessing (e.g., data cleaning, data enrichment, normalization, feature selection and/or extraction, outliers rejection, etc.) the collected data. Finally, a model learning step is performed in which a machine learning technique is applied.

Different learning approaches have been deployed in the past in smart building applications. In [26], for instance, hidden Markov models (HMMs) [27] have been used for estimating occupancy using a wireless ambient sensing system, CO_2 sensors, and a wired camera network in order to establish actual occupancy levels. The large variance in the energy consumption was found to be primarily due to the operating mode; occupants that are elected to run their AC for longer durations, at lower set points and/or throughout a larger space, consumed more energy than occupants that did not [28]. Consequently, energy reduction methods must encompass a combination of technological development, building physics, and occupants' behavior to achieve the desired performance [29]. As such, numerous studies have developed control systems and modeling methodologies to better assist occupants to play active roles in buildings. In [30], a supervised learning approach is investigated. It initially determines the common sensors to be used to estimate and classify the approximate number of people (within a range) in a room and their activities. Means to estimate occupancy include motion detection, power consumption, CO_2 concentration sensors, microphone, or door/window positions. The most useful measurements in calculating the information gains when added to the classification algorithm are then determined. Next, estimation that relies on decision tree and random forest learning algorithms is performed. The reason behind the choice of the algorithms is that they yield decision rules readable by humans, which correspond to nested if-then-else rules, where thresholds can be adjusted depending on the considered living areas. One office has been used for testing and two video cameras have been used in this approach. This highly limits the implementation of the application because of the privacy issues.

Studying occupants' activity and behavior is a key for building adaptation and energy saving, thus not limited to occupancy detection and estimation only [31–33]. The primary motivation behind studies of activity recognition is to contribute to buildings, while a comprehensive model can improve the energy performance of a building. This has been studied by previous research in this area, and large savings can be obtained with activity aware building energy management system. Such building energy management system can also warn users about activities or behaviors that adversely affect energy savings of the building. This induces an energy aware behavior that can add one-third to a building's designed energy performance [17]. Thus, the goal of this chapter is to provide a review on machine learning approaches related to activities recognition in smart buildings. Furthermore, it serves to facilitate the definitions and introduction of machine learning techniques to domain beginners and practitioners alike. Moreover, the chapter also sheds light on the various advancements made in activity recognition in smart homes using machine learning, presenting the first survey of such methods, to the best of our knowledge. Several machine learning models have been deployed over the years for activity recognition [34–36]. The process generally involves learning activity models from training data. The model learns to recognize patterns that differentiate various classes in the training data and apply this knowledge for the prediction and classification over the test data. This allows the actualization of a solution without necessarily providing domain specific knowledge. Since the problem emanates from pattern recognition or data analysis, such methods are termed data-driven. [37] identify such data-driven approaches and categorize them into generative, discriminative, and heuristic-based modeling:

1. Generative modeling: uses training data samples to form a description of the complete input space. Probabilistic models like Bayesian networks, Gaussian mixture models, and HMMs fall under this category. The underlying assumption of this model is that the training samples are representative of the entire input space/distribution and thus enough data must be available to learn the complete probabilistic representation.
2. Discriminative modeling: has the primary objective of finding a decision boundary or boundaries, rather than representing the entire input space. A basic example of this model is K-nearest neighbor (KNN) classifier, where a test point is assigned to a cluster that is at a minimum distance (the notion of distance may vary accordingly) to it. Similarly, but better performing algorithms in the same category, are decision trees and SVMs [38].
3. Heuristic-based modeling: uses a combination of both generative and discriminative models along with some heuristic information [39].

It is noteworthy that other approaches that take advantage of both generative and discriminative learning simultaneously, called hybrid generative discriminative approaches, have been proposed recently in the literature [40–43]. When training data (i.e., labeled data where the output's correct value for each instance is known) are considered, the learning approach is called supervised. Classification and regression are typical examples of supervised learning tasks. Using a set of training data grouped into classes, the goal of classification is to build a classifier to predict to which class a new observation should be assigned. Examples of classification approaches include support vector machine (SVM), decision trees, random forests, artificial neural networks, and K-nearest neighbors. Regression, on the other hand, is related to predicting a numerical value using a function built by relating outputs to inputs. Examples of regression approaches include linear regression and support vector regression. In many cases, however, the data are unlabeled and need the deployment of unsupervised learning technique to infer possible regularities (e.g., clusters) in the input space. Clustering (partitional or hierarchical) is the main example of unsupervised learning and consists of grouping observations such that intraclass and interclass similarities are maximized and minimized, respectively [44]. Partitional approaches include both centroid-based (ex. K-Means) and density-based (ex. DBSCAN) clustering. Hierarchical approaches include both divisive (i.e., top down) and agglomerative (i.e., bottom up) approaches. A compromise between supervised and unsupervised learning, called semi-supervised learning, allows to consider labeled data jointly with unlabeled data. An example of semi-supervised learning techniques is active learning which necessitates an interaction with the user to get the desired outputs for new test data. In order to avoid collecting data from scratch and disturbing the daily life of users some activity recognition approaches have been based on transfer learning. The main idea consists of transferring learned knowledge as much as possible from an existing environment, the so-called source domain, to a new target one (i.e., the environment where knowledge is applied) to reduce data collection effort. It is noteworthy that in transfer learning, feature sets, label sets as well as learning tasks in both source and target domains datasets can be different. Transfer learning approaches can be roughly classified into three groups of approaches: instance-, feature-, and parameter-based transfer techniques.

The rest of this chapter is organized as follows: Sect. 2 describes the machine learning algorithms and reviews the relevant papers in the literature pertinent to the topic at hand, Sect. 3 presents an extensive case study, and finally Sect. 4 concludes the chapter.

2 Activity Recognition in Smart Buildings

In this section, we overview the main families of approaches that have been deployed for activity (e.g., cooking, sleeping, eating, etc.) recognition in smart buildings: classification, regression, and clustering. The first two are often referred to under the umbrella of supervised learning while the latter is an unsupervised learning method. These form the two main branches of machine learning. Other derivatives and hybrid categories such as semi-supervised learning [45] and the popular deep learning methods [46] have been researched extensively recently. However, they are usually founded on one of the two main categories or even combines both of the approaches. It is noteworthy to mention that when deep learning techniques studies arise in the literature, we list them as part of the supervised learning approach.

Supervised learning refers to methodologies whereby input data has explicit labels for each of its entries or objects, depending on the nature of the pertaining data. Such data is then split into training and testing sets that are used for the learning of the parameters of the desired algorithm. Specifically, classification is presented in Sect. 2.1 and regression in Sect. 2.2. On the other hand, unsupervised learning has to be carried out without the availability of labels for the data at hand. We also review the relevant literature of the latter method applied for activity recognition in smart buildings, as appropriate. Clustering is detailed in Sect. 2.3. A complete list of the papers with the respective algorithm(s) used as well as other miscellaneous details is described in Sect. 2.4.

2.1 Classification

This section is split into two subsections whereby Sect. 2.1.1 presents general classification approaches for activity recognition and Sect. 2.1.2 expands on HMMs and their utilization in the field.

2.1.1 General Classification Approaches

Given a set of data with discrete labels or classes that may be used for training, *Classification* then refers to the correct identification of the label or class that testing data falls under. Mathematically, classification is a mapping between input data x and output label y such that:

$$y = g(x|\theta) \tag{1}$$

where $g()$ represents the classification function or algorithm, and θ is its respective parameters. The function uses the training data to approximate the parameters. The closer the approximation to the true parameters, the better the fit and hence the performance of the classification algorithm. Thus, $g()$ can also be viewed as a separator between the data points of the various classes or labels in a problem.

This approach has been critical in developing various activity recognition approaches in smart buildings. For example, [47] use ontological modeling and semantic reasoning for a real-time multisource sensor data based activity recognition system in smart homes. The algorithm first converts detected sensor activation corresponding properties into context ontologies. This constructs an activity description and then equivalency and subsumption reasoning are performed for activity recognition. Finally, semantic retrieval is used for obtaining the set of atomic activity concepts.

Hu et al. [48] present a new classification algorithm based on feature incremental random forests. Random forests are another classification algorithm that may be utilized for activity recognition. They are based on decision trees whereby the overfitting is addressed by reporting the final classification result as the mode of the various individual trees. Indeed, a decision tree approach is used for the real-time smart watch system presented in [49] for activity recognition. Incremental learning, on the other hand, refers to updating the existing model dynamically with new data or sensors instead of retraining the model from scratch and disposing of the existing one.

It is sometimes referred to by online learning in the machine learning community [50, 51]. This proposed method [48] has been tested on three different datasets and reportedly consistently outperformed other incremental learning methods. Similarly, [52] also investigate a new methodology to incorporate incremental learning for dynamic activity recognition using random forests. However, the latter is comparable to the performance of batched random forests and extremely randomized trees.

Batch learning is the opposing concept to online learning and refers to retraining the entire model as new data or data sources, such as new sensors in the case of activity recognition, become available. Online or incremental learning is usually used because it saves time and resources as well as enables real time processing for real world applications.

Gu et al. [53] introduce a classification approach based on emerging patterns that defines significant changes between different activity classes. Hence, it has the advantage of independence from the dataset used for training given that it identifies the underlying sequential patterns of an activity regardless of whether it is interleaved or concurrent. This brings us to two other important definitions in action recognition: *Concurrent* activities refer to ones in which each of the activity can be broken down into multiple ones that are carried out at the same time or simultaneously such as eating or walking. *Interleaved* activities refer to simple activities such as a wave of the hand or sleeping.

A three phased activity recognition method is proposed in [54]. Classification of the activities is carried out by four different machine learning models: random forests, K-nearest neighbors, support vector machines, and decision trees. In normal activity detection, the four models perform comparably, while the random forest approach outperforms all others in interleaved activity recognition. Support vector machine are also used for activity recognition in smart homes in [55].

Multiple classification algorithms are studied in [56] for activity recognition. These include decision tables, decision trees (C4.5), K-nearest neighbors, support vector machines, and naive Bayes. Interestingly, meta classifiers are also compared for designing the optimum classifier for the problem. These include boosting, bagging, and plurality voting. This represents the first investigation carried out to find whether combining classifiers trained on accelerometer features would result in an improved result, as claimed by the authors. Data was collected for eight different activities carried out by two individuals over different days in multiple setups and with no noise filtering. The activities were standing, walking, running, climbing up stairs, climbing down stairs, sit-ups, vacuuming, and brushing teeth. Gradient boosting, K-nearest neighbor, linear discriminant analysis, and random forests are also researched for activity recognition in smart homes in [57] as well as kernel Fisher discriminant analysis and extreme learning machine in [58].

All in all, plurality voting was found to be the optimum implementation with consistent performance across different setups [56]. It is noteworthy to mention that an accelerometer is a well-researched sensor for activity recognition. Indeed, the utility of even simple sensors has proven effectiveness such as in [59], even with an elementary classifier such as the naive Bayes [60] or with deep learning techniques such as convolutional neural networks and long short-term memory [61]. For instance, [62] use accelerometer data from 20 individuals each with five different accelerometers and a decision tree classier setup. The results suggest that the use of multiple accelerometers improves recognition.

Long short-term memory and convolutional neural networks are also used in [63], while only the latter is deployed in [64] and compared to the K-nearest neighbor and support vector machine methods. Other deep learning techniques such as recurrent neural networks are also studied in [65] (with support vector regression) and in [66] (with support vector machine, naive Bayes, and logistic recognition).

2.1.2 Hidden Markov Models

HMMs are one of the most popular methods used in the field due to the sequential nature of the problem [67, 68]. An HMM is a well-received double stochastic model that uses a compact set of features to extract underlying statistics [69]. Its structure is formed primarily from a Markov chain of latent variables with each corresponding to the conditioned observation. A Markov chain is one of the least complicated ways to model sequential patterns in time series data. It allows us to maintain generality while relaxing the independent identically distributed assumption [70].

Mathematically, an HMM is characterized by an underlying stochastic process with K hidden states that form a Markov chain. Each of the states is governed by an initial probability π, and the transition between the states at time t can be visualized with a transition matrix $B = \{b_{ii'} = P(s_t = i'|s_{t-1} = i)\}$. In each state s_t, an observation is emitted corresponding to its distribution which may be discrete or continuous. This is the observable stochastic process set (Fig. 1).

The emission matrix of the discrete observations can be denoted by $\Xi = \{\Xi_{it}(m) = P(X_t = \xi_m|s_t = i)\}$ where $[m, t, i] \in [1, M] \times [1, T] \times [1, K]$, and the set of all possible discrete observations $\xi = \{\xi_1, \ldots, \xi_m, \ldots, \xi_M\}$. On the other hand, the respective parameters of a probability distribution define the observation emission for a continuous observed symbol sequence. The Gaussian distribution is most commonly used which is defined by its mean and covariance matrix $\varkappa = (\mu, \Sigma)$ [71–73]. Consequently, a mixing matrix must be defined $C = \{c_{ij} = P(m_t = j|s_t = i)\}$ in the case of continuous HMM emission probability distribution where $j \in [1, M]$ such that M is the number of mixture components in set $L = \{m_1, \ldots, m_M\}$. Hence, a discrete or continuous HMM may be defined with the following respective parameters $\lambda = \{B, \Xi, \pi\}$ or $\{B, C, \varkappa, \pi\}$.

For the thorough explanation of the HMM algorithms to follow, we also introduce another visualization that depicts the graphical directed HMM structure as shown

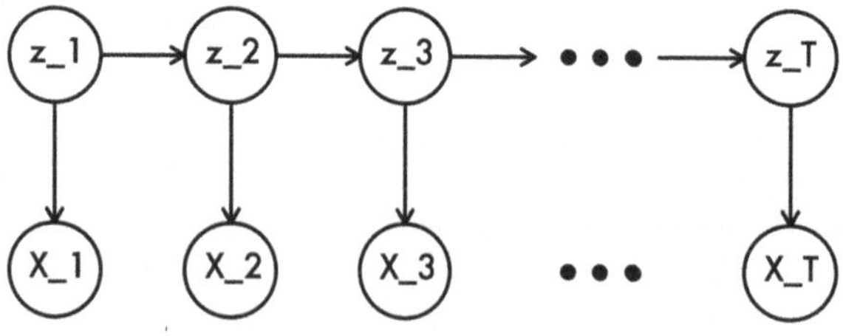

Fig. 1 A typical hidden Markov chain structure representation of a time series where z_1 denotes the first hidden state z_1 and X_1 denotes the corresponding observed state X_1. This is shown accordingly for a time series of length T

in Fig. 2. Figure 3 shows transitions then when they become trellis or lattice. Indeed, Rabiner first introduces the three classical problems of HMMs in [71] as: (1) evaluation or likelihood, (2) estimation or decoding, and (3) training or learning. These are described as follows:

1. The evaluation problem is mainly concerned with computing the probability that a particular sequential or time series data was generated by the HMM model, given both the observation sequence and the model. Mathematically, the primary objective is computing the probability $P(X|\lambda)$ of the observation sequence $X = X_1, X_2, \ldots, X_T$ with length T given an HMM model λ.
2. The decoding problem finds the optimum state sequence path $I = i_1, i_2, \ldots, i_T$ for an observation sequence X. This is mathematically $\mathbf{s}^* = \text{argmax}_{\mathbf{s}} P(\mathbf{s}|X, \lambda)$.
3. The learning problem refers to building an HMM model through finding or "learning" the right parameters to describe a particular set of observations. Formally, this is performed with maximizing the probability $P(X|\lambda)$ of the set of observation sequence X given the set of parameters determined λ. Mathematically, this is $\lambda^* = \text{argmax}_{\lambda} P(X|\lambda)$.

In the following discussion, we present the respective solutions for each of the HMM problems. We assume discrete emission observations. However, it is straightforward to extend these solutions to the HMM of continuous emission distributions given their parameters and mixing matrix. We also briefly recall the two conditional independence assumptions that allow for the tractability of the HMM algorithms [74]:

1. Given the $(t-1)$st hidden variable, the tth hidden variable is independent of all other previous variables such that:

$$P(s_t|s_{t-1}, X_{t-1}, \ldots, s_1, X_1) = P(s_t|s_{t-1}) \tag{2}$$

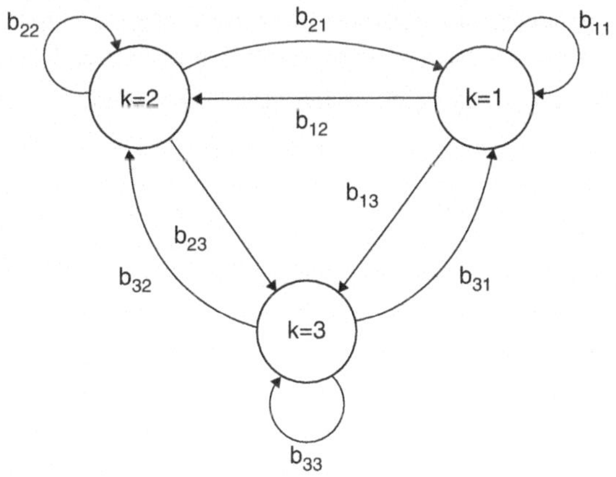

Fig. 2 An HMM transition diagram with three states

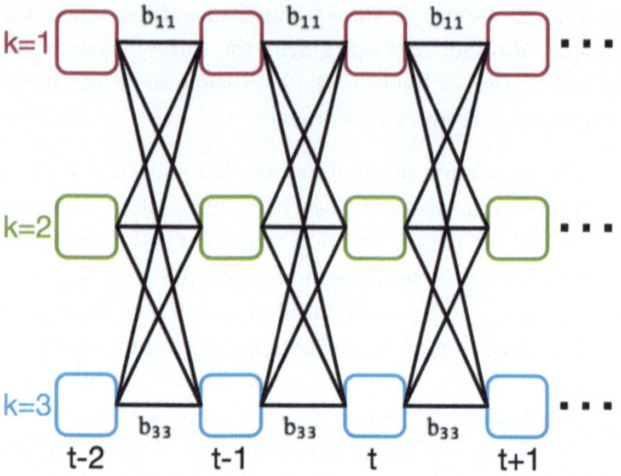

Fig. 3 Lattice or trellis HMM structure which is a representation of the hidden states

2. Given the tth hidden variable, the tth observation is independent of other variables such that:

$$P(X_t|s_T, X_T, s_{T-1}, X_{T-1}, \ldots, s_{t+1}, X_{t+1}, s_t, s_{t-1}, X_{t-1}, \ldots, s_1, X_1) \\ = P(X_t|s_t) \tag{3}$$

The first problem we address is the evaluation problem.

The forward algorithm calculates the probability of being in state s_i at time t after the corresponding partial observation sequence given the HMM model λ. This defines the forward variable $\rho_t(i) = P(X_1, X_2, \ldots, X_t, i_t = s_i|\lambda)$ which is solved recursively as follows:

1. initiate the forward probabilities with the joint probability of state s_i and the initial observation X_1: $\rho_1(i) = \pi_i \Xi_i(X_1), 1 \leqslant i \leqslant K$;
2. calculate how state $q_{i'}$ is reached at time $t+1$ from the K possible states s_i, $i = 1, 2, \ldots, K$ at time t and sum the product over all the K possible states: $\rho_{t+1}(j) = \left[\sum_{i=1}^{K} \rho_t(i) b_{ij}\right] \Xi_j(X_{t+1})$ for $t = 1, 2, \ldots, T-1, 1 \leqslant j \leqslant K$
3. Finally, compute $P(X|\lambda) = \sum_{i=1}^{K} \rho_T(i)$.

The forward algorithm has a computational complexity of K^2T which is considerably less than a naive direct calculation approach. A graphical depiction of the forward algorithm can be observed in Fig. 4.

Next, the Viterbi algorithm aims to find the most likely progression of states that generated a given observation sequence in a certain HMM. Hence, it offers the solution to the decoding problem. This involves choosing the most likely states at each time t individually. Hence, the expected number of correct separate states is maximized. This is illustrated in Fig. 5. To perform this algorithm, we need to define the following:

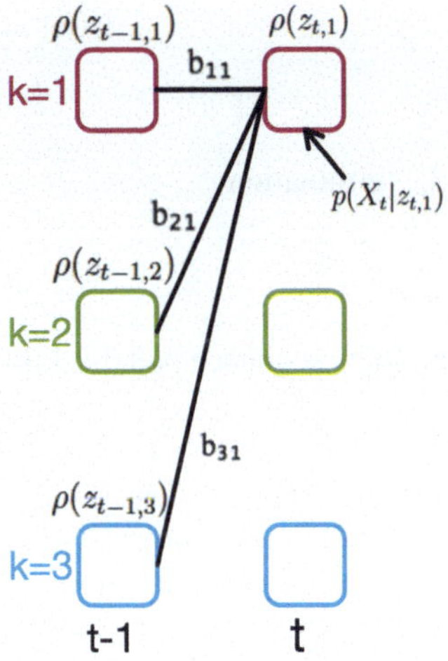

Fig. 4 Graphical representation of the evaluation of the ρ variable of the forward algorithm in an HMM lattice fragment

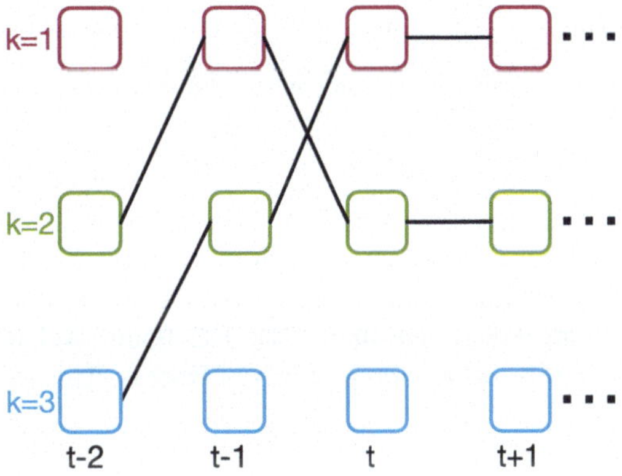

Fig. 5 Graphical representation of two probable pathways in an HMM lattice fragment. The objective of the Viterbi algorithm is to find the most likely one

$$\gamma_t(i) = P(i_t = s_i|X, \lambda) = \frac{\rho_t(i)\theta_t(i)}{p(X|\lambda)} \tag{4}$$

where $\gamma_t(i)$ is the probability of being in state s_i at time t given the observation sequence X and the HMM λ.

The main steps of the Viterbi algorithm can then be summarized as:

1. Initialization

$$\delta_1(i) = \pi_i \Xi_i(X_1), 1 \leqslant i \leqslant K \tag{5}$$

$$\psi_1(i) = 0 \tag{6}$$

2. Recursion

$$\text{For } 2 \leqslant t \leqslant T, 1 \leqslant i' \leqslant K \tag{7}$$

$$\delta_t(i') = \max_{1 \leqslant i \leqslant K} [\delta_{t-1}(i) b_{ii'}] \, \Xi_{i'}(X_t) \tag{8}$$

$$\psi_t(i') = \text{argmax}_{1 \leqslant i \leqslant K} [\delta_{t-1}(i) b_{ii'}] \tag{9}$$

3. Termination

$$P^* = \max_{1 \leqslant i \leqslant K} [\delta_T(i)] \, i_T^* = \text{argmax}_{1 \leqslant i \leqslant K} [\delta_T(i)] \tag{10}$$

4. State sequence path backtracking

$$i_t^* = \psi_{t+1}(i_{t+1}^*), \text{ for } t = T-1, T-2, \ldots, 1 \tag{11}$$

Finally, and in order to address the third HMM problem, we first describe another important algorithm. Similar to the forward algorithm, but now computing the tail probability of the partial observation from $t+1$ to the end, given that we are starting at state s_i at time t and model λ, is the backward algorithm. This has the variable $\theta_t(i) = P(X_{t+1}, X_{t+2}, \ldots, X_T, i_t = s_i | \lambda)$ and is solved as follows:

1. Compute an arbitrary initialization $\theta_T(i) = 1, 1 \leqslant i \leqslant K$;
2. $\theta_t(i) = \sum_{i'=1}^{K} b_{ii'} \Xi_{i'}(X_{t+1})$ for $t = T-1, T-2, \ldots, 1, 1 \leqslant i \leqslant K$

Figure 6 depicts the computation process of the backward algorithm in an HMM lattice structure. Together with the forward algorithm, this forms the forward-backward algorithm through consequent iteration. In the context of HMMs, the forward-backward algorithm is of extreme importance and is also known as the Baum Welch algorithm [71]. The Baum Welch algorithm is traditionally used to solve the estimation problem of HMMs. This iterative algorithm requires an initial

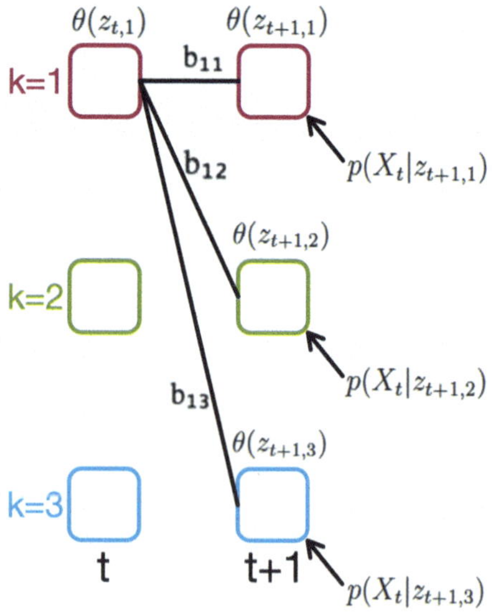

Fig. 6 Graphical representation of the evaluation of the θ variable of the backward algorithm in an HMM lattice fragment

random clustering of the data, is guaranteed to converge to more compact clusters at every step, and stops when the log-likelihood ratios no longer show significant changes [75].

In order to apply the Baum Welch algorithm, we must define

$$\varphi_t(i, i') = P(i_t = s_i, i_{t+1} = s_i'|X, \lambda) = \frac{\rho_t(i) b_{ii'} \Xi_{i'}(X_{t+1}) \theta_{t+1}(i')}{p(X|\lambda)} \tag{12}$$

where $\varphi_t(i, i')$ is the probability of path being in state s_i at time t and then transitioning at time $t + 1$ with $b_{ii'}$ to state s_i', given λ and X. $\rho_t(i)$ then considers the first observations ending at state s_i at time t, $\theta_{t+1}(i')$ the rest of the observation sequence, and $b_{ii'} \Xi_{i'}(X_{t+1})$ the transition to state $s_{i'}$ with observation X_{t+1} at time $t + 1$. Hence, $\gamma_t(i)$ may also be expressed as:

$$\gamma_t(i) = \sum_{i'=1}^{K} \varphi_t(i, i') \tag{13}$$

whereby $\sum_{t=1}^{T-1} \varphi_t(i, i')$ is the expected number of transitions made from s_i to $s_{i'}$ and $\sum_{t=1}^{T-1} \gamma_t(i)$ is the expected number of transitions made from s_i.

The general re-estimation formulas for the HMM parameters π, B, and Ξ are then:

1. $\bar{\pi}_i = \gamma_1(i), 1 \leqslant i \leqslant K$
2. $\bar{b}_{ii'} = \sum_{t=1}^{T-1} \varphi_t(i, i') / \sum_{t=1}^{T-1} \gamma_t(i)$
3. $\bar{\Xi}_{i'}(k) = \sum_{t=1, X_t=k}^{T} \gamma_t(i') / \sum_{t=1}^{T} \gamma_t(i')$

Oliver et al. [76] utilize an extension, layered HMMs to detect various activities like deskwork, phone conversations, presence, etc. The layered structure of their model makes it feasible to decouple different levels of analysis for training and inference. Each level in the hierarchy can be trained independently, with different feature vectors and time granularity. Once the system has been trained, inference can be carried out at any level of the hierarchy. One benefit of such a model is that each layer can be trained individually in isolation, and therefore the lowest layer that is most sensitive to environment noises and flickers can be retrained without touching the upper layers. HMMs and conditional random field (CRF) have been used in [77] to recognize seven different activities (leave house, toileting, showering, sleeping, preparing breakfast, preparing dinner, preparing a beverage) in a home setting. An HMM-based approach to recognize independent and joint activities among multiple residents in smart environments has been proposed in [78].

Nonetheless, HMMs suffer from some drawbacks that [79] aimed to overcome by introducing a new variant; namely, Switching Hidden Semi-Markov Model. This model supplements HMMs with a hierarchical structure to benefit from the natural hierarchy depicted by humans in activities. It also incorporates explicit state duration though the semi-HMM to address the violation of the Markovian assumption when the duration of the state is no longer geometric. The system reportedly outperforms both a traditional HMM as well as a hierarchical one.

2.2 Regression

Regression is often viewed as a variant of classification whereby the data or the variables at hand are, in contrast, of continuous nature. Specifically, *Regression* is the prediction of continuous labels given a set of labeled training data. It is also sometimes referred to as prediction and is closely related to classification. As a matter of fact, Eq. (1) can also be used to represent regression whereby $g()$ represents the regression function that is used to fit the data x to find out y. Notice that while the first assumption of the best function is a linear approximation, it is not always the case. Indeed, higher order approximations are usually used to better estimate the true distribution of the training data.

Given its nature of continuous predictions, it is not often used in the area of activity recognition due to the discrete nature of the data. Nonetheless, regression, in particular linear regression, remains one of the most traditional machine learning methods and the problem may be posed within a continuous framework for its use. For example, linear regression is used for classification of human activities in smart homes and inspires a new regression-tree-based activity forecasting algorithm in [80].

However, while linear regression is a powerful technique, it is not necessarily the most suitable in all cases. The best approach machine learning approach to be used is always dependent on the nature of the data itself. This is investigated in [81] whereby the authors argue that prior statistical analysis of the problem is imperative for choosing the best machine learning algorithm. They compare the use of random forests and linear regression finding out that the prior outperforms the latter due to the nonlinear nature of the data.

2.3 Clustering

A significant problem when tackling the activity recognition problem using supervised learning approaches is collecting ground truth information. Indeed, the large variety of possible activities makes their recognition in a supervised way challenging.

Since no labels are available in clustering, this presents an added challenge in finding homogeneous groups within the input data. The objective in such algorithms can be straightforwardly defined as: Finding homogeneous groups or clusters in data such that the intra-distance between the data points is minimized and the interdistance between the data homogeneous groups or clusters is maximized is known as *Clustering*.

The most famous clustering approach is using mixture models. Consider a set of N observation vectors $\mathcal{X} = \{\overrightarrow{\mathcal{X}}_1, \ldots, \overrightarrow{\mathcal{X}}_N\}$ represented in D-dimensional space where each vector $\overrightarrow{\mathcal{X}}_\varrho = (\mathcal{X}_{\varrho 1}, \ldots, \mathcal{X}_{\varrho D})$. If we assume that each vector $\overrightarrow{\mathcal{X}}_\varrho$ is generated from a finite mixture model with ϖ components, then the likelihood of the data is defined as:

$$p(\overrightarrow{\mathcal{X}}_\varrho|\kappa, \Lambda) = \sum_{\varsigma=1}^{\varpi} \kappa_\varsigma p(\overrightarrow{\mathcal{X}}_\varrho|\Lambda_\varsigma) \quad (14)$$

where $p(\overrightarrow{\mathcal{X}}_\varrho|\Lambda_\varsigma)$ is the mixture distribution at hand that is used to statistically model the observations or data $\mathcal{X}$, Λ_ς is the respective set of component parameters for the distribution, and κ_ς is the mixing coefficient of the mixture component ς with $\kappa = (\kappa_1, \ldots, \kappa_\varpi)$. The mixing coefficients vector follow constraints of positivity and unit summation resultant on the $\boldsymbol{\kappa}$. Each of the data observation vectors $\overrightarrow{\mathcal{X}}_\varrho$ is assigned to all of the mixture components with a responsibility or posterior probability $p(\varsigma|\overrightarrow{\mathcal{X}}_\varrho) \propto \kappa_\varsigma p(\overrightarrow{\mathcal{X}}_\varrho|\Lambda_\varsigma)$.

Clustering represents an attractive solution as it is easy to obtain unlabeled samples from routine experiments; they do not require human effort. This is also applicable for the problem at hand though more research can be invested in this particular area. For example, k-means algorithm is applied in [82] to cluster sensor readings collected from smart homes for activity recognition. Classification of non-separated activities within each cluster is then carried out by K-nearest neighbor classification approach. This also represents a system where a hybrid approach improves the overall classifier performance.

2.4 Miscellaneous

So far, we have presented papers in the literature that address the problem of activity recognition in smart buildings using supervised and unsupervised learning techniques. A summary of these papers can be observed in Table 1. On the other hand, semi-supervised learning techniques applied in [84] and [85] represent another learning approach aiming to address activity recognition issue. It exploits unlabeled data in order to improve model performance. For example, [83] introduce a method for human activity recognition that benefits from the structure and sequential properties of the training and testing data. In the training phase, a fraction of data labels has been obtained and used in a semi-supervised method for recognizing the user's activities. Label propagation has been used on a K-nearest neighbor graph to calculate the probability of the unlabeled data in each class in the training phase. These probabilities have been used to train an HMM

Table 1 A list of the papers detailed in this chapter for activity recognition in smart buildings with the respective machine learning (ML) technique utilized and the algorithm(s) used

Paper	ML technique	Algorithm(s)
[47]	Classification	Ontology modeling and semantic reasoning
[48]	Classification	Feature incremental random forests
[49]	Classification	Decision trees
[50]	Classification	Support vector machines/naive Bayes/hidden Markov models/conditional random fields
[52]	Classification	Incremental learning random forests
[53]	Classification	Emerging pattern technique
[78]	Classification	Hidden Markov models
[76]	Classification	Layered hidden Markov models
[77]	Classification	Hidden Markov models/conditional random field
[79]	Classification	Switching hidden semi-Markov model
[56]	Classification	Decision tables/decision trees (C4.5)/K-nearest neighbors/support vector machines/naive Bayes (with meta classifiers setup)
[60]	Classification	Naive Bayes
[62]	Classification	Decision trees (multiple accelerometers)
[61]	Classification	Convolutional neural networks and long short-term memory
[80]	Regression	Linear regression/regression-tree-based activity forecasting algorithm
[81]	Classification/regression	Random forests/linear regression
[82]	Clustering/classification	k-Means algorithm/K-nearest neighbors
[83]	Semi-supervised learning	K-nearest neighbors/hidden Markov model
[54]	Classification	Random forests/K-nearest neighbors/support vector machines/decision trees
[65]	Regression/classification	Support vector regression/recurrent neural network
[66]	Classification	Support vector machine/naive Bayes/logistic recognition/ recurrent neural network
[55]	Classification	Support vector machine
[58]	Classification	Kernel Fisher discriminant analysis/extreme learning machine
[57]	Classification	Gradient boosting/K-nearest neighbor/linear discriminant analysis/ random forest
[63]	Classification	Convolutional neural networks and long short-term memory
[64]	Classification	Convolutional neural networks/K-nearest neighbor/support vector machine
[59]	Classification	Naive Bayes/K-nearest neighbor/support vector machine

in a way that each of its hidden states corresponds to one class of activity. Some semi-supervised approaches have been based on active learning, also. For instance, different active learning strategies have been investigated in [86]. In particular, a dynamic k-means clustering approach has been proposed to discover unseen new activities spontaneously. These unseen activities are detected as outliers which make the clustering algorithm sensitive to the number of clusters that can increase at every iteration. The overall clustering error was recorded using an error function on the set of clusters defined as the sum of the Euclidean distances between the different data instances and the clusters centers. An objective function based on entropy is then defined to fetch the most informative data instances. The activities that were considered are cooking, sweeping, washing, and cleaning which were used for the passive learning. Three other activities, namely eating, sleeping, and talking, on the phone were left to the active learner to discover.

Some recent approaches have been based on transfer learning. In [87], for example, the authors proposed a feature-based approach to reuse learned knowledge form an original environment and tested it successfully to extract and transfer knowledge between two different smart home environments by considering only single-resident scenarios. The problem was formulated as classification task using SVM by matching the different features of the source and target environments. Two cases were considered. In the first one labeled datasets from both environments were supposed to be available. In the second one labeled data are available only in the source environment and the information from the target one is limited to sensor deployment considered as background knowledge.

Another issue refers to the features used. In any of the machine learning techniques, or any algorithm for that matter, the importance of extracted features to be used cannot be overstated. Indeed, some studies were carried out in [88, 89] to analyze the various features and their importance in activity recognition. This falls outside the scope of this chapter, but an interested reader is referred to the paper for further details.

Furthermore, in order to ensure the completeness of the activity recognition survey, it is noteworthy to mention that not all methods are dependent on machine learning techniques. For instance, [90–92] present other algorithms that do not fall under the scope of this survey. An interested reader is referred to [93] for a general reference on human action recognition.

3 Case Study

To evaluate the deployment of machine learning approaches in smart buildings in general and their potential in activity recognition, we present three recent methods for occupancy estimation that have been applied in an office H358 case study (see Chapter "Formalization of the Energy Management Problem and Related Issues"). Extensive work is currently conducted to apply these approaches for activity recognition. The proposed approaches are:

1. Estimating occupancy with a set of sensors, and possible manual labeling by an expert.
2. Estimating occupancy with a set of sensors without manual labeling but using a knowledge-based approach.
3. Estimating occupancy with a set of sensors with self-labeling by occupants (interactive learning).

These approaches depend mainly on collecting and analyzing data from non-intrusive sensors. The use of such sensors is based on the hypothesis that humans interact with their surroundings, i.e., performing some activities. It affects environmental conditions that can be in the form of CO2 concentration, moisture, temperature rise, or sound. As it is mentioned in Chapter "Formalization of the Energy Management Problem and Related Issues", different sensors exist in the H358 office, i.e., PIR motion sensors, CO2 sensors, indoor air temperature, and relative humidity sensor, pressure sensors, acoustic sensors, ultrasonic sensors, power consumption sensors, in order to define the occupancy level.

To perform the task of finding the number of occupants, a link needs to be observed between the office context and the number of occupants in it. The office context can be described as a collection of state variables, $S_t = [s_1, s_2, \ldots, s_n]_t$. This group of state variables S must characterize occupancy at each time step t.

A state variable can be presented as a feature, and therefore the features are displayed as a feature vector. Thus, the multidimensional space that includes all potential values of such a feature vector is the feature space. The underlying approach for the experiments is to formulate the classification problem as a mapping from a feature vector into a feature space that comprises several classes of occupancy. Therefore, the success of such an approach depends strongly on how useful (features which give maximum distinction between classes) the chosen features are. In this case, features are attributes from multiple sensors collected over a time interval. The selection of interval duration is highly context-dependent and has to be done according to the required granularity. The results presented here are based on an interval of $T_s = 30\,\text{min}$ (which has been referred here as one quantum).

From the large set of features discussed in [30], some of them may not be worth considering in order to achieve the target of occupancy classification. These features are the ones which, when added to the classification algorithm, make no difference to the overall output. In other words they are not useful enough for our purpose. For example, absolute humidity readings would be useless, as it is not representative of occupancy at all. Defining the most important features (sensors) is considered as a necessary study in a smart building application. It can give an essential conclusion for the required installation of the sensor in the buildings, which leads to minimizing the total cost.

Before any features are extracted for the training data, some basic preprocessing had to be done: application of an *outliers removal algorithm* and *interpolation* for non-existent data. The interpolation part is necessary for filling in missing values from the sensor data. Amayri et al. [30] concludes the most relevant features for the occupancy estimation problem in the office:

1. power consumption.
2. motion counter.
3. acoustic pressure recorded by a microphone.

These three features will be used in the three following experiments of applying machine learning techniques for occupancy estimation.

3.1 Estimating Occupancy with a Set of Sensors, and Possible Manual Labeling by an Expert

Let us start with the first experiment, where supervised learning has been deployed. Collecting the required training data has been done by counting occupancy manually using two video cameras in office H358. The average number of people visiting the office was registered every 30 min during the day.

Different supervised learning methods have been investigated (i.e., support vector machine, decision tree, random forest, linear regression). A *decision tree* based classification approach has been selected as our prediction model because it provides human-readable results that can be analyzed and easily adapted. Providing decision rules is one crucial aspect from the energy point of view to generalize the model for another similar context.

Power consumption, motion counter, and acoustic pressure are the main features for building our model. Five occupancy levels have been chosen to generate decision trees due to the maximum number of occupants met while collecting the dataset.

3.2 Resulting Occupancy Estimators

From the collected data in the office H358, a training dataset covering 11 days from 04-May-2015 to 14-May-2015 has been used. Moreover, a validation dataset is collected over 4 days from 17-May-2015 to 21-May-2015. Figure 7 shows the result obtained from the decision tree and random forest, considering the three features. It leads to occupancy estimation with an accuracy of 81.7% and an average error of 0.26 person, while random forest accuracy is 84%, and the average error is 0.26 person (Table 2).

The above results indicate that using the decision tree and random forest rules give quite a reasonable estimation of occupancy. Because of the limitation of the need to have labeled training data when deploying supervised learning, unsupervised learning based on collecting knowledge and questioning will be discussed in the next section. It will help to facilitate and generalize the occupancy estimation process.

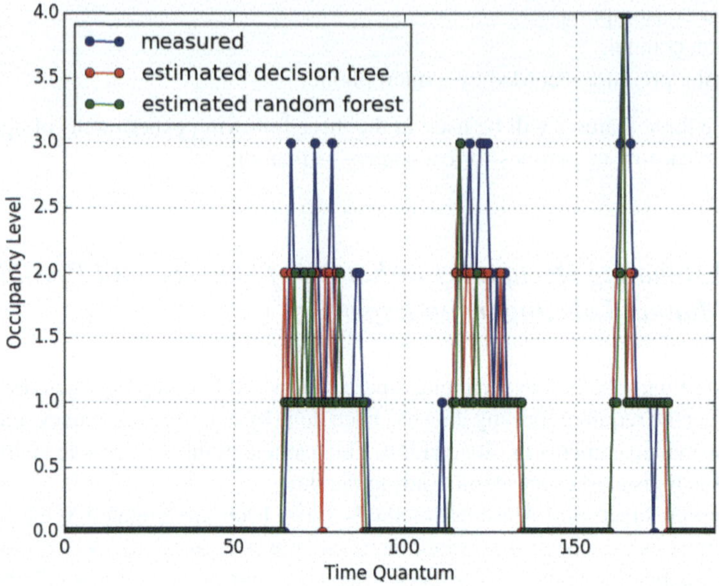

Fig. 7 Occupancy estimation from DT using three features

Table 2 Decision tree classification results after selecting main features

	Average error decision tree	Support
Class 1	0.02	132
Class 2	0.49	22
Class 3	0.88	23
Class 4	1.27	11
Class 5	1.75	4
Avg/total	0.26	192

3.3 Designing Estimators from Knowledge and Adjusting from Data

Similarly to the first approach, designing estimators from knowledge is based on sensor data and knowledge coming, respectively, from observations and questionnaires to build the estimation model. The proposed technique relies on a Bayesian Network (BN) algorithm to model human behavior with probabilistic cause-effect relations and states based on knowledge and questionnaire [94, 95].

The same case study of an office (H358) is considered as a simple and essential one-zone context with lots of sensors. Motion detection, power consumption, and acoustic pressure recorded by a microphone are used to feed this model. Collecting occupancy and activity feedbacks is very easy in the office context. Besides, there is a facility of questioning the occupants during design and validation periods of occupancy model. Unsupervised learning algorithms are used to solve problems where the solution is not known. In this case, usually, the structure is derived by clustering the sensor data based on relationships among the variables. While in the case of collecting training period, it becomes similar to supervised learning methods with the difference in prediction techniques. For each feature, different levels have been considered. For example, the power consumption values discretize in three levels: low consumption, medium consumption, and high consumption, or L, M, and H, respectively. It gives a probability table with nine values. The probability table for power consumption has been defined by proposing different questions to the office occupants. For example: when occupants are arriving and leaving the office? What is the average time for using the laptop during the working hours? According to the user answers, the conditional probabilities are either calculated or filled directly in the tables. The same process can be repeated for the recorded signal from the microphone. At the same time, two different levels have been defined for the microphone low acoustic pressure and high acoustic pressure or L and H, respectively, see Fig. 8. Three occupancy levels have been considered to generate a Bayesian Network (BN): Low, Medium, and a High number of occupants. While the probabilities table for motion counter has been suggested according to the general knowledge for three different cases, low motion, medium motion, and high motion, or L, M, and H, respectively. Figure 8 shows the results obtained from the Bayesian network for three levels and three main features. Both actual and estimated occupancy profiles have been plotted in a graph with the number of occupants and time relations (quantum time was 30 min). The accuracy achieved from the Bayesian network was 91% (the number of correctly estimated points divided by the total number of points), and the average error was 0.08 persons. Table 3 represents the average error values for each class of estimation. While "support" indicates the number of events (sensor data each quantum time) in each class, and average support indicates the sum of all events in the three classes (Fig. 9).

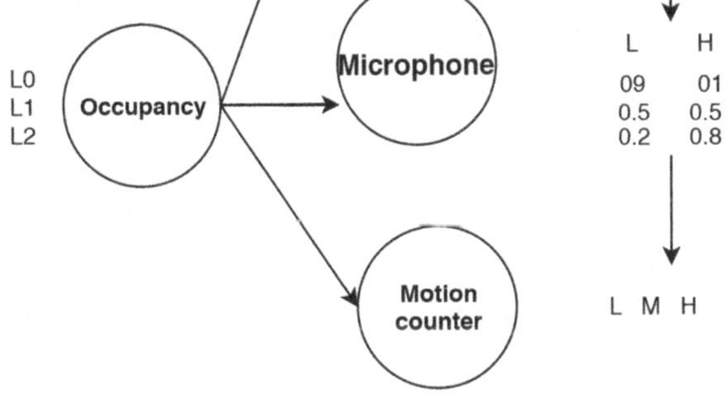

Fig. 8 Bayesian network (BN) structure of an office H358

Table 3 Bayesian network estimation results

Classes	Accuracy-BN	Average error-BN	Support
Class 1	97%	0.001	400
Class 2	84%	0.2	170
Class 3	79%	0.5	54
Average	91%	0.09	624

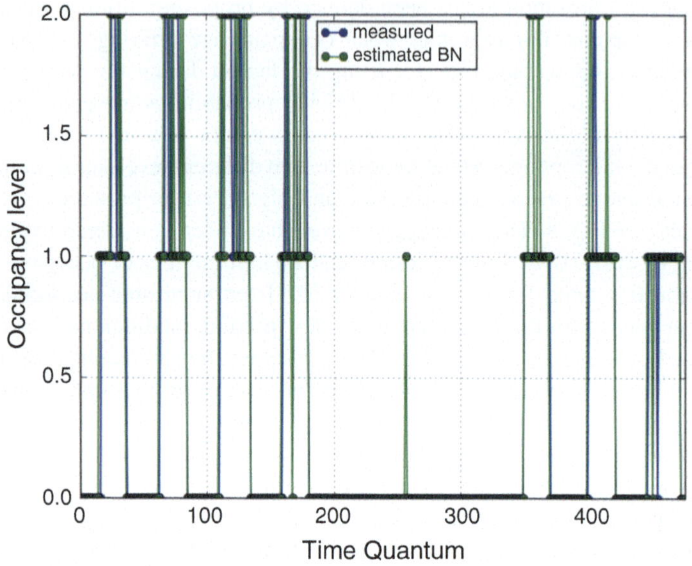

Fig. 9 Occupancy estimation from Bayesian network

Using the knowledge domain and questionnaire with data sensors in the unsupervised learning method is more flexible and open for different types of applications, with acceptable average errors for occupancy estimations. Besides, avoiding the use of video cameras has been achieved. This approach can be used widely in different contexts. Still, due to a few possibilities to validate the estimation model and poor performance in some testing period, a new innovative approach is proposed in the next section. It depends on estimating occupancy with a set of sensors, and self-labeling by occupants.

3.4 Designing Estimators from Interactive Learning

A novel way of supervised learning is analyzed to estimate the occupancy in a room where actual occupancy is interactively requested to occupants when it is the most relevant to limit the number of interactions. Occupancy estimation algorithm relies on machine learning: it uses information gathered from occupants. In this section, an interactive technique has been investigated to solve the problem of getting the required labels used in the supervised method. In practical applications, the limitation arises due to the occupant's privacy issues. Accurately estimating occupancy with a set of sensors and self-labeling by interaction with occupants are the main goals of this section.

3.4.1 The Principle of Interactive Learning

Obtaining training data is a challenging task for smart home applications in general and activity recognition in particular. Some approaches have been proposed to involve the occupants to collect informative training data. An interesting approach called interactive learning has been proposed in [96]. Interactive learning is a process involving an exchange of information with the users to collect some essential data according to the problem context. In supervised learning methods, which are widely used in a lot of applications, the problem of the required target arises in the determination of the number of occupants, i.e., the labeling issue is usually tackled using video cameras. Utilizing a camera is generally not acceptable in many places to respect the privacy of occupants. Interactive learning is an extension of supervised learning that determines the occupancy by collecting the required labeling from the occupants themselves. The problem statement of occupancy estimation has been explained in [96].

Three rules are considered to determine whether an interaction space (*ask*) is potentially useful or not:

1. The density of the neighborhood: It is the number of existing records in the neighborhood of a potential *ask*. The neighborhood is defined by the infinite distance with a radius equal to one, because of the normalization. The record is a vector of features obtained in which values are obtained from the sensors. The neighborhood can be modified according to $\epsilon \in [0, 1]$.
2. The classifier estimation error in the neighborhood of the potential *ask* leads to the concept of neighborhood quality. If the classifier estimation error is very high for a record, this record is removed from the neighborhood because of the poor quality. $E_r \in [1, 2)$ typically is an error ratio that can be adjusted. However, a value smaller than 1 means a record is considered as good. Conversely, if E_r is big, equal to 2, for instance, it means you accept error twice as big as the average error. Theoretically, E_r belongs to $[0, \infty)$, but it is limited in our experiments to 2.
3. The minimum class weight: i.e., the minimum number of records for each class. The minimum class weight, weight(class x) $< C_w$, which can be adjusted according to the problem.

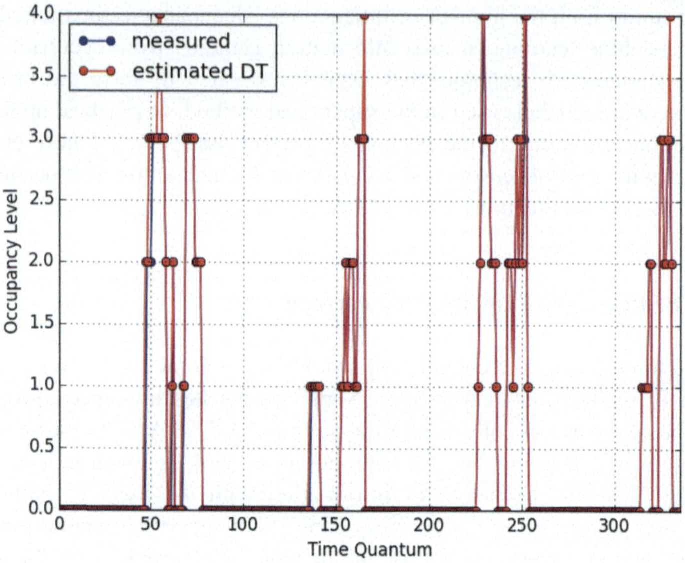

Fig. 10 Occupancy estimation from interactive learning

Table 4 Number of asks

Day	1	2	3	4	5	6	7	8	9	10
Number of *asks*	10	4	0	0	2	1	2	1	0	0

All the potential *asks* that satisfy the above three rules are asked to the occupants in order to become an additional record, possibly. The three previous rules have been checked with each new record. As a first validation, the occupant reaction has to be taken into account as a response probability whether the occupants answer or not. In a given context, the number of asks relies on the classifier used for estimation occupancy. To evaluate the interactive approach, we deploy the decision tree to compare it with the manual label approach. According to our study in [96, 97], Five occupancy levels to generate decision tree with an average error of 0.03 (see Fig. 10). Decision tree needs 21 *asks* for training data to build an acceptable estimator see the following Table 4.

Occupancy estimation using decision tree and interactive learning with an average error 0.03 person is more efficient than using decision tree and manually labeling from the video camera with an average error of 0.2 person. The precise answers to the questions can explain this improvement in occupancy estimation results. An occupant has replied to them during a training period of the decision tree. While in manually labeling from a video camera, average values of occupancy have been obtained, with some human mistakes during labeling. Probably the average error will decrease if the end-user does not feel concerned by the estimation process.

Table 5 Knowledge based vs manual labeling vs interactive learning occupancy estimation comparison

Case study office H358	Supervised learning	Interactive learning	Based knowledge learning
Average error	0.18	0.03	0.19
Requirements	Labels from video, keyboard	Frequent questions	Collecting knowledge from questioning and observation
Adaptation with high number of occupants	+	++	+/−

4 Conclusion

The Internet-of-Things (IoT) revolution has provided a variety of affordable sensors that new buildings are equipped with as well as data acquisition devices, and cloud storage. This has resulted in an unprecedented generation of raw data from sensors and smart meters. Many data mining approaches and machine learning techniques have been proposed to extract hidden knowledge from these data and then to build learning machines for a variety of applications and tasks. Activity recognition in smart buildings is one of the tasks that received a lot of attention due to its importance in energy management systems, for instance. The goal of this chapter was to review a variety of machine learning techniques that have been applied for activity recognition. Moreover, a case study and a methodology that concern occupancy estimation and that can be easily adopted for activity recognition have been presented and discussed. The results in this case study lead to the conclusion that the interactive learning approach is more efficient for occupancy estimation than the other methods taking into account the context. Two points can explain occupancy estimation improvement using interactive learning: firstly, the probability of making some human mistakes during manually labeling while using the video camera; secondly, the training period cannot be sufficient by missing some cases from the studied area. Using the ask technique considers all the events that occur when a new question is sent for each unique and different situation. This allows also to take into account the quality of the training data as deeply discussed in [98]. Interactive learning is the primary step to collect knowledge about the relations between user behavior and energy use. Moreover, its deployment allows involving occupants and increasing their awareness of energy systems. It depicts the future vision to develop energy systems, and it presents how much it is essential to put occupants in the energy process loop (Table 5).

Acknowledgments This work has been supported by the Fonds de Recherche du Quebec—Nature et Technologies (FRQNT) and the Natural Sciences and Engineering Research Council of Canada (NSERC). This work benefits from the support of the INVOLVED ANR-14-CE22-0020-01 project of the French National Research Agency. It is also supported by the French National Research Agency in the framework of the EquipEx program AmiQual4Home ANR -11-EQPX-00 and the Investissements d'avenir program (ANR-15-IDEX-02) as the cross disciplinary program Eco-SESA.

References

1. T. Labeodan, W. Zeiler, G. Boxem, Y. Zhao, Occupancy measurement in commercial office buildings for demand-driven control applications - a survey and detection system evaluation. Energy Build. **93**, 303–314 (2015)
2. W. Khalid, I.A. Shah, Q. Mehfooz, N. Irshad, W. Mahmood, Reduction in building energy requirements by modern energy conservation techniques, in *2015 Power Generation System and Renewable Energy Technologies (PGSRET)*, June 2015, pp. 1–5
3. D.-M. Petroeanu, G. Carueaeu, N.L. Carueaeu, A. Parjan, A review of the recent developments in integrating machine learning models with sensor devices in the smart buildings sector with a view to attaining enhanced sensing, energy efficiency, and optimal building management. Energies **12**(24), 4745 (2019)
4. J.L. Crowley, J. Coutaz, An ecological view of smart home technologies, in *AMI 2015 - European Conference on Ambient Intelligence, Athens*, Nov 2015
5. A. De Paola, M. Ortolani, G. Lo Re, G. Anastasi, S.K. Das, Intelligent management systems for energy efficiency in buildings: a survey. ACM Comput. Surv. **47**(1), 1–38 (2014)
6. L. Jiang, D.-Y. Liu, B. Yang, Smart home research, in *Proceedings of 2004 International Conference on Machine Learning and Cybernetics*, vol. 2 (2004), pp. 659–663
7. E. Soltanaghaei, K. Whitehouse, Walksense: classifying home occupancy states using walkway sensing, in *Proceedings of the 3rd ACM International Conference on Systems for Energy-Efficient Built Environments* (2016), pp. 167–176
8. J. Scott, A. Bernheim Brush, J. Krumm, B. Meyers, M. Hazas, S. Hodges, N. Villar, Preheat: controlling home heating using occupancy prediction, in *Proceedings of the 13th International Conference on Ubiquitous Computing* (Association for Computing Machinery, New York, 2011), pp. 281–290
9. C. Sarkar, S.N.A.U. Nambi, V. Prasad, iLTC: achieving individual comfort in shared spaces, in *Proceedings of the 2016 International Conference on Embedded Wireless Systems and Networks* (Junction Publishing, New York, 2016), pp. 65–76
10. Y. Agarwal, B. Balaji, R. Gupta, J. Lyles, M. Wei, T. Weng, Occupancy-driven energy management for smart building automation, in *Proceedings of the 2nd ACM Workshop on Embedded Sensing Systems for Energy-Efficiency in Building* (Association for Computing Machinery, New York, 2010), pp. 1–6
11. K. Basu, L. Hawarah, N. Arghira, H. Joumaa, S. Ploix, A prediction system for home appliance usage. Energy Build. **67**, 668–679 (2013)
12. H.A. Aglan, Predictive model for CO 2 generation and decay in building envelopes. J. Appl. Phys. **93**(2), 1287–1290 (2003)
13. D. Yan, W. Oabrien, T. Hong, X. Feng, H.B. Gunay, F. Tahmasebi, A. Mahdavi, Occupant behavior modeling for building performance simulation: current state and future challenges. Energy Build. **107**, 264–278 (2015)
14. K. Schakib-Ekbatan, F.Z. Cakici, M. Schweiker, A. Wagner, Does the occupant behavior match the energy concept of the building? Analysis of a German naturally ventilated office building. Build. Environ. **84**, 142–150 (2015)
15. P. Hoes, J. Hensen, M. Loomans, B. de Vries, D. Bourgeois, User behavior in whole building simulation. Energy Build. **41**(3), 295–302 (2009)
16. A. Ebadat, G. Bottegal, D. Varagnolo, B. Wahlberg, K.H. Johansson, Estimation of building occupancy levels through environmental signals deconvolution, in *Proceedings of the 5th ACM Workshop on Embedded Systems for Energy-Efficient Buildings* (Association for Computing Machinery, New York, 2013), pp. 1–8
17. M. Milenkovic, O. Amft, An opportunistic activity-sensing approach to save energy in office buildings, in *Proceedings of the Fourth International Conference on Future Energy Systems* (Association for Computing Machinery, New York, 2013), pp. 247–258
18. H. Nguyen, M. Rahmanpour, N. Manouchehri, K. Maanicshah, M. Amayri, N. Bouguila, A statistical approach for unsupervised occupancy detection and estimation in smart buildings, in

2019 IEEE International Smart Cities Conference, ISC2 2019, Casablanca, Morocco, 14–17 Oct 2019 (IEEE, Piscataway, 2019), pp. 414–419
19. D. Ankam, N. Bouguila, M. Amayri, Beta-Liouville regression and applications, in *6th International Conference on Control, Decision and Information Technologies, CoDIT 2019, Paris, France, 23–26 April 2019* (IEEE, Piscataway, 2019), pp. 1740–1745
20. N. Zamzami, M. Amayri, N. Bouguila, S. Ploix, Online clustering for estimating occupancy in an office setting, in *28th IEEE International Symposium on Industrial Electronics, ISIE 2019, Vancouver, BC, Canada, 12–14 June 2019* (IEEE, Piscataway, 2019), pp. 2195–2200
21. N. Manouchehri, J.S. Kalsi, M. Amayri, N. Bouguila, Finite two-dimensional beta mixture models: model selection and applications, in *28th IEEE International Symposium on Industrial Electronics, ISIE 2019, Vancouver, BC, Canada, 12–14 June 2019* (IEEE, Piscataway, 2019), pp. 1407–1412
22. M. Amayri, S. Ploix, F. Najar, N. Bouguila, F. Wurtz, A statistical process control chart approach for occupancy estimation in smart buildings, in *IEEE Symposium Series on Computational Intelligence, SSCI 2019, Xiamen, China, 6–9 Dec 2019* (IEEE, Piscataway, 2019), pp. 1729–1734
23. K. Padmanabh, A. Malikarjuna, S. Sen, S.P. Katru, A. Kumar, S.K. Vuppala, S. Paul, iSense: a wireless sensor network based conference room management system, in *Proceedings of the First ACM Workshop on Embedded Sensing Systems for Energy-Efficiency in Buildings* (Association for Computing Machinery, New York, 2009), pp. 37–42
24. T.A. Nguyen, M. Aiello, Beyond indoor presence monitoring with simple sensors, in *Proceedings PECCS* (2012), pp. 5–14
25. C. Miller, Z. Nagy, A. Schlueter, A review of unsupervised statistical learning and visual analytics techniques applied to performance analysis of non-residential buildings. Renew. Sust. Energ. Rev. **81**, 1365–1377 (2018)
26. B. Dong, B. Andrews, K.P. Lam, M. Höynck, R. Zhang, Y.-S. Chiou, D. Benitez, An information technology enabled sustainability test-bed (itest) for occupancy detection through an environmental sensing network. Energy Build. **42**(7), 1038–1046 (2010)
27. R. Nasfi, M. Amayri, N. Bouguila, A novel approach for modeling positive vectors with inverted Dirichlet-based hidden Markov models. Knowl. Based Syst. **192**, 105335 (2020)
28. Z. Li, Y. Jiang, Q. Wei et al., Survey on energy consumption of air conditioning in summer in a residential building in Beijing. J. Heat. Vent. Air Cond. **37**(4), 46–51 (2007)
29. A.L. Pisello, F. Asdrubali, Human-based energy retrofits in residential buildings: a cost-effective alternative to traditional physical strategies. Appl. Energy **133**, 224–235 (2014)
30. M. Amayri, A. Arora, S. Ploix, S. Bandhyopadyay, Q.-D. Ngo, V.R. Badarla, Estimating occupancy in heterogeneous sensor environment. Energy Build. **129**, 46–58 (2016)
31. D. Beymer, K. Konolige, Real-time tracking of multiple people using continuous detection, in *IEEE Frame Rate Workshop* (1999), pp. 1–8
32. O. Ardakanian, A. Bhattacharya, D. Culler, Non-intrusive techniques for establishing occupancy related energy savings in commercial buildings, in *Proceedings of the 3rd ACM International Conference on Systems for Energy-Efficient Built Environments* (Association for Computing Machinery, New York, 2016), pp. 21–30. https://doi.org/10.1145/2993422.2993574
33. R.H. Dodier, G.P. Henze, D.K. Tiller, X. Guo, Building occupancy detection through sensor belief networks. Energy Build. **38**(9), 1033–1043 (2006). http://www.sciencedirect.com/science/article/pii/S0378778806000028
34. K. Basu, V. Debusschere, S. Bacha, U. Maulik, S. Bondyopadhyay, Nonintrusive load monitoring: a temporal multilabel classification approach. IEEE Trans. Industr. Inform. **11**(1), 262–270 (2015)
35. S.T.M. Bourobou, Y. Yoo, User activity recognition in smart homes using pattern clustering applied to temporal ANN algorithm. Sensors (Basel, Switzerland) **15**(5), 11953–11971 (2015)
36. S. Daoaca, T. Hong, Occupancy schedules learning process through a data mining framework. Energy Build. **88**, 395–408 (2015)

37. L. Chen, J. Hoey, C.D. Nugent, D.J. Cook, Z. Yu, Sensor-based activity recognition. IEEE Trans. Syst. Man Cybern. Part C Appl. Rev. **42**(6), 790–808 (2012)
38. L.M. Candanedo, V. Feldheim, Accurate occupancy detection of an office room from light, temperature, humidity and CO2 measurements using statistical learning models. Energy Build. **112**, 28–39 (2016)
39. F. Pernkopf, J. Bilmes, Efficient heuristics for discriminative structure learning of Bayesian network classifiers. J. Mach. Learn. Res. **11**, 2323–2360 (2010)
40. N. Bouguila, Hybrid generative/discriminative approaches for proportional data modeling and classification. IEEE Trans. Knowl. Data Eng. **24**(12), 2184–2202 (2012)
41. N. Bouguila, O. Amayri, A discrete mixture-based kernel for SVMs: application to spam and image categorization. Inf. Process. Manag. **45**(6), 631–642 (2009)
42. N. Bouguila, Deriving kernels from generalized Dirichlet mixture models and applications. Inf. Process. Manag. **49**(1), 123–137 (2013)
43. T. Bdiri, N. Bouguila, Bayesian learning of inverted Dirichlet mixtures for SVM kernels generation. Neural Comput. Appl. **23**(5), 1443–1458 (2013)
44. A.K. Jain, M.N. Murty, P.J. Flynn, Data clustering: a review. ACM Comput. Surv. **31**(3), 264–323 (1999). https://doi.org/10.1145/331499.331504
45. X.J. Zhu, Semi-supervised learning literature survey. University of Wisconsin-Madison, Department of Computer Sciences, Tech. Rep., 2005
46. L. Deng, D. Yu, Deep learning: Methods and applications. Found. Trends Signal Process. **7**(3–4), 197–387 (2014)
47. L. Chen, C.D. Nugent, H. Wang, A knowledge-driven approach to activity recognition in smart homes. IEEE Trans. Knowl. Data Eng. **24**, 961–974 (1970). https://doi.org/10.1109/tkde.2011.51
48. C. Hu, Y. Chen, X. Peng, H. Yu, C. Gao, L. Hu, A novel feature incremental learning method for sensor-based activity recognition. IEEE Trans. Knowl. Data Eng. **31**(6), 1038–1050 (2019)
49. U. Maurer, A. Smailagic, D.P. Siewiorek, M. Deisher, Activity recognition and monitoring using multiple sensors on different body positions, in *International Workshop on Wearable and Implantable Body Sensor Networks (BSN'06)* (2006), pp. 4–116
50. N.C. Krishnan, D.J. Cook, Activity recognition on streaming sensor data. Pervasive Mob. Comput. **10**, 138–154 (2014)
51. W. Fan, N. Bouguila, Online learning of a Dirichlet process mixture of Beta-Liouville distributions via variational inference. IEEE Trans. Neural Networks Learn. Syst. **24**(11), 1850–1862 (2013)
52. C. Hu, Y. Chen, L. Hu, X. Peng, A novel random forests based class incremental learning method for activity recognition. Pattern Recognit. **78**, 277–290 (1970). https://doi.org/10.1016/j.patcog.2018.01.025
53. T. Gu, L. Wang, Z. Wu, X. Tao, J. Lu, A pattern mining approach to sensor-based human activity recognition. IEEE Trans. Knowl. Data Eng. **23**, 1359–1372 (1970). https://doi.org/10.1109/tkde.2010.184
54. M. Raeiszadeh, H. Tahayori, A. Visconti, Discovering varying patterns of normal and interleaved ADLs in smart homes. Appl. Intell. **49**(12), 4175–4188 (2019). https://doi.org/10.1007/s10489-019-01493-6
55. W. Li, B. Tan, R. Piechocki, Passive radar for opportunistic monitoring in e-health applications. IEEE J. Transl. Eng. Health Med. **6**, 1–10 (2018)
56. N. Ravi, N. Dandekar, P. Mysore, M.L. Littman, Activity recognition from accelerometer data, in *Proceedings of the 17th Conference on Innovative Applications of Artificial Intelligence - Volume 3*, ser. IAAI'05 (AAAI Press, 2005), pp. 1541–1546. http://dl.acm.org/citation.cfm?id=1620092.1620107
57. P.S. Colin Brennan, G.W. Taylor, Designing learned CO_2-based occupancy estimation in smart buildings. IET Wirel. Sensor Syst. **8**, 249–255(6) (2018)
58. Y. Tian, X. Wang, L. Chen, Z. Liu, Wearable sensor-based human activity recognition via two-layer diversity-enhanced multiclassifier recognition method. Sensors **19**(9), 2039 (2019). http://dx.doi.org/10.3390/s19092039

59. Y. Liu, L. Nie, L. Liu, D.S. Rosenblum, From action to activity: sensor-based activity recognition. Neurocomputing **181**, 108–115 (2016). Big Data Driven Intelligent Transportation Systems
60. E.M. Tapia, S.S. Intille, K. Larson, Activity recognition in the home using simple and ubiquitous sensors, in *Pervasive Computing*, ed. by A. Ferscha, F. Mattern (Springer, Berlin, 2004), pp. 158–175
61. F. Ordóñez, D. Roggen, Deep convolutional and LSTM recurrent neural networks for multimodal wearable activity recognition. Sensors **16**(1), 115 (2016)
62. L. Bao, S.S. Intille, Activity recognition from user-annotated acceleration data, in *Pervasive Computing*, ed. by A. Ferscha, F. Mattern (Springer, Berlin, 2004), pp. 1–17
63. H. Yu, G. Pan, M. Pan, C. Li, W. Jia, L. Zhang, M. Sun, A hierarchical deep fusion framework for egocentric activity recognition using a wearable hybrid sensor system. Sensors (Basel, Switzerland) **19**(3), 546 (2019). https://www.ncbi.nlm.nih.gov/pubmed/30696100
64. X. Guo, R. Su, C. Hu, X. Ye, H. Wu, K. Nakamura, A single feature for human activity recognition using two-dimensional acoustic array. Appl. Phys. Lett. **114**(21), 214101 (2019)
65. F.A. Machot, A.H. Mosa, M. Ali, K. Kyamakya, Activity recognition in sensor data streams for active and assisted living environments. IEEE Trans. Circuits Syst. Video Technol. **28**(10), 2933–2945 (2018)
66. H. Zhao, Q. Hua, H.-B. Chen, Y. Ye, H. Wang, S.X.-D. Tan, E. Tlelo-Cuautle, Thermal-sensor-based occupancy detection for smart buildings using machine-learning methods. ACM Trans. Des. Autom. Electron. Syst. **23**(4) (2018). https://doi.org/10.1145/3200904
67. D.J. Patterson, D. Fox, H. Kautz, M. Philipose, Fine-grained activity recognition by aggregating abstract object usage, in *Ninth IEEE International Symposium on Wearable Computers (ISWC'05)*, Oct 2005, pp. 44–51
68. S. Ali, N. Bouguila, Variational learning of Beta-Liouville hidden Markov models for infrared action recognition, in *The IEEE Conference on Computer Vision and Pattern Recognition (CVPR) Workshops*, June 2019
69. L.R. Rabiner, A tutorial on hidden Markov models and selected applications in speech recognition. Proc. IEEE **77**(2), 257–286 (1989)
70. C.M. Bishop, *Pattern Recognition and Machine Learning (Information Science and Statistics)* (Springer, Berlin, 2006)
71. L. Rabiner, B. Juang, An introduction to hidden Markov models. IEEE ASSP Mag. **3**(1), 4–16 (1986)
72. M. Rodriguez, C. Orrite, C. Medrano, D. Makris, One-shot learning of human activity with an MAP adapted GMM and simplex-HMM. IEEE Trans. Cybern. **47**(7), 1769–1780 (2017)
73. M. Wang, S. Abdelfattah, N. Moustafa, J. Hu, Deep gaussian mixture-hidden Markov model for classification of EEG signals. IEEE Trans. Emerg. Top. Comput. Intell. **2**(4), 278–287 (2018)
74. J. Bilmes, A gentle tutorial of the EM algorithm and its application to parameter estimation for Gaussian mixture and hidden Markov models. Int. Comput. Sci. Inst. **4**, 126 (1998)
75. J.A. Hartigan, *Clustering Algorithms*, 99th edn. (Wiley, New York, 1975)
76. N. Oliver, A. Garg, E. Horvitz, Layered representations for learning and inferring office activity from multiple sensory channels. Comput. Vis. Image Underst. **96**(2), 163–180 (2004). Special Issue on Event Detection in Video
77. T. van Kasteren, A. Noulas, G. Englebienne, B. Kröse, *Accurate Activity Recognition in a Home Setting* (Association for Computing Machinery, New York, 2008), pp. 1–9. https://doi.org/10.1145/1409635.1409637
78. G. Singla, D.J. Cook, M. Schmitter-Edgecombe, Recognizing independent and joint activities among multiple residents in smart environments. J. Ambient Intell. Humaniz. Comput. **1**(1), 57–63 (2010)
79. T.V. Duong, H.H. Bui, D.Q. Phung, S. Venkatesh, Activity recognition and abnormality detection with the switching hidden semi-Markov model, in *2005 IEEE Computer Society Conference on Computer Vision and Pattern Recognition (CVPR'05)*, vol. 1 (2005), pp. 838–845

80. B. Minor, D.J. Cook, Regression tree classification for activity prediction in smart homes, in *Proceedings of the 2014 ACM International Joint Conference on Pervasive and Ubiquitous Computing: Adjunct Publication* (Association for Computing Machinery, New York, 2014), pp. 441–450
81. A. Tsanas, A. Xifara, Accurate quantitative estimation of energy performance of residential buildings using statistical machine learning tools. Energy Build. **49**, 560–567 (1970)
82. L.G. Fahad, S.F. Tahir, M. Rajarajan, Activity recognition in smart homes using clustering based classification, in *2014 22nd International Conference on Pattern Recognition*, Aug 2014, pp. 1348–1353
83. M. Stikic, D. Larlus, B. Schiele, Multi-graph based semi-supervised learning for activity recognition, in *2009 International Symposium on Wearable Computers*, Sept 2009, pp. 85–92
84. N. Bouguila, Hybrid generative/discriminative approaches for proportional data modeling and classification. IEEE Trans. Knowl. Data Eng. **24**(12), 2184–2202 (2012)
85. B.M. Shahshahani, D.A. Landgrebe, The effect of unlabeled samples in reducing the small sample size problem and mitigating the Hughes phenomenon. IEEE Trans. Geosci. Remote Sens. **32**(5), 1087–1095 (1994)
86. H.M.S. Hossain, M.A.A.H. Khan, N. Roy, Active learning enabled activity recognition. Pervasive Mob. Comput. **38**, 312–330 (2017). Special Issue IEEE International Conference on Pervasive Computing and Communications (PerCom) 2016. http://www.sciencedirect.com/science/article/pii/S1574119216302073
87. Y. Chiang, C. Lu, J.Y. Hsu, A feature-based knowledge transfer framework for cross-environment activity recognition toward smart home applications. IEEE Trans. Hum. Mach. Syst. **47**(3), 310–322 (2017)
88. T. Huynh, B. Schiele, Analyzing features for activity recognition, in *Proceedings of the 2005 Joint Conference on Smart Objects and Ambient Intelligence: Innovative Context-aware Services: Usages and Technologies, ser. sOc-EUSAI '05* (ACM, New York, 2005), pp. 159–163
89. Q. Zhu, Z. Chen, Y.C. Soh, Smartphone-based human activity recognition in buildings using locality-constrained linear coding, in *2015 IEEE 10th Conference on Industrial Electronics and Applications (ICIEA)* (2015), pp. 214–219
90. E. Kim, S. Helal, D. Cook, Human activity recognition and pattern discovery. IEEE Pervasive Comput. **9**(1), 48–53 (2010)
91. D. Tran, A. Sorokin, Human activity recognition with metric learning, in *Computer Vision – ECCV 2008*, ed. by D. Forsyth, P. Torr, A. Zisserman (Springer, Berlin, 2008), pp. 548–561
92. N. Oliver, E. Horvitz, A. Garg, Layered representations for human activity recognition, in *Proceedings of Fourth IEEE International Conference on Multimodal Interfaces*, Oct 2002, pp. 3–8
93. O.D. Lara, M.A. Labrador, A survey on human activity recognition using wearable sensors. IEEE Commun. Surv. Tutor. **15**(3), 1192–1209 (2013)
94. M. Amayri, S. Ploix, H. Kazimi, Q. Ngo, A. Safadi, Estimating occupancy from measurements and knowledge using Bayesian network for energy management. Sensor **7**, 53932–53944 (2019)
95. M. Amayri, S. Ploix, Q.-D. Ngod, Estimating occupancy from measurement and knowledge with Bayesian networks, in *2016 International Conference on Computational Science and Computational Intelligence (CSCI)* (2016)
96. M. Amayri, S. Ploix, N. Bouguila, F. Wurtz, Estimating occupancy using interactive learning with a sensor environment: real-time experiments. IEEE Access **7**, 53932–53944 (2019)
97. M. Amayri, S. Ploix, P. Reignier, S. Bandyopadhyay, Towards interactive learning for occupancy estimation, in *ICAI'16 - International Conference on Artificial Intelligence (As Part of WORLDCOMP'16 - World Congress in Computer Science, Computer Engineering and Applied Computing)*, Las Vegas, July 2016. https://hal.archives-ouvertes.fr/hal-01407401
98. M. Amayri, S. Ploix, N. Bouguila, F. Wurtz, Database quality assessment for interactive learning: application to occupancy estimation. Energy Build. **209**, 109578 (2020)

Characterization of Energy Demand and Energy Services Using Model-Based and Data-Driven Approaches

Carlos A. Santos Silva, Manar Amayri, and Kaustav Basu

1 Introduction

The residential building sector is the second largest single consumer of final energy in Europe, accounting for 26% of the final energy consumption in 2018 and 16.6% of the primary energy [18], just behind the transportation sector (30%) and ahead of industry (25%) and services (15%). In terms of energy resources, natural gas accounted for 32%, electricity for 25%, renewable resources for 20%, and oil products 12%. In terms of end-uses, space heating accounted for 64%, followed by water heating 14.8%, lighting and appliances with 14%, and cooking with 6%.

In general, the energy efficiency in the residential sector can be improved by using more efficient energy equipment, by upgrading the building envelope characteristics, or by inducing changes in the consumer's behavior [5, 27]. The overall effects can be tracked by analyzing the trends of residential space heating intensity (energy consumption per floor area) as the largest end-use is usually space heating [27] or by analyzing the energy consumption historical time series and correlating it with the introduction of policy instruments like the building codes or appliance energy labels [5].

C. A. S. Silva (✉)
IN+ Centre for Innovation, Technology and Policy Research, LARSYS, Instituto Superior Técnico, Universidade de Lisboa, Lisboa, Portugal
e-mail: carlos.santos.silva@tecnico.pt

M. Amayri
CNRS, Grenoble, France
e-mail: manar.amayri@grenoble-inp.fr

K. Basu
Quby B.V., Amsterdam, Netherlands
e-mail: kaustav.basu@quby.com

S. Ploix et al. (eds.), *Towards Energy Smart Homes*,
https://doi.org/10.1007/978-3-030-76477-7_7

To estimate and/or measure the impact of such actions in particular buildings or households, one should be able to develop accurate dynamic models of building's energy consumption [6]. As buildings are complex systems, the energy consumption is influenced by a combination of factors, including the age and location of the building, the household size, and the penetration of appliances and electronic devices—including the type, function, dimension, quantity, and efficiency [16]. However, two households with similar characteristics and equipment will present different energy consumption, as occupants' behavior will be different. Occupants' behavior in residential buildings can be described as the occupants' presence and the consequent use of active systems—like lighting, equipment, heating and cooling systems—and the interaction with other devices passive systems as windows and blinds that influence energy consumption. Thus, knowing the occupants' behavior is a key aspect to develop accurate models of energy consumption [26].

To describe occupant's behavior, it is necessary to perform occupancy surveys and occupants' monitoring (through sensors or direct observations), which can be time consuming and intrusive [44]. Currently, the deployment of smart meters is making the information about energy use more available [15]. In fact, EU has adopted a number of initiatives aiming to improve energy consumption awareness, including the replacement of at least 80% of electricity meters with smart meters by 2020 [17]. The smart meters data allows for a temporal assessment of the electricity use, which holds the potential to reveal insights about the electricity consumption and the behavioral and technological drivers of that consumption. In this way, the accuracy of the building's energy models can be increased, enabling the simulation of impactful measures for the improvement of the energy efficiency in the residential sector [37].

Most authors categorize energy consumption models in two classes: top-down approaches, where energy consumption is estimated by means of macroeconomic variables, like income, fuel prices, or average household floor area; and bottom-up approaches that estimate the energy consumption by synthesizing the energy consumption from the consumption of individual appliances or services [30, 43]. The first type of approaches is used when there is no specific detailed data about the households under study and therefore the energy consumption is inferred from related data, while the second is preferred when data is available. Top-down models only provide us information regarding the use of a certain type of energy in a yearly time scale and are valuable to infer general variables like total energy demand forecast [43]. However, to understand clearly the dynamics of appliances' use we need to use bottom-up models, which can be categorized into two sub-classes: engineering models, which are based on physical models of the buildings and the appliances; and statistical or data-driven models, which are based on energy consumption data.

This chapter describes the state-of-the-art methods to characterize the energy consumption and energy services in residential buildings. Firstly, a review is done spanning from model-based approaches—like building thermal simulation tools—to data-driven approaches—like Non-Intrusive Load Monitoring (NILM). This study discusses the context under which each of the approaches should be followed, such as the sampling rate of data and the available data features or even the evolution of equipment's and appliances under the new IOT setting. We also discuss the integration of these approaches, like using the model-based approaches to generate data from data-driven approaches in context with scarce data or the use of data-driven models to learn model-based models and replace them in context of real-time applications where the available computational time is low. Finally, some results are presented using two novel approaches, one based on interactive learning and another using factorial hidden Markov models, to demonstrate that it is possible to achieve reasonable demand characterization models for energy services in the residential sector.

2 Engineering Models

2.1 Building Energy Simulation Models

Building Energy Simulation (BES) models are frequently used to evaluate the effect of the energy efficiency measures, since they allow to study different retrofit solutions as envelope improvement, HVAC and lighting systems improvement and operation, or occupants' behavior change [6]. They are bottom-up models.

For the past 50 years, a wide variety of building energy simulation tools have been developed and enhanced throughout the building energy simulation community [12]. These building energy simulation software have different features and various capabilities such as: general geometry modeling; definition of zonal internal loads; building envelope properties, daylight and solar radiation; infiltration, ventilation, and multi-zone airflow; renewable energy systems; electrical systems and equipment; HVAC systems; environmental emissions; economic evaluation; climate data availability, results reporting, and validation [11].

Several limitations arise related with the simulation outputs, since buildings monitoring often identifies significant gaps between the predicted and actual energy use of buildings and its thermal behavior [11]. Consequently, several techniques have been developed to support building simulation analysis, including parametric simulation, sensitivity analysis, simulation-based optimization, meta-model analysis, etc. Still, the calibration process with measurements values of building models tends to be difficult and time consuming. The amount of parameters that are uncertain and could affect the outputs of the model is normally high and difficult to identify [11].

One of the parameters that has been acknowledged to introduce more uncertainty is the occupants' behavior, as its randomness is hard to model and is influenced by multiple contextual factors [26]. Moreover, the data to support these assumptions are hard to find, as it is usually gathered through surveys, literature review, occupancy sensors and, more recently, from smart meters. To overcome this problem, typical or average profiles describing the occupants' presence are often used in energy simulations. However, the main criticism of this approach is the oversimplification, where the behavioral differences between occupants and the variability of occupants' behavior throughout the year are not considered.

2.2 *Technological Models*

Technological models assume that the energy consumption in a household is the sum of the use of different appliances. Therefore it requires the knowledge or assumption of what appliances exist in a household, and then the power consumption of each appliance and the time of use. These models require the existence of extensive databases of empirical data to support the description of each appliance. Often these models complete the bottom-up information with top-down information, like appliance ownership or appliance efficiency. As examples of use, we have [25] that present an approach to a bottom-up model at the energy service level, or [23] that estimate the heating, cooling, and domestic appliances' energy use for different climatic regions in Algeria.

2.3 *Time-of-Use-Surveys Models*

Time-of-use surveys (TUS) are surveys completed by residents, usually by keeping logbooks or diaries about the time use of activities an individual engages in during a specific time interval throughout the day. This information is extremely important to characterize the occupants' behavior and can therefore be used in building simulation models or technological models.

As examples of use of TUS, [46] analyzed the UK's time use survey 2005 to identify how social practices in the household take place in relation to the time of the day, including preparing food, washing, cleaning, washing clothes, watching TV, and using a computer. Fischer et al. [21] used the German TUS to develop a stochastic bottom-up model that generates synthetic electrical load profiles taking into consideration the seasonal occupant behavior and the correlation between the start time and the duration of an activity. More recently, [22] created models to generate a daily electricity demand profile that can be representative of a large number of Danish households using TUS.

2.4 *Using Smart Meter Data to Improve Engineering Models*

Electricity consumption data can provide useful information about the consumers and their habits. In fact, there has been an increasing use of smart meters data in current studies, namely to identify various types of consumers for short-term and midterm load forecasting, time-of-use (ToU) tariff design, and demand-side management (DSM) strategies [45]. Other studies focus only on the residential load characterization [42], or on inferring about the drivers behind the residential consumption, in terms of socio-economic status, appliances stocks, and dwellings characteristics [33]. Finally, electricity consumption disaggregation, appliances, lighting and plug load profiles distinction, as well as occupancy inference and inhabitants' routines are other uses of smart meter data [37]. Thus, smart meter data can be used to characterize the activities and equipment in building simulations and therefore used to calibrate and validate engineering models. This approach is particularly useful when it is necessary to extrapolate the energy consumption models for cities or regions based on the detailed monitoring of few households [24].

3 Data-Driven Models

Another approach to characterize energy demand characterization is to monitor in detail the energy consumption. Several load monitoring techniques can be implemented to determine the consumption and status of different appliances to understand the behavior of the different essential loads in the household. These techniques can be divided into two main types: Intrusive Load Monitoring (ILM) and Non-Intrusive Load Monitoring (NILM).

3.1 Intrusive Load Monitoring

Intrusive Load Monitoring (ILM) is a data-collection technique where measurement devices are installed at each appliance node to detect its power consumption [38] and therefore characterize in detail the household consumption. The main benefit of this technique is the accuracy of the results; however, it requires expensive and complicated installation systems [38].

The databases generated by ILM systems can be labeled in two ways: manually, which means that the appliance that is being monitored is labeled by the user; automatically, which means that the system is trained with examples from typical appliances and then recognizes the appliance that is being used. In general, manual setup ILM systems outperform automatic setup ILM systems.

3.2 Non-intrusive Load Monitoring (NILM)

Non-Intrusive Load Monitoring (NILM) technique is an alternative process, in which one single monitoring device is installed at the main distribution board at the household and an algorithm is applied to determine the energy consumption and the state of operation for each individual appliance [40]. As expected, the main advantage of NILM is the fact that only one single monitoring device is required thus lowering significantly the cost and the intrusion at the household level. The main disadvantage is the lower accuracy compared to ILM systems, in particular, those with manual labeling.

In general, any household appliance can be categorized in one of these classes [40]:

- on/off appliances, e.g., light bulbs, which have only two states: off (no consumption) and on, with a fix power demand. The duration depends on the user;
- finite-state appliances, e.g., fridges or dish-washers, which have multiple states, each one with its own power demand, and where the duration of each state is usually fixed and cyclic;
- continuously varying appliances, e.g., laptops, which have infinite states (continuous power demand) and which behavior is not cyclic;
- permanent demand appliances, e.g., routers or alarm-clocks, which are always On with a fixed power demand.

The appliances can be detected by "event-based" algorithms that detect the On/Off transitions or by "non-event-based" or "energy-based" methods that detect whether an appliance is On during the sampled duration [8].

These algorithms can use different measurements and features data, such as active and reactive power, voltage and current measurements, signal waveform and current harmonics signatures or high frequency electromagnetic interference (EMI).

The sampling rate is an important parameter in the complexity level of the disaggregation methods, as it affects not only the type of feature that can be measured but also the type of algorithm that can be used. A detailed discussion on the features and algorithms can be found in [41].

A high frequency (1 s–1 m) sample data rate allows more accuracy and detailed analysis to detect appliances loads. However, the large amount of measured data requires higher quality hardware and requires storage and processing capacity (locally or in the cloud) to run the disaggregation algorithms.

Recently the challenges in NILM approach focus in solving the disaggregation problem for the regular smart meters, which measure data at a lower frequency. The sampling rate varies between 15 min and one hour, according to the recommendation of the Energy Regulatory Authority of 2010, which states that the smart meters installed in each household must have the ability to measure the power consumption and save the actual data for at least 15-min-periods [49].

The methodologies to solve NILM problems encompass a mixture of domains. A majority of the earliest research focused on this problem from a signal processing perspective. The focus was on identifying different appliance signatures which distinguishes one appliance from another by analyzing with mathematical tools (for example, wavelet transformation) [9]. Subsequent research also considered the problem as a blind source separation task and proposed relevant techniques in that direction [32]. The details of the approaches can further be understood for the data distribution perspective.

3.2.1 NILM for High-Sapling Rate

A high-sampling rate NILM approach is generally a sampling rate of around 1 sample per second (1 Hz). In the last two decades, there has been a considerable amount of work to this effect. Each new method proposes to reduce the limitations of the previous ones both in term of signatures or applying state of the art pattern recognition techniques. The identified features are known as appliances signatures. Approaches typically consist of identifying the steady state or in some cases transient state features [51]. Subsequently, these signatures are matched with earlier learned models using a pattern recognition algorithm [9]. The drawbacks of these approaches are mainly the hardware requirement to monitor and process the information [19].

These methods do not fit well into the smart meter sampling rate, so a separate device has to be installed for training, visualization, and communication to the grid. This is a major drawback for these methods, commercially and practically speaking. The load separation at a high-sampling rate of all the appliances also raises privacy concerns as user activity can be easily detected, interpreted, and monitored [10].

3.2.2 NILM for Low Sampling Rate

At a low sampling rate (a sampling rate in the order of minutes) switching events are difficult to detect so non-event-based methods are more suited. The major issue at low sampling rate is that low energy consuming devices are difficult to be detected. However, high energy consuming appliances, such as water heater or washing machine, can still be identified with reasonable precision even at sampling rate of 15 min for example [28].

Considering the constraint of low sampling rate, the differentiation of the methods is directly dependent on the choice of algorithms. A method that partially disaggregates total household electricity usage into five load categories has been proposed at a low sampling rate in [32], where different sparse coding algorithms are compared and a Discriminative Disaggregation Sparse Coding algorithm is tested. A feature-based Support Vector Machine classifier accuracy is also mentioned but is not presented. The method of [32] is an implementation of the blind source separation problem, which aims at disaggregating mixture of sources into its individual sources. In the NILM context, the problem is undermined as there is only one mixture and a large number of sources. Another issue in using blind source separation is the assumption of no prior information about the sources. On the contrary, in this context, the sources (appliances) do have separate usage patterns which could be used. Nevertheless, blind source separation still remains a promising direction of research in this domain.

Temporal graphical models such as Hidden Markov Models also have been promisingly used in this domain as they are a classical method for sequence learning [34]. They have been successfully used in many domains, especially in speech recognition. In this context, the problem is to learn the model parameters given the set of observations as input sequence and appliances states as output. Hidden Markov Models also consider sequential patterns in consumption but in NILM problems, at a very low sampling rate it seems to have a high sensibility to training noise.

3.2.3 New Approaches Based on Machine Learning Algorithms

More recently, as in other domains, machine learning algorithms have been getting attention to solve the NILM problem [40]. Both supervised and unsupervised approaches can be used, but unsupervised methods have the advantage of not requiring a preliminary dataset to train the algorithms.

In supervised methods, the applications span from Bayesian classifiers which assume that the states of the appliances are independent although this is often not true in practice [7], to Support Vector Machine [20], Hidden Markov Models [52], or Artificial Neural Networks [47].

For unsupervised methods, most of the research is based on Hidden Markov Models, as it not required to perform event detection. This makes these algorithms suitable for low frequency samples as event detection is very difficult or not possible. In particular, Factorial Hidden Markov Model is a popular approach, as the observation for each appliance results from the output of each individual Markov model [31].

4 Case Study: Application to Residential Energy Consumption in France

In this section, we present the application of different types of methods to characterize energy demand and energy services in the residential sector using model-based and data-driven approaches for the case of France.

Between 2006 and 2008, a European project called REMODECE [13] was developed with partners from 12 EU countries that did a very detailed characterization of the electricity consumption of the residential sector in Europe. The large-scale monitoring campaign and a consumer survey around 1300 households and the study involved the collection of 6.000 questionnaires. About 11.500 single appliances were analyzed.

In the case of France, a large dataset denominated IRISE was collected, which includes the total energy consumption and particular appliances consumption in 100 households in France between 1998 and 2000. The dataset considers a broad set of electrical appliances spanning from low power-low consumption appliances such as lights to large power-large consumption appliances, such as DHW systems, HVAC systems, and wet appliances. Over the last years, this database has been used to calibrate, validate, and develop several models of the energy consumption in the residential sector in France.

After discussing the different methodological approaches to characterize the energy consumption and energy services for the residential sector, in this section we lay the foundations to discuss the applicability of each type of methods. For the engineering models, first we summarize the findings of several research papers that have been published using the IRISE dataset by type of model and then we propose a novel data-driven model to perform the characterization of energy demand, based on a new approach, the Interactive Learning [2, 3] and we compare it with standard NILM approaches for the same datasets.

4.1 Engineering Models

4.1.1 Building Energy Simulation Model

Kashif et al. [29] proposed a co-simulation environment for energy smart homes that takes into account inhabitants' dynamic and social behavior. To do that, the set-points for different controllers are adjusted using a physical building energy simulation model. To model the human behavior, the Brahms environment was used. The IRISE dataset was used to understand how inhabitants' behavior affects energy consumption. Subsequently, to model the behavior, a questionnaire was used that captured the context and the time-of-use of devices that impact the consumption.

Plessis et al. [36] proposed also a co-simulation environment using an Agent-Based Modeling (ABM) to simulate occupant behavior and a building energy simulation model that uses hybrid and differential algebraic equations to perform the dynamic thermal modeling. The "Mozart" house was modeled as it is one of the most representative houses in the French residential building stock (medium size detached house of $100\,\mathrm{m}^2$ of living surface area and an air volume of $252.15\,\mathrm{m}^3$). The FMI standard for co-simulation was used to couple the SMACH occupant behavior simulator and a building energy model built with the BuildSysPro Modelica library. Again, the IRISE dataset was used to describe the occupants' behavior.

From these examples, we can conclude that the use of building energy simulation models is mostly adequate to forecast the thermal behavior of a household given a set of different control parameters (e.g. switch on the heating or cooling system, increase the set-point temperature, close the blinds or open the windows), which has a significant impact in the energy demand of a household.

4.1.2 Technological Models

Almeida and Fonseca [1] describe the detailed monitoring campaign done under the REMODECE project and, based on that analysis, propose a technological model that describes the average energy consumption in the residential sector in Europe. From the data, they extracted the average power consumption for different appliances and the average time of use. They concluded that at the time, the electronic loads were a key contributor to the power demand and that there was a wide range of performance levels in the models available in the market. They also looked into detail into the patterns of residential lighting use, in which an increasing penetration of CFLs was being partly compensated by an increasing penetration of halogen lighting. Residential air conditioning was growing fast and was already a major contributor to summer peak demand in Mediterranean countries, as shown by the summer load curves from very hot days. Finally, based on the technological models, they were able to evaluate the potential energy savings from improving the efficiency of different appliances.

From this example, we can conclude that technological models are useful to estimate the impact of energy efficiency measures based on the replacement of appliances with low efficiencies.

4.1.3 Time-of-Use Survey Model

De Lauretis et al. [14] try to correlate the average energy and expenditure intensities of time uses of the total population as well as of income, household-composition and housing-type subgroups. To do that, they use a time-of-use survey done in France from 2009 to 2010. They find out that income is an obvious driver of energy and expenditure intensities but is revealed to influence time use as well. Household composition and housing type are also associated with substantial variations in activity patterns and in the energy and expenditure intensities of activities, even within a given income group. In conclusion, they underline the importance of household disaggregation in household energy analyses, to properly account for such disparities.

Robinson et al. [39] have proposed a structure for a new multi-agent simulation system in which occupants' presence, activity, behavior, comfort, and investments are each simulated in a coherent way using time-of-use surveys and bottom-up technological models. They suggest that this forms a robust basis for future simulations at the range of scales, from the building to the urban and beyond, with which we wish to examine occupants' impacts on sustainability and test strategies for ameliorating these impacts.

From these examples, we can conclude that time-of-use surveys are important source of information to describe the activities in the residential sector and therefore can replace detailed monitoring campaigns or questionnaires to characterize the activities in a household. Together with a technological model of the appliances that are used during each type of activity, it is possible to build a detailed disaggregated model of energy consumption.

4.2 *Data-Driven Models*

4.2.1 Intrusive Load Monitoring

The IRISE dataset was built using an Intrusive Load Monitoring approach [1, 13].

Basu et al. [8] perform a detail analysis for all the 100 households and conclude that they can be clustered in 4 main classes depending on the average load, average deferable load, area, and number of occupants.

Here, we chose to include houses with both Water Heating and Electric Heating Appliances. From the available dataset, 3 houses have been chosen: 28, 38, and 78. These houses also have other appliances with high power, such as the "Electric Cooker" or "Micro Wave Oven" that during the identification process might introduce uncertainty. Compared to the work of [8], house 28 would fall in the cluster "2," 38 would fall in cluster "1," and 78 in cluster "3."

Table 1 summarizes the main indicators regarding the consumption in each household. We can see that the houses present different profiles: house 38 has high consumption, distributed throughout the day. Houses 28 and 78 have low consumption and the consumption patterns more concentrated in specific periods (dawn for 28 and evenings and dawns for 78).

4.2.2 Non-intrusive Load Monitoring

The IRISE dataset was already used to develop Non-Intrusive Load Monitoring approaches, as describes in [8]. That work proposes a generic methodology using temporal sequence classification algorithms, based on an innovative time series distance-based approach that uses k-nearest classifier using different distance metrics (Euclidean, dynamic time warping, and temporal correlation), with 10% training and 90% testing. The results are compared with a standard NILM application based on the hidden Markov model (HMM) algorithm, using precision, recall, and F-measure, commonly used in information theory studies [48], but the proposed approach outperformed the HMM.

Table 1 Energy consumption of the houses in the dataset

House	28	38	78
Total yearly consumption (kWh)	8943	14031	8264
Average hourly consumption (kWh)	1.02	1.60	0.94
Standard deviation of hourly consumption (kWh)	1.27	1.97	1.22
Hourly peak (kW)	6.9	4.6	7.9
Water heater	16.71%	15.57%	67.19%
Electric heating	58.31%	52.25%	1.72%
Clothes drier	0.74%	2.14%	4.48%

The results for the k-nearest approach using the dataset with 10 min resolution to detect the ON event are described in Table 2. Notice that these results are not specific for houses 28,38, and 78, but for houses in the same cluster.

For this work, we asked the company WATT-IS to run the results for the chosen houses [50]. This algorithm combines the traditional NILM techniques based on "event detection"—which explores heuristics derived from the power demand, power variation, and total energy consumed for each event from different appliances—and unsupervised machine learning techniques (as there is no labeled data), like clustering to identify similar events and feature selection, to identify the most relevant data attributes. The algorithm is totally unsupervised.

The results for the NILM using the dataset with 10 min resolution are described in Table 3. We cannot directly compare these results with the ones in [8]—as those ones are for different houses within a cluster that presents similar characteristics, so the relative weight of the appliances and even the use of the appliances is different. Still we can see that the F-score is lower than the k-nearest supervised approach from [8], which demonstrates that at this sample rate, the water-heater loads, electric heating, and clothes drier, the signals can be mixed.

Finally, we also applied an algorithm based on Factorial Hidden Markov Model from [35], which was based on [31].

Table 2 Results for NILM approached based on k-nearest classifier from [8]: 10 min

House	Appliance	F-score
Cluster 2 (28) On	Water heater	94%
	Electric heating	n.a.
	Clothes drier	n.a
Cluster 1 (38) On	Water heater	91%
	Electric heating	n.a.
	Clothes drier	n.a.
Cluster 3 (78) On	Water heater	91%
	Electric heating	n.a.
	Clothes drier	39%

Table 3 Results for NILM: 10 min

House	Appliance	Precision	Recall	F-score
28 On	Water heater	32%	40%	36%
	Electric heating	66%	68%	67%
	Clothes drier	n.a.	n.a	n.a.
38 On	Water heater	66%	45%	54%
	Electric Heating	33%	75%	45%
	Clothes drier	n.a.	n.a	n.a.
78 On	Water heater	45%	31%	37%
	Electric heating	13%	1%	2%
	Clothes drier	n.a.	n.a	n.a.

Table 4 Results for factorial hidden Markov models: 10 min

House	Appliance	Precision	Recall	F-score
28 On	Water heater	93%	65%	75%
	Electric heating	80%	72%	75%
	Clothes drier	n.a.	n.a	n.a.
38 On	Water heater	91%	67%	77%
	Electric heating	55%	68%	60%
	Clothes drier	n.a.	n.a	n.a.
78 On	Water heater	93%	62%	74%
	Electric heating	n.a.	n.a.	n.a.
	Clothes drier	n.a.	n.a	n.a.

The results are described in Table 4. The results show that for water heater and electric heating, the algorithm presents higher performance and is comparable to the supervised approach presented in [8].

As a conclusion, NILM approaches can be used to disaggregate the use of the appliances, especially the ones that present a significant weight in the overall consumption. The approach presented in [8] appears to capture fairly well the water heater, but is uses a training set. The NILM proposed by WATT-IS, which does not use any labeled training data, is able to capture fairly well the appliances if the relative weight in the consumption is high, as it is the case of electric heating in houses 28 and 38 and Water heating in house 78. The NILM using Factorial Hidden Markov Model performs well, except if the relative weight in the consumption is small, like the electric heater in house 78. Overall, even with low resolution data, NILM models can be used to obtain detailed disaggregated data without resorting to ILM, in order to supply information of occupants' behavior for other models, like building simulation models or technological models.

4.3 A New Approach to Develop Data-Driven Models: Interactive Learning

In this work the deployment of Interactive Learning (IL) is used to disaggregate the appliances consumption. IL is a supervised learning methodology that involves the exchange of information with the user to collect a training dataset related to a specific context [2]. One of the advantages of IL is that useful feedback can be obtained from the end-user and increase their awareness of energy systems. This algorithm, proposed by Amayri et al. [2], has been successfully applied to estimate the occupancy in office rooms, using different sensors and avoiding the use of cameras [3]. Besides, the concept of interactive learning allows us to evaluate and improve the quality of the database [4].

In Interactive Learning, each data point is a list of features coming from sensors, which is called an "ask." In our case, the features correspond to the list of features which include the current and the previous total electricity consumption data, the hour of the day and the derivative of electricity consumption. The data point may include a label provided by the feedback of the user regarding the use of a certain appliance. The main problem in interactive learning is to determine when a candidate "ask" should be considered for collecting occupant feedback, considering the existing database. The density of the neighborhood, average error estimation, the weight of each class, and the score of the spread of the data (Qscore), are used to define the valuable time to interact with the end users. In [3, 4], the IL approach is enhanced to define the right time to question occupants when relevant, by limiting the number of interactions and maximizing the information gain. The classifier construction is part of the method, so the IL will determine what is the expected state of the appliance at the next instant.

In this work, the algorithm of IL is used as a multi-label classification model (i.e., three appliance states, on/off). The interactivity depends mainly on the methodology used to define when it is necessary to ask occupants information about the state of the appliances. It does so by limiting the number of interactions and maximizing the information's usefulness about the disaggregated appliances.

The first step for the validation is to apply IL approach with the spread rate concept [4] on IRISE case study. At this step, Human Machine Interface (HMI) interaction with end users in the IRIS houses is simulated, using as the answers of the "asks" the data labels obtained from the power consumption sensors which are connected to each appliance. Naive Bayes classifier has been applied with the interactive learning process.

In Table 5 we present the number of "asks" over 18 days and Table 6 presents the results.

Comparing IL to the NILM results, we can see that in general, the IL method outperforms the NILM for the Water Heater and Electric Heating, especially for the houses where the performance of NILM was lower (water heater On in house 38 and Electric Heating ON in house 78). Further, IL presents the advantage of identifying the clothes drier, although the accuracy of the ON detection is low (around 30%). This is due the fact that the weight of this type of appliance in the total consumption

Table 5 Number of asks each day: 10 min for one appliance

Day	1	2	3	4	5	6	7	8	9	10	11	12	13	14	15	16	17	18
Number of *asks* with 100 % replies (28)	18	3	0	0	0	1	0	0	2	1	0	0	2	1	0	1	0	0
Number of *asks* with 100% replies (38)	16	3	0	0	0	0	0	2	1	0	0	2	0	0	1	0	0	0
Number of *asks* with 100% replies (78)	20	0	1	2	0	0	0	0	0	1	0	0	0	0	1	0	0	0

Table 6 Results for Interactive learning: 10 min

House	Appliance	Precision	Recall	F-score
28 On	Water heater	85%	76%	81%
	Electric heating	50%	34%	66%
	Clothes drier	24%	33%	37%
38 On	Water heater	54%	33%	30%
	Electric heating	77%	64%	70%
	Clothes drier	35%	10%	14%
78 On	Water heater	90%	89%	87%
	Electric heating	99%	99%	99%
	Clothes drier	25%	70%	41%

is small, so there are not a lot of data to perform well the classification, compared to the other appliances.

5 Discussion and Conclusions

In this section, we summarize the findings regarding the use of the different types of approaches that can be used to characterize the energy demand and energy services in the residential sector.

The choice of the approach to use regarding the model is based on two dimensions: the objective of the study and the available data. Firstly, engineering models like building energy simulation models or technological models have to be used to estimate the energy service demand (heating and cooling needs, water heating, lighting, cooking), which are extremely difficult to obtain from data-driven models, as these are complex to measure directly. Consequently, if the objective of study is to develop an energy management system or evaluate the impact of energy management strategies, whether it refers to changing the set-points of appliances or the replacement of the appliances, it is necessary to resort to an engineering model. No data-driven model "per se" will be able to provide conclusive answers regarding the changes in consumption caused by changes in the operation of the appliances.

In any case, engineering models require in general the modeler to consider several assumptions, like, for example, the occupant's behavior in terms of schedules or set-points. In this case, time-of-use surveys provide this information, as they describe the type of activity and eventually details about the use. On its own, TUS cannot be used to characterize the energy consumption, only if coupled with an engineering model.

Finally, data-driven models can only be implemented directly to characterize energy consumption, as this is the variable which is usually measured. In case of ILM approaches, the details of the demand characterization (for example, when each appliance is operating) depends directly on the monitoring system. However, in most cases, only aggregated measurements and with low resolution are available.

In that case, NIML approaches can disaggregate information. The challenge for NILM approaches is that the data being monitored is in general low resolution data, without labeling. The comparison between three different NILM approaches shows some differences between the algorithms, but in general for this low resolution, these algorithms are able to capture the operation of the appliances that are relevant in the overall consumption (water heaters, heating systems). Even the detection of large white appliances presents a significant challenge if no labeled data is used.

To solve this challenge, we propose here the use of Interactive Learning, which is a learning method that can label data directly with the user feedback in case the available data is not enough to perform the classification. As shown in the results, after a first initial set of questions on the first days, the need to ask the user for additional information is sparse. It presents the advantages of NILM (only using aggregated data) with the advantages of ILM, which is the access to labeled data. In this way, we find IL a very promising approach to disaggregate total energy data, even for low resolution data and with the additional capability of disaggregating the use in other appliances.

The disaggregation provided by NILM models, even if the accuracy it is not very high, can replace to a certain extent the Time-of-use-surveys, as often specific activities are related to the use of specific appliances. This is the case of an electric oven (used when people are cooking) or a TV. However, this is not necessarily true for heating systems (they can operate while people are cooking, watching TV or sleeping) or water heaters with storage tanks (they can heat the water before people take their shower or when they are not at home). Again, the use of an approach like IL is less invasive than performing a detailed questionnaire regarding the use of appliances.

Finally, to characterize the energy services, it is necessary to couple the data-driven model with a technological model or energy simulation model. Take as an example the hot water service. Measuring water heating consumption without measuring the water temperature or the water consumption requires a technological model to infer the service from the data. The same applies for heating needs. Even with the temperature of the room, it is necessary to know the area of the heated space, the type of equipment to estimate the service.

From the examples described, we see that most of the works that perform detailed energy demand characterizations often use co-simulation frameworks, integrating different types of models. Thus, we believe that in the current context of large-scale deployment of smart meters in the residential sector, it becomes feasible to characterize in detail the energy consumption and energy services in the residential sector.

With the aggregated electricity consumption data, and using NILM or Interactive Learning, it is possible to disaggregated energy consumption, replacing ILM approaches. From multiples households disaggregated data it is possible to build appliances databases and generate technological models from it. This allows to identify households with poor performance equipment or inadequate uses of equipment (e.g., water heaters operating during the day and not taking advantage of lower tariffs in the evening). From disaggregate data from multiple households in the same location, e.g., neighborhood, it is possible to infer the parameters about the houses for energy simulation models. Take the example of several households that use the air-conditioning in the same building. By comparing the use of it, we can estimate infiltration or envelop losses using energy simulation models. And with good calibrated simulation models, it is possible to design smart energy management systems that learn the behavior of the user, learn the characteristics of the equipment and the buildings and optimize the energy system by minimizing consumption while providing the correct comfort level of energy service.

Acknowledgments This paper was developed under the cooperation project PESSOA2019 "INTERACTIVE BEHAVIOUR FOR URBAN FLEXIBILITY" from the Program Hubert Curien (PHC), between the Portuguese National Foundation for Science and Technology (FCT) from Portugal and CAMPUSFRANCE from France.

The contribution from Carlos Santos Silva stems from the research work of multiple doctoral students at IN+ that have contributed to understand the use of energy in the building sector and make it more efficient, namely Henrique Pombeiro, Ricardo Gomes, and Surya Pandiyan.

The contribution from Manar Amayri benefits from the support of the INVOLVED ANR-14-CE22-0020-01 project of the French National Research Agency. It is also supported by the French National Research Agency in the framework of the EquipEx program AmiQual4Home ANR -11-EQPX-00 and the Investissements d'avenir program (ANR-15-IDEX-02) as the cross disciplinary program Eco-SESA. It is also supported by the project Pessoa2019.

References

1. A. Almeida, P. Fonseca, Residential monitoring to decrease energy use and carbon emissions in Europe, in *European Council for an Energy Efficient Economy ECEEE 2007 Summer Study on Energy Efficiency, Panel 6 Products and Appliances* (2007), pp. 1–14
2. M. Amayri, S. Ploix, P. Reignie, B. Sanghamitra, Towards interactive learning for occupancy estimation, in *Proceedings of ICAI'16 - The 18th International Conference on Artificial Intelligence* (2016), pp. 1–9
3. M. Amayri, S. Ploix, N. Bouguila, F. Wurtz, Estimating occupancy using interactive learning with a sensor environment: real-time experiments. IEEE Access **7**, 53932–53944 (2019)
4. M. Amayri, S. Ploix, N. Bouguila, F. Wurtz, Database quality assessment for interactive learning: application to occupancy estimation. Energy Build. **209**, 109578 (2020). https://doi.org/10.1016/j.enbuild.2019.109578
5. E. Aydin, D. Brounen, The impact of policy on residential energy consumption. Energy **169**, 115–129 (2019). https://doi.org/10.1016/j.energy.2018.12.030
6. L. Azevedo, R. Gomes, C. Silva, Influence of model calibration and optimization techniques on the evaluation of thermal comfort and retrofit measures of a Lisbon household using building energy simulation. Adv. Build. Energy Res. 1–32 (2019). https://doi.org/10.1080/17512549.2019.1654916
7. S. Barker, M. Musthag, D. Irwin, P. Shenoy, Non-intrusive load identification for smart outlets, in *2014 IEEE International Conference on Smart Grid Communications (SmartGridComm)* (2014), pp. 548–553
8. K. Basu, V. Debusschere, A. Douzal-Chouakria, S. Bacha, Time series distance-based methods for non-intrusive load monitoring in residential buildings. Energy Build. **96**, 109–117 (2015). https://doi.org/10.1016/j.enbuild.2015.03.021
9. T. Bier, D. Benyoucef, D. Ould Abdeslam, J. Merckle, P. Klein, Smart meter systems measurements for the verification of the detection classification algorithms, in *IECON 2013 - 39th Annual Conference of the IEEE Industrial Electronics Society* (2013) pp. 5000–5005

10. B.J. Birt, G.R. Newsham, I. Beausoleil-Morrison, M.M. Armstrong, N. Saldanha, I.H. Rowlands, Disaggregating categories of electrical energy end-use from whole-house hourly data. Energy Build. **50**, 93–102 (2012). https://doi.org/10.1016/j.enbuild.2012.03.025
11. D. Coakley, P. Raftery, M. Keane, A review of methods to match building energy simulation models to measured data. Renew. Sustain. Energy Rev. **37**, 123–141 (2014). https://doi.org/10.1016/j.rser.2014.05.007. http://www.sciencedirect.com/science/article/pii/S1364032114003232
12. D.B. Crawley, J.W. Hand, M. Kummert, B.T. Griffith, Contrasting the capabilities of building energy performance simulation programs. Build. Environ. **43**(4), 661–673 (2008). https://doi.org/10.1016/j.buildenv.2006.10.027. http://www.sciencedirect.com/science/article/pii/S0360132306003234
13. A. de Almeida et al., Residential monitoring to decrease energy use and carbon emissions in Europe (2008). https://ec.europa.eu/energy/intelligent/projects/sites/iee-projects/files/projects/documents/remodece_publishable_report_en.pdf. Accessed 12 May 2020
14. S. De Lauretis, F. Ghersi, J.M. Cayla, Energy consumption and activity patterns: an analysis extended to total time and energy use for French households. Appl. Energy **206**, 634–648 (2017). https://doi.org/10.1016/j.apenergy.2017.08.180
15. F.A. Diawuo, M. Sakah, S. de la Rue du Can, P.C. Baptista, C.A. Silva, Assessment of multiple-based demand response actions for peak residential electricity reduction in Ghana. Sustain. Cities Soc. **59**, 102235 (2020). https://doi.org/10.1016/j.scs.2020.102235
16. M.M. Eskander, M. Sandoval-Reyes, C.A. Silva, S.M. Vieira, J. Sousa, Assessment of energy efficiency measures using multi-objective optimization in Portuguese households. Sustain. Cities Soc. **35**, 764–773 (2017). https://doi.org/10.1016/j.scs.2017.09.032
17. European Commission, Directive 2009/72/EC of the European parliament and of the council of 13 July 2009 concerning common rules for the internal market in electricity and repealing directive 2003/54/EC (2009). https://eur-lex.europa.eu/legal-content/EN/TXT/PDF/?uri=CELEX:32009L0072&from=EN. Accessed 07 July 2020
18. European Commission, Eurostat statistics explained: energy consumption in households (2020). https://ec.europa.eu/eurostat/statistics-explained/index.php/Energy_consumption_in_households. Accessed 07 July 2020
19. R.A.S. Fernandes, I.N. da Silva, M. Oleskovicz, Load profile identification interface for consumer online monitoring purposes in smart grids. IEEE Trans. Ind. Inf. **9**(3), 1507–1517 (2013)
20. M. Figueiredo, A. de Almeida, B. Ribeiro, Home electrical signal disaggregation for non-intrusive load monitoring (NILM) systems. Neurocomputing **96**, 66–73 (2012). https://doi.org/10.1016/j.neucom.2011.10.037
21. D. Fischer, A. Härtl, B. Wille-Haussmann, Model for electric load profiles with high time resolution for German households. Energy Build. **92**, 170–179 (2015). https://doi.org/10.1016/j.enbuild.2015.01.058
22. K. Foteinaki, R. Li, C. Rode, R.K. Andersen, Modelling household electricity load profiles based on danish time-use survey data. Energy Build. **202**, 109355 (2019). https://doi.org/10.1016/j.enbuild.2019.109355
23. R. Ghedamsi, N. Settou, A. Gouareh, A. Khamouli, N. Saifi, B. Recioui, B. Dokkar, Modeling and forecasting energy consumption for residential buildings in Algeria using bottom-up approach. Energy Build. **121**, 309–317 (2016). https://doi.org/10.1016/j.enbuild.2015.12.030
24. R. Gomes, A. Ferreira, L. Azevedo, R.C. Neto, L. Aelenei, C. Silva, Retrofit measures evaluation considering thermal comfort using building energy simulation: two Lisbon households. Adv. Build. Energy Res. **15**, 291–314 (2021). https://doi.org/10.1080/17512549.2018.1520646
25. J.P. Gouveia, P. Fortes, J. Seixas, Projections of energy services demand for residential buildings: insights from a bottom-up methodology. Energy **47**(1), 430–442 (2012). https://doi.org/10.1016/j.energy.2012.09.042. Asia-Pacific Forum on Renewable Energy 2011
26. T. Hong, S.C. Taylor-Lange, S. D'Oca, D. Yan, S.P. Corgnati, Advances in research and applications of energy-related occupant behavior in buildings. Energy Build. **116**, 694–702 (2016). https://doi.org/10.1016/j.enbuild.2015.11.052

27. International Energy Agency, Energy efficiency indicators 2020 (2020). https://www.iea.org/reports/energy-efficiency-indicators-2020. Accessed 07 July 2020
28. G. Kalogridis, C. Efthymiou, S.Z. Denic, T.A. Lewis, R. Cepeda, Privacy for smart meters: towards undetectable appliance load signatures, in *2010 First IEEE International Conference on Smart Grid Communications* (2010), pp. 232–237
29. A. Kashif, S. Ploix, J. Dugdale, X.H.B. Le, Simulating the dynamics of occupant behaviour for power management in residential buildings. Energy and Buildings **56**, 85–93 (2013). https://doi.org/10.1016/j.enbuild.2012.09.042
30. M. Kavgic, A. Mavrogianni, D. Mumovic, A. Summerfield, Z. Stevanovic, M. Djurovic-Petrovic, A review of bottom-up building stock models for energy consumption in the residential sector. Build. Environ. **45**(7), 1683–1697 (2010). https://doi.org/10.1016/j.buildenv.2010.01.021
31. J.Z. Kolter, T. Jaakkola, Approximate inference in additive factorial HMMs with application to energy disaggregation, in *Proceedings of the Fifteenth International Conference on Artificial Intelligence and Statistics* (2012), pp. 1472–1482
32. J. Kolter, S. Batra, A. Ng, Energy disaggregation via discriminative sparse coding. Adv. Neural Inf. Process. Syst. **23**(7), 1153–1161 (2010)
33. F. McLoughlin, A. Duffy, M. Conlon, Characterising domestic electricity consumption patterns by dwelling and occupant socio-economic variables: an Irish case study. Energy Build. **48**, 240–248 (2012). https://doi.org/10.1016/j.enbuild.2012.01.037
34. O. Parson, S. Ghosh, M. Weal, A. Rogers, Using hidden Markov models for iterative non-intrusive appliance monitoring, in Neural Information Processing Systems workshop on Machine Learning for Sustainability (2011). https://eprints.soton.ac.uk/272990/. Accessed 17 Dec 2011
35. O. Parson, S. Ghosh, M. Weal, A. Rogers, Non-intrusive load monitoring using prior models of general appliance types, in *Proceedings of the Twenty-Sixth AAAI Conference on Artificial Intelligence, Toronto, Ontario* (2012), pp. 356–362
36. G. Plessis, É. Amouroux, Y. Haradji, Coupling occupant behaviour with a building energy model - a FMI application, in Proceedings of the 10th International Modelica Conference March 10–12, 2014, Lund (2014), pp. 321–326
37. H.R.M.L. Pombeiro, R.A. Gomes, C.A.S. Silva, Designing an adaptive feedback platform for encouraging energy efficiency behaviors: a practical case in Portuguese households, in *2016 Future Technologies Conference (FTC)* (2016), pp. 769–776
38. A. Ridi, C. Gisler, J. Hennebert, A survey on intrusive load monitoring for appliance recognition, in *2014 22nd International Conference on Pattern Recognition* (2014), pp. 3702–3707
39. D. Robinson, U. Wilke, F. Haldi, Multi agent simulation of occupants' presence and behaviour, in *Proceedings of Building Simulation 2011: 12th Conference of International Building Performance Simulation Association, Sydney, 14–16 November* (2011), pp. 2110–2117
40. A. Ruano, A. Hernandez, J. Ureña, M. Ruano, J. Garcia, Nilm techniques for intelligent home energy management and ambient assisted living: a review. Energies **12**, 2203 (2019). https://doi.org/10.3390/en12112203
41. N. Sadeghianpourhamami, J. Ruyssinck, D. Deschrijver, T. Dhaene, C. Develder, Comprehensive feature selection for appliance classification in NILM. Energy Build. **151**, 98–106 (2017). https://doi.org/10.1016/j.enbuild.2017.06.042
42. B. Stephen, S.J. Galloway, Domestic load characterization through smart meter advance stratification. IEEE Trans. Smart Grid **3**(3), 1571–1572 (2012)
43. L.G. Swan, V.I. Ugursal, Modeling of end-use energy consumption in the residential sector: a review of modeling techniques. Renewable and Sustainable Energy Reviews **13**(8), 1819–1835 (2009). https://doi.org/10.1016/j.rser.2008.09.033
44. S. Tao, M.Y. Ru, W. Du, X. Zhu, Q.R. Zhong, B.G. Li, G.F. Shen, X.L. Pan, W.J. Meng, Y.L. Chen, H.Z. Shen, N. Lin, S. Su, S.J. Zhuo, T.B. Huang, Y. Xu, X. Yun, J.F. Liu, X.L. Wang, W.X. Liu, H.F. Cheng, D.Q. Zhu, Quantifying the rural residential energy transition in china from 1992 to 2012 through a representative national survey. Nat. Energy **3**, 567–573 (2018). https://doi.org/10.1038/s41560-018-0158-4

45. J. Torriti, Price-based demand side management: assessing the impacts of time-of-use tariffs on residential electricity demand and peak shifting in northern Italy. Energy **44**(1), 576–583 (2012). https://doi.org/10.1016/j.energy.2012.05.043
46. J. Torriti, Understanding the timing of energy demand through time use data: time of the day dependence of social practices. Energy Res. Soc. Sci. **25**, 37–47 (2017). https://doi.org/10.1016/j.erss.2016.12.004
47. M.S. Tsai, Y.H. Lin, Modern development of an adaptive non-intrusive appliance load monitoring system in electricity energy conservation. Appl. Energy **96**, 55–73 (2012). https://doi.org/10.1016/j.apenergy.2011.11.027
48. G. Tsoumakas, I. Katakis, Multi-label classification: an overview. Int. J. Data Warehousing Min. **3**(1), 13 (2007). https://doi.org/doi:10.4018/jdwm.2007070101
49. H. Wang, W. Yang, T.C.Q. Yang, An optimal load disaggregation method based on power consumption pattern for low sampling data. Sustainability **11**(251), (2019). https://doi.org/10.3390/su11010251
50. Watt Intelligent Solutions, Watt-is, your partner in energy analytics (2012). https://watt-is.com/. Accessed 18 May 2020
51. M. Zeifman, K. Roth, Nonintrusive appliance load monitoring: review and outlook. IEEE Trans. Consum. Electron. **57**(1), 76–84 (2011)
52. T. Zia, D. Bruckner, A. Zaidi, A hidden Markov model based procedure for identifying household electric loads, in *IECON 2011 - 37th Annual Conference of the IEEE Industrial Electronics Society* (2011), pp. 3218–3223

Occupant Actions Selection Strategies Based on Pareto-Optimal Schedules and Daily Schedule for Energy Management in Buildings

Monalisa Pal and Sanghamitra Bandyopadhyay

1 Introduction

Building energy management is a globally important topic of interest as it contributes to nearly 40% of global energy consumption [3, 23–25]. Architectural and technological developments [25, 28] provide solutions to decrease energy consumption while meeting the occupants' demands, such as comfort, cost, etc. However, such solutions either require the construction of new buildings (also known as the green buildings) or require installation of new devices [15, 25]. Nonetheless, the occupants' demands can be met to some extent by regulating the actions of the occupants [3, 24]. Such a solution to the building energy management problem is more beneficial as it not only helps in decreasing the occupants' demands but also can be applied to existing buildings without any extra construction or installations [17, 25]. Thus, such solutions are also known as zero-cost solutions for the building energy management problem.

A building energy management problem is, thus, characterized by occupants' actions, occupants' demands, the testbed zone (room, floor, building, neighborhood, etc.), and granularity (hourly, daily, monthly, seasonal, yearly, etc.) of data [17, 25]. The actions of the occupants, which can influence the occupants' demands, comprise opening and closing of doors, windows and window blinds, and turning on or off of electrical devices (like heater, monitor, etc.) [17, 25]. Under a given physical context (outdoor weather), these occupants' actions create various environment-building-occupant interactions through airflow and heat flow, which affect the indoor ambience, such as indoor temperature, indoor CO_2 concentration, and indoor humidity [17, 25]. These physical variables create an impact on the

M. Pal (✉) · S. Bandyopadhyay
Machine Intelligence Unit, Indian Statistical Institute, Kolkata, India
e-mail: monalisap90@gmail.com; sanghami@isical.ac.in

S. Ploix et al. (eds.), *Towards Energy Smart Homes*,
https://doi.org/10.1007/978-3-030-76477-7_8

occupants' demands like thermal discomfort [3], air quality discomfort [3], and expenses incurred due to heater operations [25]. Thus, the occupants' actions are the controllable causes that are capable of creating various desirable effects. These causal phenomena [3, 9] are the fundamental factors driving the zero-cost solution for the building energy management problem.

Due to the mathematical formalism of various physical knowledge models, it is difficult for the occupants to learn the impact of their actions on desirable effects [3]. On the one hand, machine learning approaches [18, 34] have often been adopted to simulate the occupant-building interactions. On the other hand, multi-objective optimization (MOO) approaches [3, 6, 15] are also adopted to yield occupants' actions for a preferable indoor ambience. However, merely recommending actions does not create the necessary awareness among the occupants to modify their habits towards an energy-efficient routine [3]. Towards employing this awareness in practice, various explanation generating engines [2, 25] are developed, which analyze why a recommended action is important and what happens if the recommended action is not implemented.

The MOO module generates a list of action schedules compromising among the various demands [20, 25]. However, much attention has not been provided towards the selection of the most relevant action schedule from the list of Pareto-optimal action schedules. To meet this research gap, several strategies for selecting this relevant action schedule are investigated in this work considering the data collected using various sensors in an office room at Grenoble Institute of Technology, France, where four researchers work. An MOO framework is considered for obtaining the Pareto-optimal set of schedules of occupants' actions (opening and closing of doors and windows over 12 h with hourly granularity), under a recorded physical context, while minimizing two criteria: thermal dissatisfaction and CO_2 based air quality dissatisfaction. The most relevant Pareto-optimal schedule is then chosen following the decision-making strategy presented in this work. This optimal schedule and the historical (recorded) schedule of occupants' actions can then be utilized by any explanation generating engine to describe the causal changes.

The organization of the rest of this article is as follows. In Sect. 2, a literature survey on various decision-making strategies for choosing the relevant Pareto-optimal schedule is provided. It also highlights the novelty and the importance of the proposed work for selecting relevant occupants' actions. Section 3 discusses the specifications of the various constituents involved in the problem formulation along with the operation of the proposed approach to address the building energy management problem. The proposed evolutionary framework and the occupants' actions selection strategy are analyzed with real-world data in Sect. 4. Finally, the article is concluded in Sect. 5 while mentioning the possibilities to further extend this work.

2 Brief Review on Decision-Making Strategies in Presence of Multiple Compromises of Interest

The objectives of the building energy management problem consider occupants' comfort, economy, and ecology, which are conflicting in nature. Hence, in the past decade, related problems are often formulated as multi-objective optimization problems.

Earlier studies, such as [14] and [7], are noteworthy research works where multi-objective optimization problems for physical retrofit planning are investigated. Installations of windows, roof, and solar panels are planned to minimize energy consumption while balancing the stakeholders' gain and the occupants' demands using Tchebycheff programming in [6] and mixed-integer non-linear programming in [15]. However, in these works, scalarization functions are used to transform the multi-objective problems entirely into a single objective optimization problem, which yields a single optimal solution. Thus, the need for decision-making does not arise but the entire search space is also not thoroughly explored due to these transformations.

In a related building energy management problem, the first multi-objective optimization approach (using second-generation non-dominated sorting genetic algorithm, NSGA-II [13]) has been employed in [26]. The estimated Pareto-Front provides a set of alternative solutions which is beneficial for the decision-making process. However, it specifies the requirement of expert knowledge or user preference for selecting one of the several alternatives from the estimated set of solutions [26].

Another multi-objective optimization approach (using differential evolution for multi-objective optimization, DEMO [29]) proposed the use of distance to best compromises [3] for an automated decision-making process. This approach is motivated to choose the best trade-off with equal preference to all objectives. Thus, the solution with the minimum sum of objectives (*closest* to the ideal minima or the origin of the objective space) is chosen by this approach as the recommended schedule of occupants' actions.

For allowing more control to the users, a user interface is developed in [19]. In this approach, several radial basis functions are fitted on the estimated Pareto-Front to generate the global response surface [1]. This allows the users to smoothly navigate various co-dependent sliders along each objective and the optimal solution is generated by inverse mapping and interpolation of objective functions.

When decision-making involves specifying preferences, usually a single decision-maker is considered. In the presence of multiple occupants with different preferences, multiple preferences are often managed by imposing a hierarchy or priority order among the multiple occupants [17]. A decision-making strategy using a Nash-bargaining like approach is proposed in [21] to yield the fair consensus solution concerning multiple preferences. By varying its α parameter, the kind of fairness can also be regulated to manage situations of similar occupants' preferences

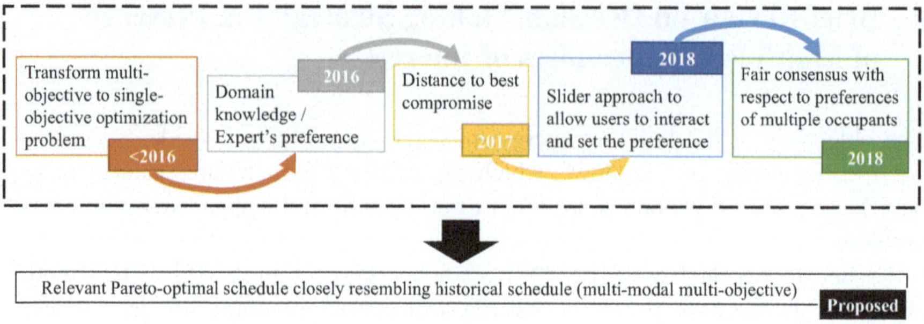

Fig. 1 Roadmap of occupants actions selection strategies

with a very diverse preference of an occupant. Thus, the fair consensus solution becomes the final chosen solution by this approach.

The above-mentioned decision-making approaches, even those considering occupants' preferences, only deal with solutions in the objective space. Moreover, the next goal in the building energy management problem is to motivate the occupants to adjust their actions as per the recommended schedule by providing explanations. Differential explanations and causal graphs suggest that occupants' actions at the time instants t and $(t + k)$ may both affect the indoor ambience at $(t + k)$, i.e., different actions can generate similar effects. Equivalently, multiple action schedules can generate nearly the same objective values (multi-modal nature of building energy management problem). Among various estimated Pareto-optimal action schedules, the schedule closely resembling the historical (recorded) schedule can be considered to be highly relevant as this provides a better alternative without highly varying the usual routine of the occupant. Thus, it will be easier to motivate the occupants to incorporate these small changes in their routine for a more energy-efficient routine. Thus, the novel contribution of this work is twofold:

1. Design a multi-modal multi-objective optimization framework that can explore both actions and effects and is capable of generating equivalent action schedules for nearly the same objective values.
2. Design an automated decision-making strategy to select (post-optimization) the relevant Pareto-optimal action schedule, which closely resembles the historical (recorded, past) schedule.

The roadmap of occupants' actions selection strategies leading to the proposed work is briefly summarized in Fig. 1.

3 Determining the Best Schedule of Occupant Actions

The underlying causal relations [3, 25] occur among various physical variables, which can be grouped into four categories as follows:

- *Occupants' actions* ($\mathcal{X}_B$): This is the set of variables that are directly controllable by the occupants. The scope of this work involves actions like opening/closing of doors ($\zeta_D(t)$) and opening/closing of windows ($\zeta_W(t)$) at time instant t.
- *Physical context* ($\mathcal{P}_B$): This is the set of variables which are not controllable by the occupants. This involves variables like outdoor temperature ($T_{out}(t)$), occupancy ($n(t)$), wind speed, humidity, etc. at time instant t.
- *Effects* ($\mathcal{F}_B$): These are the set of variables that help in quantifying various effects desired by the occupants. The scope of this work involves effects like thermal dissatisfaction ($\sigma_{temp}(t)$) and air quality dissatisfaction ($\sigma_{air}(t)$) at time instant t.
- *Intermediate variables* ($\mathcal{I}_B$): This forms the set of remaining variables. It has two distinct subsets: (1) variables which can be measured using sensors such as indoor temperature ($T_{in}(t)$) and CO_2 concentrations ($C_{in}(t)$) and (2) variables which are estimated through simulation models such as airflow ($Q(t)$) and heat flow ($\varphi(t)$).

The causal relation [3, 25] can be summarized as in Eq. (1). This represents that under a given context $\mathcal{P}_B$, the set of actions $\mathcal{X}_B$ causes the effects $\mathcal{F}_B$ through the intermediate variables $\mathcal{I}_B$.

$$\mathcal{X}_B, \mathcal{P}_B \xrightarrow{\mathcal{I}_B} \mathcal{F}_B \tag{1}$$

The goal of the building energy management problem is to generate optimal occupants' actions for desired effects (multi-objective optimization). Let the relevant optimal actions be $\mathcal{X}_B^\star$. When $\mathcal{X}_B^\star$ is used to simulate the building model under the same physical context $\mathcal{P}_B$, the obtained level of satisfaction is given by $\mathcal{F}_B^\star$ and the associated physical phenomenon is $\mathcal{I}_B^\star$ which assists in achieving $\mathcal{F}_B^\star$. Thus, in formal terms: $\mathcal{X}_B^\star, \mathcal{P}_B \xrightarrow{\mathcal{I}_B^\star} \mathcal{F}_B^\star$.

The following sub sections describe the various modules to construct the framework for achieving this above-mentioned goal.

3.1 Description of Experimental Testbed

The layout of the office room in the Grenoble Institute of Technology is shown in Fig. 2, which forms the experimental testbed. It is fitted with 27 sensors, which record indoor physical variables like humidity, temperature (T_{in} in °C), CO_2 concentration (C_{in} in ppm), motions (to assist in occupancy estimation [4, 5]), etc. These sensors also record the corridor variables like temperature (T_n), CO_2 concentration (C_n), etc. Other contextual variables like outdoor temperature (T_{out}), wind speed, illuminance, etc. are available from the weather forecast. These sensors are used to create a historical database ($\mathcal{H}_{DB}$), which provides data to tune the physical knowledge models and the reference schedule ($\tilde{\mathcal{X}}_B$) for decision-making.

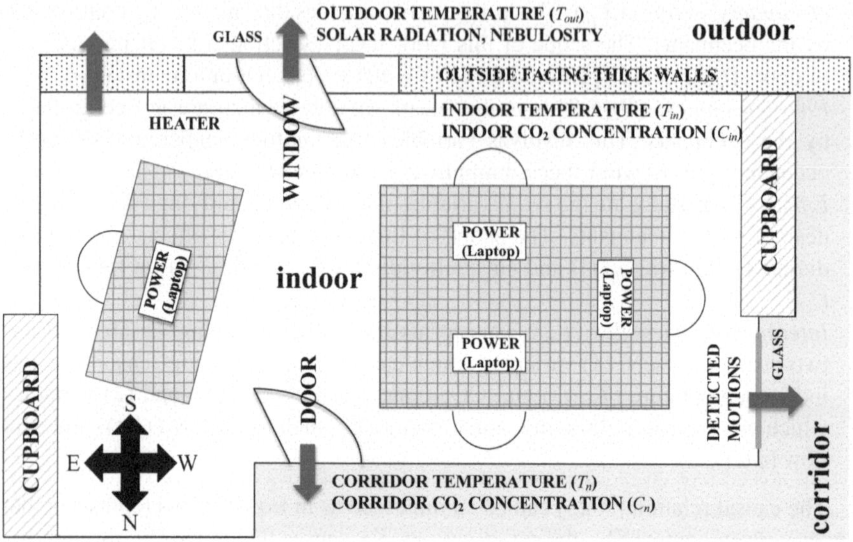

Fig. 2 Layout of the office room (testbed)

The data is available for the operational duration of the office for approximately 100 days (April 1, 2015, to July 9, 2015, 8 am to 8 pm).

3.2 *Physical Knowledge Models for Simulation*

As this study considers optimization objectives (effects of occupants' actions) as the thermal and CO_2 based air quality dissatisfaction, physical knowledge models are fitted to simulate indoor temperature ($T_{in}(t)$) and indoor CO_2 concentration ($C_{in}(t)$) for a given physical context and a given set of actions. It is interesting to note that both $T_{in}(t)$ and $C_{in}(t)$ are intermediate variables which are influenced by occupants' actions (present and past values of $\zeta_W(.)$ and $\zeta_D(.)$) and physical context and, in turn, influences the indoor comfort ($\sigma_{air}(t)$ and $\sigma_{temp}(t)$). The lack of appropriate physical knowledge models for a building zone limits the scope of the current research problem. Nonetheless, the thermal and aeraulic models of the office, whose layout is shown in Fig. 2, have been thoroughly explored in Sect. 5.1 of Chapter "Formalization of the Energy Management Problem and Related Issues", and the best-fitted models are available for further use in this experiment.

3.3 Formulation of the Optimization Problem

The concerned zero-cost building energy management problem is an optimization problem that uses the causal relation in Eq. (1) and the physical knowledge models in Sect. 5.1 of Chapter "Formalization of the Energy Management Problem and Related Issues" for optimizing $\mathcal{F}_B\ (\mathcal{X}_B \mid \mathcal{P}_B)$. For a formal definition to obtain the Pareto-optimal set of occupants' actions, this work considers the following encoding of N-dimensional solution vector ($\mathbf{X}_B$ with $N = 24$) and M-dimensional objective vector ($\mathbf{F}_B$ with $M = 2$), which are derived from occupants' actions ($\mathcal{X}_B$) and effects ($\mathcal{F}_B$), respectively.

- *Encoding of actions for the optimization module:* As the work considers an hourly granularity (counting 8 am as the first hour), the status of opening and closing of door and window at every hour helps in the construction of the 24-dimensional binary solution vector ($\mathbf{X}_B$) as shown in Eq. (2). Thus, the schedule of door and window opening/closing can easily be extracted from this encoding.

$$\mathbf{X}_B = \left[x_{B,1}, \ldots, x_{B,N}\right] = \left[\zeta_W^1, \ldots, \zeta_W^k, \ldots, \zeta_W^{12}, \zeta_D^1, \ldots, \zeta_D^k, \ldots, \zeta_D^{12}\right]$$

$$\text{where, } \zeta_W^k = \begin{cases} 1, & \text{if window is open at } k\text{th hour between 8am and 8pm} \\ 0, & \text{if window is closed at } k\text{th hour between 8am and 8pm} \end{cases}$$

$$\text{and } \zeta_D^k = \begin{cases} 1, & \text{if door is open at } k\text{th hour between 8am and 8pm} \\ 0, & \text{if door is closed at } k\text{th hour between 8am and 8pm} \end{cases} \tag{2}$$

- *Desired effects of occupants' actions forming the optimization objectives:* As this work considers minimization of thermal and CO_2 based air quality dissatisfaction, a transformation of indoor temperature (T_{in}^k) and CO_2 concentrations (C_{in}^k) to quantifiable dissatisfaction values over an entire day is required.

  At the kth hour, thermal dissatisfaction (σ_{temp}^k) and CO_2 based air quality dissatisfaction (σ_{air}^k) are computed using Eqs. (3) and (4), respectively. It is not surprising to note that when the office room is unoccupied ($n^k = 0$), the respective dissatisfaction values are zero. The indoor temperature (T_{in}^k) and CO_2 concentrations (C_{in}^k) at the kth hour are obtained using the physical knowledge models, described in Sect. 3.2.

$$\sigma_{temp}^k\left(T_{in}^k\right) = \begin{cases} \frac{21-T_{in}^k}{21-18}, & \text{if } T_{in}^k < 21 \text{ and } n^k > 0 \\ 0, & \text{if } 21 \le T_{in}^k \le 23 \text{ or } n^k = 0 \\ \frac{T_{in}^k-23}{26-23}, & \text{if } T_{in}^k > 23 \text{ and } n^k > 0 \end{cases} \tag{3}$$

$$\sigma_{air}^{k}\left(C_{in}^{k}\right)=\begin{cases}0, & \text{if } C_{in}^{k}\leq 400 \text{ or } n^{k}=0\\ \frac{C_{in}^{k}-400}{1500-400}, & \text{if } C_{in}^{k}>400 \text{ and } n^{k}>0\end{cases} \tag{4}$$

The overall thermal and CO_2 based air quality dissatisfaction over the entire day forms the two optimization objectives and is obtained in terms of the arithmetic mean over the operational duration of the office by using Eq. (5).

$$\mathbf{F_B}=\left[f_{B,1},\ldots,f_{B,M}\right]=\left[\frac{1}{12}\sum_{k=1}^{12}\sigma_{temp}^{k},\frac{1}{12}\sum_{k=1}^{12}\sigma_{air}^{k}\right] \tag{5}$$

Thus, with these definitions, the historical database and the simulation models, an evolutionary algorithm could be used to generate an estimate of the Pareto-optimal set of solutions (action schedule) from which the relevant schedule of door and window opening/closing is further selected for recommending to the occupants.

3.4 Proposed Framework to Generate Relevant Compromises

Among several multi-objective evolutionary algorithms (MOEAs), decomposition-based algorithms have superior exploration capabilities due to partitioning of the objective space into multiple sub-spaces and solving the multi-objective optimization problem within each sub-space in a collaborative manner [10]. Furthermore, integration of decomposition strategies with Pareto-dominance based strategies have huge potential to improve the estimation of the Pareto-optimal solutions [12, 31]. Moreover, only recent MOEAs, such as DE-TriM [22], take advantage of both decomposition and Pareto-dominance and are also capable of exploring the decision space to find equivalent solution vectors, which have nearly the same objective vectors. If a set of such equivalent Pareto-optimal solutions are available, the solution closely resembling the historical schedule (a schedule with minimal changes from the past schedule) can be recommended to the occupants as the first step towards learning the energy-efficient routine.

As DE-TriM was developed for real-valued solution space [22], whereas the building energy management problem has binary-valued solution space, binary reproduction operators such as those of Genetic Algorithm (single point binary crossover and bit-flip mutation) could be used instead of the operators of Differential Evolution. Thus, with some minor changes to DE-TriM, this work proposes the framework of GA-TriM (*G*enetic *A*lgorithm for *M*ulti-*M*odal *M*ulti-objective problems) to address the concerned zero-cost building energy management problem.

The steps of the GA-TriM for estimating the Pareto-optimal solutions of a multi-modal multi-objective problem are outlined in Algorithm 1. Aside from the objective functions (Eq. (5)), the population size (n_{pop}), termination condition

Algorithm 1 General Framework of GA-TriM

Input: $\mathbf{F}_B(.)$: Objectives of building energy management problem corresponding to N-dimensional decision space (lower-bounded by $\mathbf{X}_B^L$ and upper-bounded by $\mathbf{X}_B^U$) and M-dimensional objective space; n_{pop}: Population size; G_{max}: Maximum generations (termination condition); $\mathcal{W}$: n_{dir} number of M-dimensional reference vectors

Output: $\mathcal{A}_{G_{max}}$: Estimated Pareto-optimal sets; $\mathcal{A}_{\mathbf{F},G_{max}}$: Estimated Pareto-Front

1: **procedure** GATRIM($\mathbf{F}_B(.)$, n_{pop}, G_{max}, $\mathcal{W}$)
2: $\mathcal{A}_{G=1} \leftarrow$ An $n_{pop} \times N$ matrix is randomly initialized, bounded by $\mathbf{X}_B^L$ and $\mathbf{X}_B^U$
3: $\mathcal{A}_{\mathbf{F},G=1} = \{\mathbf{F}_B(\mathbf{X}_B) \mid \mathbf{X}_B \in \mathcal{A}_{G=1}\}$ (Fitness is evaluated)
4: Initialize mutation probability, $P_{mut} = 2/N$ for all candidates
5: $\mathbf{P}_{G=1}^{arr} \leftarrow$ A vector of length n_{dir} to store sub-population sizes, initialized with n_{pop}/n_{dir} for all directions
6: **for** $G = 1$ to G_{max} (until termination) **do**
7: $\mu_{P_{mut}} \leftarrow$ Mean over P_{mut} of all candidates in $\mathcal{A}_G$
8: **for** $k = 1$ to n_{dir} (for each direction) **do**
9: $\mathcal{A}_G^{sub\_k} \leftarrow$ Create a sub-population with $S_{k,G} \in \mathbf{P}_G^{arr}$ candidates from $\mathcal{A}_G$ which are closest to $\mathbf{W}_k \in \mathcal{W}$ in terms of $d2$ (Eq. (8))
10: $\mathbf{X}_B^{parent} \leftarrow$ Any $\mathbf{X}_B \in \mathcal{A}_G^{sub\_k}$ is assigned as the parent candidate
11: $P_{mut} \leftarrow N\left(\mu_{P_{mut}}, 0.1\right)$ for $\mathbf{X}_B^{parent}$ such that $P_{mut} \in (0, 1]$
12: $\mathbf{X}_B^{child} \leftarrow Reproduce\left(\mathbf{X}_B^{parent}, P_{mut}, \mathbf{W}_k, \mathcal{W}\right)$ (Algorithm 2)
13: $\mathcal{A}_G$: Append $\mathbf{X}_B^{child}$ to $\mathcal{A}_G$
14: $\mathcal{A}_{\mathbf{F},G}$: Append $\mathbf{F}_B\left(\mathbf{X}_B^{child}\right)$ to $\mathcal{A}_{\mathbf{F},G}$
15: **end for**
16: $\mathcal{A}_{G+1}$and$\mathcal{A}_{\mathbf{F},G+1}$: Create population of size n_{pop} for next generation using non-dominated sorting and SCD (as secondary criteria) on $\mathcal{A}_G$and$\mathcal{A}_{\mathbf{F},G}$
17: $\mathbf{I}_{AD}$: Find indices of the directions in $\mathcal{W}$ to which each candidate of $\mathcal{A}_{G+1}$ is closest to in terms of $d2$ (Eq. (8))
18: $\mathbf{P}_{G+1}^{arr} = Feedback\_Allocation\left(\mathbf{I}_{AD}, n_{pop}, n_{dir}\right)$ (Algorithm 3)
19: **if** G is divisible by 10 **then**
20: Re-assign $P_{mut} = 2/N$ for all candidates
21: **end if**
22: **end for**
23: Return $\mathcal{A}_{G_{max}}$ and $\mathcal{A}_{\mathbf{F},G_{max}}$
24: **end procedure**

G_{max}, and a set $\mathcal{W}$ of n_{dir} reference vectors[1] to partition the objective space are considered as inputs to GA-TriM. It estimates the Pareto-optimal set of solutions ($\mathcal{A}_{G_{max}}$), whose corresponding non-dominated set in the objective space ($\mathcal{A}_{\mathbf{F},G_{max}}$) yields the estimated Pareto-Front. GA-TriM consists of the following steps:

1. Initialization steps (Lines 2–5): In line 2, a population ($\mathcal{A}_{G=1}$) of N-dimensional solutions are randomly initialized such that each solution is a random unique vector according to the definition in Eq. (2). In line 3, the objective vector for each solution of the population is evaluated to create the matrix $\mathcal{A}_{\mathbf{F},G=1}$. In line 4, the mutation probabilities (P_{mut}) for each solution is initialized to (length of solution

[1]Reference vectors [11, 12] are obtained from http://worksupplements.droppages.com/refvecgen.

Algorithm 2 Reproduction of new candidates (in decision space)

Input: $\mathbf{X}_B^{parent}$: A parent solution vector; P_{mut}: Variable-wise mutation probability; $\mathbf{W}_k$: kth weight vector; $\mathcal{W}$: set of all weight vectors
Output: $\mathbf{X}_B^{child}$: Child solution vector

procedure REPRODUCE($\mathbf{X}_B^{parent}$, P_{mut}, $\mathbf{W}_k$, $\mathcal{W}$)
    $\mathbf{X}_{B,1}^{parent} = \mathbf{X}_B^{parent}$
    $\mathbf{W}_r \leftarrow$ Nearest neighbor of $\mathbf{W}_k$
    $\mathbf{X}_{B,2}^{parent} \leftarrow$ A random candidate associated, in terms of $d2$ (Eq. (8)), to $\mathbf{W}_r$
    if $rand(0, 1) < 0.5$ **then**
        Extract the ζ_W variables from $\mathbf{X}_{B,1}^{parent}$ and $\mathbf{X}_{B,2}^{parent}$ to yield $\mathbf{X}_1^{par}$ and $\mathbf{X}_2^{par}$
    else
        Extract the ζ_D variables from $\mathbf{X}_{B,1}^{parent}$ and $\mathbf{X}_{B,2}^{parent}$ to yield $\mathbf{X}_1^{par}$ and $\mathbf{X}_2^{par}$
    end if
    Generate $\mathbf{X}_1$ and $\mathbf{X}_2$ using binary crossover between $\mathbf{X}_1^{par}$ and $\mathbf{X}_2^{par}$
    if $rand(0, 1) < 0.5$ **then**
        $\mathbf{X}^c \leftarrow$ Flip variables of $\mathbf{X}_1$ with probability less than P_{mut}
    else
        $\mathbf{X}^c \leftarrow$ Flip variables of $\mathbf{X}_2$ with probability less than P_{mut}
    end if
    $\mathbf{X}_B^{child} \leftarrow$ Replace $\mathbf{X}_1^{par}$ in $\mathbf{X}_{B,1}^{parent}$ with $\mathbf{X}^c$
    Return $\mathbf{X}_B^{child}$
end procedure

Algorithm 3 Feedback for resource allocation to determine sub-population sizes

Input: $\mathbf{I}_{AD}$: A vector of length n_{pop} representing indices (j) of directions ($\mathbf{W}_j \in \mathcal{W}$) to which $\mathbf{X}_B \in \mathcal{A}_G$ is closest to in terms of $d2$ (Eq. (8)); n_{pop}: Population size; n_{dir}: Number of reference vectors
Output: $\mathbf{P}_{G+1}^{arr}$: A vector of length n_{dir} with updated sub-population sizes

procedure FEEDBACK_ALLOCATION($\mathbf{I}_{AD}$, n_{pop}, n_{dir})
    for $k = 1$ to n_{dir} (for each direction) **do**
        n_{sub}: Calculate the number of indices in $\mathbf{I}_{AD}$ that are equal to k
        $N_k^{share} = \frac{n_{sub} \times 100}{n_{pop}}$ (Calculate the share of population closest to kth direction)
        $S_{k,G+1} = \frac{100 - N_k^{share}}{n_{dir} - 1} \times \frac{n_{pop}}{100}$ (Larger sub-population sizes for directions with smaller shares and vice versa)
    end for
    Return $\mathbf{P}_{G+1}^{arr} = \left[\text{round}(S_{1,G+1}), \ldots, \text{round}(S_{n_{dir},G+1})\right]$
end procedure

vector/number of possible actions at each hour$)^{-1} = 1/(N/2) = 2/N$. Finally, in line 5, the size of the sub-populations associated with each reference vector is declared as $\mathbf{P}_{G=1}^{arr} = \left[S_{1,G=1}, \ldots, S_{n_{dir},G=1}\right] = \left[n_{pop}/n_{dir}, \overset{n_{dir}}{\ldots}, n_{pop}/n_{dir}\right]$.

2. Optimization loop (Lines 6–22): This is the for loop which iterates until the termination condition is met. The steps in this optimization loop are followed for any generation G. In line 7, the arithmetic mean of the mutation probabilities of all candidates in the current population is adaptively estimated as $\mu_{P_{mut}}$, which is to be used later in the hyper-parameter estimation step of the reproduction loop. The for loop in line 8–15 iterates over all the n_{dir} directions, which is involved in

the generation of a new solution. This reproduction loop is explained in the next point. At the end of line 15, there are n_{pop} old (parent) solutions and n_{dir} new (child) solutions from which only n_{pop} solutions can pass to the next generation. Hence, the elitist selection is performed in line 16.

The selection of n_{pop} compromises (trade-off solutions) is dictated by the Pareto-dominance relation where $\mathbf{X}_{B,1}$ Pareto-dominates $\mathbf{X}_{B,2}$ ($\mathbf{X}_{B,1} \prec \mathbf{X}_{B,2}$) according to Eq. (6).

$$\forall i \in \{1, \ldots, M\}, f_{B,i}(\mathbf{X}_{B,1}) \leq f_{B,i}(\mathbf{X}_{B,2}) \text{ and}$$
$$\exists j \in \{1, \ldots, M\}, f_{B,j}(\mathbf{X}_{B,1}) < f_{B,j}(\mathbf{X}_{B,2}) \tag{6}$$

Thus, when two solution vectors are compared with respect to Pareto-dominance relation, either of the three alternatives can take place: $\mathbf{X}_{B,1} \prec \mathbf{X}_{B,2}$, $\mathbf{X}_{B,2} \prec \mathbf{X}_{B,1}$ or $\mathbf{X}_{B,1}$ and $\mathbf{X}_{B,2}$ are non-dominated. Non-dominated solutions are considered to be equivalent. Moreover, the set of non-dominated solutions (which are not dominated by any solution) gives the estimation of the Pareto-optimal set. Such solutions are also known as the rank-one solutions ($\mathcal{R}_1$). The non-dominated set of solutions obtained after removing $\mathcal{R}_1$ from $\mathcal{A}_G$ yields the rank-two solutions ($\mathcal{R}_2$) and so on. Non-dominated sorting of the population $\mathcal{A}_G$ of size ($n_{pop} + n_{dir}$) partitions the population into $\mathcal{R}_1$, $\mathcal{R}_2$, ... Each rank of solutions is directly propagated to the population of the next generation $\mathcal{A}_{G+1}$ until a rank of solutions just exceeds n_{pop}. This last essential rank of solutions is further sorted based on special crowding distance (SCD) [22, 33] and only the required number of candidates are passed on to the next generation to fulfill the size requirement as shown in Fig. 3a. Special crowding distance (SCD) combines crowding distance in both objective (CDF) and decision (CDX) space according to Eq. (7) and thus, helps in maintaining diversity in both the spaces. Crowding distance of a solution is proportional to the perimeter of the hyper-rectangle formed by the neighbors of solutions [13]. It is due to the use of this SCD that the GA-TriM framework is capable of retaining solution vectors that have similar objective vectors and thus, helps GA-TriM in addressing the multi-modal multi-objective problem.

$$SCD(\mathbf{X}_B) = \begin{cases} max\,(CDX(\mathbf{X}_B), CDF(\mathbf{X}_B))\,, & \text{if } CDX(\mathbf{X}_B) > \overline{CDX(\mathbf{X}_B)} \\ & \text{or } CDF(\mathbf{X}_B) > \overline{CDF(\mathbf{X}_B)} \\ min\,(CDX(\mathbf{X}_B), CDF(\mathbf{X}_B))\,, & \text{otherwise} \end{cases} \tag{7}$$

In line 17, a vector $\mathbf{I}_{AD}$ of length n_{pop} is created whose ith element represents the index k of the direction $\mathbf{W}_k$ with which the solution vector $\mathbf{X}_{B,i}$ is associated. This association between $\mathbf{W}_k$ and $\mathbf{X}_{B,i}$ is dictated by the $d2$ distance as given by Eq. (8) and illustrated in Fig. 3b.

$$\text{Sub-space associated with } \mathbf{W}_k : \{\mathbf{F}_B(\mathbf{X}_B) \mid d2\,(\mathbf{X}_B|\mathbf{W}_k) \le d2\left(\mathbf{X}_B|\mathbf{W}_j\right)\}$$
$$\text{where, } j = \{1, 2, \ldots, n_{dir}\}, k \neq j,$$

$$d1 = \frac{\left\|\left(\mathbf{F}_B\left(\mathbf{X}_{B,i}\right)\right)^T \mathbf{W}_k\right\|}{\|\mathbf{W}_k\|} \text{ and } d2 = \left\|\mathbf{F}_B\left(\mathbf{X}_{B,i}\right) - d1\frac{\mathbf{W}_k}{\|\mathbf{W}_k\|}\right\| \tag{8}$$

The sub-population sizes in $\mathbf{P}_G^{arr}$ are readjusted, using Algorithm 3, in line 18. With the new sub-population sizes in $\mathbf{P}_{G+1}^{arr}$, the next generation takes place. However, after every small number of generations (10 generations), the mutation probabilities are re-initialized (in lines 19–21) to forget the past landscape and adapt the hyper-parameters as per the present landscape.

3. Reproduction loop (Lines 8–15): For the kth iteration of the reproduction loop, in line 9, a sub-population $\mathcal{A}_G^{sub\_k}$ is created with $S_{k,G}$ candidates from $\mathcal{A}_G$ which are nearest to $\mathbf{W}_k$. From this sub-population, any random candidate is chosen in line 10 as the parent candidate ($\mathbf{X}_B^{parent}$) for reproduction. In line 11, the mutation probability is sampled from a normal distribution whose mean was evaluated as $\mu_{P_{mut}}$ in line 7. This is the hyper-parameter estimation step. The mutation probability and the parent candidate participate in Algorithm 2 to generate the new child candidate ($\mathbf{X}_B^{child}$) in line 12. The child solution vector is appended to $\mathcal{A}_G$ and its objective vector is appended to $\mathcal{A}_{\mathbf{F},G}$ in lines 13 and 14, respectively. In line 12, the reproduction loop calls Algorithm 2 where another parent ($\mathbf{X}_{B,2}^{parent}$) is randomly chosen from the sub-population associated with a neighboring direction. The two parents engage in the if-else tree of Algorithm 2 to create $\mathbf{X}_B^{child}$ as exemplified in Fig. 3c.
4. Termination (Line 23): At the end of G_{max} generations, the evolved population of solution vectors is returned in $\mathcal{A}_{G_{max}}$ and its associated set of objective vectors is returned in $\mathcal{A}_{\mathbf{F},G_{max}}$.

As mentioned, line 18 of Algorithm 1 calls Algorithm 3 for the re-adjustment of sub-population sizes. Algorithm 3 calculates n_{sub} as the number of candidates associated with direction k. Then, it evaluates the percentage of the total population associated with a direction in N_k^{share}. Negatively correlating the sub-population size ($S_{k,G+1}$) with the percentage of associated candidates (N_k^{share}), the sub-population size ($S_{k,G+1}$) is re-evaluated to allow more exploration (diverse candidates lead to global search) of less explored sub-spaces and to allow more exploitation (less diverse candidates lead to local search) of already explored areas. Thus, instead of forming sub-population with associated candidates, a predetermined number of candidates form sub-populations by borrowing or donating candidates from neighboring sub-spaces in line 9 of Algorithm 1. The feedback of sub-population size is explained through the working of Algorithm 3 using an example in Fig. 4.

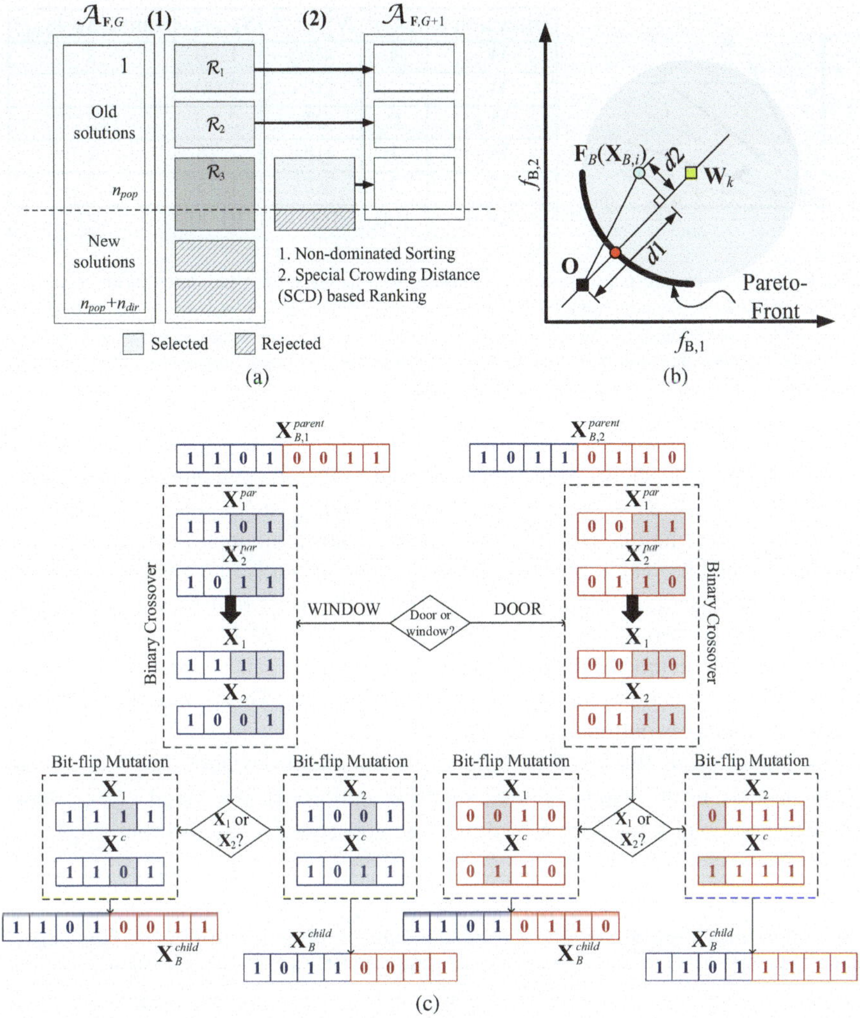

Fig. 3 (**a**) Elitist selection strategy of GA-TriM, (**b**) Association of $\mathbf{X}_{B,i}$ to $\mathbf{W}_k$ is related to $d2$, (**c**) Illustration for creation of $\mathbf{X}_B^{child}$ (true dimension of solution vector is ignored in the example)

3.5 Proposed Schedule Selection Approach

Once the estimated Pareto-optimal set ($\mathcal{A}_{G_{max}}$) and Pareto-Front ($\mathcal{A}_{\mathbf{F},G_{max}}$) are obtained, one of relevant action schedule (solution vector) is to be chosen. From the literature survey, it can be seen that most of the contemporary decision-making approaches [19, 21, 26] are based on the users' preference. Only the method of distance to best compromises, used in [3, 25], is an automated decision-making approach.

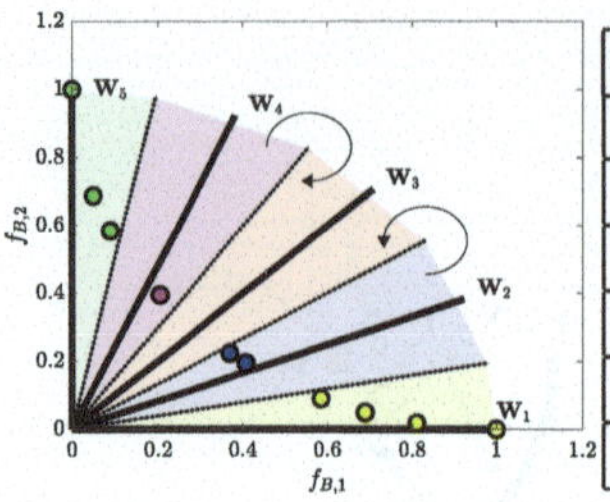

Direction k	n_{sub}	N_k^{share}	$S_{k,G+1}$	round($S_{k,G+1}$)
$\mathbf{W}_1$	4	40%	1.50	2
$\mathbf{W}_2$	2	20%	2.00	2
$\mathbf{W}_3$	0	0%	2.50	2
$\mathbf{W}_4$	1	10%	2.25	2
$\mathbf{W}_5$	3	30%	1.75	2
Total	100	100%	10.00	10

Fig. 4 Feedback on sub-population sizes in order to allocate more candidates (with respect to number of associated candidates) for exploring less explored areas (e.g. sub-space associated with $\mathbf{W}_3$) and less candidates for exploring more explored areas (e.g. sub-space associated with $\mathbf{W}_1$). The goal is to allocate candidates such that the size of all the sub-populations, associated with n_{dir} directions, evolves to n_{pop}/n_{dir} for uniform diversity

Recent literature shows an incremental mindset/incremental motivational framework [16, 27, 30] as a powerful teaching method. Thus, a subset of estimated Pareto-optimal solutions can be chosen, which shows minimal changes from the historical schedule. The motivation for such an approach is to convey the message to the occupants: "Even with little variations in the usual routine, a large amount of change in comfort can be attained."Once this subset of Pareto-optimal solutions is obtained, the method of distance to best compromises can be applied to generate the most relevant action schedule for a recommendation. This proposal of schedule selection strategy is outlined in Algorithm 4.

In Algorithm 4, lines 2–5 determine the minimum deviation (Δ_{min}^{sch}) over all the schedules in the Pareto-optimal set from the historical schedule ($\tilde{X}_B$). Next step (line 6) obtains the subset of those solutions (pruned Pareto-Front, $\mathcal{A}_{\mathbf{F},G_{max}}^{sch}$) which can generate Δ_{min}^{sch}. The objective vector ($\mathbf{F}_B^{\star}$) showing minimum net global dissatisfaction [3, 25] among those in $\mathcal{A}_{\mathbf{F},G_{max}}^{sch}$ is obtained in line 7 and its corresponding solution vector (action schedule) is obtained in line 8. This action schedule ($\mathbf{X}_B^{\star}$) is finally returned as the estimate of the most relevant Pareto-optimal schedule of actions in line 9 from Algorithm 4. In this work, the evaluation of net global dissatisfaction considers the origin of the objective space as the reference objective vector ($\mathbf{F}_B^{ref}$).

In the next section, the proposed action schedule selection approach is validated using real-world data.

4 Discussion on Various Schedule Selection Strategies

The proposed approach (GA-TriM) is implemented on a computer having 8 GB RAM and 2.20 GHz Intel Core i7 processor using Python 3.4. The data collected

Algorithm 4 Choosing relevant solution vector from estimated Pareto-optimal set

Input: $\mathcal{A}_{G_{max}}$: Estimated Pareto-optimal set; $\mathbf{F}_B^{ref}$: Reference objective vector; $\tilde{\mathbf{X}}_B$: Historical schedule from $\mathcal{H}_{DB}$
Output: $\mathbf{X}_B^{\star}$: Relevant solution vector

```
procedure SCHEDULE_SELECT(A_{G_max}, A_{F,G_max}, X̃_B)
    for i = 1 to n_pop (for each solution vector in A_{G_max}) do
        ΔX_{B,i} = Σ_{j=1}^{N} |x̃_{B,j} − x_{B,i,j}| (Net deviation of ith solution from historical schedule)
    end for
    Δ^{sch}_{min} ← Minimum deviation in schedule over all ΔX_{B,i}
    A^{sch}_{F,G_max} = {F_B(X_B) | ΔX_B = Δ^{sch}_{min}} (Obtain the subset of objective vectors for those solution vectors which generate Δ^{sch}_{min})
    F*_B = argmin Σ_{j=1}^{M} |f^{ref}_{B,j} − f_{B,j}| over all F_B ∈ A^{sch}_{F,G_max} (Find best objective vector from pruned Pareto-Front using distance to best compromises [3])
    X*_B = arg(F*_B) (Obtain the corresponding solution vector)
    Return X*_B
end procedure
```

in the database $\mathcal{H}_{DB}$ from Grenoble Institute of Technology helps in analyzing the efficacy of the proposed approach.

In the first experiment, the effectiveness of GA-TriM as a multi-objective evolutionary algorithm is assessed. For this purpose, the estimated Pareto-Front from GA-TriM is compared with those obtained from other optimization algorithms in terms of purity metric.

Purity metric [8] reveals the proportion of non-dominated solutions contributed by the ith non-dominated solution set ($\mathcal{A}_{\mathbf{F},i}$) towards the union of K_{PF} sets of non-dominated solutions ($\mathcal{A}_{\mathbf{F}}^{\star}$ = non-dominated set with respect to $\cup_{i=1}^{K_{PF}} \mathcal{A}_{\mathbf{F},i}$). The purity metric ($PM$) for the ith approximation of Pareto-Front is formally evaluated using Eq. (9).

$$PM\left(\mathcal{A}_{\mathbf{F},i}, \mathcal{A}_{\mathbf{F}}^{\star}\right) = \frac{\left|\mathcal{A}_{\mathbf{F},i} \cap \mathcal{A}_{\mathbf{F}}^{\star}\right|}{\left|\mathcal{A}_{\mathbf{F},i}\right|}, \text{ for } i = 1, 2, \ldots, K_{PF} \quad (9)$$

GA-TriM is compared with those MOEAs which have already established their effectiveness in addressing the concerned building energy management problem. Along with these MOEAs, a recent decomposition and dominance based MOEA (known as NAEMO [31]) is also used in the comparison. The specifications of these MOEAs, used in this experiment, are mentioned in Table 1. It should be noted that a generation of GA-TriM is over n_{dir} directions, not over n_{pop} candidates. Hence, G_{max} is different for GA-TriM. However, for a fair comparison, the number of evaluations of objective functions is kept constant at 6000 for all the MOEAs.

Comparison of the abilities of MOEAs in obtaining a better Pareto-optimal approximation is assessed in Table 2 in terms of purity metric for 15 days. Moreover, for a fair comparison, 20 candidates are randomly sampled from $\mathcal{A}_{G_{max}}$ (output of GA-TriM) such that the final population size of all the algorithms is equal.

Table 1 Specifications of different MOEAs compared with GA-TriM for solving the MOO problem of building energy management

GA-TriM	$n_{dir} = 10, n_{pop} = 10 \times n_{dir} = 100, G_{max} = 6000/n_{dir} = 600$
DEMO [3]	DE/rand/1/bin, crossover rate (CR) = 0.8, scale factor (F^{DE}) $\in [0, 2], n_{pop} = 20, G_{max} = 300$
NSGA-II [13, 23, 26]	Single point binary crossover, binary tournament selection, bit-flip mutation (mutation probability = 1/24), $n_{pop} = 20, G_{max} = 300$
AGE-II [25, 32]	Single point binary crossover, binary tournament selection, bit-flip mutation (mutation probability = 1/24), degree of additive approximation (ϵ) = 0.01, $n_{pop} = 20, G_{max} = 300$
NAEMO [31]	$n_{pop} = n_{dir} = 20, G_{max} = 300$, remaining parameters are set as specified in [31]

Table 2 Purity metric (rank of algorithm) to compare the quality (number and Pareto-optimality) of the solution vectors estimated by various MOEAs where best and second-best performing values are highlighted in dark and light shades of gray, respectively

Sl. No.	Date	AGE-II [25]	DEMO [3]	GA-TriM	NAEMO [31]	NSGA-II [23]
1	01-Apr-2015	0.3333 (4)	0.2222 (5)	1.0000 (1)	0.8333 (3)	0.8500 (2)
2	08-Apr-2015	0.0000 (4)	0.6667 (3)	0.7500 (2)	0.0000 (4)	0.8889 (1)
3	10-Apr-2015	0.0000 (2)	1.0000 (1)	0.0000 (2)	0.0000 (2)	0.0000 (2)
4	15-Apr-2015	0.0000 (2)	0.0000 (2)	1.0000 (1)	1.0000 (1)	1.0000 (1)
5	20-Apr-2015	0.0000 (3)	0.0000 (3)	1.0000 (1)	1.0000 (1)	0.9474 (2)
6	30-Apr-2015	0.0000 (2)	0.0000 (2)	1.0000 (1)	1.0000 (1)	1.0000 (1)
7	06-May-2015	0.0000 (2)	0.0000 (2)	1.0000 (1)	1.0000 (1)	1.0000 (1)
8	11-May-2015	0.0000 (2)	0.0000 (2)	1.0000 (1)	1.0000 (1)	1.0000 (1)
9	20-May-2015	0.0000 (2)	0.0000 (2)	1.0000 (1)	1.0000 (1)	1.0000 (1)
10	26-May-2015	0.0000 (3)	0.0000 (3)	0.6667 (1)	0.5000 (2)	0.6667 (1)
11	01-Jun-2015	0.0000 (2)	1.0000 (1)	0.0000 (2)	0.0000 (2)	0.0000 (2)
12	10-Jun-2015	0.0000 (2)	0.0000 (2)	1.0000 (1)	1.0000 (1)	1.0000 (1)
13	19-Jun-2015	0.0000 (2)	0.0000 (2)	1.0000 (1)	1.0000 (1)	1.0000 (1)
14	30-Jun-2015	0.0000 (3)	1.0000 (1)	0.6667 (2)	1.0000 (1)	0.6667 (2)
15	07-Jul-2015	0.0000 (2)	0.0000 (2)	1.0000 (1)	1.0000 (1)	1.0000 (1)
Average ranks		2.47	2.20	1.27	1.47	1.33

The following insights are noted from the observations in Table 2:

- In 11 out of 15 cases, GA-TriM has a purity metric of 1, which implies that all the candidates from GA-TriM are superior over the non-dominated solutions generated by other MOEAs.
- In several cases, NSGA-II, NAEMO, and GA-TriM tie in the resulting purity values. Thus, these algorithms have nearly similar capabilities to approximate the Pareto-optimal solution set. The poor performance of DEMO [3] can be attributed to its non-elitist nature. For AGE-II [32], the additive approximations render several objective vectors (not equivalent in terms of Pareto-dominance) to be equivalent.

- In terms of purity metric, the overall rank of GA-TriM is the highest over 15 days, followed by NSGA-II.

In the second experiment, two automated decision-making approaches (new/distance to best compromises in pruned Pareto-Front versus old/distance to best compromises in estimated Pareto-Front) are compared for 15 days. Comparison is done in the objective space in terms of net global dissatisfaction attained when the recommended routine (new/old) is followed versus when the usual routine is followed. Comparison is also done in the decision space in terms of the percentage of net deviation required from the usual routine to follow the recommended routine (new/old). These results are mentioned in Table 3.

The following insights are obtained from the observations in Table 3:

- On all the days, the net global dissatisfaction can be reduced using the GA-TriM recommended schedule from the usual schedule.
- Using the existing schedule selection strategy (distance to best compromises [3, 25]), although a lower net dissatisfaction can be obtained, the numerical value is very close to that obtained by the proposed schedule selection strategy. Moreover, this net dissatisfaction can be obtained with much less disturbance to the original schedule of the occupant.

  Pareto-Fronts estimated by GA-TriM are also visualized in Fig. 5, for 2 days, to compare the position of various compromises of interest.

Table 3 Comparison of proposed decision-making strategy (New) with existing decision-making strategy (Old) for recommending the occupants actions schedule where best and second-best performing values are highlighted in dark and light shades of gray, respectively

Sl. No.	Date	Net global dissatisfaction			Net deviation (%) in schedule	
		Usual	GA-TriM (New)	GA-TriM) (Old)	GA-TriM (New)	GA-TriM (Old)
1	01-Apr-2015	0.3912	0.3605	0.3603	25.00	33.33
2	08-Apr-2015	0.7300	0.7172	0.7126	33.33	41.67
3	10-Apr-2015	0.2062	0.1955	0.1955	41.67	45.83
4	15-Apr-2015	0.5066	0.4151	0.4150	58.33	62.50
5	20-Apr-2015	0.5976	0.5952	0.5948	25.00	41.67
6	30-Apr-2015	0.3717	0.3652	0.3652	37.50	45.83
7	06-May-2015	0.3506	0.2243	0.2242	45.83	50.00
8	11-May-2015	0.6536	0.5761	0.5759	29.17	45.83
9	20-May-2015	0.3967	0.3930	0.3930	42.00	42.00
10	26-May-2015	0.5038	0.4660	0.4660	54.17	62.50
11	01-Jun-2015	1.3041	1.1661	1.1661	79.17	87.50
12	10-Jun-2015	1.1021	1.1013	1.1013	29.17	29.17
13	19-Jun-2015	0.7829	0.7828	0.7828	4.17	8.33
14	30-Jun-2015	1.9972	1.9963	1.9963	25.00	45.83
15	07-Jul-2015	3.2075	3.1834	3.1834	45.83	58.33

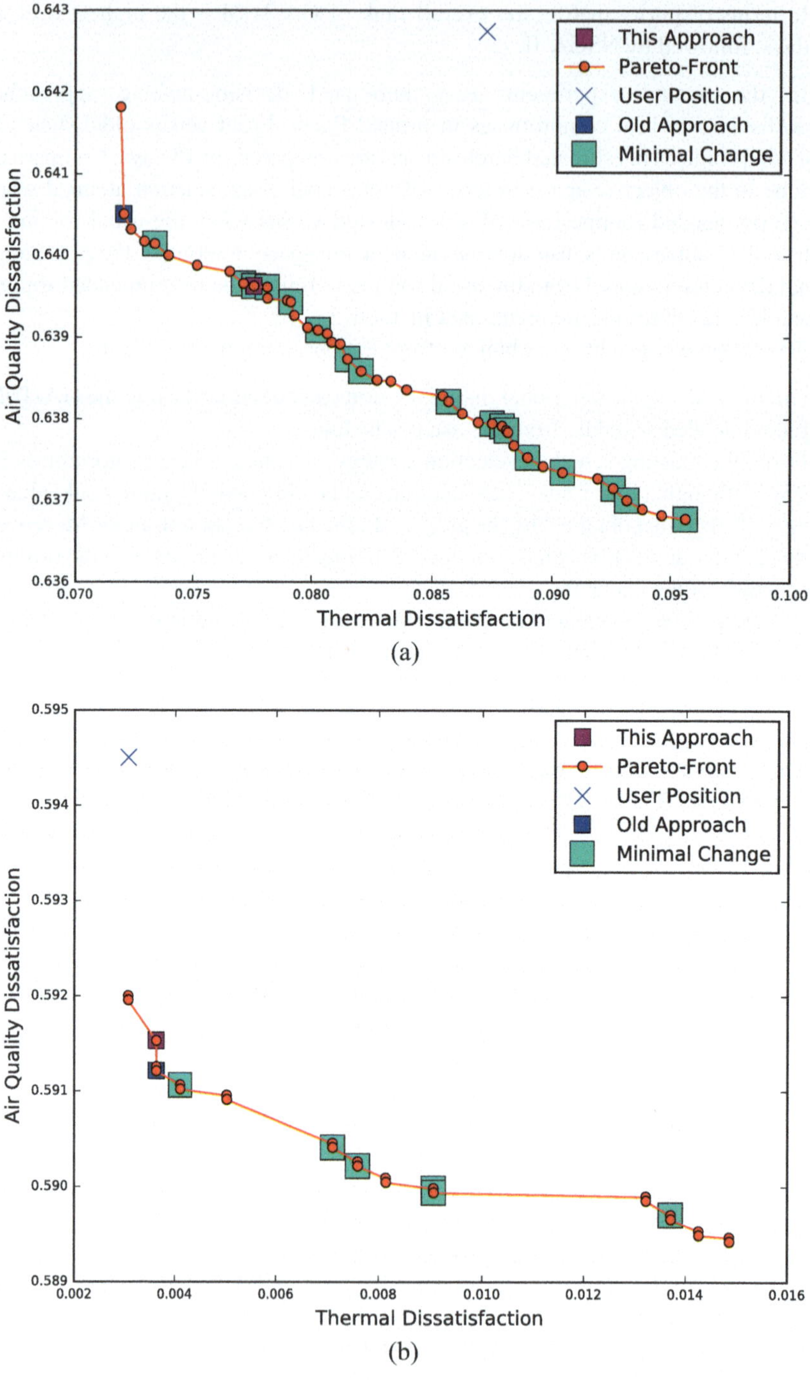

Fig. 5 Estimated Pareto-Fronts using real-world data showing equivalent objective vectors are distributed over the entire Pareto-optimal surface where cyan square objective vectors correspond to action schedules having equal deviation from the historical schedule, blue square and violet

- On several days (such as on 30-Apr-2015, 01-Jun-2015, etc.), both the strategies yield similar net global dissatisfaction. Yet the proposed strategy can be considered to be better as it attains this value with fewer changes. This result is possible only because GA-TriM facilitates discovering the equivalent solution vectors for the same objective vectors.
- The net dissatisfaction values attained after June 2015 (summer), are higher than those obtained before. Even with high percentage of changes in action schedule (such as on 01-Jun-2015), the net dissatisfaction value cannot be minimized any further. This situation necessitates the use of HVAC in the office room, as has been discussed in [24, 25].

5 Conclusion and Scope of Further Research

As building energy management is a global concern, this research domain is of vital importance. Several existing works have considered the optimization of the impact of occupants' actions so that occupants' dissatisfaction is reduced by enforcing an energy-efficient schedule of opening/closing of doors/windows. However, multi-objective optimization algorithms result in a set of trade-off solutions. From this solution set, only one relevant solution can be implemented for practical application.

Selection strategies for choosing this relevant occupants' action schedule are known as decision-making. The present work discusses various decision-making approaches and notices that existing schedule selection strategies are based on preferences in the objective space. Moreover, when an optimal action schedule is directly recommended to occupants, the necessary awareness is not created to modify their habits. However, it will be easier to convince the occupants to introduce small changes in their action schedule. With this motivation, a schedule selection strategy is proposed in this work which creates a subset from the estimated Pareto-optimal solution set by selecting action schedules that have the minimum deviation from the historical schedule. Then, the schedule selection strategy continues over this subset of solutions in the objective space as done in earlier works. However, to find various equivalent solutions over the entire Pareto-optimal surface it is necessary to design a multi-objective evolutionary algorithm that is capable of exploring the solution space (in addition to the objective space) such that multiple action schedules for the same objective vector can be found. This is the multi-modal nature of the building energy management problem. Thus, a genetic algorithm for

Fig. 5 (continued) square correspond to relevant solutions using old decision-making approach [3] and this proposed decision-making approach, respectively. Additionally, user position (blue cross) marks the objective vector resulting from the historical routine of the occupant. (**a**) 08-April-2015. (**b**) 20-April-2015 multi-modal multi-objective problems (GA-TriM) is designed in this work which helps in addressing the concerned the building energy management problem.

The result demonstrates the superiority of GA-TriM in comparison to other algorithms that have been previously used for building energy management. Moreover, the proposed schedule selection strategy also demonstrates that nearly the same optimal value of objectives (net global dissatisfaction) can be attained with fewer changes in the action schedule. However, the present framework suffers from the following limitations, and future work can be carried out to overcome each of these limitations:

1. Lack of data and physical knowledge models of various physical parameters such as humidity, pollutants, etc. create a bottleneck to expand the current framework. Moreover, every new zone requires new models to be fitted. Machine learning models (such as neural networks, logistic regression, etc.) or time-series processing methods may be used to bypass these simulation models. However, research in this aspect is still in its infancy.
2. Seasonal variations in outdoor weather heavily influence the indoor ambience. However, due to the unavailability of data for random extreme fluctuations (like natural disasters), it cannot be concluded whether the existing framework can operate effectively in such situations. Data acquisition and rigorous investigations in these aspects are extremely important in this era of global warming.

Acknowledgments The authors would like to thank Prof. Stéphane Ploix for allowing us to utilize the data acquired at Grenoble Institute of Technology, France, for building energy management associated research studies.

This work is partially supported by the Indian Statistical Institute, Kolkata, and J. C. Bose Fellowship (SB/SJ/JCB-033/2016) of the Department of Science and Technology, Government of India.

References

1. P. Alotto, M. Gaggero, G. Molinari, M. Nervi, A "design of experiment" and statistical approach to enhance the "generalised response surface" method in the optimisation of multiminima problems. IEEE Trans. Magn. **33**(2), 1896–1899 (1997). https://doi.org/10.1109/20.582657
2. A.A. Alyafi, V.B. Nguyen, Y. Laurillau, P. Reignier, S. Ploix, G. Calvary, J. Coutaz, M. Pal, J.P. Guilbaud, From usable to incentive-building energy management systems. Model. Using Context **2**(Issue 1), 1–30 (2018). https://doi.org/10.21494/ISTE.OP.2018.0302
3. A. Alzouhri Alyafi, M. Pal, S. Ploix, P. Reignier, S. Bandyopadhyay, Differential explanations for energy management in buildings, in *IEEE Technically Sponsored SAI Computing Conference* (2017), pp. 507–516
4. M. Amayri, S. Ploix, S. Bandyopadhyay, Estimating occupancy in an office setting, in *Sustainable Human–Building Ecosystems* (2015), pp. 72–80
5. M. Amayri, A. Arora, S. Ploix, S. Bandhyopadyay, Q.D. Ngo, V.R. Badarla, Estimating occupancy in heterogeneous sensor environment. Energy Build. **129**, 46–58 (2016)
6. E. Asadi, M.G. Da Silva, C.H. Antunes, L. Dias, Multi-objective optimization for building retrofit strategies: a model and an application. Energy Build. **44**, 81–87 (2012)

7. E. Asadi, M.G. da Silva, C.H. Antunes, L. Dias, L. Glicksman, Multi-objective optimization for building retrofit: a model using genetic algorithm and artificial neural network and an application. Energy Build. **81**, 444–456 (2014)
8. S. Bandyopadhyay, S.K. Pal, B. Aruna, Multiobjective GAs, quantitative indices, and pattern classification. IEEE Trans. Syst. Man Cybern. B Cybern. B **34**(5), 2088–2099 (2004)
9. B. Bredeweg, K.D. Forbus, Qualitative modeling in education. AI Mag. **24**(4), 35–46 (2003). http://www.aaai.org/ojs/index.php/aimagazine/article/view/1729
10. C.A.C. Coello, Recent results and open problems in evolutionary multiobjective optimization, in *Theory and Practice of Natural Computing*, ed. by C. Martín-Vide, R. Neruda, M.A. Vega-Rodríguez (Springer International Publishing, Cham, 2017), pp. 3–21
11. I. Das, J.E. Dennis, Normal-boundary intersection: a new method for generating the pareto surface in nonlinear multicriteria optimization problems. SIAM J. Optim. **8**(3), 631–657 (1998)
12. K. Deb, H. Jain, An evolutionary many-objective optimization algorithm using reference-point-based nondominated sorting approach, part I: solving problems with box constraints. IEEE Trans. Evol. Comput. **18**(4), 577–601 (2014)
13. K. Deb, S. Agrawal, A. Pratap, T. Meyarivan, A fast elitist non-dominated sorting genetic algorithm for multi-objective optimization: NSGA-II, in *Parallel Problem Solving from Nature PPSN VI. PPSN 2000*, ed. by M. Schoenauer et al. Lecture Notes in Computer Science, vol. 1917 (Springer, Berlin, 2000), pp. 849–858
14. C. Diakaki, E. Grigoroudis, D. Kolokotsa, Towards a multi-objective optimization approach for improving energy efficiency in buildings. Energy Build. **40**(9), 1747–1754 (2008)
15. Y. Fan, X. Xia, A multi-objective optimization model for building envelope retrofit planning. Energy Procedia **75**, 1299–1304 (2015)
16. E.A. Gunderson, N.S. Sorhagen, S.J. Gripshover, C.S. Dweck, S. Goldin-Meadow, S.C. Levine, Parent praise to toddlers predicts fourth grade academic achievement via children's incremental mindsets. Dev. Psychol. **54**(3), 397 (2018)
17. T. Hong, S.C. Taylor-Lange, S. D'Oca, D. Yan, S.P. Corgnati, Advances in research and applications of energy-related occupant behavior in buildings. Energy Build. **116**, 694–702 (2016)
18. J. Langevin, J. Wen, P.L. Gurian, Modeling thermal comfort holistically: Bayesian estimation of thermal sensation, acceptability, and preference distributions for office building occupants. Build. Environ. **69**, 206–226 (2013)
19. Y. Laurillau, V.B. Nguyen, J. Coutaz, G. Calvary, N. Mandran, F. Camara, R. Balzarini, The TOP-slider for multi-criteria decision making by non-specialists, in *Proceedings of the 10th Nordic Conference on Human-Computer Interaction* (ACM, New York, 2018), pp. 642–653
20. A.T. Nguyen, S. Reiter, P. Rigo, A review on simulation-based optimization methods applied to building performance analysis. Appl. Energy **113**, 1043–1058 (2014). https://doi.org/10.1016/j.apenergy.2013.08.061. http://www.sciencedirect.com/science/article/pii/S0306261913007058
21. M. Pal, S. Bandyopadhyay, Consensus of subjective preferences of multiple occupants for building energy management, in *2018 IEEE Symposium Series on Computational Intelligence (SSCI)*, pp. 1815–1822 (2018). https://doi.org/10.1109/SSCI.2018.8628670
22. M. Pal, S. Bandyopadhyay, Differential evolution for multi-modal multi-objective problems, in *Proceedings of the Genetic and Evolutionary Computation Conference Companion, GECCO '19* (ACM, New York, 2019), pp. 1399–1406. http://doi.acm.org/10.1145/3319619.3326862
23. M. Pal, R. Sengupta, S. Bandyopadhyay, A.A. Alyafi, S. Ploix, P. Reignier, S. Saha, Analysis of optimizers to regulate occupant's actions for building energy management, in *2017 Ninth International Conference on Advances in Pattern Recognition (ICAPR)* (2017), pp. 1–6. https://doi.org/10.1109/ICAPR.2017.8593024
24. M. Pal, A.A. Alyafi, S. Bandyopadhyay, S. Ploix, P. Reignier, Enhancing comfort of occupants in energy buildings, in *Operations Research and Optimization. FOTA 2016*, ed. by S. Kar, U. Maulik, X. Li. Springer Proceedings in Mathematics & Statistics, vol. 225 (Springer, Singapore, 2018), pp. 133–144

25. M. Pal, A.A. Alyafi, S. Ploix, P. Reignier, S. Bandyopadhyay, Unmasking the causal relationships latent in the interplay between occupant's actions and indoor ambience: a building energy management outlook. Appl. Energy **238**, 1452–1470 (2019). https://doi.org/10.1016/j.apenergy.2019.01.118. http://www.sciencedirect.com/science/article/pii/S0306261919301047
26. S. Papadopoulos, E. Azar, Optimizing HVAC operation in commercial buildings: a genetic algorithm multi-objective optimization framework, in *Proceedings of the 2016 Winter Simulation Conference* (IEEE Press, Piscataway, 2016), pp. 1725–1735
27. D. Park, E.A. Gunderson, E. Tsukayama, S.C. Levine, S.L. Beilock, Young children's motivational frameworks and math achievement: relation to teacher-reported instructional practices, but not teacher theory of intelligence. J. Educ. Psychol. **108**(3), 300 (2016)
28. S. Ramakrishnan, X. Wang, J. Sanjayan, J. Wilson, Thermal performance assessment of phase change material integrated cementitious composites in buildings: experimental and numerical approach. Appl. Energy **207**, 654–664 (2017). https://doi.org/10.1016/j.apenergy.2017.05.144. http://www.sciencedirect.com/science/article/pii/S030626191730689X. Transformative Innovations for a Sustainable Future – Part II
29. T. Robič, B. Filipič, DEMO: differential evolution for multiobjective optimization, in *Evolutionary Multi-Criterion Optimization* (Springer, Berlin, 2005), pp. 520–533
30. A.A. Ryazanov, N.J. Christenfeld, Incremental mindsets and the reduced forgiveness of chronic failures. J. Exp. Soc. Psychol. **76**, 33–41 (2018). https://doi.org/10.1016/j.jesp.2017.12.003. http://www.sciencedirect.com/science/article/pii/S0022103117301658
31. R. Sengupta, M. Pal, S. Saha, S. Bandyopadhyay, NAEMO: neighborhood-sensitive archived evolutionary many-objective optimization algorithm. Swarm Evol. Comput. **46**, 201–218 (2019). https://doi.org/10.1016/j.swevo.2018.12.002. http://www.sciencedirect.com/science/article/pii/S2210650218304231
32. M. Wagner, F. Neumann, A fast approximation-guided evolutionary multi-objective algorithm, in *Proceedings of the 15th Annual Conference on Genetic and Evolutionary Computation* (ACM, New York, 2013), pp. 687–694
33. C. Yue, B. Qu, J. Liang, A multiobjective particle swarm optimizer using ring topology for solving multimodal multiobjective problems. IEEE Trans. Evol. Comput. **22**(5), 805–817 (2018). https://doi.org/10.1109/TEVC.2017.2754271
34. J. Zhao, B. Lasternas, K.P. Lam, R. Yun, V. Loftness, Occupant behavior and schedule modeling for building energy simulation through office appliance power consumption data mining. Energy Build. **82**, 341–355 (2014)

Generation of Optimal Strategies for Energy Management of Living Area Depicted by Thousands of Constraints

Quoc-Dung Ngo, Hussein Joumaa, and Mireille Jacomino

1 Introduction

Nowadays, the residential energy consumption is about 27.2% of the EU final energy consumption, representing the second largest consuming sector after transport [1]. Therefore, energy consumption reduction in building has become an important challenge for researchers. A lot of Building Energy Management Systems (BEMS) have been proposed [2–5] aiming at minimizing the daily electrical consumption while maintaining a satisfactory level of comfort for occupants. Modern dwelling systems may be complex in terms of number of appliances, including production and storage means but also in terms of applications such as monitoring and model-based estimation energy management. Several approaches are proposed in order to deal with the problem of generation of optimal strategies for energy management. In recent years, sophisticated methods, formalisms, and tools have been developed for different applications in order to better master dwelling energy consumption such as:

- G-homeTech, MACES, A Mathematical Programming Language (AMPL) for energy management. Actually, an energy management plan for instance proposed

Q.-D. Ngo
Faculty of Information Technology, Posts and Telecommunications Institute of Technology, Ho Chi Minh City, Vietnam
e-mail: dungnq@ptit.edu.vn

H. Joumaa (✉)
Capgemini DEMS, Montbonnot-Saint-Martin, France
e-mail: Hussein.Joumaa@capgemini.com

M. Jacomino
Université Grenoble Alpes, Grenoble INP Ense3, Laboratoire G-Scop, Grenoble Cedex 1, France
e-mail: Mireille.Jacomino@grenoble-inp.fr

S. Ploix et al. (eds.), *Towards Energy Smart Homes*,
https://doi.org/10.1007/978-3-030-76477-7_9

by G-homeTech aims at proposing to occupants the best configurations of appliances for next 24 h in order to optimize the compromise comfort/cost. To deal with thousands of variables and constraints problems in reasonable time, a mixed-integer linear programming (MILP) solver is used. Therefore, an acausal linear problem is required for this application.
- Matlab/Simulink, Modelica for simulation or parameter estimation. These tasks aim at simulating the dwelling behavior during a given time or at calculating parameters required for physically explicit models. In these applications, Matlab/Simulink, Modelica can handle causal models including both linear and non-linear ones.

However, these tools require for each a special formalism to be executable. It also means that a complex model developed in Matlab/Simulink for a simulation purpose cannot be reused in G-homeTech for example. This major drawback implies a rewriting process manually of models for each target application that represents time-consuming and error-prone. Therefore, tools to handle and transform models are required. This chapter presents different approaches proposed to generate an optimal strategy for energy management system and ends by presenting a solution to handle and transform models. Section 2 presents the regular/centralized solving approach based on mixed-integer linear programming (MILP) optimization. Section 3 presents an overview of works based on model predictive control (MPC). In Sect. 4, a solution for automatic model generation for energy management is presented.

2 Principle of Regular/Centralized Solving Approach

Components like rooms and appliances are elements of the structural representation of dwellings, but functional representation is more relevant for global energy management because it points out the role of each element. The concept of service is proposed in order to incorporate the functional representation of these components in the energy management solving process.

2.1 The Concept of Service

2.1.1 Definition of Service

A functional entity, a service SRV_i, is defined by:

- A set of supporting components and appliances of the dwelling
- A time period T_i where the service may occur: for instance, the time period in which a washing machine may consume power to do a specific washing

- A set of actions $U_i(T_i)$ that may modify the achievement of the service: it may be set points or controlled variables
- A set of available observations $Y_i(T_i)$ that provide information about an actual behavior: it has to include consumed, stored, or provided powers because energy is focused.
- A set of modeling constraints $K_{i,j}\ (U_i(T_i), Y_i(T_i)) = 0, \forall j$ that depict the links between actions and observations: these constraints depend on the supporting components and appliances
- A set of operational constraints $K'_{i,k}\ (U_i(T_i), Y_i(T_i)) \diamond 0, \forall k$[1] that depict the operational limits of the dwelling: these constraints depend on the supporting components and appliances
- A service performance indicator $S_i\ (U_i(T_i), Y_i(T_i))$: it may be an occupant comfort level indicator, a quantity of stored energy or a cost (energy or environmental cost for instance).

2.1.2 Type of Services

Dwellings with appliances aim at providing comfort to inhabitants as a final aim. Services can then be decomposed into three kinds:

- The end-user services that provide directly comfort to inhabitants
- The intermediate services that manage energy storage
- The support services that produce electrical power to intermediate and end-user services

Support services usually deal with electric power supplying thanks to conversion from a primary energy to electricity. *Fuel-cells-based generators*, *photovoltaic power suppliers*, and *grid power suppliers* belong to this class. Intermediate services are generally achieved by electrochemical batteries. Among the end-user services, the well-known services such as *cloth washing*, *water heating*, *specific room heating*, *cooking in oven*, and *lighting* can be found.

2.1.3 Service Qualification

Let us assume a given time range for anticipating the energy needs (typically 24 h). A service is qualified as *permanent* if its energetic consumption/production/storage covers the whole time range of energy assignment plan; otherwise, the service is named *temporary service*. Table 1 gives some examples of services according to this classification.

The services can also be classified according to the way their behavior can be modified.

[1] $\diamond$ stands for a comparison operator.

Table 1 Examples of temporary and permanent services

	Temporary services	Permanent services
Support services	Photovoltaic panels	Power provider
Intermediate services	–	Storage
End-user services	Washing	Room heating

Whatever the service is, an end-user, an intermediate, or a support service, it can be modifiable or not. A service is qualified as *modifiable by an energy management system* if the energy management system is capable to modify its behavior (the starting time for example).

There are different ways of modifying services. Sometimes, modifiable services can be considered as continuously modifiable such as the temperature set points in *room heating services* or the shift of a washing. Some other services may be modified discretely such as the interruption of a *washing service*. The different ways of modifying services can be combined: for instance, a washing service can be considered both as interruptible and as continuously shiftable. A service modelled as discretely modifiable contains discrete decision variables in its model, whereas a continuously modifiable service contains continuous decision variables. Of course, a service may contain both discrete and continuous decision variables.

A service can also be characterized by the way it is known by an automation system. The consumed or produced power may be observable or not. Moreover, for end-user services, the impact of a service on the inhabitant comfort may be known or not.

Obviously, a service can be taken into account by an energy management system if it is at least observable. Some services are indirectly observable. Indeed, all not observable services can be gathered into a virtual non-modifiable service whose consumption/production is deduced from a global power meter measurement and from the consumption/production of all observable services. In addition, a service can be taken into account for long-term scheduling if it is predictable. In the same way as for observable services, all the unpredictable services can be gathered into a global non-modifiable predictable service. A service can be managed by an automation system if it is observable and modifiable. Moreover, it can be long-term managed if it is predictable and modifiable.

2.2 *Principle of Control Mechanism*

An important issue in dwelling energy management problems is the uncertainties in the model data. For instance, solar radiation, outdoor temperature, or services requested by inhabitants may not be predicted with accuracy. In order to solve this issue, a three-layer architecture is proposed: a local layer, a reactive layer, and an anticipative layer.

The *anticipative layer* is responsible for scheduling end-user, intermediate, and support services taking into account predicted events and costs in order to avoid as much as possible the use of the *reactive layer*. The prediction procedure forecasts various information about future user requests but also about available power resources and costs. Therefore, it uses information from predictable services and manages continuously modifiable and shiftable services. This layer has slow dynamics (e.g., a 1 h sampling time) comparing to other layers and includes predictive models with learning mechanisms, including models dealing with inhabitant behaviors. This layer also contains a predictive control mechanism that schedules energy consumption and production of end-user services several hours in advance. This layer computes plans according to available predictions. The sampling period of the anticipative layer is denoted as Δ. This layer relies on the most abstract models.

The *reactive layer* has been detailed in [6]. Its objective is to manage adjustments of energy assignment in order to follow up a plan computed by the upper *anticipative layer* in spite of unpredicted events and perturbations. Therefore, this layer manages modifiable services and uses information from observable services (comfort for end-user services and power for others). This layer is responsible for decision-making in case of violation of predefined constraints dealing with energy and inhabitant comfort expectations: it performs arbitrations between services. The set points determined by the plan computed by the upper *anticipative layer* are dynamically adjusted in order to avoid user dissatisfaction. The control actions may be dichotomic in enabling/disabling services or more gradual in adjusting set points such as reducing temperature set point in room heating services or delaying a temporary service. Actions of the reactive layer have to remain transparent for the plan computed by the anticipative layer: it can be considered as a fast dynamic unbalancing system taking into account actual dwelling state, including unpredicted disturbances, to satisfy energy, comfort, and cost constraints. If the current state is too far from the computed plan, the anticipative layer has to re-compute it.

The *local layer* is composed of devices together with their existing local control systems generally embedded into appliances by manufacturers. It is responsible for adjusting device controls in order to reach given set points in spite of perturbations. This layer abstracts devices and services for upper layers: fast dynamics are hidden by the controllers of this level. This layer is considered as embedded into devices.

2.3 *Modeling and Solving Approach*

2.3.1 Modeling Services

The model of a service can be decomposed into two aspects: the modeling of the behaviors with operational constraints, which depends on the types of involved models, and the modeling of the service performances, which depends on the types of service. Whatever the type of model it is, it has to be defined all over a time horizon $K \times \Delta$ for anticipative problem solving composed of K sampling periods lasting Δ each.

2.3.2 Modeling Behavior of Services

In order to model the behavior of different kinds of services, three different types of models have been used: discrete events are modelled by finite-state machines, continuous behaviors are modelled by differential equations, and mixed discrete and continuous evolutions are modelled by hybrid models that combine the two previous ones. In this chapter, we present the case of finite-state machines. Other models are presented in [5].

Finite-State Machines (FSM)

A finite-state machine dedicated to a service, denoted SRV, is composed of a finite number of states $\{\mathcal{L}_m; m \in \{1, \ldots, M\}\}$ and a set of transitions between those states $\{\mathcal{T}_{p,q} \in \{0, 1\}; (p, q) \in S \subset \{1, \ldots, M\}^2\}$. Each state $\mathcal{L}_m$ of a service SRV is linked to a phase characterized by a maximal power production $P_m > 0$ or consumption $P_m < 0$.

A transition triggers a state change. It is described by a condition that has to be satisfied to be enabled. The condition can be a change of a state variable measured by a sensor, a decision of the anticipative mechanism, or an elapsed time for phase transition. If it exists a transition between the state $\mathcal{L}_m$ and $\mathcal{L}_{m'}$, then $\mathcal{T}_{m,m'} = 1$, otherwise $\mathcal{T}_{m,m'} = 0$. An action can be associated to each state: it may be a modification of a set point or an on/off switching. As an example, let us consider a washing service.

The service provided by a washing machine may be modelled by a FSM with 4 states: the first state is the *stand-by* state $\mathcal{L}_1$ with a maximal power of $P_1 = -5W$ (it is negative because it deals with consumed power). The transition toward the next state is triggered by the anticipative mechanism. The second state is the *water heating* state $\mathcal{L}_2$ with $P_2 = -2400W$. The transition to the next state is triggered after τ_2 time units. The next state corresponds to the *washing* characterized by $\mathcal{P}_3 = -500W$. And finally, after a given duration τ_3 depending on the type of washing (i.e., the type of requested service), the spin-drying state is reached with $\mathcal{P}_3 = -1000W$. After a given duration τ_4, the *stand-by* state is finally recovered. Considering that the initial state is $\mathcal{L}_1$, this behavior can be formalized by

$$\begin{cases} (state = \mathcal{L}_1) \wedge (t = t_{start}) & \rightarrow state = \mathcal{L}_2 \\ (state = \mathcal{L}_2) \wedge (t = t_{start+\tau_2}) & \rightarrow state = \mathcal{L}_3 \\ (state = \mathcal{L}_3) \wedge (t = t_{start+\tau_2+\tau_3}) & \rightarrow state = \mathcal{L}_4 \\ (state = \mathcal{L}_4) \wedge (t = t_{start+\tau_2+\tau_3+\tau_4}) & \rightarrow state = \mathcal{L}_1. \end{cases} \tag{1}$$

2.3.3 Modeling the Performance of Services

Depending on the type of service, the quality of the service achievement may be assessed in different ways. End-user services provide comfort to inhabitants, intermediate services provide autonomy, and support services provide power that can be assessed by its cost and its impact on the environment. In order to evaluate these qualities, different types of criteria have been introduced. In this chapter, the case of end-user services is presented. Other types of services are treated in [5].

End-User Services

The global function of comfort is very complex to compute. This function not only depends on the satisfaction regarding each service (heating, cooking, washing,...) taken on its own but also on psychological complex factors. Let us try to specify how is this global satisfaction function.

Let σ be the global function of comfort or the global function of satisfaction in a living space. Leaving implicit psychological factors, it can be stated: $\sigma = \sigma(\sigma_1, \ldots, \sigma_n)$, where σ_i represents the satisfaction related to a service SRV_i.

Each satisfaction function σ or σ_i takes a value in the interval [0, 1] with 0 the limits of acceptability and 1 the ideal value. The global satisfaction function has to satisfy:

1. $\sigma = \sigma(\sigma_1, \ldots, \sigma_n)$.
2. If $\gamma > 0$ is a satisfaction increase $\sigma_i + \gamma \leq 1$ of an end-user service, $\sigma(\sigma_1, \ldots, \sigma_i + \gamma, \ldots, \sigma_n) \geq \sigma(\sigma_1, \ldots, \sigma_i, \ldots, \sigma_n)$, it can be translated by $\frac{\delta\sigma(\sigma_1,\ldots,\sigma_n)}{\delta\sigma_i} \geq 0, \forall i$.
3. The global marginal satisfactions regarding the different services are possibly different: possibly, $\frac{\delta}{\delta\sigma_i}\sigma \neq \frac{\delta}{\delta\sigma_j}\sigma$ if $i \neq j$.
4. The bound conditions lead to: $s(0, \ldots, 0) = 0$ and $s(1, \ldots, 1) = 1$.

The global satisfaction function is complex, but it can be studied in a particular validity domain: $\forall i, \sigma_i \in \left[\sigma_{min}^i, 1\right]$ and $\sigma \in [\sigma_{min}, 1]$. Let σ^* be the middle point of a validity domain: $\sigma^* = \frac{1}{2}(1 + \sigma_{min})$. σ^* can be studied:

$$\sigma(\sigma_1, \ldots, \sigma_n) \approx \sigma(\sigma^*, \ldots \sigma^*) + \sum_i \left.\frac{\delta\sigma(\sigma_1, \ldots, \sigma_n)}{\delta\sigma_i}\right|_{\sigma_i=\sigma^*\forall i} (\sigma_i - \sigma^*). \tag{2}$$

Considering the both bound condition, it yields

$$\begin{aligned} 1 &= \sigma(1, \ldots, 1) \\ \sigma_{min} &= \sigma\left(\sigma_{min}^1, \ldots, \sigma_{min}^n\right). \end{aligned} \tag{3}$$

With (2) and (3), it comes out:

$$\sum_i \left. \frac{\delta\sigma(\sigma_1, \ldots, \sigma_n)}{\delta\sigma_i} \right|_{\sigma_i = \sigma^* \forall i} = 1$$

$$\sigma(\sigma^*, \ldots, \sigma^*) = \sigma^*.$$

With other terms, if $\forall i, \sigma_i \in \left[\sigma_{min}^i, 1\right]$ and $\sigma \in [\sigma_{min}, 1]$, we can present the global average satisfaction function with

$$\sigma(\sigma_1, \ldots, \sigma_n) \approx \sum_i a_i \sigma_i \text{with} \forall i, a_i > 0 \textbf{ and} \sum_i a_i = 1 \tag{4}$$

or, using dissatisfaction $D = 1 - \sigma$ instead of satisfaction:

$$D(D_1, \ldots, D_n) \approx \sum_i a_i D_i \text{with} \forall i, a_i > 0 \textbf{ and} \sum_i a_i = 1. \tag{5}$$

Let us now consider indicators to assess the performance of some services. In the next, indicators have to be considered as proposals, but alternative indicators coming from further researches could also be used and then they can be reformulated with a MILP formalism.

Generally speaking, modifiable permanent services use to control a physical variable: the user satisfaction depends on the difference between an expected value and an actual one. Let us consider for example the temperature of a room heating service. A dwelling can usually be split into several heating services related to rooms (or thermal zones) assumed to be independent.

Let us consider the comfort standard 7730 [7] for thermal comfort assessment. According to the comfort standard 7730 [7], three qualitative categories of thermal comfort can be distinguished: A, B, and C. In each category, typical value ranges for temperature, air speed, and humidity of a thermal zone that depends on the type of environment have been proposed [7]: office, room,... These categories are based on an aggregated criterion named predictive mean vote (PMV) modeling the deviation from a neutral ambience.

The absolute value of this PMV is an interesting index to evaluate the quality of a HVAC service related to a thermal zone because it can be easily transformed into a MILP formalization. In order to simplify the evaluation of the PMV, typical values for humidity and air speed are used. Therefore, only the ambient temperature corresponding to the neutral value of PMV (PMV $= 0$) is dynamically concerned. Under this assumption, an ideal temperature T_{opt} is obtained. Depending on the environment, an acceptable temperature range coming from the standard leads to an interval $[T_{min}, T_{max}]$. For instance, in an individual office in category A, with typical air speed and humidity conditions, the neutral temperature is $T_{opt} = 22°$ and the acceptable range is $[21°, 23°]$.

Therefore, considering the HVAC service $SRV(i)$, the discomfort criterion $D(i, k)$, which in our case is more usable than comfort criterion, is modelled by the following formula where assumptions are depicted by two parameters a_1 and

a_2:

$$D(i) = \begin{cases} \dfrac{f(i) - f_{opt}(i)}{f_{max}(i) - f_{opt}(i)} \text{ if } f(i) > f_{opt}(i) \\ \dfrac{f_{opt}(i) - f(i)}{f_{opt}(i) - f_{min}(i)} \text{ if } f(i) \leq f_{opt}(i). \end{cases} \quad (6)$$

Generally speaking, modifiable temporary end-user services do not aim at controlling a physical variable. Temporary services such as washing are expected by inhabitants to finish at a given time. Therefore, the quality of achievement of a temporary service depends on the amount of time it is shifted. Therefore, in the same way as for permanent services, a user dissatisfaction criterion for a service $SRV(i)$ is defined by

$$D(i,k) = |PMV(T_{in}(i,k))| = \begin{cases} a_1 \times \dfrac{(T_{opt} - T_{in}(i,k))}{T_{opt} - T_{Min}} & \text{if } T_{in}(i,k) \leq T_{opt} \\ a_2 \times \dfrac{(T_{in}(i,k) - T_{opt})}{T_{Max} - T_{opt}} & \text{if } T_{in}(i,k) > T_{opt} \\ , \end{cases} \quad (7)$$

where f_{opt} stands for the requested end time and f_{min} and f_{max} stand, respectively, for the minimum and maximum acceptable end times.

2.3.4 Formulation of the Anticipative Problem as a Linear Problem

Formulation of the energy management problem contains both behavioral models with discrete and continuous variables, differential equation, and finite-state models, and quality models with non linearities such as in the PMV model. In order to get mixed-linear problems that can be solved by well-known efficient algorithms, transformations have to be done. The ones that have been used are summarized in [5].

2.3.5 Formalizing Time Shifting

Temporary services are modelled by finite-state machines. The consumption of a state can be shifted such as task in scheduling problems. The starting and ending times of services can be synchronized to an anticipative period [8]. It leads to a discrete-time formulation of the problem. However, this approach is both a restriction of the solution space and an approximation because the length of a time service has to be a multiple of Δ. The general case has been considered here.

In the scientific literature, continuous-time formulations of scheduling problems exist [9–11]. However, these results concern scheduling problems with disjunctive

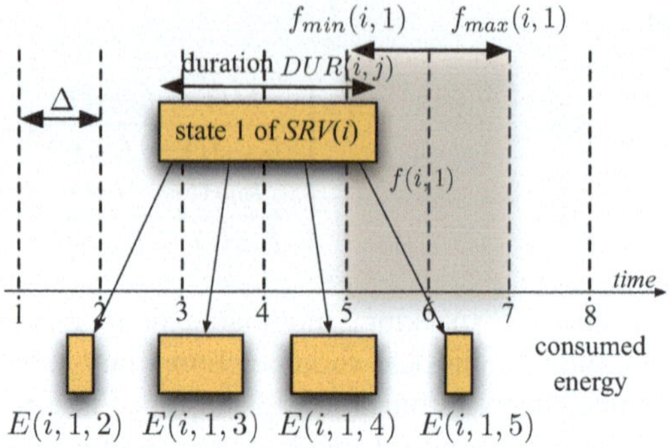

Fig. 1 Shift of temporary services

resource constraints. Instead of computing the starting time of tasks, the aim is to determine the sequence of tasks on shared resources. In energy management problems, the matter is not restricted to determine such sequence because several services can be achieved at the same time.

Figure 1 presents an example of a state in temporary services. Temporary services can be continuously shifted. Let $DUR(i, j)$, $f(i, j)$, and $p(i, j)$ be, respectively, the duration of the state j of service $SRV(i)$, the ending time, and the power related to the service $SRV(i)$ during the state j. $f(i, j)$ is defined according to inhabitant comfort models: they correspond to extrema in the comfort models presented in Sect. 2.3.3.

According to [12], the potential consumption/production duration (effective duration if positive) $d(i, j, k)$ of a service $SRV(i)$ in state j during a sampling period $[k\Delta, (k+1)\Delta]$ is given by (see Fig. 1)

$$d(i, j, k) = min(f(i, j), (k+1)\Delta) - max(f(i, j) - DUR(i, j), k\Delta). \tag{8}$$

Therefore, the consumption/production energy $E(i, j, k)$ of the service $SRV(i)$ in state j during a sampling period $[k\Delta, (k+1)\Delta]$ is given by

$$E(i, j, k) = \begin{cases} d(i, j, k)p(i, j) & \text{if } d(i, j, k) > 0 \\ 0 & \text{otherwise} \end{cases}, \tag{9}$$

where $d(i, j, k)$ stands for the duration of the state j of the service i during the period k. It is null if the state j of service i does not intersect the anticipative period k. $p(i, j)$ is the power consumed during the state j of the service i.

$E(i, j, k)$ can be modelled using a binary variable: $\delta_{t0}(i, j, k) = (d(i, j, k) \geq 0)$ and a semi-continuous variable: $z_{t0}(i, j, k) = \delta_{t0}(i, j, k)d(i, j, k)$. It leads to the following inequalities:

$$d(i, j, k) \leq \delta_{t0}(i, j, k)K\Delta \tag{10}$$

$$d(i, j, k) > (\delta_{t0}(i, j, k) - 1)\, K\Delta \tag{11}$$

$$E(i, j, k) = z_{t_0}(i, j, k)p(i, j) \tag{12}$$

$$z_{t_0}(i, j, k) \leq \delta_{t0}(i, j, k)K\Delta \tag{13}$$

$$z_{t_0}(i, j, k) \geq -\delta_{t0}(i, j, k)K\Delta \tag{14}$$

$$z_{t_0}(i, j, k) \leq d(i, j, k) + (1 - \delta_{t0}(i, j, k))\, K\Delta \tag{15}$$

$$z_{t_0}(i, j, k) \geq d(i, j, k) - (1 - \delta_{t0}(i, j, k))\, K\Delta, \tag{16}$$

where $z_{t_0}(i, j, k)$ is an abstract semi-continuous variable.

But the model still contains non-linear functions min and max in the expression of $d(i, j, k)$. Therefore, Eq. (8) has to be transformed into a mixed-linear form. Let us introduce two binary variables $\delta_{t1}(i, j, k)$ and $\delta_{t2}(i, j, k)$ defined by

$$\delta_{t1}(i, j, k) = (f(i, j) - k\Delta \geq 0)$$

$$\delta_{t2}(i, j, k) = (f(i, j) - DUR(i, j) - k\Delta \geq 0)\,.$$

Using transformations, it yields

$$f(i, j) - k\Delta \leq \delta_{t1}(i, j, k)K\Delta \tag{17}$$

$$f(i, j) - k\Delta \geq (\delta_{t1}(i, j, k) - 1)\, K\Delta \tag{18}$$

$$f(i, j) - DUR(i, j) - k\Delta \leq \delta_{t2}(i, j, k)K\Delta \tag{19}$$

$$f(i, j) - DUR(i, j) - k\Delta \leq (\delta_{t2}(i, j, k) - 1)\, K\Delta. \tag{20}$$

Therefore, min and max of Eq. (8) become

$$f_{min}(i, j, k) = \delta_{t1}(i, j, k+1)(k+1)\Delta + (1 - \delta_{t1}(i, j, k+1))\, f(i, j) \tag{21}$$

$$s_{max}(i, j, k) - \delta_{t2}(i, j, k)(f(i, j) \quad DUR(i, j)) + (1 - \delta_{t2}(i, j, k))\, k\Delta \tag{22}$$

with $min(f(i, j), (k+1)\Delta) = f_{min}(i, j, k)$ and $max(f(i, j) - DUR(i, j), k\Delta) = s_{max}(i, j, k)$.

The duration $d(i, j, k)$ can then be evaluated:

$$d(i, j, k) = f_{min}(i, j, k) - s_{max}(i, j, k). \tag{23}$$

Equations (10)–(23) model the time shifting of a temporary service.

Some services have been modelled by mixed-integer linear form. Other services can be modelled in the same way. Anticipative control in dwelling energy management can be formulated then as a multi-criteria mixed-linear programming problem represented by a set of constraints and optimization criteria. The regular solving approach is presented in [5].

3 Model Predictive Control for Energy Management

Approaches based on model predictive control (MPC) become a significant part of solutions proposed to design efficient energy management systems in building. This is due to its ability to compute near-optimal solution while handling multi-variable character and respecting the constraints of the problem.

Among these works, there is a group based on the stochastic model predictive control as in [13], where a stochastic control has been applied to building heating control. The objective of the controller is to reduce the internal air temperatures in the morning and overnight, in anticipation of solar gains during the day in order to reduce the energy consumption and improve the comfort of the occupants.

Borrelli et al. [14] propose a MPC-based controller for the thermal energy storage in building. A dual-stage optimization is used in order to tackle complexity and feasibility issues: the first stage operates using heuristic rules, while the second stage optimally controls based on a periodic moving window blocking strategy that is used in order to reduce the computational time associated with the resulting non-linear constrained optimization. The approach has been tested on the campus of the University of California. The results of the simulation showed that, using the MPC-based controller, the daily electricity bill can be reduced of 24.5% compared to the manual control sequence.

Collazos et al. [15] proposed a management of polygeneration systems with predictive techniques. A model-based predictive controller has been developed using a mixed-linear and integer programming model to define the optimal management strategy of micro-cogeneration systems in building applications. The model includes the balance of the hot-water storage as well as the heat accumulation in the building envelope.

Negenborn in [16] proposes a model predictive controller (MPC) that uses mixed-logical framework to model and control the energy flows in buildings. This is one among the few works, based on model predictive control, that focuses on energy flows including power generation, energy storage capabilities, and the possibility of energy exchange with an external energy supplier. Most studies based on MPC are exclusively centered on thermal aspects. Other shortcomings included non-linear (bilinear) characteristics are scarcely explicitly handled as in [14]. Non-linear heat emission characteristics of Fan Coil Units are generally linearized around a functioning point. Other similar works are in [15, 17–19].

Lamoudi in [20] mentions that most studies show at least one of the following drawbacks:

- Focus is exclusively centered on thermal aspects.
- Non-linear (bilinear) characteristics are scarcely explicitly handled.
- Non-linear heat emission characteristics of Fan Coil Units are generally linearized around a functioning point.

Lamoudi in [20–23] proposes a solution based on a distributed model predictive control framework to overcome these drawbacks. The designed solution explicitly handles the non-linear features of the system model, and the comfort variables in the solution are addressed while meeting specific power limitations for each energy type (power from the grid, local solar production, etc.). This chapter does not detail this solution here.

4 Automatic Model Generation for Energy Management

4.1 Introduction

Automatic model generation becomes a promising approach to handle and transform models required for energy management systems. In this perspective, gPROMS [24] or General Algebraic Modeling System (GAMS) [25] is developed allowing the user to concentrate on the modeling problem by forgetting the application formalism requirement. Once the core model is defined, the system takes care of the time-consuming details of the specific machine and system software implementation to transform this model into different formalisms. Nevertheless, such an automatic transformation cannot always be performed in dwelling energy management applications. Indeed, it is not always sure to get an acausal linear model for energy management application from an initial non-linear model for example if there are non-linearizable terms inside.

The model construction of a whole dwelling is not a trivial task. Moreover, dwelling model is not really stable due to equipment (add/remove) changes. It means that dwelling model construction also needs to be flexible. Therefore, it is not a good solution to build a whole dwelling model at once but to compose step-by-step equipment models to get the final model afterwards. When there is a change, it needs just to add/remove the corresponding equipment model into/from the final model.

Based on the MDE approach, this section presents a method consisting in automatically manipulating equipment models to get a dwelling final model, the so-called pivot model, and then automatically projecting this pivot model into target application models.

4.2 Problem Formulation

In this section, different key concepts aiming at composing a generic dwelling model, based on the Model-Driven Engineering (MDE) approach, are presented with the the help of an illustrative example.

4.2.1 Concept of Model Transformation

Let us consider resistor modelled by

$$C_0 : U_1 = R_1 \times I_1. \tag{24}$$

This simple model may be used by a designer into different optimization problems, adding information like lower and upper bounds of variables and an objective function for instance:

$$C_0 : U_1 = R_1 \times I_1 \tag{25}$$

$$C_1 : 0 \leq R_1 \leq 5 \tag{26}$$

$$C_2 : 0 \leq U_1 \leq 4 \tag{27}$$

$$Objective : Max I_1. \tag{28}$$

In spite of its simplicity, if another resistor R_2 is added in parallel, the whole system model has to be rewritten as

$$C_0 : U_1 = R_1 \times I_1 \tag{29}$$

$$C_1 : U_2 = R_2 \times I_2 \tag{30}$$

$$C_2 : I_{total} = I_1 + I_2 \tag{31}$$

$$C_3 : 0 \leq R_1 \leq 5 \tag{32}$$

$$C_4 : 0 \leq R_2 \leq 3 \tag{33}$$

$$C_5 : 0 \leq U_1 \leq 4 \tag{34}$$

$$C_6 : 0 \leq U_2 \leq 4 \tag{35}$$

$$C_7 : U_1 = U_2 \tag{36}$$

$$Objective : Max I_{total}. \tag{37}$$

Although the rewriting process is not time-consuming for this example, it becomes a tough work for complex systems that contain hundreds of variables and constraints. In addition, model may also have to be rewritten depending on the target application. For instance, some optimization algorithms require a causal ordering (simulated annealing), and some others require linearization (mixed-integer linear programming). Therefore, two difficulties to be dealt are:

- A model must be composed of elements that can be reused.
- A model has to be transformable.

To handle model transformation in optimization problems, the concept of pivot model is used. Actually, a pivot model is a high-level application-independent description that can be transformed into target application formats. This section is aiming at automatizing the model rewriting processes.

In the computer science literature, model rewriting processes are usually managed using the concept of Model-Driven Engineering (MDE).

4.3 Concept of MDE

Basically, the Model-Driven Engineering (MDE) approaches aim at separating models based on company know-how and those related to software implementations in order to maintain the sustainability of the company know-how in spite of the changes of development environment [25]. To do this, it is necessary first to define Platform-Independent Models (PIM), i.e., pivot model, technically independent from execution platform. It enables the generation of a set of Platform-Specific Models (PSM) afterwards. Based on the MDE approach, the problem can be decomposed into 2 abstraction levels. The two concepts of PSM and PIM are corresponding, respectively, to the levels M0 and M1. Shortly, the signification of each level is:

- Level M0 (PSM) is the real system that contains executable object.
- Level M1 (PIM) is the model that represents the system.

Then, the main objective of this approach is to be able to perform transformations to generate different models related to the levels. There are two types of transformations of models:

- Transformation model to code (PIM to PSM)
- Transformation model to model (PIM to PIM)

Generally speaking, the transformation model to code can be viewed as a special case of model-to-model transformation. Basically, a transformation model to model is performed with the help of transformation rules that consist in transforming a set of input models to targeted models.

4.4 Concept of Pivot Model

Thanks to this architecture and based on the MDE concept, a pivot model can be considered as a PIM (level M1) and the PSM can be associated to a specific optimization model format. Basically, PIM is supposed to be available initially, and then PIM to PSM or PIM to PIM transformations have to be computed by applying transformation processes. Generally speaking, the PIM construction is built from elementary models, denoted as EM, that describe element parts in the system. An elementary model EM, in the field of optimization, is associated with a subspace of a vector space defined on $\mathbb{R}^n$. It is considered that integer set $\mathbb{N}$ is a specialization of the real number space $\mathbb{R}$: $\mathbb{N} \subset \mathbb{R}$ and that True, False are modelled by a binary set $\{0,1\}$, which is a specialization of $\mathbb{N}$. An element model representing an element in a given mode is defined as

Definition 1 EM : mode(EM) $\leftrightarrow \mathcal{V}_{\mathcal{S}} \in \mathcal{E}; \mathcal{E} \subset dom(\mathcal{S}) \subset \mathbb{R}^n$
with:

- $\mathcal{V}_{\mathcal{S}} = \{v_0, \ldots, v_{n-1}\}$ is a set of variables, respectively, related to the tuple of symbols $\mathcal{S} = \{symbol_0, \ldots, symbol_{n-1}\}$.
- $dom(\mathcal{S}) = dom(symbol_0) \times \ldots \times dom(symbol_{n-1})$ is the set of value domains of symbols corresponding to variables.
- Mode is generally and implicitly ok (except in diagnosis analysis) for normal behavior.
- The subspace $\mathcal{E}$ is defined by a set of n_j constraints $\mathcal{K}$ defined over $\mathbb{R}^n$.

$$\mathcal{K} = \{\mathcal{K}_j(\mathcal{S}) \diamond 0; \forall j\} \tag{38}$$

where $\diamond$ stands for a comparison operator.

The notion of element has to be clarified. Let us consider a dual-flow ventilation system; it is composed of two speed variation control devices and two electric drives associated with the extraction and insufflated air. An element of this ventilation system model can be a variation control device model or an electric drive level but at a higher level of consideration. By the way, the whole ventilation system can also be seen as an element of the dwelling system depending on the point of view. Therefore, the more the elements are decomposed, the more they can be reused.

Actually, a pivot model is composed step by step by adding required element models (*EM*s). This solution facilitates the pivot model construction for system designer because instead of building a unique model containing all the needed constraints, the designer can compose components by composing different *EM*s. Moreover, these blocs can possibly be reused afterwards to get bigger ones and so on before they are used for building a pivot model. A pivot system model PM = $(\mathcal{K}_{\sum}(\mathcal{S}_{\sum}, dom(\mathcal{S}_{\sum}))$ is an union of elementary models EM$_i$ plus connection constraints K$_j$.

In this chapter, two application models that are used the most in BEMS are considered: simulation model and energy management model using the MILP formalism. Let us consider the pivot system model $PM = (\mathcal{K}_{\sum}(\mathcal{S}_{\sum}, dom(\mathcal{S}_{\sum}))$; constraints can be decomposed into equality constraints, denoted as $\mathcal{K}_{\sum}^{=}$, and inequality constraints, denoted as $\mathcal{K}_{\sum}^{\nabla}$. A model is said to be simulable if it exists a function φ: $\mathcal{S}_{\sum}^{in} \rightarrow \mathcal{S}_{\sum}^{out}$ such as $\mathcal{K}_{\sum}^{=}(\mathcal{S}_{\sum}) \leftrightarrow \varphi(\mathcal{S}_{\sum}^{in}) = \mathcal{S}_{\sum}^{out}$ on $dom(\mathcal{S}_{\sum})$ with $(\mathcal{S}_{\sum}^{in}, \mathcal{S}_{\sum}^{out})$ is a partition of $\mathcal{S}_{\sum}$.

Transforming PM into simulation model, denoted as PM_{SA}, consists in selecting and projecting $\mathcal{K}_{\sum}^{=}(\mathcal{S}_{\sum})$ into $\varphi(\mathcal{S}_{\sum}^{in}) = \mathcal{S}_{\sum}^{out}$, and a causal ordering has to be performed. It requires usually to set values of some variables that will become parameters and input variables. The Dulmage–Mendelsohn algorithm [26] is generally used providing $\mathcal{K}_{\Sigma}^{=}(\mathcal{S}_{\Sigma})$.

A MILP model is defined as $PM_{MILP} = (\mathcal{K}_{\Sigma}^{MILP}(\mathcal{S}_{\Sigma}^{MILP}, dom(\mathcal{S}_{\Sigma}^{MILP}))$ with $\mathcal{K}_{\Sigma}^{MILP}(\mathcal{S}_{\Sigma}^{MILP}$ linear). Transforming PM into PM_{MILP} relies on linearization transformation that may require the introduction of new variables, The specification of the domains of some variables, and the introduction of new constraints. A MILP model also contains criteria to be optimized expressed by a linear function of $\mathcal{S}_{\Sigma}$.

Different transformation processes for composing a pivot model then project this pivot model into simulation, and MILP formalism is detailed in Sect. 4.5.

4.5 *Transformation Process Principles*

This section gives an overview on different transformation process principles by using an illustrative example. A transformation is composed of two main steps. The first one aims at manipulating element models to build a pivot model. Then, different projection processes are applied to obtain target application models. The simulation and energy management using MILP formalism models are shown in this section.

4.5.1 Composition Process

This sub-section focuses on how a pivot model is built. Thus, the most important step to build a pivot model is the composition of different element models. The objective of composition is to encourage the reusability of element models and make the pivot model construction more modular. A composition can be applied for a set of element models, a set of compositions of element models, a set of compositions of compositions, and so on. Moreover, recursive compositions can be performed unlimitedly to get bigger compositions. To illustrate this point, consider now an electric circuit as presented by Fig. 2.

The system presented in Fig. 2 is composed of two blocs of four electrical resistors R_1, R_2, R_3, and R_4. Independently of any formalism, the construction of

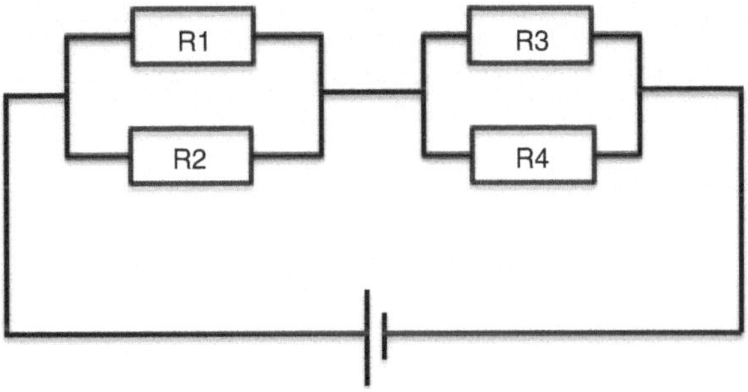

Fig. 2 Example of electric circuit

such a pivot model can be done by composing first a bloc of 2 parallel resistors. Then, the pivot model is built by duplicating this bloc and connecting the whole system.

When composing step by step the pivot model, there are two remaining problems that have to be considered. The first one consists in specializing all resistors with the corresponding values, and the second one consists in establishing different connections between element models.

To deal with the first problem, each element model (*EM*) is necessarily specialized before being used in a composition. The specialization concept presented in [27] is well suited to this problem. It makes an element more specific by adding some supplementary information like a prefix or a type. According to the definition given in Part 1, the specialization of an *EM* consists only in adding a distinct prefix to symbol representing a variable each time it is used. For instance, $R_1.U$ is not the same as $R_2.U$ and so on. An *EM* could be specialized as many times as desired. The more specialized an *EM* is during a composition process, the more specific it is. For instance, $bloc_1.R_1.U$ is not the same as $bloc_2.R_1.U$. Nevertheless, a set of specialized *EM*s cannot form a composition without connection between them. Indeed, two specialized *EM*s, for instance resistor R_1 and resistor R_2, require explicitly the following connecting equations, which is a common concept in [27]:

$$R_1.U = R_2.U$$

$$I_{total} = \frac{R_1.U}{R_1.R} + \frac{R_2.U}{R_2.R}.$$

These connecting equations are added into the compositions. The pivot model of this system could be:

- The parallel bloc composition with the dots "." represents suffixes of prefixes.

$$C_0 : R_1.U = R_1.R \times R_1.I \tag{39}$$

$$C_1 : R_2.U = R_2.R \times R_2.I \tag{40}$$

$$C_2 : R_1.U = R_2.U \tag{41}$$

$$C_3 : I_{total} = \frac{R_1.U}{R_1.R} + \frac{R_2.U}{R_2.R}. \tag{42}$$

- By duplicating the parallel bloc composition above twice and by adding connecting equations can be established the final circuit. The system pivot model is thus built:

$$C_0 : bloc_1.R_1.U = bloc_1.R_1.R \times bloc_1.R_1.I \tag{43}$$

$$C_1 : bloc_1.R_2.U = bloc_1.R_2.R \times bloc_1.R_2.I \tag{44}$$

$$C_2 : bloc_1.R_1.U = bloc_1.R_2.U \tag{45}$$

$$C_3 : bloc_1.I_{total} = \frac{bloc_1.R_1.U}{bloc_1.R_1.R} + \frac{bloc_1.R_2.U}{bloc_1.R_2.R} \tag{46}$$

$$C_4 : bloc_2.R_1.U = bloc_2.R_1.R \times bloc_2.R_1.I \tag{47}$$

$$C_5 : bloc_2.R_2.U = bloc_2.R_2.R \times bloc_2.R_2.I \tag{48}$$

$$C_6 : bloc_2.R_1.U = bloc_2.R_2.U \tag{49}$$

$$C_7 : bloc_2.I_{total} = \frac{bloc_2.R_1.U}{bloc_2.R_1.R} + \frac{bloc_2.R_2.U}{bloc_2.R_2.R} \tag{50}$$

$$C_8 : bloc_1.I_{total} = bloc_2.I_{total} \tag{51}$$

$$C_9 : U_{total} = bloc_1.R_1.U + bloc_2.R_1.U. \tag{52}$$

To recapitulate the above pivot model construction, the resistor model (24) is first used two times to create two different resistors. Then, a bloc of two parallel resistors is created by adding connecting equations. Finally, the pivot model is built by duplicating this parallel bloc and adding new connecting equations. This pivot model can automatically be generated if these three steps are defined in a recipe. The notion of recipe is also the main idea of this chapter aiming at automatizing the generation between models. In other words, an expert has to list all steps required for:

- Generating a pivot model from different element models
- Generating a pivot model into different target application models

Each step is considered as a transformation rule that is implemented and put into a common rule set. This rule set contains all rules that are required for these two points above. Then, expert can create recipes that call gradually implemented rules from the rule set to get desired formalisms. However, the equation manipulation to create such a pivot model is not a trivial task. To automatize the prefix adding, connecting equation adding, or constraints duplication, an engine that handles symbolic constraints is strongly required.

In the recent decades, symbolic computation or computer algebra [28, 29] has become an important research area of mathematics and computer scientists aiming at developing tools for solving symbolical equations. The capabilities of major general-purpose Computer Algebra Systems (CAS) are presented in [30, 31]. Moreover, among the mathematical features of a CAS, there are transformations allowing to manipulate and optimize symbolic computations in order to automatically generate optimization code [32].

The GIAC/XCAS CAS [33] has been developed to solve a wide variety of symbolic problems and was awarded the 3rd price at the Trophées du Libre 2007 in the scientific software category http://www-fourier.ujf-grenoble.fr/~parisse/giac.html. This CAS has been used for symbolic manipulation of all the constraints in the pivot model. With GIAC/XCAS, each constraint is treated as a n-ary tree equation as

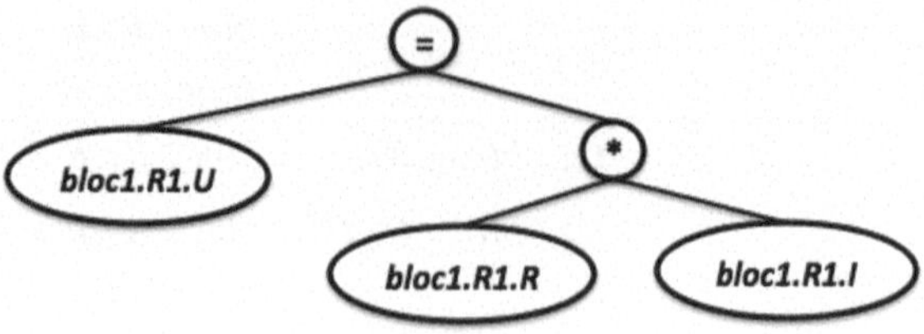

Fig. 3 Equation (43) n-ary tree representation

presented in Fig. 3. This tool is also the core of the implemented program allowing to carry out all the manipulations and transformations symbolically.

Finally, the set of required manipulations for composing a pivot model is, respectively, summarized as follows:

- Specialization of *EM* by adding prefixes
- Adding connecting equations

4.5.2 Projection Process

Once a pivot model is composed, the next step consists in applying different projection processes to get desired formalisms. These projection processes can always be detailed in recipes to automatize the transformation between models. This sub-section shows different steps to get simulation and energy management using MILP formalisms. These processes are summarized in Fig. 4.

Energy management application using MILP formalism requires an acausal linear model containing 3 kinds of variables: binary, continuous, and integer, and 2 kinds of constraints: equality and inequality. It means that MILP transformation process aims at projecting dwelling pivot model including appliance phenomena physics, occupant requirement, and energy flows into MILP formalism as follows:

- Transform all constraints into equality and inequality constraints.
- Linearize all non-linear terms.

Based on the platform PREDIS/MHI model that is detailed in Sect. 4.6, the first point is necessarily performed to transform all ordinary differential equations (ODE) and logical constraints into equality and inequality constraints. In this study case, an approximation of ODE time discretization is shown instead of the exact transformation solution. This approximation consists in developing the derivative variable into

$$\frac{dv_i}{dt} = \frac{v_i(t+1) - v_i(t)}{t} \tag{53}$$

with $v_i \in \mathcal{V}_S$ and t is the predefined time step.

This time discretization transformation of all ODEs, $\frac{dv_i}{dt} = f(\mathcal{V}_S)$, is performed symbolically as presented by Fig. 5.

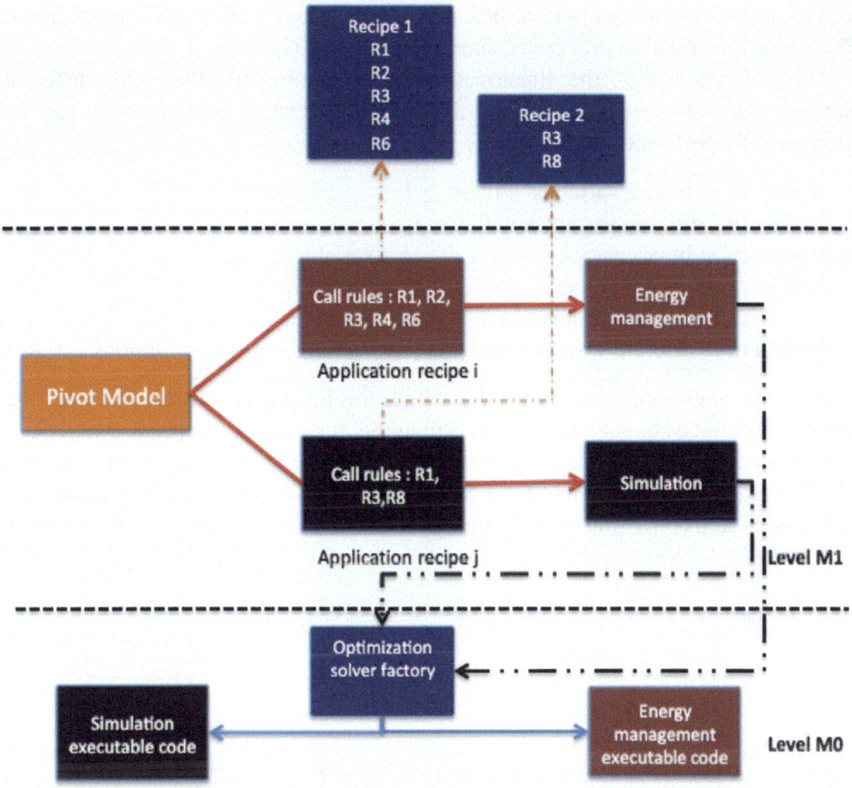

Fig. 4 Example of two projection processes

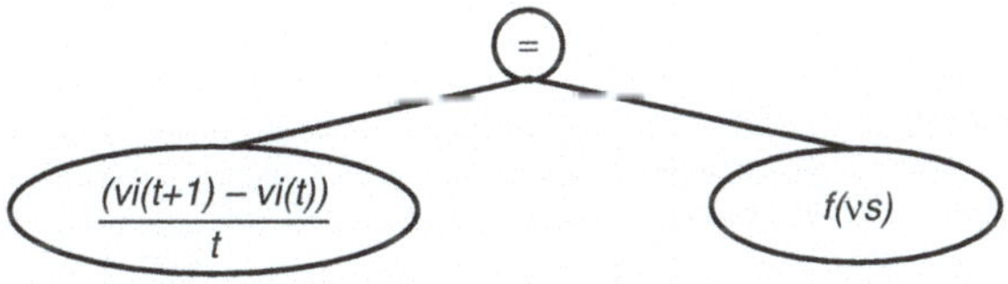

Fig. 5 Time discretization pattern

The main idea of this transformation is the same for logical constraint transformation, and it can be found in [5]. Once this step is completed, the pivot model contains equality and inequality constraints. The next step to do consists in searching and linearizing all non-linear terms.

The difference between non-linear terms is based on the nature of variables and/or the nature of functions that contain variables. Indeed, product of two discrete variables cannot be linearized in the same way than a production of two continuous variables or a cosine function for example. To linearize the pivot model, it is preferable to sort out all the non-linear terms in different kinds of non-linearity first. Then each kind of non-linearity is linearized by corresponding rules. It means that recipes, rules, and rule set have to be easily extended to cover all eventual changes. The whole linearization process is summarized in Fig. 6.

This schema makes the linearization process automatically using different patterns that were presented in [5, 34] and linearization patterns that can be automatically performed are:

- Product of m binary variables with $m \geq 1$
- Product of l discrete variables with $l \geq 1$
- Product of m binary variables and l discrete variables
- Product of m binary variables and 1 continuous variable
- Product of l discrete variables and 1 continuous variable
- Product of m discrete variables and l discrete variables and 1 continuous variable

However, there are some terms for which the linearization process cannot be automatized, but a human intervention is required, for instance the product of n ($n \geq 1$) continuous variables. Indeed, it does not exist a linearization pattern for this type of non-linearity to be performed directly. Linearizing such non-linearity requires a preliminary step consisting in discretizing the domain of $n - 1$ continuous variables

Fig. 6 Linearization process

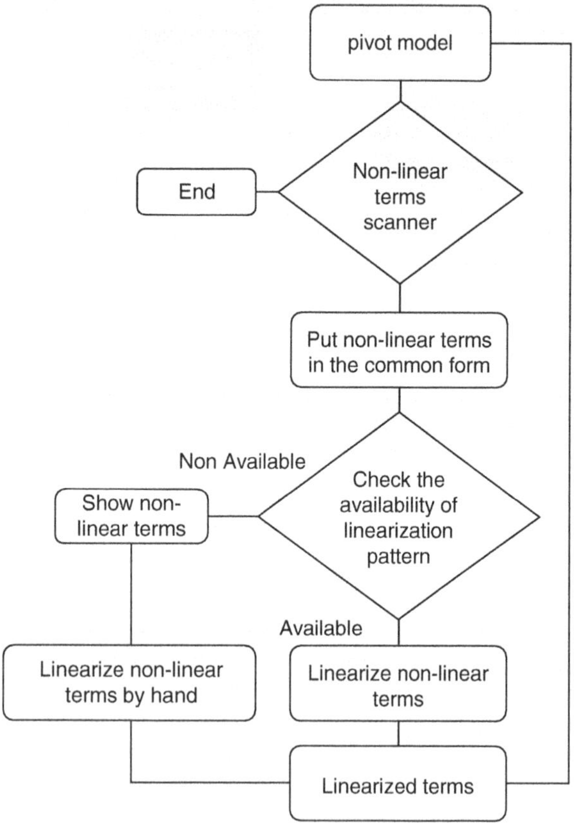

into sets of discrete values. Then, the pattern of discrete and continuous product can be used to get a linear term. Discretization also means approximation to realistic values; therefore, the choice of discrete values impacts strongly on final results, and this step cannot be automatically performed by system. Only expert who masters his dwelling system can take good values for linearization process afterwards.

Let us linearize the circuit system (25) by discretizing for instance the resistor into $R = \{3, 4\}$. Then, the discrete and continuous linearization patterns can be used by introducing a new variable, denoted as Z, representing the product $R \times I$ with

$$Z = R \times I = (\delta_1 \times v_1 + \delta_2 \times v_2) \times I \tag{54}$$

with δ_i is a binary variable that takes value $\{0, 1\}$. Actually, the goal is to select the best value among those of R to maximize or minimize the objective function. Equation (54) can be factorized as

$$Z = \delta_1 \times v_1 \times I + \delta_2 \times v_2 \times I \tag{55}$$

with v_1 and v_2 parameters. There are two binary and continuous products to be linearized. Let us linearize for instance the first binary and continuous product term: $\delta_1 \times v_1 \times I$. The corresponding pattern implies to create a new continuous variable, denoted as $Z^{'}$ with 4 new constraints delimiting the bounds of $Z^{'}$ as follows:

$$Z^{'} \leq \delta_1 \times v_1 \times \overline{I} \tag{56}$$

$$Z^{'} \geq \delta_1 \times v_1 \times \underline{I} \tag{57}$$

$$Z^{'} \leq (I - \underline{I} \times (1 - \delta_1)) \times v_1 \tag{58}$$

$$Z^{'} \geq (I - \overline{I} \times (1 - \delta_1)) \times v_1 \tag{59}$$

with $\underline{I}$ and $\overline{I}$ are, respectively, the lower and upper bounds of the continuous variable I. The second binary and continuous production $\delta_2 \times v_2 \times I$ is linearized in the same way. Once all of the non-linear terms are linearized, the MILP model formalism is obtained.

Regarding the simulation model formalism, the required projection aims at making the pivot model simulable. First, a model is simulable if and only if it is a just-determined model structurally meaning that the number of variables is equal to the number of equality constraints; therefore, variables could have exactly one solution. Dulmage–Mendelsohn algorithm [26] is used for checking these criteria. However, a pivot model is not necessarily just determined, but it could also be:

- Under-determined structurally meaning that there are fewer equality constraints than variables. In this case, variables can have no solution or infinitely many solutions making the pivot model non-simulable. This kind of model could be only used by energy management application.

- Over-determined structurally meaning that there are more equality constraints than variables. In this case, variables can have no solution or only one solution. This kind of model could be used only by diagnosis application for which Analytical Redundancy Relations (ARR) for fault detection are performed.

Normally, a correct simulation model gives only a just-determined set, while other sets are empty. The equality constraints can be reorganized according to the upper-triangular just-determined part of the incidence matrix of equality constraints. The presence of an under-determined set or of an over-determined part means that the whole model cannot be simulated and it is necessary to recheck the element models.

Usually, in building energy management, the reorganized matrix is strictly upper triangular with no bloc on the diagonal but sometimes blocs may appear. In this case, the projection cannot be fully automatized because there is no general process to solve implicit systems of non-linear equations.

It is important to note that only equality constraints are taken into account for generating the simulation model formalism. It means that a preliminary step to extract equality constraints from logical operator constraints, ODE, and inequality constraints. Then, if the simulability criteria of this set of equality constraints are verified, the next step consists in making this pivot model causal. In other words, this pivot model $\mathcal{S}_{\sum}$ needs to be separated into $\mathcal{S}_{\sum}^{in}$ and $\mathcal{S}_{\sum}^{out}$. Therefore, causal ordering process is necessary to be performed. After this step, a resolution sequence of variables is obtained to provide to the simulator.

Let us consider an academic example consisting in simulating the first parallel bloc described by the equations

$$C_0 : R_1.U = R_1.R \times R_1.I \tag{60}$$

$$C_1 : R_2.U = R_2.R \times R_2.I \tag{61}$$

$$C_2 : R_1.I = \frac{R_1.U}{R_1.R} \tag{62}$$

$$C_3 : R_2.I = \frac{R_2.U}{R_2.R} \tag{63}$$

$$C_4 : R_1.U = R_2.U = 220\,\mathrm{V} \tag{64}$$

$$C_5 : R_1.R \in [100, 200]ohm \tag{65}$$

$$C_6 : I_{total} = R_1.I + R_2.I. \tag{66}$$

The objective is to maximize the variable I_{total}. This model is simulable because it is just determined and the reorganized incidence matrix is given by Table 2.

The constraint order computation is $[C_5, C_2, C_0, C_4, C_3, C_1, C_6]$.

In order to automatize the whole transformation process, a software architecture is proposed in Fig. 7.

Table 2 Obtained result from Dulmage–Mendelsohn computation

	I total	$R_2.I$	$R_2.R$	$R_2.U$	$R_1.U$	$R_1.I$	$R_1.R$
C_6	1	1	0	0	0	1	0
C_1	0	1	1	1	0	0	0
C_3	0	1	1	1	0	0	0
C_4	0	0	0	1	1	0	0
C_0	0	0	0	0	1	1	1
C_2	0	0	0	0	1	1	1
C_5	0	0	0	0	0	0	1

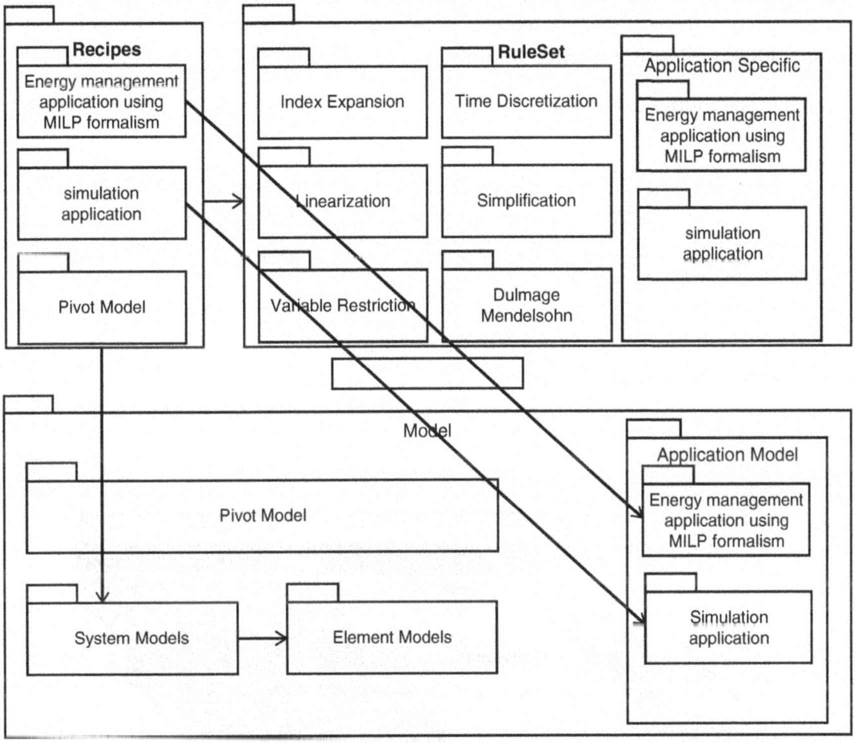

Fig. 7 Software architecture

In this section, the different steps to transform a pivot model into two target application models:

- Energy management model using MILP formalism is obtained after all of the non-linear terms are linearized as shown by equations.
- Simulation model is obtained with Dulmage–Mendelsohn computation.

The next section is used for illustrating the application of this proposed method on model transformation of the platform PREDIS/MHI.

4.6 Application on PREDIS/MHI

This section aims at presenting an implementation of this proposed method to deal with energy management problem of the platform PREDIS/MHI located in Grenoble, France. The "Monitoring and Habitat Intelligent" PREDIS platform is a research platform for company and academic researchers working on energy management. This platform is an office low-consumption building highly instrumented where most of the energy flows are measured using different sensor technologies. The structure of this platform is given by Fig. 8. For the sake of clarity, this section focuses on the classroom zone that is equipped with computers for students and a heating and ventilation system containing:

- An air treatment unit model:

$$AirFlow = coef \times Q_{Air} \tag{67}$$

$$P_{airTreatementUnit} = P_{ventilation} + P_{heating} \tag{68}$$

- A thermal balance model:

$$Phi_{Total} = Phi_{Sun} + P_{heating} + Phi_{Occup} \tag{69}$$

- A thermal comfort model distinguishing whether there is someone or not in the classroom:

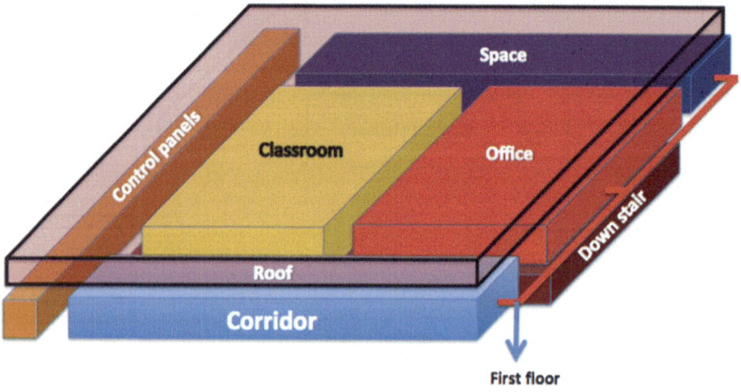

Fig. 8 An overview of PREDIS/MHI platform

$$\begin{aligned}&If\ presence = 1 : \\ &T_{felt} < T_{pref} \Rightarrow sigma_{incomfort} = \\ &1/(T_{pref} - T_{max}) \times T_{felt} - T_{pref}/(T_{pref} - T_{max})\end{aligned} \tag{70}$$

$$\begin{aligned}&T_{felt} >= T_{pref} \Rightarrow sigma_{incomfort} = \\ &1/(T_{max} - T_{pref}) \times T_{felt} - T_{pref}/(T_{max} - T_{pref})\end{aligned} \tag{71}$$

$$\begin{aligned}&If\ presence = 0 : sigma_{incomfort} = 0 \\ &T_{felt} <= T_{max_{absence}} \\ &T_{felt} >= T_{min_{absence}}\end{aligned} \tag{72}$$

- A thermal zone model:

$$\begin{aligned}R_{Ventilation} =& 1/((1 - efficiency) \times Cp_{Air} \\ &\times rho_{Air} \times AirFlow)\end{aligned} \tag{73}$$

$$\begin{aligned}R_{Eq} =& 1/(1/(R_{Ventilation} + R_w) \\ &+ \sum(1/R[neighborhood]))\end{aligned} \tag{74}$$

$$\begin{aligned}\frac{d}{dt}T_w =& -1/(R_{Eq} \times C_w) \times T_w + 1/((R_{Ventilation} \\ &+ R_w) \times C_w) \times T_{out} + \sum(T[neighborhood]/ \\ &(R[neighborhood] \times C_w)) + R_{Ventilation} \times Phi_{total}/ \\ &(C_w \times (R_{Ventilation} + R_w))\end{aligned} \tag{75}$$

$$\begin{aligned}T_{In} =& R_{Ventilation} \times T_w/(R_{Ventilation} + R_w) \\ &+ R_w/(R_{Ventilation} + R_w) \times T_{Out} \\ &+ R_{Ventilation} \times R_{Eq} \times Phi_{total}/(R_{Ventilation} + R_w)\end{aligned} \tag{76}$$

- A CO_2 comfort model:

$$sigma_{CO_2} = (C_{CO_2} - C_{fav})/(C_{max} - C_{fav}) \tag{77}$$

- A CO_2 zone model:

$$\begin{aligned}\frac{d}{dt}C_{In_{CO_2}} =& Q_{Breath} \times occupancy \\ &\times (C_{Breath} - C_{In_{CO_2}})/Vol_{Zone} \\ &+ AirFlow \times (C_{Out_{CO2}} - C_{In_{CO_2}})/Vol_{Zone}\end{aligned} \tag{78}$$

- Finally, the total power consumption model:

$$P_{total} = P_{airTreatementUnit} + P_{lighting} + P_{computer} \tag{79}$$

$$Total_{cost} = P_{total} \times PricePerKwh \tag{80}$$

These models describe only the physical phenomena of PREDIS/MHI. In this section, the goal is to build up the pivot model of the classroom and then to project it into MILP formalism and simulation model. To automatically generate these different models, the following recipe is used:

- Pivot model composition recipe of the classroom is realized in 3 steps:
  - Compose the CO_2 system:
    - Specialize: CO_2 comfort with prefix: CO_2Comfort
    - Specialize: CO_2 zone with prefix: CO_2Zone
    - Connect: CO_2Comfort.C_{CO_2} = CO_2Zone.C_{InCO_2}
  - Compose the thermal system:
    - Specialize: thermal comfort with prefix: thermalComfort
    - Specialize: thermal zone with prefix: thermalZone
    - Connect: thermalComfort.T_{felt} = thermalZone.T_{In}
  - Compose the final pivot model:
    - Specialize: CO_2 system with prefix: CO_2System
    - Specialize: thermal system with prefix: thermalSystem
    - Specialize: power consumption with prefix: powerConsumption
    - Specialize: thermal balance with prefix: thermalBalance
    - Specialize: air treatment unit with prefix: airTreatmentUnit
    - Connect: airTreatmentUnit.$AirFlow$ = thermalSystem.$AirFlow$
    - Connect: airTreatmentUnit.$P_{airTreatementUnit}$ = powerConsumption. $P_{airTreatementUnit}$
    - connect: thermalBalance.$P_{heating}$ = airTreatmentUnit.$P_{heating}$

Initially, element models of PREDIS/MHI are represented in textual description files. Thanks to GIAC/XCAS CAS symbolic mathematical system [33], each constraint is represented as a n-ary equation tree. Different variables inside this constraint are detected and memorized under different distinct symbols. After the parsing process, an *EM* is represented by a set of n-ary equation trees that facilitate the different manipulations and projection afterwards. Considering now the representation of the CO_2 zone model, its n-ary equation tree representation is given by Fig. 9.

At this stage, constraints are specialized with a given prefix, and the new connecting equation is also added to compose the pivot model. Each projection rule has been defined in the common rule set, and these steps are also found in the corresponding application recipe. Let us develop some specific rules to illustrate how the symbolical transformation is performed: time discretization, linearization for getting a MILP formalism.

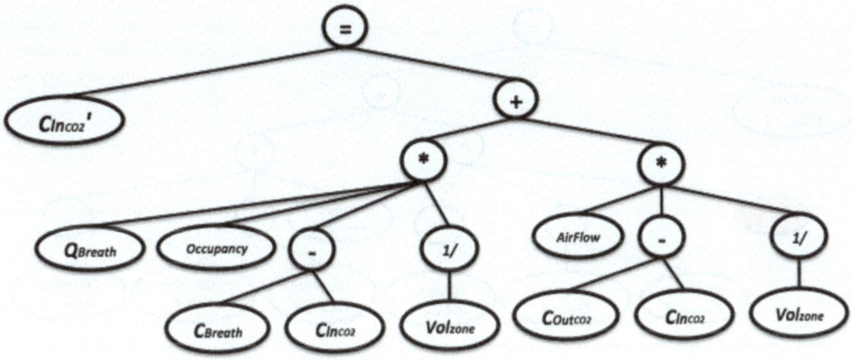

Fig. 9 CO_2 zone model n-ary representation

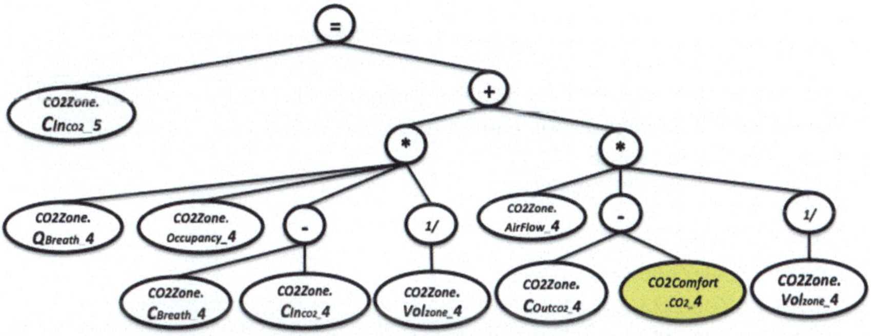

Fig. 10 CO_2 zone model after the ODE transformation process

To provide a discretized linear model to MILP solver, the required projection is the time discretization. Based on a daily period plan, the time is discretized into 24 sampling steps of 1 h. It means that there is one best appliance and envelope configuration for each hour. To process it, the time discretization multiplies 24 times each constraint of pivot model with time index ranging from 0 to 23. The ODE implementation of CO_2 zone at 5th time step is given by Fig. 10.

The next important projection rule consists in linearizing all the non-linear terms inside constraints of pivot model. First, all of the non-linear terms are detected by a checker process. Once they are detected, the nature of each term is checked before being linearized. Let us linearize the binary–continuous product:

$CO_2Zone.Q_{Breath} \times CO_2Zone.occupancy$ in the CO_2 zone model where *occupancy* is 0 whenever there is nobody or 1 otherwise. In this case, a temporal variable, denoted as z, is used for replacing the considered term in the corresponding constraint as given by Fig. 11.

In order to keep the same meaning of the non-linear term, pivot model adds four new constraints resulting of this binary–continuous product linearization pattern transformation: $z\_4 \leq CO_2Zone.occupancy\_4 \times \overline{CO_2Zone.Q_{Breath}\_4}$

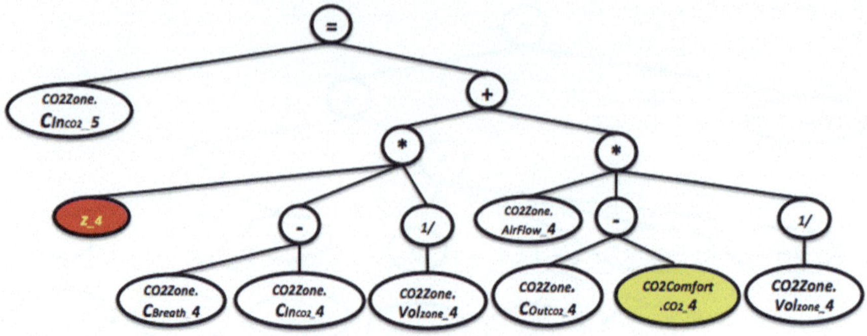

Fig. 11 CO_2 zone binary model after the first linearization process

$z\_4 \geq CO_2Zone.occupancy\_4 \times \overline{CO_2Zone.Q_{Breath}\_4}$
$z\_4 \leq CO_2Zone.Q_{Breath}\_4\text{-}(1\text{-}CO_2Zone.occupancy\_4) \times \underline{CO_2Zone.Q_{Breath}\_4}$
$z\_4 \geq CO_2Zone.Q_{Breath}\_4\text{-}(1\text{-}CO_2Zone.occupancy\_4) \times \overline{CO_2Zone.Q_{Breath}\_4}$.

The result of this linearization pattern represents exactly the considered binary–continuous product because:

- If $occupancy\_4 = 1$:
  $z\_4 \leq sup(\text{CO}_2\text{Zone}.Q_{Breath}\_4)$
  $z\_4 \geq inf(\text{CO}_2\text{Zone}.Q_{Breath}\_4)$
  $z\_4 \leq \text{CO}_2\text{Zone}.Q_{Breath}\_4$
  $z\_4 \geq \text{CO}_2\text{Zone}.Q_{Breath}\_4$.
  In this case, the first two constraints are always true so they can be eliminated. The last two constraints make it possible to take into account the real values of $\text{CO}_2\text{Zone}.Q_{Breath}\_4$.
- If $occupancy\_4 = 0$:
  $z\_4 \leq 0$
  $z\_4 \geq 0$
  when there is nobody in the classroom, it means that the Q_{Breath} is equal to 0, too.

Once all the non-linear terms are linearized, the optimized anticipative model is obtained. After the computation with MILP solver (IBM ILOG CPLEX 12.3), the result obtained within 2 min for next 24 h is given by Fig. 12.

4.6.1 Transformation of the PREDIS/MHI Pivot Model to the Fast Simulated Annealing Optimization Model

In order to handle the simulated annealing process, a causal ordering has to be performed in order to get a problem looking like:

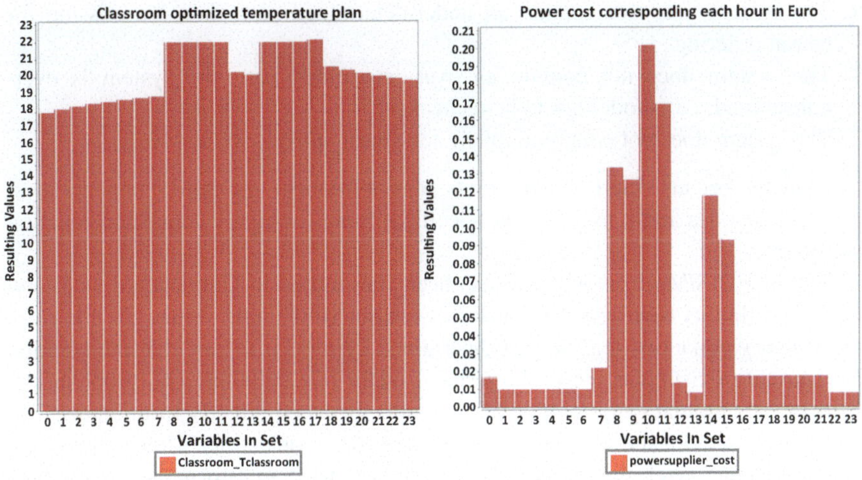

Fig. 12 Results of classroom temperature plan and power cost corresponding for next 24 h generated by MILP solver

$$Y = f(X); X \in R^m, Y \in R^n$$

$$X \diamond 0$$

$$Y \diamond 0,$$

where $\diamond$ stands for comparative operators.

Therefore, transforming the pivot non-linear model of PREDIS/MHI into the SA application starts by distinguishing equality constraints from inequalities. Then, equalities have to be reorganized to be solved. A Dulmage–Mendelsohn decomposition [26] has been done in the same way as it is done in Modelica [27]. It reorganized an incidence matrix into an upper block triangular matrix using the Hopcroft–Karp bipartite maximum matching search algorithm, which is $O((|V|+|E|)^{3.5})$, where V and E are, respectively, variables and equality constraints. Then, the presence of an under-determined set is searched to check whether the problem can be solved or not. Finally, the presence of an over-determined part is also searched: it should be empty; otherwise, contradictions may occur between over-determined variables. Whenever it has been checked that the under- and over-determined sets are empty, the equality constraints can be reorganized according to the upper-triangular just-determined part of the incidence matrix of equality constraints. Generally, in building energy management, the reorganized matrix is strictly upper triangular with no block on the diagonal but sometimes blocks may appear. In this case, the transformation cannot be fully automatized because there is no general process to solve implicit systems of non-linear equations. Generally speaking, the transformation can be fully automatized whether:

1. The system does not contain an under-determined part: data are missing for causal ordering.
2. The system does not contain an over-determined part: the system is over-constrained, i.e., model has to be rechecked.
3. The system does not contain implicit non-linear subsystem to solve.

Actually, because the first two points are not frequent, the third one is the most problematic ones and it may involve specific solving for highly connected equation subsystems.

The PREDIS/MHI pivot model has been transmitted to Dulmage–Mendelsohn algorithm. Inputs correspond to variables restricted to single values. In this case, the equality equations lead to just-determined system. It means that the solution for causality exists and the problem has exactly one solution. The problem can be resolved automatically: output values can be deduced directly from inputs. The inputs will be adjusted by SA to satisfy inequality constraints while minimizing objective. Giac symbolic mathematical system is used to reformulate constraints and solve them in order to carry out the SA process.

Simulated annealing uses this model as a simulation problem to optimize a part of its inputs according to an objective computed iteratively. A part of inputs is imposed as parameters and others as degrees of freedom to be optimized. The simulated annealing algorithm chooses the new value for each degree of freedom randomly. A variable called temperature is updated for each iteration of the program, and it decreases exponentially. The optimization process keeps in memory the chosen values of degrees of freedom from the last iteration. If the new values improve the objective, these new values replace the old ones in memory. If the new values worsen the objective, they can replace or not the old values in the memory according to the results of this condition:

$$proba > exp(-\delta/temperature)$$

where:

$proba$: a random value generated by random method.

δ: the difference between the old and new objective values.

$temperature$: a variable that decreases exponentially during the evolution of the algorithm.

To avoid repetitions in values instantiation, a tabu list is added in the algorithm. For each random choice, the tabu list is checked before validation of this new value.

The simulated annealing optimization supports the interactions with occupants. The pivot model used in the MILP optimization and the SA optimization is the same, but the differences are in variables considering as degrees of freedom for solvers. In MILP optimization, degrees of freedom are fixed by expert when the transformation from pivot model to applicative model is done. In SA optimization, occupants choose the variables that they want to change according to their personal requirements interactively. The rest of the variables are for the results of MILP optimization, and the initial values of their variables are taken from MILP optimization

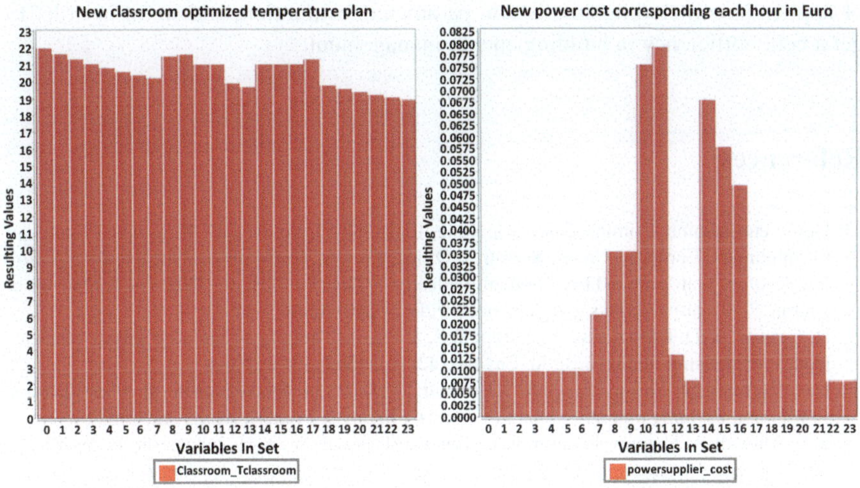

Fig. 13 Results of new classroom temperature plan and new power cost correspondingly generated by SA solver

results. The requirements of occupants can be expressed as additional constraints on variables.

Consider now the SA computation within few seconds for the new temperature set points. The new optimized plan is given by Fig. 13.

5 Conclusion

This chapter presented (Sect. 2) a global model-based anticipative energy management system to generate an optimal strategy in order to harvest the maximum energy. This model has been adapted and applied to the CANOPEA building prototype proposed by the Rhône-Alpes team for Solar Decathlon Europe 2012 contest [35]. The BEMS of CANOPEA building prototype was able to compute energy management strategies and to propose advices to occupants.

Approaches based on model predictive control (MPC) proposed to design efficient energy management systems in building have also been presented in Sect. 3. These approaches have the ability to compute near-optimal solution while handling multi-variable character and respecting the constraints of the problem.

Finally, a model transformation methodology based on MDE approach is proposed aiming to automatically generate application models in building energy management. The core specifications to transform a pivot model into application models are defined. A prototyped software has been developed for PREDIS/MHI platform to validate the proposed approach. It has been shown that the proposed approach can be advantageous in BEMS problem where two kinds of optimizations are presented: an initial global MILP optimization and several SA fast optimizations to support interaction with occupants. It is under development to handle other kinds of applications including simulation, parameters estimation, and so on in order to get a better efficiency in building energy management.

References

1. Union européenne. Commission européenne, EUROSTAT, Statistical Office of the European Communities, European Union. Statistical Office, *Energy, Transport and Environment Statistics*. Energy, Transport and Environment Indicators, Data (Publications Office of the European Union, 2019). https://books.google.fr/books?id=GUMhzAEACAAJ
2. H. Doukas, K.D. Patlitzianas, K. Iatropoulos, J. Psarras, Intelligent building energy management system using rule sets. Build. Environ. **42**(10), 3562–3569 (2007)
3. G. Levermore, *Building Energy Management Systems: An Application to Heating, Natural Ventilation, Lighting and Occupant Satisfaction* (Taylor & Francis, New York, 2002)
4. S.A. Ahmadi, I. Shames, F. Scotton, L. Huang, H. Sandberg, K.H. Johansson, B. Wahlberg, *2012 Seventh International Conference on Towards more Efficient Building Energy Management Systems, Knowledge, Information and Creativity Support Systems (KICSS)* (2012), pp. 118–125
5. D.L. Ha, H. Joumaa, S. Ploix, M. Jacomino, An optimal approach for electrical management problem in dwellings. Energy Build. **45**, 1–14 (2012)
6. S. Abras, S. Ploix, S. Pesty, M. Jacomino, A multi-agent home automation system for power management, in *Informatics in Control, Automation and Robotics III* (Springer, Berlin, 2006)
7. AFNOR, Ergonomie des ambiances thermiques, Détermination analytique et interprétation du confort thermique par le calcul des indices PMV et PDD et par des critère de confort thermique local; Norme europèenne, norme française (2006)
8. S. Ha, H. Jung, O. Yong, Method to analyze user behavior in home environment. Personal Ubiquitous Comput. **10**, 110–121 (2006)
9. J.M. Pinto, I.E. Grossmann, A continuous time mixed integer linear programming model for short term scheduling of multistage batch plants. Ind. Eng. Chem. Res. **34**, 3037–3051 (1995)
10. J.M. Pinto, I.E. Grossmann, Assignment and sequencing models for the scheduling of process systems. Ann. Oper. Res. **81**, 433–466 (1998)
11. P.M. Castro, I.E. Grossmann, An efficient MIPL model for the short-term scheduling of single stage batch plants. Comput. Chem. Eng. **30**, 1003–1018 (2006)
12. P. Esquirol, P. Lopez, *L'ordonnancement*, Collection Gestion Série Production et techniques quantitatives appliquées à la gestion (Economica, Paris, 1999)
13. A.M. Nygard-Ferguson, Predictive Thermal Control of Building Systems, PhD Thesis, Ecole Polytechnique Federale de Lausanne, 1990
14. Y.M. Borrelli, F. Hencey, B.B. Packard, Model predictive control of thermal energy storage in building cooling systems, in *Proceedings of the 48th IEEE Conference on Decision and Control* (2009)
15. A. Collazos, F. Maréchal, C. Gähler, Predictive optimal management method for the control of polygeneration systems. Comput. Chem. Eng. **33**, 1584–1592 (2009)
16. R.R. Negenborn, M. Houwing, B. De Schutter, J. Hellendoorn, Model predictive control for residential energy resources using a mixed-logical dynamic model, in *IEEE International Conference on Networking, Sensing and Control* (2009)
17. S. Yuan, R. Perez, Multiple-zone ventilation and temperature control of a single-duct HVAV system using model predictive strategy. Energy Build. **38**, 1248–1261 (2006)
18. M. Gwerder, J. Tödtli, System for controlling room temperature in building using predictive control, Patent (2009)
19. F. Oldewurtel, D. Gyalistras, M. Gwerder, C.N. Jones, A. Parisio, V. Stauch, B. Lehmann, M. Morari, Increasing energy efficiency in building climate control using weather forecasts and model predictive control, in *10th REHVA World Congress CLIMA* (2010)

20. M.Y. Lamoudi, M. Alamir, P. Béguery, Model predictive control for energy management in buildings part 1: zone model predictive control, in *IFAC Conference on Nonlinear Model Predictive Control* (2012)
21. M.Y. Lamoudi, M. Alamir, P. Béguery, Unified NMPC for multi-variable control in smart buildings, in *IFAC 18th World Congress* (2011)
22. M.Y. Lamoudi, P. Béguery, M. Alamir, Use of simulation for the validation of a predictive control strategy, in *12th International IBPSA Conference* (2011)
23. M.Y. Lamoudi, M. Alamir, P. Béguery, Model predictive control for energy management in buildings part 2: Distributed model predictive control, in *IFAC Conference on Nonlinear Model Predictive Control* (2012)
24. PSE. gPROMS (2011). http://www.psenterprise.com/gproms/
25. M.R. Bussieck, A. Meeraus, General algebraic modeling system (GAMS), in *Modeling Languages in Mathematical Optimization* (Springer, Boston, 2004), pp. 137–157
26. A.L. Dulmage, N.S. Mendelsohn, Coverings of bipartite graphs. Can. J. Math. **10**, 517–534 (1958)
27. Modelica, Modelica® – A Unified Object-Oriented Language for Physical Systems Modeling, Technical Report, Volume 3.2 (2010)
28. J. Von Zur Gathen, J. Gerhard, *Modern Computer Algebra* (Cambridge University Press, Cambridge, 2003)
29. F. Winkler, Advances and Problems in Algebraic Computation, in *Contributions to General Algebra 12 (Proc. AAA'58, Vienna)*, Verlag Johannes Heyn, Klagenfurt (Austria) (2000)
30. W.S. Brown, A.C. Hearn, Applications of symbolic algebraic computation. Comput. Phys. Commun. **17**(1–2), 207–215 (1979)
31. M. Wester, A review of CAS mathematical capabilities. Comput. Algebra Nederland **13**, 41–48 (1994)
32. A. Dall'Osso, Computer algebra systems as mathematical optimizing compilers, Sci. Comput. Program. **59**(3), 250–273 (2006)
33. P. Bernard, Giac/Xcas, a free computer algebra system. Technical Report, University of Grenoble (2008)
34. G.D. Oliveira, M. Jacomino, D.L. Ha, S. Ploix, Optimal power control for smart homes, in *18th IFAC World Congress* (Elsevier, Amsterdam, 2011)
35. Y.H. Said, S. Ploix, S. Galmiche, B. Lechat, A. Djellouli, T. Scheid, S. Bergeon, X. Brunotte, A global model based energy management system applied to the CANOPEA building, in *Proceedings of BS 2013: 13th Conference of the International Building Performance Simulation Association* (2013)

Distributed and Self-learning Approaches for Energy Management

Hussein Joumaa, Khoder Jneid, and Mireille Jacomino

1 Principle of Distributed Solving Approach

The role of a power management system is to adapt the power consumption to the available power resources and vice versa, taking into account inhabitant comfort criteria. It has to reach a compromise between the priorities of the inhabitant in terms of comfort and in terms of cost while satisfying technological constraints of devices. The problem is distributed by nature. It deals with heterogeneous multi-services having different types of models and divergent preferences, it is evolutive. The centralized optimization is not able to take into account this kind of service representation which forces its resolution to be distributed as well. The distributed artificial intelligence approaches or Multi-agent approaches have been used to manage these services. The interesting capabilities in terms of openness and adaptability, while solving problems, of such systems motivate for solving energy management problems using the Multi-Agent Systems (MAS).

The first MAS approach for energy distribution has been presented in [12] and [19]. Kok et al. [14] put forward a market-based control concept for the supply and demand matching (SDM) in electricity networks. It aims to propose a Multi-Agent System for the electronic market. Its purpose is to control tasks in future

H. Joumaa (✉)
Capgemini DEMS, Montbonnot-Saint-Martin, France
e-mail: Hussein.Joumaa@capgemini.com

K. Jneid
eLichens Grenoble, Laboratoire G-Scop, Grenoble Cedex 1, France
e-mail: khoder.jneid@elichens.com

M. Jacomino
Université Grenoble Alpes, Grenoble INP Ense3, Laboratoire G-Scop, Grenoble Cedex 1, France
e-mail: Mireille.Jacomino@grenoble-inp.fr

S. Ploix et al. (eds.), *Towards Energy Smart Homes*,
https://doi.org/10.1007/978-3-030-76477-7_10

electricity network which is expected to develop into a network of networks with a vast number of system parts communicating and coordinating with each other. The developments of solutions based on Multi-Agent Systems, well suited to solve spatially distributed and opened problems, permit to imagine an intelligent Multi-Agent Home Automation system. The multi-agent approaches have some advantages but, generally speaking, cannot ensure an optimal solution of the energy management problem contrary to the centralized approaches. An agent-based architecture, MAHAS (Multi-Agent Home Automation System) was proposed in [2–4, 6, 7, 13] in order to solve the energy management problem. The architecture proposes an agent modeling decomposed into two complementary mechanisms: an anticipative mechanism which computes, through a negotiation protocol (solving strategy) between agents, a solution for power management problem, and a reactive mechanism that protects from constraint violation.

Multi-agent approaches allow the agents to cooperate and coordinate their actions in order to find an acceptable solution for power management. As the verification and the control of the emergent behavior of multi-agent systems is extremely complex, the performance of the centralized system is when possible better than that obtained with distributed systems. On the other hand, the distributed system based on Multi-Agent System techniques does offer advantages over the centralized approach: its openness, its scalability, and its capability to manage diversity.

2 Principle of Mixed Solving Approach

An alternative approach, noted mixed approaches, proposes a formulation to combine the centralized solving approach [11] for energy management problem in homes with a multi-agent solving system. This approach uses the multi-agent systems by integrating the agentified-equipment models in the global solution of the problem that is provided by a centralized way. The proposed system is a mixed centralized/distributed approach [1, 5, 7, 10, 13] for the resolution of global energy management problem.

These solutions integrate mainly three parts (Fig. 1):

- The regular services: the services that can be integrated directly into a centralized energy management problem without the limits cited before. For example, these services consist of the ones having a mixed integer linear model that can be integrated directly into the energy management problem in [15].
- The singular services (agents): All the other services having models that cannot be integrated into the global solving process. For example in the energy management system [15], all singular services that do not have a mixed integer linear model and should communicate with the solver to give their *energetic profiles* can be considered singular services or agents. The solving process and the orchestration between regular and singular services use a specific protocol of

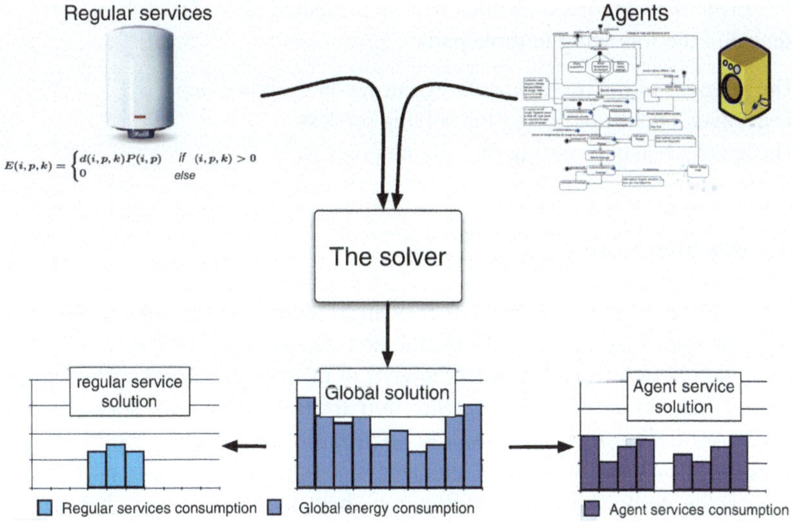

Fig. 1 Global architecture of the mixed solving system

communication to incorporate agents and regular services into a main system, in order to compute a global energy consumption plan. An example of this orchestration strategy is presented in the following parts.

- The solver is usually the "orchestrator" between regular and agent services. In many works, the solver consists of a regular solver with the ability to communicate with agents. The solver integrates the information sent by the agent's local solvers with regular service models in order to generate a global problem to solve.

2.1 The Solving Strategy in Mixed Solving Approach

The system is composed of regular services having a linear model and agents that use several different models that are not explicit for the mixed solver.

There is only one communication needed between the regular services and the solver. At the beginning of the solving process, the solver receives the linear model from the regular services. The models are used all along the solving process.

In the case of agents, a lot of communications are needed with the solver. Each exchange between the agents and the solver is considered as a step in the solving process. In each step, an intermediate problem is created by the solver then computed. The solver decides which information is needed to be sent to the agents in the next step. The agent takes into account the information sent by the solver and sends energetic profiles that are the exchange data from the singular services to the solver. The energetic profiles will be presented later. The solving process is presented in the following in three parts:

- The progress of the problem solving during one solving step.
- The solver's behavior during the solving process.
- The agent's behavior during the solving process.

2.1.1 One Step Solving

Figure 2 presents the information exchanged between the solver, the regular services, and the agent services during the first step in the solving process.

First, the solver receives the linear models of the regular services. This operation is the initialization of the problem. Once initialization is done, the solver computes the *relevance indicator*. It is an indicator which aims at directing the local solving problem for the agents. When this indicator is computed, it is sent to all agents.

The agents do not have any information about the environment, but they have the ability to solve their own local problem. When agents receive the relevance indicator, they compute their solutions taking into account this indicator serving as information about their environment. They obtain several solutions, which are called *energetic profiles*. It is the consumption for the concerned agent for each period of the optimization horizon. All these profiles are sent back to the solver which in turn includes them in the problem to be solved at this step together in the global problem with all the services.

Next, the solver begins a new step by computing the relevance indicator taking into account the received *energetic profiles* sent by the agents in order to improve the global solution at each step in the solving process.

2.1.2 Solver's Role

The solver has two tasks to do in each step. In order to formulate these tasks, we introduce some notations:

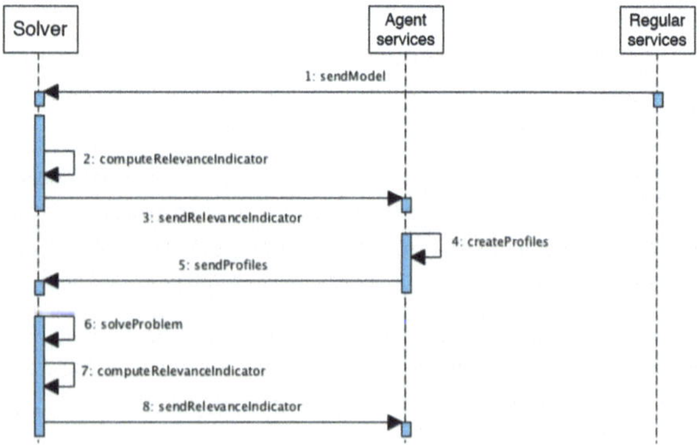

Fig. 2 Solving process during one step

- k is the index of anticipative period
- $\mathcal{S}$ is the set of services
- $\mathcal{S}^L$ is the set of *regular services*
- $\mathcal{S}^D$ is the set of *singular services*
- S is a service included in $\mathcal{S}$
- E_k^{max} is the available energy during the period k before any optimization
- $E_k^j(S)$ is the consumed energy by the *regular service* $S \in \mathcal{S}^L$ during the period k in the step j
- $E_k^j(S, i, \mathbb{P}_k)$ is the consumed energy by the *agent service* $S \in \mathcal{S}^D$ during the period k for the i^e profile in the step j
- C_k is the cost of energy during the period k
- $\upsilon(S)$ is the characteristic of inhabitant request for the service S
- $D(\upsilon(S))$ is the dissatisfaction of the *regular service* $S \in \mathcal{S}^L$
- $D(\upsilon(S), i, \mathbb{P}_k)$ is the dissatisfaction of the *agent service* $S \in \mathcal{S}^D$ for the i^e profile
- $\mathbb{P}_{k,\forall k}$ is the relevance indicator for the current step of resolution

Optimization Problem

At each step, the solver computes a linear problem to find a solution. The *regular services* models are represented in [9]. This problem is extended by including *agent services*. Some equations are added to take into account the *agent services*. The solver has to choose the most pertinent energetic profiles amount those given by the agents. A new set of variables for each *agent service* is introduced (see Eq. (1)). $\zeta_i(S)$ is a binary variable whose value is 1 if the profile i of the *agent service* S is chosen by the solver, 0 otherwise. Combined with Eq. (2), ensure the solver to keep only one profile for each *agent service* in the solution.

$$\zeta_i(S) \in \{0, 1\}, \forall i \tag{1}$$

$$\sum_i \zeta_i(S) = 1 \tag{2}$$

The criterion to minimize is modified and becomes a two parts criterion (3). The first part concerns the *regular services* and the second part the *agent services*. They are designed on the same scheme to have a standardized criterion. This scheme splits into two influences:

- The influence on the cost: the global energy cost must be minimized.
- The influence on the inhabitants: the dissatisfaction of the inhabitants must be minimized.

$$J_{iter} = \sum_{S \in \mathcal{S}^L} \left(\sum_k C_k E_k(S, \theta(S)) + \lambda \times D(\upsilon(S, \theta(S))) \right)$$

$$+ \sum_{S \in \mathcal{S}^D} \sum_i \zeta_i \left(\sum_k C_k E_k(S, i, \mathbb{P}_k) + \lambda \times D(\upsilon(S), i, \mathbb{P}_k) \right) \tag{3}$$

Those influences can be found in both *regular services* part and *agent services* part. But there is a fundamental difference between these two parts, and it is symbolized by the sum on the index i in the *agent services* part. The solver keeps for each *service agents* only one profile. For each profile, the solver receives one consumption plan and an associated dissatisfaction. The sum in the criterion with binary variables forces to keep only one profile for each agent for the minimization.

Figure 3 shows the complexity of the problem to be solved at each step. Each *service agent* provides n profiles, if there are m *singular services*, then there are n^m different solutions. But the solver has to minimize the criterion to keep one.

Relevance Indicator

The relevance indicator is computed during each solving step to direct the local solving process of service agents for the next step. After the solving step j, the relevance indicator is computed with Eq. (4). The purpose of this approach is to

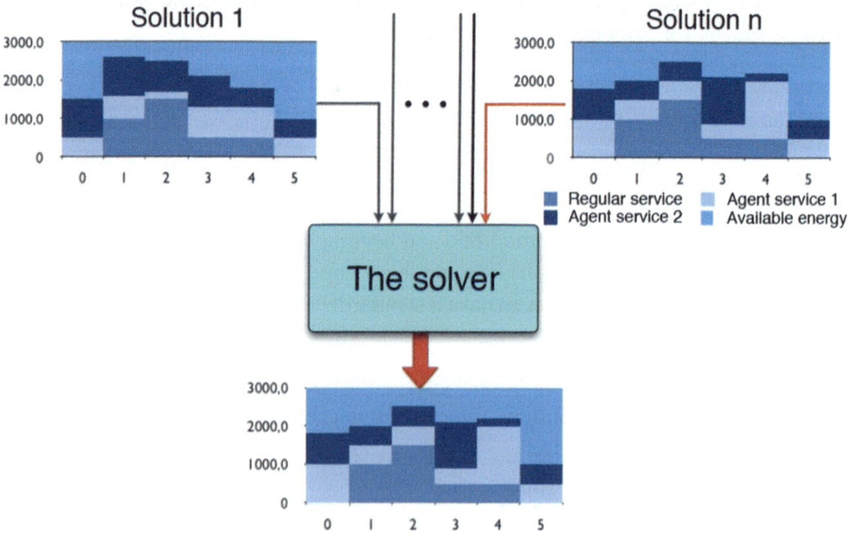

Fig. 3 Solution found by the solver

share the information about the energy consumption and price between solver and service agents. The service agents integrate the received information in their local solving process of the step j+1. This indicator is high when the consumed energy is important or/and when the energy is expensive. This indicator aims to improve the current solution. During the first step, the consumption of the *agent services* is null.

$$\mathbb{P}_k^j = \frac{1 + E_k^{max}}{1 + E_k^{max} - \sum_{S \in \mathcal{S}^L} E_i^j(S)} C_k \tag{4}$$

2.1.3 Role of the Agents

An agent is dedicated to a specific entity whose behavioral model cannot be conveniently linearized and then taken into account directly by the centralized solver. In this part, the algorithm used by agents in order to provide energy planning is explained using an example of a washing machine service agent.

The washing machine service agent has its internal state model. The states are shown by Fig. 4. They consist of:

- Some behavioral states like heating, prewash, washing, and spin-drying
- Two states representing the beginning and the end of the service
- Some states denoting *wait i* that represents the waiting time between behavioral states
- Some states modeling the interruption within each state, denoted *interrupted state*

The standard behavior of the washing machine service is given by the state sequence scenario [start, heating, prewash, washing, spy-drying, end]. The other states are only visited when the service agent tries to find some neighboring

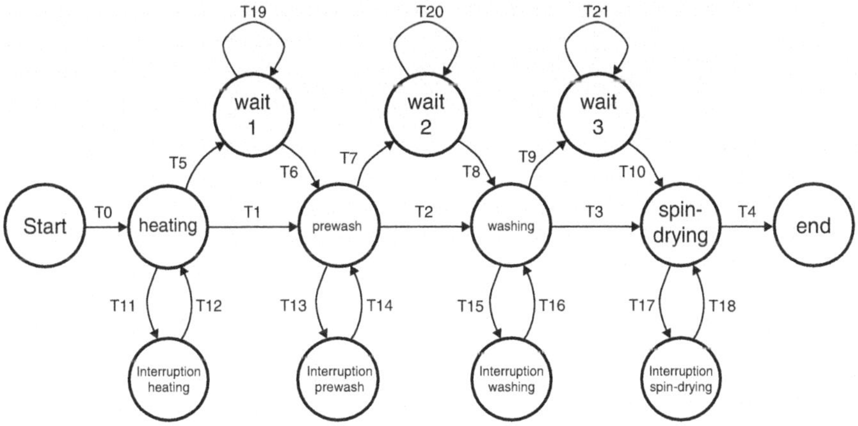

Fig. 4 State model of the washing machine service agent

energetic profiles in order to respond to some criteria sent by the solver that coordinates the agents to deal with global optimization of energy consumption.

Each visit to an *interrupted state* has a fixed time period $\tau_{interrupted}$. It is possible to visit the *interrupted state* more than once in order to increase the interruption time in a state. For example, in the state sequence scenario [start, heating, interrupted heating, heating, interrupted heating, prewash, washing, spy-drying, end], the time spent in the interrupted heating state is $2 * \tau_{interruption}$.

A behavioral profile is the state sequence scenario with the date of each state visit. The behavioral profile is characterized by:

- The starting time of the service
- The number of visits to each *interrupted state* and the number of visits for each *wait i* state
- The date of each visit to *interrupted states* and *wait i* states.

These characteristics are denoted in the following *parameters of behavioral profile*. It is interesting to note that a behavioral profile is computed in order to be converted into an energetic profile. The energetic profile consists of the energy consumed by the service in each period of the anticipative horizon. The energetic profile is then sent to the solver.

The agent satisfaction is computed according to the energetic profile. The satisfaction depends on the number of visited interrupted states and also on the effective ending time regarding its expected value by the occupants. The increase in the number of interruptions affects the agent satisfaction.

2.1.4 Agent Solving Algorithm

The agent solving algorithm is presented in Fig. 5.

First, the agent receives the relevance indicator. The relevance indicator consists of information about the penalization and the energy price during the anticipative horizon. The agent receives also the chosen energetical profile at step j.

The first step in the algorithm is to normalize the values of relevance indicators (5). The goal of this step is to obtain $RI_{k(normalized)}$ that can be used in the computation of CA_k, the agent coefficient. It is composed of both the information received from the solver and on the local satisfaction computed by the agent.

$$RI_{k(normalized)} = RI_k / Max(RI_k) \tag{5}$$

where $Max(RI_k)$ is the maximum value of relevance indicator associated with energetic profiles received for the period p.

The second step consists of the computation of the agent coefficient CA_k. The CA_k merges the information about the penalization, the energy price, and the agent dissatisfaction denoted I_k (6).

$$CA_k = RI_k + \lambda * I_k \tag{6}$$

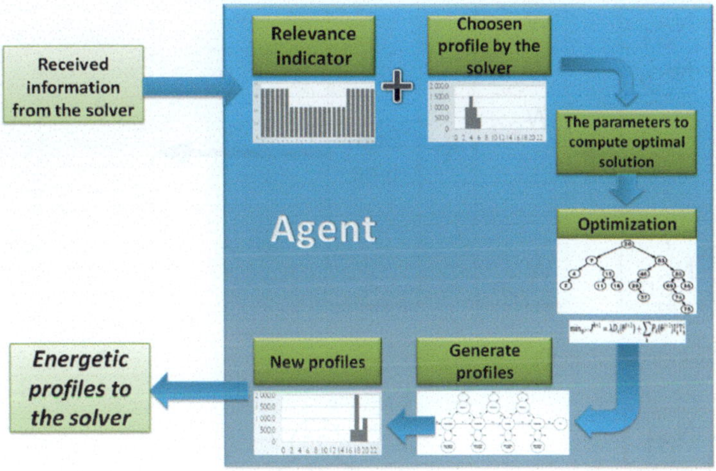

Fig. 5 Solving algorithm in the agent

In order to generate an energetic profile, the first step is to compute the behavioral profile. The parameters of the behavioral profile are listed above. The first one is the starting time of the service. We begin by finding the best intervals over 6 periods in the 24 h horizon according to the values CA_k. For each interval j we compute X_j (7).

$$X_j = \left(\sum_{k \in [j, j+6]} CA_k \right) / 6 \tag{7}$$

The minimum of the list X_j is denoted $X_{j_{min}}$.

Then we try to find the intervals having no significant difference with $X_{j_{min}}$. We denote L_{min}, the list:

$$L_{min} = \{k / 1 - (X_{j_{min}} / X_k) < 0.1\} \tag{8}$$

The interval χ with the maximum variance in L_{min} is chosen for the optimization. The starting time of the service corresponds to the starting time of the chosen interval χ.

The parameters of the optimization are presented in Fig. 6 where N_{Si} is the number of interruptions in the state Si. W_{Si} is a value to select the time for interruption within the state.

A branch and bound optimization is performed on this parameter (Fig. 7) within the chosen interval χ. Each agent solves the optimization problem with this function. It represents the minimization of the energetic cost and dissatisfaction

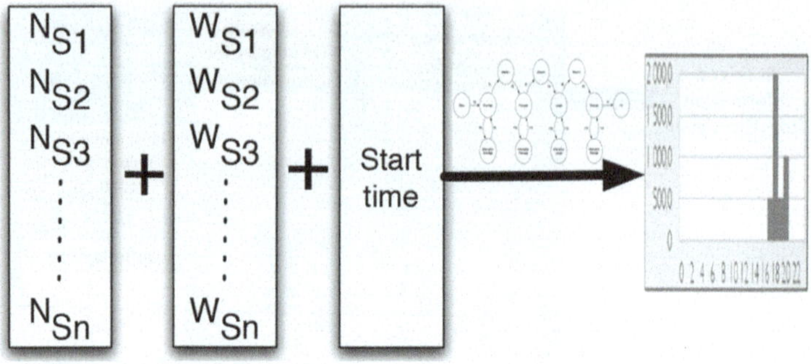

Fig. 6 Parameters of a profile

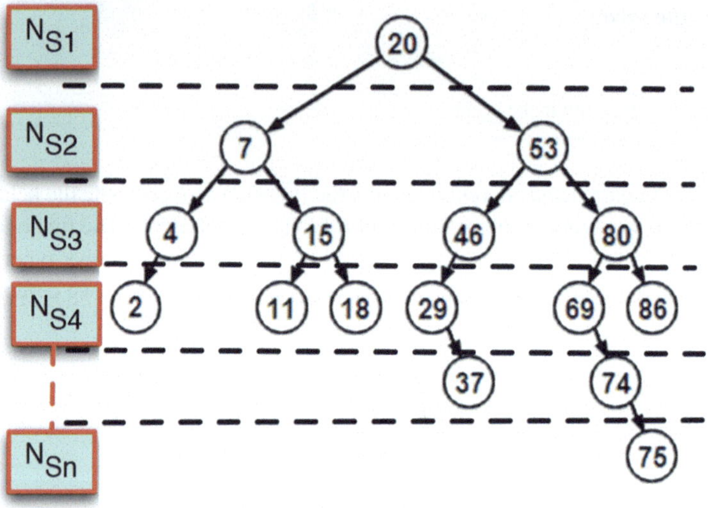

Fig. 7 Optimization using branch and bound

from a local point of view. The function to be minimized (9) is similar to the one presented for the solver.

$$min_{\theta^{j+1}} J^{k+1} = \sum E_k(\theta^{j+1})\mathbb{P}_k^j \mathbb{T}_k^j + \lambda D_i(\theta^{j+1})) \tag{9}$$

θ^{j+1} represents the parameters of the user that define the usage conditions. The function is composed of two parts: the first one is the influence of the energetic cost and the second one is the influence of the satisfaction of the agent.

The result of this optimization is a list of parameters required to generate the behavioral profile (parameters of behavioral profile). Then, the energetic profile can be computed and sent to the solver to be integrated in the global problem solving.

2.2 Results of the Implementation

The implemented system consists of five components (Fig. 8):

- The classical regular solver used in [9]
- The global solver including regular services and agent services
- The broker agent is a communication component that receives all the local problems from service agents and constructs one global service agent problem. This problem is then sent to the global solver. The broker receives also the relevance indicator from the solver and dispatches the information to service agents
- The service agents with the capabilities to solve a local problem.

The system is tested by using two service agents and some regular services. Figure 9 presents simulation results. The first part of the figure presents the anticipated consumption of an agent in the next 24 h. The second part presents the anticipated consumption of a regular service. The system delivers also the computed temperature setpoints for the next 24 h.

The system has been tested in order to identify the relation between the different parameters. The goal of the presented tests is to improve the solution and to decrease the execution time of the system. Figure 10 presents the relation between the criterion values (the criterion is the function to be minimized J [Eq. (3)]) and the number of iterations. For 30 profiles sent by the agents, the value of the criterion is not dependent on the number of iterations. In other cases, the value of criterion has an optimum for 15 iterations. Figure 11 presents the variation of the time of execution with the number of iterations. The time of execution increases slowly for 18 profiles around 15 iterations. Figure 12 presents the relation between the

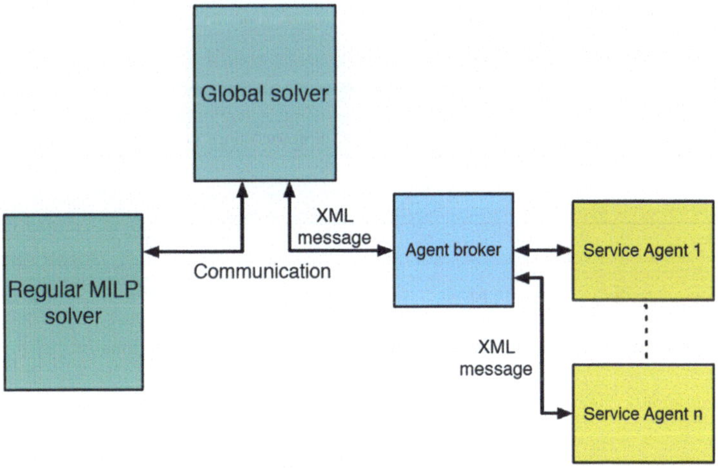

Fig. 8 The components of the mixed solving system

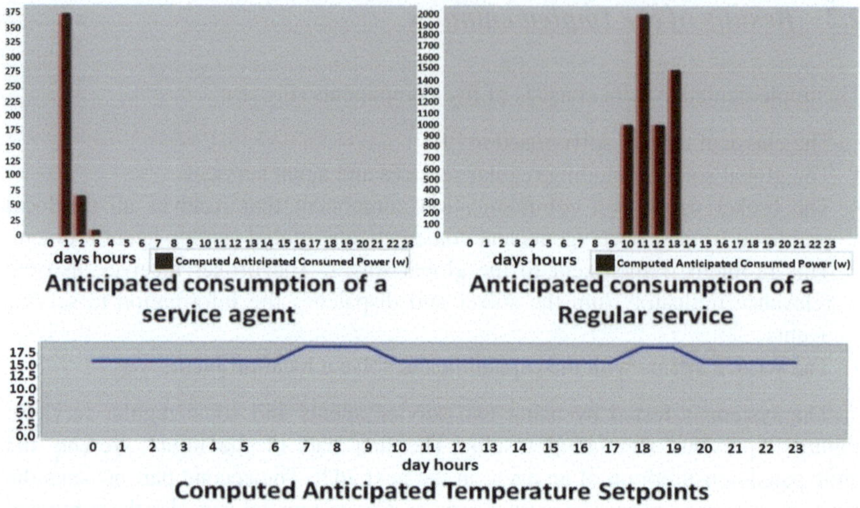

Fig. 9 Anticipating regular and service agents

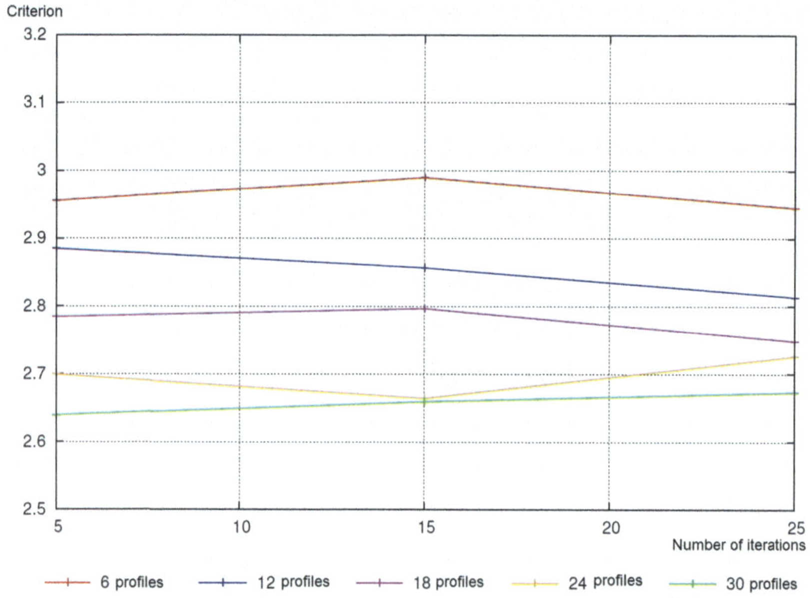

Fig. 10 The relation between the criterion values and the number of iterations

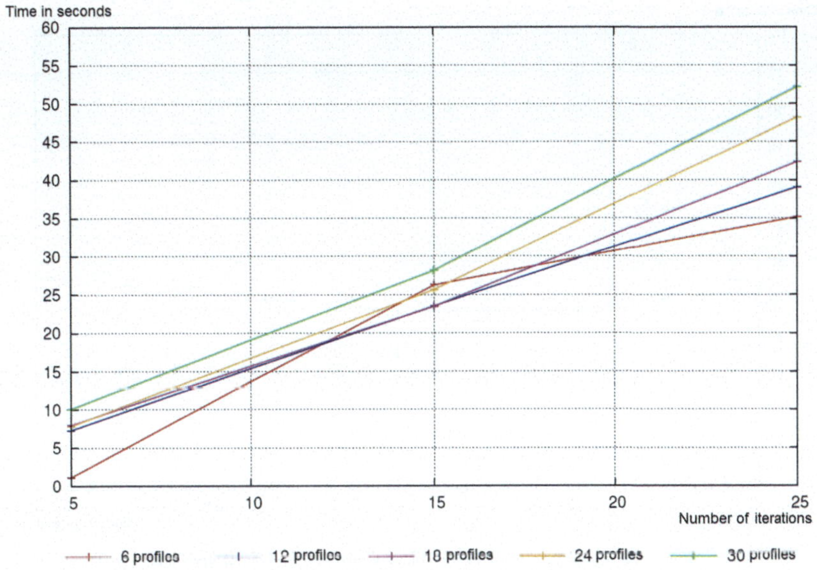

Fig. 11 The variation of the time of execution with the number of iterations

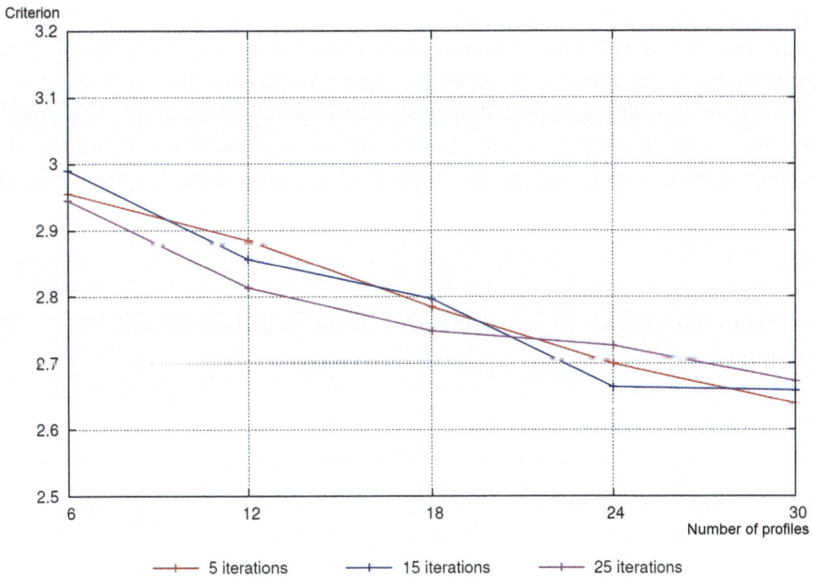

Fig. 12 The number of profiles sent by the solver and the value of the criterion

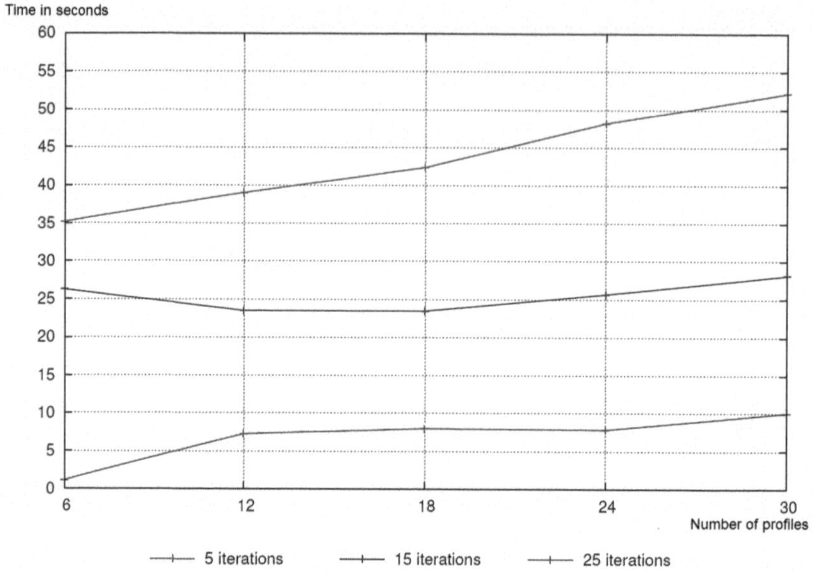

Fig. 13 The relation between the number of profiles and the time of execution

number of profiles sent by the solver and the value of the criterion. The value of optimization criterion is better when the number of profiles is higher. These results are not related to the number of iterations. Figure 13 presents the relation between the number of profiles sent by the agents and the time of execution. The time of execution is higher when the number of profiles sent by the solver is higher. This is related to the treatment time of the information received needed in the solver and the agents.

3 Principle of Solving Approach Using Reinforcement Learning

3.1 Reinforcement Learning

Reinforcement learning (RL) is a subcategory of machine learning that involves learning by interaction [20]. Two main players exist in this kind of learning: the agent and the environment. The agent is the learner and decision maker at the same time. The environment is the thing or system the agent interacts with. It generates rewards that represent an evaluative feedback for the agent. A reward is a scalar value that represents the quality of the action taken by the agent at a specific state; such as +1 for rewarding a good action and −1 for penalizing a bad action.

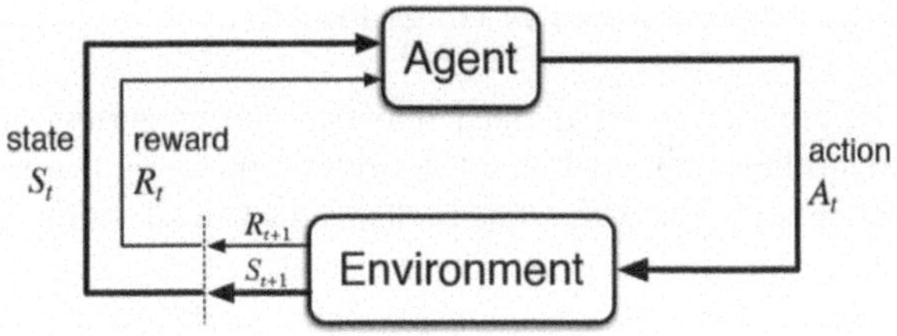

Fig. 14 Agent–environment interaction

The agent learns through trial and error and then a great number of interactions with the environment is needed to learn the optimal actions. The agent–environment interaction is illustrated in Fig. 14. At each time step t, the agent observes the state S_t that includes a reward R_t then takes an action a_t. The environment will move then to a new state S_{t+1}, and the agent will receive an evaluative feedback of the action taken, reward R_{t+1}. The agent's goal is to learn the optimal policy π^* that maximizes the cumulative or discounted reward at a given state. The optimal policy specifies the best action to take at a specific state. The discounted reward (return) is computed as follows:

$$r_{t+1} + \gamma r_{t+2} + \gamma^2 r_{t+3} + \ldots = E\left[\sum_{k=0}^{\infty} \gamma^k r_{t+k+1}\right] \tag{10}$$

$\gamma \in [0, 1]$ is called a discount factor.

3.1.1 Markov Decision Process

RL problems are modeled as a Markov decision process (MDP), since the transition from state (s) to a new state (s′) after executing action (a) depends only on the current state and not the history of the passed states. A MDP can be described using these four concepts: States, actions, probabilities, and rewards:

- S represents the state space.
- A represents the action space.
- $p(s_{t+1}|s_t,a_t)$ represents the state transition probability.
- $q(s_{t+1}|s_t,a_t)$ represents the probability governing the reward received when moving to a new state after an action's execution.

MDP results an optimal policy that maps between states and actions and thus guides the agent through time to maximize the reward. Having a complete knowledge of the environmental model (S, A, p, q), the problem can be solved using dynamic programming to obtain an optimal policy.

However, the environmental model is not fully observable, especially p and q, and then the optimal policy cannot be obtained using traditional dynamic programming. A solution to compute the optimal policy is through estimating the missing model (model-based RL) or through learning directly the policy (model-free RL).

3.1.2 Model-Based vs Model-Free RL

RL algorithms are divided into two subcategories: model-based and model-free algorithms. In model-based algorithms, the agent learns a model of the environment by observing how the state is changed when an action is taken and thus learning the state transition function. When a model of the environment is learned, it could be coupled then with a planning algorithm to obtain the optimal policy such as explicit-explore and exploit [17].
However, model-free algorithms do not need a model of the environment, and they learn the policy through trial and error aiming to have an approximation of the optimal policy. Most of the used RL algorithms belong to the model-free algorithms such as: Q-learning [17] and SARSA [18]. Model-free algorithms are mainly used in the literature because they are less computationally expensive and do not require estimating an environmental model that could be a complex task.

3.1.3 Q-learning

Q-learning is widely used in RL and is a model-free method. Q-learning outputs a Q-table mapping states to actions, Q = S X A. The policy followed by the agent in order to take an action is determined by this table. The Q-table is updated at each step the agent takes an action (a) that generates a reward (r), as follows:

$$Q(s_t, a_t) = (1 - \alpha)Q(s_t, a_t) + \alpha[r + \gamma \max_{a_t} Q(s_{t+1}, a_t) - Q(s_t, a_t)] \quad (11)$$

$\alpha \in [0, 1]$ represents the learning rate and $\gamma \in [0, 1]$ is called the discount factor. The discount factor represents the importance given for future rewards. A discount factor equals to 0 leads to a greedy agent that considers only immediate rewards whereas a value equals to 1 leads to an agent that considers a long-term reward. The discount factor specifies how much the agent should exploit and explore.

3.1.4 Deep Reinforcement Learning (DRL)

Tabular methods such as Q-learning, SARSA, face complexity problems when dealing with high dimensional state and action spaces. The Q-table for example will become very large and infinite with continuous state and action spaces. Thus, tabular methods are not scalable in complex problems. A way to overcome this issue is to use deep reinforcement learning where a deep neural network is used as a function approximator to replace for example the Q-table and then approximation the policy. Deep reinforcement learning has been used widely recently in atari games such as Alpha GO and in chess game where they have showed outstanding results.

3.2 *Optimizing HVAC Systems Using RL*

Reinforcement learning has shown interesting results in robotics and playing video games. Recently, it has been used in the domain of building energy management, especially on Heating, Ventilation, and Air Conditioning systems (HVAC) but just inside simulated environments. In this section, we present the literature review that has been done so far in this area.

3.2.1 Learning a Simple Thermostat Controller

In [8], the authors first develop a physical model for a room equipped with an HVAC system. Then, they develop a prediction model that can predict if the room will be occupied or not. Using the physical model with the occupancy prediction model, they use reinforcement learning to learn a zone thermostat controller that has four actions:

- Heat_{on}: turns on heating
- Heat_{off} turns off heating
- Cool_{on} turns on cooling
- Cool_{off} turns off cooling

Whereas the state space depends on:

- T_{in}: room indoor temperature
- tto: time to occupancy using the occupancy model
- T_{out}: outdoor temperature

Concerning the rewards, they are designed according to certain scenarios. The rewards are scalar values that represent the quality of the action. Here are some rewards for some scenarios:

1. (Room.occupied = false) (Action = Heat_{on}) (rt > sp || rt < sp); R = 0
2. (Action = Heat_{off}) (rt = sp); R = 1
3. ...

sp is the setpoint temperature and thus the agent should take optimal actions that drive the room temperature to this setpoint. The state and action spaces are kept small so that the problem can be tractable using simple reinforcement learning like tabular methods. Even though only three state variables are used, the state space can go very large, for instance the outdoor temperature can have very high range. Q-learning is used by the authors to learn the thermostat controller that leads to 10% of energy saving over the initial controller. The energy minimization was in the heating energy because of the limitations of the data used.

3.2.2 Zone Air Flow Controller

In [22], the authors study a variable-air-volume (VAV) system, which is simulated in the software EnergyPlus. In VAV system, a building is divided into many zones, and every zone is equipped with a VAV box which is responsible of regulating the air flow inside the zone to meet a specific temperature setpoint. The zone temperature at a specific time step depends only on the previous zone temperature and some environmental parameters as well as the conditioned air input coming out from the HVAC system, the VAV box specifically. Therefore, the HVAC control can be treated as a Markov decision process (MDP) and hence the authors used DRL to control the VAV box.

Control Actions The building is divided into n zones and every zone has a local VAV box. The VAV box provides the conditioned air to the zones with a specific flow rate. The control actions are then related to regulating the air flow rate in every VAV box and thus having an independent control for every zone. The control action could be discretized into many values such as F = (f1, f2, f3, f4, ..., fm). For n zones then, we will have an action space that has n^m possible actions. Obviously, the action space highly increases with the increase of the number of zone and the number of actions and taking more time for the agent to converge. As a result, the authors use a DRL algorithm to regulate the air flow inside every zone.

System States The agent observes the environment first to decide then what action to take. The environmental or system states are considered to be: zone temperature, outdoor temperature, equipment power, occupants activities, solar irradiance intensity, time, and some forecast values of outdoor temperature. By considering forecast values of outdoor temperature, the DRL agent will have the ability to take a proactive control since it can learn the weather trend or pattern.

Reward Function The DRL agent aims in this study at regulating the VAV box airflow of every zone to meet the desired temperature while minimizing the energy cost. The reward function at time step t is as follows:

$$r_t = -\cos t(a_{t-1}, s_{t-1}) - \lambda \sum_i ([T_t^i - \bar{T}_t^i]_+ + [\underline{T}_t^i - T_t^i]_+) \quad (12)$$

$\bar{T}_t^i$ is the desired upper bound temperature of zone i at time step t, and $\underline{T}_t^i$ is the desired lower bound temperature. The reward function includes the energy cost and the temperature violation. λ represents the weight of the temperature violation. The reward described here is negative. Therefore, the DRL algorithm aims at maximizing the reward and thus minimizing the energy cost and the temperature violation by having a good control of VAV box air flow. The results of this study achieve 19% of energy reduction over the EnergyPlus baseline for an area with five zones.

3.2.3 Cooling Optimization of a Simulated Data Center

In [16], the authors use deep reinforcement learning to optimize a simulated data center in EnergyPlus where the objective is to reduce the power consumption while maintaining the thermal comfort. The data center consists of two zones (server rooms). Each zone is equipped with an HVAC system that consists of:

- A mixing box that exchanges air between outdoor and indoor.
- A variable air volume (VAV) fan that takes in the outdoor air at a specific flow rate.
- Many coils that have a common setpoint temperature to meet and are responsible for lowering or cooling the temperature so that the zone temperature meets the zone setpoint.

The optimization of the data center is considered as a deep reinforcement learning that is composed of three main components: state, action, and space.
State: The state space contains the following:

- Outdoor air temperature in [−20 °C, 50 °C]
- Zone air temperature (zone 1 and zone 2) in [−20 °C, 50 °C]
- Total electric demand power in [0 W, 1 GW]
- Non-HVAC electric demand power in [0 W, 1 GW]
- HVAC electric demand power in [0 W, 1 GW]

Action: The actions considered to optimize are the VAV fan flow rate and the common coils' setpoint temperature of each zone.
Reward: The aim of the DRL model is to reduce the power consumption of the data center while maintaining the thermal comfort. Thus, the reward function should reflect this goal by penalizing a temperature violation and high power consumption.

$$r_t = -P_t - \lambda \sum_{i=1}^{z} ([T_t^l - \bar{T}_t^l]_+ + [\underline{T}_t^l - T_t^l]_+) \tag{13}$$

P_t is the total power consumption at time t and z is the number of zones.

The DRL controller shows 22% reduction of total electric demand power. The work is applied on a simulated data center, and the next step is thus to test the model on a real system.

3.3 Discussions

In the domain of HVAC systems, reinforcement learning has been used mainly in simulated environments. The reason behind this is that reinforcement learning requires a lot of interactions with the environment since it learns through trial and error. The agent needs to explore the environment in order to learn the optimal policy of taking an action at a specific state. Hence, learning the optimal policy online (in real environment) can be computationally expensive and can lead to taking very dangerous and bad actions during learning, which result in high energy consumption and bad comfort. As a result, accurate simulators are needed so that the RL agent learns offline a good initial policy that can be tuned later and used in a real system. Another solution would be to use model-based reinforcement learning where the agent learns at first a model of the environment and thus can learn not to take critical actions. However, learning environmental models (the physics of the building: indoor temperature evolution, air quality evolution, power consumption, ...) requires a lot of representative data (yearly dataset) that may not be available.

References

1. S. Abras, S. Ploix, S. Pesty, M. Jacomino, A multi-agent home automation system for power management, in *Informatics in Control, Automation and Robotics III* (2006)
2. S. Abras, S. Ploix, S. Pesty, M. Jacomino, A Multi-agent design for a home automation system dedicated to power management, in *IFIP International Federation for Information Processing, Volume 247, Artificial Intelligence and Innovations 2007: From Theory to Applications*, ed. by C. Boukis, L. Pnevmatikakis, L. Polymenakos (Springer, Boston, 2007), pp. 233–241
3. S. Abras, S. Ploix, S. Pesty, M. Jacomino, A multi-agent design for a home automation system dedicated to power management, in *International Federation for Information Processing Digital Library: Artificial Intelligence and Innovations 2007: From Theory to Applications* (2007)
4. S. Abras, S. Ploix, S. Pesty, M. Jacomino, An anticipation mechanism for power management in a smart home using multi-agent systems, in *Information and Communication Technologies: From Theory to Applications, 2008. ICTTA 2008*
5. S. Abras, S. Ploix, S. Pesty, M. Jacomino, Advantages of MAS for the resolution of a power management problem in smart homes, in *8th International Conference on Practical Applications of Agents and Multi-Agent Systems, PAAMS'2010* (2010)
6. S. Abras, S. Ploix, S. Pesty, M. Jacomino, Advantages of MAS for the resolution of a power management problem in smart homes, in *Advances in Practical Applications of Agents and Multiagent Systems* (2010)
7. S. Abras, S. Ploix, S. Pesty, Managing power in a smart home using multi-agent systems, in *Housing, Housing Costs and Mortgages: Trends, Impact and Prediction* (2010)
8. E. Barrett, S. Linder, Autonomous HVAC control, a reinforcement learning approach, in *Joint European Conference on Machine Learning and Knowledge Discovery in Databases* (Springer, Cham, 2015)
9. G. De Oliveira, M. Jacomino, D.L. Ha, S. Ploix, Optimal power control for smart homes, in *18th IFAC World Congress, 2011* (2011)
10. A.M. Elmahaiawy, N. Elfishawy, M.N. El-Dien, Anticipation the consumed electrical power in smart home using evolutionary algorithms, in *MCIT 2010 Conference* (2010)
11. S. Ha, H. Jung, Y. Oh, Method to analyze user behavior in home environment. Personal Ubiquitous Comput. **10**, 110–121 (2006)
12. N. Jennings, The ARCHON system and its applications, in *2nd International Conference on Cooperating Knowledge Based Systems (CKBS-94)*, Keele (1994), pp. 13–29
13. H. Joumaa, S. Ploix, S. Abras, G. De Oliveira, A MAS integrated into Home Automation system, for the resolution of power management problem in smart homes, in *1st Conference and Exhibition Impact of Integrated Clean Energy on the Future of the Mediterranean Environment* (2011)

14. J.K. Kok, C.J. Warmer, I.G. Kamphuis, PowerMatcher: multiagent control in the electricity infrastructure, in *Proceedings of the Fourth International Joint Conference on Autonomous Agents and Multiagent Systems* (2005)
15. H.D. Long, J. Hussein, P. Stephane, M. Jacomino, An optimal approach for electrical management problem in dwellings. Energy Build. **45**, 1–340 (2012)
16. T. Moriyama et al., Reinforcement learning testbed for power-consumption optimization, in *Asian Simulation Conference* (Springer, Singapore, 2018)
17. S. Ray, P. Tadepalli, Model-based reinforcement learning, in *Encyclopedia of Machine Learning* (Springer, Boston, 2010), pp. 690–693
18. G.A. Rummery, M. Niranjan, *On-Line Q-Learning Using Connectionist Systems*, vol. 37 (University of Cambridge, Cambridge, 1994)
19. O. Sousa, Z. Vale, C. Ramos, J. Neves, Object-oriented agents in power distribution automation, in *2000 10th Mediterranean Electrotechnical Conference. Information Technology and Electrotechnology for the Mediterranean Countries. Proceedings. MeleCon 2000 (Cat. No. 00CH37099)* (2000)
20. R.S. Sutton, A.G. Barto, *Reinforcement Learning: An Introduction* (The MIT Press, Cambridge, 2011)
21. C.J.C.H. Watkins, Learning from Delayed Rewards (1989)
22. T. Wei, Y. Wang, Q. Zhu, Deep reinforcement learning for building HVAC control, in *Proceedings of the 54th Annual Design Automation Conference 2017* (2017)

Model Predictive Control Based on Stochastic Grey-Box Models

Christian Ankerstjerne Thilker, Rune Grønborg Junker, Peder Bacher, John Bagterp Jørgensen, and Henrik Madsen

Acronyms

ACF	Autocorrelation function
AIC	Akaike's information criterion
AR	Autoregressive
CCF	Cross-correlation function
CDEKF	Continuous–discrete extended Kalman filter
CTSM-R	Continuous-time stochastic modelling in R
EKF	Extended Kalman filter
EV	Electrical vehicle
FF	Flexibility function
FI	Flexibility index
HMM	Hidden Markov model
KF	Kalman filter
LRT	Likelihood-ratio test
LS	Least squares
MLE	Maximum likelihood method
MPC	Model predictive control
PI	Proportional integral
PID	Proportional integral derivative

C. A. Thilker (✉) · R. G. Junker · P. Bacher · J. B. Jørgensen
Technical University of Denmark, Lyngby, Denmark
e-mail: chant@dtu.dk; rung@dtu.dk; pbac@dtu.dk; jbjo@dtu.dk

H. Madsen
Technical University of Denmark, Lyngby, Denmark

Norwegian University of Science and Technology, Trondheim, Norway
e-mail: hmad@dtu.dk; henrik.madsen@ntnu.no

S. Ploix et al. (eds.), *Towards Energy Smart Homes*,
https://doi.org/10.1007/978-3-030-76477-7_11

PV	Photo-voltaic cells
SDE	Stochastic differential equation
SOC	State of charge
RMS	Root mean square

1 Introduction

If the proliferation of renewable energy sources is to be continued, solutions for the related problems have to be implemented. The problems include, but are not limited to, mismatch in generation and load, voltage deviation, congestion, and demand ramps. While historically these problems were manageable through control of the generation, this will not be an option in the future, as a majority of power generation will be coming from intermittent renewable energy sources. On the other hand, recent advances and adoption of digital solutions and smart devices present new opportunities for smart energy *demand* [33, 44, 49, 50], by utilising the inherent *energy flexibility* [25]. With buildings accounting for around 40% of energy demand [31], they have been identified as key assets in this context [34]. However, to actually make buildings smart and unlock the inherent energy flexibility, suitable methods for controlling them have to be employed. While smart buildings seek to resolve high-level problems, there remains the dilemma that the buildings themselves are subject to decentralised and independent control and given over to the controllers commissioned by the building owners. To deal with this, it has been proposed to use a two-level control hierarchy [12, 17, 58] in which the upper level consists of controllers that formulate price signals. The price signals are then sent to the lower-level controllers that are controlling energy flexible systems such as smart buildings. The objective of the lower-level controllers is to minimise costs that, if the price signals are formulated correctly, also solve the grid problems [19, 29]. A generalisation of this hierarchical setup of nested controllers is described as the smart-energy operating system (SE-OS) in [36, 44, 45].

For smart control of buildings, both one-way and two-way communication setups are used—often referred to as indirect and direct control, respectively. The simplest and most resilient setup is achieved by one-way communication where a price signal is sent to a group of buildings in a certain part of the grid. In the paper [12], it is shown how consumers, which are sensitive to varying prices, can be used to control the electricity load using a one-way price signal. Estimation of the price response is based on data measurable at grid level, removing the need to install sensors and communication devices between each individual consumer and the price-generating entity.

A sizeable list of examples of smart control of buildings from EBC's Annex 67 project can be found in [24] with details on the central control strategies presented in [56]. The potential of the energy flexibility of buildings was thoroughly investigated in this project, and it was found that by applying suitable control methods, it was found that the suitable application of control methods exposed enormous potential for energy flexibility in buildings. Another important project in regards to this is the SmartNet project, which, together with the CITIES project, demonstrated the potential of automatic energy flexible control for a number of buildings with an indoor swimming pool [4, 36].

In this chapter, special focus will be put on how to formulate extended and advanced disturbance models in such a way that short-term forecasts are well described by them—the most important disturbances to consider for smart control of buildings are weather-related.

The models will be formulated as stochastic grey-box models. This modelling framework bridges the gap between physical and statistical modelling that makes it possible to combine knowledge from physics and statistics in an optimal way. The grey-box models will be formulated using *discretely observed stochastic differential equations* written down as *continuous–discrete-time stochastic state-space models*. In statistics, such models are also called *continuous state-space Hidden Markov models*.

Grey-box models are typically rather simple models in terms of physics, but they are formulated with emphasis on the stochastic part of the models. This implies that we will be able to use rigorous statistical model techniques and that the models enable for an efficient use of online sensors for control and forecasting. This modelling framework has been used to describe the thermal dynamics of buildings [5, 9, 26, 27, 40] and energy systems in many control-oriented projects [16].

First, we will describe the grey-box modelling framework in Sect. 2. Next, we shall describe some examples of grey-box models for buildings and smart building-related components in Sect. 4. This includes models for heat pumps, stationary, and mobile batteries (EVs).

The states of a building, e.g., the indoor air temperature, are heavily influenced by the weather conditions, and a special focus of this chapter is to establish rather simple stochastic models for the most important weather variables. In relation to control, the weather acts as a disturbance, and in order to obtain the best possible controllers, it is important that the controllers are able to take advantage of short-term forecasts of the disturbances. Models for the most relevant disturbance variables for control of buildings are described in Sect. 5.

The theory for model predictive control is outlined in Sect. 6. A special emphasis is put on how models for predicting the weather variables can be integrated into the concepts of model predictive control, and this is the topic of Sect. 7.

2 Grey-Box Models

The models in this chapter are based on the *grey-box* modelling framework. This framework is typically based on a non-linear model with a partial theoretical structure and some unknown parts derived from data. Consequently, the grey-box framework bridges the gap between models based on first principle (*white-box* models) and models based solely on data (*black-box* models) (Fig. 1).

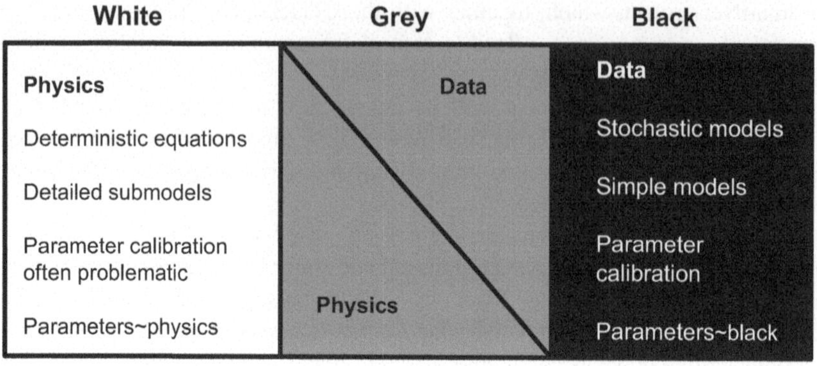

Fig. 1 Grey-box modelling bridges the gap between white- and black-box modelling

Grey-box models are formulated as a state-space model where the dynamics of the states is described in continuous time by a set of stochastic differential equations (SDEs) (*system equations*). The discrete-time observations are related to the states by a set of static equations (*observation equations*). Hence, a grey-box is formulated as continuous–discrete-time stochastic state-space model in the form

$$\mathrm{d}\boldsymbol{x}(t) = \underbrace{f(\boldsymbol{x}(t), \boldsymbol{u}(t), \boldsymbol{d}(t), t)\mathrm{d}t}_{\text{Drift}} + \underbrace{g(\boldsymbol{x}(t), \boldsymbol{u}(t), \boldsymbol{d}(t), t)\mathrm{d}\boldsymbol{\omega}(t)}_{\text{Diffusion}} , \tag{1}$$

$$\boldsymbol{y}_k = h(\boldsymbol{x}(t_k)) + \boldsymbol{v}_k , \qquad \boldsymbol{v}_k \sim N(\boldsymbol{0}, \mathsf{R_V}) , \tag{2}$$

where $\boldsymbol{x}$ is the system vector, $\boldsymbol{\omega}$ is a standard Wiener process (also often called a Brownian motion), and f and g are the drift and diffusion functions, respectively. h is the observation function and $\boldsymbol{v}_k$ is the observation noise. The drift function is the deterministic part of the SDE, whereas the diffusion function describes all the uncertainties not properly described in the drift.

If the system in (1)–(2) is linear, the model is written as

$$\mathrm{d}\boldsymbol{x}(t) = (\mathsf{A}\boldsymbol{x}(t) + \mathsf{B}\boldsymbol{u}(t) + \mathsf{E}\boldsymbol{d}(t))\,\mathrm{d}t + \Sigma\mathrm{d}\boldsymbol{\omega}(t) , \tag{3}$$

$$\boldsymbol{y}_k = \mathsf{C}\boldsymbol{x}(t_k) + \boldsymbol{v}_k , \qquad \boldsymbol{v}_k \sim N(\boldsymbol{0}, \mathsf{R_V}) , \tag{4}$$

where A, B, E, C, and Σ are matrices governing the state evolution, input, disturbance, observation, and noise, respectively.

Modelling physical systems using SDEs provides a natural method to represent the phenomenon as it evolves in continuous time. In contrast to discrete-time models, prior physical knowledge about the system can rather easily be included, and the estimated parameters do not depend on the sampling time.

There are many reasons for introducing the system noise (the diffusion term):

- *Modelling approximations*. For example, the dynamics, as described by the drift term, might be an approximation to the true system.
- *Unrecognised and unmodeled inputs*. Some variables that are not considered, such as wind speed, may affect the system.
- *Noise in measurements of input variables*. In such cases the measured input signals are regarded as the actual input to the system, and the deviation from the true input is described by the noise term.

In the observation equation, a noise term is also introduced. The reason for this noise term is:

- *Noise in measurements of output variables*. The sensors that measure the output signals are affected by noise and drift.

It seems reasonable to assume that the system noise and the measurement noise are independent.

This chapter focuses on simple grey-box models, describing the heat dynamics of a building and related components such as a heat pump and batteries (stationary and mobile). The main purpose is to describe the dynamics of the building and relevant components. In particular we shall focus on how the heat dynamics are affected by outdoor climate.

2.1 A Simple Linear Grey-Box Model

Let us consider a simple second-order grey-box model for the thermal dynamics of a building ([40]). Here, the so-called RC formulation is used and the thermal capacity is lumped into two states, and each of these states has an associated thermal mass.

$$\begin{bmatrix} dT_m \\ dT_i \end{bmatrix} = \begin{bmatrix} \frac{-1}{r_i c_m} & \frac{1}{r_i c_m} \\ \frac{1}{r_i c_i} & -\left(\frac{1}{r_a c_i} + \frac{1}{r_i c_i}\right) \end{bmatrix} \begin{bmatrix} T_m \\ T_i \end{bmatrix} dt$$
$$+ \begin{bmatrix} 0 & 0 & A_w p/c_m \\ 1/(r_a c_i) & 1/c_i & A_w(1-p)/c_i \end{bmatrix} \begin{bmatrix} T_a \\ \phi_h \\ \phi_s \end{bmatrix} dt + \begin{bmatrix} d\omega_m(t) \\ d\omega_i(t) \end{bmatrix} . \quad (5)$$

$$\boldsymbol{T}_r(t) = \begin{bmatrix} 0 & 1 \end{bmatrix} \boldsymbol{T}(t) + \boldsymbol{e}(t) . \quad (6)$$

The states of the model are given by the temperature T_m of a large heat accumulating medium with the heat capacity c_m and by the temperature T_i of the room air and possibly the inner part of the walls with the capacity c_i. The term r_i is the resistance against heat transfer between the room air and the large heat accumulating medium, while r_a is the resistance against heat transfer from the room air to the ambient air with the temperature T_a.

The input energy is supplied by the electrical heaters ϕ_h and the solar radiation that penetrates through the windows facing south $A_w\phi_s$, where A_w is the effective window area. The effective window area is the window area corrected for shade effects and absorption and reflection by the triple glazed windows. Note that only the indoor air temperature is measured.

This model has been identified in [40]. It is concluded that for the considered building, this second-order model provides a good description of all the variations in the data since the residuals are white noise. In that paper, it is described how the parameters are estimated using a maximum likelihood method, and furthermore, it was concluded that all the solar radiation is influencing the indoor air since the assumption $p = 0$ seems reasonable (the parameter p was not significant).

3 Identification of Grey-Box Models

Formulating suitable grey-box models is an iterative process in which physical considerations are combined with information obtained by statistical observations. The typical starting point is to formulate the mathematical equations governing the most important physical dynamics. These equations are then used as the initial model. Next, the parameters of the model are estimated, and finally, the model is used to generate residuals. These residuals are key to the model validation step, and if it is concluded that the residuals still show systematic behaviour, then the residuals are analysed in order to identify how the model can be improved and extended.

3.1 *Initial Model Structure Identification*

Typically, the initial model order, i.e., the number of state equations, and the dominating structure of the model are determined by physics. However, also statistical methods are useful. For instance, it is well known that the autocorrelation and partial autocorrelation function contain important information about the order of (linear) models. The following step-by-step guide summarises the procedure of formulating grey-box models:

1. Make a drawing of the physical system that includes the various methods for heat transfer (conductive, convective, and radiation).
2. Write down the mass and energy balance equations for the system.
3. Determine the *causality* of the system. Which time series data can be considered as input and which as output? For instance, for a building with feedback or controlled internal air temperature, the output could be the heat consumption, whereas for a building with no feedback, e.g., when the heating signal is determined by a PRBS signal, the internal air temperature could be the output.

4. Evaluate if any *non-linear* phenomena must be taken into account explicitly in the initial phase (later on statistical methods can be used for identifying non-linear phenomena). Such phenomena could be significantly influenced by wind speed, complicated glass construction, humidity, influence from rainfall, etc. Some non-linear effects can be described by a transformation of the input variables. In [54] the non-linear effect of solar radiation is described in a grey-box model using spline basis functions.
5. Evaluate if any *non-stationary* phenomena must be taken into account explicitly in the initial phase (later on statistical methods can be used for identifying non-stationary phenomena). Examples of such phenomena could be the fermentation of a new concrete building, moisture in the construction, opening of windows and doors, etc. For control applications, slowly varying non-stationary phenomena can be handled by considering adaptive and recursive methods [1, 38].

3.2 Estimation of Model Parameters

Typically, the model parameters are estimated either using the least squares method (LS) or the maximum likelihood estimation (MLE) method. The advantage of the MLE method is that this method also allows for estimating the parameters related to the noise term. Here, we briefly introduce the MLE method for estimating parameters in grey-box models. The method is described in detail in [30].

Given a sequence of measurements $\mathcal{Y}_N = \{\boldsymbol{Y}_1, \boldsymbol{Y}_2, \cdots, \boldsymbol{Y}_N\}$, the likelihood function is the joint probability density of all the observations but considered as a function of the unknown parameters. Thus, the likelihood function can be written as the product of the one-step ahead conditional densities:

$$\mathcal{L}(\boldsymbol{\theta}|\mathcal{Y}_N, \mathcal{U}_N) = \prod_{k=1}^{N} p(\boldsymbol{Y}_k|\boldsymbol{\theta}, \mathcal{Y}_{k-1}, \mathcal{U}_k) p(\boldsymbol{X}_0|\boldsymbol{\theta}) , \tag{7}$$

where $p(\boldsymbol{Y}_k|\boldsymbol{\theta}, \mathcal{Y}_{k-1}, \mathcal{U}_k)$ is the probability of observing $\boldsymbol{Y}_k$ given the previous observations, inputs, and set of parameters $\boldsymbol{\theta}$. This is the so-called exact likelihood function that contains a parameterisation of the density associated with the initial state $\boldsymbol{X}_0$.

Since the systems are assumed to be driven by Wiener processes for which the increments are Gaussian, the one-step ahead density for linear systems is also Gaussian. For most non-linear systems, this is still a reasonable assumption, and this assumption can be checked—see, e.g., [6].

In the Gaussian case, the conditional density is completely characterised by the conditional mean (the prediction) and the conditional covariance. By introducing the one-step prediction error (also called the *innovation error* or *residuals*)

$$\boldsymbol{\epsilon}_{k|k-1} = \boldsymbol{Y}_k - \hat{\boldsymbol{Y}}_{k|k-1} , \tag{8}$$

and the associated covariance, $\mathsf{R}_{k|k-1} = \mathrm{Var}(\boldsymbol{Y}_k|\mathcal{Y}_{k-1}, \boldsymbol{\theta})$, the likelihood function can be written as

$$\mathcal{L}(\boldsymbol{\theta}; \mathcal{Y}_N, \mathcal{U}_N) = p(\mathcal{Y}_N|\mathcal{U}_N, \boldsymbol{\theta}) \tag{9}$$

$$= \left(\prod_{k=1}^{N} \frac{\exp\left(-\frac{1}{2}\boldsymbol{\epsilon}_{k|k-1}^{\top}\mathsf{R}_{k|k-1}^{-1}\boldsymbol{\epsilon}_{k|k-1}\right)}{\sqrt{\det\left(\mathsf{R}_{k|k-1}\right)\left(\sqrt{2\pi}\right)^{\mathrm{L}}}} \right) p(\boldsymbol{X}_0|\boldsymbol{\theta}) \,, \tag{10}$$

where L is the dimension of the observation space. Using the logarithm, we obtain the log-likelihood function

$$l(\boldsymbol{\theta}; \mathcal{Y}_N, \mathcal{U}_N) = -\frac{1}{2}\sum_{k=1}^{N}\left(\boldsymbol{\epsilon}_{k|k-1}^{\top}\mathsf{R}_{k|k-1}^{-1}\boldsymbol{\epsilon}_{k|k-1} + \log\left(\det\left(\mathsf{R}_{k|k-1}\right)(2\pi)^{\frac{\mathrm{L}}{2}}\right)\right)$$
$$+ \log(p(\boldsymbol{X}_0|\boldsymbol{\theta})).$$

The parameter estimates are found by maximising the log-likelihood function

$$\hat{\boldsymbol{\theta}} = \arg\max_{\boldsymbol{\theta}}\left\{l(\boldsymbol{\theta}; \mathcal{Y}_N, \mathcal{U}_N)\right\}. \tag{11}$$

The corresponding value of the log-likelihood is the observed maximum likelihood value given the available data set.

For linear models, the conditional mean and covariance are calculated using an ordinary Kalman filter, while for non-linear models, an extended Kalman filter is used. See [30] for further details.

3.3 Uncertainty of Parameter Estimates

Uncertainty of parameter estimates is an essential output of any statistical parameter estimation scheme. This uncertainty lies in the facilitation of subsequent statistical tests. For the software implementation used here [28, 39], an estimate of the uncertainty of the parameter estimates is obtained by using the fact that by the central limit theorem the ML estimator is asymptotically Gaussian with mean θ and covariance:

$$\Sigma_{\hat{\theta}} = \mathsf{H}^{-1}, \tag{12}$$

where the matrix H is given by

$$h_{ij} = -E\left\{\frac{\partial^2}{\partial\theta_i\partial\theta_j}\left(l(\theta|\mathcal{Y}_N)\right)\right\}, i, j = 1, \ldots, p\,.$$

An approximation to H can be obtained from

$$h_{ij} \approx -\left(\frac{\partial^2}{\partial\theta_i\partial\theta_j}\left(l(\theta|\mathcal{Y}_N)\right)\right)\Big|_{\theta=\hat{\theta}}, i, j = 1, \ldots, p\,,$$

which is simply the Hessian evaluated at the maximum of the log-likelihood function. To obtain a measure of the uncertainty of the individual parameter estimates, the covariance matrix is decomposed as

$$\Sigma_{\hat{\theta}} = \sigma_{\hat{\theta}} \mathsf{R} \sigma_{\hat{\theta}}\,, \tag{13}$$

into $\sigma_{\hat{\theta}}$, which is a diagonal matrix of the standard deviations of the parameter estimates, and R, which is the corresponding correlation matrix.

3.4 Selection of Model Structure

Basically, the two main categories of problems related to the order of the model are:

1. **Model too simple:** A common problem is that *the residuals* for a given model are *autocorrelated*. In this case the model needs to be extended (for grey-box models, more states are needed). Another common problem is that the residuals are *cross-correlated* with some explanatory variables (e.g., large residuals for large wind speeds). In this case, this (or these) explanatory variable needs to be included in the model.
2. **Model too large:** A common problem is that some of the *parameters* are *insignificant*. In order to ensure a reliable estimation of the performance, the amount of parameters must be reduced by removing insignificant parameters.

3.5 Model Validation

If the residuals from a given modelling step show systematic variation, then the model is too simple and it can be improved. Consequently, model validation is a very important step in model building.

The following methodologies can be used in relation for model validation:

1. **Test for white noise residuals.**
   Typically, the autocorrelation function (ACF) of the residuals is used here. If a test for white noise residuals fails, then the model must be extended by extending the model order, which for grey-box models is the number of states.
2. **Test for dependency with inputs.**

Plot residuals against the inputs to see if any dependency exists. The cross-correlation function (CCF) (see [38, p. 230]) can be used to identify linear dependencies that have to be added to the model.

3. **Test for parameter significance.**
   See the next section on model validation. Here, it is mentioned that if a parameter is found to be insignificant, then in general this parameter should be removed from the model and the parameters of the reduced model are estimated.
4. **Check for correlation between parameters.**
   Most software for parameter estimation provides a correlation matrix of the estimated parameters. A numerically very high (say larger than .98) correlation between two parameter estimates indicates that one of these two parameters should be either excluded from the model or fixed to some physically assumed values.

3.6 Comparison of Models

1. **Test between (nested) models.**
   If two models are nested, i.e., the smaller model (B) can be found just by removing parts of a larger model (A), then the *likelihood-ratio test (LRT)* is very useful. The LRT value is given as $D = 2(\log L(A) - \log L(B))$, where $\log L(A)$ is the logarithm of the likelihood function for model A. For grey-box modelling, the asymptotic test principles based on Wilks' Theorem are used. Given that the model can be reduced to model B, the quantity D is according to Wilks' Theorem asymptotically $\chi^2(k-m)$ distributed, where k and m are the number of parameters in models A and B, respectively. For large values of D, it is concluded that the best model is the larger model. See, e.g., [42] for further details.
   In CTSM-R, the value $\log L$ is found using summary().
2. **Comparison between (non-nested) models**. If two models are non-nested, then methods based on *information criteria* can be used—see page 174 in [38]. This consists of computing an information criterion, such as the AIC or BIC:

$$AIC = 2k - 2\log L(A), \quad BIC = 2\log(N) - 2\log L(A)\ .$$

The preferred model is then simply found as the model with the lowest information criteria. Alternatively, and preferably, when a lot of data are available, cross-validation can be used [8]. In its simplest form, this procedure can be summarised as:

a. Split data into two parts, $\mathcal{Y}_{\text{Train}}$ and $\mathcal{Y}_{\text{Validate}}$. A typical split is 80% for $\mathcal{Y}_{\text{Train}}$ and 20% for $\mathcal{Y}_{\text{Validate}}$.
b. Estimate model parameters using only the data contained in $\mathcal{Y}_{\text{Train}}$. With these model parameters, compute the one-step residuals for $\mathcal{Y}_{\text{Validate}}$.

c. Using these residuals, evaluate either the likelihood function (9) or the sum of squared residuals (RMS).
d. The model with the highest likelihood or lowest RMS is preferred.

4 Smart Building-Related Models

This section presents multiple models for a building where the model of the building itself is the same, but the heating system models and control strategies are different. The first model uses conventional electrical heaters (radiators) to supply heat to the room air. The second model uses a heat pump: a compressor heats water, which then flows into pipes based under the floor. This ground-sourced heating is very efficient electricity-wise due to the compressor and is thus an attractive solution for heating.

The models are formulated using stochastic differential equations (SDEs), and the stochastic model for the building is closely related to the simple model introduced in Sect. 2.1.

4.1 *The Heat Pump Model*

Halvgaard et al. [20] describe a model for a building with a heat pump that is reused in this chapter. The model includes the same two important states as the model in Sect. 2.1: the room air temperature and the floor medium temperature. Additionally, it includes the temperature of the water connected to the heat pump. That makes the system states $\boldsymbol{x}(t) = \left[T_r(t), T_f(t), T_w(t)\right]^T$. Table 1 lists and describes all variables in the model, and Fig. 2 shows an illustration of the smart building model and the directions of the heat dynamics.

Regarding the disturbances of the model, two elements are of high importance: the solar radiation and ambient air temperature. The solar radiation generally plays a double role regarding the heating of buildings: it directly enters a building through, e.g., windows, but it also highly influences the ambient air temperature, which in

Table 1 Description of the variables in the heat pump model

Variable	Unit	Description
T_r	°C	The room air temperature
T_f	°C	The temperature of the floor medium
T_w	°C	The temperature of the water in the compressor and pipes
T_a	°C	The ambient air temperature
W_c	W	The energy delivered to the compressor of the heat pump
ϕ_s	W/m^2	The solar radiation entering the building

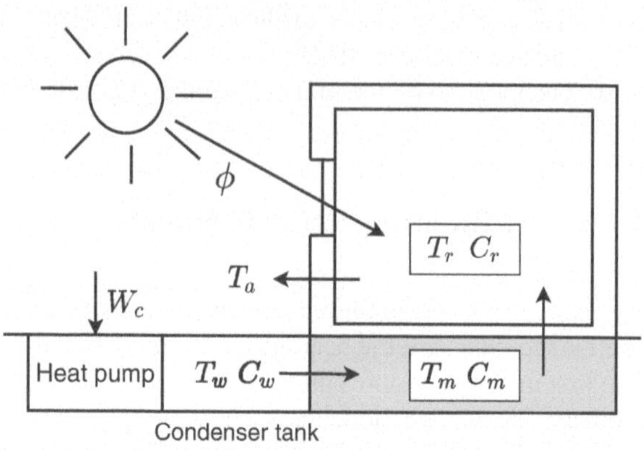

Fig. 2 An illustration of the smart house model. Each state has an associated heat capacity and temperature. The arrows indicate the positive direction of energy flow

turn affects the building. Later sections will describe and model these dynamics and cross-correlations.

Based on Fig. 2, the equations below describe the overall building dynamics

$$\begin{aligned} C_r dT_r(t) &= \left(Q_{fr}(t) - Q_{ra}(t) + A_w p \phi_s(t)\right) dt + \sigma_r d\omega_r(t) , \\ C_f dT_f(t) &= \left(Q_{wf}(t) - Q_{fr}(t) + A_w (1-p) \phi_s(t)\right) dt + \sigma_f d\omega_f(t) , \\ C_w dT_w(t) &= \left(\eta W_c(t) - Q_{wf}(t)\right) dt + \sigma_w d\omega_w(t) , \end{aligned} \tag{14}$$

where C_r, C_f, and C_w are heat capacities for the room air, floor, and water, respectively. ω_r, ω_r, and ω_w are Wiener processes for each state, and σ_r, σ_f, and σ_w are noise constants. The heat flows are given by

$$\begin{aligned} Q_{ra}(t) &= r_{ra}^{-1} \left(T_r(t) - T_a(t)\right) , \\ Q_{fr}(t) &= r_{fr}^{-1} \left(T_f(t) - T_r(t)\right) , \\ Q_{wf}(t) &= r_{wf}^{-1} \left(T_w(t) - T_f(t)\right) . \end{aligned} \tag{15}$$

We can write the set of SDEs in linear form as in (3)

$$\begin{aligned} \mathsf{A} &= \begin{bmatrix} -\frac{1}{r_{fr}C_r} - \frac{1}{r_{ra}C_r} & \frac{1}{r_{fr}C_r} & 0 \\ \frac{1}{r_{fr}C_f} & -\frac{1}{r_{wf}C_f} - \frac{1}{r_{fr}C_f} & \frac{1}{r_{wf}C_f} \\ 0 & \frac{1}{r_{wf}C_w} & -\frac{1}{r_{wf}C_w} \end{bmatrix}, \quad \mathsf{B} = \begin{bmatrix} 0 \\ 0 \\ \frac{\eta}{C_w} \end{bmatrix}, \\ \mathsf{E} &= \begin{bmatrix} \frac{1}{r_{ra}C_r} & A_w \frac{(1-p)}{C_r} \\ 0 & A_w \frac{p}{C_f} \\ 0 & 0 \end{bmatrix}, \quad \mathsf{C} = \begin{bmatrix} 1 \, 0 \, 0 \end{bmatrix} , \end{aligned} \tag{16}$$

with the variables $\boldsymbol{x}(t) = \left[T_r(t), T_f(t), T_w(t)\right]^T$, $u(t) = W_c(t)$, $\boldsymbol{d}(t) = [T_a(t), \phi_s(t)]^T$. Table 2 lists and briefly describes all the parameters in the model.

4.2 The Electrical Heater Model

The model using electrical heaters is almost identical to the one introduced in Sect. 2.1, with the exception of p, which is significant. Otherwise the parameters for this model are the same as used for the heat pump model. By separating the input and disturbance variables in the previous introduced model, we obtain the following linear second-order system:

$$\begin{aligned} \mathsf{A} &= \begin{bmatrix} -\frac{1}{r_{fr}C_r} - \frac{1}{r_{ra}C_r} & \frac{1}{r_{fr}C_r} \\ \frac{1}{r_{fr}C_f} & -\frac{1}{r_{fr}C_f} \end{bmatrix}, \quad \mathsf{B} = \begin{bmatrix} \frac{1}{C_w} \\ 0 \end{bmatrix}, \\ \mathsf{E} &= \begin{bmatrix} \frac{1}{r_{ra}C_r} & A_w\frac{(1-p)}{C_r} \\ 0 & A_w\frac{p}{C_f} \end{bmatrix}, \quad \mathsf{C} = \begin{bmatrix} 1 & 0 \end{bmatrix}, \end{aligned} \tag{17}$$

with the variable $\boldsymbol{x}(t) = \left[T_r(t), T_f(t)\right]^T$, $u(t) = W_c(t)$, $\boldsymbol{d}(t) = [T_a(t), \phi_s(t)]^T$.

4.3 Buildings with Stationary Batteries and Electrical Vehicles

In the very near future, electrical vehicles (EVs) will be in almost every household, and it is believed by many that future smart buildings will include stationary

Table 2 The values used in the model for a single smart home in (14) and (17)

Parameter	Value	Unit	Description
C_r	810	kJ/°C	Heat capacity constant for the room air
C_f	3315	kJ/°C	Heat capacity constant for the floor
C_w	836	kJ/°C	Heat capacity constant of the water in the pipes
r_{ra}	0.036	kJ/(°C h)	Resistance against heat transfer between the room air and the ambient air
r_{fr}	0.0016	kJ/(°C h)	Resistance against heat transfer between the floor and the room air
r_{wf}	0.036	kJ/(°C h)	Resistance against heat transfer between the water and the floor
p	0.1		The fraction of energy from the solar energy into the room air
η	3		The heat pump coefficient of performance
A_w	2.9		The effective window area

batteries [21, 57, 63]. The latter has the purpose of storing electricity harvested from photo-voltaic cells (PVs) and to buy and sell electricity from the market when the price is low and high, respectively. Adding a stationary battery and potentially an EV also greatly increases the *flexibility* of a building. We follow the modelling approach as in [63]. For more extended state-space models for batteries, see [7, 62]. We shall not, however, use or demonstrate these models in this chapter.

The fundamental differential equation governing the state of charge (SOC) of an (very simplified) integrating battery has the form

$$\dot{\gamma}(t) = \frac{V(t)}{Q} i(t) , \tag{18}$$

where $\gamma \in [0, 1]$ is the SOC (0 is discharged, while 1 is fully charged), V is the voltage, Q is the total battery capacity, and i is the current. We can rewrite this as the *power* flowing in and out of the battery

$$\dot{\gamma}(t) = \frac{1}{Q} \left(\eta^+ P^+(t) - \eta^- P^-(t)\right) , \tag{19}$$

where P^+ and P^- are the power flow in and out of the battery and η^+ and η^- are the respective efficiency constants. The corresponding SDE formulation is

$$\mathrm{d}\gamma(t) = \left(\eta^+ P^+(t) - \eta^- P^-(t)\right) \mathrm{d}t + \sigma_\gamma \mathrm{d}\omega(t) . \tag{20}$$

If the smart building is equipped with both a stationary battery and an EV, then we need a description of both

$$\mathrm{d}\gamma_{ev}(t) = \left(\eta_{ev}^+ P_{ev}^+(t) - \eta_{ev}^- P_{ev}^-(t)\right) \mathrm{d}t + \sigma_{ev} \mathrm{d}\omega_{ev}(t) , \tag{21a}$$

$$\mathrm{d}\gamma_{bat}(t) = \left(\eta_{bat}^+ P_{bat}^+(t) - \eta_{bat}^- P_{bat}^-(t)\right) \mathrm{d}t + \sigma_{bat} \mathrm{d}\omega_{bat}(t) , \tag{21b}$$

where γ_{ev} and γ_{bat} are the EV and stationary battery SOC, respectively. P_{bat}^+ and P_{bat}^- are the bought and sold electricity from the market, and P_{ev}^+ and P_{ev}^- are charging and discharging the EV battery, respectively.

The two batteries thus add two additional states to the smart building state-space model. To write the model in state-space form requires some assumptions. First, we assign the EV usage as a disturbance variable, $d_{ev}(t) = P_{d,ev}^-(t)$. Second, we assume that all the electricity generated by PVs and bought from the market go directly to the stationary battery. The solar radiation is thus also a disturbance, $d_{bat}(t) = \phi_s(t)$. The input variable for the battery is $u_{bat}(t) = \left[P_{bat}^+(t), P_{bat}^-(t), W_c(t), P_{ev}^+(t)\right]^T$. Writing the state-space formulation for the *entire* smart building including batteries, the state-space variables become $\boldsymbol{x}(t) = \left[T_r(t), T_f(t), T_w(t), \gamma_{ev}(t), \gamma_{bat}(t)\right]^T$, $\boldsymbol{u}(t) = \left[P_{bat}^+(t), P_{bat}^-(t), W_c(t), P_{ev}^+(t)\right]^T$, and $\boldsymbol{d}(t) = \left[T_a(t), \phi_s(t), P_{ev}^-(t), \phi_s(t)\right]^T$. Similarly, the continuous-time linear system likewise is

$$\mathsf{A} = \begin{bmatrix} -\frac{1}{r_{fr}C_r} - \frac{1}{r_{ra}C_r} & \frac{1}{r_{fr}C_r} & 0 & 0\,0 \\ \frac{1}{r_{fr}C_f} & -\frac{1}{r_{wf}C_f} - \frac{1}{r_{fr}C_f} & \frac{1}{r_{wf}C_f} & 0\,0 \\ 0 & \frac{1}{r_{wf}C_w} & -\frac{1}{r_{wf}C_w} & 0\,0 \\ 0 & 0 & 0 & 0\,0 \\ 0 & 0 & 0 & 0\,0 \end{bmatrix}, \tag{22a}$$

$$\mathsf{B} = \begin{bmatrix} 0 & 0 & 0 & 0 \\ 0 & 0 & 0 & 0 \\ \frac{\eta}{C_w} & 0 & 0 & 0 \\ 0 & \frac{\eta_{ev}}{Q_{ev}} & 0 & 0 \\ -\frac{\eta_{bat}}{Q_{bat}} & -\frac{\eta_{bat}}{Q_{bat}} & \frac{\eta_{bat}}{Q_{bat}} & -\frac{\eta_{bat}}{Q_{bat}} \end{bmatrix}, \quad \mathsf{C} = \begin{bmatrix} 1\,0\,0\,0\,0 \\ 0\,0\,0\,1\,0 \\ 0\,0\,0\,0\,1 \end{bmatrix}, \tag{22b}$$

$$\mathsf{E} = \begin{bmatrix} \frac{1}{r_{ra}C_r} & \frac{(1-p)}{C_r} & 0 & 0 \\ 0 & \frac{p}{C_f} & 0 & 0 \\ 0 & 0 & 0 & 0 \\ 0 & 0 & -\frac{\eta_{ev}}{Q_{ev}} & 0 \\ 0 & 0 & 0 & \frac{\eta_{bat}}{Q_{bat}}\eta_{pv}n_{pv} \end{bmatrix}. \tag{22c}$$

5 Disturbance Modelling

In this section, we shall use the well-documented meteorological data presented in [3] as the foundation for the grey-box models for the disturbances. Table 3 lists and describes each attribute of the data, which is collected from two weather stations in Værløse and Taastrup in Denmark. Samples are taken hourly from January 1, 1967 to December 31, 1973. The cloud cover is measured on the so-called okta scale. An okta is an integer in the range from 0 to 9, where 0 is completely clear skies, gradually gets more cloudy up till 8 that is fully overcast. Okta 9 is the class of non-observable cloud cover conditions, e.g., in foggy weather or heavy snow fall. We thus denote the okta state space by

$$C = \{0, 1, \ldots, 8, 9\}. \tag{23}$$

Table 3 Facts about the data and they are measured

Attribute	Notation	Unit	Measurement method
Cloud cover	$\{c', c, \kappa, Z_\kappa\}$	okta	Measured once every hour
Diffuse radiation	I_D	W/m^2	The average of 6 observations within an hour
Direct radiation	I_N	W/m^2	The average of 6 observations within an hour
Net radiation	R_n	W/m^2	The average of 6 observations within an hour
Ambient air temperature	T_a	°C	The average of 6 observations within an hour

5.1 Cloud Cover

The type, height, and amount of cloud cover have enormous influence on energy levels and balances of the lower atmosphere, i.e., the local weather close to Earth's surface. The variations of the solar radiation are mainly due to the absorption of energy by the molecules of the clouds. For example, in case of a heavy cloud cover, much less solar radiation gets through the atmosphere down to the surface. In a control context of a smart building that has PVs and is able to harvest energy from the sun, it is crucial to know the amount of solar radiation available. The cloud cover is undoubtedly the single most important factor in this case. The cloud cover also has a big impact on the air temperature in the lower atmospheric layers. When the rays from the sun hit the Earth's surface, a certain fraction gets absorbed and heats up the soil that, in turn, heats up the air. A good model for the cloud cover is thus a crucial element of a disturbance model for describing the local weather.

5.1.1 Discrete State-Space Cloud Cover Model

Figure 3 shows a plot of the cloud cover data from March. The overall dynamics seems fluctuating and is to some extend random. However, the cloud cover seems to spend more time and be more stable at both the ends of the scale (around okta 0 and 8).

Due to the discrete measure of the cloud cover, it is tempting to opt for a discrete state-space model to describe the cloud cover. An example of this is a continuous-time Markov model, see, e.g., [53]. The literature does describe successful models of this kind using both homogeneous and in-homogeneous models to describe the diurnal behavioural variation of the cloud cover in Denmark [41, 43]. The results confirm that the cloud cover is more stable in the clear sky and overcast states, with greater fluctuation in the middle states.

This model very well describes the probabilistic dynamics of the cloud cover. It also supplies estimates of the future expected value through the *Kolmogorov forward equations*. But the rest of the disturbance models will be formulated as continuous state-space models and rely on SDEs. Therefore, in a combined disturbance model for the smart building and in an MPC framework, a SDE describing the cloud cover

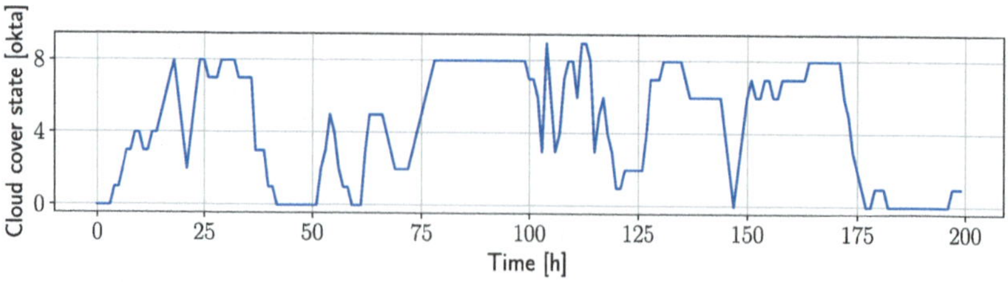

Fig. 3 A sample visualisation of the cloud cover data in March

becomes more convenient. For this reason, we shall now formulate, estimate, and validate a SDE-based model for the cloud cover.

5.1.2 Continuous State-Space Model Based on Stochastic Differential Equations

Recall the typical form of SDEs, explained in Sect. 4

$$x(t) = f(x(t), t)\mathrm{d}t + g(x(t), t)\mathrm{d}\omega(t) . \tag{24}$$

The analysis from the results of the Markov models in, e.g., [43] show that the cloud cover is governed by very special dynamics. It turns out that the process is less likely to move when it is in the end points (okta 0 and 8) and more likely for middle oktas. Formulating a SDE with these dynamics is not a trivial process. A first observation is that the cloud cover state space has boundaries. It is thus important to ensure that the SDE does not allow the process to go outside the boundaries. Let $\mathcal{K} = [0, 1]$ be the set of real numbers from 0 to 1, and let $\kappa \in \mathcal{K}$ denote the cloud cover state on *some* normalised scale. Starting with the diffusion, g, a very appropriate function can be

$$g(\kappa(t), t) = \sigma_\kappa \kappa(t) (1 - \kappa(t)) , \tag{25}$$

where σ_κ is a constant (Fig. 4). This choice ensures that the diffusion goes to zero in both ends of the okta scale and is also largest in the middle—which is desirable in order to make the process stay at either end longer time while ensuring that the middle states are more transient. In the grey-box modelling framework, we should assume some structure on the drift function, but it can be useful to also use some flexible functions such as the Legendre polynomials to allow the data to freely form them to easily maximise the likelihood function. The work in this chapter uses the following *mean reverting process*:

$$\mathrm{d}\kappa(t) = \theta(\kappa(t))(\mu(\kappa(t)) - \kappa(t))\,\mathrm{d}t + \sigma_\kappa \kappa(t)(1 - \kappa(t))\mathrm{d}\omega(t) , \tag{26}$$

where μ is the mean value and θ is the reversion speed. This SDE has a state-dependent diffusion term that has some nice modelling features, but it also has some pretty significant disadvantages from an estimation and simulation standpoint. The next part will detail these disadvantages.

5.1.3 Transformation into a State-Independent Diffusion Process

State-dependent diffusion terms in SDEs give rise to problems in estimation and simulation [46]. Two of the more influential problems are:

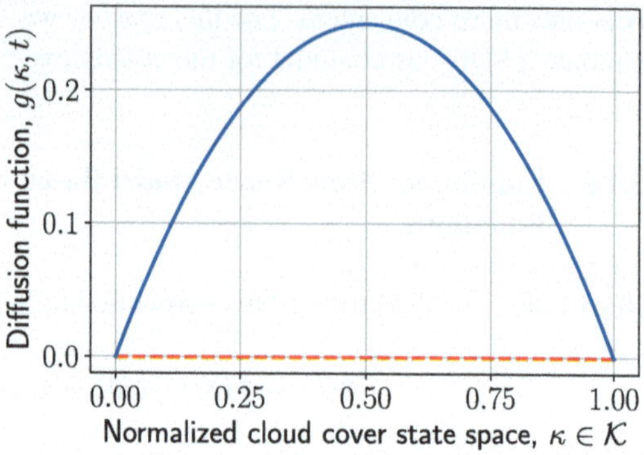

Fig. 4 The diffusion function in (25) with $\sigma_\kappa = 1$

- *Simulation* from a SDE with state-dependent diffusion can have slower convergence rate and then require more computational power.
- *Predictions* using methods like the extended Kalman filter (EKF) can be wrong and even illegal if they go outside of the bounds of the domain due to the linearisation.

A popular solution to this problem in the literature is to use the so-called *Lamperti transformation* [48].

The Lamperti transformation heavily relies on the result from stochastic calculus called Ito's lemma. Informally speaking, Ito's lemma corresponds to the chain rule for stochastic calculus: given a process, X, and a function, $\psi(X)$, Ito's lemma states the derivative of the function as a stochastic process $Z = \psi(X)$. That way, we can alternatively view the lemma as the equivalence of two processes, X and Z, by a closed formula using a transformation ψ. Consider a strictly positive process: by taking the natural logarithm of the process, we create a new process that lives on the entire real line. To obtain the differential equation governing this process, usually we would use the chain rule. But for a stochastic process, we need Ito's lemma. The special case where we choose a function that results in a constant diffusion term, we call the Lamperti transformation. Let us start by stating Ito's very famous lemma [22] (using some simplifying notation).

Lemma 1 (Ito's Lemma) *Let X be an Ito's process in the form*

$$\mathrm{d}X = f(X, t)\mathrm{d}t + g(X, t)\mathrm{d}\omega.$$

Let the function $\psi(X, t) \in C^2(\mathbb{R} \times [0, \infty))$, and then the process

$$Z = \psi(X, t)$$

is an Ito's process. Furthermore, Z is governed by the process

$$\mathrm{d}Z = \left(\frac{\partial \psi}{\partial t}(X, t) + f(X, t)\frac{\partial \psi}{\partial X}(X, t) + \frac{1}{2}\frac{\partial^2 \psi}{\partial X^2}(X, t)g(X, t)^2\right)\mathrm{d}t + \frac{\partial \psi}{\partial X}(X, t)g(X, t)\mathrm{d}\omega.$$

By choosing $\partial\psi/\partial X(X,t)$ to be equal to $1/g(X,t)$, the diffusion term becomes exactly a unit for the Ito-transformed process, Z. The following theorem states this result [22, 46].

Theorem 1 (Lamperti Transformation) *Let X be an Ito's process as in Lemma 1. Define the function*

$$\psi(X,t) = \int \frac{1}{g(x,t)} \mathrm{d}x \bigg|_{x=X}$$

. If $\psi(X,t)$ is bijective onto $\mathbb{R}$, then Z has a unit diffusion term and has the following process:

$$\mathrm{d}Z = \left(\frac{\partial \psi}{\partial t}(\psi^{-1}(Z,t),t) + \frac{f(\psi^{-1}(Z,t),t)}{g(\psi^{-1}(Z,t),t)} - \frac{1}{2}\frac{\partial g}{\partial X}(\psi^{-1}(Z,t),t) \right) \mathrm{d}t + \mathrm{d}\omega.$$

Applying the Lamperti transformation on the specific SDE in (26) (except for leaving a constant on the diffusion term), the Lamperti-transformed process, $Z_\kappa = \psi(\kappa,t)$, is

$$\begin{aligned} Z_\kappa = \psi(\kappa,t) &= \int \frac{1}{x(1-x)} \mathrm{d}x \bigg|_{x=\kappa} \\ &= \log(\kappa) - \log(1-\kappa) = \log\left(\frac{\kappa}{1-\kappa}\right). \end{aligned} \tag{27}$$

For $\kappa \in \mathcal{K}$, the process Z_κ is in all of the real numbers, $\mathbb{R}$. The inverse of ψ is

$$\kappa = \psi^{-1}(Z_\kappa, t) = \frac{\exp(Z_\kappa)}{1+\exp(Z_\kappa)}, \tag{28}$$

Using Theorem 1, the *state-independent* Lamperti-transformed process becomes

$$\begin{aligned} \mathrm{d}Z_\kappa &= \left(0 + \frac{f(\kappa,t)}{\kappa(1-\kappa)} + \kappa - \frac{1}{2}\right)\mathrm{d}t + \sigma_\psi \mathrm{d}\omega \\ &= \left(\frac{f\left(\frac{\exp(Z_\kappa)}{\exp(Z_\kappa)+1}, t\right)}{\frac{\exp(Z_\kappa)}{\exp(Z_\kappa)+1}\left(1 - \frac{\exp(Z_\kappa)}{\exp(Z_\kappa)+1}\right)} + \frac{\exp(Z_\kappa)}{\exp(Z_\kappa)+1} - \frac{1}{2} \right) \mathrm{d}t + \sigma_\psi \mathrm{d}\omega \\ \mathrm{d}Z_\kappa &= f_\psi(Z_\kappa)\mathrm{d}t + \sigma_\psi \mathrm{d}\omega\,. \end{aligned} \tag{29}$$

To get an intuition of how the Lamperti transformation works, Fig. 5 shows an example of a SDE with simple linear drift while having the diffuse function as in (25), together with the Lamperti-transformed drift. For the Lamperti-transformed process, the drift will make sure that it always stays around 0 and does not go towards $\pm\infty$. But the most interesting feature of the Lamperti drift is perhaps that

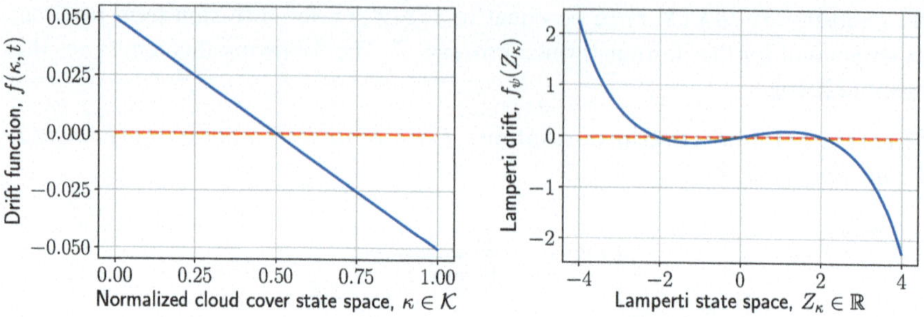

Fig. 5 A simple linear drift function (left) and its Lamperti-transformed equivalent (right) from the process $\mathrm{d}X = 0.1 \cdot (0.5 - X)\mathrm{d}t + X(1 - X)\mathrm{d}\omega$

there is a stable stationary point at $Z_k = 0$ driving the process towards zero. The explanation should be found in the shape of the diffusion function that is largest in the middle and goes to zero in the ends. Around $Z_k = 0$, the diffusion is the dominating force and has zero mean—i.e., is expected to keep the process around zero due to the zero mean (at least expectation-wise). The variance, however, will make sure to drive the process away from zero. But when the process gets too far out, the drift will dominate again and force it towards zero.

Due to the advantages of dealing with a state-independent SDE, all estimation, simulation, and prediction happen in the Lamperti domain in (29) and are subsequently transformed back to the original cloud cover domain by ψ^{-1}. We now turn to estimate parameters in (26).

5.1.4 Estimation of Parameters Embedded in the SDE

As previously mentioned, we need to base the choice on the drift function of the SDE on physical properties of the process we attempt to model. We choose to use the following SDE:

$$\mathrm{d}\kappa = \theta\sqrt{\kappa(1-\kappa)}\left(\frac{\exp(P_7(\kappa))}{1+\exp(P_7(\kappa))} - \kappa\right)\mathrm{d}t + \sigma\kappa(1-\kappa)\mathrm{d}\omega\,, \tag{30}$$

where $P_7(\kappa)$ is a linear combination of the first seven Legendre polynomials. The model in (30) has a very complex mean value. The intuition is that it allows the mean value to move rather freely in the range from 0 to 1, depending on the cloud cover state. The reverting-speed term $\sqrt{\kappa(1-\kappa)}$ may seem like an over-complication in the model. But previous cloud cover modelling attempts suggest that the process spends more time in the ends of the okta scale. The term $\sqrt{\kappa(1-\kappa)}$ makes the drift smaller at the ends of the scale and therefore intuitively makes the process stay there for longer before reverting back to the middle.

Before we are able to estimate parameters in the SDE in (26), we require a transformation of the data, $\zeta : C \rightarrow \mathcal{K}$, due to the diffusion term (since it requires the process to be in the interval $\mathcal{K} = [0, 1]$). First, we let okta 9 be missing observations in the data, such that the observable okta state space (and the state space we should model) is $\{0, 1, \ldots, 8\}$. Choosing a good transformation is not straightforward. We cannot simply divide the okta state space by 8. Doing so implies that okta 0 in C corresponds to 0 in $\mathcal{K}$ and likewise okta 8 corresponds to 1 in $\mathcal{K}$. But the drift and diffusion of the SDE equal zero for $\kappa = 0$ and $\kappa = 1$ and the SDE gets stuck. Furthermore, the Lamperti transformation is not well defined for these values.

In the discrete okta state space, C, the distance between each pair of neighbouring states is the same. However, in the continuous state space $\mathcal{K}$, this is definitely not given. In fact, the definition of the oktas [3] indicates that the end points of okta pairs $\{0, 1\}, \{7, 8\} \in C$ are more alike compared to the rest of oktas. Thus by moving the end points of the okta scale closer together as in Fig. 6 before dividing by 8, we might obtain a good transformation that behaves well. We thus choose the following transformation for the cloud cover to get it into $\mathcal{K}$:

$$\kappa = \zeta(c) = c/8\,, \quad c \in C\,,\ \kappa \in \mathcal{K}\,, \tag{31}$$

where the oktas 0 and 8 have been perturbed according to Fig. 6. But how much should we move the end points as to get the best model? To answer this question, we use *Akaike's information criterion* (AIC) as a measure to compare estimated models. We perform a small grid search for the positions of oktas 0 and 8 around the points 0.5 and 7.5 and choose the model that performs best in terms of the AIC value.

To estimate the parameters in the SDE, we apply ML estimation using the continuous–discrete extended Kalman filter (CDEKF), see, e.g., [10, 23]. We use the CDEKF to predict from the SDE. Let $\boldsymbol{\theta}$ denote the set of parameters in the model. Given $\boldsymbol{\theta}$, we use the CDEKF to calculate the 1-step prediction and variance of the state, $\hat{\kappa}_{k|k-1}(\boldsymbol{\theta})$ and $R_{k|k-1}(\boldsymbol{\theta})$. Let $\epsilon_k(\boldsymbol{\theta}) = \hat{\kappa}_{k|k-1}(\boldsymbol{\theta}) - \kappa_k$ be the prediction error of the state using $\boldsymbol{\theta}$, and then the ML estimate is given by (9).

Table 4 shows the result of the grid search and suggests that the perturbation (okta0, okta8) $= (0.6, 7.4)$ is by far the best choice in terms of AIC—not surprisingly, since we expected the end points to behave more like their neighbours.

Figure 7 shows histograms of the long-term distributions of the data and model 3 using the state transformation in Fig. 6 with the values (okta0, okta8) $= (0.6, 7.4)$.

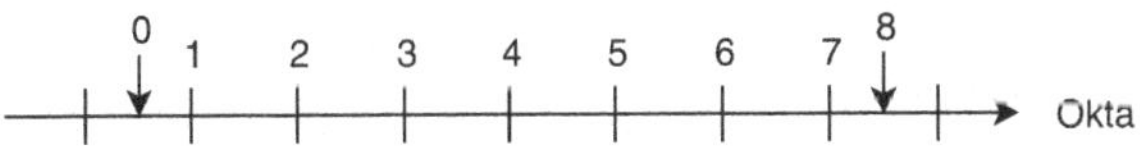

Fig. 6 An improved cloud cover transformation, ζ. Due to oktas 0 and 1, and okta 7 and 8, supposedly being more alike compared to the other oktas, we propose the following transformation

Table 4 The AIC value for each combination of end points in the grid search using the model in (30)

Okta 0 \ Okta 8	7.40	7.45	7.50	7.55
0.50	21079	21432	22436	22557
0.55	21330	21591	21680	26515
0.60	20442	22817	23294	24994

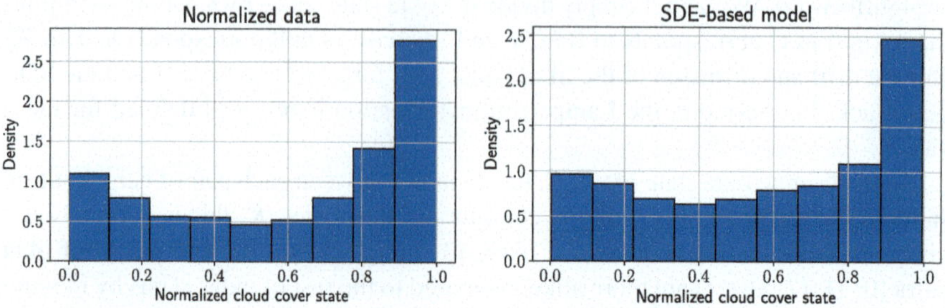

Fig. 7 The long-term distribution of the data (left) and model 3 (right) using the improved state transformation in (31) with oktas 0 and 8 moved to 0.6 and 7.4, respectively

Fig. 8 The autocorrelation function of the 1-step prediction residuals of model using the okta transformation as in (31)

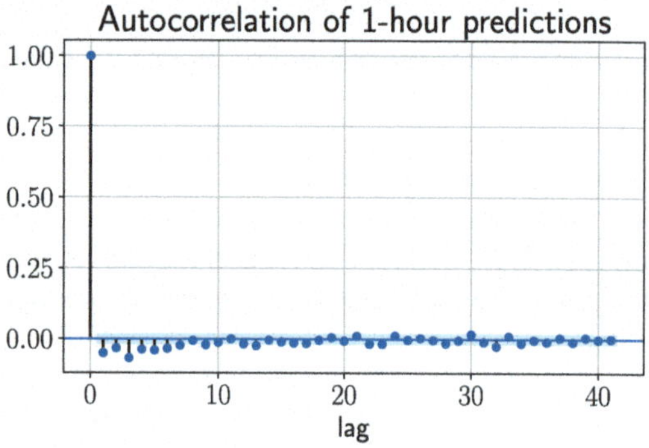

Table 5 Parameter estimates for the model in (30) with the locations of oktas 0 and 8 moved to 0.60 and 7.40. $\hat{p}_i, i = 1, \ldots 7$, is the parameter for the ith Legendre polynomial

Parameter	$\hat{p}_1$	$\hat{p}_2$	$\hat{p}_3$	$\hat{p}_4$	$\hat{p}_5$	$\hat{p}_6$	$\hat{p}_7$	$\hat{\theta}$	$\hat{\sigma}$
Estimate	–53.1	14.6	–42.3	8.8	–58.1	–30.3	–45.7	0.187	0.835
Std. Err.	2.13	0.831	1.715	0.434	2.273	1.329	1.873	0.003	0.012

Even though the distributions are not identical, the model mimics the overall pattern very well. The autocorrelation function in Fig. 8 is also close to zero as desired.

Table 5 shows the parameter estimates for the model. It can be hard to interpret the model by simply looking at the parameters since the Legendre polynomial parameters do not make much sense by themselves. Instead, we show the drift and Lamperti drift functions of the cloud cover state in Fig. 9 to give an intuition of the model. It truly has a very non-trivial and complex shape. It is partly the mean value function, $\exp(P_7(\kappa))/(1 + \exp(P_7(\kappa)))$, and the reverting-speed function, $\theta\sqrt{\kappa(1-\kappa)}$, that makes this possible. The former allows the drift function to make

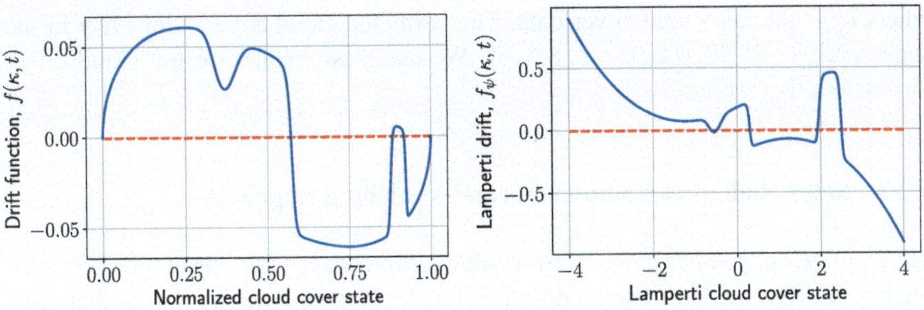

Fig. 9 Left: The drift function of model 3 in (30) using the state transformation in (31). Right: The same drift function, but in the *Lamperti domain*. The dashed line simply indicates the zero line

sudden changes from positive to negative and vice versa (which happens around $\kappa = 0.6$ and $\kappa = 0.9$). The ladder makes sure that the drift goes to zero in both ends and gives it the overall bending shape. The very sharp bend seen in the drift around $\kappa = 0.9$ surely seems odd and out of place. But it has a crucial role in the long-term distribution. It creates a stable stationary point for the process and therefore makes the process stay in the overcast states for more time. This is especially visible in the long-term distribution by the larger density in the overcast state compared to the clear skies states.

The Lamperti drift is harder to interpret, as it lives in the logistic domain. But it illustrates how the Lamperti process behaves and how the state dependence affects it.

5.2 *Solar Radiation*

Now that we have established a model describing the cloud cover dynamics based on SDEs, we move on to describe the next component of an advanced disturbance model: the solar radiation. It is responsible for some of the fast heating dynamics influencing the indoor air temperature and is thus an important disturbance.

5.2.1 Modification of the Cloud Cover Data

The data introduced in Sect. 5.1 are averages of multiple observations within an hour, except for the cloud cover. The cloud cover value taken at time t_k is thus not representative for the cloud cover in the interval $[t_k, t_{k+1}]$. To obtain better estimates for such cloud cover values, we use the average of the cloud cover at time t_k and t_{k+1}

$$c_{k+1} = (c'_{k+1} + c'_k)/2 \,, \tag{32}$$

where c'_k is the raw cloud cover data. The corrected cloud cover values live in the state space $c_k \in \{0, 0.5, 1, \ldots, 8.5, 9\}$. We use these values for the cloud cover throughout this section.

5.2.2 Solar Radiation Components and Modelling Approach

The term *global solar radiation* covers all the short-wave radiation at the surface of the Earth. These are high-energy rays that transfer large amounts of energy that turns into heat when absorbed by objects or electricity by the PVs. Previous modelling attempts range from simple polynomial fits to complex black-box neural networks, see, e.g., [15, 59, 60, 65]. The global solar radiation consists of two components: *diffuse* and *direct* radiation. The direct radiation is all short-wave radiation travelling undisturbed to the Earth's surface. The diffuse radiation is all short-wave radiation that is reflected from molecules in the atmosphere. The fundamental relationship is

$$\phi_s(t) = I_N(t)\sin(\alpha(t)) + I_D(t)\,, \tag{33}$$

where ϕ_s, I_N, and I_D are the global, direct, and diffuse radiation, respectively, and α is the solar elevation angle. That is, obtaining models for each component gives a model for the global radiation. Lambert–Beer's law gives an analytical expression for the intensity of the radiation when it arrives at the Earth's surface

$$I_\lambda = I_{0\lambda} e^{-\int \mu_\lambda(s)\mathrm{d}s}\,, \tag{34}$$

where $I_{0\lambda}$ is the initial intensity and μ_λ is the attenuation of the medium the ray travels in. The integral in (34) is hard to evaluate since $\kappa_{a\lambda}$ is difficult to estimate for the atmosphere due to its very non-uniform density and dependence on the solar elevation.

We shall employ a more data-driven approach, namely kernel regression [64]. It estimates the conditional expectation, $E(Y|X = x)$, of a variable. In our case, we estimate the expected solar radiation given the cloud cover okta, $c \in C$, and the solar elevation angle, $\alpha(t)$. Local linear regression has an advantage over *constant* regression in that it generally induces less bias in the ends of the support. Figure 10 shows the direct radiation for some selected oktas (the diffuse radiation is omitted). It shows that for larger oktas, the data seem more scattered—and behave very poorly for okta 8.

Figure 11 shows the result of applying local linear kernel regression on the direct and diffuse radiation data for each okta. The conditional expectations are in line with the physical properties of the radiation types: The direct radiation is highest when there is close to zero clouds, while the diffuse radiation tops for a certain presence of clouds.

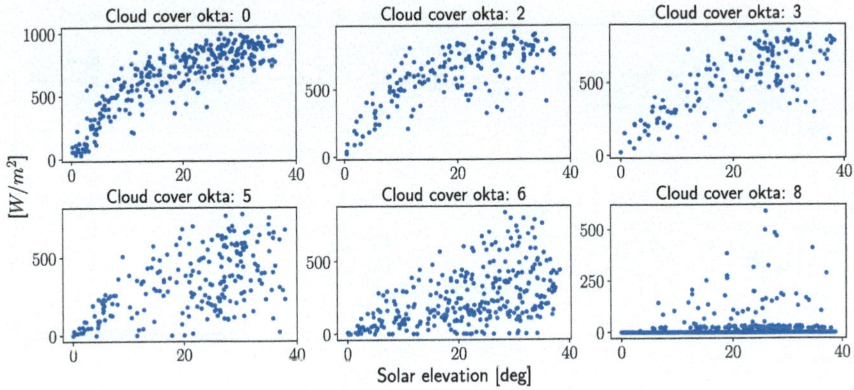

Fig. 10 The direct radiation for some example oktas: 0, 2, 3, 5, 6, and 8

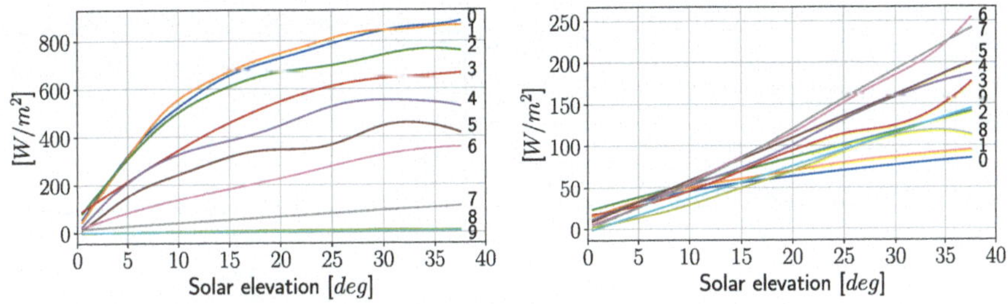

Fig. 11 The results from the kernel regression applied on the direct and diffuse radiation. For simplicity, only half of the okta values are shown

5.2.3 Describing the Deviation and Autocorrelation

Now that we have described the conditional expectation, we move on to describe the deviation and the potential autocorrelation left in the residuals. Figure 12 shows the residuals of the kernel regression applied to the direct radiation (we omit the diffuse radiation due to space limitation, but the behaviour is the same). It also suggests a rough linear increase in the standard deviation is the case (the same is true for the diffuse radiation). That is, we employ the following model for the standard deviation for each okta:

$$\sigma^{(c)}(t) = \beta_0^{(c)} + \beta_1^{(c)}\alpha(t) , \tag{35}$$

for both the direct and diffuse radiations. $\beta_0^{(c)}$ and $\beta_1^{(c)}$ are constant parameters for each $c \in C$. Let $\epsilon_k^{(c)} = \hat{y}_k^{(c)} - y_k^{(c)}$ be the residual, and let $\sqrt{R_k} = \sigma_k^{(c)}$ be the standard deviation at time t_k, $k = 1, 2 \ldots N$, for okta c. The above sets of parameters, $\{\beta_0^{(c)}, \beta_1^{(c)}\}$, can be estimated using ML estimation as in (9) using a numerical solver. The parameter estimates can be found in [61].

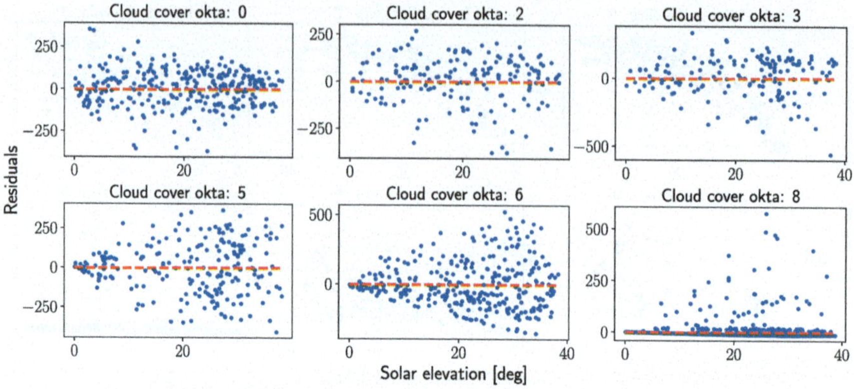

Fig. 12 The residuals for the direct radiation

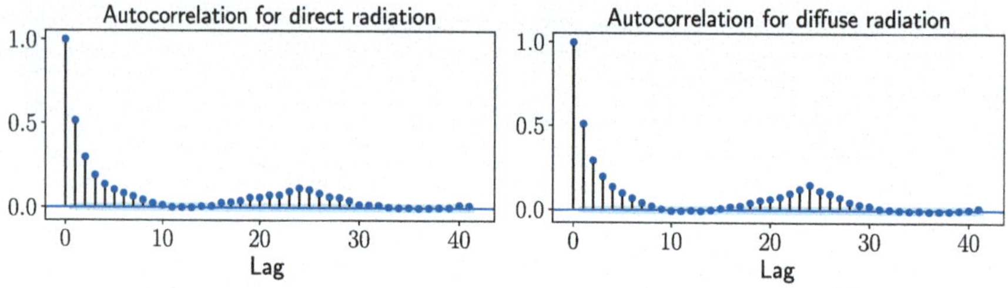

Fig. 13 The ACF for the direct and diffuse radiations

The cloud cover is not the single cause of variation in the solar radiation. For example, vapour, dust, ozone, and other particles give rise to *autocorrelation*. Let

$$e(t) = \frac{\hat{\epsilon}^{(c)}(t)}{\hat{\beta}_0^{(c)} + \hat{\beta}_1^{(c)}\alpha(t)} \tag{36}$$

be the standardised residuals of the direct and diffuse data (i.e., two processes). Note that each of the processes is standard normal distributed and independent of the cloud cover. We consider the two processes as a *multivariate* time series with the variable, $\boldsymbol{e}_k = \left[e_{N,k}, e_{D,k}\right]^T$, with missing observations (during night-time). See [38] for how to deal with missing observations in a time series. Figure 13 now shows the autocorrelation in $\boldsymbol{e}_k$. The fast exponential decay in the first few lags suggests that a first-order *autoregressive* (AR) model is necessary

$$\begin{aligned} \boldsymbol{e}_k &= \Phi \boldsymbol{e}_{k-1} + \boldsymbol{\epsilon}_k \, , \\ \boldsymbol{a}_k &= \boldsymbol{e}_k + \boldsymbol{\epsilon}_{a,k} \, , \end{aligned} \tag{37}$$

where Φ is the AR coefficients and $\epsilon_k \sim N(\mathbf{0}, \mathsf{Q_e})$ and $\epsilon_{a,k} \sim N(\mathbf{0}, \mathsf{R_a})$ are the process and observation noise, respectively. $\boldsymbol{e_k}$ is thus the noise process driven by its previous values and a noise term, ϵ_k. The observation equation (ref equation) is a white noise process that encumbers the observations with noise, $\epsilon_{a,k}$, which we also need to estimate. We estimate the parameters by applying ML estimation and use the Kalman filter to estimate the covariance matrices for the noise terms. The results become

$$\Phi = \begin{bmatrix} \underset{(0.013)}{0.609} & \underset{(0.009)}{0.109} \\ 0 & \underset{(0.010)}{0.675} \end{bmatrix}, \quad \mathsf{R_a} = \begin{bmatrix} \underset{(0.011)}{0.160} & 0 \\ 0 & \underset{(0.019)}{0.162} \end{bmatrix}, \quad \mathsf{Q_e} = \begin{bmatrix} \underset{(0.019)}{0.466} & \underset{(0.005)}{0.160} \\ \underset{(0.005)}{0.160} & \underset{(0.016)}{0.456} \end{bmatrix} \tag{38}$$

with the standard errors in parentheses beneath the estimate.

5.3 Net Radiation

The net radiation itself is not directly important for describing the heat dynamics of a building. But it is an important meteorological variable that heavily influences the ambient air temperature. The model for the ambient air temperature thus requires a model for the net radiation. The net radiation, also known as the net flux, is the balance of the total energy at the boundary of the atmosphere. It is simply the sum of the total outgoing and incoming energy of the atmosphere. A negative net radiation corresponds to more energy leaving the atmosphere and vice versa. In general, the net radiation, R_n, is given by the analytical formula [2]

$$R_n = (1 - \alpha_g)\psi_s + L_u + L_d \,, \tag{39}$$

where α_g is the albedo fraction and L_u and L_d are the upward and downward components of the long-wave radiation. The albedo is the fraction of global solar radiation that is reflected on the Earth's surface into space, and $(1 - \alpha_g)$ is therefore the fraction of global radiation that is partly absorbed and that becomes long-wave radiation.

Figure 14 shows the net radiation for a single March, which makes the dependence on the time of the day clear. A simple and convenient model to use for our purpose is the following, suggested by [37], which relies on the current cloud cover and the global radiation

$$R_n(c(t), \phi_s(t), t) = K_c + k_c\phi_s(t) + k\alpha(t)^2 + \epsilon(t) \,, \tag{40}$$

where K_c and k_c are constants that are dependent on the present cloud cover and α is the solar elevation. The model in (40) is linear in its parameters. This makes linear least squares estimation useful to fit the parameters

Fig. 14 The net radiation during March

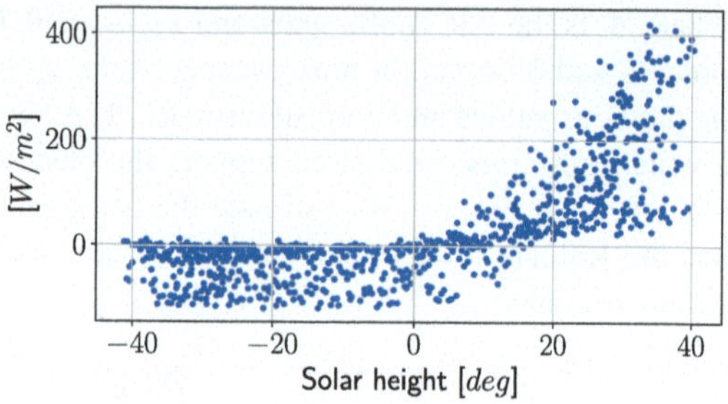

$$\min_{x} \|\mathsf{A}x - b\|_2^2 \,, \tag{41}$$

having the unique solution

$$\hat{x}_{LS} = (\mathsf{A}^T \mathsf{A})^{-1} (\mathsf{A}^T b) \,. \tag{42}$$

A and b in (41) have the forms

$$\mathsf{A} = \begin{bmatrix} e_{c_0}^T & e_{c_0}^T \phi_s(0) & h(0)^2 \\ e_{c_1}^T & e_{c_1}^T \phi_s(1) & h(1)^2 \\ & \vdots & \\ e_{c_i}^T & e_{c_i}^T \phi_s(i) & h(i)^2 \\ & \vdots & \\ e_{c_N}^T & e_{c_N}^T \phi_s(N) & h(N)^2 \end{bmatrix}, \qquad b = \begin{bmatrix} R_n(0) \\ R_n(1) \\ \vdots \\ R_n(i) \\ \vdots \\ R_n(N) \end{bmatrix}, \tag{43}$$

where N is the total number of observations and e_i is a vector of zeros with a one in the ith entry. c_i is the cloud cover index of the ith observation—recall the modified cloud cover data state space is $\{0, 0.5, \ldots, 8.5, 9\}$. x is thus the parameters, $x = (K_0, K_{0.5}, K_1, \ldots, K_9, k_0, k_{0.5}, k_1, \ldots, k_9, k)$. Table 6 shows the estimated parameters in (40). The increasing trend in constant net radiation, $\hat{K}_c$, with the increasing amount of cloud cover indicates that the clouds ‘contain’ the net radiation (net energy flux) within the atmosphere. This can be recognised by the phenomenon that colder nights typically appear when the skies are completely clear.

5.4 *Ambient Air Temperature*

The only missing piece in the puzzle now in the advanced disturbance model is the ambient air temperature. While the net radiation describes the net flux at the boundary of the atmosphere, the following fundamental relationship describes the heat fluxes: *close to Earth’s surface*

Table 6 The estimated parameters of the model in (40). Even though we have not imposed any regularisation or other things that tie the estimates together, the values are somewhat nicely distributed

Cloud cover okta	$\hat{K}_c$	$\hat{k}_c$	$\hat{\sigma}_\varepsilon$	$\hat{k}$
0	−69.2	0.549	40.1	0.0418
0.5	−68.2	0.551	38.5	0.0418
1	−74.1	0584	38.3	0.0418
1.5	−74.9	0.576	37.7	0.0418
2	−74.1	0.571	39.7	0.0418
2.5	−75.5	0.566	45.2	0.0418
3	−73.6	0.565	48.4	0.0418
3.5	−71.5	0.566	54.4	0.0418
4	−73.6	0.589	58.1	0.0418
4.5	−67.5	0.606	57.9	0.0418
5	−69.5	0.661	59.8	0.0418
5.5	−63.6	0.695	60.7	0.0418
6	−56.5	0.717	59.7	0.0418
6.5	−46.5	0.699	56.7	0.0418
7	−29.5	0.595	49.4	0.0418
7.5	−13.5	0.461	43.5	0.0418
8	2.5	0.150	31.8	0.0418

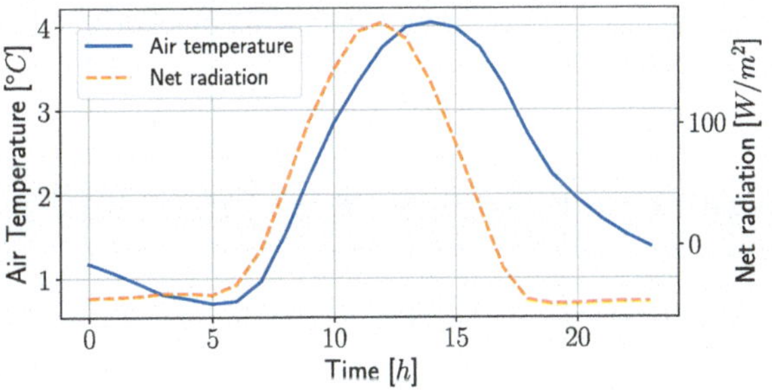

Fig. 15 The hourly mean values of the ambient air temperature and the net radiation during a day. The delayed response from the ambient air temperature suggests that a dynamic model is necessary

$$R_n = L_f + S_f + G_f \ , \tag{44}$$

where R_n is the net radiation described in Sect. 5.3 and G_f is the soil heat flux. L_f and S_f are the latent and sensible heat fluxes. The latent heat flux is heat gradients related to absorbed or released heat due to phase changes by matter—e.g., when water evaporates, it absorbs heat in order to decrease the molecule density. The sensible heat flux is all energy required to change the temperature of matter without phase changes taking place. The latent and sensible heat fluxes thus relate to the gradients of the air temperature. To get an idea of the kind of model needed to describe the air temperature, Fig. 15 shows the diurnal mean value variations of the net radiation and ambient air temperature. It supports the fact that the net radiation is an important explanatory variable to describe the air temperature. It further suggests that a dynamical model is needed due to the time lag between the peak values of 3–4 h.

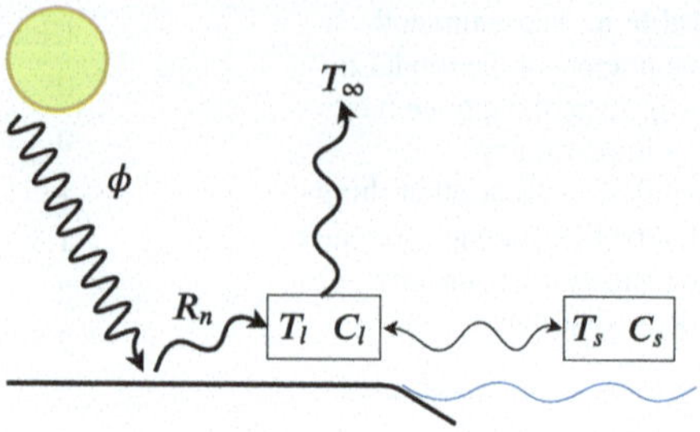

Fig. 16 An illustration of the dynamical model describing the ambient air temperature. The states are indicated by the squares containing the heat capacities

The atmospheric air directly above the Earth's surface has a relatively small heat capacity, making it quick to respond to level changes in the net radiation. Water in contrast has a very large heat capacity. The temperature of the seas thus highly regulates the temperature of the air above it. The air masses above sea and land interact due to climatic motions, and the sea consequently regulates the land air temperature. For instance, it is well known that the largest annual temperature difference occurs in the middle of large continents. Hence, the level of regulation by the sea depends on geographical location and local climate. Using the above knowledge about the behaviour and balances of the air temperatures, we are ready to formulate the stochastic dynamical model describing the ambient air temperature above land

$$C_w \mathrm{d}T_w(t) = \left(\frac{1}{R_{wl}}(T_l(t) - T_w(t))\right)\mathrm{d}t + \sigma_w \mathrm{d}\omega_w(t)\,, \tag{45a}$$

$$C_l \mathrm{d}T_l(t) = \left(\frac{1}{R_{wl}}(T_w(t) - T_l(t)) + \frac{1}{R_{l\infty}}(T_\infty - T_l(t)) + R_n(t)\right)\mathrm{d}t + \sigma_l \mathrm{d}\omega_l(t)\,, \tag{45b}$$

$$T(t_k) = T_l(t_k) + v_k, \quad v_k \sim N(0, R_v)\,, \tag{45c}$$

where C_w and C_l are the heat capacities for sea and land, R_{wl} and $R_{l\infty}$ are the resistances against the heat flows between the states, ω_w and ω_l are the standard Wiener processes, and v_k is the observation noise. The model is also illustrated in Fig. 16: The solar radiation influences the net radiation, which in turn acts as an input to the land air temperature. The land air temperature interacts with the sea temperature and a constant outflow of energy, T_∞, to counteract heat inputs and to ensure stability of the model. Equation (45) thus uses the sea temperature as a *hidden state* to describe the land air temperature.

Since this is a continuous-time model, we use the continuous–discrete Kalman filter to calculate one-step predictions and estimate the observation variance to compute one-step predictions and filter the estimates. Let $\hat{x}_{k|k-1}(\boldsymbol{\theta})$ and $R_k(\boldsymbol{\theta})$ be the one-step prediction and observation variance for x_k at time t_k, calculated using

Table 7 The parameter estimates in (45) and the corresponding standard errors

Parameter	$\hat{C}_w$	$\hat{C}_l$	$\hat{R}_{wl}$	$\hat{R}_{l\infty}$	$\hat{\sigma}_w$	$\hat{\sigma}_l$
Estimate	534.56	58.99	0.0145	0.1017	0.913	0.0003
Std. Error	8.44	0.49	0.0001	0.0026	0.0123	0.0001

a given set of parameters $\boldsymbol{\theta} = (C_w, C_l, R_{wl}, R_{l\infty}, \sigma_w, \sigma_l)$. Then, the ML estimate is the solution to the problem in (9).

Using the data from Højbakkegård, Table 7 shows the estimation results for the model in (45). As expected, the heat capacity for the sea is much larger compared to the land air. Also note the very small process noise for the land air temperature, ω_l, compared to the sea temperature. This indicates that it is primarily the sea temperature that drives the land air temperature.

6 Model-Based Predictive Control

The previous sections focused on establishing statistically determined dynamical models for the smart building and the most important disturbances. This section shows the potential benefits of using the advanced disturbance models for forecasting. We start by introducing model predictive control (MPC) and deriving the optimisation problem involved with computing the optimal control. Furthermore, we discuss how to incorporate and use the given disturbance forecasts in the MPC algorithm. Lastly, we present a more classical method for handling disturbances in an MPC setup, where the disturbances are not modelled but instead an integrator is introduced to estimate the current disturbances. Even though the method provides offset-free control, we discuss why it is not ideal when dealing with very fast dynamics (as with the solar radiation).

6.1 Constrained Model Predictive Control

Many variations of MPC exist and have gained high popularity for control purposes due to the framework's superiority over non-predictive control schemes such as PI/PID control [35] and its simplicity. In general, the MPC framework is given by the following (*Bolza*) problem:

$$J(\hat{\boldsymbol{x}}_{k|k}, \{\hat{\boldsymbol{d}}_{k+i|k}\}_{l\in N}) = \min_{\boldsymbol{u}} \int_{t_k}^{t_{k+N_p}} \ell(\boldsymbol{x}(\tau), \boldsymbol{u}(\tau), \boldsymbol{d}(\tau))\mathrm{d}\tau + \ell_b(\boldsymbol{x}(t_{k+N_p})), \tag{46a}$$

$$s.t. \quad \boldsymbol{x}(t_k) = \hat{\boldsymbol{x}}_{k|k}\,, \tag{46b}$$

$$\boldsymbol{d}(t) = \hat{\boldsymbol{d}}_{k+i|k},\ t \in [t_{k+i}, t_{k+i+1}[\ , \tag{46c}$$

$$\mathrm{d}\boldsymbol{x}(t) = f(\boldsymbol{x}(t), \boldsymbol{u}(t), \boldsymbol{d}(t))\mathrm{d}t\ , \tag{46d}$$

$$\boldsymbol{x}(t) \in \mathcal{X}(t)\ ,\ \boldsymbol{u}(t) \in \mathcal{U}(t)\ , \tag{46e}$$

where $\mathcal{N} = \{0, 1, \dots, N_p - 1\}$ is the control times and N_p is the prediction horizon. J is the cost function, $\boldsymbol{x}$ is the system, $\boldsymbol{u}$ is the input, and $\mathcal{X}(t)$ and $\mathcal{U}(t)$ are the allowed sets for $\boldsymbol{x}$ and $\boldsymbol{u}$. $\hat{\boldsymbol{x}}_{k|k}$ is the filtered estimate of $\boldsymbol{x}$ at time t_k and acts as the initial condition. $\{\hat{\boldsymbol{d}}_{k+i|k}\}_{i\in\mathcal{N}}$ is the sequence of disturbances, which in general comes from outside of the MPC framework. In this text, we get it from the separate disturbance model developed in the previous sections. ℓ_b is a cost on $\boldsymbol{x}$ on the boundary of the time domain sometimes called a *cost-to-go term.*

The cost function in (46a) involves evaluation of an integral. In practice though, a computer can only deal with discrete time. Consequently, the problem in (46) is typically reformulated as a discrete problem (in the case of a linear system)

$$J(\hat{\boldsymbol{x}}_{k|k}, \{\hat{\boldsymbol{d}}_{k+i|k}\}_{i\in\mathcal{N}}) = \min_{\hat{\boldsymbol{u}}_k} \sum_{i\in\mathcal{N}} \left[\ell_k(\hat{\boldsymbol{x}}_{k+i+1}, \hat{\boldsymbol{u}}_{k+i}, \hat{\boldsymbol{d}}_{k+i})\right] + \ell_{N_p}(\hat{\boldsymbol{x}}_{k+N_p})\ , \tag{47a}$$

$$s.t.\quad \hat{\boldsymbol{x}}_k = \hat{\boldsymbol{x}}_{k|k}\ , \tag{47b}$$

$$\hat{\boldsymbol{d}}_{k+i} = \hat{\boldsymbol{d}}_{k+i|k}\ , \tag{47c}$$

$$\hat{\boldsymbol{x}}_{k+i+1} = \mathsf{A}_\mathsf{d}\hat{\boldsymbol{x}}_{k+i} + \mathsf{B}_\mathsf{d}\hat{\boldsymbol{u}}_{k+i} + \mathsf{E}_\mathsf{d}\hat{\boldsymbol{d}}_{k+i}\ , \tag{47d}$$

$$\hat{\boldsymbol{x}}_{k+i+1} \in \mathcal{X}_{k+i+1}\ ,\ \hat{\boldsymbol{u}}_{k+i} \in \mathcal{U}_{k+i}\ , \tag{47e}$$

$$i \in \mathcal{N}\ , \tag{47f}$$

where the subscript d in (47d) indicates that the matrices are discretised. We obtain such a discrete system using, e.g., *zero-order hold*. That is, we assume that the input variable is constant during each preferably small time sample $\boldsymbol{u}(t) = \boldsymbol{u}_k$, for $t \in [t_k, t_{k+1}]$, $k \in \mathcal{N}$. (47d) describes the dynamics of the system and provides the so-called *Kalman predictions* given by the recursion. The disturbances, $\{\hat{\boldsymbol{d}}_{k+i|k}\}_{i\in\mathcal{N}}$, are again obtained from the separate disturbance model. We let the cost-to-go term be zero $\ell_{N_p}(\hat{\boldsymbol{x}}_{k+N_p}) = 0$. But it can be very important to include in some cases. For example, when batteries are included, the controller will try to sell all stored electricity (which we do not immediately want) since it minimises the cost. Unless we include a cost-to-go term, that weights the value of the electricity left in the battery [57].

The cost function is of crucial importance in terms of defining the behaviour of the controller. It is important that it minimises a term that reflects the desired behaviour and ensures stability. The latter is usually not a problem when dealing with systems of slow dynamics such as the temperature of a building. Often the cost function is minimising some distance between a control variable and a set point. For building climate control, the control variable can be the room air temperature and the set point can be the desired temperature. Two common examples of cost functions are the following:

$$\text{Quadratic cost} \qquad \boldsymbol{x}^T \mathsf{Q} \boldsymbol{x} + \boldsymbol{u}^T \mathsf{R} \boldsymbol{u} \,, \tag{48a}$$

$$\text{Economic (linear) cost} \qquad \boldsymbol{c}^T \boldsymbol{u} \,. \tag{48b}$$

The quadratic cost function typically minimises a relative weighting between the variables and can provide a trade-off between the input and the regulation of the system. The linear cost function measures an amount of some resource. In temperature regulation of a building, it is often the energy consumption or price. But in general, the resource is an abstract size and can also measure the CO_2 emission from electricity generation or even a generic penalty signal manually designed to force a certain behaviour. For our purpose, we use *economic MPC*, while also softening the constraints

$$J(\hat{\boldsymbol{x}}_{k|k}, \{\hat{\boldsymbol{d}}_{k+i|k}\}_{i \in \mathcal{N}}) = \min_{\hat{\boldsymbol{u}}_k, \hat{\boldsymbol{s}}_k} \sum_{i \in \mathcal{N}} \boldsymbol{c}_{k+i} \hat{\boldsymbol{u}}_{k+i} + \sum_{i \in \mathcal{N}^+} \boldsymbol{\rho}_{k+i} \hat{\boldsymbol{s}}_{k+i} \,, \tag{49a}$$

$$s.t. \quad \hat{\boldsymbol{x}}_k = \hat{\boldsymbol{x}}_{k|k} \,, \tag{49b}$$

$$\hat{\boldsymbol{d}}_{k+i} = \hat{\boldsymbol{d}}_{k+i|k} \,, \qquad i \in \mathcal{N} \,, \tag{49c}$$

$$\hat{\boldsymbol{x}}_{k+i+1} = \mathsf{A_d} \hat{\boldsymbol{x}}_{k+i} + \mathsf{B_d} \hat{\boldsymbol{u}}_{k+i} + \mathsf{E_d} \hat{\boldsymbol{d}}_{k+i} \,, \qquad i \in \mathcal{N} \,, \tag{49d}$$

$$\hat{\boldsymbol{y}}_{k+i} = \mathsf{C_d} \hat{\boldsymbol{x}}_{k+i} \,, \qquad i \in \mathcal{N}^+ \,, \tag{49e}$$

$$\hat{\boldsymbol{y}}_{k+i} - \hat{\boldsymbol{s}}_{k+i} \le \boldsymbol{y}_{max,k+i} \,, \qquad i \in \mathcal{N}^+ \,, \tag{49f}$$

$$\boldsymbol{y}_{min,k+i} \le \hat{\boldsymbol{y}}_{k+i} + \hat{\boldsymbol{s}}_{k+i} \,, \qquad i \in \mathcal{N}^+ \,, \tag{49g}$$

$$\Delta \boldsymbol{u}_{min,k+i} \le \Delta \hat{\boldsymbol{u}}_{k+i} \le \Delta \boldsymbol{u}_{max,k+i} \,, \qquad i \in \mathcal{N} \,, \tag{49h}$$

$$\boldsymbol{u}_{min,k+i} \le \hat{\boldsymbol{u}}_{k+i} \le \boldsymbol{u}_{max,k+i} \,, \qquad i \in \mathcal{N} \,, \tag{49i}$$

$$\boldsymbol{0} \le \hat{\boldsymbol{s}}_{k+i} \,, \qquad i \in \mathcal{N}^+ \,, \tag{49j}$$

where $\mathcal{N}^+ = \{1, 2, \ldots, N_p\}$, $\Delta\hat{\boldsymbol{u}}_{min,i}$ and $\Delta\hat{\boldsymbol{u}}_{max,i}$ are the minimum and maximum allowed changes of input, and $\hat{\boldsymbol{s}}_i$ and $\boldsymbol{\rho}_i$ are the slack variable and slack penalty, respectively. $\boldsymbol{u}_{max,i}$ and $\boldsymbol{u}_{min,i}$ are the upper and lower constraints on the input, and $\hat{\boldsymbol{y}}_{max,i}$ and $\hat{\boldsymbol{y}}_{min,i}$ are the upper and lower constraints on the observed variables.

The slack variable has the purpose of *softening* the constraints. That is, it allows the solution to move outside of the constraints without making the problem infeasible—but at a cost!

6.1.1 Rewriting the State Equations of the Optimisation Problem

We are now familiar with the objective of MPC and what the purpose of the constraints is, but it is not directly clear how to write out the optimisation problem such that we can implement it. First, we need to recognise that the variables of the optimisation problem are the input and the slack variables, $[\boldsymbol{u}_k, \boldsymbol{s}_k]$. That is, we need to write each constraint in (49) as an equation using $\boldsymbol{u}_k$ and $\boldsymbol{s}_k$. To do this, we use the Kalman predictions of the system to obtain a matrix expression for the states for all prediction times in $\mathcal{N}$. Writing out the observed system using the Kalman predictions is

$$\begin{aligned} \hat{\boldsymbol{y}}_k = \mathsf{C_d}\hat{\boldsymbol{x}}_k &= \mathsf{C_d}(\mathsf{A_d}\hat{\boldsymbol{x}}_{k-1} + \mathsf{B_d}\hat{\boldsymbol{u}}_{k-1} + \mathsf{E_d}\hat{\boldsymbol{d}}_{k-1}) , \\ &= \mathsf{C_d}\mathsf{A_d}\hat{\boldsymbol{x}}_{k-1} + \mathsf{C_d}\mathsf{B_d}\hat{\boldsymbol{u}}_{k-1} + \mathsf{C_d}\mathsf{E_d}\hat{\boldsymbol{d}}_{k-1} . \end{aligned} \tag{50}$$

The state development for $\hat{\boldsymbol{x}}_k$ is again given by the Kalman predictions, where $\hat{\boldsymbol{x}}_{k-1} = \mathsf{A_d}\hat{\boldsymbol{x}}_{k-2} + \mathsf{B_d}\hat{\boldsymbol{u}}_{k-2} + \mathsf{E_d}\hat{\boldsymbol{d}}_{k-2}$. Inserting this into (50) yields

$$\begin{aligned} \hat{\boldsymbol{y}}_k &= \mathsf{C_d}\mathsf{A_d}\left(\mathsf{A_d}\hat{\boldsymbol{x}}_{k-2} + \mathsf{B_d}\hat{\boldsymbol{u}}_{k-2} + \mathsf{E_d}\hat{\boldsymbol{d}}_{k-2}\right) + \mathsf{C_d}\mathsf{B_d}\hat{\boldsymbol{u}}_{k-1} + \mathsf{C_d}\mathsf{E_d}\hat{\boldsymbol{d}}_{k-1} , \\ &= \mathsf{C_d}{\mathsf{A_d}}^2\hat{\boldsymbol{x}}_{k-2} + \mathsf{C_d}(\mathsf{A_d}\mathsf{B_d}\hat{\boldsymbol{u}}_{k-2} + \mathsf{B_d}\hat{\boldsymbol{u}}_{k-1}) + \mathsf{C_d}(\mathsf{A_d}\mathsf{E_d}\hat{\boldsymbol{d}}_{k-2} + \mathsf{E_d}\hat{\boldsymbol{d}}_{k-1}) . \end{aligned}$$

Continuing this approach until an initial state is reached (and shifting the time to start at t_k and end at t_{k+Np}), the result is

$$\hat{\boldsymbol{y}}_{k+N_p} = \mathsf{C_d}{\mathsf{A_d}}^k\hat{\boldsymbol{x}}_k + \mathsf{C_d}\sum_{i=0}^{N_p-1}{\mathsf{A_d}}^i\mathsf{B_d}\hat{\boldsymbol{u}}_{k+N_p-1-i} + \mathsf{C_d}\sum_{i=0}^{k-1}{\mathsf{A_d}}^i\mathsf{E_d}\hat{\boldsymbol{d}}_{k+N_p-1-i} . \tag{51}$$

Let $\hat{\boldsymbol{Y}}_{k+1}$ be a vector containing the predictions N_p steps ahead starting from t_{k+1}, $\hat{Y}_{k+1} = [\hat{\boldsymbol{y}}_{k+1}^T, \hat{\boldsymbol{y}}_{k+2}^T, \ldots, \hat{\boldsymbol{y}}_{k+N_p}^T]^T$. Then (51) shows how to formulate an expression for $\hat{\boldsymbol{y}}_{k+i}$ for $i \in \mathcal{N}^+$ using a convenient matrix-vector notation

$$\boxed{\hat{\boldsymbol{Y}}_{k+1} = \Phi\hat{\boldsymbol{x}}_k + \Gamma\boldsymbol{U}_k + \Pi\hat{\boldsymbol{D}}_k}, \tag{52}$$

where

$$\Phi = \begin{bmatrix} \mathrm{C_dA_d} \\ \mathrm{C_dA_d}^2 \\ \mathrm{C_dA_d}^3 \\ \vdots \\ \mathrm{C_dA_d}^{N_p} \end{bmatrix}, \quad \Gamma = \begin{bmatrix} \mathrm{C_dB_d} & 0 & \cdots & \cdots & 0 \\ \mathrm{C_dA_dB_d} & \mathrm{C_dB_d} & \ddots & & 0 \\ \mathrm{C_dA_d}^2\mathrm{B_d} & CAB & \mathrm{C_dB_d} & \ddots & \vdots \\ \vdots & \vdots & & \ddots & 0 \\ \mathrm{C_dA_d}^{N_p-1}\mathrm{B_d} & \mathrm{C_dA_d}^{N_p-2}\mathrm{B_d} & \cdots & \cdots & \mathrm{C_dB_d} \end{bmatrix},$$

$$\boldsymbol{U}_k = \begin{bmatrix} \boldsymbol{u}_k \\ \boldsymbol{u}_{k+1} \\ \boldsymbol{u}_{k+2} \\ \vdots \\ \boldsymbol{u}_{k+N_p-1} \end{bmatrix}, \quad \Pi = \begin{bmatrix} \mathrm{C_dE_d} & 0 & \cdots & \cdots & 0 \\ \mathrm{C_dA_dE_d} & \mathrm{C_dE_d} & \ddots & & 0 \\ \mathrm{C_dA_d}^2\mathrm{E_d} & \mathrm{C_dA_dE_d} & \mathrm{C_dE_d} & \ddots & \vdots \\ \vdots & \vdots & & \ddots & 0 \\ \mathrm{C_dA_d}^{N_p-1}E & \mathrm{C_dA_d}^{N_p-2}\mathrm{E_d} & \cdots & \cdots & \mathrm{C_dE_d} \end{bmatrix},$$

$$\hat{\boldsymbol{D}}_k = \begin{bmatrix} \hat{\boldsymbol{d}}_k \\ \hat{\boldsymbol{d}}_{k+1} \\ \hat{\boldsymbol{d}}_{k+2} \\ \vdots \\ \hat{\boldsymbol{d}}_{k+N_p-1} \end{bmatrix}. \tag{53}$$

6.1.2 Rewriting the Constraints in the Optimisation Problem

Now that we have an expression for $\hat{\boldsymbol{y}}_{k+i}$, $i \in \mathcal{N}^+$, we are able to eliminate the dependence on $\hat{\boldsymbol{y}}_{k+i}$ in the constraints. Starting with (49f)

$$\begin{aligned} & \hat{\boldsymbol{Y}}_k - \boldsymbol{S}_{k+1} \leq \boldsymbol{Y}_{max} , \\ \Longrightarrow \quad & \Phi\hat{\boldsymbol{x}}_k + \Gamma\boldsymbol{U}_k + \Pi\hat{\boldsymbol{D}}_k - \boldsymbol{S}_{k+1} \leq \boldsymbol{Y}_{max} , \\ \Longrightarrow \quad & \Gamma\boldsymbol{U}_k - \boldsymbol{S}_{k+1} \leq \boldsymbol{Y}_{max} - \Phi\hat{\boldsymbol{x}}_k - \Pi\hat{\boldsymbol{D}}_k , \end{aligned}$$

where $\boldsymbol{S}_{k+1} = \{s_{k+i}\}_{i \in \mathcal{N}^+}$ is a vector with the slack variables. We can do the same thing with the lower constraint for $\hat{\boldsymbol{y}}_{k+i}$,

$$-\Gamma\boldsymbol{U}_k - \boldsymbol{S}_{k+1} \leq -\boldsymbol{Y}_{min} + \Phi\hat{\boldsymbol{x}}_k + \Pi\hat{\boldsymbol{D}}_k . \tag{54}$$

To rewrite (49h), we need the following transcription:

$$\begin{bmatrix} \Delta \boldsymbol{u}_{min} + \boldsymbol{u}_{-1} \\ \Delta \boldsymbol{u}_{min} \\ \vdots \\ \Delta \boldsymbol{u}_{min} \end{bmatrix} \leq \begin{bmatrix} \mathsf{I} & 0 & \cdots & & 0 \\ -\mathsf{I} & \mathsf{I} & 0 & & \\ 0 & -\mathsf{I} & \mathsf{I} & 0 & \vdots \\ \vdots & \ddots & \ddots & \ddots & \ddots \\ & & 0 & -\mathsf{I} & \mathsf{I} & 0 \\ 0 & \cdots & & 0 & -\mathsf{I} & \mathsf{I} \end{bmatrix} \boldsymbol{U}_k \leq \begin{bmatrix} \Delta \boldsymbol{u}_{max} + \boldsymbol{u}_{-1} \\ \Delta \boldsymbol{u}_{max} \\ \vdots \\ \Delta \boldsymbol{u}_{max} \end{bmatrix}, \tag{55}$$

where $\boldsymbol{u}_{-1}$ is the input given to the system at time t_{k-1}, and I is the identity matrix with the same size as the length of $\boldsymbol{u}_k$. Denoting the matrix in (55) by Λ, then the constraint in (49h) is

$$\begin{aligned} -\Lambda \boldsymbol{U}_k &\leq -\Delta \boldsymbol{U}_{min} , \\ \Lambda \boldsymbol{U}_k &\leq \Delta \boldsymbol{U}_{max} . \end{aligned} \tag{56}$$

(49i) is straightforward in the sense that it requires no further notational introduction

$$\begin{aligned} -\boldsymbol{U}_k &\leq -\boldsymbol{U}_{min} , \\ \boldsymbol{U}_k &\leq \boldsymbol{U}_{max} . \end{aligned} \tag{57}$$

Finally, we demand the slack variables to be non-negative, $-\boldsymbol{S}_{k+1} \leq 0$. We are now able to write the problem in (49) as an expression of the input and slack variables

$$\begin{aligned} J = \min_{\boldsymbol{U}_k, \boldsymbol{S}_{k+1}} \quad & \begin{bmatrix} \boldsymbol{C}_u^T & \boldsymbol{P}_s^T \end{bmatrix} \begin{bmatrix} \boldsymbol{U}_k \\ \boldsymbol{S}_{k+1} \end{bmatrix} \\ s.t. \quad & \begin{bmatrix} -\Gamma & -\mathsf{I} \\ \Gamma & -\mathsf{I} \\ -\mathsf{I} & 0 \\ \mathsf{I} & 0 \\ -\Lambda & 0 \\ \Lambda & 0 \\ 0 & -\mathsf{I} \end{bmatrix} \begin{bmatrix} \boldsymbol{U}_k \\ \boldsymbol{S}_{k+1} \end{bmatrix} \leq \begin{bmatrix} -\boldsymbol{Y}_{min} + \Phi \hat{\boldsymbol{x}}_k + \Pi \hat{\boldsymbol{D}}_k \\ \boldsymbol{Y}_{max} - \Phi \hat{\boldsymbol{x}}_k - \Pi \hat{\boldsymbol{D}}_k \\ -\boldsymbol{U}_{min} \\ \boldsymbol{U}_{max} \\ -\Delta \boldsymbol{U}_{min} \\ \Delta \boldsymbol{U}_{max} \\ \boldsymbol{0} \end{bmatrix}, \end{aligned} \tag{58}$$

where $\boldsymbol{C}_u = [\boldsymbol{c}_k^T, \ldots, \boldsymbol{c}_{k+N_p-1}^T]^T$ and $\boldsymbol{P}_s = [\boldsymbol{\rho}_{k+1}^T, \ldots, \boldsymbol{\rho}_{k+N_p}^T]^T$ are the electricity costs and the slack variable penalty, respectively. We have now written the optimisation problem in (49) as a constrained *linear program* that gives us the optimal input, $\hat{\boldsymbol{U}}_k^* = [\hat{\boldsymbol{u}}_k^{*T}, \ldots, \hat{\boldsymbol{u}}_{k+N_p-1}^{*T}]^T$, that minimises the cost J based on initial conditions for the system, $\hat{\boldsymbol{x}}_{k|k}$, and disturbance forecasts, $\{\hat{\boldsymbol{d}}_{k+i}\}_{i \in \mathcal{N}}$. This is referred to as *optimal control*. Figure 17 displays the overall MPC framework and how the elements interact. Very often, for systems governed by uncertainty, it is necessary to use a *moving horizon scheme*, where only the current optimal input

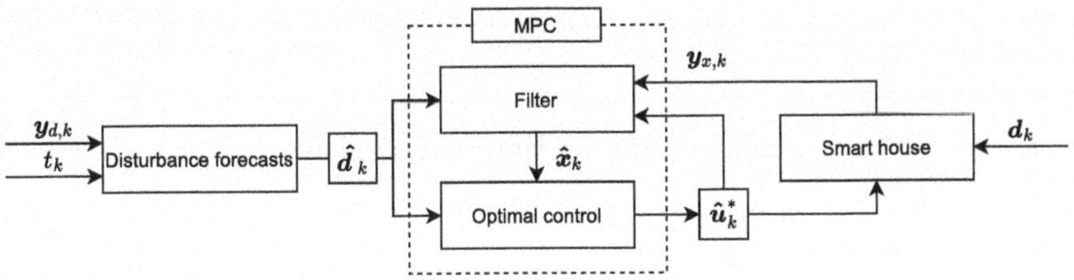

Fig. 17 The MPC framework for the smart building and how the disturbance model is incorporated

is applied to the system, u_k^*. When arriving at the next time step, t_{k+1}, the optimal control problem is computed again and the current input is applied. This is known as *closed-loop feedback control* and ensures stability of the system as the controller can account for unforeseen uncertainty in the system between time steps. As this is a linear in-equality constrained problem, a closed-form solution to (58) does not exist in general. We therefore use numerical optimisation to find the unique solution that exists due to convexity as long as the problem is well posed.

6.2 *Offset-Free Control Without Separate Disturbance Model*

In more conventional MPC setups where separately modelled disturbance models are not feasible, there exist ways to deal with unforeseen disturbances. In practice, parameter uncertainties, lack of model accuracy, and non-modelled disturbances all usually necessitate some kind of action; otherwise, *offsets* can arise. For example, if the disturbances act with a constant (or slowly varying) force, we can obtain a non-zero distance between the system state and the desired set point. The literature suggests multiple ways to deal with this [51, 52, 55], and this is still an active research area. A popular method (among others) is known as the *augmented disturbance model*. In practice, two variants are widely used and well studied: the *input* and *output* disturbance models.

Morari and Stephanopoulos [47] derive some important concepts and results regarding disturbance modelling for continuous-time systems in the deterministic case. Consider a continuous-time linear state-space system of the form

$$\begin{aligned} \mathrm{d}\boldsymbol{x}(t) &= (\mathsf{A}\boldsymbol{x}(t) + \mathsf{B}\boldsymbol{u}(t))\,\mathrm{d}t\,, \\ y_k &= \mathsf{C}\boldsymbol{x}(t_k)\,, \end{aligned} \tag{59}$$

where $\boldsymbol{x} \in \mathbb{R}^n$, $\mathsf{A} \in \mathbb{R}^{n\times n}$, $\boldsymbol{u} \in \mathbb{R}^{n_u}$, $\mathsf{B} \in \mathbb{R}^{n\times n_u}$, and $\mathsf{C} \in \mathbb{R}^{n_y}$. Note that the disturbances are not a part of the model in the first place. The augmented disturbance model approach assumes that the disturbances act on the system as integrated white noise; that is, we can add an *integrator* as an independent state, $\boldsymbol{\eta} \in \mathbb{R}^{n_d}$ in the system by

$$\begin{aligned} \mathrm{d}\boldsymbol{x}(t) &= \Big(\mathsf{A}\boldsymbol{x}(t) + \mathsf{B}\boldsymbol{u}(t) + \bar{\mathsf{B}}\boldsymbol{\eta}(t)\Big)\,\mathrm{d}t\ , \\ \mathrm{d}\boldsymbol{\eta}(t) &= \bar{\Sigma}\mathrm{d}\bar{\boldsymbol{\omega}}(t)\ , \\ \boldsymbol{y}_k &= \mathsf{C}\boldsymbol{x}(t_k) + \bar{\mathsf{C}}\boldsymbol{\eta}(t_k)\ , \end{aligned} \tag{60}$$

where $\bar{\mathsf{B}} \in \mathbb{R}^{n\times n_d}$ and $\bar{\mathsf{C}} \in \mathbb{R}^{n\times n_d}$ are the disturbances on the input and output (hence the name). We shall assume $C_d = \mathbf{0}$ in the rest of this section. The case when the disturbances only act on the system is called *input disturbances* because it acts as an input on the system. We can augment the disturbance and obtain the *augmented* system

$$\begin{aligned} \mathrm{d}\begin{bmatrix}\boldsymbol{x}(t)\\ \boldsymbol{\eta}(t)\end{bmatrix} &= \left(\begin{bmatrix}\mathsf{A} & \bar{\mathsf{B}}\\ 0 & 0\end{bmatrix}\begin{bmatrix}\boldsymbol{x}(t)\\ \boldsymbol{\eta}(t)\end{bmatrix} + \begin{bmatrix}\mathsf{B}\\ 0\end{bmatrix}\boldsymbol{u}(t)\right)\mathrm{d}t + \begin{bmatrix}0\\ \bar{\Sigma}\end{bmatrix}\mathrm{d}\bar{\boldsymbol{\omega}}(t)\ , \\ \boldsymbol{y}_k &= \begin{bmatrix}\mathsf{C} & 0\end{bmatrix}\begin{bmatrix}\boldsymbol{x}(t)\\ \boldsymbol{\eta}(t)\end{bmatrix}\ . \end{aligned} \tag{61}$$

Note that we are (obviously) not able to influence the disturbances and (in general) know nothing about them. However, it is crucial for us to *estimate* them in order to obtain offset-free control. *Observability* is an important concept that relates to whether we are able to estimate all states in the given system, $\boldsymbol{x}$, based on the observed information we have, $\boldsymbol{y}$. We say that the system (C, A) in (59) is observable if

$$\operatorname{rank}\begin{bmatrix}\mathsf{CA}\\ \mathsf{CA}^2\\ \vdots\\ \mathsf{CA}^{n-1}\end{bmatrix} = n\ . \tag{62}$$

In general, we are able to estimate all states in a system if and only if it is observable. That is, we need to make sure that the augmented system in (61) is observable—otherwise we cannot estimate the disturbances and in turn not obtain offset-free control. The system $\left(\begin{bmatrix}\mathsf{C} & 0\end{bmatrix}, \begin{bmatrix}\mathsf{A} & \bar{\mathsf{B}}\\ \mathsf{C} & 0\end{bmatrix}\right)$ is observable if and only if the following requirements are fulfilled [47]:

1. The system (C, A) is observable.
2. $\operatorname{rank}\begin{bmatrix}\mathsf{A} & \bar{\mathsf{B}}\\ \mathsf{C} & 0\end{bmatrix} = n + n_d$.

This implies that we are able to insert at most n_y (the number of independently observed variables) integrators into the system while ensuring observability of the augmented system. We can estimate the disturbance states simply by using the Kalman filter or Luenberger observer [32] (treating them as any other hidden state). This method also supplies disturbance forecasts by computing the predictions supplied by the system. It is easy to see that it corresponds to zero-order disturbance forecasts also called *persistent forecasts*, see, e.g., [38, p. 333] and [11]. For this reason, the integrator approach works best when the disturbance dynamics are slow—and not very well for faster dynamics such as the solar radiation for a smart building. We will show this in the next section.

7 Predictive Control with Embedded Disturbance Models

We now combine the individual weather models in the previous sections into a combined disturbance model framework. Ultimately, we want to show that by modelling the disturbances, we can obtain more accurate control than using, e.g., augmented integrators. The advanced disturbance model should return a vector, $\boldsymbol{d}(t) = \left[d_{T_a}(t), d_{\phi_s}(t)\right]^T$, containing the solar radiation and ambient air temperature. Writing up all equations for the individual disturbances gives the following complete description:

$$
\begin{aligned}
&\text{Cloud cover model} \begin{cases} \mathrm{d}Z_\kappa = f_\psi(Z_\kappa)\mathrm{d}t + \sigma_\psi \mathrm{d}\omega_\kappa \\ c = \zeta^{-1}(\psi^{-1}(Z_\kappa)) \end{cases} \\
&\text{Solar radiation model} \begin{cases} \phi_s = I_N(c,t) + I_D(c,t) \end{cases} \\
&\text{Net radiation model} \begin{cases} R_n = R_n(c, \phi_s, t) \end{cases} \\
&\text{Air temperature model} \begin{cases} \mathrm{d}T_w = f_{T_w}(T_l, T_w)\mathrm{d}t + \sigma_w \mathrm{d}\omega_w \\ \mathrm{d}T_l = f_{T_l}(T_l, T_w, R_n)\mathrm{d}t + \sigma_l \mathrm{d}\omega_l \end{cases} \\
&\text{Observations} \begin{cases} d_{\phi_s} = \phi_s + v_{\phi_s}, \quad v_{\phi_s} \sim N(0, R_{\phi_s}) \\ d_{T_a} = T_l + v_{T_a}, \quad v_{T_a} \sim N(0, R_{T_a}) \\ \boldsymbol{d} = \left[d_{T_a}, d_{\phi_s}\right]^T. \end{cases}
\end{aligned}
\tag{63}
$$

Since the disturbance model in (63) is based on SDEs, we are able to use the CDEKF to compute *certainty equivalent* Kalman predictions—which MPC requires. This procedure requires numerical solutions of coupled differential equations, which in turn requires initial conditions preferably from observations coming from the building site in order to ensure accuracy. Due to the one-way coupling of the individual weather models, the computation of the predictions becomes much easier, as it can be split into smaller and simpler calculations:

1. Compute the cloud cover predictions $\{\hat{c}_{k+i|k}\}_{i\in\mathcal{N}}$.
2. Compute the solar radiation predictions $\{\hat{\phi}_{s,k+i|k}\}_{i\in\mathcal{N}}$.
3. Compute the net radiation predictions $\{\hat{R}_{n,k+i|k}\}_{i\in\mathcal{N}}$.
4. Compute the ambient air temperature predictions $\{\hat{T}_{a,k+i|k}\}_{i\in\mathcal{N}}$.

7.1 Comparison of Advanced Disturbance Forecasts and Persistent Forecasts

All the necessary elements are now introduced for us to demonstrate how to control the room air temperature of a smart building presented in Sect. 4 using the advanced disturbance forecasts from (63). Additionally, we want to show that the great effort put into modelling the disturbances actually improves the quality of the smart building regulation. To do this, we compare the advanced disturbance forecasts with a more typical and conventional kind of offset-free control that is explained in Sect. 6.2. In this text, we use *persistent forecasts*, which are often used as a reference model for weather and energy forecasting models. It uses the following constant predictions:

$$\hat{\boldsymbol{d}}_{k+i|k} = \hat{\boldsymbol{d}}_{k|k}\,, \quad i \in \mathcal{N}\,. \tag{64}$$

That is, we assume that all future disturbances equal the disturbance at the present time, t_k, and we assume that we actually observe them.

Figure 18 shows the persistent and advanced disturbance forecasts using a prediction horizon of 96 h. The simplicity of the persistent forecasts becomes very

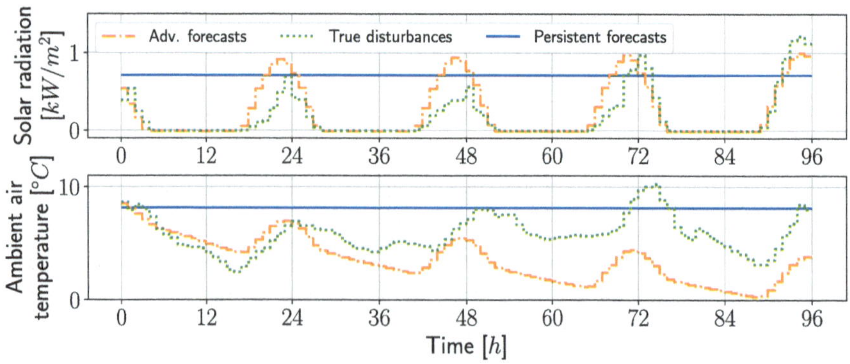

Fig. 18 The persistent forecasts against the advanced disturbance forecasts. The latter is computed by integrating (63) forward using the current observations as initial conditions

visible compared to the complex dynamics of the true disturbances. The advanced disturbance forecasts are of course most accurate in a short future time span due to the initial conditions. They then drift towards some stationary dynamics, highly dictated by the stationary points of the cloud cover that can be seen in Fig. 9. We therefore cannot hope to accurately forecast the disturbances 96 hours into the future using these methods—instead they give *expected disturbance values*. For this reason, the literature normally uses meteorological forecasts. They are, however, less accurate for short-term predictions. [66] suggest that in practice, the advanced disturbance forecasts work best 4–10 h into the prediction horizon and from that point on meteorological forecasts in general perform better. The latter is based on large systems of differential equations and is calculated using very powerful computers. In practice, it is believed that a combination of short- and long-term forecasts will be the best solution.

8 Simulation Results

As previously mentioned, we use data from March as the true disturbances acting on the smart building. This gives us 7 months of data to simulate control of the smart building using the two forecasting schemes. In this section, we show the results of controlling the smart buildings presented in Sect. 4. Furthermore, we present the results where the heat pump is combined with both electrical heaters and air conditioners (for cooling). Recall that the heat pump is a factor 3 more efficient compared to the electrical heaters, which makes it economically attractive and interesting to combine.

All simulations in this section use a prediction horizon of 96 h, a time sample of 1 hour, and temperature constraints $T_{r,min}$=20 C° and $T_{r,max}$=24 C° (which are softened). We use the slack penalty suggested by [57], $\rho_k = 5000$. The electricity price is taken from Nord Pool and is the average over all March data and equals to $c_k = 0.27 \cdot 10^{-3}$ EUR/Wh.

Figure 19 shows a 15-day sample of the 7 months of simulation for two smart buildings: one using electrical heaters and one using a heat pump. The smart building equipped with electrical heaters acts faster and is therefore more capable of adjusting to sudden changes from the disturbances. This is borne out by the electrical heaters that operate at a level that sets the room air temperature to the lower constraint to minimise costs—except when the sun shines and additional heat is not needed. The differences between two forecasting schemes are not greatly visible from this sample, however, due to the effect of the control feedback every hour.

The solution for the smart building equipped with a heat pump looks much different. The overall dynamics are much slower. In contrast to the case with the electrical heaters, the advanced disturbance forecasts seem to enable the controller

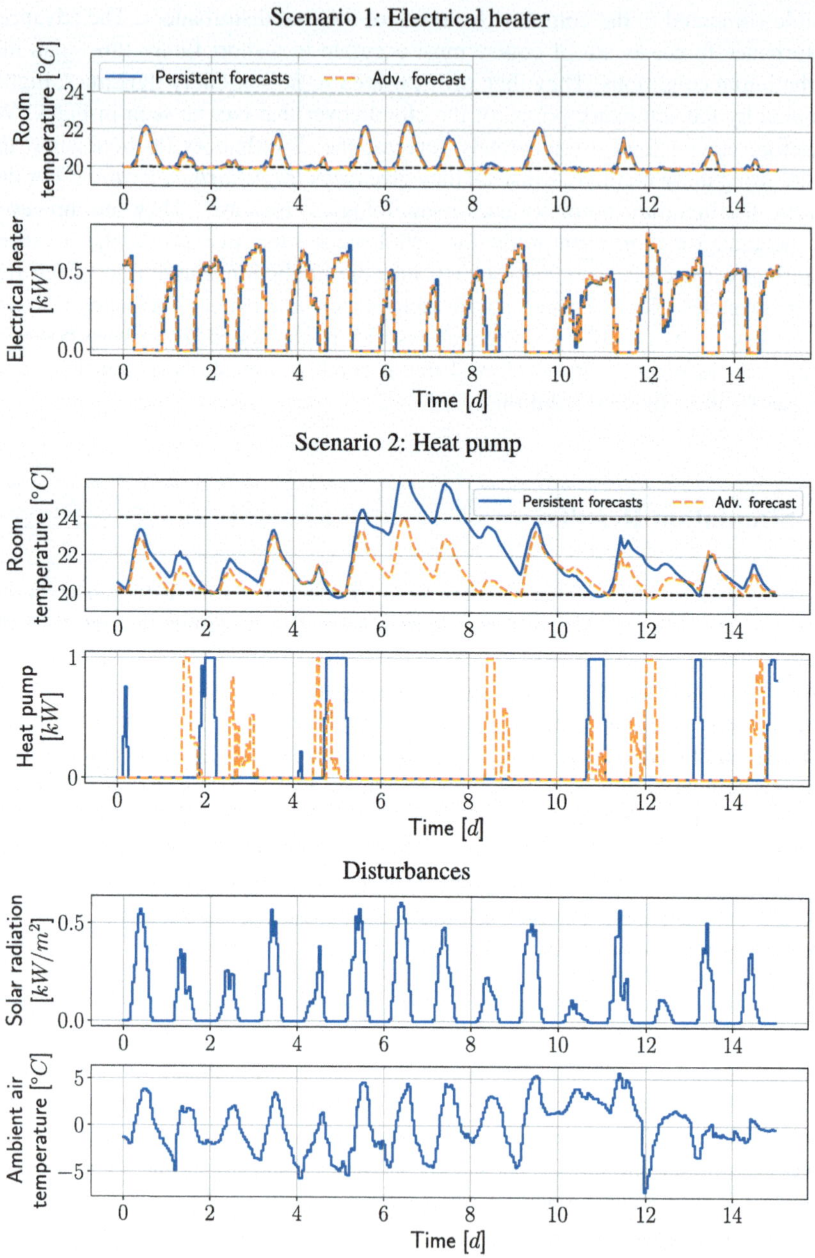

Fig. 19 A 15-day sample of the total 7 months of simulation. It shows the indoor air temperature and the heat input for the two scenarios at the same point in the time series of simulation. The black dashed lines are the constraints

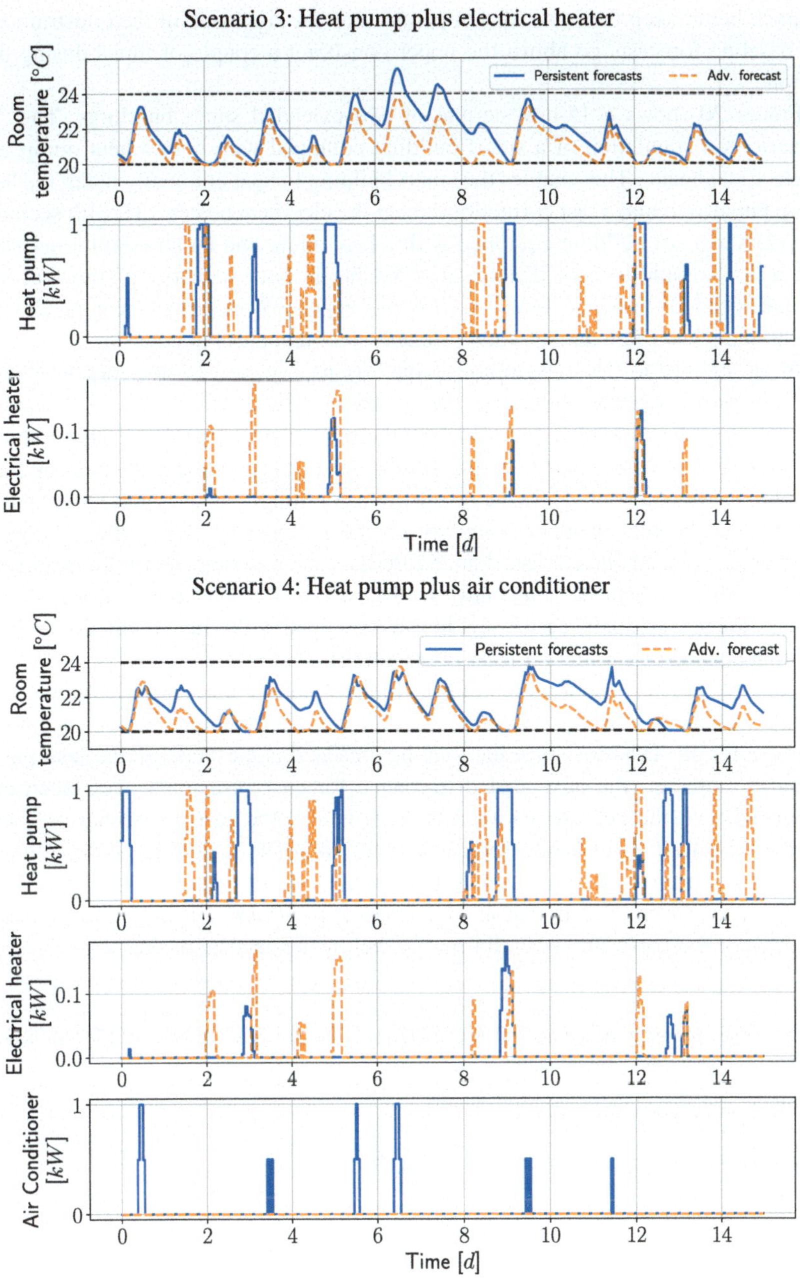

Fig. 20 A 15-day sample of the total 7 months of simulation. It shows the indoor air temperature and the heat input for the two scenarios where the heat pump is combined with faster heating inputs

to much better keep the room air temperature on the right side of the constraints—the baseline forecasts go above the upper constraint a couple of times during this sample.

Figure 20 shows a 15-day sample of two extended smart buildings. The 3rd scenario is a simulation of a smart building equipped with both a heat pump and an electrical heater. This enables the smart building to heat efficiently using the heat pump but also to make fast corrections using the electrical heaters. The 4th scenario considers a smart building equipped with a heat pump and an air conditioner such that it is also able to cool if necessary. Visible in both scenarios is the fact that the advanced disturbance forecasts use the expensive electrical heaters and air conditioners less often and are therefore able to obtain cheaper control.

In an attempt to draw asymptotic conclusions, we turn to consider how well the forecasting schemes minimise the actual cost function in (49) of the entire simulation of the 7 months, as this is what the solutions are based on. Table 8 shows the constraint violations of the entire simulations corresponding to the second term in (49a). Additionally, it shows the results for a controller that uses perfect forecasts: this gives a theoretical upper boundary on the performance using the settings in this chapter. The advanced disturbance forecasts seem to outperform the persistent forecasts in all scenarios. Especially in the case of the heat pump alone: this is perhaps the most realistic case—that houses equipped with a heat pump do not have addition heating or cooling (at least in Denmark).

Looking at the cost term in (49a), Table 9 shows the total electricity cost for all scenarios. It is obvious that the cost for the electrical heaters is almost identical for all scenarios since the total heat needed is the same. In the heat pump scenario, however, the advanced disturbance forecasts use much less electricity compared with the persistent forecasts. This is also the case for scenarios 3 and 4—the advanced disturbance forecasts seem to offer a significant decrease in electricity consumption and in general are very close to the perfect forecasts. This is also visible from the simulation samples in Figs. 19 and 20 where the advanced disturbance forecasts almost at all times lie below the persistent forecasts.

Table 8 The constraint violations (the second term in the cost function in (49a)) for all heating strategies for each forecasting scheme

Constraint violation of the control simulations			
Forecasting method	Persistent	Advanced disturbances	Perfect
Scenario 1: Electrical heater	48.5	39.6	25.11
Scenario 2: Heat pump	157.9	12.3	1.7
Scenario 3: Heat pump plus electrical heater	48.0	6.7	1.2
Scenario 4: Heat pump plus AC	4.4	2.4	0

9 Hierarchical Control

In the next section, we will first illustrate how the controllers described in the previous sections can be considered as the low-level controllers of a multi-level or hierarchical control setup for solving grid or ancillary service problems in future smart energy systems. Subsequently, we shall briefly outline how these principles can be generalised to multi-level and hierarchical control problems. This section will also outline how to establish a connection between the multi-level control problems and conventional electricity markets.

9.1 *Two-level Control for Utilising Energy Flexibility*

In the previous sections, it has been shown how to develop controllers for controlling smart buildings according to forecasts of prices, weather conditions, and indoor climate requirements. In this section, it will be explained how to leverage this by generating prices that are used indirectly to control the demand of the smart buildings. The basic concept is illustrated by Fig. 21, where a smart building, from an external perspective, takes an input (price) and gives an output (demand). Analysed in this way, a model, termed the Flexibility Function, can be developed that predicts demand as a dynamic function of price. The Flexibility Function could be any dynamic model. In [18], a linear model (finite impulse response model) is suggested, but in [19], it is shown that a grey-box model using stochastic differential equations is more appropriate.

Table 9 The electricity price in EUR (the first term in the cost function in (49a)) for all heating strategies for each forecasting scheme

Electricity cost of the simulations			
Forecasting method	Persistent	Advanced disturbances	Perfect
Scenario 1: Electrical heater	303.2	302.2	302.0
Scenario 2: Heat pump	117.3	110.4	107.7
Scenario 3: Heat pump plus electrical heater	113.0	108.2	107.5
Scenario 4: Heat pump plus AC	117.9	108.3	107.5

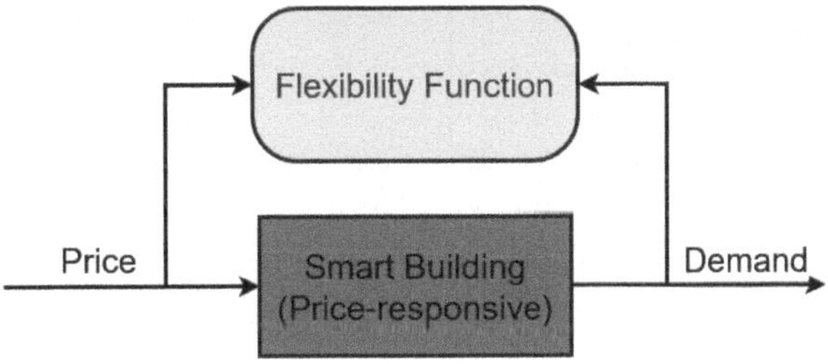

Fig. 21 The demand of a smart building can be predicted as a function of prices

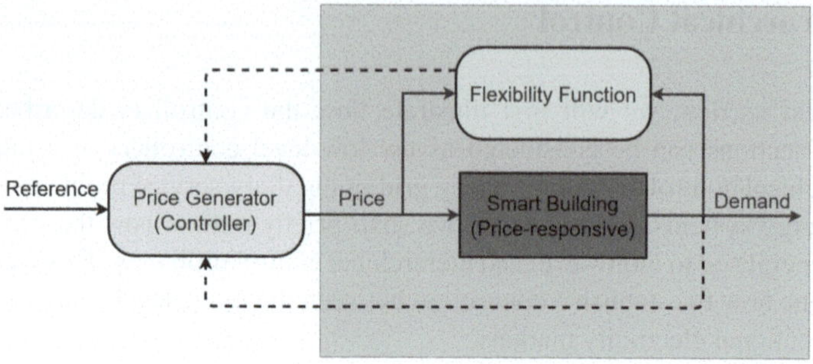

Fig. 22 Using a Flexibility Function to generate price signals and demand as control feedback

Once a Flexibility Function has been estimated, a second controller can be formulated where the objective is to control demand according to some criteria, and the decision variable is the price. As shown in Fig. 22, the Flexibility Function can be used to generate prices according to some reference. Notice how the demand acts as the feedback to the controller, closing the loop.

If FF is the Flexibility Function that takes prices as input and gives expected demand as output, while r_l is a reference load, then a naive upper-level optimisation problem can be written as

$$\min_{\boldsymbol{C}_u} \quad (\mathrm{FF}(\boldsymbol{C}_u) - r_l)^2. \tag{65}$$

Obviously, it might be necessary to impose limits on how much the price can change, requirements on the average value, and a more sophisticated optimisation problem than the minimum variance formulation as discussed in [17]. Combining this optimisation problem with the one presented in (58) reveals how the price signal, $\boldsymbol{C}_u$, couples the two in an elegant fashion

$$\min_{\boldsymbol{C}_u} \quad (\mathrm{FF}(\boldsymbol{C}_u) - r_l)^2 \qquad \text{Upperlevel}$$

$$\min_{U_k, S_{k+1}} \quad \begin{bmatrix} \boldsymbol{C}_u{}^T & \boldsymbol{P}_s^T \end{bmatrix} \begin{bmatrix} \boldsymbol{U}_k \\ \boldsymbol{S}_{k+1} \end{bmatrix} \qquad \text{Lowerlevel}$$

$$s.t. \quad \begin{bmatrix} -\Gamma & -\mathrm{I} \\ \Gamma & -\mathrm{I} \\ -\mathrm{I} & 0 \\ \mathrm{I} & 0 \\ -\Lambda & 0 \\ \Lambda & 0 \\ 0 & -\mathrm{I} \end{bmatrix} \begin{bmatrix} \boldsymbol{U}_k \\ \boldsymbol{S}_{k+1} \end{bmatrix} \leq \begin{bmatrix} -\boldsymbol{Y}_{min} + \Phi\hat{\boldsymbol{x}}_k + \Pi\hat{\boldsymbol{D}}_k \\ \boldsymbol{Y}_{max} - \Phi\hat{\boldsymbol{x}}_k - \Pi\hat{\boldsymbol{D}}_k \\ -\boldsymbol{U}_{min} \\ \boldsymbol{U}_{max} \\ -\Delta\boldsymbol{U}_{min} \\ \Delta\boldsymbol{U}_{max} \\ \boldsymbol{0} \end{bmatrix}.$$

Notice how the two optimisation problems are solved independently from each other, thus preserving autonomy and privacy for the building owners while simultaneously allowing an aggregator to utilise the energy flexibility. In practice, there are going to be a lot of smart buildings for each aggregator that all have independent control problems. This method scales well to this case since the computational burden for the upper level remains constant—with the Flexibility Function simply representing the aggregated response from the smart buildings.

In [29], it is shown how the Flexibility Function can be used to generate a Flexibility Index for a building.

9.2 Multi-Level Control and Markets

Ultimately, the purpose of the future smart-energy system is to establish a connection between the controllers operating at local scales and high-level markets operating at large scales. Essentially, a spectrum of all relevant spatial aggregation levels (building, district, city, region, country, etc.) has to be considered. At the same time, control or market solutions must ensure that the power system is balanced at all future temporal scales. Consequently, data-intelligent solutions for operating flexible electrical energy systems have to be implemented on all spatial and temporal scales.

To address these issues, several solutions have been proposed in the literature in recent years. These major solutions are transactive energy, peer-to-peer, and control-based solutions, as described in [13].

Traditionally power systems are operated by sending bids to a market. However, in order to balance the systems on all relevant horizons, several markets are needed. Examples are day-ahead, intra-day, balancing, and regulation markets. The bids are typically static consisting of a volume and duration. However, we believe that we need a disruption related to principles for activating low-level flexibility.

Given all the bids, the so-called supply and demand curves for all the operated horizons can be found. Mathematically, these supply and demand curves are static and deterministic. Merit order dispatch is then used to optimise the cost of generation. However, if the production is from wind or solar power, then the supply curve must be stochastic, and the demand flexibility has to be described dynamically, by the introduced Flexibility Function. Consequently, we need to introduce new digitised markets that are dynamic and stochastic. And instead of using a large number of markets for different purposes (frequency, voltage, congestion, etc.) and on different horizons, we will suggest to use concepts based on the Flexibility Function and stochastic control theory, exactly as described in the previous section for the two-level case. We call this a Smart-Energy Operating System (SE-OS) [36, 44, 45].

If we zoom out in space and time, i.e., consider the load in a very large area on a horizon of days, or maybe next day, then both the dynamics and stochasticity can be eliminated, and hence, we can use conventional market principles as illustrated in

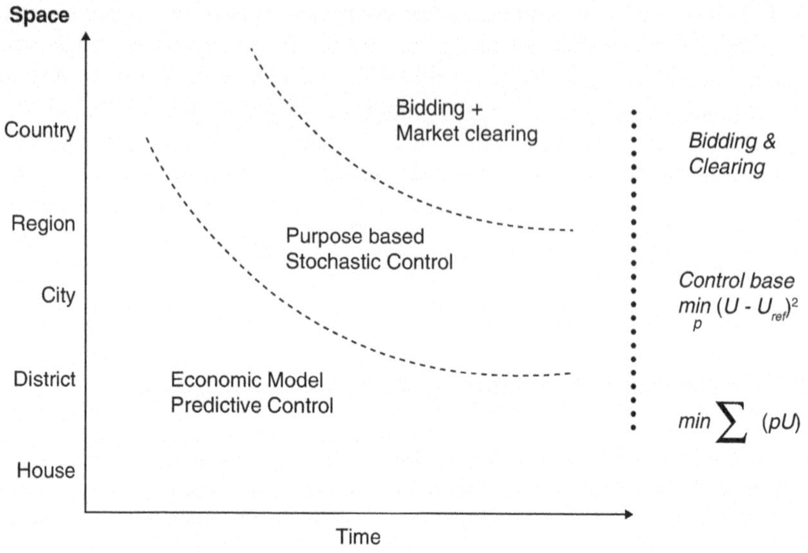

Fig. 23 Hierarchical control and markets

Fig. 23. If we zoom in on higher temporal and spatial resolutions (like for instance a house), the dynamics and stochasticity become important, and consequently, we will suggest to use the control-based methods for the flexibility as discussed in this chapter.

The total setup consists of a combination of all these options, and the best option depends on the zoom level. The conclusion is that we need new future digitised refined market principles, which operate as a hierarchy of conventional market-based bidding and clearing on the higher levels and control-based approaches on the lower level—see Fig. 23.

All these principles for forecasting, control, and optimisation are included in the so-called Smart-Energy Operating System (SE-OS), which is used to develop, implement, and test solutions (layers: data, models, optimisation, control, communication) for operating flexible electrical energy systems at all scales. See [14, 36, 44, 45] for further information.

10 Summary

In this chapter, we have presented methods for modelling relevant for the control of smart buildings. Specifically, we have introduced the grey-box modelling framework, and we have used this modelling framework to establish models for a building—as well as for some of the most important weather-related disturbances—namely cloud cover, solar radiation, net radiation, and ambient air temperature.

Most importantly, the grey-box principle bridges the gap between models based on first principles (physics) and models based on information obtained from the data (statistics).

Further, methods for model development are suggested. For parameter estimation, we suggest using the maximum likelihood method as this method allows for an integrated estimation of parameters related to the embedded description of the stochastic part.

Having models for the buildings and disturbance models as stochastic differential equations enables and promotes the use of model predictive control (MPC) as the regulation scheme for the indoor air temperature. MPC is widely described and used in the literature for building climate control problems. We introduced and formulated the mathematical optimisation problem involved with MPC and showed how to numerically compute the optimal control solution. We explained the problem of dealing with disturbances in control and showed how to incorporate them—both by simple means (using an augmented integrator) and by embedding the advanced disturbance models to supply forecasts. The last section presented simulation-based results of MPC applied to the presented smart building models using different heating strategies. The obtained results strongly suggest that the use of sophisticated disturbances models over conventional methods to supply weather forecasts can improve the building climate control.

Lastly, we have briefly explained how energy flexibility can be leveraged through price-based control, by utilising a two-level framework in which prices are generated by a controller to actuate the energy flexibility of the smart buildings. These principles are generalised to multi-level controllers for solving all types of ancillary and balancing service problems in future weather and data-driven energy systems.

Acknowledgments The authors are financed by a number of projects related to smart buildings, smart grids, and energy systems. Consequently, we would like to acknowledge support related to the following projects: *Centre for IT-Intelligent Energy Systems (IFD[1] CITIES)*, *Flexible Energy Denmark (IFD FED)*, *Digitally supported Smart District Heating (IFD HEAT 4.0)*, *Research Centre on Zero Emission Neighbourhoods in Smart Cities (FME[2]-ZEN)*, *Sustainable Energy Plus Neighbourhoods (H2020 syn.ikia)*, *Energy balancing and resilience solutions to unlock the flexibility and increase market options for distribution grid (H2020 ebalance-plus)*, *Demand response for energy cooperatives (H2020 FLEXCoop)*, *The value of end-use flexibility in the future Norwegian energy system (FlexBuild)*, *IEA EBC Annex 67: Energy Flexible Buildings*, *IEA EBC Annex 71: Building energy performance assessment based on in-situ measurements*, *IEA EBC Annex 81: Data-Driven Smart Buildings*, and *IEA EBC Annex 82: Energy Flexible Buildings Towards Resilient Low Carbon Energy Systems*.[3] Finally, we would like to thank all the partners of this long list of projects for many very inspiring discussions.

[1]Innovation Fund Denmark.

[2]Norwegian Centres for Environment-friendly Energy Research.

[3]The participations in IEA Annex activities are supported by EUDP: a support scheme for development and testing of energy systems solutions under the Danish Energy Agency.

References

1. M. Alam et al., Applying extended Kalman filters to adaptive thermal modelling in homes. Adv. Build. Energy Res. **12**(1), 48–65 (2018). ISSN: 17562201, 17512549. https://doi.org/10.1080/17512549.2017.1325398
2. N. An, S. Hemmati, Y.-J. Cui, Assessment of the methods for determining net radiation at different time-scales of meteorological variables. J. Rock Mech. Geotech. Eng. **9**(2), 239–246 (2017)
3. B. Andersen et al., Meteorological Data for Design of Building and Installation: A Reference Year. Technical Report. 66. Technical University of Denmark. Thermal Insulation Laboratory (1977)
4. A.G. Azar et al., SmartNet: Results of Pilot B (2019)
5. P. Bacher, H. Madsen, Identifying suitable models for the heat dynamics of buildings. Energy Build. **43**(7), 1511–1522 (2011)
6. J. Bak et al., Goodness of fit of stochastic differential equations, in *Symposium i anvendt statistik*. Conference 01 January 1999 (1999)
7. B.S. Bhangu et al., Nonlinear observers for predicting state-of-charge and state-of-health of lead-acid batteries for hybrid-electric vehicles. IEEE Trans. Vehic. Technol. **54**(3), 783–794 (2005)
8. C.M. Bishop, *Pattern Recognition and Machine Learning*. Springer, New York (2006)
9. O.M. Brastein et al., Parameter estimation for externally simulated thermal network models. Energy Build. **191**, 200–210 (2019). ISSN: 18726178, 03787788. https://doi.org/10.1016/j.enbuild.2019.03.018
10. N.L. Brok, H. Madsen, J.B. Jørgensen, Nonlinear model predictive control for stochastic differential equation systems. IFAC-PapersOnLine, in *6th IFAC Conference on Nonlinear Model Predictive Control NMPC 2018*, vol. 51(20) (2018), pp. 430–435. ISSN: 2405-8963. http://www.sciencedirect.com/science/article/pii/S2405896318327290
11. C.F.M. Coimbra, J. Kleissl, R. Marquez, Overview of solar-forecasting methods and a metric for accuracy evaluation, in *Solar Energy Forecasting and Resource Assessment*, chap. 8, ed. by J. Kleissl (Academic, Boston, 2013), pp. 171–194
12. O. Corradi et al., Controlling electricity consumption by forecasting its response to varying prices. IEEE Trans. Power Syst. **28**(1), 421–429 (2013). ISSN: 15580679, 08858950. https://doi.org/10.1109/TPWRS.2012.2197027
13. G. De Zotti et al., Ancillary services 4.0: a top-to-bottom control-based approach for solving ancillary services problems in smart grids. IEEE Access **6**, 11694–11706 (2018a). ISSN: 21693536. https://doi.org/10.1109/ACCESS.2018.2805330
14. G. De Zotti et al., Consumers' flexibility estimation at the TSO level for balancing services. IEEE Trans. Power Syst. **34**(3), 1918–1930 (2018b). ISSN: 15580679, 08858950. https://doi.org/10.1109/TPWRS.2018.2885933
15. J.S.G. Ehnberg, M.H.J. Bollen, Simulation of global solar radiation based on cloud observations. Solar Energy, in *ISES Solar World Congress 2003*, vol. 78(2) (2005), pp. 157–162. ISSN: 0038-092X. https://doi.org/10.1016/j.solener2004.08.016
16. C. Ghiaus, A. Chicinas, C. Inard, Grey-box identification of air-handling unit elements. Control Eng. Pract. **15**(4), 421–433 (2007). ISSN: 18736939, 09670661. https://doi.org/10.1016/jconengprac.2006.08.005
17. J.R. Grønborg, Characterisation and Integration of Energy Flexibility through Stochastic Modelling and Control (2019)
18. J.R. Grønborg et al., Characterizing the energy flexibility of buildings and districts. Appl. Energy **225**, 175–182 (2018)
19. J.R. Grønborg et al., Stochastic nonlinear modelling and application of price-based energy flexibility. Appl. Energy **275**, 115096 (2020). https://doi.org/10.1016/j.apenergy.2020.115096
20. R. Halvgaard et al., Economic model predictive control for building climate control in a smart grid, in *2012 IEEE PES Innovative Smart Grid Technologies (ISGT)* (2012a), pp. 1–6

21. R. Halvgaard et al., Electric vehicle charge planning using Economic Model Predictive Control, in *2012 IEEE International Electric Vehicle Conference* (2012b)
22. S. Iacus, *Simulation and Inference for Stochastic Differential Equations: With R Examples*, vol. 1 (2008). https://doi.org/10.1007/978-0-387-75839-8
23. A.H. Jazwinski, *Stochastic Processes and Filtering Theory*. Mathematics in Science and Engineering (Academic, New York, 1970)
24. S.Ø. Jensen et al., Examples of Energy Flexibility in Buildings (2019a)
25. S.Ø. Jensen, J. Parker, A.J. Marszal-Pomianowska, Principles of Energy Flexible Buildings (2019b)
26. M.J. Jimenez, H. Madsen, Models for describing the thermal characteristics of building components. Build. Environ. **43**(2), 152–162 (2008). ISSN: 1873684x, 03601323. https://doi.org/10.1016/j.buildenv.2006.10.029
27. M.J. Jimenez, H. Madsen, K.K. Andersen, "Identification of the main thermal characteristics of building components using MATLAB. Build. Environ. **43**(2), 170–180 (2008). ISSN: 1873684x, 03601323. https://doi.org/10.1016/j.buildenv.2006.10.030
28. R. Juhl, J.K. Møller, H. Madsen, *CTSMR - Continuous Time Stochastic Modelling in R* (2016). arXiv:1606.00242 [stat.CO]
29. R.G. Junker, R. Relan, H. Madsen, Designing individual penalty signals for improved energy flexibility utilisation. IFAC-PapersOnLine **52**(4), 123–128 (2019). https://doi.org/10.1016/j.ifacol.2019.08.166
30. N.R. Kristensen, H. Madsen, S.B. Jørgensen, "Parameter estimation in stochastic grey-box models. Automatica **40**(2), 225–237 (2004). ISSN: 0005-1098
31. K.B. Lindberg, Impact of Zero Energy Buildings on the Power System. Doctoral Thesis. 2017–35. NTNU (2017)
32. D.G. Luenberger, An introduction to observers. IEEE Trans. Autom. Control **16**(6), 596–602 (1971)
33. H. Lund, E. Münster, Integrated energy systems and local energy markets. Energy Policy **34**(10), 1152–1160 (2006)
34. Z. Ma, B.N. Jørgensen, J. Parker, Stakeholders' Perspectives on Energy Flexible Buildings (2019)
35. J.M. Maciejowski, *Predictive Control with Constraints* (Prentice Hall, London, 2002)
36. C. Madina et al., Technologies and protocols: the experience of the three smartNet pilots, in *TSO–DSO Interactions and Ancillary Services in Electricity Transmission and Distribution Networks*, ed. by G. Migliavacca (2019), pp. 141–183. https://doi.org/10.1007/978-3-030-29203-4
37. H. Madsen, Statistically Determined Dynamical Models for Climate Processes. PhD Thesis. Technical University of Denmark (1985)
38. H. Madsen, *Time Series Analysis* (Chapman & Hall, Boca Raton, 2008)
39. H. Madsen, CTSM-R. http://henrikmadsen.org/software/ctsm-r/. Accessed May 2020
40. H. Madsen, J. Holst, Estimation of continuous-time models for the heat dynamics of a building. Energy Build. **22**, 67–79 (1995)
41. H. Madsen, P. Thyregod, Inhomogenous Markov models for the variations in cloud cover, in *International Conference on Statistical Climatology*, vol. 3 (1986), pp. 71–76. http://imsc.pacificclimate.org/proceedings/3IMSC.pdf.
42. H. Madsen, P. Thyregod, *An Introduction to General and Generalized Linear Models* (Chapman & Hall, Boca Raton, 2011)
43. H. Madsen, H. Spliid, P. Thyregod, Markov models in discrete and continuous time for hourly observations of cloud cover. J. Clim. Appl. Climatol. **24**, 629–639 (1985)
44. H. Madsen et al., Control of electricity loads in future electric energy systems, in *Handbook of Clean Energy Systems*, ed. by A.J. Conejo, E. Dahlquist, J. Yan (Wiley, New York, 2015)
45. H. Madsen, J. Parvizi, P. Bacher, Smart-energy operating-system - a framework for implementing flexible electric energy systems in smart cities, in SUSTAIN 2016, DTU, November 2016 (2016)

46. J. Møller, H. Madsen, From State Dependent Diffusion to Constant Diffusion in Stochastic Differential Equations by the Lamperti Transform (2010)
47. M. Morari, G. Stephanopoulos, Minimizing unobservability in inferential control schemes. Int. J. Control **31**(2), 367–377 (1980)
48. J.N. Nielsen, H. Madsen, Applying the EKF to stochastic differential equations with level effects, in *Automatica (Oxford)* (2000). ISSN: 18732836, 00051098. https://doi.org/10.1016/S0005-1098(00)00128-X
49. N. O'Connell et al., Benefits and challenges of electrical demand response: a critical review. Renew. Sustain. Energy Rev. **39**, 686–699 (2014). ISSN: 18790690, 13640321. https://doi.org/10.1016/j.rser.2014.07
50. M. O'Malley et al., Energy systems integration: defining and describing the value proposition. Contract **303**, 275–300 (2016)
51. G. Pannocchia, J. Rawlings, Disturbance models for offset-free model-predictive control. AIChE J. **49**, 426–437 (2003). https://doi.org/10.1002/aic.690490213
52. G. Pannocchia, M. Gabiccini, A. Artoni, Offset-free MPC explained: novelties, subtleties, and applications. IFAC-PapersOnLine **48**(23), 342–351 (2015)
53. M.A. Pinsky, S. Karlin, Continuous time Markov chains, in *An Introduction to Stochastic Modeling*, ed. by M.A. Pinsky, S. Karlin, 4th edn. (Academic, Boston, 2011), pp. 277–346. ISBN: 978-0-12-381416-6
54. C. Rasmussen et al., Semi-parametric modelling of sun position dependent solar gain using B-splines in grey-box models. Solar Energy **195**, 249–258 (2020). ISSN: 14711257, 0038092x. https://doi.org/10.1016/j.solener.2019.11.023
55. J. Rossiter, Model-based Predictive Control - A Practical Approach (2003)
56. A.Q. Santos, B.N. Jørgensen, Control Strategies and Algorithms for Obtaining Energy Flexibility in Buildings (2019)
57. H.A. Schløijter et al., Economic model predictive control for energy systems in smart homes, in *2019 IEEE Conference on Control Technology and Applications (CCTA)* (2019)
58. F.C. Schweppe et al., Homeostatic utility control. IEEE Trans. Power Appar. Syst. **99**, 1151–1163 (1980)
59. M.A. Shamim et al., An improved technique for global solar radiation estimation using numerical weather prediction. J. Atmos. Solar-Terrestrial Phys. **129**, 13–22 (2015). ISSN: 1364-6826. http://www.sciencedirect.com/science/article/pii/S1364682615000590
60. V. Sharma et al., Short term solar irradiance forecasting using a mixed wavelet neural network. Renew. Energy **90**, 481–492 (2016). ISSN: 0960-1481. http://www.sciencedirect.com/science/article/pii/S0960148116300209
61. C.A. Thilker, Optimization for Smart Energy Systems". Can be found at https://www.findit.dtu.dk. MA Thesis. Department of Applied Mathematics and Computer Science: Technical University of Denmark (2020)
62. T.O. Ting et al., State-space battery modeling for smart battery management system, in Lecture Notes Notes in Engineering and Computer Science, vol. 2210 (2014)
63. O. Tremblay, L.-A. Dessaint, A.-I. Dekkiche, A generic battery model for the dynamic simulation of hybrid electric vehicles, in *2007 IEEE Vehicle Power and Propulsion Conference* (2007), pp. 284–289
64. L. Wasserman, *All of Nonparametric Statistics*. Springer Texts in Statistics (Springer, Berlin, Heidelberg, 2006)
65. D. Yang, P. Jirutitijaroen, W.M. Walsh, Hourly solar irradiance time series forecasting using cloud cover index. Solar Energy **86**(12), 3531–3543. Solar Resources (2012). ISSN: 0038-092X. http://www.sciencedirect.com/science/article/pii/S0038092X12003039
66. Y. Zhang, V.I. Hanby, Short-term prediction of weather parameters using online weather forecasts, in *Proceedings: Building Simulation 2007* (2007)

Explanations Generation with Knowledge Models

Amr Alzouhri Alyafi, Patrick Reignier, and Stephane Ploix

1 Problem Statement and General Solving Principles

This work tackles the problem of generating causal explanations from a physical model-based EMS (as an example of knowledge models), to involve occupants in the loop with their EMS.

Due to the complexity and mathematical formalism of the knowledge models, they are not suitable for interactions with inhabitants: the intrinsic knowledge they contain is not directly intelligible. Human beings are trained, since elementary school, to think and to understand the world through causal representations like: what happens, when does it happen, what affects it, and what does it affect [1]. The equations of the EMS are elaborated by experts whose goal is to predict the evolution of physical variables, not to model causality. These equations, which are not designed to produce explanations, do not represent causal relationships explicitly. As a result, causal relationships cannot be determined automatically from the analysis of the equations. This causal knowledge is hidden inside the equations and should be extracted to be made explicit.

A. A. Alyafi (✉)
Univ. Grenoble Alpes, CNRS UMR5269, G2Elab, Grenoble, France
e-mail: amr.alyafi@gmail.com

P. Reignier
Univ. Grenoble Alpes, CNRS UMR5217, Grenoble INP, LIG Lab/APTIKAL Team, Grenoble, France
e-mail: patrick.reignier@imag.fr

S. Ploix
G-SCOP lab, Grenoble Institute of Technology, University of Grenoble Alps, CNRS UMR5272, Grenoble, France
e-mail: stephane.ploix@grenoble-inp.fr

S. Ploix et al. (eds.), *Towards Energy Smart Homes*,
https://doi.org/10.1007/978-3-030-76477-7_12

Causality from sensory data is difficult to model mathematically. Effects can be directly observed, but causal relationships cannot. Considering phenomena as events, a cause (C) always precedes the observation of an effect (E), but an effect (E) observed after (C) and correlated with it does not necessarily mean that (C) is the cause of (E). The "car allergic to vanilla ice cream" scenario illustrates this case [2]: a man used to buy ice cream after dinner for his family. He complained to General Motors that every time he bought vanilla ice cream, he had difficulties in starting the car engine (other ice cream flavors were fine). General Motors engineers finally found that the cause of the problem was vapor lock. Actually, it took less time to buy vanilla ice cream than for other flavors because of a dedicated counter. As a result, the engine remained too hot for the vapor lock to dissipate. The co-occurrence of buying vanilla ice cream and the car not starting did not mean that buying vanilla ice cream was the cause of the car failure.

As illustrated in Fig. 1, the co-occurrence (with a potential time delay dt) of two phenomena calls for several interpretations: precedence only (Fig. 1a), direct causal relationship (Fig. 1b), and consequences of a third phenomenon that may be outside of perception (Fig. 1c). For instance, having a flu may first cause fever and then coughing. Ignoring the existence of viruses may lead to the belief that the fever is the cause of coughing.

The work focuses on the generation of explanations about energy impact of user actions. Explanations occur in different ways [3] and for different reasons. One of the main motivation for having explanations is to be able to behave in a better way if similar events or scenarios arise in future [4]. Explanations usually rely on causal relationships. There are at least four kinds of causal explanations: common cause, common effect, linear causal chains, and causal homeostasis (cyclic causal relationships) [5]. According to [6], explanations are ubiquitous, come in a variety of forms and formats, and are used for a variety of purposes. Still, the common feature about most explanations is their limitation. For most natural phenomena and many artificial ones, the full set of relations to be explained is complex and far beyond the grasp of any one individual.

Cooperation between inhabitants and EMSs can be built using explanations as a powerful and intuitive tool to transfer knowledge with expert systems. This work proposes two approaches to generate causal explanations from EMSs: differential

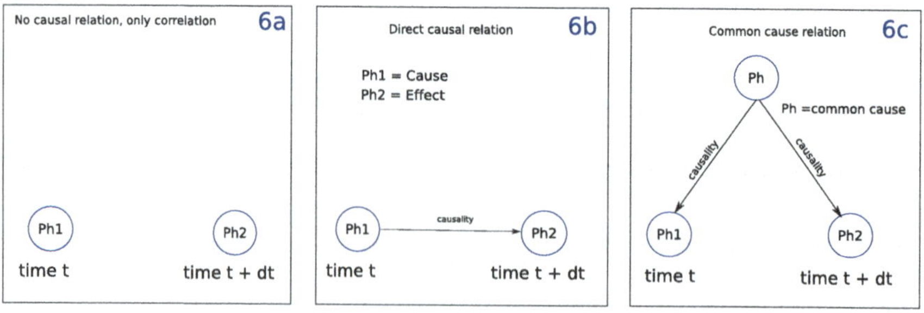

Fig. 1 Causal relationships between co-occurrent phenomena

explanations and direct explanations. These two approaches rely on a model-based EMS (knowledge model).

The next section will describe the generation of explanations using the described energy models and data collected from the previous chapter (Chap. 4).

2 Generating Explanations

From the model-based EMS (presented in Chap. 4), Figs. from 2, 3, 4, and 5 present different outputs of the system for May 5, 2015 throughout the day from 8 a.m. to 8 p.m. (normal working hours for the office). In Fig. 2, it can be seen on the left the different window actions registered on that day and the recommended actions generated by the system (window-opening-best); on the right, there are the same actions but for the door opening. Figure 3 represents the different simulated intermediate variables (heat flow on the left and air flow on the right). Figure 4 presents the solar radiation on the left and the estimated occupancy on the right. Figure 5 left presents the different inside temperature, outside temperature, corridor temperature, and the best temperature simulated when the occupant follows the recommended actions. The right part is associated to the air quality (CO_2 concentrations).

From this first simple example (office with only one thermal zone), we can already observe:

1. The difficulty for the user, to understand his/her environment. He/She needs to know and correlate the different variables present in the environment to understand how they are impacting his/her comfort criteria (here, the effects).
2. The importance of the occupant's actions. The simulated effects in Fig. 5 show how two simple actions like opening the door and window have a considerable impact on the comfort criteria.

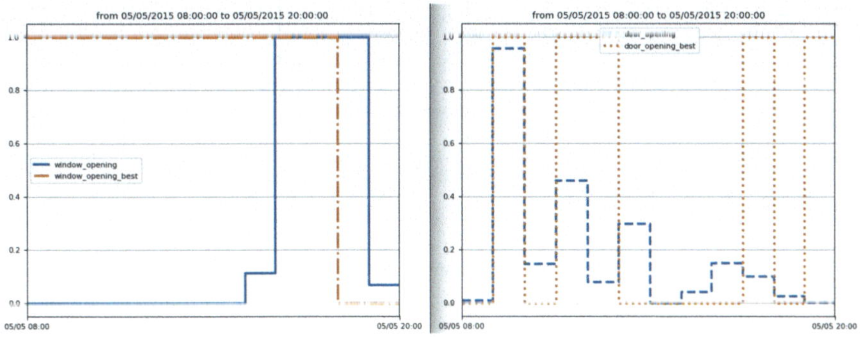

Fig. 2 Window (left)/door (right) opening (measured and recommended)

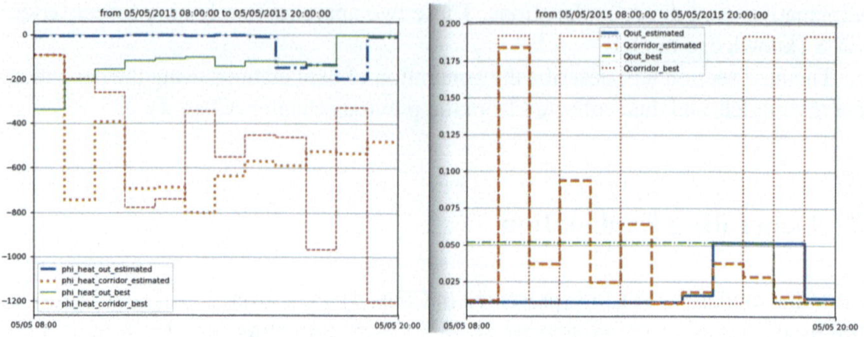

Fig. 3 Different estimations of the heat flow (left) and air flow (right)

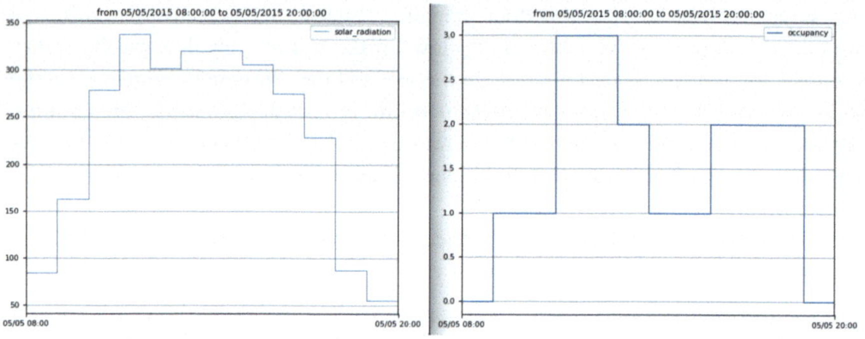

Fig. 4 Solar radiation (left) and estimated occupancy (right)

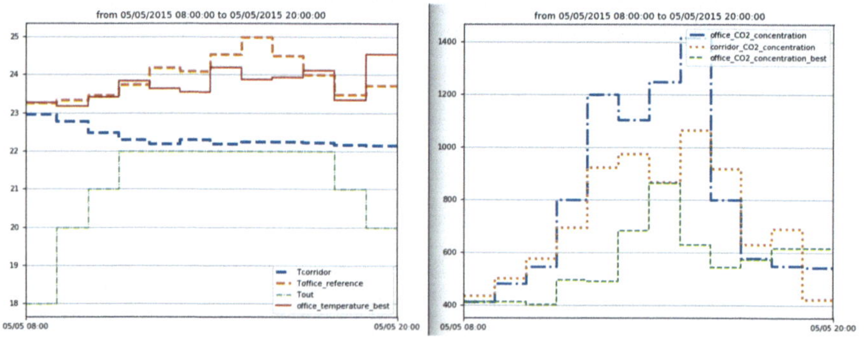

Fig. 5 Outside, corridor, inside, best (with recommended actions) temperature (left) and corridor, inside, recommended air quality (right)

3. The cognitive dissonance problem (between occupant's goals and actions). Inhabitants may act in contradiction to their goals (their comfort criteria, cost, ...) because they do not understand the impact of their actions. In this example, the occupant's goal is to maintain a good level of air quality, yet he/she closed the window in the morning contrary to what he/she should have done to reach his/her objective.

This shows why it is very important for occupants to understand the impact of their actions, and here explanations can be very helpful in doing that.

The next section describes differential explanations and how they are generated.

2.1 Differential Explanations

Differential explanations are constructed by analyzing the difference between two scenarios. Scenarios can be measured, simulated, or imagined; they can be set in the past, present, or future. For example, the occupants can compare a past day's actions with an imaginary plan of actions and learn the impact of those actions on their comfort criteria. The scenario can also be a comparison between what the user has done and the recommended actions generated by the EMS to see what he can gain if he applied the recommended actions, as in Fig. 7. This comparison can also be done between recommended actions for a future day and what the inhabitant likes to do, or simply any plans of actions for that day. This comparison with the recommended actions is important as it could play a role in persuading occupants to change their behavior and follow the recommended one. The comparison includes the set of actions, the intermediate variables, and the effects. Intermediate and effects variables emphasize the consequences of the difference between the user's actions and the recommended actions for instance and form the basis for the explanations.

Effect variables are impacted by changes in the actions and also by the context variables. To clearly understand the difference between occupants' different actions, scenarios that are compared must have a similar context (as defined in the next paragraph) to visualize the actions' effects. Otherwise, the different effects may be related to the difference in the context variables and not the occupants' actions. For instance, opening a window in summer or in winter does not have the same consequences.

To reduce the risk of false causal relationships or circular explanations, the available variables are classified into four groups: actions, context, intermediate, and effects variables, as shown in Fig. 6. From actions to effects corresponds to the natural tendency of causal relations between the groups of variables depending on their role. This is considered as expert abstract knowledge [7].

More precisely, the different groups are:

1. Context group: It contains all the uncontrollable variables that the system needs to take into consideration like the outside temperature T_{out}, the temperatures in the neighbor zones T_{cor}, the number of occupants, etc.

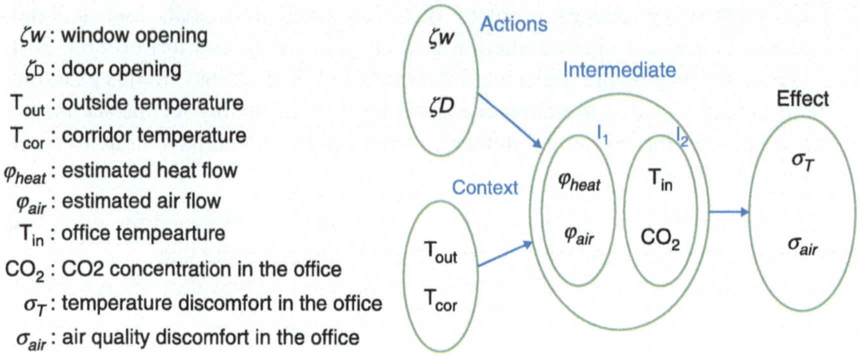

Fig. 6 General schema to generate explanations

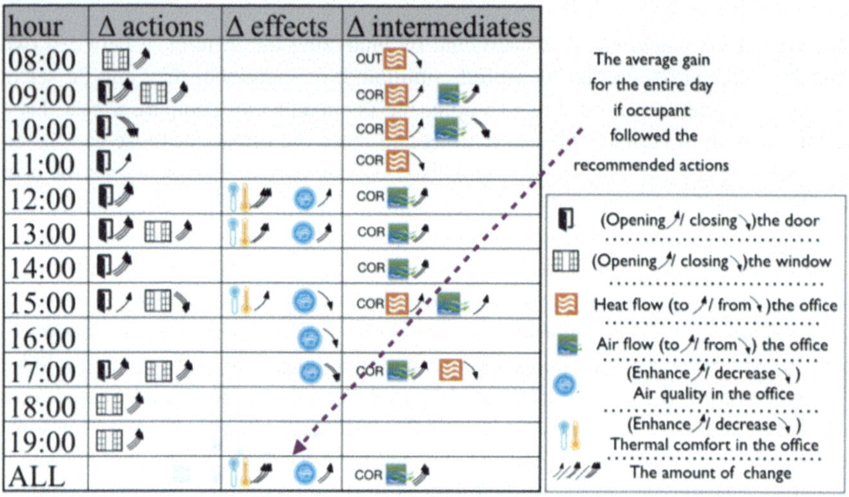

Fig. 7 Differential explanations through difference between the historical scenario and a Pareto-optimal scenario for 5 May 2015. A door or window with an uprising arrow means it should have been opened longer for the corresponding period of time

2. Action group: It contains all the different actions that the system can propose to the occupants to enhance their comfort levels (like opening the window ζ_w and opening the door ζ_D).
3. Effect group: It contains the variables that will be directly experienced by the occupants like thermal dissatisfaction σ_T and air quality dissatisfaction σ_{air}.
4. Intermediate group: It contains different sub-groups and different levels for multiple intermediary variables. These variables are either measured like the indoor temperature T_{in} and the indoor CO_2 concentration C_{in} or estimated like the heat flow φ_{heat} and the air flow φ_{air}.

In Fig. 7, the differential explanations are illustrated in a table where the first column represents the difference of occupants' actions with what the occupant should have done according to the recommended plan, as shown in Eq. (1).

$$A_k^\star - \tilde{A}_k = \Delta A_k, \tag{1}$$

where $A_k^\star$ represents the action calculated by the energy model (or any other scenario) at instant k, while $\tilde{A}_k$ represents a measured occupant's action at the same instant. The variable, k, can take any integer value representing hours in the day. In this example, $k = 8, \ldots, 20$ because it focuses on the time where occupants are potentially present (daytime period). At 8 a.m., for instance, the inhabitant should have opened the window longer during this time slot. At 4 p.m., the user behaves according to best scenario.

The second column presents the effects (Fig. 6), like in the thermal comfort and the air quality. This is given by Eq. (2).

$$E_k^\star - \tilde{E}_k = \Delta E_k, \tag{2}$$

where $E_k^\star$ represents the calculated effect by the system at instant k, while $\tilde{E}_k$ represents the measured effect. The right-hand side of Eqs. (1) and (2) denotes the difference in actions of the occupants and the resulting difference in the effect at the k-th instant, respectively. This differential explanation does not explicitly expose causal relationships but only differences of actions and effects between two different scenarios.

To add more information, the intermediate variables are added in the fourth column of Fig. 7. Those variables are explicitly extracted from the system.

The last row, labeled *ALL*, represents the overall gain or loss in the comfort criteria throughout the day, to generate a small summary and give the inhabitant an indicator of their enhancement in general, for the entire day if they follow the recommendations.

When computing the differential explanation, it is necessary to transform quantitative variable values into qualitative ones for a better understanding by the occupants and to define the qualitative distance. For instance, telling the occupant that closing the door at 2 p.m. will cause a large decrease in the airflow and that he will obtain a significant decrease in the air quality level is easier to understand than telling him/her that a difference in airflow of 30% will lead to a difference in CO_2 concentration of 400 ppm. The transformation from quantitative to qualitative data here is done by dividing the value domain of a variable into 7 sub-domains (3 positive, 3 negative, and 1 no-change levels). Those levels were chosen from human feelings according to their impact on the occupants.

The levels for variations in thermal dissatisfaction are determined with the help of an expert:

$$\Pi^{T}_{-0.25,-0.15,-0.05,0.05,0.15,0.25}\Big(\Delta\sigma_T^k(T_{in})\Big).$$

The levels for the variations in air quality dissatisfaction are given by

$$\Pi^{C_{CO_2}}_{-0.2,-0.1,-0.05,0.05,0.1,0.2}\Big(\Delta\sigma_{air}^k(C_{in})\Big).$$

The levels for the variations in the opening of the door and the window are given by

$$\Pi^{opening}_{-0.7,-0.5,-0.2,0.2,0.5,0.7}(\Delta\zeta_D)$$

$$\Pi^{opening}_{-0.7,-0.5,-0.2,0.2,0.5,0.7}(\Delta\zeta_w).$$

The arguments of each of these discretization functions describe the difference of the measured quantity with the proposed optimal value of the quantity.

Except for the no-change level, where arrows are omitted, 1 to 3 arrows have been used to represent the associated sign of variation (arrows direction) and intensity (number of arrows). For instance, in Fig. 7, the logo of window with three adjacent upward arrows means that the occupant should have opened the window for a much longer period of time during the corresponding time period. Algorithm 1 presents the different steps needed to obtain the differential explanations.

It can be seen that the differential explanations are much easier to understand than the analysis of the 13 plotted curves (Figs. 2, 3, 4 and 5) where the inhabitant has to correlate the different actions, effects, and intermediate variables. With a differential explanation, it is easy for an occupant to identify the actions that need to be modified, and monitor the difference gained with respect to different criteria while at the same time using the intermediate variables as elements of understanding.

Algorithm 1 Tabulating differential explanations

Require:
Scenario 1: $\tilde{\mathcal{A}}, C$.
Scenario 2: $\mathcal{A}^{\star}, C$.
Ensure: $\mathcal{T}$: table for differential explanations
Use physical model to get $\tilde{\mathcal{I}}$ and $\tilde{\mathcal{S}}$: $\tilde{\mathcal{A}}, C \xrightarrow{\tilde{\mathcal{I}}} \tilde{\mathcal{S}}$
Use physical model to get $\mathcal{I}^{\star}$ and $\mathcal{S}^{\star}$: $\mathcal{A}^{\star}, C \xrightarrow{\mathcal{I}^{\star}} \mathcal{S}^{\star}$
for $k = t_{start}$ to t_{end} **do**
  $row = k - t_{start} + 1$
  $\mathcal{T}_{row,1} \leftarrow k$
  Obtain $\zeta^{\star,k}$ from $\mathcal{A}^{\star}$ Different actions at instance k
  Obtain $\tilde{\zeta}^k$ from $\tilde{\mathcal{A}}$
  Calculate $\Delta\zeta^k = \zeta^{\star,k}$ - $\tilde{\zeta}^k$
  Obtain $\sigma^{\star,k}$ from $\mathcal{S}^{\star}$ obtain user satisfaction
  Obtain $\tilde{\sigma}^k$ from $\tilde{\mathcal{S}}$
  $\Delta\sigma^k = \sigma^{\star,k}$ - $\tilde{\sigma}^k$
  Obtain $Q^{\star,k}$ from $\mathcal{I}^{\star}$
  Obtain $\tilde{Q}^k$ from $\tilde{\mathcal{I}}$
  $\Delta Q^k = Q^{\star,k}$ - $\tilde{Q}^k$
  $\mathcal{T}_{row,2} \leftarrow$ Qualitative transformation of $\Delta\zeta^k$
  $\mathcal{T}_{row,3} \leftarrow$ Qualitative transformation of $\Delta\sigma^k$
  $\mathcal{T}_{row,4} \leftarrow$ Qualitative transformation of ΔQ^k
end for
$\mathcal{T}_{(row+1),3} \leftarrow$ Qualitative transformation of average values of $\Delta\sigma$
$\mathcal{T}_{(row+1),4} \leftarrow$ Qualitative transformation of average values of ΔQ

2.2 Differential Explanations with Contextual Causality

The differential explanation is yielding a list of behavior modifications (opens the door for a longer period, for instance) with a list of impacts. However, there are limitations with such descriptions. In particular, there is not a direct link between an action modification and its corresponding impact. There is no explicit causal relationships between actions and effects. Buildings have inertia, i.e., energy dynamically stored in their structure. This inertia causes a delay and has a smoothing effect on different changes in the building preventing a rapid degradation or augmentation in temperature. Inertia is also present in the room volume for the CO_2 concentration. Thus, an impact may not be on the same time slot as the action: occupant actions might have a delayed impact.

In Fig. 8, closing the door at 10am does not have an immediate impact, but it does have a strong impact on the air quality at 12 p.m.: it is a calculated delayed impact.

Second, not all the proposed action modifications have the same importance; some of them have a limited impact and could be skipped if necessary (the inhabitant might not want, for instance, to interrupt his/her current activity to close the window). But some of them should be followed because of their high impact on the selected criteria (like the previous door example having a strong impact on the air quality).

To evaluate the impact of the ith action at the jth quantum time, i.e., $\mathcal{A}_i^j$, the difference between the following two scenarios needs to be calculated: (1) a scenario (the recommended one or any other scenario for that day) ($\mathcal{A}^\star$) and (2) a second scenario (like a measured one) ($\hat{\mathcal{A}}^j$). But when calculating this difference (differential explanation), we first replace the ith action at the jth quantum time

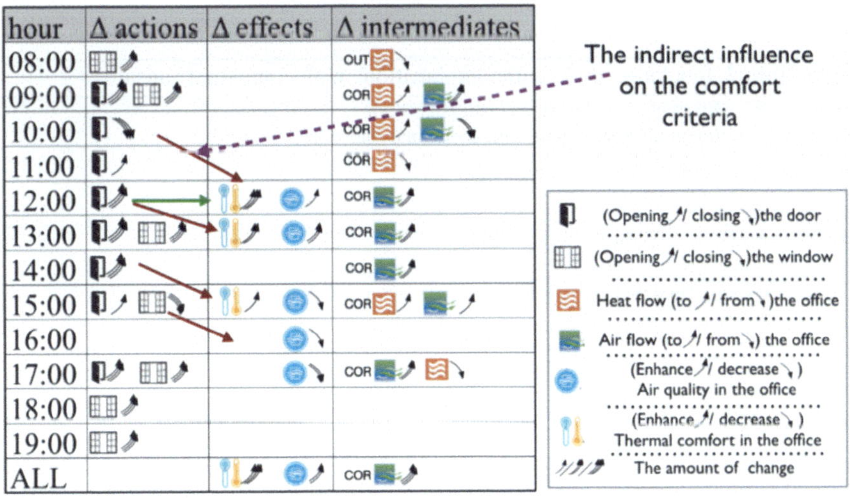

Fig. 8 Differential explanations with contextual causality

($\mathcal{A}_i^{\star j}$) by the action on the same slot from the second scenario (like what the occupant has done) ($\tilde{\mathcal{A}}_i^j$) as shown in Eq. (3). Both scenarios are simulated using the physical model of the office. The difference between the effects indicates the impact of not performing the first scenario action $\mathcal{A}_i^{\star j}$.

$$\hat{\mathcal{A}}^j = \left\{\mathcal{A}_i^{\star k}; \forall k \neq j\right\} \cup \left\{\tilde{\mathcal{A}}_i^j\right\}$$
$$\text{where,} \mathcal{A}_i \in \{\zeta_W, \zeta_D\}. \tag{3}$$

It is interesting to note that when the differences between these two scenarios are considered, Eq. (4) follows, i.e., the difference in actions is zero for all time slots except the jth time slot and at the kth time slot, and the difference is identical to the difference between the actual and recommended scenarios. Hence, by considering change in actions between these two scenarios, the change in the ith action at the jth hour can be isolated and its effects can be investigated.

$$\Delta\mathcal{A}_i^j = \mathcal{A}_i^{j\star} - \hat{\mathcal{A}}_i^j = \begin{cases} 0, & \forall k \neq j \\ \mathcal{A}_i^{\star j} - \tilde{\mathcal{A}}_i^j, & k = j. \end{cases} \tag{4}$$

Using Algorithm 2, the impact of the ith action at the jth hour ($\mathcal{A}_i^j$) can be obtained. For a complete list of impacts, Algorithm 2 has to be repeated for every ith kind of action ($\mathcal{A}_i$) and for every jth time slot. For instance, opening the door between 12 a.m. and 1 p.m. not only impacts the air quality and thermal comfort in the same time slot but also impacts the air quality and thermal comfort in the succeeding time slot (1–2 p.m.). This is also an example of a common cause leading to multiple effects.

Algorithm 2 Tabulating differential explanations with contextual causality

Require: $\mathcal{T}_0$: Differential explanations from Algorithm 1
Ensure: $\mathcal{T}$: differential explanations with contextual causalities

1: From $\mathcal{T}_0$ get: $\tilde{\mathcal{A}}, C \xrightarrow{\tilde{I}} \tilde{\mathcal{S}}$
2: From $\mathcal{T}_0$ get: $\mathcal{A}^\star, C \xrightarrow{I^\star} \mathcal{S}^\star$
3: **for** $j = t_{start}$ to t_{end} **do**
4:     **for** $k = t_{start}$ to t_{end} **do**
5:         **if** $k = j$ **then** then $\mathcal{A}^{\star,j} = \tilde{\mathcal{A}}^k$
6:         **end if**
7:         apply differential explanations Algorithm 1 to obtain $\mathcal{T}_k$
8:         Compare between $\mathcal{T}_k$ and $\mathcal{T}_0$ if there is any difference insert an arrow between the between the $\mathcal{A}^{\star,j}$ and the different satisfaction
9:     **end for**
10: **end for**

2.3 *Model Fragment*

The effect variables are caused through intermediate variables like air flow and heat flow.

Using the equations of the energy models, it is possible to generate cause and effect relations between actions and final effects, but the causality between the different levels of the intermediate variables and the final effects are indiscernible because their changes cannot be monitored with the energy models [8], as shown in Fig. 9.

Integrating relations between the intermediate and final variables is important to provide occupants with complete explanations. To overcome this difficulty, we propose the model fragment, inspired from GARP3.[1] GARP3 is a workbench that allows modelers to build, simulate, and inspect qualitative models of systems behavior [9].

A model fragment represent potential causalities as well as impossible ones. It is generic expert knowledge based on physical laws. For instance, heat flow may have an influence on air temperature but not on CO_2 concentration, as shown in Fig. 10.

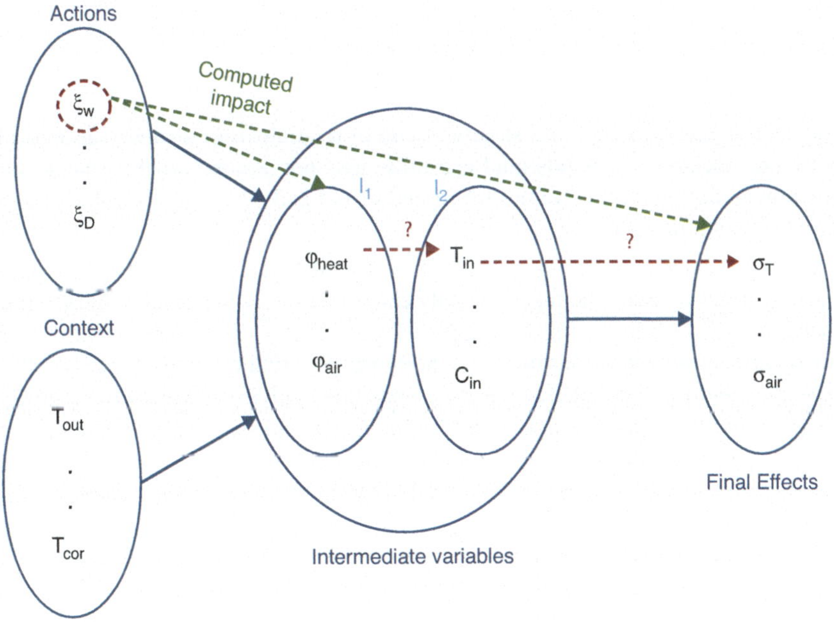

Fig. 9 Undetected causality between intermediate variables and effects

[1] https://ivi.fnwi.uva.nl/tcs/QRgroup/QRM//software/.

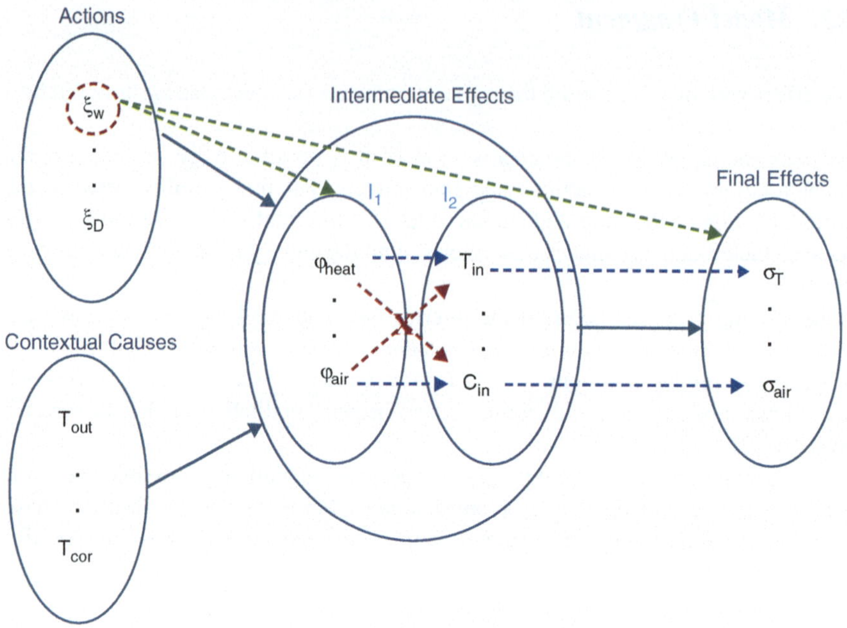

Fig. 10 The representation of the potential and forbidden causalities

Potential causality (pc) is a structural causal relation from a cause variable v1 to a target variable v2. A potential causality does not assume anything about the direction of variation of the values for v1 and v2 (v1 and v2 are labels with domains dom(v1) and dom(v2)).

It is represented as: $v_1 \xrightarrow{pc} v_2$ or $v_1 \nrightarrow^{pc} v_2$ for the forbidden ones. The forbidden causality helps to avoid the false causality caused by the co-occurrence of different events.

Conditional potential causality: A conditional potential causality is activated by a specific condition modelled as a logic proposition applying to values of variables. For example, the causality link between the heater and the inside temperature is correct only when the heater is ON; when the heater is OFF, it is a forbidden causality even if a co-occurrence appeared between the heater temperature and the inside temperature.

By integrating calculated causalities and potential ones, a full causal graph for the whole system can be done. Part of this diagram is represented in Fig. 11 where five categories of nodes appear, viz. actions (red), context (yellow), air flow (blue), heat flow (orange), and effects (green). It can be seen that action nodes have several outward edges and several paths from actions eventually leading to some effects. For example, opening the door between 9 am and 10 am (ζ_D^9) not only leads to thermal comfort (σ_{temp}^9) through heat flow from the corridor (ϕ_n^9) but also leads to air-quality-based comfort between 11 a.m. and 12 p.m. (σ_{air}^{11}).

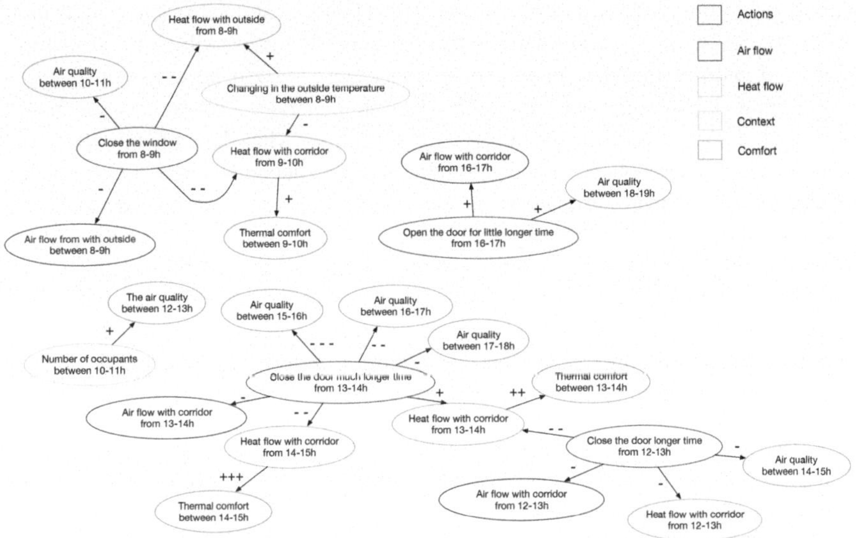

Fig. 11 Causal graph

Thus, differential explanations allow the occupants to have an explanation based on the cause–effect relations of their actions, and they may decide to change their routines or learn from their historical actions. Chapter six presents how the generated explanations can be transformed into natural language to be shared with occupants. The next section describes the second form of the causal explanations: "direct explanations," based on Bayesian networks. It will also present an initial proof of concept.

3 Direct Explanations

As we have seen in the previous section, a differential explanation is a relative explanation that highlights the consequences of a behavior change in relation to a behavior (scenario) taken as a reference. We will now describe the construction of an explanation no longer based on this reference scenario but establishing causal relations between the actions (and no longer the changes in action) and the physical quantities: the direct explanation.

Direct explanations are based on Bayesian networks learned from many different simulations generated by the EMS for a day. These different simulations with the model fragments (presented earlier) are used as an input for a Bayesian search algorithm to learn the Bayesian network structure.

Bayesian networks are a member of a vast class of models, ones that can be used to describe nested, acyclic statistical models of virtually any kind of nonpathological joint probability distribution [10]. Their signature characteristic is their ability to encode *directional* relations that can represent cause–effect relationships, compared to other graphical models that cannot, e.g., Markov networks [11].

Learning a Bayesian network from data in general involves two sub-tasks: learning the structure of the system (i.e., determining what depends on what) and learning the parameters (i.e., the strength of these dependencies). Learning the parameters for a given structure from a complete data set is trivial (the observed frequencies are optimal with respect to the maximum likelihood estimation [12]). We will focus on the structure learning aspect, which is a central for explanations.

The Bayesian search structure learning algorithm is one of the earliest algorithms. It was introduced in [13] and was refined somewhat in [14]. It follows essentially a hill climbing procedure (guided by a scoring heuristic) with random restarts. The algorithm has the following parameters:

Max Parent: count limits the number of parents that a node can have because the size of conditional probability tables of a node grows exponentially by the number of the node's parents.

Iterations: sets the number of restarts of the algorithm. Generally, the algorithm is searching through a hyper-exponential search space, and its goal can be compared to searching for a needle in a haystack. Restarts allow for probing more areas of the search space and increase the chance of finding a structure that will fit the data better. The number of iterations gives an idea of how long the algorithm will take when the number of iterations is large.

Sample size: is a factor in the score calculation, representing the inertia of the current parameters when introducing new data.

Seed: (default 0) is the initial random number seed used in the random number generator. A seed equal to zero (the default) makes the random number generator really random by starting it with the current value of the processor clock.

Link Probability: (default 0.1) is a parameter used when generating a random starting network at the outset of each of the iterations. It essentially influences the connectivity of the starting network.

Max Time: (seconds) (default 0, which means no time limit) sets a limit on the runtime of the algorithm. It is a good idea to set a limit for any sizable data set so as to have the algorithm terminate within a reasonable amount of time.

Accuracy: as a scoring function (default OFF). When checked, the algorithm will use the classification accuracy as the scoring function in search for the optimal graph.

The algorithm produces an acyclic directed graph that gives the maximum score. The score is proportional to the probability of the data providing the structure, which, assuming that the same prior probability was assigned to any structure, is proportional to the likelihood of the structure given the data. The algorithm allows the injection of expert knowledge in the form of potential/forbidden causalities (model fragments). This is helpful in organizing the variables and eliminating the correlation between variables and extracting causalities.

To apply the Bayesian search algorithm in the case study, presented in Fig. 12, 300 simulations were obtained from a genetic algorithm used by the EMS to optimize actions and find the recommended actions. This might have had an effect on the learning of the structure as the simulations are not completely random. They are oriented by the genetic algorithm that searches the best set of actions according to occupants' preferences (more details about the genetic algorithm used are in Appendix 1). The model fragments are presented in the form of potential causalities, as in Fig. 13. For instance, the heat flow cannot be the cause for the change in the CO_2 concentration.

Direct explanations is a set of probability cause–effect relationships between the variables in general. However, direct explanations cannot present the cause–effect relationships for each action at any time. They also cannot represent the delayed impact of each action on the different criteria. Another limitation is that for any 1 day the algorithm cannot learn the impact of the contextual variables on the comfort criteria because there is not enough variation in the context variables to detect their causality. The inertia in the buildings limits the learning of the Bayesian network, as the impact of the action is delayed by the inertia, and the search algorithm cannot learn that. One possible solution that we must investigate is to use the dynamic Bayesian network to overcome this difficulty.

The next section describes the field studies that had been done to validate the explanations and their utility.

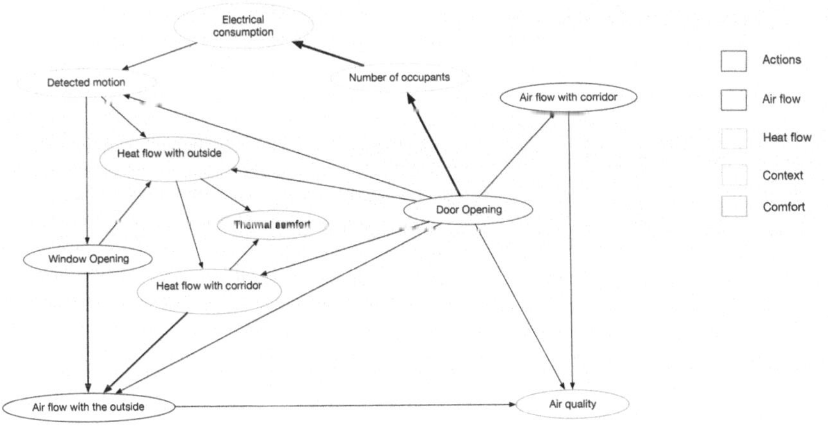

Fig. 12 Direct explanations

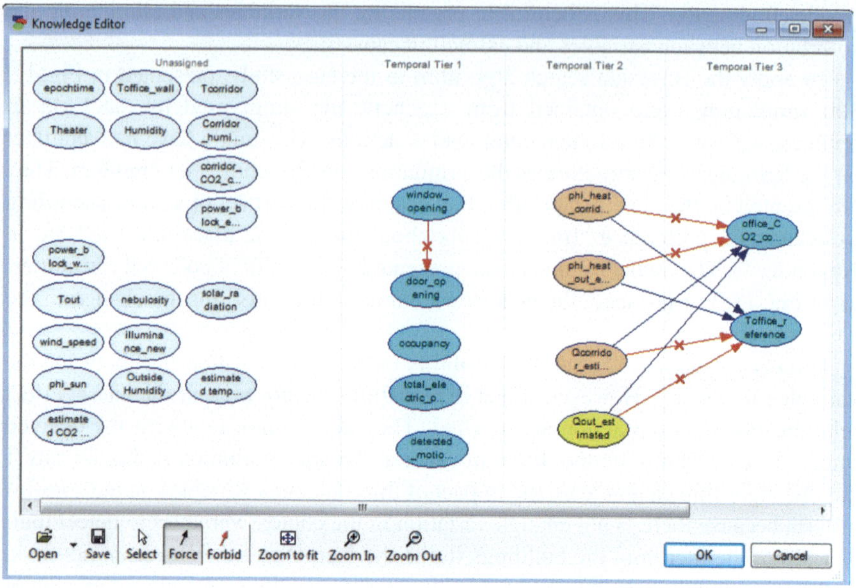

Fig. 13 Model fragment in the form of potential causalities (figure is realized using the BayesFusion (http://www.bayesfusion.com) program)

4 Validation Scenario for the Generated Explanations

4.1 Context and Goals

Introduction to the validation scenario:

> Homes are complex systems where different phenomena are present. Occupants generally would like their comfort criteria (like thermal comfort and air quality) to be at the optimum level without increasing their consumption or energy bill. Yet, within these complex systems, occupants have difficulties in determining the optimal set of actions they should perform within a specified context or being able to estimate the impact of their actions. For this reason, this work proposes an assisting tool ('the explanation generator') to help occupants to understand the impact of their actions and the cause-effect relations between different variables within different contexts.

The objective is to measure how the causal explanations proposed in this work might assist occupants to better understand their homes. For that, three criteria need to be satisfied. Are the generated explanations:

1. Intelligible?
2. Credible?
3. Easy for the user to understand them?

4.2 Method

The method aims to evaluate if the proposed tool to generate explanations can help the users to understand their energy systems or not. To do that, the first step (a training phase) immerses the participant in the scenario: an office (the case study) with an energy system. This allows us to evaluate their initial knowledge of the system and to aid them in thinking deeper into the problem. The generated explanations are provided to the participants. Then, they are given the second task to evaluate the utility of those explanations. Finally, a registered semi-structured interview takes place with each participant to get their feedback about:

- If the participant finds the explanations intuitive or not
- If the explanations are clear or not
- If the participant is ready to adopt them or not
- The form of explanations
- Other comments

Their feedback is analyzed to determine their understanding of the different phenomena.

4.3 Participants

The 10 participants were between 18 and 65 years old, 5 women and 5 men. The participants were from different backgrounds (scientific and non-scientific), none of them from the domain of the research and never worked on the problem of energy management. They were also volunteers and were not paid for their participation.

4.4 Independent Variables

In order to get a valid comparison, all the tests were done in an office in the G-SCOP laboratory considering the same day: 05/05/2015 (chosen randomly from the set of days available in our sensors database). All the participants were asked to perform the same actions: opening of the window and door. The measured variables were the proposed programs for the opening of door/window throughout the day. The results were evaluated using a physical model and evaluated by how much they improved the comfort of occupants.

The participants were interviewed by a researcher from the human and social sciences (Hélène Haller[2]) not directly involved in this research. All interviews were recorded via two microphones and then analyzed.

[2]helene.haller@umrpacte.fr.

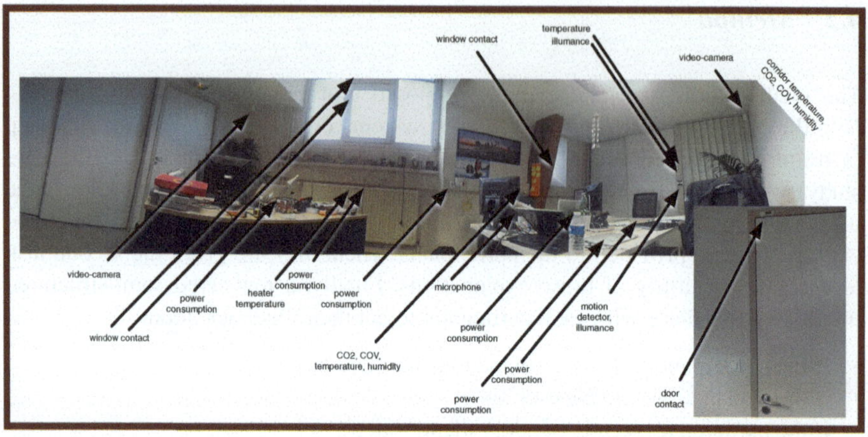

Fig. 14 Office with sensors: the only possible actions are window and door openings during summer

4.5 *Tasks*

The participants were asked to perform two tasks. In Task 1 (T1), the participants were instructed to look at data from the 5th of May 2015 with the context variables for the case study. Then they were asked to determine the best window/door position to enhance their comfort. Task 2 (T2) was repeating the first task after having been given the system's explanations. Then, a short semi-structured interview with each participant was done to see if they had any preference or any comments on the explanations.

4.6 *Scenario of the Interview*

This section describes the exact introductory speech and questions presented to the different participants:

> We thank you for agreeing to participate in this experiment. It is part of a doctoral work on the realization of an energy management system. This system is based on the generation of explanations, containing tips for users. Today, we wish to observe whether the proposed explanations are understandable and acceptable for you.
>
> For this, we offer a scenario, during which we will ask you several questions. (Give Fig. 14—office and sensors.) The questions we will ask you are related to this office, located in the laboratory G-SCOP. This office is equipped with different sensors. The only possible actions on this room are to open or close the window and the door. Now imagine that we are May 5th 2015. (Give image Fig. 15—office registered sensor data for the 5th of May 2015.) Here are the data provided by the sensors installed in the office, data relating to the air quality, the number of occupants, the outside temperature (context), the corridor temperature, and the office temperature at 8 am.

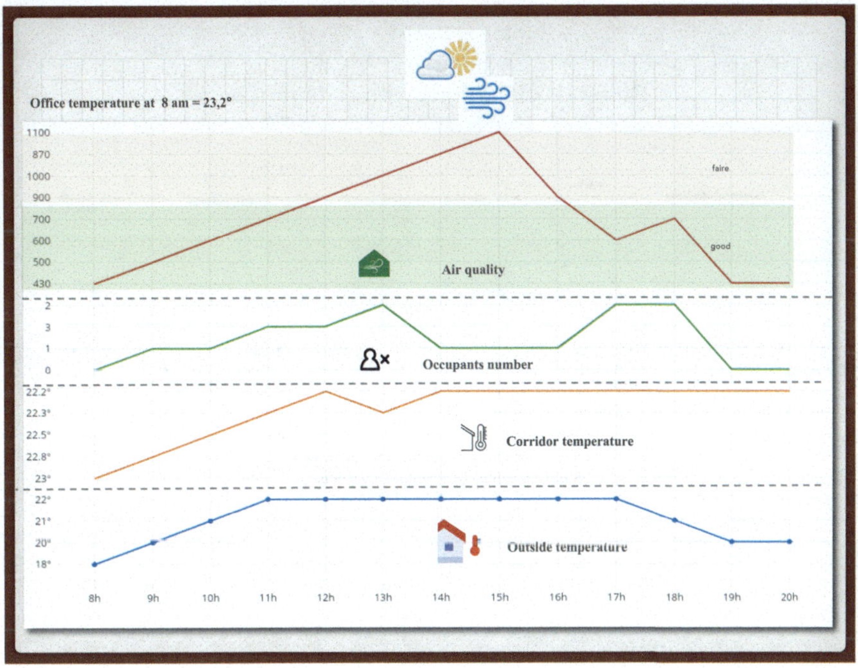

Fig. 15 Office registered sensor data for the 5th of May 2015

Given these data, what are in your opinion the best sequences of actions (opening door/window) to get the maximum comfort in this office? Generally, comfort is defined by a temperature between 21° and 23 °C and a CO_2 concentration as low as possible. I will now show you how schematically the generation of direct explanations works. (Give picture Fig. 16—model for direct explanations.) I will let you read it.

Given this pattern, what would you do to get the maximum comfort in this office? I remind you that comfort is generally defined by a temperature between 21 °C and 23 °C and a CO_2 concentration as low as possible.

- Do you find that the direct explanations I showed you were logical or not? Why?
- Did those explanations help you better understand how to get maximum comfort in the office? Why?
- Do you find this type of explanation (direct explanations) intuitive/understandable for you?
- Do you think that these explanations could be better presented? If yes, how?

(Give picture Fig. 17—Recommended solution.) Here are the actions recommended by the system for maximum comfort. Do you understand these recommendations? I will now show you a differentiated causal explanation. (Give picture Fig. 18—Differential explanations.) Between 8 am and 9 am, if you left the window and the door open longer, there would have been a light heat input from the outside and from the corridor, which will improve your thermal comfort in the 9h–10h time slot and improve the air quality in the time slot 10h–11h.

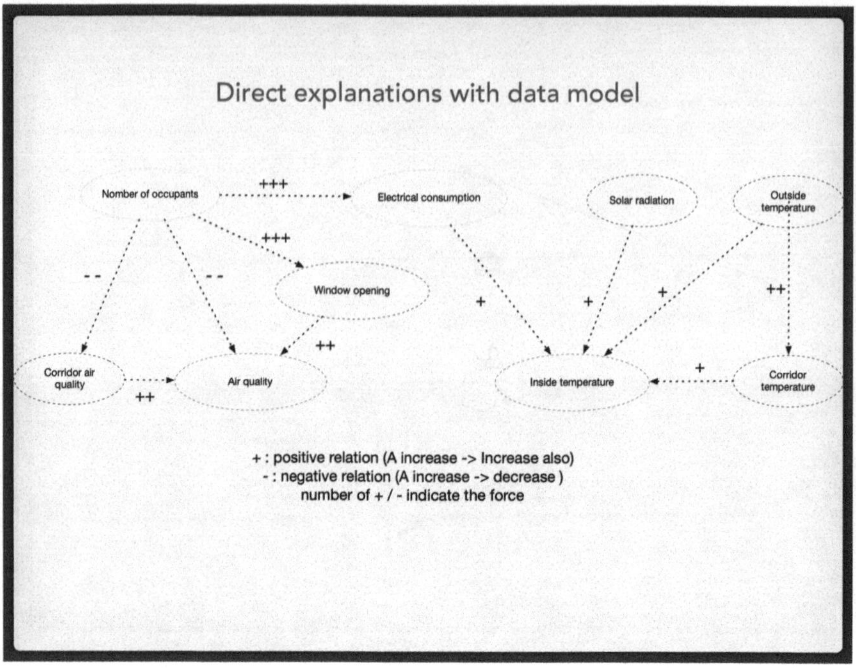

Fig. 16 Direct explanations

- Do you find these explanations logical? Why?
- Have these explanations helped you better understand how to get maximum comfort in the office? Why?
- Do you find this type of explanation (causal explanations) intuitive?
- Do you think that these explanations could be better presented?
- Have explanations and understandings encouraged you to follow the recommendations of the system?
- Have these explanations increased confidence in the system?
- Could you give an estimate of the improvement of your understanding of the environment following these explanations?

Before concluding this interview, I would like to ask you two questions:

- By comparing the two types of explanations (direct explanation/differentiated causal explanation), which one do you find the best? Why?
- If you had these explanations at home, would you use them? If yes, how often? In what occasion?

4.7 Results

In the first task (T1), participants in general were a little bit lost and tried to imagine scenarios that were far from what the system would recommend them to do (the best solution). In general, they based their decisions on their habitual activities or certain

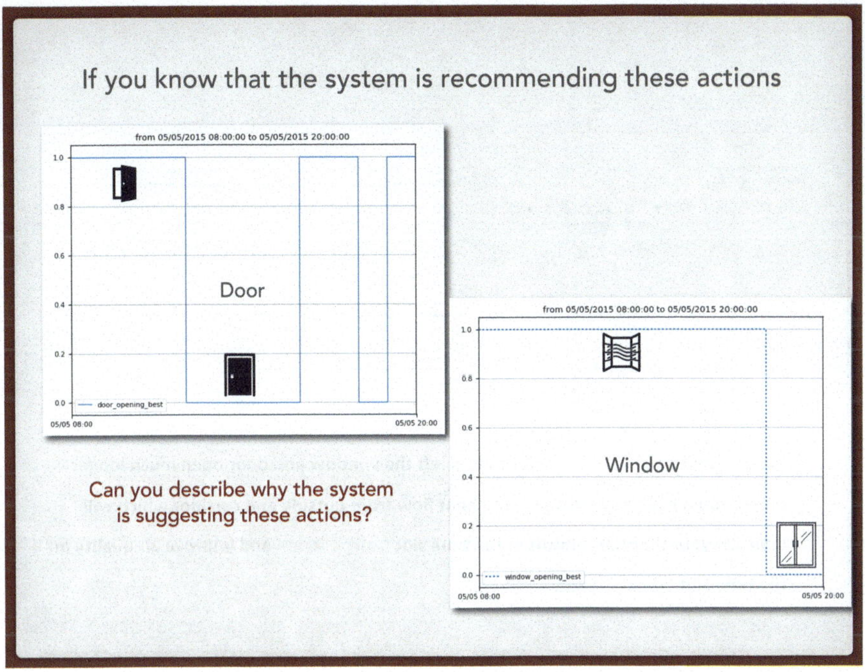

Fig. 17 Recommended solution

beliefs (like "we should open the window each morning"), except for one participant who tried to analyze the context data and seemed to have a better understanding. After showing the direct explanations, participants were asked to complete the task (T2). A clear improvement was noticed, as they (from what they reported) started to know what were the variables affecting the comfort variables (inside temperature, air quality). So they had started to realize the relation between actions and resulting effects.

Then, the differential explanations were introduced with the optimal solutions. Finally, the participants were asked to compare the differential and direct explanations and give their opinions.

Participants in general did appreciate the explanations, and most of them repeated the same words "I learned new things" or "I didn't know that before" or "It confirmed what I thought." Around 10% of the participants preferred the direct explanations, others preferred the differential explanations and also liked the natural language form of the explanations. In remarks, they said in general that "heat flow" is not clear for them and asked for an easier term to replace it.

Participants said that they would like to get the explanations and system recommendations, even if they would not follow them all the time. They will consult the explanations more when they suspect that something is not right (like

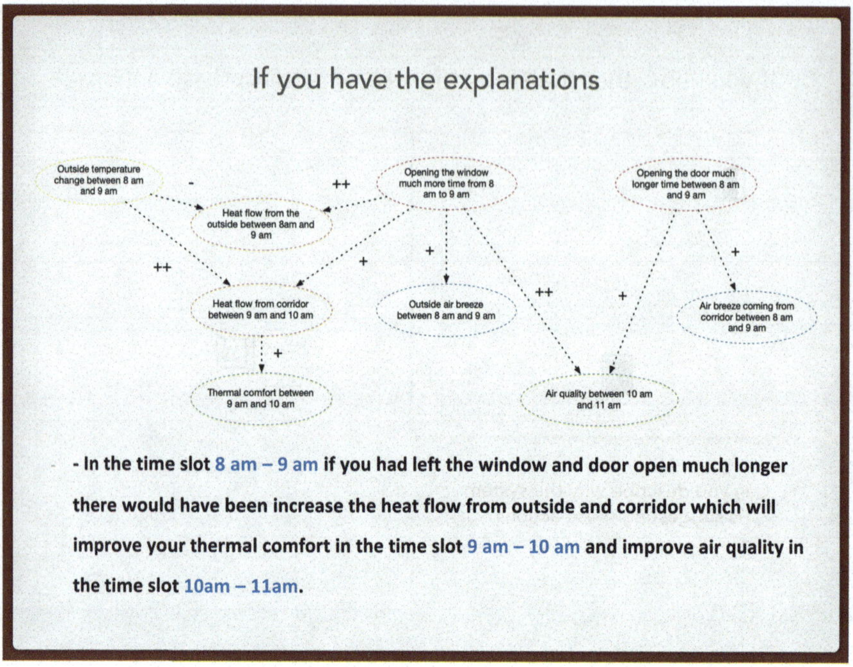

Fig. 18 Differential explanations

to checking it when they feel cold even when the heater is ON), or, in the case of pollution periods, to know what to do. One participant said that she will check it each day to confirm the air quality for her children.

5 Second Validation

A second field study by the IIHM team at the LIG laboratory was done to evaluate the UI and the causal explanations that were part of it after being integrated [15].

This study was done with a different interviewer, different buildings, and with different participants from the previous validation study. 13 participants from different backgrounds, age, and sex were chosen. For the explanation part, all of the participants understood well the explanations provided by the e-consultant, and most of them found the explanations well formulated; results are presented in the following table.

	Yes		NO	
	Value	Percentage %	Value	Percentage %
Could you understand the explanations provided by the system?	13	100	0	0
Would you formulate the explanations differently?	2	15.3	11	84.7
Do you find the explanations useful?	13	100	0	0
Do you think that explanations are necessary to understand how the e-consultant works?	8	61.5	5	38.5
Would you find it useful to provide the system with explanations regarding your behavior?	10	76.9	3	23.1

Only two participants declared they would formulate explanations differently. Nevertheless, they used different terms to express their understanding of the purpose of explanations such as: "reasons" (Participant 1), "explain (ations)" (Participants 2, 5, 7, 11, 12, and 14), the "Why" (Participant 3), "consequences" (Participants 4 and 14), "motivator" (Participant 6), and "utility" (Participant 8). All participants declared they found the explanations useful, and a majority (8/13) found them necessary in order to understand how the e-consultant works. In addition to usefulness, explanations appeared to contribute generally in a positive way regarding the differences as highlighted by the verbatim report below (some of them are in French and translated into English):

1. Answer:
   «C'est exactement ça, on te demande de faire des trucs. Du coup, par défaut, ça m'énerve parce que je n'ai pas envie de faire des trucs ; et, du coup, mais il m'explique ; je vais t'expliquer pourquoi quoi. Donc, ça, c'est bien. Du coup, quand il m'explique comme ça ; moi, après, je comprends et je dis OK . . . Et ben, c'est directement lié aux motivations quoi. Donc, si je suis motivé par la raison, si je n'en ai rien à foutre . . . C'est pour ça que j'imaginerais la possibilité de lui dire mes motivations. S'il sait exactement la température que je préfère, ça, c'est parfait. Du coup, je n'ai pas trop à m'en occuper ». (Participant 2)
   English translation:
   "That's exactly it, you're asked to do things. So, by default, it annoys me because I do not want to do things; and suddenly, but he explains to me; It explain why. So, that's good. So, when it explains me like that; me, after, I understand and I say OK . . . Well, it's directly related to what motivations. So, if I am motivated by reason, if I do not give a fuck . . . That's why I would imagine the possibility of telling him my motives. If he knows exactly the temperature that I prefer, that's perfect. So, I do not have too much to take care of it" ». (Participant 2)
2. Answer:
   «Oui, parce que la première fois, tu as envie de savoir pourquoi. Est-ce que c'est par rapport à ce que tu penses ; est-ce que ça confirme tes attentes ? Des fois, c'est une autre raison. Voilà, c'est pour conforter l'utilisateur ». (Participant 5)
   English translation:

«Yes, because the first time, you want to know why. Is it in relation to what you think; does that confirm your expectations? Sometimes, that's another reason. That's it to comfort the user »(Participant 5)

3. «This is, I guess, some information that makes you motivated. Why should I do this now? You will have your answer ». (Participant 6)
4. Answer:
   «En fait, ça te donne une explication … L'explication qu'il y a derrière l'action, derrière son conseil … parce qu'en fait, te dire : ferme la fenêtre, ouvre la fenêtre, mais si tu ne sais pas pourquoi tu le fais, ça peut te … A un moment, tu peux te dire : pourquoi je le fais ; tu peux t'arrêter mais, quand tu vois une explication, en plus qui est plausible, qui tient la route, tu vas te dire : je le fais ». (Participant 11)
   English translation:
   «In fact, it gives you an explanation … The explanation behind the action, behind his advice … because in fact, tell you: close the window, open the window, but if you do not know why you can do it, it can … At a certain moment, you can say to yourself: why I do it; you can stop but, when you see an explanation, besides which is plausible, who holds the road, you will say to you: I do it ». (Participant 11)

These two field studies demonstrate the importance of explanations for occupants. They present the explanations' utility in allowing the occupants to understand how the environment is functioning and why the e-consultant is recommending different actions at different times.

6 Conclusion

This chapter presents the generation of explanations with the use of knowledge models. Knowledge models can be any type of model that can provide simulation between input variables and output ones, like physical models or linear regression models [16]. This chapter described with a real case study the difficulty in understanding these types of systems, and at the same time why it is very important for occupants to understand the impact of their actions.

Then, it describes the different steps to obtain differential/direct explanations, and how they can help the occupants to understand the impact of their actions. It presents how it is possible to explore the implicit causality in the knowledge-based EMS and render it explicit through the differential explanations with contextual causality. This chapter also presents the model fragments concept to allow the injection of the expert knowledge and help determining the causalities. It also describes direct explanations and their limitations.

Finally, it presents the field studies to validate the direct/differential explanations' usefulness and acceptance by occupants.

References

1. B. Bredeweg, K.D. Forbus, Qualitative modeling in education. AI Mag. **24**(4), 35–35 (2003)
2. K. Tregoe, Help! My car is allergic to vanilla ice cream; a study in problem solving. The complete case (2018)
3. F.C. Keil, Explanation and understanding. Annu. Rev. Psychol. **57**, 227–254 (2006)
4. F. Heider, *The Psychology of Interpersonal Relations* (Psychology Press, London, 1982)
5. E. Sober, Parsimony, likelihood, and the principle of the common cause. Philos. Sci. **54**(3), 465-469 (1987)
6. R.A. Wilson, K. Frank, The shadows and shallows of explanation. Minds Mach. **8**(1), 137–159 (1998)
7. J.B. Tenenbaum, C. Kemp, T.L. Griffiths, N.D Goodman, American association for the advancement of science: how to grow a mind: statistics, structure, and abstraction. Science **331**, 1279–1285 (2011)
8. L. Scanu, S. Ploix, E. Wurtz, P. Bernaud, Model Tuning Approach for Energy Management of Office and Apartment Settings, Proceedings of the 15th IBPSA Conference (Building Simulation 2017)
9. B. Bredeweg, et al. Garp3: a new workbench for qualitative reasoning and modelling, in *Proceedings of the Fourth International Conference on Knowledge Capture* (2007)
10. P.M. Kevin, Dynamic Bayesian Networks: Representation, Inference and Learning, PhD Thesis, 2002
11. M.K. Slifka, J.L. Whitton, Clinical implications of dysregulated cytokine production. J. Mol. Med. **78**, 74–80 (2000). https://doi.org/10.1007/s001090000086
12. D. Cameron, Theoretical debates in feminist linguistics: Questions of sex and gender, in ed. by R. Wodak, *Gender and Discourse* (Sage Publications, London, 1997), pp. 99–119
13. G.F. Cooper, E. Herskovits, A Bayesian method for the induction of probabilistic networks from data. Mach. Learn. **9**(4), 309-347 (1992)
14. D. Heckerman, D. Geiger, D.M. Chickering, Learning Bayesian networks: the combination of knowledge and statistical data. Mach. Learn. **20**(3), 197–243 (1995)
15. Y. Laurillau,, V.B. Nguyen, J. Coutaz,, G. Calvary,, N. Mandran,, F. Camara, R. Balzarini, The TOP-slider for multi-criteria decision making by non-specialists, in *Proceedings of the 10th Nordic Conference on Human-Computer Interaction* (2018), pp. 642–653
16. L. Scanu, *Towards Archetypes of Self-Tuned Models for Connected* (2017)

The Mondrian User Interface Pattern: Inspiring Eco-responsibility in Homes

Yann Laurillau, Joëlle Coutaz, Gaëlle Calvary, and Van Bao

1 Introduction

Domestic energy consumption represents approximately one-third of the worldwide total energy consumption, with a projected increase of up to 40% by 2040 [1]. This increase in energy demand has motivated a large body of research, using a diverse collection of approaches to reduce domestic energy use. One possible approach is purely technical, as illustrated by autonomous smart energy management systems. For example, using sensing and machine learning, such systems can predict occupancy and adjust temperature accordingly [2]. At the other extreme, the task of reducing energy use is delegated to building occupants, by making them aware of the problem and providing them with quantitative information about their consumption, or by using popular media hammering home the message that global warming is a critical societal challenge.

Between these two extremes—full machine autonomy vs. full human responsibility, the current popular approach is to support a strong positive collaboration between the system and residents. Typically, based on users' preferences and the physical characteristics of the habitat, the system is able to optimize cost and comfort. This means that the system takes decisions on the users' behalf and has control of the home, at least temporarily [3]. However, by recommending users to perform eco-responsible actions along with explanations, the system can improve the sense of agency [4]. In this case, the research question becomes "*How to support Human-System collaborative interaction in the context of eco-responsibility*"?

In this chapter, we present the *Mondrian User Interface (Mondrian UI) as a UI design pattern* that can help system developers to structure and populate

Y. Laurillau (✉) · J. Coutaz · G. Calvary · V. Bao
Université Grenoble Alpes, CNRS, Grenoble INP Institute of Engineering Univ. Grenoble Alpes, Grenoble, France
e-mail: Yann.Laurillau@univ-grenoble-alpes.fr

S. Ploix et al. (eds.), *Towards Energy Smart Homes*,
https://doi.org/10.1007/978-3-030-76477-7_13

the interactive components of their system in order to inspire eco-responsibility. In the following, we synthesize the key lessons drawn from the state of the art used to inform the design of the Mondrian UI pattern. This includes results from research on domestic environments, on properties of ambient displays, as well as on informational representations relevant to eco-responsibility. We then give a detailed description of the design pattern followed by an illustrative example with the user interface of an e-coach smart energy management system [5].

2 Domestic Environment and Design Implications

Previous research on domestic environments has been conducted for diverse purposes including household communication [6, 7], time management and planning, technology control [8], as well as energy use in everyday life [9]. With regard to energy, it has been demonstrated that few people are aware of their consumption [10, 11]. To address this problem, many forms of eco-feedback [12] as well as theoretical work on persuasive technologies have been developed [13–18], seeking to change human behavior without coercion or deception. However, as demonstrated by Erickson et al. in their longitudinal experiment with a web portal aimed at supporting reductions of electricity consumption, people "*keep forgetting*" or "*do not have time*" to use the system [19].

Thus, households need to be provided with "*a low-cost way to modify existing behaviors*" [8] that can be integrated in their life style and routines. In turn, routines evolve periodically and opportunistically, typically when an exception occurs [20]. In their ethnographic study, Davidoff et al. observe that ordinary life is punctuated by key moments, such as leaving for work, that require focusing attention on demanding activities. In addition, users' knowledge about technical and persuasive systems improves over time, which in turn may result in loss of interest [21, 22].

These observations call for a UI solution based on *calm technology*, notably the concept of *ambient display* that can support *different levels of users' attention investment* [4]. In particular, when users are in a hurry, it is essential that *information be glanceable*, possibly *catching the eye* in the case where an urgent decision is required. On the other hand, when people are willing to invest more attention in the system, a *zoomable UI* is more relevant provided that information is "*meaningful and contextually appropriate*" [23]. As interest and engagement have to be maintained, *aesthetics* has a role to play [24].

3 Ambient Displays and Aesthetics

Ambient displays "*move to the center of attention only when appropriate and desirable*" [25], while the use of aesthetics and lifelike forms are promising techniques to inspire positive changes in human behavior [26]. The combination

The use of ambient and artistic displays. (With permission of Paay et al. [28])

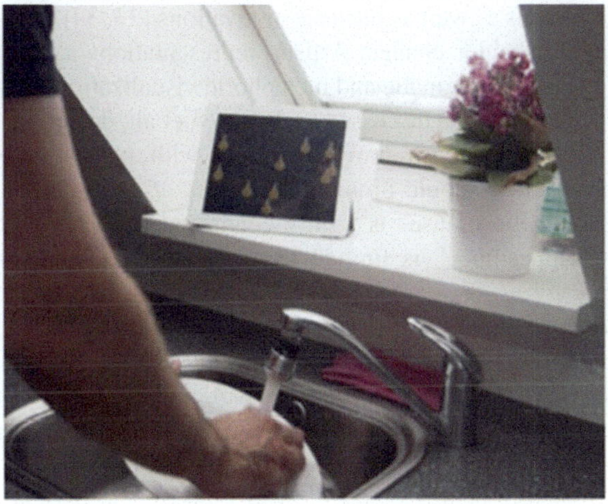

of ambient displays with aesthetics is believed to raise at-a-glance awareness [23], to increase inhabitants' engagement and to promote intrinsic motivations [26, 27]. Figure 1 shows examples of ambient displays using aesthetic content for eco-feedback and eco-responsibility.

3.1 Design for the Periphery

As a *calm technology*, ambient displays seamlessly provide information in a non-intrusive manner in the periphery of an occupant's attention [26, 29]. According to [30], calm technology provides background information that does not continuously force the user to actively pay attention. Jafarinaimi [26] and Ferscha [25] observe that calm technology allows users to interact with the system when they desire, rather than passively receive pushed information from the system. Paay et al. [28] found that the "ambiency" of real-time eco-feedback keeps inhabitants in context, moving their attention from the periphery to focus, then possibly to actions aimed at reducing energy use. This focus change relates to Fogg's notion of a trigger, an important factor of his theoretical model of behavior change [13].

3.2 Aesthetic Representations

In Human-Computer Interaction, aesthetics and enjoyment are considered to be essential to the user's experience [31]. Aesthetics provide motivational affordances and improves system attractiveness, and is a key dimension of several persuasive design spaces [11, 27] and design principles [32]. Aesthetic values inspire positive emotions as well as intrinsic motivations [33, 34], which in turn may foster sustainable behavior change. Artistic representations also avoid negative reinforcements that some pragmatic and metaphoric visualization techniques may convey [27, 34].

However, as discussed by Petersen et al., designing aesthetic interactive systems is not limited to providing users with pleasing visual effects [35]. Aesthetic interaction should promote *aesthetics of use* as well as *aesthetics of appearance*. "Aesthetic of use" is tightly connected to the context of use whereas "aesthetic of appearance" is limited to superficial beauty. Aesthetic interaction "*is about triggering imagination, it is thought-provoking and encourages people to think differently about the encountered interactive systems, what they do and how they might be used differently to serve differentiated goals*" [35]. Finally, in the context of domestic use, aesthetic displays participate to the personal decoration of the interior.

Aesthetic representations fall within three categories: abstract representations, metaphorical representations, and informative art.

3.2.1 Abstract Representations

Abstract representations are intended to raise curiosity and, from there, to increase awareness about the system behavior [26]. Abstraction can bundle large data sets into synthetized at-a-glance information. For instance, in Fig. 2, the particles that fall from the top of the screen at different speeds, sizes, and colors express the electricity consumption of an office building in real time [21]. On the other hand, People Garden message board is a flower garden that informs users about their social environment [36]: each user is represented by an abstract flower whose shape, color, and position evolve according to time and the number of messages posted by this person. Abstraction may also be used to hide personal information [37].

Rodgers & Bartram [23] consider abstraction as a suitable solution for designing in the periphery. They have explored different data representations via abstract geometric shapes for a variety of contexts of energy consumption (see Fig. 3-

Fig. 2 Particles as an abstract representation of electricity consumption. (Image taken from [21])

Fig. 3 (top) Energy consumption through pinwheels visualization (with permission of Rodgers et al.), (bottom) users' activities reflected as globular objects. (With permission of Nakajima et al. [38])

top). Similarly, Nakajima & Lehdonvirta [38] present users' activities as the transformation of globular objects (see Fig. 3-bottom).

3.2.2 Metaphorical Representations

Metaphorical representations "*refer to the understanding of one idea, or conceptual domain, in terms of another*" [27]. They are often used to visualize data in a pro-environmental manner, using nature-inspired elements such as trees, animals, earth, and forest. For instance, Fig. 4 (left) shows a landscape to reflect electricity usage in the home [34]. The more the appliances are turned on, the more nature-based elements (here, animals and flowers) appear in the landscape.

3.2.3 Informative Art

Informative art consists of augmenting artworks, such as paintings and posters, with additional information [30, 40]. For instance, in order to motivate users to walk

Fig. 4 (left) A Landscape metaphorical representation to illustrate household consumption (with permission of Nisi et al. [34]), (right) Nuage vert as a metaphorical representation of a city's energy consumption. (With permission of Evans et al. [39])

Fig. 5 (left) Mona Lisa's face reflects user's walking activities (with permission of Nakajima et al. [38]), (middle) Mondrian-inspired prototype illustrates email-traffic (with permission of Redström et al. [30]), (right) earthquake activities inspired from Richard Long's artworks. (With permission of Holmquist and Skog [41])

more, the number of steps that users have walked is reflected on Mona Lisa's face [38]. As shown in Fig. 5 (left), Mona Lisa looks older if the user has not walked enough. As another example, the representation of email traffic of Fig. 5 (center) is inspired from the paintings of Dutch artist Piet Mondrian where each rectangle represents the email activity of a person [30]. With Stone Garden (Fig. 5 right), earthquakes are represented as types of stone [41]. The size and position of a stone represent the magnitude as well as the latitude/longitude of an earthquake. This representation looks like the artwork of the famous English artist Richard Long.

Although aesthetic representations play an important role in inspiring eco-responsibility, in some situations, users want to "*see the numbers*" [23]. The combination of aesthetics with pragmatic representations is one way to satisfy this need.

3.3 Combining Aesthetics with Pragmatic Representations

Whereas aesthetic representations are primarily thought-provoking and support at-a-glance sensemaking, pragmatic representations provide "*concrete quantitative information*" using traditional scientific visualization techniques such as bar charts and scatter plots. By contrast with aesthetics representations, pragmatic representations are expected to be explicit and unambiguous. A number of studies have shown that, although occupants understand the mapping of their energy use into artistic illustrations, they also need clear quantitative information for a deeper understanding [23, 34]. Spark [37] and PowerViz [28] illustrate the combined use of aesthetic and pragmatic representations.

In Fig. 6, the Spark system uses shapes of different colors and sizes to represent users' daily activities complemented with a quantitative chart-based representation.

Similarly, PowerViz combines a metaphorical representation with analytic tools to help inhabitants to understand their domestic consumption at the appliance level [28]. As shown in Fig. 7, PowerViz consists of a *screen saver, a usage history*, and an *appliance usage.* The screen saver uses a metaphorical visualization to "*create ongoing engagement with the system while giving a playful overview of total household energy consumption*" [28]. The other two displays are intended for in-depth analysis of energy consumption over time as well as for comparing the energy consumption of individual appliances. Comparative analysis may allow the detection of "greedy" appliances whose usage may then be reduced.

The combined use of aesthetic and pragmatic representations raises the following question: "*How to provide an easy transition between the representations?*" In PowerViz, switching from the metaphorical view to the pragmatic views is performed by touching the screen. Although touching involves minimal motor effort, it is not sufficient to support the change at the cognitive level. We address this question with the concept of Zoomable UI coupled with Focus+context techniques.

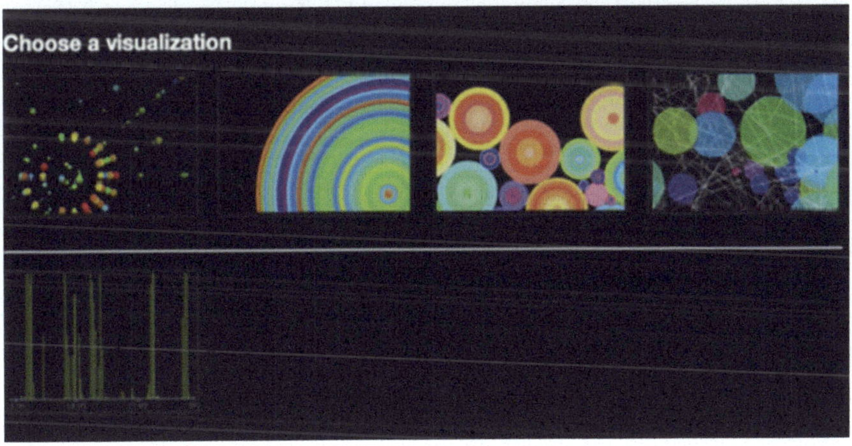

Fig. 6 Spark: Art-based visualizations of people activities using various types of shapes (top, left to right): Spiral, Rings, Bucket, Pollock, (bottom) users can also track their activities as a bar chart. (With permission of Fan et al. [37])

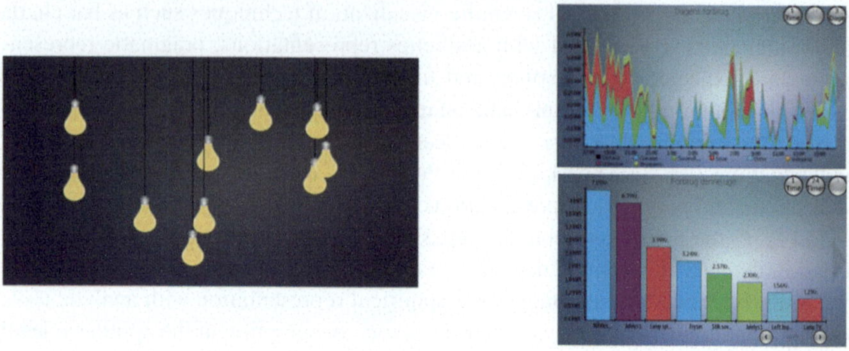

Fig. 7 (left) PowerViz: light bulbs metaphorical real-time representation of the domestic power usage at the appliance level, (right) screens for in-depth analysis: (from top to bottom) Usage History and Appliance usage. (With permission of Paay et al. [28])

4 Zoomable User Interface and Focus+Context Techniques

Zoomable user interfaces allow users to change the scale of graphical contents in order to see more or less details. Zoomable user interfaces are known to provide an easy transition between multiple levels of details. They are therefore good candidates to satisfy "*easy transition to more in-depth information,*" a criterion that Mankoff et al. recommend for evaluating the usability and effectiveness of ambient information systems [42].

However, zooming introduces a temporal separation, leaving the user to assimilate and remember the relations between views [43]. Zooming is thus not sufficient to support smooth transitions between radically distinct representations as exemplified by PowerViz when it comes to switching between the metaphorical and the pragmatic representations. Focus+context techniques, on the other hand, suppress temporal separation.

4.1 Focus+Context Techniques

Focus+Context "*integrates the [visual] focus and its [visual] context into a single display where all parts are concurrently visible: The focus is displayed seamlessly within its surrounding context.*" [43]. This technique eliminates the temporal separation used by zooming and, at the same time, minimizes the spatial separation by displaying the focused item within its context. "*By presenting all regions in a single coherent display, focus+context systems aim to decrease the short term memory load associated with assimilating distinct views of a system, and thus potentially improve user ability to comprehend and manipulate the information*" [43].

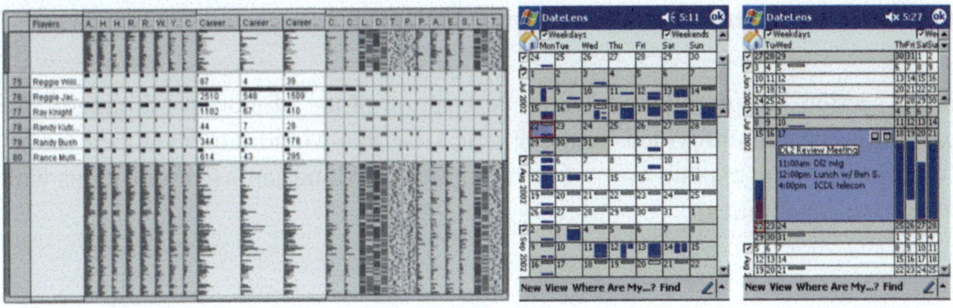

Fig. 8 DateLens [44] explores the Focus+Context approach for a calendar display. (With permission of Cockburn et al. [43])

Fisheye lens, a popular focus+context technique, combines entirety and details in a single view using various forms of visual distortion. Visual distortion must be designed carefully as it may impede legibility and interpretation [43]. Tablelens [45] and DateLens [44], illustrated in Fig. 8, both use a fisheye geometric transformation that preserves the rectangular format of all the regions of the display.

TableLens provides a condensed overview of large datasets by displaying rows and columns as rectangular bars. Users can expand a specific block of information based on the selected row and column. The zoomed block reveals additional data values. Similarly, DateLens (Fig. 8-right) uses a fisheye Focus+Context approach for displaying calendars on small-size screens.

4.2 Semantic Zoom

Focus+Context results in magnifying the focused region while shrinking the other areas. Semantic zoom is generally used to adapt the presentation of data items at different scale levels based on the available space. DateLens, for example, uses three zoom factors simultaneously; information for the day of interest is fully available as text in a focused large rectangular area. The days of the week that form the close temporal context of the day of interest show the number of meetings as colored vertical bars displayed in smaller vertical rectangles, while the other days, further away from the temporal context, are denoted as a day number in even smaller rectangles (see Fig. 8). A major difficulty for system designers is first to identify the information users need at each level of details, and then to decide how this information is represented at different scales.

The Mondrian UI pattern, presented next, brings together the key results discussed above into a holistic interaction framework that system developers can use to integrate multiple forms of eco-responsibility techniques.

5 The Mondrian User Interface Pattern

In this section, we provide the overall description of the Mondrian UI pattern. We explain the benefits provided by the artistic style that Mondrian developed for his Compositions, and explain how we have augmented Mondrian's Compositions with interactive capabilities.

5.1 Overall Description

As Fig. 9 shows, the Mondrian UI pattern organizes the user interface of a system as a focus+context tiled display, where each tile can be zoomed in and zoomed out to support multilevel representations of the same semantic information while maintaining informational context accessible. In order to minimize attentional

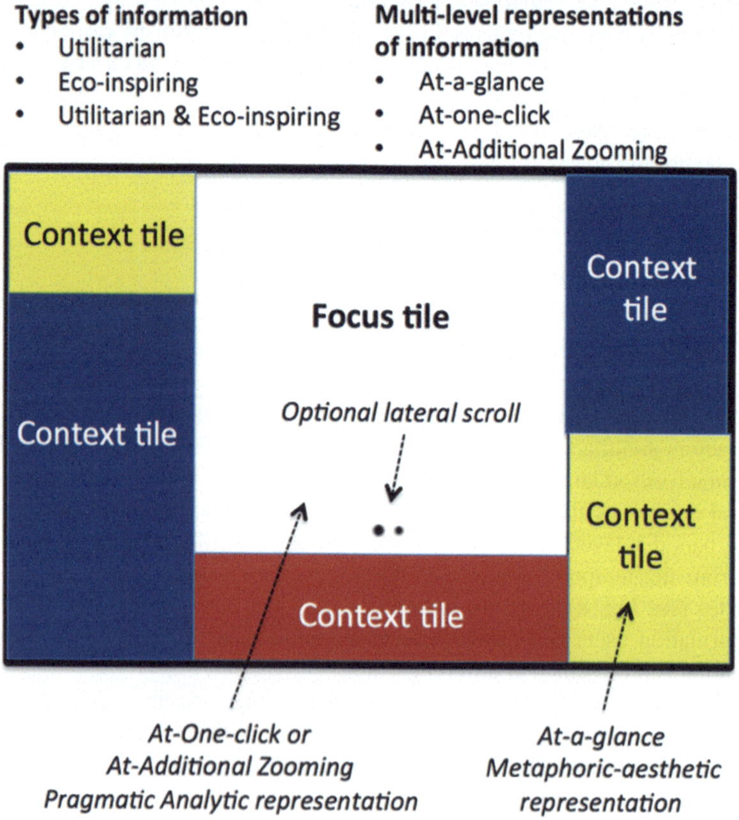

Fig. 9 The overall structure and interactive principles of the Mondrian UI pattern

effort, the top-level content of a tile is intended to communicate information at a glance while the other levels provide analytic details on demand with minimalist one-click interaction.

Each tile of the Mondrian UI is populated with a specific type of content: eco-inspiring content and/or utilitarian content. Eco-inspiring content includes spatiotemporal information about energy use, recommendations and explanations, social incentives such as intra- and inter-social comparisons and gaming. Utilities are services that family members have the habit of consulting frequently such as the weather forecast, stock market, time, or the local traffic conditions. In addition to semantic zoom, tiles may include horizontal scrolling as a mechanism to switch between sets of multilevel representations. This can be used to switch from a set of pragmatic representations to a set of artistic and metaphoric expressions of the same semantic information.

In order to be unavoidable without being intrusive, the Mondrian UI pattern has been devised as an always-on ambient display running on tablets located in a socializing space of the home such as the kitchen or the entrance hall [6]. As an element of the interior decoration, the ambient display has to be aesthetically pleasing and attractive. This requirement for aesthetics has motivated the use of Mondrian's abstract Compositions to structure the display.

5.2 *Mondrian's Abstract Compositions and Their Benefits*

As illustrated by the Composition shown in Fig. 10 (left), Mondrian uses black, or gray, lines to divide the canvas into rectangles of different sizes, and paints the rectangles in primary colors. For the De Stijl avant-garde movement, Mondrian's Compositions define a new aesthetic language. From our perspective, the Compositions bring additional benefits:

- Familiarity with tiled screens as used by popular systems such as Windows applications (see illustration in Fig. 10, right).

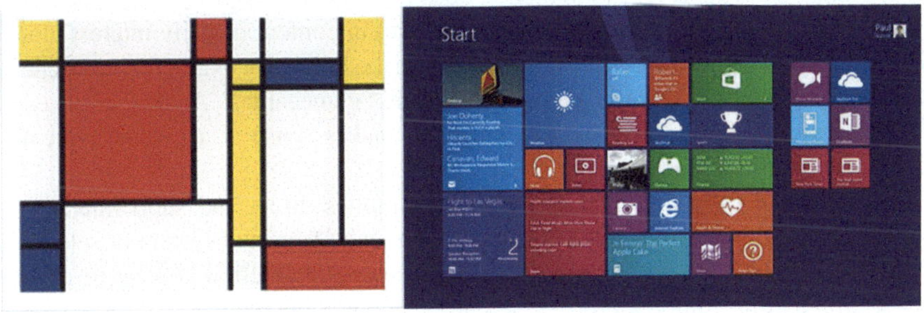

Fig. 10 (left) An example of Mondrian's composition, (right) Windows 8 screen consists of tiles, each tile referring to one block of functions ((c) Microsoft)

- Easy mapping between the rectangles and information: rectangles can be populated with specific types of content and primary colors can serve to convey content types.
- Direct application of the focus+context technique, as Mondrian's rectangles are of different sizes.
- Extensibility and support for iterative and incremental development: rectangles can be added or removed in phase with the development process of the system. Alternatively, some rectangles can be populated with content while others can be left as empty "pure Mondrian" tiles for further increments. In addition, the design of the content of a particular rectangle can be modified iteratively without affecting the others.

The next paragraph shows how Mondrian's Compositions can be augmented with interactive capabilities.

5.3 Mondrian's Compositions Augmented with Multilevel Interaction

As discussed above, people's attention investment depends on the context of use [46]. Users may feel too busy to interact with the system, they may forget to use it, or may have lost interest. On the other hand, when mentally available, they may want to explore the system in depth, to learn and understand, or to find the cause of some unexpected system behavior. We propose to augment Mondrian's Compositions with multilevel interaction that enables users to transition freely between their levels of attention. Multilevel interaction combines focus+context and semantic zoom techniques with several levels of physical involvement.

5.3.1 Three Levels of Interaction: At-A-Glance, At-One-Click, At-Additional Zoom

At-a-glance UI. This level of interaction does not require any physical contact with the display, except looking at the screen. Utilitarian content of daily interest along with a well-thought out location of the display in the home, for example, the kitchen, makes the display physically unavoidable, thus "glanceable." At this level, eco-inspiring content is represented in a way that makes sense at a glance, typically using metaphorical representations.

At-One-click UI. This level of interaction implies direct and short interaction with the system, typically accepting an eco-challenge from the system or selecting another tile related to eco-information. This class of interaction aims at involving the user in a more complex process beyond glancing at the display.

At-Additional-zoom UI. At the next level of interaction, the UI enables the user to explore the system facilities in depth by zooming in and out while supporting transitioning between the Mondrian's tiles.

5.3.2 Transitioning Between the Mondrian's Tiles

Figure 11 illustrates the two types of transition: (a) transitions from ambient to focus and (b) transitions within a focused view.

Figure 11a-left shows an ambient screen composed of 6 tiles. By tapping a tile—in this example Tile 1, the tile of interest is magnified while the other tiles are shrunk (cf. 11-a-center). Focus is now on Tile 1 and its content is semantically zoomed in while the content of the other tiles are zoomed out. As illustrated in Fig. 11a-right, moving the focus from Tile 1 to Tile 3 is performed in one single action by tapping the "shrunk Tile 3" shown in 11-a-center. As the focus migrates from tile to tile, the geometric reconfiguration of the tiles as well as the coloring used by the focus+context technique preserves the Mondrian's style. By doing so, we support both *aesthetics of use* and *aesthetics of appearance*.

Some tiles, such as Tile 1 in Fig. 11b, may offer more than one level of focus+context interaction. Within such tiles, the same interactive behavior applies as described above. For example, focused Tile 1 is zoomed in as 4 sub-tiles denoted as *a, b, c, d*. Tapping sub-tile *a* brings the focus on *1-a* (center), and from there, tapping *b* brings the focus to *1-b* (right).

We now illustrate the application of the Mondrian UI pattern with the user interface of an e-coach system.

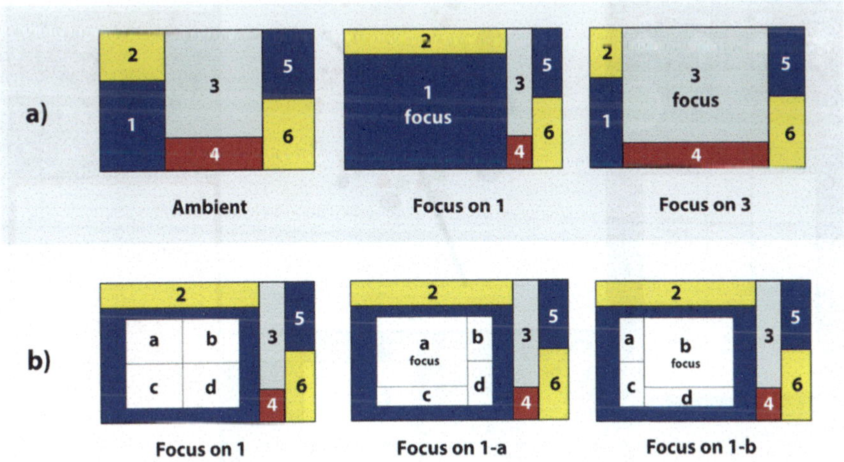

Fig. 11 (**a**) Transitions from the ambient screen (left) to focused Tile 1 (center) and from Tile 1 (center) to Tile 3 (right), (**b**) transitions within focused Tile 1 (left) to focused Tile 1-a (center) and from Tile 1-a (center) to Tile 1-b (left)

6 The E-coach Mondrian User Interface

The user interface described in this section is the front-end of a cooperative e-coach management system for domestic use. At the time of this writing, this user interface serves as a proof of concept for the Mondrian UI pattern. Some tiles have been implemented and validated with end-users in controlled experiments [47], but deployment in real-world settings has yet to be performed.

6.1 *Overall Structure of the e-Coach UI*

As Fig. 12 shows, the overall structure of the e-coach UI is a combination of utilitarian and eco-inspiring tiles whose colors and interactive behavior comply with the Mondrian UI pattern. Utilities include the weather forecast and the current date. Eco-information is expressed as spatial, temporal, or social incentives, as well as in terms of explainable recommendations and human control. To serve as an ambient display, the tiles of the home screen are intended to make sense at a glance.

In the following, we describe the content of the tiles as users move their focus of interest.

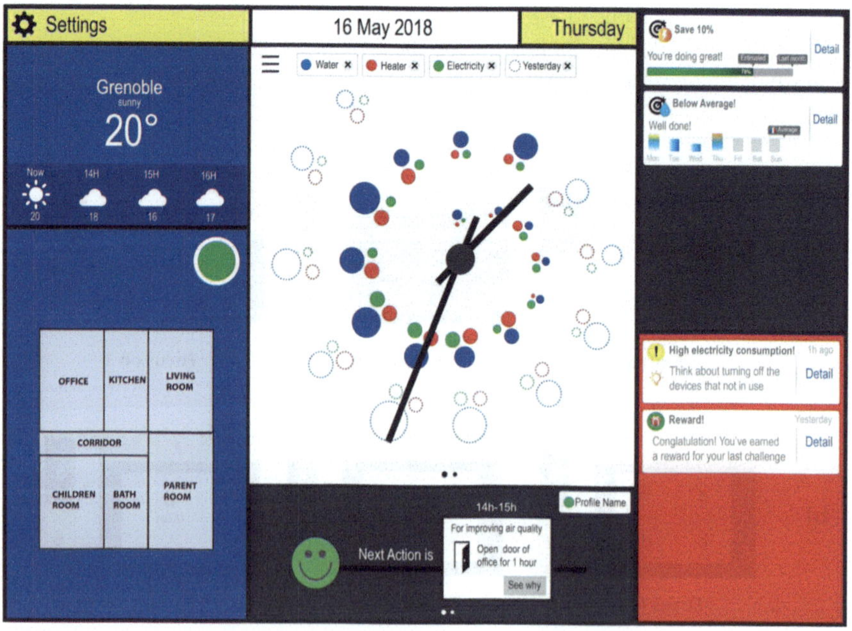

Fig. 12 The home screen of the e-coach is an "at-a-glance ambient display" that applies the Mondrian UI pattern

6.2 *Spatial Eco-information*

Spatial eco-information relates to the home as a set of places. A place is a space where specific types of activities and interactions take place. As an initial prototype design, we have assimilated places to rooms, although in real-world settings, a physical room may include distinct places. Spatial eco-information is displayed on the representation of an abstracted floor plan whose rectangles comply with the Mondrian style.

6.2.1 At-A-Glance Spatial Eco-information

As shown in Fig. 13, a circle is used to represent the overall status of the habitat. The color and animation of the circle are inspired from the semantics of traffic lights: green when everything is fine (Go! You can leave for work, as illustrated in Fig. 12), yellow as a warning signal (there may be something to check before leaving for work), or, as illustrated in Fig. 13, blinking red to force attention (Stop! Action is required). The rooms/places of the floor plan that need to be checked display an icon that denotes the level of urgency and cause of the problem.

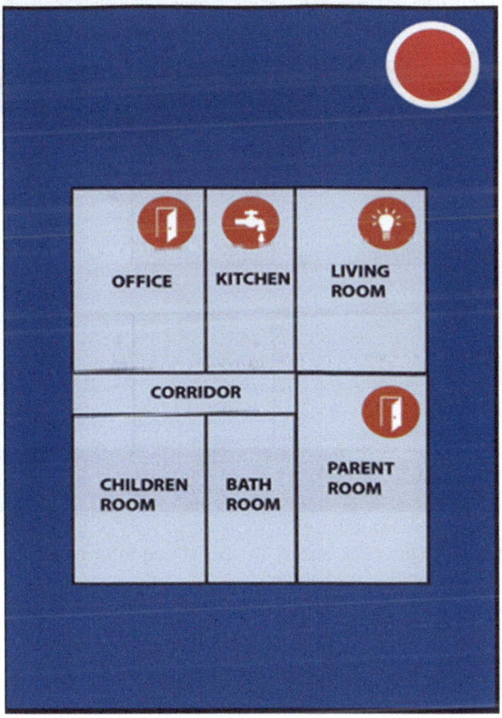

Fig. 13 The blinking red circle denotes the existence of problem(s): icons represent the issues that need human attention (e.g., water is running in the kitchen where no one is present)

Note that the tile of the weather forecast is purposefully close to the spatial representation of the habitat to augment the chance that people check the overall status of their home frequently.

6.2.2 At-One-Click Spatial Eco-representation

Tapping the spatial eco-information tile of Fig. 13 results in the display presented in Fig. 14. The inhabitant has now access to additional information to analyze the situation, to understand the impact of the household behavior, and from there, to decide to behave differently. In this example, the outdoor of the office is opened. Although it is currently 1:35 pm and sunny, the outside temperature is 8 °C, while that of the office is 22 °C. Consequently, the air flows towards the exterior resulting in energy loss. If we refer to the Habit Alteration Model [16] and to the Dual Process theory, the repetition of this situation should break the habit of leaving doors inadvertently opened, and to form new eco-responsible habits.

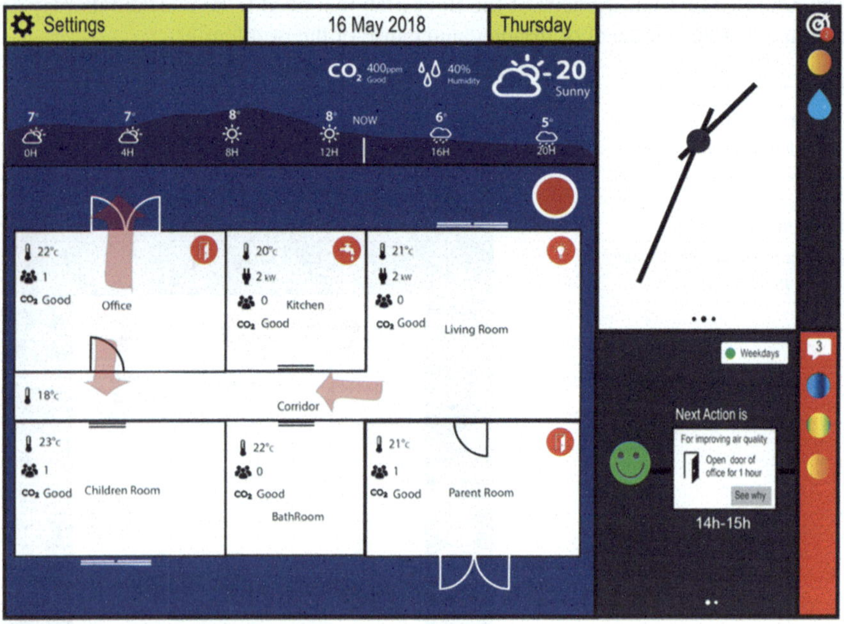

Fig. 14 One-click access to spatial eco-information of the home. Arrows show how airflows circulate within the habitat

6.2.3 Additional Zoom-In of Spatial Eco-information

By zooming the tile of a room/place one level down, the inhabitant obtains the temporal evolution of the ambient conditions of the room. Figure 15 shows the display that results when tapping the office tile of Fig. 14.

Whereas "one-click eco-visualization" shows the value of the key indicators in real time, additional zoom-in allows inhabitants to monitor the indicators of a room through time. In this situation, a chronogram shows the "heart beat" of the room at the top of the zoomed-in tile. In the example of Fig. 15, the chronogram shows that state changes are concentrated in the morning. The red vertical ruler allows the exploration of the states across the time lines so that causalities between the states can be detected. In the example, at 8:05 am, the temperature in the office was 22 °C, the office was occupied by 2 persons, and the level of CO_2 was fine as the window was left opened ajar for a while early in the morning.

Whereas "additional zoom-in of room tiles" allows access to information that is primarily space-based, eco-information discussed next is primarily time-based.

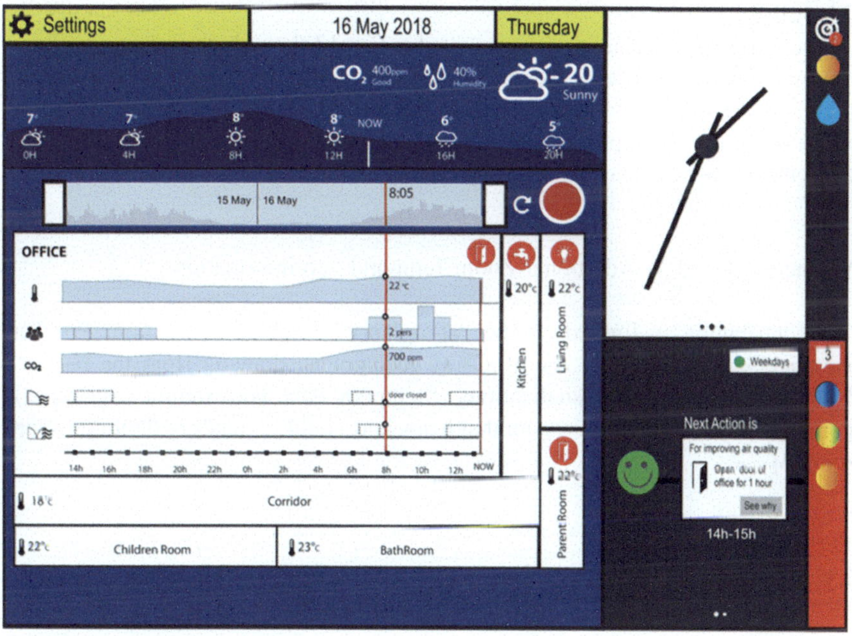

Fig. 15 Additional zoom-in of the office to support analytic tasks. The tiles of the other rooms display overview indicators

6.3 Temporal Eco-information Combined with Utilitarian Information

Our representation of temporal eco-information is based on the popular clock metaphor. Clock-based visualizations are often used in persuasive interactive systems such as eForecast [48] and Clock Cast [49] to indicate when energy price is cheap. Similarly, EnergyAwareClock [50] uses a clock to display domestic 24-h electricity consumption. The utilitarian function of a clock is also a key factor for using this metaphor.

6.3.1 At-A-Glance Temporal Eco-information

As shown in Fig. 12, the clock tile occupies most of the screen real estate to promote its utilitarian function. An analog clock is augmented with an abstract representation of the hourly consumptions of the household on a 24-h time period. We have used a spiral-based layout, as this technique is appropriate for the analysis of cyclic data [51, 52].

The spiral-based visualization presented in Fig. 12 shows both real-time and past consumptions of electricity, water and heater, each one associated with a colored circle. For ambient purpose, this visualization provides an at-a-glance overview and progress of the consumptions in a way that does not overload the interface and that keeps the aesthetics pleasant.

At-One-Click and At-Additional-Zoom Temporal Eco-information

Our contribution is limited to the "at-a-glance" level of interaction. For "at-one-click" and "additional zoom-in," we recommend drawing from the research developed on interactive visualizations of temporal data. For example, Activelec is an interactive visualization system that helps households to analyze their consumption [53].

6.4 Social Eco-information

At the time of this writing, the prototype covers social eco-information in a very limited way. This is illustrated in Fig. 12 with the tiles at the top-right corner of the display. These tiles include a combination of goal setting, challenges, and social comparison. In most theories of behavior change, goal setting and challenges are effective incentives for triggering eco-responsibility [12]. In addition, it is believed that goal setting is even more effective when coupled with feedback about self-monitoring as well as related to social and normative comparisons [54–56].

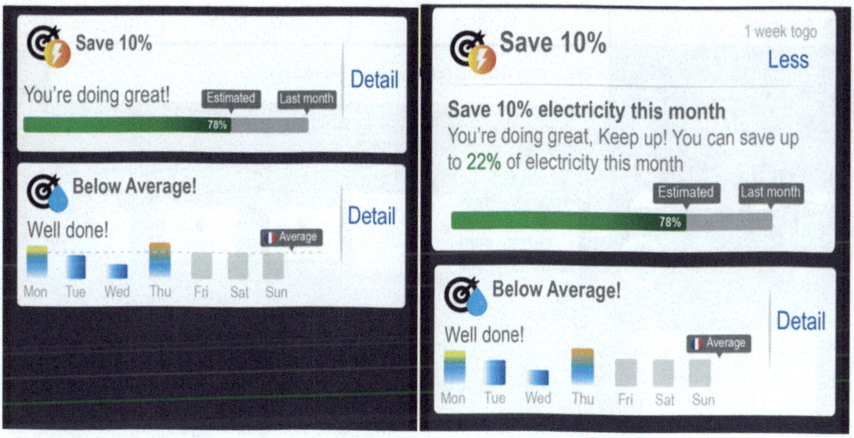

Fig. 16 (left) Social eco-information as goal setting and social comparison, (right) the corresponding at-one-click view

As illustrated in Fig. 16-left, occupants can track their progression at a glance and have an idea about how their goal compares with that of others. Figure 16-right shows an example of "at-one-click" social eco-information.

So far, we have described the tiles of the Mondrian UI with examples of eco-information that mainly serves as eco-feedback. We need now to address human control and collaboration with the e-coach.

6.5 *Human Control and Collaboration*

Informed by users' preferences in terms of thermal comfort, air quality, and cost, the e-coach system generates recommendations along with their explanations. The tile with a black background at the bottom center of the Mondrian UI is dedicated to this human-system collaboration.

6.5.1 At-A-Glance Human Control and Recommendations

As shown in Fig. 12, the collaboration tile tells the occupant which action to perform next so that the ambient comfort of the home can be maintained in accordance with the household preferences. In this example, the e-coach recommends to open the door of the office for 1 h between 14:00 and 15:00. If needed, a "See why" button tells why this action is appropriate. The green happy status-man indicates that, so far, all recommended actions have been performed and that the system can maintain the ambient comfort as requested.

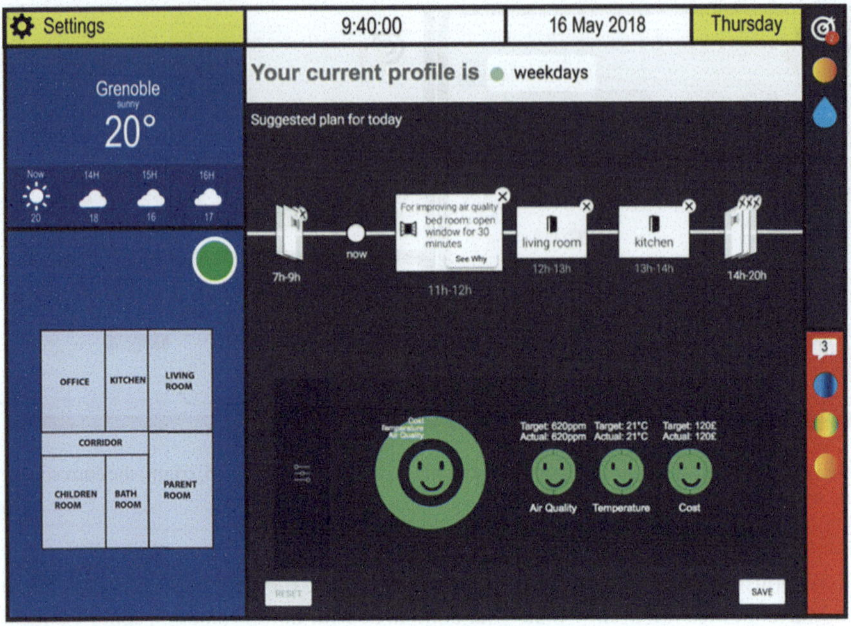

Fig. 17 At-one-click human control and recommendations. The action plan consists of a sequence of actions recommended to users in order to maintain comfort in conformity with their preferences

6.5.2 At-One-Click Human Control and Recommendations

Figure 17 shows the display when the focus moves to the "Human Control and Recommendations" tile. Recommendations are presented as a sequence of actions, such as opening and closing windows, that the system is unable to perform on human's behalf, but that should be performed so that ambient comfort is maintained in conformity with the users' preferences. This action plan, however, is not coercive: it is browsable and editable (recommended actions can be suppressed and the system will note this); the plan may be executed partially by the users, or even completely ignored. The status man is updated accordingly to show how far the occupant is, or will be, from optimal behavior.

6.5.3 Additional Zoom-In of the Human Control and Recommendations

Additional zooming allows users to express their preferences in terms of cost, thermal comfort, and air quality, using the TOP-sliders shown in Fig. 18. Many studies show that people want to stay in control and be involved in the housing management process. Giving the control to occupants is key to inspire eco-responsibility.

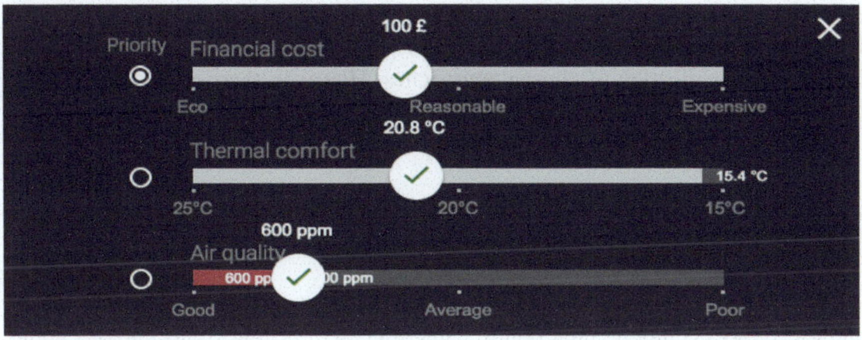

Fig. 18 The TOP-Sliders allow users to find the best tradeoff between conflicting criteria, financial cost, thermal comfort, and air quality

The TOP-Sliders help occupants to find optimal compromises between tightly coupled and conflicting criteria such as cost, temperature, and air quality. The precise details of the interactive behavior of the TOP-sliders, along with their usability, are described in [47]. Once the user has found the best compromise, the e-coach generates the appropriate action plan, which in turn can be edited, possibly leading users to modify their preferences (or just leave things as they are), and with time, learn and move towards eco-responsible behavior.

The combined and simultaneous availability of eco-information, TOP-Sliders, and explainable and editable action plan at multiple levels of interaction is one way "*to support Human-System collaborative interaction in the context of eco-responsibility.*"

7 Conclusion

Our contribution is the conceptual Mondrian user interface aiming at supporting long-term user interaction to accompany a behavior change and supporting multiple contexts of use. The design rationale relies on applying ambient/artistic approaches in an always-on display, on the combination of pragmatic and artistic representations, and on semantic zoom techniques to provide a multilevel user interaction. As a proof of concept, we instantiated this concept for INVOLVED project's e-coach engine [5], aiming at supporting end-users in promoting sustainable behavior in energy in residential context. Household contexts introduce various constraints such as appliance placement, visibility, aesthetic choices, and interactive affordances [23]. Besides, home settings include issues related to how occupants are willing to interact with the smart system, and how to effectively design user interface that adapt to these constraints and complexities.

We emphasize that the purpose of this study is an exploration of practical design approaches and the identification of very basic interactional bricks rather than an evaluative study. We conducted a literature review about current persuasive system for energy in household. We present our design solution and some interactional bricks with justification of chosen approaches. However, there exist some aspects that could be improved.

7.1 Personalization

As an e-coach system for household, it needs to adapt to different contexts, purposes, and especially people. It is relevant to one criticism of persuasive technology about how designers define what is “good” or “bad” for users [54]. Because there is no “one-size-fit-all” solution, hence the needs to personalize the system in terms of functionalities, design elements are obvious. For instance, in our case, rather than Mondrian style, alternative modes could be given for customization purpose. Besides, we could imagine artistic styles as items that can only be unlocked for usage when occupants accomplished certain tasks and challenges.

7.2 Long-Term Study

As behavior change is a long-term and complex process, the study must involve a longitudinal evaluation in order to measure the persuasive aspect, and more importantly in our case, how chosen design elements affect the change. Therefore, future works include a longitudinal study of whether persuasive interaction respecting on user values actually promoted desired change in energy consuming behaviors. Currently, a long-term evaluation is out of the scope of this work but constitutes a mandatory perspective.

References

1. U.S. Energy Information Administration. International Energy Outlook (2013). https://www.eia.gov/pressroom/presentations/sieminski_07252013.pdf
2. R. Yang, M.W. Newman, J. Forlizzi, Making sustainability sustainable: challenges in the design of eco-interaction technologies, in Proceedings of the SIGCHI Conference on Human Factors in Computing System (CHI 2014) (ACM, 2014), pp. 823–832
3. R.H. Jensen, et al., HeatDial: beyond user scheduling in eco-interaction, in Proceedings of the 9th Nordic Conference on Human-Computer Interaction - NordiCHI ’16 (Gothenburg, Sweden, 2016), pp. 1–10
4. H. Limerick, D. Coyle, J.W. Moore, The experience of agency in human-computer interactions: a review. Front. Hum. Neurosci. **8**, 643 (2014). https://doi.org/10.3389/fnhum.2014.00643
5. A.A. Alyafi et al., From usable to incentive-building energy management systems. Modélisation et Utilisation du Contexte **2**, 1 (2018). https://doi.org/10.21494/ISTE.OP.2018.0302

6. A. Crabtree, The Social Organization of Communication in Domestic Settings, in Proceedings of the 8th Conference of the International Institute of Ethnomethodology and Conversation Analysis, Manchester (2003)
7. E.D. Mynatt, J. Rowan, A. Jacobs, S. Graighill, Digital family portraits: supporting peace of mind for extended family members, in Proceedings of the SIGCHI Conference on Human Factors in Computing Systems, March 2001 (CHI'01), pp. 333–340. https://doi.org/10.1145/365024.365126
8. S. Davidoff, M.K. Lee, C. Yiu, J. Zimmerman, D. Anind, Principles of Smart Home Control, in Proceedings of the 8th international conference on Ubiquitous Computing (Ubicomp'06), September 2006, pp. 19–34, https://doi.org/10.1007/11853565_2
9. J. Pierce, et al., Home, habits, and energy: examining domestic interactions and energy consumption, in Proceedings of the SIGCHI Conference on Human Factors in Computing Systems (New York, NY, 2010), pp. 1985–1994
10. P.B. Crabb, Effective control of energy-depleting behavior. Am. Psychol. **47**(6), 815–816 (1992). https://doi.org/10.1037/0003-066X.47.6.815
11. J. Pierce, et al., Energy aware dwelling: a critical survey of interaction design for eco-visualizations, in Proceedings of the 20th Australasian Conference on Computer-Human Interaction Designing for Habitus and Habitat - OZCHI '08 (Cairns, Australia, 2008), p. 1
12. J. Froehlich, et al., The design of eco-feedback technology, in Proceedings of the 28th International Conference on Human Factors in Computing Systems - CHI '10 (Atlanta, GA, 2010), p. 1999
13. B. Fogg, A behavior model for persuasive design, in Proceedings of the 4th International Conference on Persuasive Technology - Persuasive '09 (Claremont, CA, 2009), p. 1
14. B. Fogg, Captology: The study of computers as persuasive technologies, in CHI '97 Extended Abstracts on Human Factors in Computing Systems (New York, NY, 1997), pp. 129–129
15. V.B. Nguyen, Designing Persuasive Interactive Systems: energy as an application domain, PhD thesis, Grenoble-Alpes University, 2019
16. C. Pinder et al., Digital behaviour change interventions to break and form habits. ACM Trans. Comput.-Hum. Interact. **25**(3), 1–66 (2018). https://doi.org/10.1145/3196830
17. J.O. Prochaska, W.F. Velicer, The transtheoretical model of health behavior change. Am. J. Health Promot. **12**(1), 38–48 (1997). https://doi.org/10.4278/0890-1171-12.1.38
18. E. Shove, *Changing Human Behaviour and Lifestyle: A Challenge for Sustainable Consumption? The Ecological Economics of Consumption* (Edward Elgar Publishing, 2004)
19. T. Erickson, M. Li, Y. Kim, A. Deshpande, S. Sahu, T. Chao, P. Sukaviriya, M. Naphade, The Dubuque electricity portal: evaluation of a city-scale residential electricity consumption feedback system, in Proceedings of the SIGCHI Conference on Human Factors in Computing Systems (CHI'13), April 2013, pp. 1203–1212, https://doi.org/10.1145/2470654.2466155
20. R. Yang, M.W. Newman, Learning from a learning thermostat: lessons for intelligent systems in the home, in Proceedings of the 2013 ACM International Joint Conference on Pervasive and Ubiquitous Computing (UbiComp 2013) (ACM, 2013), pp. 93–102. https://doi.org/10.1145/2493432.2493489
21. J. Coutaz et al., "Will the last one out, please turn off the lights": promoting energy awareness in public areas of office buildings, in *Ambient Intelligence*, ed. by A. Kameas, K. Stathis, (Springer International Publishing), pp. 20–36
22. T. Schwartz, G. Stevens, T. Jakobi, S. Denef, L. Ramirez, V. Wulf, D. Randall, What people do with consumption feedback: a long-term living lab study of a home energy management system. Interact. Comput. **27**(6), 551–576 (2015). https://doi.org/10.1093/iwc/iwu009
23. J. Rodgers, L. Bartram, Exploring ambient and artistic visualization for residential energy use feedback. IEEE Trans. Vis. Comput. Graph. **17**(12), 2489–2497 (2011). https://doi.org/10.1109/TVCG.2011.196
24. L. Hallnäs, J. Redström, From use to presence: on the expressions and aesthetics of everyday computational things. ACM Trans. Comput.-Hum. Interact. **9**(2), 106–124 (2002). https://doi.org/10.1145/543434.543441

25. A. Ferscha, Informative art display metaphors. Universal access in human-computer interaction, in *Ambient Interaction*, ed. by C. Stephanidis, (Springer, Berlin Heidelberg, 2007), pp. 82–92
26. N. Jafarinaimi, et al., Breakaway: an ambient display designed to change human behavior, in CHI '05 Extended Abstracts on Human Factors in Computing Systems - CHI '05 (Portland, OR, 2005), p. 1945
27. W.-C. Fang, J.Y.-J. Hsu, Design concerns of persuasive feedback system, in Proceedings of the 7th AAAI Conference on Visual Representations and Reasoning (2010), pp. 20–25
28. J. Paay, et al., Design of an appliance level eco-feedback display for domestic electricity consumption, in Proceedings of the 26th Australian Computer-Human Interaction Conference on Designing Futures the Future of Design - OzCHI '14 (Sydney, NSW, Australia, 2014), pp. 332–341
29. M. Shipworth, *Motivating Home Energy Action: A Handbook of What Works* (Australian Greenhouse Office, 2000)
30. J. Redström, et al., Informative art: using amplified artworks as information displays, in Proceedings of DARE 2000 on Designing Augmented Reality Environments - DARE '00 (Elsinore, Denmark, 2000), pp. 103–114
31. D.A. Norman, *The Design of Everyday Things* (Basic Books, 2002)
32. B.J. Fogg, Persuasive technology: using computers to change what we think and do. Ubiquity. 2002. https://doi.org/10.1145/764008.763957
33. T.G. Holmes, Eco-visualization: combining art and technology to reduce energy consumption, in Proceedings of the 6th ACM SIGCHI Conference on Creativity & Cognition - C&C '07 (Washington, DC, 2007), p. 153
34. V. Nisi, et al., SINAIS from Fanal: design and evaluation of an art-inspired eco-feedback system, in Proceedings of the Biannual Conference of the Italian Chapter of SIGCHI on - CHItaly '13 (Trento, Italy, 2013), pp. 1–10
35. M.G. Petersen, et al., Aesthetic interaction: a pragmatist's aesthetics of interactive systems, Proceedings of the 5th Conference on Designing Interactive Systems: Processes, Practices, Methods, and Techniques (New York, NY, 2004)
36. R. Xiong, J. Donath, PeopleGarden: creating data portraits for users, in Proceedings of the 12th Annual ACM Symposium on User Interface Software and Technology (New York, NY, 1999), pp. 37–4
37. C. Fan, et al., A spark of activity: exploring informative art as visualization for physical activity, in Proceedings of the 2012 ACM Conference on Ubiquitous Computing - UbiComp'12 (Pittsburgh, PA, 2012), p. 81
38. T. Nakajima, V. Lehdonvirta, Designing motivation using persuasive ambient mirrors. Pers. Ubiquit. Comput. **17**(1), 107–126 (2013). https://doi.org/10.1007/s00779-011-0469-y
39. H. Evans, et al., Nuage Vert. July–September 2009, pp. 13–15, vol. 16 IEEE MultiMedia.XX (2009)
40. T. Skog, et al., Bringing computer graphics to everyday environments with informative art, in ACM SIGGRAPH 2002 Conference Abstracts and Applications on XX - SIGGRAPH '02 (San Antonio, TX, 2002), p. 153
41. L.E. Holmquist, T. Skog, Informative art: information visualization in everyday environments, in Proceedings of the 1st International Conference on Computer Graphics and Interactive Techniques in Australasia and South East Asia (2003), pp. 229–235
42. J. Mankoff, et al., Heuristic evaluation of ambient displays, in Proceedings of the SIGCHI Conference on Human Factors in Computing Systems (New York, NY, 2003), pp. 169–176
43. A. Cockburn et al., A review of overview+detail, zooming, and focus+context interfaces. ACM Comput. Surv. **41**(1), 1–31 (2008). https://doi.org/10.1145/1456650.1456652
44. B.B. Bederson et al., DateLens: a fisheye calendar interface for PDAs. ACM Trans. Comput.-Hum. Interact. **11**(1), 90–119 (2004). https://doi.org/10.1145/972648.972652
45. J. Lamping, et al., A focus+context technique based on hyperbolic geometry for visualising large hierarchies, in Proceedings of the SIGCHI Conference on Human Factors in Computing Systems (New York, NY, 1995), pp. 401–408

46. A. Blackwell, T.R.G. Green, Investment of attention as an analytic approach to cognitive dimensions, in 11th Annual Workshop of the Psychology of Programming Interst Group (PPIG-11) (1999), pp. 24–35
47. Y. Laurillau, et al., The top-slider for multi-criteria decision making by non-specialists, in Proceedings of NordiCHI'18 (Oslo, 2018), pp. 169–176. https://doi.org/10.1145/3240167.3240185
48. J. Kjeldskov, et al., Eco-forecasting for domestic electricity use, in Proceedings of the 33rd Annual ACM Conference on Human Factors in Computing Systems - CHI '15 (Seoul, Republic of Korea, 2015), pp. 1985–1988
49. M.K. Rasmussen, et al., Exploring the flexibility of everyday practices for shifting energy consumption through clockcast, in Proceedings of the 29th Australian Conference on Computer-Human Interaction - OZCHI '17 (Brisbane, QLD, Australia, 2017), pp. 296–306
50. L. Broms, et al., *The Energy AWARE Clock: Incorporating Electricity Use in the Social Interactions of Everyday Life*. (2009)
51. P. Dragicevic, S. Huot, SpiraClock: a continuous and non-intrusive display for upcoming events, in CHI '02 Extended Abstracts on Human Factors in Computer Systems, Minneapolis, MN (ACM Press, 2002), pp. 604–605, https://doi.org/10.1145/506443.506505
52. C. Tominski, H. Schumann, Enhanced interactive spiral display, in Proceedings of the Annual SIGRAD Conference, Special Theme: Interactivity (2008), pp. 53–56
53. J. Wambecke, G.-P. Bonneau, R. Blanch, R. Vergne, Activelec: an interaction-based visualization system to analyze household electrivity consumption, in Workshop Vis in Practice—Visualization Solutions in the Wild, IEEE VIS 2017, Oct 2017, Phoenix, (2017), 4p
54. H. Brynjarsdottir, et al., Sustainably unpersuaded: how persuasion narrows our vision of sustainability, in Proceedings of the 2012 ACM Annual Conference on Human Factors in Computing Systems - CHI '12 (Austin, TX, 2012), p. 947
55. S. Kuznetsov, E. Paulos, UpStream: motivating water conservation with low-cost water flow sensing and persuasive displays, in Proceedings of the 28th International Conference on Human Factors in Computing Systems - CHI '10 (Atlanta, GA, 2010), p. 1851
56. R. Orji, et al., Towards personality-driven persuasive health games and gamified systems, in Proceedings of the 2017 CHI Conference on Human Factors in Computing Systems - CHI '17 (Denver, CO, 2017), pp. 1015–1027

Faults and Failures in Smart Buildings: A New Tool for Diagnosis

Houda Najeh, Mahendra Pratap Singh, and Stephane Ploix

1 Introduction

Building systems are becoming more and more efficient as well as deliver useful building services that make occupants comfortable by providing them thermal comfort, air quality, ... They are an important provider of technology systems as well as they include HVAC systems, sophisticated controllers, energy management systems, and a large number of sensors. However, they are vulnerable to various faults, failures, and various events that could cause a discrepancy in building performance and consequently a discomfort to occupants. Fault diagnosis and maintenance of a whole building system is a complex task to perform. Available building fault detection and diagnosis tools are only capable of performing fault detection using behavioral constraints analysis. Singh et al. [53] proposes to use heterogeneous tests with validity constraints in the context of building fault diagnosis, but the proposed approach assumes that the sensors are reliable.

In a building system, a universal model, i.e., valid whatever the context, is difficult to set up . Different parts, such as the building envelop, the use and behavior of occupants, the devices for energy management, and various appliances interact, and it is difficult to model this interaction and the model is in most cases rather good ⊥°C. Diagnosis reasoning must differ in different scenarios, e.g., fault detection and diagnosis approaches should be different for normal working days and a vacation period. However, there are contextual models with limited validity. The problem

H. Najeh (✉) · S. Ploix
G-SCOP lab, Grenoble Institute of Technology, University of Grenoble Alps, CNRS UMR5272, Grenoble, France
e-mail: houda.najeh@grenoble-inp.fr; stephane.ploix@grenoble-inp.fr

M. P. Singh
Fraunhofer Institute for Systems and Innovation Research ISI, Karlsruhe, Germany
e-mail: mahendra.singh@isi.fraunhofer.de

S. Ploix et al. (eds.), *Towards Energy Smart Homes*,
https://doi.org/10.1007/978-3-030-76477-7_14

is that the validity is measured with potentially faulty sensors. If these sensors are faulty, the diagnostic result is not guaranteed, and there is a need for an indicator to assess a level of validity of a test. Nevertheless, the diagnostic result is calculated from a set of tests, each one defined by its level of validity. The question that arises is how to conclude in terms of diagnosis and how to take into account the level of validity in the diagnosis.

A test relies on several variables and that all the data should be available along a given time period. However, the data gaps are the main sensor fault in buildings and performing test could be an issue. Sensor values are not uniformly sampled, and the delay depends on the measured value and the type of sensor. The problem is to decide from which delay the sensor becomes faulty?

The objective of this work is to highlight these challenges as well as to provide a strategy about how to solve them. This chapter is motivated by new challenges as well as new solutions for diagnosis in building are proposed. A state of the art about diagnosis in buildings is presented in Sect. 2. In Sect. 3, the major challenges that lie in this problem domain are identified. Section 4 discusses the need for new services for diagnosis in building. Sections 5 and 6 present respectively the application example and the proposed methodologies for diagnostic reasoning. Finally, a conclusion is presented in Sect. 7.

2 Fault Diagnosis in Buildings: State of the Art

2.1 Faults in Buildings

It is difficult to obtain detailed information on energy consumption in buildings since it requires more detailed monitoring and measurements than what is usually available. However, the energy consumption yearly report [58] shows how the total energy consumption can be divided into different end-uses in buildings (Table 1). The statistics show that the energy end-uses of commercial building are as follows: lighting 20.2%, spacing heating 16.0%, spacing cooling 14.5%, ventilation 9.1%, refrigeration 6.6%, and other end-uses 33.6% [57, 58]. The faults in the system can occur in connection with each of these end-uses. This figure reveals how costly a fault could be in terms of its energy use. Studies show that 25–45% of energy HVAC energy consumptions are wasted due to faults, i.e., to a difference between the characteristic observed on the device and the reference characteristic when it is out of specification, including improper control logic and strategy, malfunction of controllers and controlled devices, etc. [1]. In addition, the waste of energy is also due to a non-optimal control. In fact, a number of studies [35, 39] indicated that optimal control strategies can reduce the energy waste and improve the overall building energy efficiency.

Table 1 Commercial energy end-use spilt

Source	Percentage
Lighting	20%
Space heating	16%
Space cooling	15%
Ventilation	9%
Refrigeration	7%
Water heating	4%
Electronics	4%
Computers	4%
Cooking	1%
Other	15%
Adjust to SEDS	5%

Common faults in buildings are:

Breakdown
The assets within buildings may be classified as parts of several systems according to the services they provide to owners. For example, HVAC systems provide hot water and heat and electrical systems provide electric power and possibly heat. However, each building system breaks down over time, and regular maintenance, repairs, and renewals are required to keep a building in working order. For example, items such as light fixtures and control panels in electrical part in buildings are subject to breakdown. In addition, the HVAC system in buildings are responsible for the heating, domestic hot water, ventilation, and cooling of a building. It consists of assets such as pumps, filters, boilers, fans, and air-conditioning equipment. These assets tend to have short to medium service lives, breaking down due to regular wear.

Misusage
Another important type of faults in buildings is misusage, but it has not yet received much attention in the scientific literature. Usage in general refers to the function of serving or using something. We must think and conceive quality, not only from techniques and standards but also from the user as a human and social being.

When users occupy a building, they appropriate these three entities: envelope, equipment, and internal organization, and adjust them as much as possible to their own comfort level. Occupants interact with the environment around them. One of the main criteria of the comfort of occupants is the control of these interactions. In general, users prefer simple and modular equipment and systems. All stages of the project life influence the final quality of the building. Therefore, even if the use of a building is only effective during its operational phase, it is important to ensure that certain good practices are followed during each stage of the building's life in order to guarantee good quality of use.

The notion of quality of use is a factor that must be integrated when planning the building. About 65% of the discomfort encountered during the follow-up is due to errors made during the programming and design phases. The evolution of home automation in buildings certainly requires a new grip on the part of the user, but it must not be forgotten that this is the building that must adapt to the needs of the user and not the other way around.

Misuse of the building may have effects harmful to:

- The comfort of the user
- The health of the user
- The durability of the building
- The environment with overconsumption linked to poor control of the equipment

The bad uses are almost the same in any residential dwelling. The consideration of human behavior is essential in the application of diagnosis in buildings. Let us take the following examples: frequent door opening and use of an important number of appliances.

Human Mistake

Human mistake is another important type of faults in buildings which has not received an adequate level of attention. Beyond the construction literature, it is common ground that human error, not technology, predominates in failures of all types [22, 48]. Examples of such faults are HVAC left on when space is unoccupied.

Human mistake is a very known fault type in diagnosis domain. A lot of studies like [5] show the role of early detection of human errors in building projects.

Wrong Configuration

Further investigation shows that faults due to the wrong configuration are also a typical fault in new buildings, which has not received an adequate level of attention [38, 55]. Examples of such faults are: wrongly configured building equipment, where the setting of the equipment is wrong and misplaced or wrongly wired sensors and actuators.

Data Failure

Data failure is another important type of faults in buildings. The applications for sensor technology are increasing rapidly. Sensors are currently being used for applications in buildings. Sensors are continually being developed with advanced capabilities, such as more reliable data extracting. These sensors can also be used to better control the building but also to estimate occupant practices essential for energy consumption by estimating the number of occupants per area and their metabolic contribution, their activities, and their routines [2]. With the cost and size of sensors becoming cheaper and smaller at a fast rate, it has been forecasted that sensors in the near future will be installed in dense arrays to eventually monitor the entire built environment [17]. There is currently a gap between modern sensing technologies and their application and applicability in the field for monitoring the performance of buildings. Research and experimental validation tests are required to assess the limitations, challenges, and performance of installing new sensor technologies to monitor certain aspects of concrete structures [51]. The concept of healthy sensors is known in the literature. Authors like [33] assumed that there are two groups of sensors: sensors that correctly measure structural responses (termed as "reference sensors") and failed (or uncertain) sensors.

Roth et al. [49] concluded that typical faults in commercial buildings consist of 13 types of faults. The annual impact of each of them in terms of energy consumption is presented in Fig. 1.

All the types of faults mentioned above are faults of the type "normal faults," i.e., easy to reveal. On the other hand, there are other insidious types of faults. For example, the absence of noise in the ventilation system does not imply no fault, but no fault has been revealed.

A few studies like [25] defined the concept of insidious faults. In this study, authors show that rising damp refers to ground water seeping up through the footings and base walling of houses due to the absence of damp proof courses, or these being poorly edited or dislodged. It is an insidious fault that can be difficult to address, without understanding scientific concepts such as capillary action of water, drainage, and hydrostatic pressure of ground water.

breakdown

misuage

human mistake

Fault	Fault type	By %	Annual energy [quards]
Duct leakage	Air Distribution	30%	0.3
HVAC left on when space unoccupied	HVAC	20%	0.2
Light left on when space unoccupied	Lightning	18%	0.18
Airflow not balanced	Air Distribution	7%	0.07
Improper refrigerant change	Refregreation circuit	7%	0.07
Dampers not working properly	Air Distribution	6%	0.055
Insufficient evaporator airflow	Air Distribution	4%	0.035
Improper controls setup / commissioning	Controls	2%	0.023
Control components failure or degradation	Controls	2%	0.023
Software programming error	Controls	1%	0.012
Improper controls hardware installation	Controls	1%	0.010
Air-flowed condenser fooling	Refrigration circuit	1%	0.008
Valve leakage	Waterside Issues	1%	0.007
Total	–	100%	1

+ data failure

+ wrong configuration (wrongly configured building equipment)

Fig. 1 The annual impact of faults in terms of energy consumption [49]

2.2 Overview of General Diagnosis Methods

Over recent years, FDD became an appealing area of research for building researchers. Various methodologies and tools have been developed to identify the faults in buildings to track the whole building performance.

Plenty of published research and survey papers are available to classify the building diagnostic techniques [27, 55]. Lately, building's faults and failures are covered at a more granular level with an impact analysis in the terms of energy consumption and financial consequences [18, 49]. Hybrid diagnosis approaches have shown an improved result over the conventional model-based diagnosis approaches [15, 32].

Katipamula and Brambley [27] presents a detailed review of fault detection and diagnosis techniques in buildings.

In August 1996 (Revised in 2001), International energy agency (IEA) published Annex-25, "Building optimization and fault diagnosis source book" [12, 24]. This work is considered as a beginning of fault detection and diagnosis in smart buildings. The aim of this work is to highlight the major faults that affect HVAC systems and controllers. In 2002, a technical report called NBCIP1 was published by Iowa Energy Center and United States Environmental Protection Agency (USEPA) [4]. The report articulates 67 case studies with 110 field studies for buildings. The aim of this work is to highlight the main source of faults in buildings coming from humans, software, and hardware. In more recent works, the Automatic Building Commissioning Analysis Tool (ABCAT), and Whole Building Diagnostician (WBD) developed by Texas A& M University and Pacific Northwest National Laboratory (PNNL) have been developed as new tools for identifying the whole building level faults [7, 29]. Recently, the Lawrence Berkeley National Laboratory and Simulation a model-based diagnostic tool has been developed [6].

In general, all the major approaches that have been used for building diagnoses are quantitative (model-based), qualitative (rule-based), or signal-based methods.

2.2.1 Building Fault Diagnosis Using Model-Based Techniques

Model-based diagnosis (MBD) uses an explicit model of the system under diagnosis. It can be qualitative or quantitative models. In general, all the model-based diagnosis approaches consist of three important stages: symptom generation, symptom evaluation, and fault isolation.

Quantitative model-based approaches are based on physical models and require detailed mathematical relations among all the operating variables with the characteristic of all components within the system. Mostly, these models are in form of a differential equation or state-space model and presume to have additional knowledge of the normal operation of system under the investigation. Unlike, the quantitative model-based diagnosis, *qualitative model-based* uses qualitative reasoning or knowledge-based information to conclude whether system or its components are in the faulty or normal state.

Dexter and Ngo [12] presented a fuzzy model to diagnose several faults in the air handling unit. Through comparing the outputs of the fuzzy model with those of the reference model, the faults that occurred in the air handling unit can be diagnosed. Norford et al. [44] developed a physical model to detect commonly occurred faults in the air handling unit. Castro [9] presented a physical model to detect the faults in the chillers. Wang and Chen [59] also presented the model-based strategy to diagnose the sensor faults in the chilling plant system. Yu et al. [63] presented a virtual model to estimate the supply airflow rate in the rooftop air-conditioning units. Employing the mass balance and energy balance, the physical residues can be calculated by comparing the outputs of the models with real measurements. Besides the physical diagnosis models, the gray-box [26] and black-box [3] models have also been developed to diagnose the chiller faults. Generally, the model-based methods [62] have been most widely developed in the HVAC systems. The application of the model-based FDD method relies on the accurate mathematical physical models.

Qualitative model-based approach uses a set of rules to diagnose the system abnormality. For example, [19] proposed a fault diagnosis of air-conditioning systems based on a qualitative bond graph. The main privilege of model-based techniques is that they require only a knowledge of normal operation and a reasoning method based on consistency. The model-based diagnosis has been developed by two communities: fault detection and isolation (FDI) community in the field of automatic control and Logical Diagnosis (DX) in the field of artificial intelligence (AI).

> Model-based fault detection and isolation methods rely on an analytical model, derived from a physical relation. In connection with buildings, it is really impossible to develop a complete physical-model matching accurately the reality for a whole building system. The modeling of various phenomenon like heat transfer from facade or unplanned occupancy is a challenge. In addition, they believe only in behavioral constraints and assumed to be true in all circumstances. However, universally valid behavioral models i.e. valid whatever the contexts are difficult to set up.

> Model-based fault diagnosis and isolation techniques assumes that model represents the reality of building operation independately of the current context and any fault can be detected by measuring the physical variables and checking the consistency with a reference model. A physical variable is a potentially observable element of information about the actual state of a building system. Nevertheless, universal models are difficult to set up. Erroneous all-context models might lead to invalid diagnoses [45].

These approaches are relevant for data failure, human mistake, and breakdown type faults, but they augmented a full analytical model.

2.2.2 Building Fault Diagnosis Using Rule-Based Techniques

Knowledge-Based FDD methods require a sufficient amount of historic data. These methods use methods from artificial intelligence to extract the knowledge based from the historic data reflecting the relationship between system variables. The behavior of the system is monitored in real-time and is compared with the knowledge base to detect possible deviations and make fault diagnosis decisions. Depending on the knowledge extraction process in this category, the methods could be divided into qualitative methods [23, 40] and quantitative methods [34, 64]. Some of the most popular qualitative knowledge-based FDD methods are those that are based on expert systems. This approach basically evaluates real-time data according to a set of rules, which are derived from the knowledge of an expert human operator.

There are a number of papers that discuss expert system applications for fault diagnosis of specific systems. Initial attempts at the application of expert systems for fault diagnosis can be found in [10, 43]. The objectives of this expert system were twofold. First, the system classifies the reasons for the observed problem as an operator error, equipment failure, or system disturbance. Second, the expert system offers prescriptive remedies to restore the process to normal operation.

In parallel, a contemporary group of researchers also focused on qualitative models for fault diagnosis analysis. In buildings, rule-based qualitative models are used to diagnose faults in air handling units or other parts of HVAC [21, 28, 50]. With a set of rules, the faults that occurred in the air handling unit can be diagnosed successfully. Also, rule-based diagnosis methods are also adopted in the literature to manage the whole building [13]. In this work, authors present an intelligent decision support model using rule sets based on a typical building energy management system. In addition, the model's impact on the energy consumption and indoor quality of a typical office building in Greece is presented. The model can control how the building operational data deviates from the settings as well as carry out diagnosis of internal conditions and optimize building's energy operation. In this context, the integrated "decision support model" can contribute to the management of the daily energy operations of a typical building, related to the energy consumption, by incorporating the following requirements in the best possible way: the guarantee of the desirable levels of living quality in all building's rooms and the necessity for energy savings.

> Qualitative models are not enough to cover all the possible actions by following rules. Moreover, tests derived from rules are challenged by their validity. For instance, testing indoor temperature without validating door or window position might lead to a false alarm.

> In addition, the application of rule-based FDD methods depends on the rules constructed. For example, for testing a HVAC system using a set of rules, if the rules are not detailed enough, the diagnosis efficiency may be limited.

These approaches are relevant to human mistake and breakdown type faults.

2.2.3 Building Fault Diagnosis Using Signal-Based Techniques

Signal-based FDD methods mainly use signals, which are obtained from measurements for diagnostics [20]. The algorithms within this category derive symptoms of a healthy system as an output of the symptom analysis and the knowledge of the system, which are at disposal. When a system is faulty, symptoms that appear in the measured signal differ from those of healthy systems. Typically, these methods analyze signals in either time domain or frequency domain. However, there are also methods in this category that use both time and frequency domains. The difference between signal-based and rule-based diagnostic methods is explained by the fact that signal-based methods are based on signal processing techniques, whereas rule-based methods are based on rules coded in the form of if-then-else statements.

As a new FDD method, recently, the data-driven methods have been paid more attention in HVAC field. The data-driven method such as principal component analysis [14], neural network [65], etc.... never need to build the accurate mathematical physical models or detailed experience rules.

Authors in [30] presented a general regression neural network in the air handling unit. It can be used to diagnose the abrupt and performance degradation faults. Wang and Chen [59] developed a detection model based on neural network in the variable air volume systems. The neural network can be used to diagnose the faults of outdoor air, supply air, and return air flow rate sensors after training using operation data. A fault detection and diagnosis strategy using combined neural networks and subtractive clustering analysis is presented in [16].

Actually, the data-driven FDD methods usually take advantage of the intrinsic relations among the various data. Through calculating the deep intrinsic mathematic relations of the variables, the normal and abnormal operation can be distinguished. When faults occur, the intrinsic relations among variables will be broken, which is different from that under normal conditions.

These methods are relevant for data failure and might detect fault signature for breakdown type fault.

Most signal-based fault detection and isolation techniques in the literature are interested only by the following known fault types: drift, outliers and bias. The occurrence of data gap faults has also not been given an adequate span of attention in the academia.

3 Diagnosis in Buildings: New Challenges

Current work highlights the following key challenges in building fault diagnosis:

- Complexity
- Modeling difficulty
- No universal model
- Unreliable sensors in buildings

Following sub-sections explain these issues and proposed methodology in detail.

3.1 Complexity

The first challenge is complexity in testing a whole building system using both rule and pure model-based tests. Buildings are complex systems, and the relations among the different sub-systems are intricated.

Buildings are becoming more complex because of the higher expectations of users as well as the ongoing integration of many technologies. They are equipped with HVAC systems, sensors, building automation system, and supervisory controllers.

According to [31], the problems in building complex systems often arise in the interfaces between the various elements such as hardware, software, and human components.

Complexity is defined as uncertainty in [8, 60], ambiguity in [36], variability in [46], and dynamism in [37, 61], which are caused by changes in organizational and technological project environments. Changes may result from either the stochastic nature of the environment or a lack of information and knowledge about the project environment.

3.2 No Universal Model

In connection with buildings, it is tough to develop a physical model that matches precisely the reality. The various phenomenon like heat transfer from facade or unplanned occupancy are challenging jobs to model. Clarke et al. [11] show that models simulate reality within $+/-\ 1°C$ (well enough). The problem is the inputs. The IEA EBC Annex 58-project "Reliable Building Energy Performance Characterisation Based on Full Scale Dynamic Measurements" [47] is developing the necessary knowledge and tools to achieve reliable in situ dynamic testing and data analysis methods that can be used to characterize the actual thermal performance and energy efficiency of building components and whole buildings. For identifying systems, ARX-models are one of the standard tools. ARX model structure is a linear difference equation which relates the current output at time t to a finite number of past outputs and inputs. The main problem when applying ARX-models is the inputs which are obtained from potentially faulty sensors.

Singh [52] proposed the concept of contextual model, i.e., a model valid under specific contexts. These local contexts define the validity constraints. For example, for testing the indoor temperature, we should take into account several factors such as occupancy, the door and window positions, and weather conditions. The validity is measured with possibly faulty sensors. In fact, these sensors are subject to bias, outliers, or could be missed. The problem is how to conclude about a test that can be valid or not knowing that validity can only be tested with possibly faulty sensors? This is a challenge.

3.3 Unreliable Sensors in Buildings

In buildings, an important amount of data is available from sensors. Sensor values are not necessarily uniformly sampled. While after pre-processing the sensors report values regularly, reality shows that quite many values are missing. The gaps that as a result exist are sometimes too small to be visible on a graph.

4 Need for New Services for Diagnosis in Buildings

This section discusses the need for new services for diagnosis in buildings.

4.1 Need for Testing in Specific Context Under the Hypothesis of Fault Modeling

In the domain of fault diagnosis, a symptom is defined as a measurable change in the behavior of a system from its normal behavior i.e. an indication of fault. Conventional model or rule-based behavioral tests are used to generate only symptoms. These models appear in the behavioral constraints and it is assumed that the behavioral test could be applied to any situation without taking into account different contexts. These tests can be more or less valid due to the difficulty of getting good datasets in building and to sensor aging. If the validity is not taken into account during the conception of tests, false symptoms may be produced and consequently a false diagnosis. *The aim of this work is to explicitly take into account the validity of the tests to make the diagnostic decision.*

However, a model valid for all context is difficult to design and the validity of a test result is always questioned in fault diagnosis. In addition, the validity is measured with potentially faulty sensors. In fact, sensors are subject to different kinds of faults. The data can be biased, subject to outliers or missing.

The test of the space of validity consists of observation points given either by the sensors or by an expert. If the sensors are in operating mode, then the performance guarantee is assured. Otherwise, there is no guarantee. The question is: Is the test space always covered or not? *Hence the need for an indicator to assess the level of validity for each test.*

An example of contextual test considering the building thermal performance test is given in Fig. 2.

Let's consider a range-based test that checks the indoor temperature T_{in} for the building shown in Fig. 2 is estimating and testing the indoor temperature with a behavioral constraint i.e. T_{in} lies between the maximum temperature (T_{max}) and minimum temperature (T_{min}). This model-based thermal test only considers the behavioral constraint and evaluates symptoms under the predetermined thermal

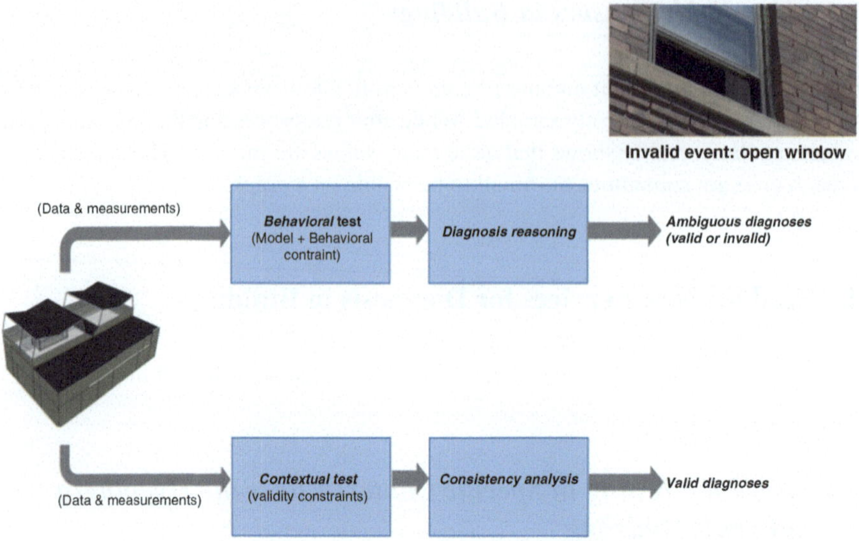

Fig. 2 Behavioral and contextual test for diagnosis

bounds for a specific building. However, no validity constraints are integrated with the test. Figure 2 illustrates the significance of behavioral and contextual tests along with an example of an invalid event, i.e., open window. In this case, the following validity constraints are needed to be combined with behavioral constraints:

- Testing indoor temperature without *verifying occupancy level for all times* might lead to a false alarm.
- *The door and window position need to be verified for all times* because these inputs are not easy to model.
- Similarly, *outdoor weather condition needs to be verified for all times.*

These validity constraints are difficult to model and due to the lack of knowledge about the validity. A pure model-based test might lead to an ambiguous test result. In conclusion, model validity is another kind of knowledge about the behavior. In order to launch a valid diagnosis analysis, each test needs to satisfy the validity constraints V and behavioral constraints B simultaneously (see Table 2).

4.2 *Need for Indicators to Assess a Level of Validity of a Test and a Confidence Level for Global Diagnosis*

A test is performed in a period of time considering behavioral and validity constraints. Validity constraints evaluate whether the tests can be performed or not. However, the validity is measured with potentially faulty sensors. If the sensors are

Table 2 Table of validity

B	V	Conclusion
Satisfied $\forall t$	Satisfied $\forall t$	Normal behavior
Satisfied $\forall t$	Non-satisfied	Invalid
Non-satisfied	Satisfied $\forall t$	Abnormality
Non-satisfied	Non-satisfied	Invalid
Satisfied $\exists t$	Satisfied $\exists t$	Normal behavior
Satisfied $\exists t$	Non-satisfied	Invalid
Non-satisfied	Satisfied $\exists t$	Abnormality
Non-satisfied	Non-satisfied	Invalid

in OK state, the test result is always guaranteed. If the sensors are faulty, there is no longer guarantee and the question that arises is how to evaluate the level of validity of a test in the presence of sensor faults? *Hence, there is a need for an indicator to assess a level of validity for each test.*

The diagnostic result is calculated from a set of tests, each defined by its level of validity. The problem is how to evaluate the confidence level of diagnoses in the presence of partially valid tests? *Hence, there is a need for a confidence level for global diagnosis.*

4.3 Need to Know the Periods of Good Operation of Sensors

After receiving signals from a sensor, these signals need to be processed. An acceptable and accurate process of these signals requires:

1. Full knowledge regarding the operation of the sensors and the nature of signals:
   In order to be able to use signals' information correctly, the operation of a sensor, and the nature of signals they produce should be well understood. By having this knowledge, we are able to choose the right tools for the acquisition of data from the sensor. For instance, if the sensor produces a time varying signal where the information is embedded in its frequency signatures, then a frequency counter and possibly a frequency analyzer are needed.
2. Posteriori knowledge regarding the received signals:
   A posteriori knowledge about the received signals is important in order to assure that the data will be interpreted correctly and that the right device is used in the measurement process. We need to have a good understanding of what is expected from the sensor and system. The measured value can be significantly different from the real measurand.
   A posteriori knowledge is dependent on experience.
3. Information about the dynamic and static characteristics of the sensing systems:
   The characteristics of a sensor can be classified into two groups: static and dynamic. Understanding the dynamic and static characteristics behaviors is imperative for mapping the output versus the input of a system (measurand).

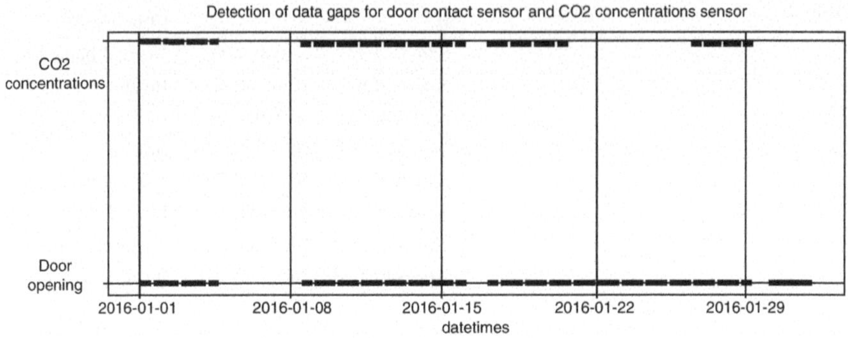

Fig. 3 Time of recordings for door contact sensor and CO2 concentrations sensor

With the increasing number of sensor devices, as well as sensor data types, the acquisition of the sensor data samples becomes time and energy consuming, which is undesirable on low power wearable devices.

Many values are missing. Figure 3 shows the evolution of raw sensor measurements for two different sensors: a door contact sensor and a CO_2 concentration sensor.

Sensor values are not necessarily uniformly sampled. There are no regularly delayed data for a variable. Delays depend not only on the type of sensor but also on the measured values. The question that arises is from which delay can we say that the sensor becomes faulty? Hence, the necessity for automatic thresholding for data gap detection for heterogeneous sensors in instrumented buildings.

5 Application Example

In this section, a case study is presented to discuss the challenges in building diagnosis.

5.1 Presentation of the Platform

The test bed is a classroom in the University of Southern Denmark (see Fig. 4).

The classroom is equipped with about 40 sensors that measure the indoor temperature, the CO_2 concentrations, the airflow, the heat from the radiators, the rpm of the ventilation system, the number of occupants, the illuminance, the corridor temperature, and the corridor CO_2 concentrations and many more. Two types of heating are installed in the platform: hydraulic heating and forced air heating from the ventilation system.

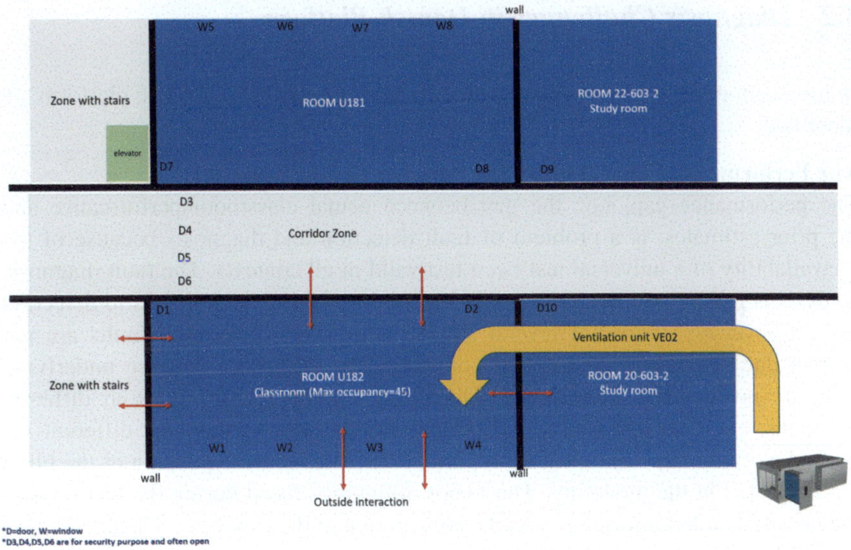

Fig. 4 Test bed

The ventilation systems at OU44 are of the type VAV (Variable Air Volume) which serves three floors called the ground floor, living room, and 1st floor. Living room and ground floor are divided into zones with one or more VAV dampers. On the first floor, there is an office area where the offices are provided with VAV dampers and common exhaust with a pressure holding damper which regulates the total extraction in relation to the total supplied air. Each AHU unit has an exhaust fan, outside and supply airflow measuring stations, mixing box, pre-filter, final filter, heating hot water coil, chilled water coil, and supply fan. A centrifugal fan provided by NK Industri (NKI) is installed in the ventilation unit. NKI climate control units are integrated with either centrifugal fans, axial fans, or chamber fans.

The fan type is suitable for installations, with changes in air performance and energy consumption. At the start of the HVAC system, the main damper is opened. The air flow for each fan is calculated in CTS (Clear-to-send) programs based on pressure measurements at the input rings for supply and return fan, respectively. Fan electricity consumption (absorbed electrical power) should be provided in the CTS system with a continuous exercising. Air flow sensors are installed inside the air flow measuring stations in the Schneider BMS system to record the air flow rate for three ventilation fan units. Accuracy of the airflow measuring satiation is marked as ±2% at 6000 feet per minute and ±0.5% at 2000 feet per minute. A graphical user interface sMAP 2.0 is a plotting engine to display the raw data. In order to deal with missing data and ambiguity in measurements, the measured data is re-sampled with the one minute sample period.

5.2 Diagnosis Challenges in Danish Platform

In this section, the major challenges that lie in diagnosis in the Danish platform are identified.

The Performance Gap is Due to the Absence of Universal Test
The performance gap, i.e., the gap between actual classroom performance and the prior estimates, is a problem of fault detection and diagnosis because of the unavailability of a universal test i.e. a test valid in all contexts. The fault diagnosis analysis is generated from the modeled behavior of the system thanks to detection tests. Conversely, there are several situations in which diagnosed faults are not correct due to change in the local context of the classroom because underlying tests are not context independent. Diagnosis reasoning must differ in different scenarios, e.g., fault detection and diagnosis approaches should be different for normal working days and a vacation period. Variables like the position of the blind are neglected in the modeling. The major difficulties faced during the test include the lack of detailed information on the constitution of the classroom, the uncertainty about occupant use, and behavior. For example, for testing the indoor temperature, several factors are linked with each other, and it is necessary to model the airflow through the windows and through the corridor, the airflow from the ventilation system, the weather station, the thermal conduction of walls. These local contexts are hard to model and lead to invalid diagnosis results. So, it is difficult to build a universal test.

The Contextual Test Facilitates Testing
The contextual test [53] consists of testing the behavior only in a particular context. For example, for testing the indoor temperature in the classroom, the test is made only in the following contexts: absence, door and window closed, and an outdoor temperature between a determined upper and lower limits. The contextual test combining different events is based on validity constraints [53] for a test. The validity is measured with potentially faulty sensors. Let us consider the example of the indoor temperature test where the behavioral constraint is defined by a behavioral constraint $B_\tau(X_{\mathbb{T}}) \in \mathcal{B}_\tau : T_{in} \in [-3, 12°C], \forall t \in \mathbb{T}$ and a validity constraint defined by $V_\tau(X_{\mathbb{T}}) \in \mathcal{V}_\tau : T_{out} \in [T_{out}^{min}, T_{out}^{max}] \wedge \zeta_D = 0 \wedge \zeta_W = 0, \forall t \in \mathbb{T}$ where $\mathbb{T}$, T_{in}, T_{out}, ζ_D, and ζ_W represent respectively the time period, the indoor temperature, the outdoor temperature, and the positions of the door and the window. Figure 5 shows the door position over 23 hours in June 3rd, 2016 from $t = 00{:}00$ to $t = 23{:}00$.

The door contact sensor always shows that the door is closed, but at a defined time ($t = 9$ h), the door is open. The question that arises is how to conclude that this test is valid or not knowing that the door contact sensor is faulty?

Missing Data
Missing data is the most interesting fault type for a building system. To make a test, the first step is to identify the periods when the datasets are complete. Figure 6

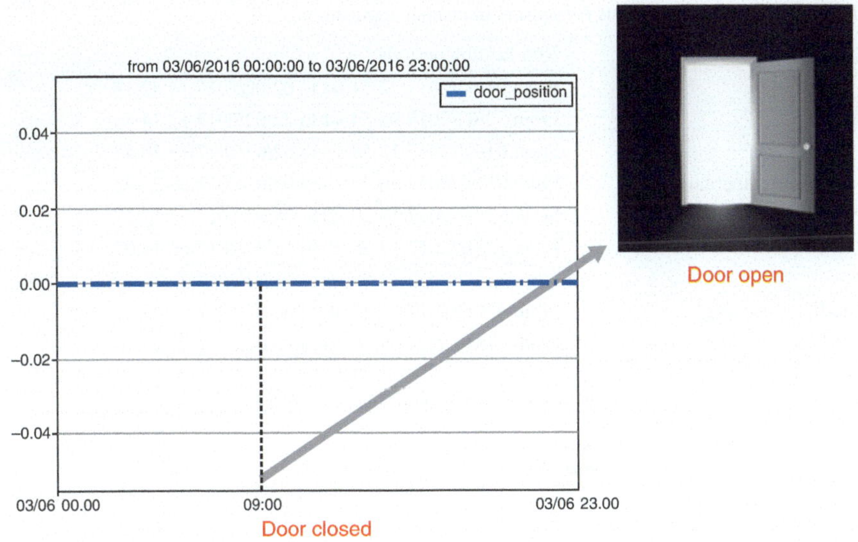

Fig. 5 Door position over 23 h

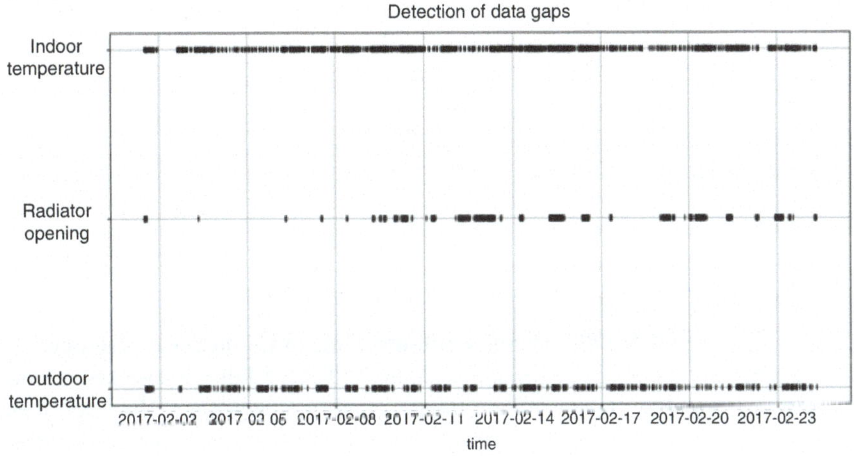

Fig. 6 Detection of data gaps

shows respectively the detection of data gaps for the indoor temperature sensor, the radiator opening sensor, and the outdoor temperature sensor.

An algorithm based on statistical approaches [42] is used for this purpose. The idea is to determine the non-healthy periods for each sensor then make the intersection of the periods and distinguish the healthy periods from the non-healthy ones. Table 3 shows the non-healthy periods for only these three sensors.

Table 3 Non-healthy periods for sensors in Danish application

Sensor	Non-healthy periods
Temperature	From 01/02/2017 22:51:04 to 02/02/2017 14:42:18
	From 22/02/2017 08:54:44 to 22/02/2017 17:14:43
	From 01/02/2017 17:54:18 to 02/02/2017 15:39:45
Outdoor temperature	From 02/02/2017 16:49:25 to 03/02/2017 08:25:52
	From 07/02/2017 17:29:20 to 08/02/2017 08:12:17
	From 12/02/2017 17:26:20 to 13/02/2017 08:16:07
	From 01/02/2017 13:04:40 to 03/02/2017 07:36:12
Radiator opening	From 03/02/2017 07:53:48 to 06/02/2017 07:45:11
	From 17/02/2017 08:30:16 to 19/02/2017 00:01:46

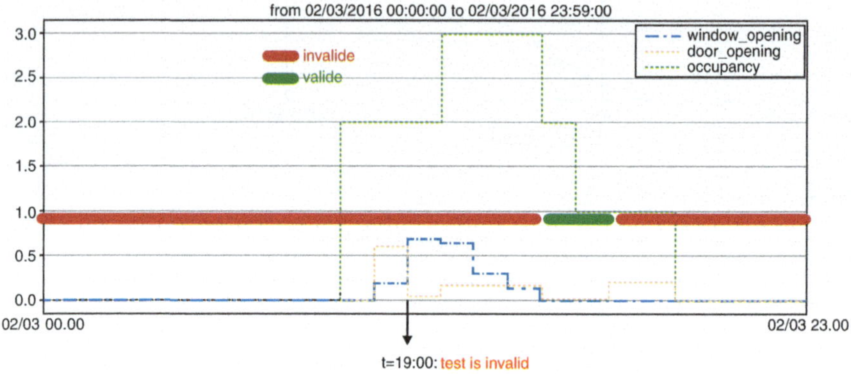

Fig. 7 24 h validity test

The intersection of the non-healthy periods for all the sensors installed in the platform is from February 9th, 2017, to February 15th, 2017.

Good Behavior and Validity Require Infinite Time to Confirm Consistency
A single instant is enough to reveal an anomaly if the validity is checked. Good behavior and validity require infinite time to confirm consistency. However, it is difficult to test over an infinite time because the data are unavailable for long periods of time. To overcome this problem, the tests are performed over a finite time, and if the conclusion of the test is ok over a finite time, then it is also ok over an infinite time. Another problem is related to testing over a continuous period of time. Let us go back to the temperature test where the validity constraint is defined by $V_\tau(X_{\mathbb{T}}) \in \mathcal{V}_\tau : T_{out}(t) \in [T_{out}^{min}, T_{out}^{max}] \wedge \zeta_D(t) = 0 \wedge \zeta_W(t) = 0, \forall t \in \mathbb{T}$. Figure 7 shows the satisfaction of the validity constraint over a continuous period of 24 h on March 2, 2016, from $t = 00{:}00$ to $t = 23{:}59$.

The test is invalid, for example, at $t = 19$ h because the validity constraint is not satisfied. Also, we note that the validity is discontinuous. There are invalid periods designated by a red color and other invalid periods designated by a green color. Therefore, testing over a continuous period always results in the conclusion of an invalid test. To overcome this problem, testing over discontinuous periods makes testing easier. The test period consists only of the instants where the validity constraint is checked.

6 Diagnostic Analysis in Danish Application

6.1 Design of Partial Valid Tests

Let us consider the following tests in the framework of the Danish application:

Test 1: Damper
This test verifies if the damper is faulty or not. $Test_1$ generates test results about the presence/absence of fault in the damper using a set of rules.

$$Test1 = \begin{cases} B(X_{\mathbb{T}}) \in \mathcal{B}_{\tau} \wedge V(X_{\mathbb{T}}) \in \mathcal{V}_{\tau} \rightarrow ok(\Sigma) \\ B(X_{\mathbb{T}}) \notin \mathcal{B}_{\tau} \wedge V(X_{\mathbb{T}}) \in \mathcal{V}_{\tau} \rightarrow \neg ok(\Sigma) \end{cases}$$

Since it is difficult to establish a linear law between the concentration of CO2 and the position of damper, we tried to define a law of control experimentally (see Fig. 8).

This test is always valid. This test is defined by a behavioral constraint $B_{\tau}(X_{\mathbb{T}}) \in \mathcal{B}_{\tau}$ with $\mathcal{B}_{\tau} : CO_2(t) \geq 800 \wedge damper_{position}(t) \geq 50, \forall t \in \mathbb{T}$

The bunch of data required for the test of behavioral constraint is $damper_{position(\mathbb{T})}$ and $CO_{2(\mathbb{T})}$. They are a sequence of similar intervals of damper positions and CO_2 concentrations measured respectively by damper position sensor and CO_2 concentrations sensor installed in Denmark application.

This test is also defined by a test support. The possible fault explanations for this test in case of inconsistency are ¬ ok(damper) ∨ ¬ ok(damper controller) ∨ ¬ ok(damper sensor) ∨ ¬ ok(CO2 concentrations sensor)

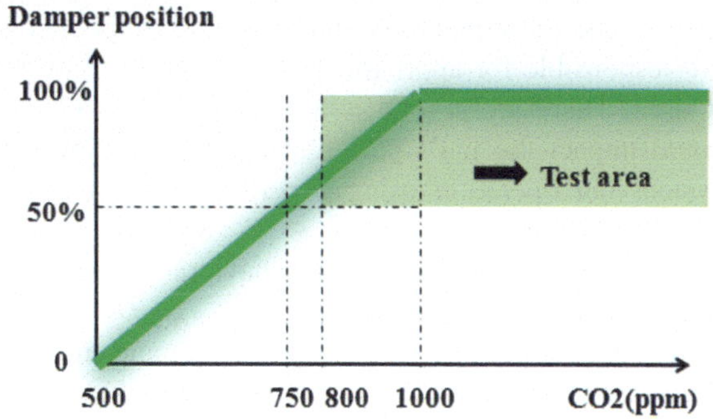

Fig. 8 Law of control between the concentration of CO2 and the position of damper

Test 2: Efficiency of the Heater Exchanger

This test verifies the efficiency of the heater exchanger. This test is defined by a validity constraint $V_\tau(X_\mathbb{T}) \in \mathcal{V}_\tau$ with $\mathcal{V}_\tau$: rpm$(t) > 0 \wedge$ airflow$(t) > 0, \forall t \in \mathbb{T}$. This test is also defined by a behavioral constraint $B_\tau(X_\mathbb{T}) \in \mathcal{B}_\tau$ with $\mathcal{B}_\tau$: efficiency$(t) \geq 70\%, \forall \in \mathbb{T}$ with efficiency is the efficiency of the heating exchanger and it is calculated by:

$$\text{efficiency}(t) = \frac{\Delta_{T1} - \Delta_{T2}}{100}, \forall t \in \mathbb{T} \quad (1)$$

with

$$\Delta_{T1} = T_{Suply\ air} - T_{intake\ air} \quad (2)$$

$$\Delta_{T2} = T_{exhaust\ air} - T_{intake\ air} \quad (3)$$

The bunch of data required for the test of behavioral constraint are $T_{Suply\ air(\mathbb{T})}$, $T_{intake\ air(\mathbb{T})}$, and $T_{exhaust\ air(\mathbb{T})}$. They are sequences of similar intervals of supply air temperature, intake air temperature, and exhaust air temperature measured respectively by supply air temperature sensor, intake air temperature sensor, and exhaust air temperature sensor. The bunch of data required for the test of validity constraint is $rpm_{(\mathbb{T})}$ and $airflow_{(\mathbb{T})}$. They are a sequence of similar intervals of rotation speed per minute of the fan and airflow measured respectively by the rotation speed sensor and airflow sensor. The rpm is used to model that the ventilation system is in mode ON.

This test is also defined by a test support. The possible fault explanations for this test in case of inconsistency are ¬ ok (heater exchanger) ∨¬ok(temperature controller) ∨¬ ok(supply air temperature sensor) ∨¬ ok(intake air temperature sensor) ∨¬ok(exhaust air temperature sensor) ∨¬ ok(rpm sensor) ∨¬ ok(airflow sensor).

Test 3: Performance of the Fan [54]

Ventilation fans are an important component of any mechanically ventilated building. Poor fan performance could significantly affect the whole building performance metrics and more precisely and according to the literature. Air Handling Unit fans are responsible for approximately 40% of all electricity consumption in a HVAC system. There are several issues such as dirty blades and mechanical wear that could impact the fan's performance. This test evaluates the building ventilation system fan operation using a performance curve [56]. The test is defined by a validity constraint $V_\tau(X_\mathbb{T}) \in \mathcal{V}_\tau$ with $\mathcal{V}_\tau$: $electricity(t) > 0, \forall t \in \mathbb{T}$. The test is also defined by a behavioral constraint $B_\tau(X_\mathbb{T}) \in \mathcal{B}_\tau$ with $\mathcal{B}_\tau$: $P_{expected}$ - $DN^+ < P_{measured} < P_{expected}$ + DN^-, DN^+ and DN^- are upper and lower value of performance design number and can be obtained from the ventilation commissioning team ($DN^+ = 30\%$ and $DN^- = 30\%$), $P_{expected}$ and $P_{measured}$ represent receptively the expected and the measured power consumption. A set of fan performance curve (FPC), provided by NK Industry is used for the modeling purpose.Expected fan performance is modeled with the help of manufacturer data and compared against the real-time fan performance. Two data-driven models are developed and implemented. The first model is used to compute expected total fan pressure at a given airflow rate while second is a Support Vector Regression (SVR) model, to predict the fan efficiency. The performance monitoring of the ventilation fan unit is determined in terms of expected and actual fan energy consumption.

To estimate expected fan energy consumption of ventilation system fan using airflow measurement, it is necessary to model total fan pressure in the terms of measured airflow. A second model requires to estimate expected fan efficiency followed by the expected fan energy consumption [54].

This test is also defined by a test support. The possible fault explanations for this test in case of inconsistency are $\neg$ ok(fan) $\vee\neg$ ok(electricity meter sensor)$\vee\neg$ ok(energy consumption sensor)$\vee\neg$ ok(airflow sensor).

Test 4: Indoor Air Quality

Here is an example of a range-based test that verifies the indoor air quality (i.e., CO_2 concentrations) range in the classroom in the University of Southern Denmark. $Test_4$ generates test results for the deviation of indoor air quality performance. Possible fault explanations for this test combine all the major components that potentially affect the office air quality performance. For example, a faulty ventilation system or an important number of occupants could be responsible for the poor air quality performance. Possible fault explanations include sensor level fault. For instance: ok(indoor CO_2 sensor) $\rightarrow obs(C_{in}) = C_{in}$ where ok signifies the non-faulty behavior of CO_2 concentrations sensor and *obs* stands for an observed value. Indoor air quality test is given as:

Test 4 is always valid. This test is also defined by a behavioral constraint $B(X_{\mathbb{T}}) \in \mathcal{B}_\tau$ with $\mathcal{B}_\tau = C_{in}(t) \in [C_{min}, C_{max}], \forall t \in \mathbb{T}$ where C_{min}, C_{max} represent respectively the lower and the upper values for CO_2 in fault free case of sensor in the month of March, 2016 (C_{min}=390 ppm, C_{max}=1828.6 ppm).

The bunch of data required for the behavioral constraint is $C_{in(\mathbb{T})}$. It contains the measurements of CO_2 concentrations during the valid time period $\mathbb{T}$. These measurements are collected from CO_2 concentrations installed in the classroom

This test is also defined by a test support. The possible fault explanations for Test 4 in case of inconsistency are: $\neg$ ok (CO_2 concentration sensor) $\vee$ $\neg$ ok (ventilation system) $\vee$ $\neg$ ok(damper).

6.2 *Diagnosis Reasoning for Danish Application*

This section deals with performing diagnosis in the Danish platform with partially valid tests. In the following, 3 methods for diagnostic analysis are discussed: visual diagnostic analysis, diagnostic analysis by [53], and the proposed diagnostic analysis [41].

6.2.1 Visual Diagnostic Analysis

This section discusses the visual diagnostic analysis. We have simulated a fault scenario that we ask the reader to guess. Figures 9, 10, 11, and 12 show respectively the CO_2 concentrations as a function of the damper position, the efficiency of the heat exchanger, the fan power, and the air quality.

The tests 1, 2, and 4 are inconsistent because the behavioral constraint is not satisfied during the valid time period. The third test is consistent because the

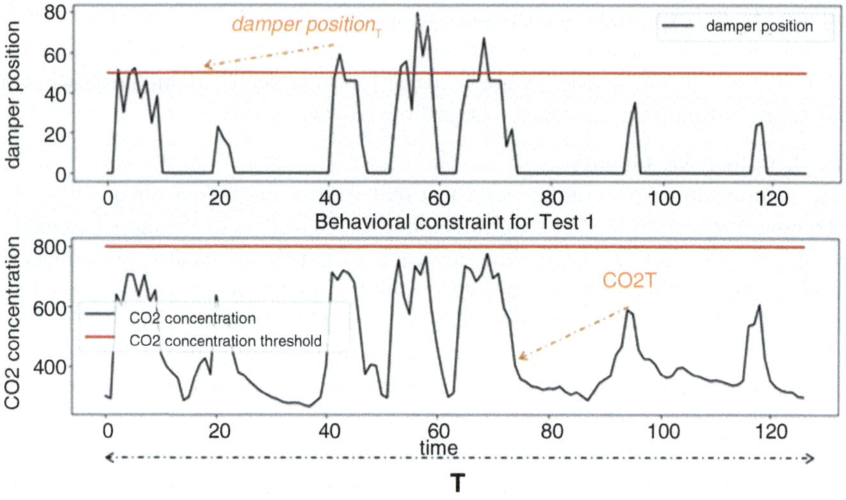

Fig. 9 Behavioral constraint satisfaction for Test 1

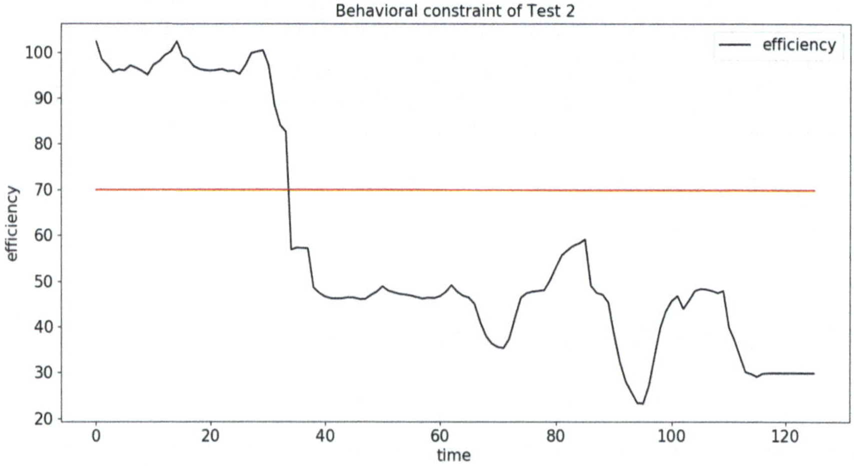

Fig. 10 Behavioral constraint satisfaction for Test 2

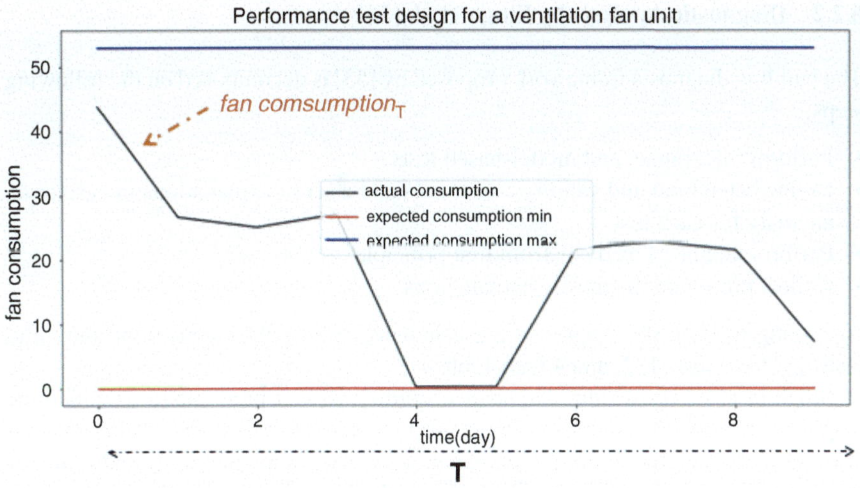

Fig. 11 Behavioral constraint satisfaction for Test 3

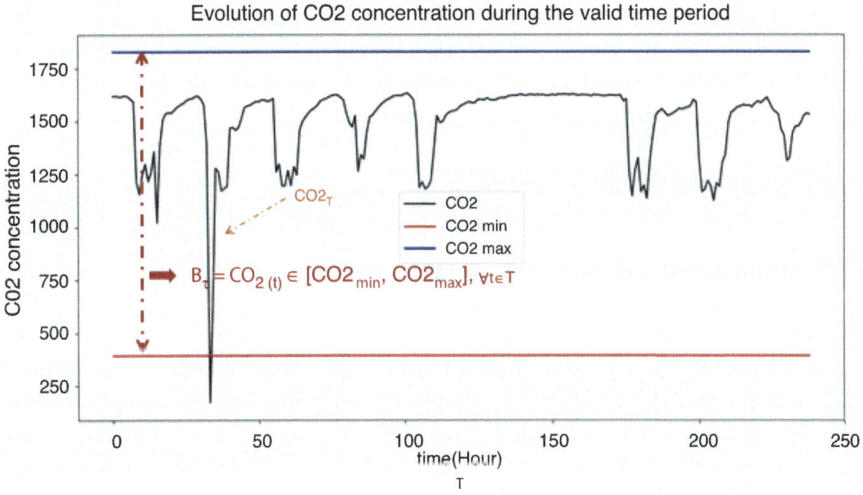

Fig. 12 Behavioral constraint satisfaction for Test 4

behavioral constraint is satisfied during the valid time period. The question that arises is: What is the simulated fault? We conclude that it is difficult to conclude on the source of default.

6.2.2 Diagnostic Analysis by Singh et al. [53]

The building diagnosis framework proposed by [53] is decomposed on the following steps:

- Perform rule, range, and model-based tests
- Define behavioral and validity constraints and a set of explanations in case of an anomaly for each test
- Perform diagnosis analysis from first principle
- Collect minimum diagnostic explanations

According to [53], the diagnoses are calculated from only inconsistent and valid tests, i.e. from tests 1, 2, and 4 (see Table 4)

Singh et al. [53] uses the diagnosis according to first principle to calculate the diagnoses because the supposed faults are not necessarily revealed and because it allows to calculate the minimum explanations at the component level and it allows the detection of multiple faults. With the method proposed by [53], we obtain 27 diagnoses, but none of which is right. In this work, only diagnoses number 0, 6, 8, 12, 26, and 27 are presented (see Table 5).

The simulated fault is an offset on the CO_2 concentrations sensor, a bias on the air temperature supply sensor, and an offset on the rpm sensor.The obtained result is inexact. The diagnoses are calculated from the 3 tests which are Test1, Test2, and Test4. In this case, a bias is applied to the rpm sensor which intervenes in the validity of test 2. So, with [53], the diagnoses are calculated by 3 tests, the validity of one of which is measured by a faulty sensor.

6.2.3 Proposed Diagnostic Analysis [41]

The methodology proposed by [41] assumes that the validity is measured with faulty sensors. In the beginning, for each inconsistent test, a level of completeness (i.e., a level of validity) is calculated using a partitioning approach (see Table 7). The diagnostic result is calculated from a set of tests, each one defined by its completeness level. To compute the confidence level of a global diagnosis deduced from a set of tests, in which some have a completeness level lower than 1, we are going to adapt a method based on fuzzy logic reasoning. Table 6 summarizes the two cases for fuzzy logic reasoning.

Table 4 Diagnosis analysis by [53]

Tests	Consistent/inconsistent	Valid/invalid
Test1	Inconsistent	Valid
Test2	Inconsistent	Valid
Test3	Consistent	Valid
Test4	Inconsistent	Valid

Table 5 Diagnosis analysis by [53]

Diagnostics found	Results according to the proposed approach	Fault probability
D0	Heater exchanger AND damper	100%
D6	Heater exchanger AND CO_2 concentrations sensor	100%
D8	Air temperature supply sensor AND CO_2 concentrations sensor	100%
D12	Airflow sensor AND damper	100%
D26	Fan AND rpm sensor AND damper	83.33%
D27	Fan AND Airflow sensor AND damper position sensor	83.33%

Table 6 Confidence level with fuzzy logic reasoning

	∃ negative tests	∃ doubtful tests mostly negative	Conclusion	Confidence level
Case 1	Yes	Yes/no	Diagnostics calculated from safe negative tests	1
Case 2	No	Yes	Diagnostics calculated from doubtful tests mostly negative	Max (degrees of belonging to Not Ok of doubtful tests mostly negative)

Table 7 Results using partitioning approach

Tests	Consistent/inconsistent	Completeness level using partitioning approach	Membership level to False
Test 1	Inconsistent	1	Negative
Test 2	Inconsistent	0.5	0.75 (doubtful mostly negative)
Test 3	Consistent	–	–
Test 4	Inconsistent	1	Negative

In this case of calculation of completeness level using partitioning approach, tests 1 and 4 are negative and test 2 is doubtful mostly negative because $\mu(Test_2 = False) = 0.75 \geq 0.5$. Table 7 summarizes these results.

With the proposed methodology, we obtain 4 diagnoses of which 1 is correct (see Table 8).

The simulated fault is an offset on the CO_2 concentrations sensor, a bias on the air temperature supply sensor, and an offset on the rpm sensor. The obtained result is inexact. The diagnoses are calculated from only 2 tests which are Test 1 and Test 4. In fact, a bias is applied to the rpm sensor which intervenes in the validity of test 2.

Table 8 Proposed diagnostic analysis [41]

Diagnostics found	Results according To the proposed approach	Fault probability
D0	CO_2 concentrations sensor	100%
D1	Damper	100%
D2	Fan AND temperature controller	83.33%
D3	Fan AND damper position sensor	83.33%

Table 9 Comparison between different diagnosis methods

	Visual diagnostic analysis	Diagnostic analysis by Singh et al. [53]	Proposed diagnostic analysis
Remarks	No formal tests	Reliable sensors	Validity measured by potentially faulty sensors
Conclusion	No idea about the simulated fault	Diagnostics calculated by three tests	Diagnostics calculated by two tests

Table 9 summarizes a comparison between the different diagnosis methods.

7 Conclusion

Accurate diagnosis of faults in buildings presents a significant building operation cost saving opportunity. With the needs of new retrofitting and building intelligence solutions, diagnosis has become significantly more important and, thus, requires new approaches to be designed, as well as a further enhancement of the existing ones. In this chapter, we have summarized the efforts in the area of fault detection and diagnosis for smart buildings with the aim of identifying the gaps and challenges that have not yet been given an adequate span of attention. This also highlights the scope of our research and has yielded an initial framework for addressing these issues.

The majority of existing building fault diagnosis techniques rely on behavioral knowledge. Model-based fault diagnosis and isolation techniques (FDI) assume that the model represents the reality of building operation independently of the current context and any fault can be detected by measuring the physical variables and checking the consistency with a reference model. A physical variable is a potentially observable element of information about the actual state of a building system. Nevertheless, behavioral models always valid in any context are difficult to set up. Erroneous all-context models might lead to invalid diagnoses. This is a challenge.

Complexity in testing a whole building system using both rule and pure model-based test, insidious faults, and unreliable sensors are also challenging one.

Thus, this work on building diagnosis differs from previous approaches since we are proposing some solutions for modeling, complexity, testing in a specific context taking into account that the validity is measured with possibly faulty sensors, confidence level for diagnosis, and unreliable instrumentation in buildings.

Acknowledgments This work is supported by the French National Research Agency in the framework of the "Investissements d'avenir" Eco SESA program (ANR-15-IDEX-02) and by the ADEME in the framework of the COMEPOS project.

References

1. B. Akinci, J. Garrett, Ö. Akin, Identification of functional requirements and possible approaches for self-configuring intelligent building systems. Final Report Submitted to Steven Bushby, NIST, 100, 20899–8631 (2011)
2. M. Amayri, A. Arora, S. Ploix, S. Bandhyopadyay, Q.-D. Ngo, V.R. Badarla, Estimating occupancy in heterogeneous sensor environment. Energy Build. **129**, 46–58 (2016)
3. K.K. Andersen, T.A. Reddy, The error in variables (EIV) regression approach as a means of identifying unbiased physical parameter estimates: application to chiller performance data. HVAC&R Res. **8**(3), 295–309 (2002)
4. M.M. Ardehali, T.F. Smith, Literature review to identify existing case studies of controls related energy inefficiencies in buildings. Prepared for the National Building Controls Information Program. Technical Report: ME-TFS-01-007. Iowa City, Iowa. Department of Mechanical and Industrial Engineering, the University of Iowa (2002)
5. A. Atkinson, Human error in the management of building projects. Construct. Manage. Econ. **16**(3), 339–349 (1998)
6. M. Bonvini, M.A. Piette, M. Wetter, J. Granderson, M.D. Sohn, Bridging the gap between simulation and the real world an application to FDD, in *Proceedings of 2014 ACEE Summer Study* (2014)
7. J.D. Bynum, D.E. Claridge, J.M. Curtin, Development and testing of an automated building commissioning analysis tool (ABCAT). Energy Build. **55**, 607–617 (2012)
8. D.J. Campbell, Task complexity: a review and analysis. Acad. Manage. Rev. **13**(1), 40–52 (1988)
9. N.S. Castro, Performance evaluation of a reciprocating chiller using experimental data and model predictions for fault detection and diagnosis/discussion. ASHRAE Trans. **108**, 889 (2002)
10. D. Chester, D. Lamb, P. Dhurjati, Rule-based computer alarm analysis in chemical process plants, in Proceedings of the Seventh Annual Conference on Computer Technology (1984), pp. 22–29
11. J.A. Clarke, J. Cockroft, S. Conner, J.W. Hand, N.J. Kelly, R. Moore, T. O'Brien, P. Strachan, Simulation-assisted control in building energy management systems. Energy Build. **34**(9), 933–940 (2002)
12. A.L. Dexter, D. Ngo, Fault diagnosis in air-conditioning systems: a multi-step fuzzy model-based approach. HVAC&R Res. **7**(1), 83–102 (2001)
13. H. Doukas, K.D. Patlitzianas, K. Iatropoulos, J. Psarras, Intelligent building energy management system using rule sets. Build. Environ. **42**(10), 3562–3569 (2007)
14. Z. Du, X. Jin, L. Wu, Fault detection and diagnosis based on improved PCA with JAA method in VAV systems. Build. Environ. **42**(9), 3221–3232 (2007)
15. Z. Du, X. Jin, Y. Yang, Fault diagnosis for temperature, flow rate and pressure sensors in VAV systems using wavelet neural network. Appl. Energy **86**(9), 1624–1631 (2009)

16. Z. Du, B. Fan, X. Jin, J. Chi, Fault detection and diagnosis for buildings and HVAC systems using combined neural networks and subtractive clustering analysis. Build. Environ. **73**, 1–11 (2014)
17. D. Estrin, L. Girod, G. Pottie, M. Srivastava, Instrumenting the world with wireless sensor networks, in *2001 IEEE International Conference on Acoustics, Speech, and Signal Processing. Proceedings (Cat. No. 01CH37221)* (2001), pp. 2033–2036
18. H. Friedman, E. Crowe, E. Sibley, M. Effinger, The Building Performance Tracking Handbook (California Commissioning Collaborative, Berkeley, 2011)
19. C. Ghiaus, Fault diagnosis of air conditioning systems based on qualitative bond graph. Energy Build. **30**(3), 221–232 (1999)
20. A. Giantomassi, F. Ferracuti, S. Iarlori, G. Ippoliti, S. Longhi, Signal based fault detection and diagnosis for rotating electrical machines: issues and solutions, in *Complex System Modelling and Control Through Intelligent Soft Computations* (Springer, New York, 2015), pp. 275–309
21. A.S. Glass, P. Gruber, M. Roos, J. Todtli, Qualitative model-based fault detection in air-handling units. IEEE Control Syst. **15**(4), 11–22 (1995)
22. Health and Safety Commission, ACSNI study group on human factors (1993)
23. E.J. Henley, Application of expert systems to fault diagnosis (1984)
24. A. Hyvärinen, E. Oja, Simple neuron models for independent component analysis. Int. J. Neural Syst. **7**(06), 671–687 (1996)
25. B.D. Ilozor, M.I. Okoroh, C.E. Egbu et al., Understanding residential house defects in Australia from the state of Victoria. Build. Environ. **39**(3), 327–337 (2004)
26. Y. Jia, T.A. Reddy, Characteristic physical parameter approach to modeling chillers suitable for fault detection, diagnosis and evaluation. J. Solar Energy Eng. **125**(3), 258–265 (2003)
27. S. Katipamula, M.R. Brambley, Methods for fault detection, diagnostics, and prognostics for building systems-a review, part I. HVAC&R Res. **11**(1), 3–25 (2005)
28. S. Katipamula, R.G. Pratt, D.P. Chassin, Z.T. Taylor, K. Gowri, M.R. Brambley, Automated fault detection and diagnostics for outdoor-air ventilation systems and economizers: methodology and results from field testing. Trans.-Am. Soc. Heat. Refrig. Air Condition. Eng. **105**, 555–567 (1999)
29. S. Katipamula, M.R. Brambley, N. Bauman, R.G. Pratt, Enhancing building operations through automated diagnostics: field test results (2003)
30. W.-Y. Lee, J.M. House, N.-H. Kyong, Subsystem level fault diagnosis of a building's air-handling unit using general regression neural networks. Appl. Energy **77**(2), 153–170 (2004)
31. N.G. Leveson, Software engineering: stretching the limits of complexity. Commun. ACM **40**(2), 129–132 (1997)
32. S. Li, J. Wen, A model-based fault detection and diagnostic methodology based on PCA method and wavelet transform. Energy Build. **68**, 63–71 (2014)
33. X. Li, D. Huang, Z. Sun, A routing protocol for balancing energy consumption in heterogeneous wireless sensor networks, in *International Conference on Mobile Ad-Hoc and Sensor Networks* (Springer, New York, 2007), pp. 79–88
34. M.-D. Ma, D.S.-H. Wong, S.-S. Jang, S.-T. Tseng, Fault detection based on statistical multivariate analysis and microarray visualization. IEEE Trans. Ind. Inf. **6**(1), 18–24 (2010)
35. P. May-Ostendorp, G.P. Henze, C.D. Corbin, B. Rajagopalan, C. Felsmann, Model-predictive control of mixed-mode buildings with rule extraction. Build. Environ. **46**(2), 428–437 (2011)
36. J.D. McKeen, T. Guimaraes, J.C. Wetherbe, The relationship between user participation and user satisfaction: an investigation of four contingency factors. MIS Quart. **18**(4), 427–451 (1994)
37. M.H. Meyer, K.F. Curley, An applied framework for classifying the complexity of knowledge-based systems. Mis Quart. **15**, 455–472 (1991)
38. J. Mickens, M. Szummer, D. Narayanan, Snitch: interactive decision trees for troubleshooting misconfigurations, in *Proceedings of the 2007 Workshop on Tackling Computer Systems Problems with Machine Learning Techniques* (2007)
39. P.-D. Moroşan, R. Bourdais, D. Dumur, J. Buisson, Building temperature regulation using a distributed model predictive control. Energy Build. **42**(9), 1445–1452 (2010)

40. S.A. Mostafa, M.S. Ahmad, M.A. Mohammed, O.I. Obaid, Implementing an expert diagnostic assistance system for car failure and malfunction. Int. J. Comput. Sci. Iss. **9**(2), 1 (2012)
41. H. Najeh, Diagnostic du système bâtiment: nouveaux défis. PhD thesis (2019)
42. H. Najeh, M.P. Singh, S. Ploix, K. Chabir, M.N. Abdelkrim, Automatic thresholding for sensor data gap detection using statistical approach, in *Sustainability in Energy and Buildings* (Springer, New York, 2020), pp. 455–467
43. K. Niida, S. Kobayashi, T. Umeda, J. Itoh, A. Ichikawa, Some expert system experiments in process engineering. Chem. Eng. Res. Des. **64**, 372–380 (1986)
44. L.K. Norford, J.A. Wright, R.A. Buswell, D. Luo, C.J. Klaassen, A. Suby, Demonstration of fault detection and diagnosis methods for air-handling units. HVAC&R Res. **8**(1), 41–71 (2002)
45. S. Ploix, M. Désinde, S. Touaf, Automatic design of detection tests in complex dynamic systems. IFAC Proc. Vol. **38**(1), 478–483 (2005)
46. P.M.A. Ribbers, K.-C. Schoo, Program management and complexity of ERP implementations. Eng. Manage. J. **14**(2), 45–52 (2002)
47. S. Roels, P. Bacher, G. Bauwens, H. Madsen, M.J. Jiménez, Characterising the actual thermal performance of buildings: current results of common exercises performed in the framework of the IEA EBC annex 58-project. Energy Proc. **78**, 3282–3287 (2015)
48. R.S. Rollings, M.P. Rollings, Pavement failures: oversights, omissions and wishful thinking. J. Perform. Construct. Fac. **5**(4), 271–286 (1991)
49. K.W. Roth, D. Westphalen, M.Y. Feng, P. Llana, L. Quartararo, Energy impact of commercial building controls and performance diagnostics: market characterization, energy impact of building faults and energy savings potential. Prepared by TAIX LLC for the US Department of Energy. November. 412 pp (Table 2–1) (2005)
50. J. Schein, S.T. Bushby, N.S. Castro, J.M. House. A rule-based fault detection method for air handling units. Energy Build. **38**(12), 1485–1492 (2006)
51. G. Simkin, J. Ingham, Digital buildings: using sensors to monitor the performance of concrete buildings during the Christchurch earthquake rebuild, in *Concrete Innovation Conference* (2014)
52. M. Singh, Improving Building Operational Performance with Reactive Management Embedding Diagnosis Capabilities. PhD Thesis, University of Grenoble Alpes (2017)
53. M. Singh, N.T. Kien, H. Najeh, S. Ploix, A. Caucheteux, Advancing building fault diagnosis using the concept of contextual and heterogeneous test. Energies **12**(13), 2510 (2019)
54. M. Singh, M. Jradi, H.R. Shaker, Monitoring and evaluation of building ventilation system fans operation using performance curves. Energy Built Environ. **1**(3), 307–318 (2020)
55. H.R. Shaker, N. Mohamed, S. Lazarova-Molnar, Fault detection and diagnosis for smart buildings: state of the art, trends and challenges, in *2016 3rd MEC International Conference on Big Data and Smart City (ICBDSC)* (2016), pp. 1–7
56. J. Stein, M.M. Hydeman, Development and testing of the characteristic curve fan model, in *ASHRAE Transactions*, vol. 110 (2004), pp. 347–356
57. S. Treado, Y. Chen, Saving building energy through advanced control strategies. Energies **6**(9), 4769–4785 (2013)
58. USDOE, Commercial energy end-use splits, by fuel type (Quadrillion Btu), in *Building Energy Data Book* (CTCN, Copenhagen, 2010)
59. S. Wang, Y. Chen, Fault-tolerant control for outdoor ventilation air flow rate in buildings based on neural network. Build. Environ. **37**(7), 691–704 (2002)
60. T.M. Williams, The need for new paradigms for complex projects. Int. J. Project Manage. **17**(5), 269–273 (1999)
61. R.E. Wood, Task complexity: definition of the construct. Organ. Behav. Hum. Decis. Process. **37**(1), 60–82 (1986)
62. B. Yu, D.H.C. Van Paassen, S. Riahy, General modeling for model-based FDD on building HVAC system. Simul. Pract. Theor. **9**(6–8), 387–397 (2002)
63. D. Yu, H. Li, M. Yang, A virtual supply airflow rate meter for rooftop air-conditioning units. Build. Environ. **46**(6), 1292–1302 (2011)

64. Y. Zhang, C. Bingham, M. Gallimore et al., Fault detection and diagnosis based on extensions of PCA. Adv. Milit. Technol. **8**(2), 27–41 (2013)
65. Y. Zhu, X. Jin, Z. Du, Fault diagnosis for sensors in air handling unit based on neural network pre-processed by wavelet and fractal. Energy Build. **44**, 7–16 (2012)

Analyzing Load Profiles in Commercial Buildings Using Smart Meter Data

Srinka Basu, Kakuli Mishra, and Ujjwal Maulik

1 Introduction

Buildings consume majority of global electricity produced. With the advent of advanced metering infrastructure (AMI) [34] in smart grids and buildings, mining information from the growing electricity consumption data has opened many new challenges. Mining the electricity consumption of buildings has a great potential to solve the problems like load forecasting, clustering [14], classification, anomaly detection [5], demand side management programs [11, 25]. The electricity consumption in buildings also gives an idea about the behavioral usage of consumers in the buildings [14], their occupancy/in-occupancy states [2], based on which the load shifting operation can be decided.

However, not all the historic data stored are equally important. Hence, to extract useful information from the existing time series data sets, several compact time series representation techniques have been proposed in literature. The time series discretization into symbols is one such dimensionality reduction technique that has allowed efficient manipulations based on the applications [22].

Existing literature on segmentation and symbolic representation breaks the time series into pre-determined fixed length segments. One of the limitation of fixed length segmentation is that it might fail to identify the critical breakpoints that fall within a window, as a result, a single window might contain two very different patterns. In addition, determining the window size might be difficult for real life applications. Figure 1 is a time series sample which has been segmented with a fixed length of size 10. However, the rising and falling trend patterns in each segment

S. Basu (✉)
Department of Engineering and Technological Studies, University of Kalyani, Kalyani, India

K. Mishra · U. Maulik
Department of Computer Science and Engineering, Jadavpur University, Kolkata, India

S. Ploix et al. (eds.), *Towards Energy Smart Homes*,
https://doi.org/10.1007/978-3-030-76477-7_15

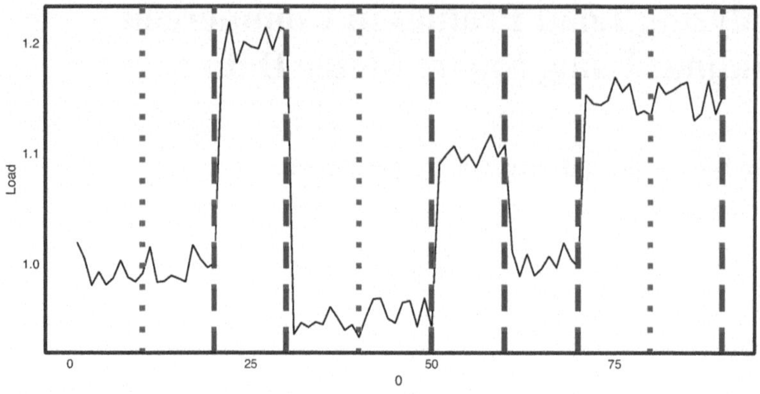

Fig. 1 Sample time series data showing the limitations of fixed window length. The time series has fixed window length of 10; the *red dotted* and the *blue dashed* lines denote the varying window length

could not be detected as a result of the fixed length windows. The ideal breakpoints for capturing the trend pattern in the time series has been shown in *blue dashed lines*.

Given the limitations of the existing segmentation techniques, in this work we propose an automatic segmentation and symbolic representation based approach for analysis of load profiles in commercial buildings. The major contributions of this work are given below:

- We first propose an automatic segmentation of the original building load data into varying length time windows where two successive segments demonstrate statistically significant difference in load patterns. We use a piecewise polynomial regression to model a segment with an initial window length decided on the basis of periodicities present in the time series data obtained from partial autocorrelation function (PACF). Successively a local search based approach that minimizes Bayesian information criterion (BIC) is used to decide the length of a segment. The proposed approach breaks a time series in piecewise non-linear statistically significant segments and thus can also be used for break-point identification.
- We next propose a symbolic representation of time series by encoding each window obtained from the segmentation process based on average load during the time indicated by the window and shape of the load pattern. The symbolic representation of a segment by the mean usage, the leading coefficient of the fitted polynomial, intercept of the fitted polynomial, and the degree of the best fit polynomial allows dimensionality/numerosity reduction of the original time series while preserving the characteristics of each window.
- We then propose a measure based on intercluster and intracluster similarity of the time series, to benchmark the symbolic representation of time series focusing on their ability to cluster multiple time series.

- We finally cluster the buildings on the basis of their symbolic representation and propose a data analysis method to identify the most frequently occurring patterns (MFOP) within building and across buildings.

The article has been organized into seven sections. First we discuss the literature in Sect. 2, subsequently the proposed method in Sect. 3, experimental setup in Sect. 4, results and analysis in Sect. 5, and finally with the conclusions and future work in Sect. 6.

2 Literature Survey

Due to growing surge in building load consumption data, researchers in smart grid area have emphasized on dimensionality reduction techniques for time series data mining [32]. Dimensionality reduction of time series can be carried out in time domain or after transforming the time series data in other domain, for example, frequency domain. One of the popular time domain based dimensionality reduction technique is the piecewise aggregate approximation (PAA) [16], which captures the mean of the segments. The segmentation step is extensively used for similarity search while mining large time series [23]. The length of the window in case of segmentation can either be fixed or computed. The advantage of using a fixed window length is that it reduces the time complexity for the data mining operations because no analysis needs to be done to determine the window length. Fixed window length can very easily discover the motif and discord in time series [24], if chosen correctly. Storing the means of the segments in PAA is rather a simplified representation of the time series data, but we cannot interpret the shape or shift in the window by the mean values.

Following PAA, Keogh et al. proposed a varying window length based segmentation in [17]. Varying length windows can efficiently capture the motif, discord in the time series. Authors in [5] have evaluated the window length using the classification and regression tree algorithm (CART). The authors in [1] review different supervised and unsupervised techniques to identify the breakpoints for segmentation. Window lengths have also been determined using the polynomial approximation of time series. The piecewise linear approximation (PLA) approach originally proposed in [3] and has been used in [19] refers to the approximation of time series of length n with K straight lines. Authors in [38] have proposed a variant of PLA with an aim to reduce the number of lines required to generate the time series by adjusting each line segment to approximate the maximum number of stream points. In the context of data mining, the PLA has been used in change point detection [29], estimate the trend of patent publishing [6], similarity measure [20], time series indexing [17]. Three major dynamic time series segmentation techniques used in PLA are sliding window, top-down, and bottom-up [13]. In case of sliding window approach, the window length is grown until it exceeds some error bound [18]; however, it gives poor results in case when abrupt changes are present in the data [31]. In literature, the breakpoints in case of sliding window based segmentation have been decided based on some error calculation which is often not well approximated [9, 21]. In case of top-down approach, a time series is recursively partitioned until some criteria is met [18] and in the bottom-up approach, smaller length subsequences are repeatedly merged until some criteria is met [18]. The performance of both top-down and bottom-up is dependent on selection of the error criteria [13].

One limitation of PLA is that due to its simple linear representation, the curves in the time series are discarded. The piecewise polynomial approximation (PPA) is a non-linear representation of the time series with polynomials of any arbitrary degree. In [9] authors have proposed an online segmentation of time series using PPA. The segment length is decided on the basis of the coefficients of orthogonal polynomials. Deciding the segment length on the basis of polynomial coefficients in [9] cannot capture the correlated power readings. Other than time series segmentation, PPA has been used in a wide variety of data mining applications, like the authors in [8] have used PPA for classification of time series. Not only in real valued time series data, the polynomial based features have been used for motion based activity recognition in image data sets [4]. Unlike [9], we propose an automatic segmentation approach in addition to symbolic representation technique that captures the features of the time series.

As this paper aims to segment the time series data based on piecewise polynomial approximation and represent the time series data symbolically for further analysis, below we discuss some of the existing symbolic representation techniques and analyses on time series data. Applications of symbolic representation include motif discovery [27, 30], discord discovery [22, 33], clustering [10, 24], classification [15], all-pairs-similarity-search[35]. Symbolic Aggregate Approximation (SAX) [22] converts the PAA transformed time series data to a symbolic representation by using the mean values of each segment. The PAA transformed data follows a Gaussian distribution as a result of which an equiprobable set of breakpoints are defined that are represented as symbols based on the region to which it falls [22]. As SAX becomes a smoothed representation, the extended SAX in [23] is an improvement which uses two other symbols besides mean: minimum and maximum. The minimum and the maximum values improve the preciseness in the symbolic representation. Authors in [37] propose an improvement in SAX by integrating the standard deviation in addition to the mean and confirm both the highest classification accuracy and the highest dimensionality reduction ratio with respect to the existing SAX techniques. Some of the symbolic representation techniques follow the trend based approaches [36] where the authors compute the trend distance factor and trend shape factor for each segment. However, the trend based approaches cannot capture the shift in the pattern. Authors in [12] have proposed a symbolic representation of time series segments where they capture the Harr wavelet coefficients and other defined key points of the segment. The authors claim that the advantage of their method [12] over others is that they have proposed a parameter free approach for symbolic representation. But analysis of the segments on Haar wavelet transformed data is a computationally intensive process.

Based on analysis of the symbolic representation of time series, in [24], the authors have identified the weekday and weekend motifs by extracting the most frequently occurring patterns using a visualization tool called suffix tree. No major contribution has been done on any of the symbolic representation or the analysis part in [24]. Authors in [5] have proposed a methodology to identify the infrequent time series patterns from a set of building data.

Most of the proposed symbolic representation techniques in literature have used fixed length segments and have analyzed the performance by computing the distance between time series based on their proposed symbolically represented data. Analyzing the performance of the proposed symbolic representation only on the basis distance measures does not add to the application area of the work. Moreover, the symbolic representation varies with the change in segment length. In this work, we decide the segment length on the basis of correlation, hence all the autocorrelated values being placed in a window add a relationship among the load values in the window, which is an additional information integrated in the segmentation step. The proposed symbolic representation can capture the average, peak/off-peak, shift, and shape of the segment. In case of DSM policies, the symbolic representation assists in analyzing the common patterns, properties of the pattern, analyzing the time during which the load pattern is common in an area of interest. However, in literature, authors have focused only on capturing the motif and discord from the symbolically represented data, but we aim to integrate the above analyses on the proposed symbolically represented data. Motifs are basically the most frequently occurring patterns in a time series data. Most of the works in literature have obtained motif on a single building level, however in this work, we define the most frequently occurring patterns as the patterns which not only occur at single building level but also can be present in multiple buildings.

3 Proposed Method

Given the short comings of the fixed length partitions in finding the right patterns from a time series and the limitations of the existing methods for symbolic representation of time series to group such patterns having similar load profile, peak-off-peak time, and average consumption, we propose a new approach to load profile analysis as discussed in this section. The proposed method first partitions a given time series into variable length segments and successively finds the symbolic representations of the respective segments. The symbolic representations can further be used to group the segments having similar load profile, peak-off-peak time, and average consumption that in-turn can help to mine useful insight about the building load profiles like, the most frequently occurring pattern in a given building, the time duration when similar consumption pattern is exhibited across buildings, and many more.

3.1 Segmentation

The objective of the segmentation approach is to partition a given time series into non-overlapping, contiguous segments such that:

(a) all the data points lying in a segment demonstrate a particular pattern,
(b) two successive segments of a time series exhibit two different patterns, and
(c) the boundary points of two successive segments are considered as break points showing a sharp change in pattern.

To meet the objectives, we propose an approach based on piecewise polynomial fitting discussed as below.

- Given a time series, a segment is determined using a sliding window protocol where initial window size is determined by a parameter named as $init\_window$. The parameter $init\_window$ is considered to be a small value sufficient enough to fit a polynomial of low degree.
- The window size is gradually increased in every incremental step to include additional data points until the breakpoint criteria are satisfied or the window size reaches a maximum window length.
- In every incremental step, we find the best fitted polynomial by minimizing the Bayesian Information Criteria (BIC).
- Subsequently, from all possible windows, the polynomial having the minimum BIC value is chosen as the best fitted polynomial that can represent the segment.
- The maximum window size is decided based on the periodicity of a stationary time series. For any given time series, we may find the maximum window using the partial autocorrelation factor (PACF) test on the differenced time series as differencing ensures removal of trend, seasonality from the data and PACF gives the maximum lag up to which the differenced values are autocorrelated irrespective of the previous lags.
- The process of segmentation is continued from the data point of time series which lies immediately after the currently obtained segment.

The procedure to obtain the maximum window length using PACF lag is discussed in Algorithm 1. The maximum window length, lag in Algorithm 1, is approximated by finding the time t at which the PACF value exceeds the 95% confidence band. If no value beyond t exceeds the $PACF$ at t, then $lag = t$ else the differencing is repeated as shown in step 1d. The output of Algorithm 1 is the maximum window length, lag which is chosen as the maximum window length throughout the time series T.

The procedure to find the best fitted polynomial is discussed in Algorithm 2. In Algorithm 2, the BIC is computed as:

$$BIC(T_s) = k \log n - 2(L(\theta)) \tag{1}$$

Algorithm 1 Find maximum window length: findLag(T)

Input: Time series data T
Output: Maximum length lag of time series window.

1. **for** $i = 1 : |T|$ **do**
    a. Obtain a differenced time series $\tilde{T}$ of lag i.
    b. Compute PACF for the differenced time series and get the PACF plot.
    c. $pa_c \leftarrow$ PACF exceeding the confidence band; $t \leftarrow$ time at which pa_c is obtained.
    d. **if** $PACF[t+1 : |T|] < pa_c$ **then**
        lag=t
    else
        continue
    end
end
2. return (lag)

Algorithm 2 Best Fit polynomial: bestFit(T_s)

Input: Time series segment T_s
Output: Degree of polynomial deg and bic.

1. $n \leftarrow \lfloor length(T_s)/\ 2 \rfloor$
2. **for** $i = 1 : n$ **do**
    a. Fit a polynomial regression line of degree i.
    b. $B[i] \leftarrow$ BIC value of the fitted polynomial regression line.
    c. $deg[i] \leftarrow$ i
end
3. $bic \leftarrow \arg\min(B)$
4. $deg \leftarrow \arg\min(deg)$
5. return (bic, deg)

where n is length of segment T_s, k is the number of parameters estimated by the polynomial, $L(\theta)$ is the likelihood of all the parameters θ of polynomial [26].

The breakpoint criteria are determined using two conditions:

(a) The degree of the best fitted polynomial in the incremented window should be the same as the degree of the best fitted polynomial of the original window. If there is a change in the degree of the best fitted polynomial, then the breakpoint criteria are satisfied assuming that there is a change in the load consumption pattern.
(b) Simultaneously, if the BIC value of the fitted polynomial in two successive windows are greater than given threshold ϵ, then it is assumed that the break point lies between the previous window and the current window as a result of which the break-point criteria is satisfied when the fraction of change in BIC over the average BIC information is greater than ϵ.

The details of segmentation process are discussed in Algorithm 3 which initially computes the maximum window length using the given Algorithm 1. Segmentation

Algorithm 3 Segmentation: $segment(T)$

Input: A time series T
Output: Segmented time series, $\hat{T}$

1. $l_{max} = findLag(T)$
2. $l_{min} = |init\_window|, i = l_{min}$
3. $deg, bic = bestFit(init\_window)$
4. **while** $i <= |T| - l_{min}$ **do**
    a. $store\_b = NULL$
    b. **for** j in $i + 1 : l_{max}$ **do**
        i. $init\_window$=append $T[j]$ to $init\_window$
        ii. $new_{deg}, new_{bic} = bestFit(init\_window)$
        iii. $store\_b[j] = new_{bic}$
        iv. $\alpha = abs(new_{bic} - bic)/mean(store\_b)$
        v. **if** $new_{deg} == deg$ and $\alpha < \epsilon$ **then**
            continue
        else
            A. $init\_window = T[j : (j + l_{min})]$
            B. $deg, bic = bestFit(init\_window)$
            C. $breakpoint \leftarrow j$
        end
        vi. $\hat{T}[i] \leftarrow init\_window$
    end
    c. $i = j$
end
5. return ($\hat{T}$)

process starts by computing the best fit polynomial on $init\_window$. l_min in step 2 is the initial window length or the length of $init\_window$. In step 3, we compute the degree and BIC for $init\_window$ shown in Algorithm 2. To avoid the over fitting issue on each window, we continue to append values to $init\_window$ given in step 4(b)i, based on the two conditions. The threshold α is computed as the absolute difference between the original BIC of the $init\_window$ to that of the BIC obtained after an append of an element to $init\_window$ divided by the mean BIC value. We continue this operation until the entire time series is segmented.

A Toy Example of Segmentation Process We explain the segmentation process with an example. Let us assume a time series T and its difference time series $\tilde{T}$ as shown in Fig. 2. The PACF plot obtained through PACF test as shown in Fig. 2c has a confidence band. The last value that exceeds the confidence band is at lag 34, that is up to lag 34, the values are correlated. Hence using Algorithm 1, we obtain the maximum window length as 34. Let us also assume that the length of $init\_window$ parameter value is 6. In the first iteration, initially the segment length is assumed as $T[6] = \{0.90, 0.8, 0.91, 0.91, 0.90, 0.7\}$. The best fitted polynomial for $T[6]$ is found to have polynomial degree 3 with BIC value as -8.9. By appending

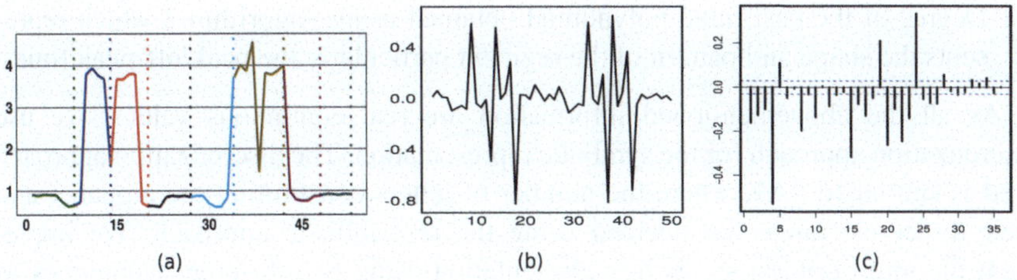

Fig. 2 An example of segmentation procedure. (**a**) shows the segments with different colors. The length of the segment is varying. (**b**) shows the differenced time series, and (**c**) shows the PACF plot

an additional data point to $T[6]$, we obtain $T[7]$ as described in step 4(b)i of Algorithm 3. The best fitted polynomial of $T[7]$ has degree 3 and BIC -8.11. Assuming the $\epsilon = 0.2$, we get $\alpha = abs(-8.11 + 8.9/ - 8.51) = 0.09$. As the condition 4(b)v of Algorithm 3 is satisfied, we continue appending the next data point to $T[7]$ and obtain the new window $T[8]$. Degree of the best fitted polynomial in $T[8]$ is 4 with BIC -12.23. As this violates condition 4(b)v of Algorithm 3, $T[7]$ is assumed to be the first segment and the process is restarted from the 8th data point. By repeating the above process until condition 4 of Algorithm 3 is satisfied, we obtain 8 segments as shown in Fig. 2a. The lengths of the segments obtained are 7, 6, 6, 8, 7, 8, 6, 6. As shown in Fig. 2, every two successive segments represent two different patterns unlike the fixed length partitions.

3.2 Symbolic Representation

The objective of the proposed method is to encode pattern based information of a segment along with peak/off-peak information and average consumption value. However, the existing symbolic representation primarily uses the average value of the segment in SAX or some additional statistics about the maximum or minimum values of the segment like in E-SAX. In PLA, only the trend information of the segment can be captured using the slope. To the best of authors' knowledge, none of the existing methods can capture information about the pattern of a segment along with the peak/off-peak information and average value. In order to meet the objective, we use a minimum of four criteria for symbolic representation as discussed below:

(a) Average value of the segment.
(b) Leading coefficient (LC) of the best fitted polynomial obtained using Algorithm 2 which gives relative information about the maximum amplitude.
(c) Intercept of the best fitted polynomial obtained using Algorithm 2 which gives a relative information about the shift.

(d) Degree of the best fitted polynomial obtained using Algorithm 2 which represents the shape and pattern of the segment particularly the peak/off-peak times.

As all the abovementioned information are real continuous values, we use discretization approach for the symbolic representation. The discretization approach used is similar to SAX where the number of letters constituting the alphabet and their respective range are decided using the probabilistic approach. We implement the multi-criteria symbolic representation using two different techniques as described below:

(1) **Sequential encoding:** This is a sequential encoding approach where in every step, a criterion is considered eventually forming a hierarchical group. For the symbolic representation, the mean of the segment is converted to symbols using the equiprobability rule in SAX technique. However, for degree, LC, and intercept, we follow a histogram based approach where the bin sizes are obtained using Sturges algorithm [28]. The rare values falling beyond $\mu \pm 3\sigma$, where μ is the mean and σ is the standard deviation, are discarded. The number of alphabet and their respective range corresponds to number of bins in the histogram.

We illustrate the symbols obtained from the sequential encoding approach in Fig. 3.

At level 1, we show the window patterns with same SAX symbol. The windows in group *a* and group *b* clearly show that they do not have segregated patterns inside. This signifies that mean of the window cannot carry all the information. In the second level, we group on the basis of LC symbols and obtain a group *ac*, where *a* is the SAX symbol and *c* is the LC symbol. In the

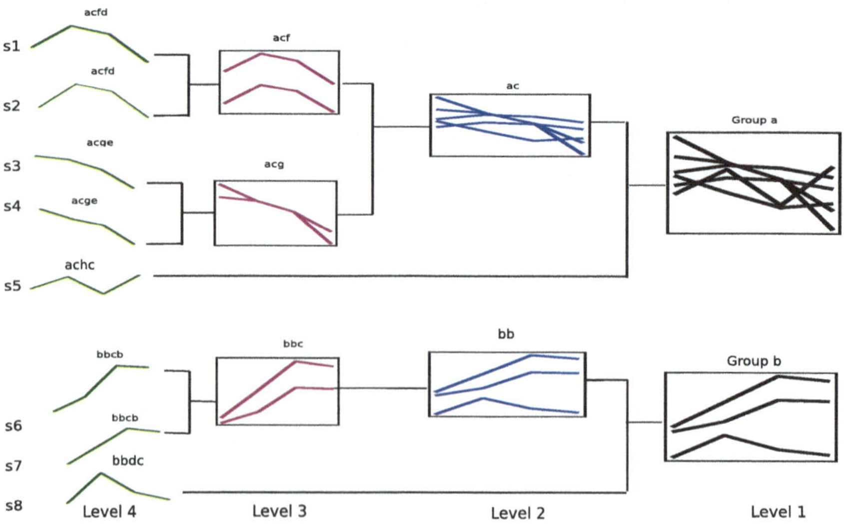

Fig. 3 An example of the sequential encoding

second level, we obtain a distinct pattern in the windows unlike first level. In the 3rd level, we group on the basis of intercept symbols and obtain *acf* and *acg* and *bbc*, where *f*, *g*, *c* are the intercept symbols, respectively. The window pattern of symbol *acf* has a rising pattern followed by a fall, whereas in case of window pattern with intercept symbol *acg*, we get a continuous falling pattern. At level 4, the patterns are segregated based on degree of the polynomial fit like the symbol *acfd* has degree *d* and has two windows in it. Symbols *achc* and *bbdc* have unique pattern and have been segregated from the rest. Hence, the example clearly shows that at each level, we obtain better patterns in the groups formed by the symbols.

(2) **Simultaneous encoding:** All the four criteria are considered simultaneously, that is, the degree, LC, intercept, and the mean of all the segments across all the buildings are considered for formation of respective probability density function. The advantage of this method is that the alphabet range remains comparable across all the groups.

A Toy Example of Symbolic Representation In Fig. 4, we illustrate the segmentation process for first five segments of the time series shown in Fig. 2. The segments are formed based on the four criteria mentioned in Sect. 3.2. For the first segment, shown in color green, the value of the average is −0.59, LC is 0.0008, intercept is 0.49, degree is 3, and it gets the symbol *bccf*. For second symbol shown in color blue, the value of average is 0.70, LC is −0.05, intercept is 18, degree is 4, and it gets the symbol *dcce*. For the third segment, shown in color red, the value of average is −0.78, LC is 0.0009, intercept is −5.94, degree is 4, and it gets the symbol, *dcae*. For the fourth segment, shown in color black, the value of average is −0.63, LC is −0.001, intercept is −3.67, degree is 3, and it gets the symbol *bcef*. For the fifth segment, shown in color skyblue, the value of average is −0.63, LC is 0.001, intercept is 1.04, degree is 3, and it gets the symbol *bccf*.

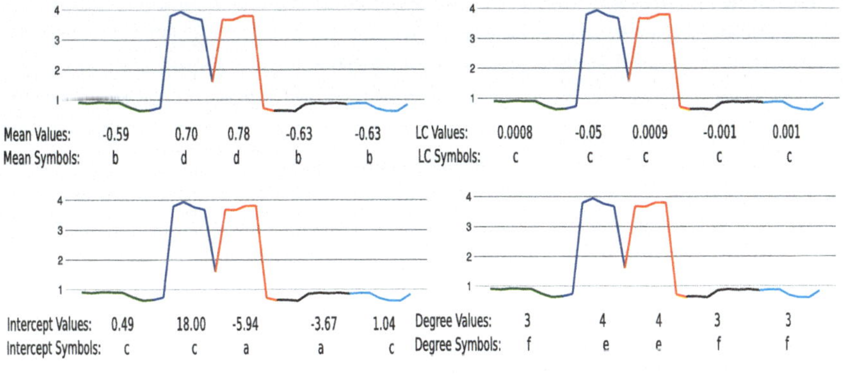

Fig. 4 An example of symbolic representation in a sample time series with five segments colored in green, blue, red, black, and sky blue, respectively

3.3 Time Complexity Analysis

Complexity of Algorithm 3 is dominated by two factors: time to compute the maximum window length l_{max} and the time to find the best fitted polynomial for every segment. For the segmentation process, the time taken to fit a k degree polynomial in a segment of size s is $O(k^2 \times s \times \beta)$, where $\beta = l_{max} - l_{min}$. To obtain all the segments of a given time series of length n, the segmentation time required is $O(k^2 \times n \times \beta)$. In our case, the maximum allowed degree $k_{max} = l_{max}/2$; therefore, the worst case time complexity of the segmentation process is $O(l_{max}^3 \times n \times \beta)$. As $l_{max} \ll n$, segmentation process is a linear time process.

For the symbolic representation process, assume there are w segments obtained after the segmentation technique. For each symbol, if the alphabet size is a, the time required for symbolic representation in both the techniques, sequential encoding and the simultaneous encoding, is $O(wa)$. The maximum value of $w = n/l_{max}$, where n is the length of time series. Hence, the total complexity for symbolic representation is $(n/l_{max}) \times a$. As $a/l_{max} << n$ and is a constant, hence symbolic representation takes linear time.

4 Experimental Setup

To evaluate the performance of the proposed method we carry out two sets of experiments. We first assess the accuracy of the proposed method in identifying a set of artificially planted pattern. Planted patterns are the artificially generated patterns placed within a random time series data. Successively we assess the performance of the proposed method in clustering similar segments through a comparative analysis on the data set discussed in the following section using the parameters mentioned below.

The value of $init\_window$ is based on the data frequency. In our experiments, the parameter $init\_window$ is kept as 6. We did not choose a lesser value because in case of commercial buildings, a minimum 6 h data are required to identify a pattern in the load values. The ϵ value is kept 0.2, which gave good set of results. The proposed symbolic representation technique has been compared with Extended SAX [23] and Piecewise Linear approximation (PLA). The fixed length window for E-SAX has been chosen as 8. For PLA, we used the proposed segmentation technique to segment the time series data. The symbols used in PLA are mean, LC, and intercept. The ϵ in PLA is same as that in our experiments, that is 0.2.

4.1 Data Description

The experiments are performed on four different data sets collected by Commission of Energy Regulation, Ireland (CER-IRISH) each containing half-hourly readings of small and medium enterprise (SME) buildings. The buildings have been categorized into industrial, office, retail, and enterprise professional (EP). For our experiments we used 40 industrial buildings, 60 office buildings, 40 retail buildings, and 40 EP buildings, each building having 6 month data during the year 2009. In addition to the power readings, the data set contains information about the number of employees in the building, the working hours, and the electrical devices used. Some of buildings have missing values which are replaced by moving averages.

4.2 Quality Measure

To measure the performance of the symbols in clustering, we propose a quality measure as described below:

Tightness Measure (T) Let $C = \{c_1, c_2, \ldots c_m\}$ where each c_i, for $i = 1$ to m represents all the time series segments having the symbolic representation as c_i. Let $F(c_i)$ be a matrix that represents the pairwise intracluster distance between all the elements of c_i and $G(c_i)$ be the matrix representing the pairwise intercluster distance between the elements of c_i and all the elements that belong to $(C - c_i)$. This is given as

$$F(c_i) = \Big[DTW(t_p, t_q) \Big]_{\forall (t_p, t_q) \in c_i} \tag{2}$$

$$G(c_i) = \Big[DTW(t_p, t_q) \Big]_{t_p \in c_i, t_q \in C - c_i} \tag{3}$$

Let min and max functions be defined on a matrix as: $min(.)$ returns the minimum value of a matrix and $max(.)$ returns the maximum value of a matrix. Tightness is computed as:

$$T = \sum_{\forall c_i \in C} \frac{(max(F(c_i)) - min(F(c_i)))^2}{(max(G(c_i)) + min(G(c_i)) + \Gamma)^2} \tag{4}$$

where T is the tightness. The power factor 2 introduces a penalty on the higher values of distance over the lower values of distance. To avoid the divide by zero error, we introduce parameter Γ. The value of Γ is zero in our experiments as because there do not exist any undefined condition in our case. The generic family of function to measure the tightness given as $\ddot{T}$ can be written as:

$$\ddot{T} = \left(\frac{(\alpha \times max(F)) - ((1 - \alpha) \times min(F))}{(\alpha \times min(G)) + ((1 - \alpha) \times max(G))} \right)^{\eta} \tag{5}$$

The above $\ddot{T}$ converges to Dunn score [7] when $\alpha = 1$ and $\eta = -1$. Dunn score is biased toward partitions having exactly one good cluster while the rest of the clusters may be of inferior quality. In our proposed measure, the quality of all the clusters is considered, unlike the Dunn score. The other descriptive statistics like mean and standard deviation can be used in place of maximum and minimum distances; however, for simplicity we chose only the maximum and minimum value.

Consider an example of the intracluster and intercluster matrices obtained from the group $acfd$ in Fig. 3, formed by $s1$ and $s2$. The intracluster matrix $F(acfd)$ is a 2×2 matrix and the intercluster matrix $G(acfd)$ is a 2×6 matrix where the rows correspond to the segments $s1$ and $s2$, respectively, and the columns corresponds to $s3$, $s4$, $s5$, $s6$, $s7$, and $s8$. Each cell (i, j) of the matrices contains pairwise DTW distance between the two time series segments corresponding to the ith row and the jth column, respectively.

5 Results and Analysis

In this section, we first report the results of the accuracy of the proposed method in identifying the planted patterns. We subsequently discuss the results of the comparative analysis, followed by the application of the proposed method for analysis of MFOP and DSM.

5.1 Planted Patterns

To estimate the accuracy of the proposed symbolic representation in case of planted patterns, we integrate repeated patterns into five random time series data. The detected patterns obtained have been illustrated in Fig. 5.

The length of the planted patterns ranges from 6 to 12. For example, in time series (a) in Fig. 5, all the planted patterns are of same length but in time series (c), the planted patterns are of different length; however, the proposed symbolic representation successfully captures all the varying length patterns.

5.2 Comparative Analysis

Table 1 shows the results of the clustering obtained in case of E-SAX and PLA and the proposed method measured in terms of the tightness measure T and Dunn score. The segments obtained using SAX technique fall into as many clusters as the size of the alphabet. Due to the nature of the encoding used in SAX, no representative pattern is observed in any of the groups. As a result SAX is not directly considered

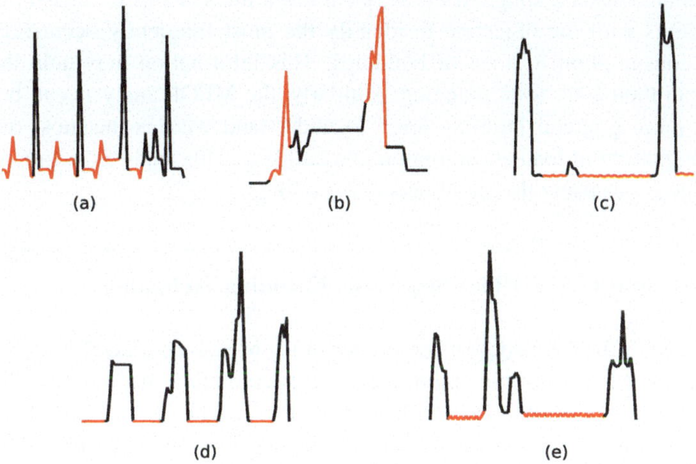

Fig. 5 Planted patterns in five different time series data are shown in red: subfigures (**a**)–(**e**) show the five different time series along with the planted patterns

Table 1 The table shows the comparison of the proposed symbolic representation technique with others. The best values of the tightness measure T and Dunn score are marked with boldface

	T			Dunn score		
Type	E-SAX	PLA	Proposed	E-SAX	PLA	Proposed
Industrial	0.94	0.41	**0.19**	0.004	**0.0002**	0.0001
Office	0.49	0.68	**0.26**	0.0002	0.0002	**0.001**
Retail	0.34	0.51	**0.28**	0.007	**0.001**	**0.001**
EP	0.32	**0.29**	0.35	0.0001	**0.0002**	0.0001

for the clustering of the segments. As shown in the Table 1, the proposed symbolic representation outperforms other methods in 3 out of 4 data sets. Though, PLA is found to outperform other methods in 3 out of 4 data sets in terms of Dunn score, the values obtained using Dunn Score are very low and performance of the proposed method is very close to the performance of PLA. It is also important to note that, the Dunn score is biased toward partitions having fewer good clusters while the rest of the clusters may be of inferior quality while the proposed measure considers the quality of all the clusters.

5.3 *Most Frequently Occurring Pattern (MFOP)*

The MFOP is described as the symbol with highest frequency of occurrence. We analyze the most frequently occurring pattern (MFOP) at two levels: segment level and building level. At segment level, we analyze the MFOP for sequential encoding and simultaneous encoding. At the segment level, the analysis is carried out across all buildings with the objective to identify the most frequently occurring pattern of load consumption in a set of buildings. This information may help the utility supplier in their long term planning. Similarly, the MFOP analysis can be carried out at a more granular building level, to understand what is the most frequently occurring pattern of load consumption in a building. This analysis is carried out on every time series using the simultaneous encoding.

5.3.1 Analysis of MFOP for Sequential Encoding Technique

Figure 6 shows the frequency of occurrence of the MFOP obtained for each data set.

We analyzed the symbol *edad* for EP type buildings because of its highest frequency of occurrence. We studied the survey data of the CER-IRISH and discovered some common properties of the buildings with pattern *edad*. The selected buildings have maximum 50 employees. Most of the buildings with pattern *edad* have 8–12 working hours during weekdays from 9 a.m. to 5 p.m. or 10 a.m. to 6 p.m.

With reference to the MFOP given in Fig. 6, we obtain two groups common in all data sets, that is *edac* and *fcac*. In Fig. 7, we illustrate the common groups and analyze the common patterns. In case of group *edac*, the office and retail type buildings have the matching peaks. The mean of the industrial and EP buildings lies in the range of 0 to 2.00, that is the similar range out of all four data sets. As shown in the Fig. 7, in case of group *fcac*, the patterns of office and EP type buildings are similar with a sharp rise at the end of the segment.

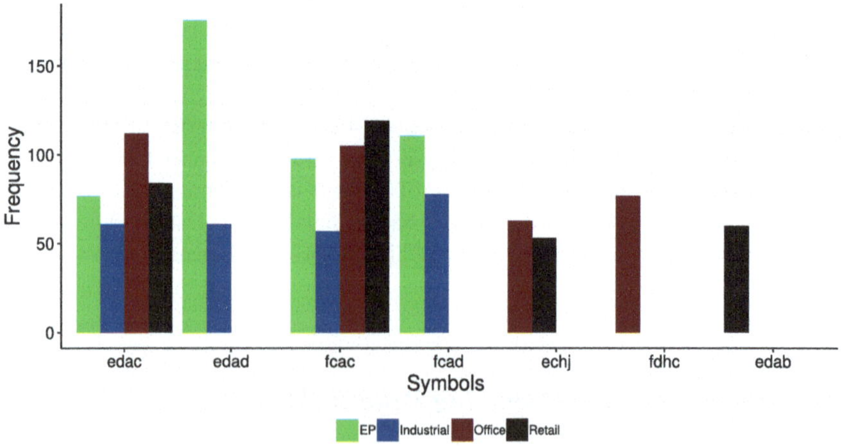

Fig. 6 Frequency of occurrences of top 4 most frequently occurring patterns for all data sets

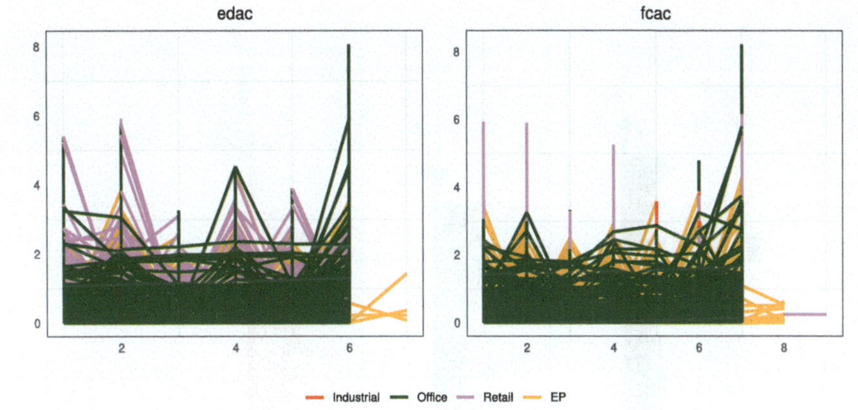

Fig. 7 Segments belonging to the common groups representing the MFOP "edac" and "fcac" for all four data sets industrial, office, retail, and EP buildings in the sequential encoding approach

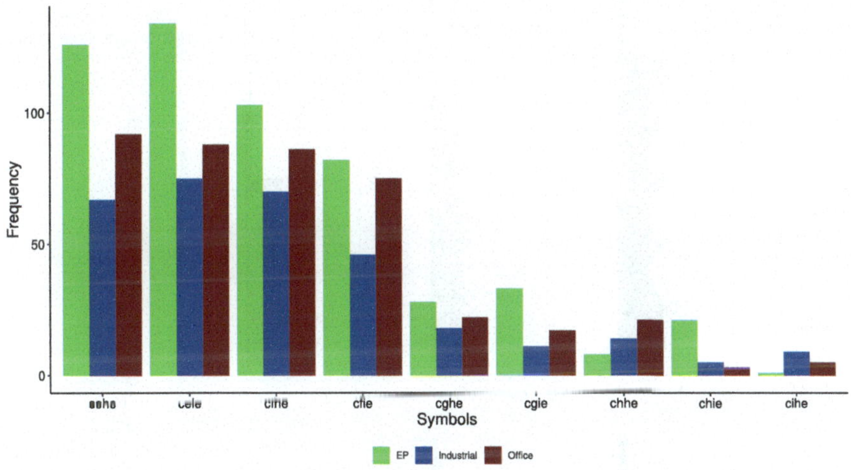

Fig. 8 Frequency of occurrences of each encoding for EP, industrial, and retail data sets

5.3.2 Analysis of MFOP for Simultaneous Encoding Technique

In Fig. 8, we plot the frequency of occurrence of the MFOP in respective data sets. For groups *cehe*, *ceie*, *cfhe*, *cfie*, EP has the maximum frequency. All the symbols have large frequency of occurrence except the last three: *chhe*, *chie*, *cihe*.

We illustrate the patterns of the common groups in Fig. 9. Although mean and degree of all the groups shown in Fig. 9 lie in the same range, the LC or intercept is varying. The groups *cehe* and *ceie* have varying range of the LC, that is in *cehe*, LC ranges from -0.00414 to -1.04729, while in *ceie*, the LC ranges from 0.00087 to 1.99564. There exists a total of 100 unique buildings with patterns *cehe* and 80

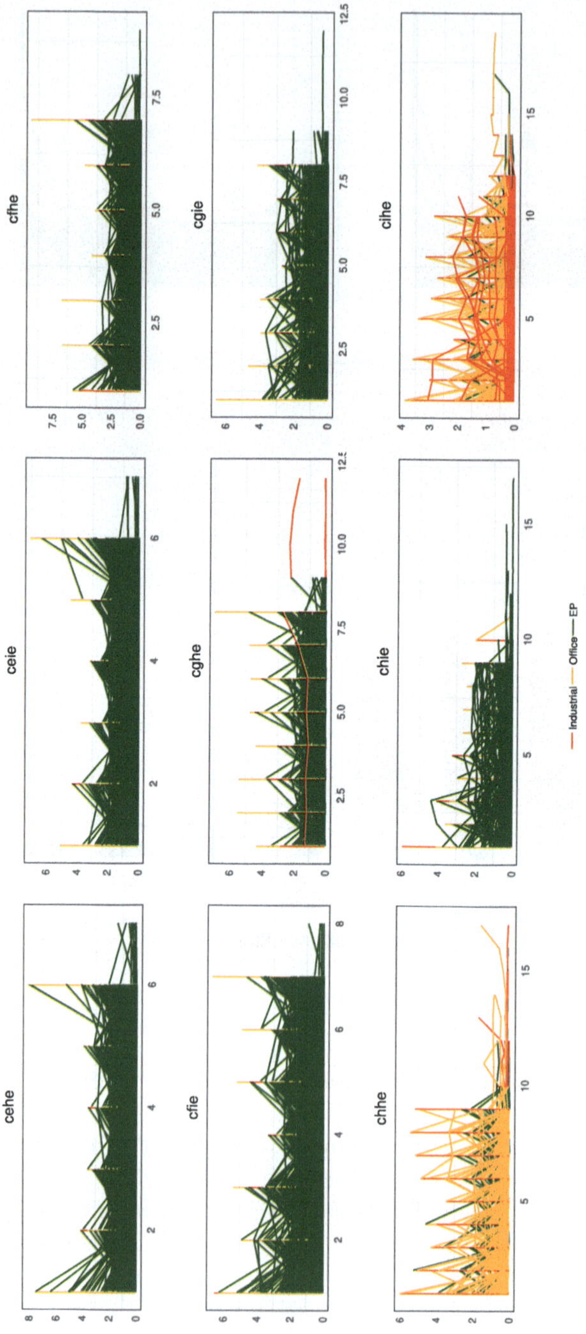

Fig. 9 Load patterns corresponding to the MFOP common in industrial, office, and EP buildings

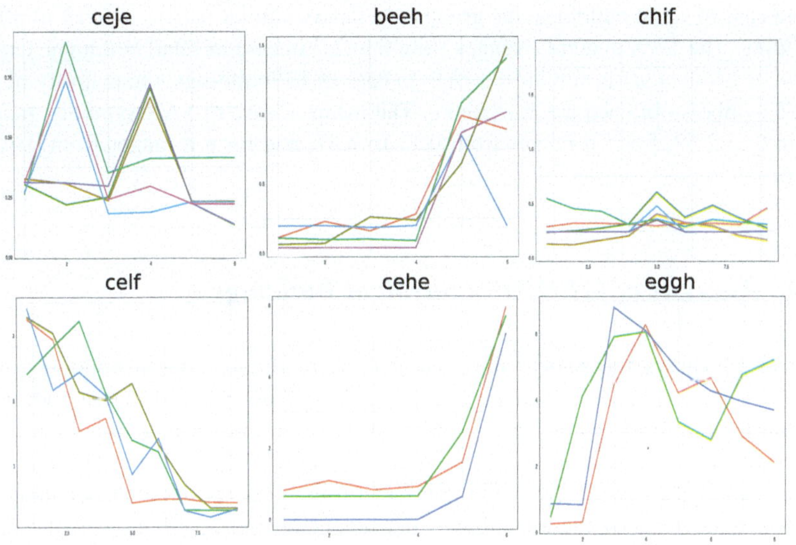

Fig. 10 Representative of groups formed by sequential encoding technique. Given groups are: *ceje*—industrial buildings, *beeh* and *chif*—office buildings, *celf*—retail buildings, *cehe* and *eggh*—EP buildings

buildings with pattern $ceie$, $cfhe$, $cfie$. In all the groups, and all data sets, more than 50% of the buildings have 1–5 employees. The working hours for the buildings ranges from approximately 8–10 h on weekdays.

We illustrate the representative patterns of the groups formed by simultaneous encoding in Fig. 10. Each group captures distinct characteristics of the load patterns.

The group $ceje$ captures the patterns with two peaks. In case of group $ceje$, majority of the segments falls into time $5:00$ a.m. $10:00$ a.m. The mean electricity consumption is lower during that time. On analyzing the buildings in group $ceje$, we discovered that those buildings mostly use heating and cooling appliances during times of low electricity use, to adjust the temperature of the building. Hence the electricity consumed between $5:00$ a.m.–$10:00$ a.m. is mostly due to the use of heating and cooling appliances. Majority of windows in case of industrial buildings falls into c, d, and h symbols. The mean of the windows in c SAX symbol ranges from 0 to 2.73, for d the mean ranges from 0.19 to 3.8, and h has the windows with highest power readings with mean ranging from 1.54 to 26.64.

The group $beeh$ for office data set has a constant pattern at the beginning of the window, and then a rising trend is observed. The group $chif$ for office building has two peaks at the end of the window. The frequently occurring SAX symbol here are b, c, and d. Here, the range of mean in case of SAX symbols b ranges from 0 to 1.9, for c, the mean ranges from 0 to 3.17, for d, mean ranges from 0.63 to 4.94.

In case of retail buildings, the group *celf* have a similar pattern to *beeh* of office building. The SAX symbol *c* ranges from 0 to 2.7 in case of retail buildings. Group *cehe* of EP data set is similar to *cehf*. In case of EP buildings, the majority of the SAX symbols obtained are *c*, *d*, and *e*. The mean in case of SAX symbol *c* ranges from 0 to 2.59, for *d* it ranges from 0.13 to 3.87, and for *e* it ranges from 0.61 to 5.16.

5.4 Analysis of MFOP of Individual Buildings

Figure 11 shows the MFOP obtained from 7 randomly chosen buildings. The symbols *ccbe* and *dcce* and *bcae* are obtained from three different industrial buildings. *ccbbe* identifies the constant power consumption, *dcce*, *bcae* identifies the wavy pattern. The next two *bcae* in the second row are for the office and retail type buildings respectively. It is important to note that as the alphabet range of each symbol used in the encoding is dependent on the distribution of the data,

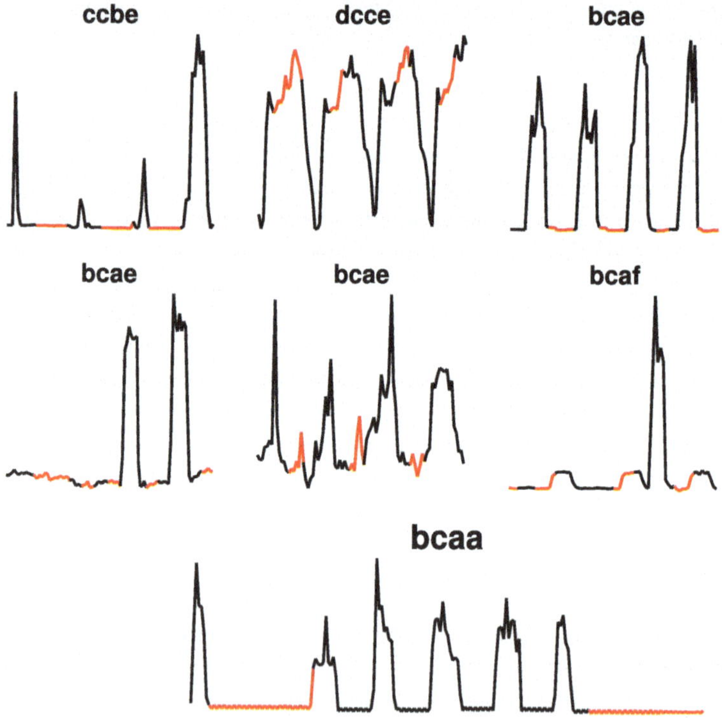

Fig. 11 MFOP obtained from 7 randomly chosen buildings. The red parts in the plots mark the segments with the given symbols. The symbols from top left are—*ccbe*, *dcce*, *bcae*: industrial data set, *bcae*: office data set, *bcae*: retail data set, *bcaf*: retail data set, *bcaa*: retail data set

same encoding may represent two different patterns in two different data sets. For example, the encoding *bcae* shows different patterns in office and retail data sets. The symbol *bcaf* is obtained from a retail type building. The pattern *bcaa* obtained from retail type building shows that the proposed symbolic representation identifies the pattern which exists for the longest period of time.

5.5 Applications Toward Demand Side Management (DSM)

Here, we discuss how the discovered MFOPs obtained from sequential encoding, simultaneous encoding, and building level can be helpful in DSM operations.

- **Sequential encoding**
  Identifying the common hours of usage of a symbol type can help the utility providers to manage the loads and in load shifting operations in case of peak hours. In case of the common groups detected like *edac*, *fcac*, and *echj* given in Fig. 6, we extract the hours where it has the highest frequency of occurrence shown in Table 2. The group *edac*, which occurs in all the data sets, shown in Fig. 6 has common hours, 07 : 00–12 : 00 in industrial and retail data sets. Similarly, the group *echj* has its occurrence from 00 : 00–5 : 00 in both office and retail data sets. This common interval existing in multiple building signifies some common behavior being captured in the group and that time can be opted for load shifting in case of peak hours of usage.
  Similarly the group *fcac* has common hours, 20 : 00 : 00–02 : 00 : 00 in case of office and EP data sets. In case of group *edad*, the pattern does not start at the same time but the duration of occurrence is same and also there is a coincidence of 3 h, that is from 05 : 00 : 00 to 08 : 00 : 00 a.m. Hence, the hour of coincidence can be considered as the peak hour for the utility providers.
- **Simultaneous Encoding**
  In Table 3 we summarize the time range of the groups common in industrial, office, and EP data sets obtained from simultaneous encoding technique. For each of the given groups in Table 3, the time range of windows is either common or there exists an hour of phase shift. For example, as shown in Table 3, for

Table 2 Table showing the similarity in time usage obtained from sequential encoding

Group	Data set	Time-From	Time-To
edac	Industrial	07:00:00	00:00:00
	Retail	07:00:00	00:00:00
fcac	Office	20:00:00	02:00:00
	EP	20:00:00	02:00:00
edad	EP	03:00:00	08:00:00
	Industrial	05:00:00	10:00:00
echj	Office	00:00:00	05:00:00
	Retail	00:00:00	05:00:00

Table 3 Table showing the similarity in time usage obtained from simultaneous encoding

Groups	Data set	Time-From	Time-To
cehe	Industrial	00:00:00	05:00:00
	EP	00:00:00	05:00:00
cgie	Industrial	00:00:00	07:00:00
	Office	00:00:00	07:00:00
chhe	Industrial	22:00:00	06:00:00
	Office	21:00:00	05:00:00
chie	Industrial	14:00:00	22:00:00
	Office	14:00:00	22:00:00

groups *cehe*, *cgie*, and *chie*, the time duration of the load pattern is common for the given data sets. However, in case of group *chhe*, the industrial and office buildings have an hour of phase shift. Discovery of these coincident groups with same time duration will help in load shifting operations.

- **Building level**

  With respect to the patterns obtained at building level analysis shown in Fig. 11, we discuss some of the applications toward DSM below:

  - **Locating the peaks**: Identifying the peaks at regular intervals can help in load balancing. For example in *dcce* in Fig. 11 for the industrial buildings, the peaks are synchronized in time and have same patterns, which makes it easier for utility providers to maintain the demand–supply equilibrium.
  - **Energy saving**: The symbol *bcae* in Fig. 11 for three different buildings has different patterns. A peak exists in the third *bcae*, that is for the retail type building, however, no peak in the first *bcae*, that is for the industrial data set. With an aim to energy saving policy, identifying the buildings with lower load consumption in segments can act as an ideal building, that is the *bcae* in the industrial data set. The discovery of ideal building can be used as an example for others to manage their respective loads thus helping in energy saving policy.

6 Summary and Conclusions

In this article, we developed an automatic segmentation and symbolic representation technique for time series to analyze the load profiles obtained from the smart meter data in commercial buildings. The limitations of the existing literature are that it breaks the time series into pre-determined fixed length segments and represents the segments based on the statistical measures of the data which fails to capture information about the pattern of a segment along with the peak/off-peak information and average value. We develop a piecewise polynomial regression model to segment a time series that can also be used for break-point identification. We next develop a symbolic representation of time series by encoding each segment by the mean value, the leading coefficient of the fitted polynomial, intercept of the fitted polynomial, and the degree of the best fit polynomial. The encoding allows dimensionality reduction of the original time series while preserving each unique pattern. We finally propose a measure to quantify the performance of any symbolic representation based method in clustering time series segments. The proposed method emphasizes on the quality of each cluster formed, unlike the common clustering indexes. One of the limitations of the proposed tightness measure, however, is that it is sensitive to outliers. In case of outliers, the $max(F(c_i))$ will increase and as a result the numerator value will be very large. Suppose $max(F(c_i)) - min(F(c_i)) \approx \gamma$. Now, even if the two different clusters are dissimilar, the denominator too will be very large, that is, $max(G(c_i)) + min(G(c_i)) \approx \gamma$, hence $Tightness \approx 1$, which signifies a poor clustering.

The performance of the proposed method evaluated for plated pattern identification and clustering is found to be an improvement over the existing methods. The developed method is applied for the analysis of the load profiles in commercial buildings and applied for demand side management. The results show that the common groups discovered across the buildings exhibit common load consumption pattern which can help in aggregated DSM applications.

The current study also opens many important questions for future analysis, like:

- What would be a data driven approach to learn the various parameters involved in the proposed segmentation and symbolic representation based approach?
- What would be the data driven approach for symbolic representation of univariate times series that optimize multiple criteria like dimensionality reduction, information loss while preserving the unique patterns?
- How can the information about the similarity in load profile across consumers be exploited for better energy management and generation of renewable energy?

Research attempting to address these questions would further enrich the field of smart energy.

References

1. S. Aminikhanghahi, D.J. Cook, A survey of methods for time series change point detection. Knowl. Inf. Syst. **51**(2), 339–367 (2017)
2. Z.D. Belafi, T. Hong, A. Reith, A library of building occupant behaviour models represented in a standardised schema. Energy Eff. **12**(3), 637–651 (2019)
3. R. Bellman, B. Kotkin, On the approximation of curves by line segments using dynamic programming. ii. Technical Report, Rand Corp Santa Monica Calif (1962)
4. U. Blanke, B. Schiele, M. Kreil, P. Lukowicz, B. Sick, T. Gruber, All for one or one for all? combining heterogeneous features for activity spotting, in *2010 8th IEEE International Conference on Pervasive Computing and Communications Workshops (PERCOM Workshops)* (IEEE, Piscataway, 2010), pp 18–24
5. A. Capozzoli, M.S. Piscitelli, S. Brandi, D. Grassi, G. Chicco, Automated load pattern learning and anomaly detection for enhancing energy management in smart buildings. Energy **157**, 336–352 (2018)

6. H. Chen, G. Zhang, D. Zhu, J. Lu, A patent time series processing component for technology intelligence by trend identification functionality. Neural Comput. Appl. **26**(2), 345–353 (2015)
7. J.C. Dunn, A fuzzy relative of the isodata process and its use in detecting compact well-separated clusters. J. Cyb. **3**, 32–57 (1973)
8. D. Fisch, T. Gruber, B. Sick, Swiftrule: Mining comprehensible classification rules for time series analysis. IEEE Trans. Knowl. Data Eng. **23**(5), 774–787 (2010)
9. E. Fuchs, T. Gruber, J. Nitschke, B. Sick, Online segmentation of time series based on polynomial least-squares approximations. IEEE Trans. Pattern Analy. Mach. Intell. **32**(12), 2232–2245 (2010)
10. N.A. Funde, M.M. Dhabu, A. Paramasivam, P.S. Deshpande, Motif-based association rule mining and clustering technique for determining energy usage patterns for smart meter data. Sustainable Cities Soc. **46**, 101415 (2019)
11. J.A. Gomez, M.F. Anjos, Power capacity profile estimation for building heating and cooling in demand-side management. Appl. Energy **191**, 492–501 (2017)
12. X. He, C. Shao, Y. Xiong, A non-parametric symbolic approximate representation for long time series. Pattern Analy. Appl. **19**, 111–127 (2014). https://doi.org/10.1007/s10044-014-0395-5
13. B. Horst, K. Abraham, *Data Mining in Time Series Databases*, vol. 57. World Scientific, Singapore (2004)
14. F. Iglesias, W. Kastner, Analysis of similarity measures in times series clustering for the discovery of building energy patterns. Energies **6**(2), 579–597 (2013)
15. B. Kalluri, A. Kamilaris, S. Kondepudi, H.W. Kua, K.W. Tham, Applicability of using time series subsequences to study office plug load appliances. Energy Build. **127**, 399–410 (2016)
16. E. Keogh, K. Chakrabarti, M. Pazzani, S. Mehrotra, Dimensionality reduction for fast similarity search in large time series databases. Knowl. Inf. Syst. **3**(3), 263–286 (2001)
17. E. Keogh, K. Chakrabarti, M. Pazzani, S. Mehrotra, Locally adaptive dimensionality reduction for indexing large time series databases, in *Proceedings of the 2001 ACM SIGMOD International Conference on Management of Data*, vol. 30 (ACM, New Yrok, 2001), pp. 151–162
18. E. Keogh, S. Chu, D. Hart, M. Pazzani, Segmenting time series: A survey and novel approach, in *Data Mining in Time Series Databases* (World Scientific, Singapore, 2004), pp 1–21
19. R. Li, T.P. Tian, S. Sclaroff, Simultaneous learning of nonlinear manifold and dynamical models for high-dimensional time series, in *2007 IEEE 11th International Conference on Computer Vision* (IEEE, Piscataway, 2007), pp 1–8
20. H. Li, C. Guo, W. Qiu, Similarity measure based on piecewise linear approximation and derivative dynamic time warping for time series mining. Expert Syst. Appl. **38**(12), 14732–14743 (2011)
21. G. Li, Z. Cai, X. Kang, Z. Wu, Y. Wang, ESPSA: A prediction-based algorithm for streaming time series segmentation. Expert Syst. Appl. **41**(14), 6098–6105 (2014)
22. J. Lin, E. Keogh, S. Lonardi, B. Chiu, A symbolic representation of time series, with implications for streaming algorithms, in *Proceedings of the 8th ACM SIGMOD Workshop on Research Issues in Data Mining and Knowledge Discovery* (ACM, New York, 2003), pp 2–11
23. B. Lkhagva, Y. Suzuki, K. Kawagoe, Extended SAX: Extension of symbolic aggregate approximation for financial time series data representation. DEWS2006 4A-i8 7 (2006)
24. C. Miller, Z. Nagy, A. Schlueter, Automated daily pattern filtering of measured building performance data. Autom. Constr. **49**, 1–17 (2015)
25. S. Nan, M. Zhou, G. Li, Optimal residential community demand response scheduling in smart grid. Appl. Energy **210**, 1280–1289 (2018)
26. D. Posada, T.R. Buckley, Model selection and model averaging in phylogenetics: advantages of Akaike information criterion and Bayesian approaches over likelihood ratio tests. Syst. Biol. **53**(5), 793–808 (2004)
27. P. Senin, J. Lin, X. Wang, T. Oates, S. Gandhi, A.P. Boedihardjo, C. Chen, S. Frankenstein, Grammarviz 3.0: Interactive discovery of variable-length time series patterns. ACM Trans. Knowl. Discovery Data **12**(1), 10 (2018)
28. H.A. Sturges, The choice of a class interval. J. Am. Statist. Assoc. **21**(153), 65–66 (1926)

29. N. Sugiura, R. Ogden, Testing change-points with linear trend. Commun. Stat. Simul. Comput. **23**(2), 287–322 (1994)
30. S. Torkamani, V. Lohweg, Survey on time series motif discovery. Wiley Interdiscipl. Rev. Data Mining Knowl. Discovery **7**(2), e1199 (2017)
31. C. Wang, X.S. Wang, Supporting content-based searches on time series via approximation, in *Proceedings. 12th International Conference on Scientific and Statistica Database Management* (IEEE, Piscataway, 2000), pp. 69–81
32. Y. Wang, Q. Chen, C. Kang, M. Zhang, K. Wang, Y. Zhao, Load profiling and its application to demand response: a review. Tsinghua Sci. Technol. **20**(2), 117–129 (2015)
33. X. Wang, J. Lin, N. Patel, M. Braun, Exact variable-length anomaly detection algorithm for univariate and multivariate time series. Data Mining Knowl. Discovery **32**(6), 1806–1844 (2018)
34. L. Wenpeng, Advanced metering infrastructure. Southern Power Syst. Technol. **3**(2), 6–10 (2009)
35. C.C.M. Yeh, Y. Zhu, L. Ulanova, N. Begum, Y. Ding, H.A. Dau, D.F. Silva, A. Mueen, E. Keogh, Matrix profile i: all pairs similarity joins for time series: a unifying view that includes motifs, discords and shapelets, in *2016 IEEE 16th International Conference on Data Mining (ICDM)* (IEEE, Piscataway, 2016), pp 1317–1322
36. Y. Yu, Y. Zhu, D. Wan, H. Liu, Q. Zhao, A novel symbolic aggregate approximation for time series, in *International Conference on Ubiquitous Information Management and Communication* (Springer, Berlin, 2019), pp 805–822
37. C.T. Zan, H. Yamana, An improved symbolic aggregate approximation distance measure based on its statistical features, in *Proceedings of the 18th International Conference on Information Integration and Web-Based Applications and Services* (ACM, New York, 2016), pp 72–80
38. H. Zhao, Z. Dong, T. Li, X. Wang, C. Pang, Segmenting time series with connected lines under maximum error bound. Inf. Sci. **345**, 1–8 (2016)

A Modern Approach to Include Representative Behaviour Models in Energy Simulations

Ayesha Kashif, Stephane Ploix, and Julie Dugdale

1 Introduction

Energy simulations take into account inhabitants' behaviour in modelling energy trends because their decisions and actions have a strong impact on the energy consumption. These are benchmarked at the office buildings using controlled activity profiles and predefined scenarios. However, at home this behaviour is quite complex and difficult to predict as compared to at work. This requires dynamic (reactive, deliberative, social) inhabitants' behaviour to be taken into account to fully understand its potential effect on energy consumption and modelling energy trends. In this chapter, a methodology to generate and validate inhabitants' dynamic behaviour model is presented. In this methodology, a concept of fine-tuning parameters is proposed where simulated consumption curves, as inhabitants' behaviour, are mapped to actual consumption curves with model fitting methods. The resulting models exhibit behaviour closer to what could possibly happen in inhabitants' daily life. This will provide an opportunity to analyse more realistic impact of these behaviours on the energy consumption patterns.

Energy management approaches of living places put emphasis on modelling and simulating various physical factors related to energy consumption e.g. thermal performance of insulation, energy used by heating and cooling, and other electrical appliances, outdoor environment, and energy efficient appliances. Modelling represents system elements and their interactions whereas simulation helps to analyse responses of the system to some change which in real life might not be possible.

A. Kashif (✉) · S. Ploix
G-SCOP lab, Grenoble Institute of Technology, University of Grenoble Alps, CNRS UMR5272, Grenoble, France
e-mail: stephane.ploix@grenoble-inp.fr

J. Dugdale
Grenoble Informatics Laboratory, Grenoble, France

S. Ploix et al. (eds.), *Towards Energy Smart Homes*,
https://doi.org/10.1007/978-3-030-76477-7_16

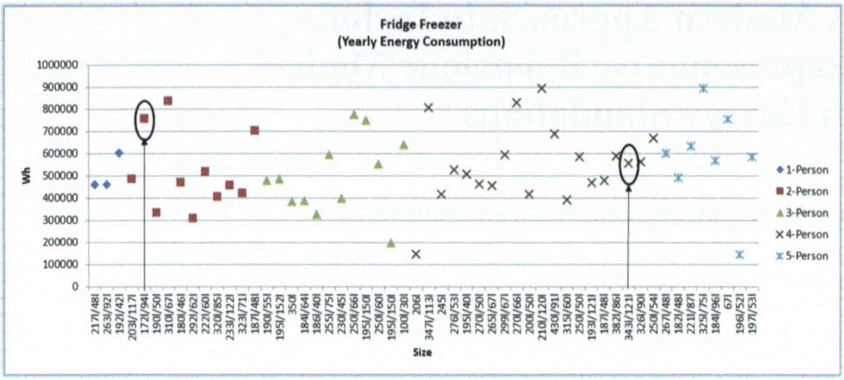

Fig. 1 Fridge freezer consumption patterns from the Irise dataset

In this section, the focus is not only on the physical aspects of the building and appliances but also on the inhabitants' dynamic behaviour because learning ecological behaviours and temperance will empower the energy simulations to understand energy consumption trends and reduce energy waste.

In order to understand how inhabitants' behaviour impacts energy consumption, the results from an analysis performed on different appliances in the Irise dataset[1] are presented. They help to assess the sensitivity of these appliances to inhabitants' behaviour. The results of the analysis performed on the fridge freezer are presented in Fig. 1. The x-axis shows the size of the fridge freezer in each house and the y-axis shows the energy consumption. Each point in the graph corresponds to the energy consumption of a fridge freezer over the period of a year along with the number of persons in each house represented by different colours. In some cases, the energy consumption depends upon the size of the fridge freezer and the number of people in the house, but in others it does not. An example of where the energy consumption does not depend upon the number of people in the house nor on the size of the fridge freezer, is shown with an oval. This shows that the energy consumption of the fridge freezer does not necessarily depend upon the number of people in the house nor on the size of the appliance. Instead, it depends on how the inhabitants use the appliance, i.e. their behaviours. This analysis also provides a good justification that simple presence/absence profiles are insufficient in order to model the household behaviour, especially for cold appliances. Figure 2 shows the yearly energy consumption of 12 place setting dishwashers for all houses in Irise that have a dishwasher. The box plot shows that the fluctuation of energy consumption among these houses is irrespective of the number of people inside the house. The example of one of the extreme cases is the *1 person* house where

[1]This is part of the European Residential Monitoring to Decrease Energy Use and Carbon Emissions (REMODECE) project. It contains energy consumption data, for each appliance from 98 French houses, recorded at every 10 min, over a one year period.

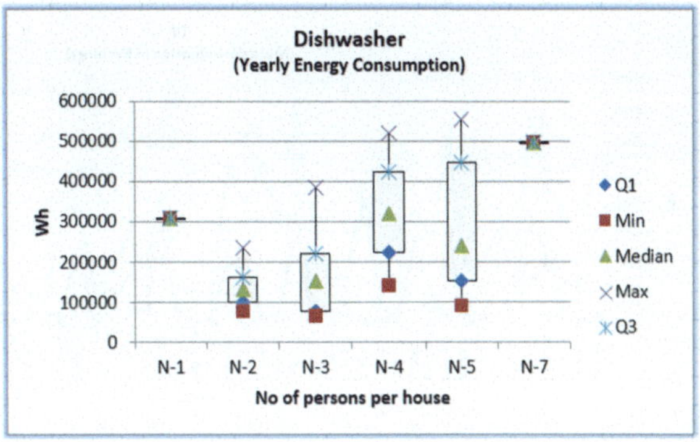

Fig. 2 Dishwasher yearly energy consumption for all houses in Irise dataset

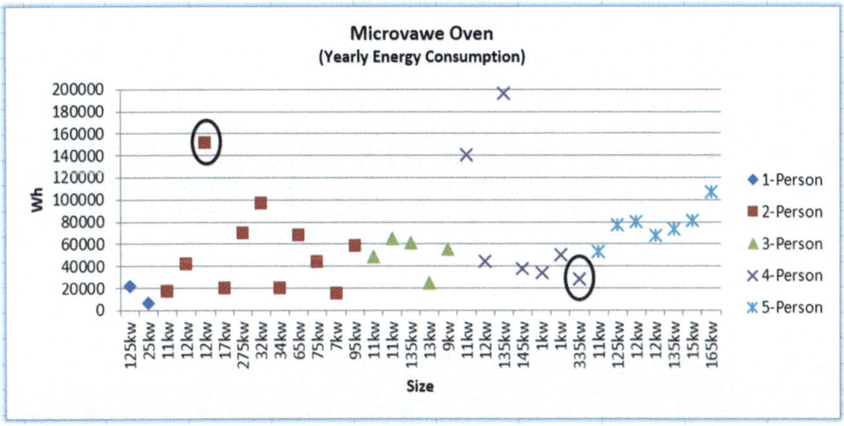

Fig. 3 Microwave oven yearly energy consumption for all houses in Irise dataset

the dishwasher consumes more than the *5 person* houses (the "Median" value) irrespective of the fact that both have a *12 place* setting dishwasher. Figure 3 shows the yearly energy consumption of the microwave ovens of different sizes. In this example, also no strict correlation between the number of persons in the house and the size of appliance with the consumption is found. The ovals show the case where a 12kw microwave in a *2 person* house is consuming more than a 335kw microwave in a *4 person* house. There could be certain reasons e.g. the inhabitants in a *2 person* house eat ready meals at home most of the time and those in the *5 person* house use the standard cooker every time they want to eat. Similarly, covering the food while warming up, the duration for which the food is warmed up, etc. impacts the overall consumption. These factors, however, belong to inhabitants' behaviour rather than the size of the appliance or the number of persons in the house and hence are important to be considered in energy simulations and demand predictions. There are certain environmental parameters that impact the inhabitants' behaviour regarding energy consumption. These include seasons, day type (weekday, weekend), day time

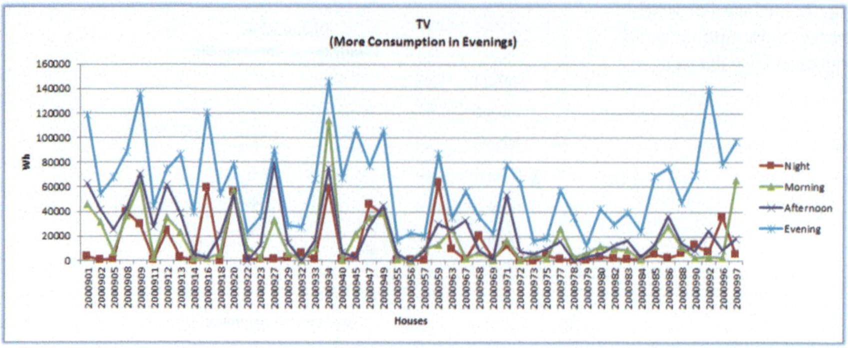

(a)

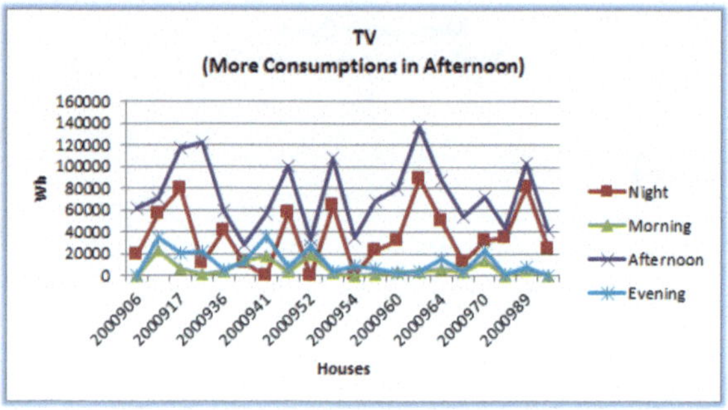

(b)

Fig. 4 Consumption of the TV for all the houses from the Irise dataset. (**a**) Higher consumption in the evening. (**b**) Higher consumption in the afternoon

(morning, afternoon, evening, night), and weather conditions (sun, rain, etc.). These parameters are also somehow influenced by the behaviour of occupants. Figure 4 shows the energy consumption of the TV in all houses in the Irise database. There is more consumption in the evening (Fig. 4a) as mostly people are at home and like to watch TV during this period. There are however some houses where the inhabitants watch TV mostly in the afternoon (Fig. 4b). In these houses, the second most probable time to watch TV is at night. This could be due to the fact that in these houses most of the family members stay at home, perhaps because they are elderly and retired or housewives or kids watching cartoons, etc.The consumption of the water heater in Fig. 5a is more in the evening as compared to other periods of the day. This is because during these periods inhabitants are mostly at home and interact more with thermostat settings or windows, etc. Figure 5b shows another case where consumption is more at nights rather than in the evenings.Figure 6a shows an example of a washing machine where there is significantly more consumption on weekends than weekdays. Conversely, the houses in Fig. 6b do not have a significant difference between the energy consumption on weekdays and weekends.

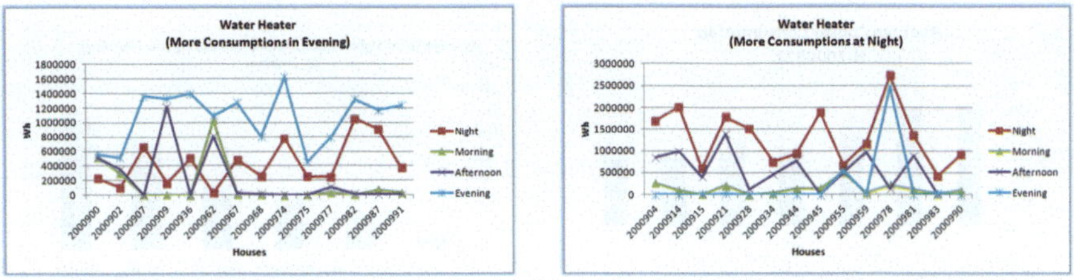

(a) (b)

Fig. 5 Water heater consumption for all the houses from the Irise dataset. (**a**) Higher consumption in the evening. (**b**) Higher consumption at night

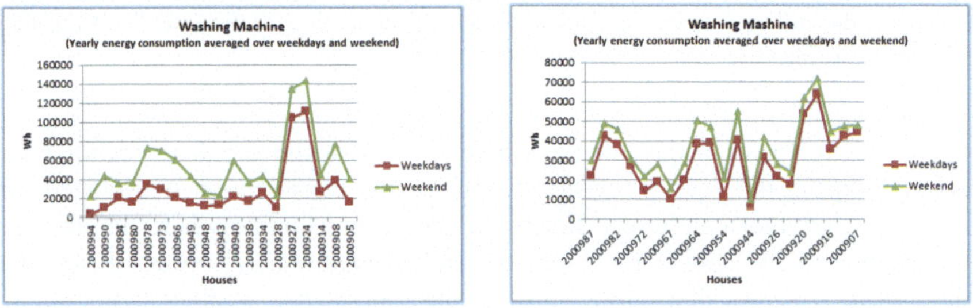

(a) (b)

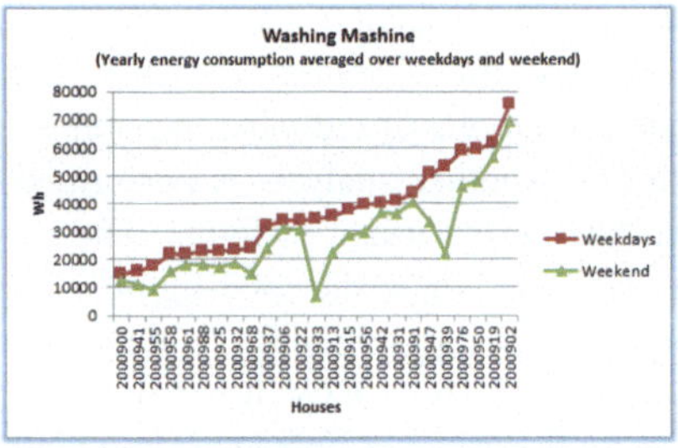

(c)

Fig. 6 Washing machine consumption of the averaged over weekdays and weekends. (**a**) Significant difference in consumption. (**b**) Small difference in consumption. (**c**) Higher consumption on weekdays

Figure 6c shows the houses where the washing machine is used more on weekdays than on weekends. Thus high variability is found in the inhabitants' behaviours regarding the weekdays and weekend consumptions. In addition to the parameters discussed above, the weather is another important factor that affects inhabitants' way of interacting with some appliances. For example, if the weather is good it may influence the inhabitants' desire to eat out. This behaviour could vary from one family to another based on their norms, culture, region, etc. In order to see the impact of weather on cooking behaviour, an analysis is performed on the houses in the Irise database. In this analysis, the consumption of the electric cooker (hotplate+oven)

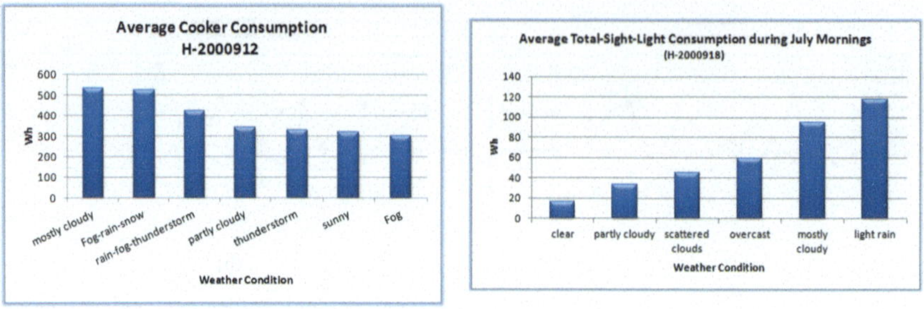

(a) (b)

Fig. 7 Impact of weather conditions on appliances' consumption behaviour. (**a**) Cooker consumption during different weather conditions for a house in the Irise database. (**b**) Total-Lighting consumption during different weather conditions

is summed up for each day for the whole year. Also the weather condition for each day during the year is registered. Finally, the consumption is averaged for each of the weather conditions. Figure 7a shows an example where the average consumption of the cooker for different weather conditions is averaged over the whole year. It shows that during most of the times when weather is not sunny the consumption is higher compared to when it is sunny. There could be certain reasons behind this consumption behaviour of this family, e.g. the tendency to eat out when the weather is good, or the inhabitants are eating cold food (salads, etc.). In literature weather is found to be one of the most important and influencing factors on energy consumption [1]. Another analysis is performed to find the impact of different weather conditions on the usage of lights. The experiment is performed on a house in the Irise database where the total-lighting consumption is summed for each morning during the period of a month. Then the consumption against each weather condition is summed up. The results shown in Fig. 7b clearly depict that as the weather is getting worse the usage of lights is significantly increased.

Including behaviour in energy control and management is currently focused on either static profiles or predictive models (sensor based inhabitants' occupancy detection). However, current approaches are also based on single user interactions with the environment and do not include reactive/deliberative decision making or complex human behaviours. The purpose of this research is to capture the behaviour that not only represents a simple presence or absence of an inhabitant in an environment but also represents a realistic interaction of the human with the environment. This means that the dynamic, reactive, deliberative, and social behaviour of inhabitants must also be taken into account in order to fully understand its possible effect on energy consumption. Such an approach considers the inhabitants as reactive, intelligent agents instead of simply "fixed metabolic heat generators passively experiencing the indoor environment" [23].

2 Inclusion of Occupants' Behaviour in Buildings Energy Management

The literature suggests that behaviour strongly influences energy consumption patterns and is an important factor for energy waste reduction in buildings. Certain programs that can improve consumption based behavioural efficiency have been found to be quite effective [2]. Azar and Menassa [3] observed significant sensitivity levels taking into account the behavioural parameters that vary with the building size and weather conditions. Building energy simulation tools are used to evaluate building designs, energy efficiency, demands, human comfort, emissions, and associated costs during design stages and performance predictions. The existing simulation tools exhibit significant differences in predicted and simulated energy consumptions. This is due to the fact that factors influencing energy consumptions in buildings, (i) outdoor/indoor climate, (ii) building characteristics and (iii) inhabitants' behaviour are poorly understood and included only with standard basic assumptions. The role of inhabitants' behaviour clearly indicates our inability to properly model inhabitants' complex behaviour, taking into account the reactive and deliberative mechanisms and to better quantify uncertainties in energy efficiency predictions.

Repetitive inhabitants' actions are included in same simulation tools (e.g. DAYSIM) as intelligent algorithms [4] and [5], but they are not representative of the actual behavioural variations. Inclusion of inhabitants' behaviour within energy simulations is discussed in literature across two dimensions: (i) behaviour models based on statistical algorithms [6] and (ii) predefined fixed schedule based behaviour models [7]. The statistical behavioural models are based on stochastic processes with probabilities of control events, but fixed schedules refer to deterministic, predictable and repeatable behaviours. This is an important limitation in these simulation tools that restrict us to achieve more accurate energy estimates and predictions. The inclusion of a probabilistic discomfort model in addition to a stochastic behavioural model [8] often results in more realistic simulations, but occupancy model with only presence and absence profiles is still a challenge.

The approaches used for energy simulations in buildings do not include the social interactions of occupants inside the building. The importance of social interactions in human societies as explained by [9] is that as compared to the linear systems where the properties of the whole system are a simple aggregation of its parts, the human societies are different. Human societies are rather complex as their behaviour cannot be determined by partitioning it and understanding the behaviour separately for each part. This complexity is due to the non-linear interactions between the people where the exchange of knowledge and materials affects the recipient's behaviour. Thus in case of humans, the behaviour emerges from the actions of its units.

Recently, the multi-agent systems (MAS) are being used in the domain of energy management within buildings. For example, a MAS approach is used in monitoring and controlling the Heating, Ventilation and Air Conditioning (HVAC) system and lighting in office buildings [10]. In smart homes, the approach has also been used for the anticipatory and reactive control of HVAC and lighting [11]. Likewise, an agent based control system was used for the optimization of a simulated residential water heating system [12]. The prediction of the mobility patterns and device usage of inhabitants has been done in the MAVHome project in order to satisfy the tradeoff between cost and comfort [13]. Abras and his colleagues [14] gave the control of appliances and sources to the software agents that are used in a home automation system. Liao and Barooah [15] developed a multi-agent system to predict and simulate the occupancy at room and zone level in commercial buildings.

Lopes et al. [16] proposed that the behavioural aspects should be included with a multidisciplinary approach that integrates engineering and social science together. They also found that the energy based behaviours are highly dynamic that can cause inconsistencies, so need to be properly investigated. Surveys have been designed to identify the wasteful behaviours and indicated that between different dwellings, energy efficient behaviours account for 51%, 37% and 11% of variance in heat, electricity and water consumption, respectively [24]. Hoes et al. [17] reported that occupant's comfort is negatively affected if they are not given the control over environment e.g. thermostat, windows, blinds fans, etc. Thus, the occupants should be provided with control over their environment to increase their comfort levels [18]. Liu et al. [19] Environmental factors and variations lead to conscious or unconscious adaptive behaviours, such as restraining physical activity levels, changing or adjusting clothes, opening or closing windows, adjusting thermostat, etc.

3 Multi-Agent Based Approach for Dynamic Behaviour Model Generation and Validation

The approaches mostly used for taking into account human behaviour in building simulations are based on purely stochastic methods. They are based on already established statistical models and hence are more reliable. These processes take the presence and/or activity profiles and reproduce the consumption behaviour according to what is real or measured. They also enable the prediction of such profile patterns for different appliances to help analyse the power demand or generate load profiles in advance.

Another approach for including occupant's behaviour in realistic building energy simulations is to not only consider behaviour as fixed profiles but to also model the reactive, deliberative, social and group phenomenon taking place in real situations. It also involves the stochastic process but mixed with an artificial intelligence based multi-agent human behaviour model. The key advantages of this approach over purely statistical approach are that it helps in finding out: (i) how to identify the high energy impacting behaviours, (ii) how the complex i.e. reactive, deliberative, social and group behaviour can be co-simulated with the thermal model of the

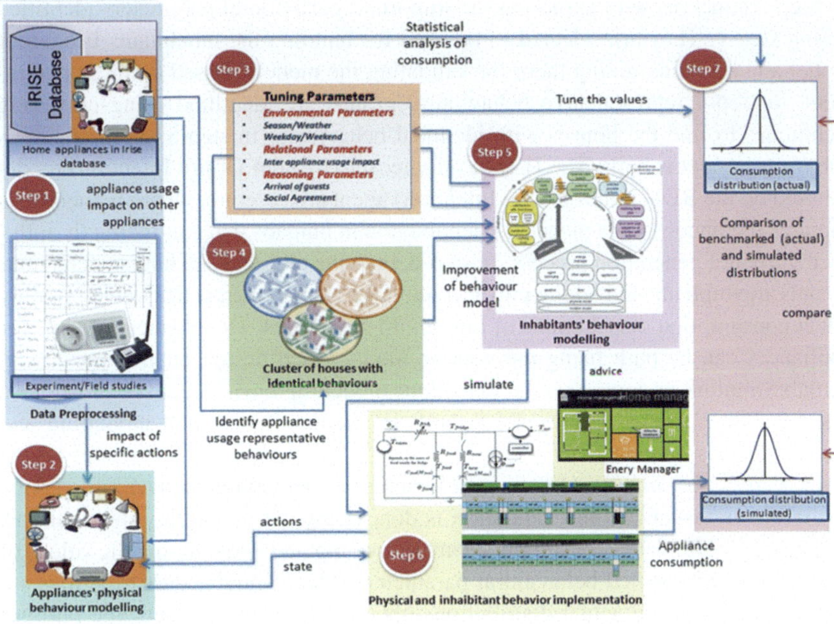

Fig. 8 Methodology to validate behaviour model

building and physical models of appliances in residential buildings, (iii) how can the complex behaviour models be validated to ensure its representativeness and (iv) how to validate BEMS with building system and inhabitants. The proposed dynamic behaviour generation and validation methodology is shown in Fig. 8. Step-1 involves data analysis and data pre-processing. An analysis of the Irise energy consumption dataset is performed to find the energy consumption behaviour of households. The data in Irise is further complemented with some additional information in order to understand the effect of certain other parameters on the energy consumption behaviour of households. This information includes the day of the week (i.e. weekend or weekdays), holidays, the state of the weather and the parallel usage of other appliances. The pre-processing step also involves the field studies and experiments where both the consumptions and the activities behind these consumption patterns are recorded. In order to further elaborate certain consumption patterns, the reasons behind certain activities are also recorded.

In step-2, the physical model of the building envelop and/or the appliance is constructed. The important inputs for this step include the data about the household activities and the data about the consumption of appliance. The impact of the usage of one appliance on another is also important to construct the model of appliances. This impact can be analysed both from the Irise database and the field studies. However, the impact of specific actions on appliance consumption is analysed through field studies.

Step-3 concerns with exploring the important behavioural parameters identified during step-1. They are selected to be used for building the inhabitants behaviour model and later fine-tuning them for validating the models. Based on these parameters, in step-4 representative behavioural groups are identified using clustering techniques to find the houses with identical behaviours. In step-5 the inhabitant's behaviour model is built. The H-BDI (Homeostasis-Belief Desire Intention) model is based on the BDI (Belief Desire Intention) agent architecture, where homeostasis captures the unconscious physical phenomenon in humans that can indirectly influence the belief generation process. In step-6, both the physical and human behaviour models are implemented taking into account different parameters that could possibly affect consumption distributions of household appliances. The physical models for appliances can be built using the generic model based design environments, e.g. matlab/Simulink or some specific behaviour modelling environments e.g. Brahms. The Inhabitant's behaviour model is built in Brahms multi-agent modelling and simulation environment.

In step-7, the impact of household energy consumption behaviours on the appliances is visualized and validation is done using the probability distributions. Likewise, the simulated consumptions are also computed while tuning the values of the previously identified behavioural parameters. These simulated distributions are then compared to the actual distributions obtained from the dataset. The purpose of this comparison is to see how close the proposed behaviour model and scenario implemented in Brahms (multiagent system) is to reality. The process of tuning the parameters continues until the actual and simulated error is significantly reduced. Besides the interaction with appliances, the co-simulations also involve the building envelop models and the Building Energy Management Systems (BEMS). These co-simulations help to identify and validate the role of BEMS while the inhabitants are also part of the whole process.

3.1 Data Collection, Pre-Processing and Analysis (Step 1)

In order to co-simulate the inhabitants' behaviour with the physical model of the appliance, both the consumption of the appliance and the actions behind these consumption patterns are required. However, the benchmarked Irise dataset only contains information about the consumption of electrical appliances as shown in Fig. 9. It does not include any information about the activities behind those consumption patterns. The question is to how the energy consuming behaviours could be deduced from consumption that would be used later for building behavioural models. In order to identify the energy consuming behaviour of inhabitants, Irise database is pre-processed to identify the behaviour of inhabitants based on critical parameters as consumption behaviours based on seasons, weekend, weekdays, holidays, and the impact of usage of one appliance over another. The sections below provide the detail of how the pre-processing is done to complement Irise

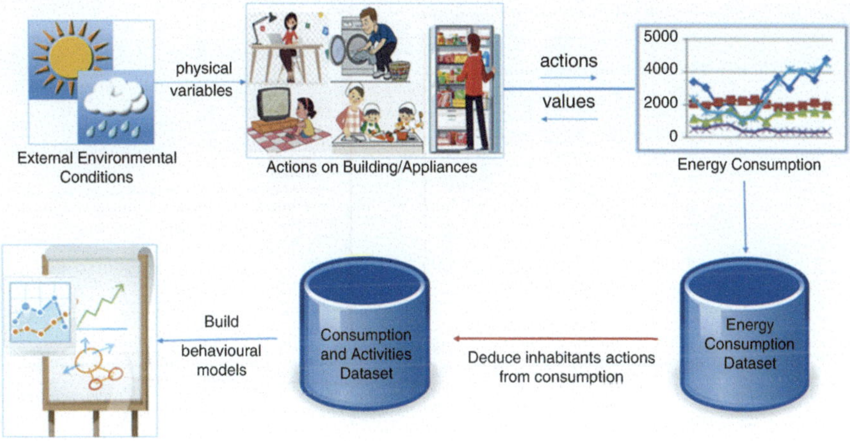

Fig. 9 Complementing the dataset

database with the extra information as, weather, appliance correlation patterns and inhabitants' behaviour, etc.

3.1.1 Case Study of a Fridge

Since, it is critical to complement the structural discrepancy of missing activities information against energy consumption trends in the Irise dataset we performed certain experiment on fridge freezer. The goal was to find energy consumption patterns associated with behavioural actions. This includes identifying the reasons behind certain activities and to link these to the consumption data in the Irise database. These behaviours are then mapped to data in the Irise database in order to provide heuristic rules that will be used in the co-simulator. The experiments were very carefully designed to model the impact of an action on the fridge freezer cycles to predict (i) when the current fridge cycle shall end, (ii) what will be the length of the next fridge cycles, (iii) how many cycles it will take to reach a stable cycle period and (iv) duration of stable cycle. Firstly the cycles of an empty fridge are modelled against controlled experimental conditions and then with food having different characteristics as (i) different quantity, (ii) different temperature and (iii) covered/uncovered was added to the fridge at different fridge cycle positions, e.g. start, middle and end of cycle periods.

3.1.2 Fridge Freezer On-Cycle Durations Computation

The fridge freezer needs a pre-processing step as compared to other appliances, because it consumes power in continuous (on and off) cycles, whereas, other

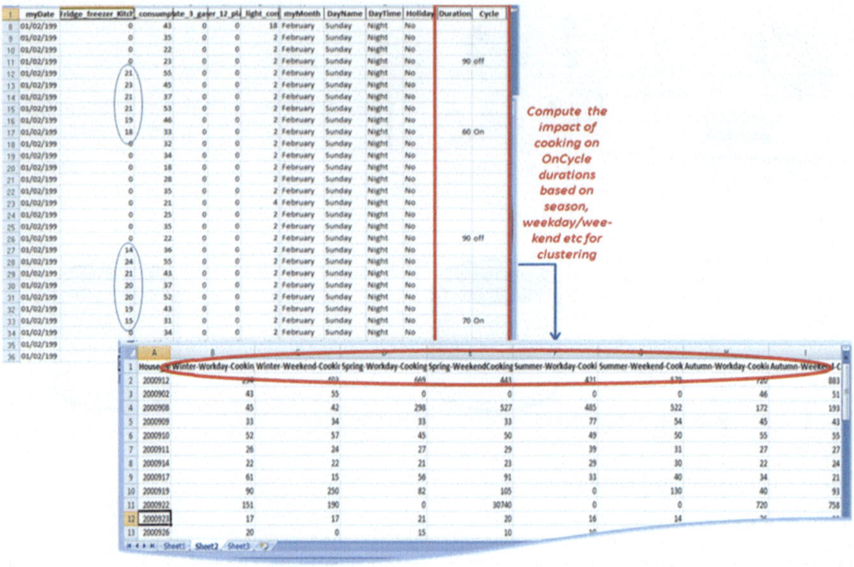

Fig. 10 Fridge on-cycle durations

appliances consume only when they are turned on. Thus, cycles of the fridge freezer need to be computed from its consumption. In Fig. 10 a snapshot of the data file from the Irise database shows the consumption of the fridge every 10 min time stamp. However, these consumption values are not very meaningful in their present form because the compressor works in continuous cycles. Thus, it is important to extract "on" and "off" cycle durations from the consumption. The flowchart in Fig. 12a explains the process of how the on and off cycles are actually computed from the Irise database. A list of selected houses is made where both the electric cooker and the fridge freezer are in the kitchen. A new field "Duration" is added to each house table in the Irise database, where the values in minutes for on and off cycles are stored. Similarly, "Cycle" field tags the computed duration with text "On" and "Off" if the duration corresponds to the "on" or "off" fridge cycles, respectively (Fig. 10). The on-cycles include the consumption values for the fridge that are above 3Wh during 10 min, whereas, when the fridge consumption is less than or equal to 3Wh, this consumption is added to the off-cycle. The reason for putting the values below or equal to 3Wh in the off-cycle is that in some houses the consumption of the fridge never goes to 0 when the compressor is off, but remains at some small value e.g. 1, 2, 3 or occasionally 4 and 5Wh. If this fact is neglected during the computation of the cycle duration, the compressor cycle will never come back to the off state. Further, the difference between the times where the values are either below (off-cycle) or above (on-cycle) 3Wh is used to compute cycle durations, respectively.

3.1.3 Impact of Seasons, Day Type and Cooking Activity

In this analysis, the global impact of the different parameters e.g. cooking activity, seasons, day types on both the fridge freezer and fridge consumption cycles is considered. This is achieved by clustering the houses in the Irise database that identify the similarities and differences that exist in the behaviour of inhabitants regarding the usage of appliances. However, it is necessary to pre-process the data for clustering by extracting the information about all the other parameters that impact the consumption. The pre-processing is done not only to complement the Irise database with additional information but also to organize the information in a meaningful way to be input to a clustering algorithm. Since, one of most important factors that impact the consumption of the fridge is the cooking activity, the houses in Irise database with both a cooker and a fridge are selected for pre-processing. Only those houses where the fridge is located in the kitchen are selected. This selection is made because the impact of the cooking activity on the fridge cycles is not only due to the interactions with the fridge but also due to the temperature change in the kitchen. Thus, the houses with fridges located in other areas e.g. living room or utility room are not included in the experiments.

3.1.4 How the Impact of Cooking Activity on Fridge On-Cycles Is Computed

In order to find the impact of the cooking activity on the consumption of the fridge, the on-cycle duration is computed when the cooker in turned on. The impact of the cooking activity on the fridge cycles is not only considered for the cycles where cooker was on but also on the subsequent fridge cycles as well. There are multiple reasons for this, e.g. the temperature increases in the kitchen affecting the fridge, the inhabitants interact with the fridge often more during the cooking activity, the inhabitants can put warm food inside the fridge, etc. This means that the fridge cycles after the cooker has been turned off must be taken into account. Hence, the fridge consumption cycles are considered to be impacted by the cooking activity until they become normal or stable.

Different trends have been observed in the fridge on-cycle durations during the cooking activity. Figure 11 shows that as a result of the cooking activity, the on-cycle duration increased compared to the previous on-cycle. Then, the subsequent on-cycle also increased showing an increasing trend in on-cycle durations. Then it started to decrease before increasing again. The decision about which cycles should be considered as being impacted by the cooking activity based on different trends is explained through an example in Table 1. Before explaining the example, the flowchart to compute the impact of the cooking activities on the fridge cycle durations is shown in Fig. 12a. It starts by taking as input all the houses where both the fridge and the electric cooker are in the kitchen. Then for every month, for each house, the "normal" compressor cycle durations are computed. "Normal" compressor cycles are those that are not influenced by the cooking activity or some

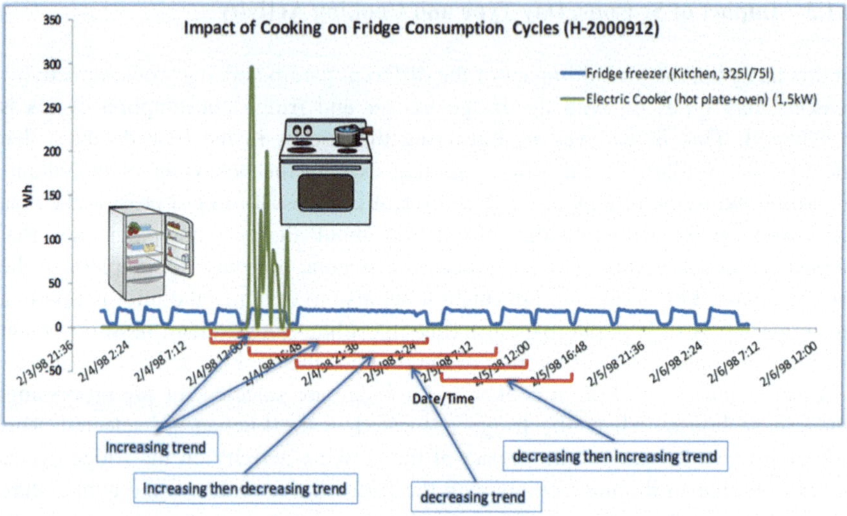

Fig. 11 Different trends identified in on-cycle durations

other activity that affects the fridge consumption. These are the cycles where the fridge is assumed to behave in the standard way and are assumed to lie in the first quartile of data. A list of SQL queries is prepared to compute the fridge on-cycle durations based on different criteria i.e. seasons and day types. This step is important as the fridge cycle durations are impacted by not only the cooking activity but also the season and day type (weekday/weekend).

3.1.5 Heuristic Approach to Compute Fridge Consumption During Cooking Activity

The important variables used in this algorithm are cVal, Dc, Dp and myFlag. The variable cVal is a pointer that scrolls down in the "OnCycle" field. This field contains the on-cycle durations of the fridge (Fig. 12b). The pointer stores the current value of the on-cycle in the OnCycle field. The Dc and Dp variables correspond to current and previous differences, computed from three consecutive fridge on-cycle durations. These variables identify the increasing or decreasing trends in the fridge cycles. The myFlag [0,1,2] variable is computed based on the Dc and Dp values to see whether the impact of cooking activity on the subsequent fridge cycles should be included or not. There could be an increasing trend (myFlag=0), decreasing trend (myFlag=1), increasing then decreasing trend (myFlag=1) (the decision criteria are the same in the last two cases, so myFlag is given the same value) and decreasing then increasing trend (myFlag=2). These values are further used with cutoff criteria (i.e. whether the normal OnCycle is reached) to decide the cooking impact on the next on-cycles.

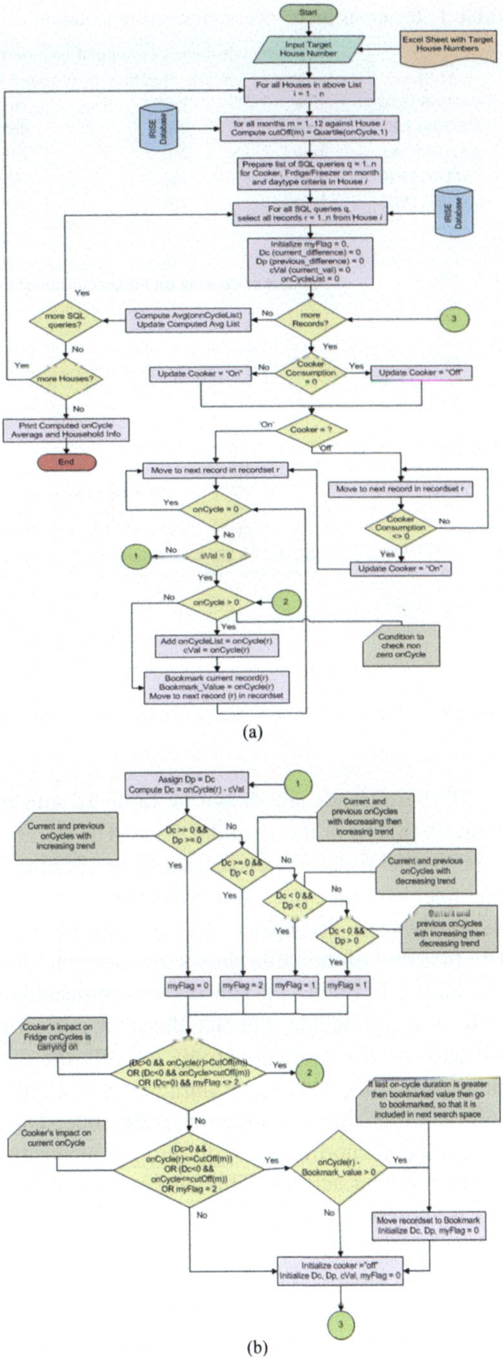

Fig. 12 Flowchart for average on-cycle duration computation based on the cooking activity

Table 1 Iterations for on-cycle duration computation

on-cycle duration computation based on cooking activity					
Date/Time	Cooker_Impact	onCycle_Duration	Dc (Current_Duration)	DP (Previous_Duration)	myFlag
2/4/1998 15:40	true	190	0	0	0
2/5/1998 3:20	true	650	460	0	0
2/5/1998 8:40	true	240	-410	460	1
2/5/1998 12:10	true	160	-80	-410	2
2/5/1998 15:30	false	180	20	-80	2

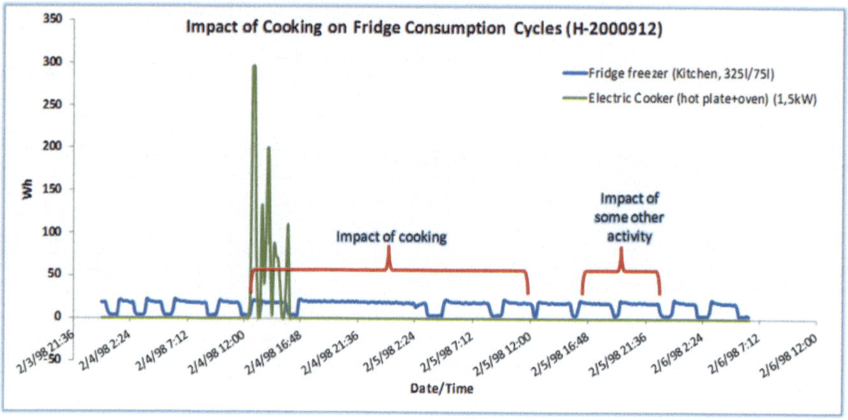

Fig. 13 Impact of cooking activities on fridge cycle durations

All these trends are shown in Table 1, with the help of an example taken from house 2000912 in the Irise database. The iterations show for how long the cooker impacts the fridge on-cycle durations. Figure 13 shows the graph of the same example; here on-cycles that are included under the impact of cooking activity can be clearly seen. The normal on-cycle duration computed for this house is 140 min. The first row of the table shows the on-cycle duration (190 min) where the cooker was turned on (Table 1). This cycle is impacted by the cooking activity.

In order to decide whether the next cycle (650 min) should be considered as impacted by the cooking activity as well, the difference between the current cycle duration and the next cycle duration is computed. If the current difference (Dc = 460) is larger than the previous difference (Dp = 0) and the on-cycle duration is larger than the normal cycle duration, it means that the current on-cycle is impacted by the cooking activity. This shows an increasing trend in the on-cycle duration and is represented by myflag = 0. The next on-cycle duration is 240 min and in order to decide whether this cycle has to be considered under the impact of cooking activity the same process is repeated, i.e. the difference between current cycle and the previous cycle is computed. The current difference (Dc= −410) is less than the previous difference (Dp = 460). The trend between the three consecutive cycles is increasing then decreasing, thus myFlag = 1. Although the current on-cycle duration has decreased, it needs to be compared with the normal on-cycle duration (cut off criteria). Since the current cycle duration is greater than the normal on-cycle duration, it is considered to be impacted by the cooking activity. The next on-cycle duration is 160 min that again shows a decreasing trend. Now the three consecutive on-cycles have a decreasing trend, and the current on-cycle duration is greater than the normal cycle. Thus, it is included under the impact of cooking activity. The next on-cycle duration is 180 min, and the trend between three consecutive cycles is decreasing and then increasing (myflag = 2). This cycle will not be considered under the impact of cooking activity. This is because once the cycle durations gradually decrease and then increase again, it is assumed that the inhabitants have performed some activity other than cooking that caused the cycles to become larger. Thus, these cycles are not considered to be impacted by the cooking activity. If there is no further cooking impact on the on-cycles, then the pointer cVal returns to the previous bookmarked on-cycle record i.e. the previous on-cycle duration it has stored. The cooker variable is set to "Off" and all other pointers are initialized to 0. The process is repeated until all the on-cycle durations for the current SQL query are computed and averaged. The process will then start for the next SQL query for the same house, until all the queries have been run. It will then move to the next house in the given list of houses. Figure 10 shows that final output of the above process, giving average fridge cycle durations based on seasons, day type and the cooking activity.

3.2 *Physical Behaviour Modelling (Step 2)*

Physical behaviour modelling is building the models for the building envelope or the appliances inside the building.

3.2.1 Building Envelop Modelling

The thermal model of the building used in this work is for the house called the SIMBAD-MOZART as was built in Matlab/Simulink by CSTB (Centre Scientifique et Technique du Bâtiment). SIMBAD-MOZART calculates the temperature in each zone by taking into account various input variables. Some of the most important variables, in Fig. 29, include the power of all the different appliances present in the zone, the position of the blinds e.g. open/closed, number of occupants in the zone, respiration flow rate, weather data, artificial lighting and ventilation. The impact of window states (opened/closed) is also taken into account through ventilation, i.e. the air mass flow between the inside and outside of the building.

3.2.2 Appliance's Behaviour Modelling

Based on the case study of the fridge, the consumption data along with the experimental results help in building a physical model for the fridge. Figure 14

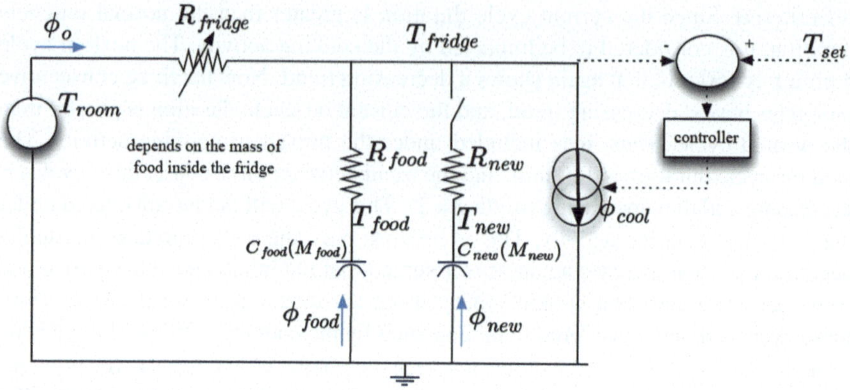

Fig. 14 Fridge behaviour model

Table 2 Variables used in the fridge model

Variable	Description
$T_{fridge}(k)$	Inside fridge temperature during reactive time k, => $T_{fridge}(k)\epsilon[T_{min}; T_{max}]$
$T_{set}(k)$	Setpoint temperature
$T_{room}(k)$	Ambient room temperature
M_{food}	Food quantity
C_{food}	$M_{food}C_p$, capacity of what is inside fridge
M_{new}	Quantity of a new food
T_{new}	New food temperature
R_{food}	Resistivity to heat exchange between food and fridge
C_{new}	$M_{new}C_p$, capacity of a new food added to fridge
R_{new}	Resistivity to heat exchange between new food and fridge
R_{fridge}	$R_{open} + \zeta(R_{close} - R_{open})$ resistivity for heat exchange between inside fridge and room
σ	Dead zone: represents the temperature zone where compressor stops the refrigeration cycles

below shows the model where the description of different variables used in the model is given in Table 2.

The fridge controller provides the cooling power Φ_{cool} that maintains the setpoint temperature. Φ_0 is the impact of heating power of the room on fridge. Its impact depends on resistance of the fridge R_{fridge}. Φ_{food} and Φ_{new} are the heating power coming from the food already present in the fridge and the newly introduced food, respectively. Their effect on the fridge temperature depends upon their capacity and mass as well as the corresponding resistance. The heat pump is an important element in modelling fridge cycles; let ρ be the performance factor of the heat pump that yields $C_{elec} = \rho\Phi_{cool}$. The fridge controller is made to follow the following criteria:

$$T_{fridge}(t) - T_{set}(t) < -\sigma \rightarrow \zeta(t+dt) = 0 \tag{1}$$

$$-\sigma \geq T_{fridge}(t) - T_{set}(t) \leq -\sigma \rightarrow \zeta(t+dt) = \zeta(t) \tag{2}$$

$$T_{fridge}(t) - T_{set}(t) > -\sigma \rightarrow \zeta(t+dt) = 1 \tag{3}$$

$$\Phi_{Cool}(t) = \zeta(t)\Phi_{Cool} \tag{4}$$

We have modelled three major events for the fridge: (a) permanent mode, where the fridge operates following the normal refrigeration cycles, (b) temporary mode when the fridge door is opened and closed; as a result, heat is exchanged and the inside temperature rises to impact the instantaneous refrigeration cycles and (c) temporary mode when food is introduced in the fridge.

(a) The model for the permanent state or normal cycles is proposed as below:

$$\frac{d}{dt}[T_{fridge}] = \left[-\frac{1}{R_{fridge}C_{food}}\right][T_{fridge}] + \left[\frac{-\rho\phi_{cool}}{C_{food}} \; \frac{1}{R_{fridge}C_{food}}\right]\begin{bmatrix}\xi \\ T_{new}\end{bmatrix} \tag{5}$$

$$T_{fridge}(0) = T_{fridge}^{init} \tag{6}$$

The model of the permanent state (1st order) is obtained when $T_{new} = T_{fridge}$.

(b) The model for the temporary mode follows that of the permanent state, but with a change in the resistance of the fridge as below:

$$R_{fridge} = R_{open} + \xi(R_{close} - R_{open}) \tag{7}$$

(c) The model for the mode when new food is introduced is proposed as below:

$$\frac{d}{dt}\begin{bmatrix}T_{fridge} \\ T_{new}\end{bmatrix} = \begin{bmatrix}-\frac{R_{new}+R_{fridge}}{R_{new}R_{fridge}C_{food}} & \frac{1}{R_{new}C_{food}} \\ \frac{1}{R_{new}C_{new}} & -\frac{1}{R_{new}C_{new}}\end{bmatrix}\begin{bmatrix}T_{fridge} \\ T_{new}\end{bmatrix} + \begin{bmatrix}\frac{-\rho\Phi_{Cool}}{C_{food}} & \frac{1}{R_{fridge}C_{food}} \\ 0 & 0\end{bmatrix}\begin{bmatrix}\xi \\ T_{room}\end{bmatrix} \tag{8}$$

with

$$T_{fridge}(t) - T_{set}(t) < -\sigma \rightarrow \zeta(t+dt) = 1 \tag{9}$$

$$-\sigma \geq T_{fridge}(t) - T_{set}(t) \leq -\sigma \rightarrow \zeta(t+dt) = \zeta(t) \tag{10}$$

$$T_{fridge}(t) - T_{set}(t) > -\sigma \rightarrow \zeta(t+dt) = 0 \tag{11}$$

$$\zeta(0) = 0 \tag{12}$$

$$\Phi_{Cool}(t) = \zeta(t)\Phi_{Cool} \tag{13}$$

We make the following assumptions in modelling the fridge:

(i) Opening the door modifies R_{fridge}.
(ii) Removing food from the fridge is assumed to have a very small impact (except for the door opening).
(iii) Adding food sets a new value to T_{new} (the temperature of the food) and parameters like C_{new} and R_{new} may be adjusted depending on the food.

3.3 Tune Parameters of Inhabitant's Behaviour Models (Step 3)

Based on the experiments and field studies, identify the important parameters that impact the consumption behaviours of inhabitants.

Weekend and Weekday Cooking Probabilities This defines the probability that the family cooks more during weekends or weekdays. While cooking, the agents interact more with the fridge. If a higher probability is assigned to weekend cooking, then the family will interact more with the fridge during weekends compared to weekdays when they may eat out or use the food they have already cooked during weekends.

Weather This defines and controls the perception by agents about the outside weather. It means that if the weather is good, e.g. sunny and warm, then the family might prefer to eat out.

Communication Based Agreement/Disagreement over Cooking or Dining Out This involves the social interaction between agents where they agree or disagree on dining out or cooking at home. The purpose of introducing this parameter is to show how the social interactions of agents are interesting to include in simulations in order to make them closer to reality.

Guests This is a random parameter that increases the interactions with the fridge resulting in larger fridge cycles, hence large energy consumptions.

3.4 Clustering Houses with Similar Behaviours (Step 4)

The impact of the cooking activity on the consumption of the fridge due to the inhabitants' behaviour varies with different seasons and day types (weekdays and weekends). The stacked chart for average on-cycle durations for the fridge in all the houses in the Irise database, where both the cooker and fridge are located in the kitchen, is shown in Fig. 15. On the x-axis there is the season (one month from each season is taken), day type and whether it is cooking activity or not; on the y-axis there is the fridge on-cycle durations in minutes. This graph shows that when

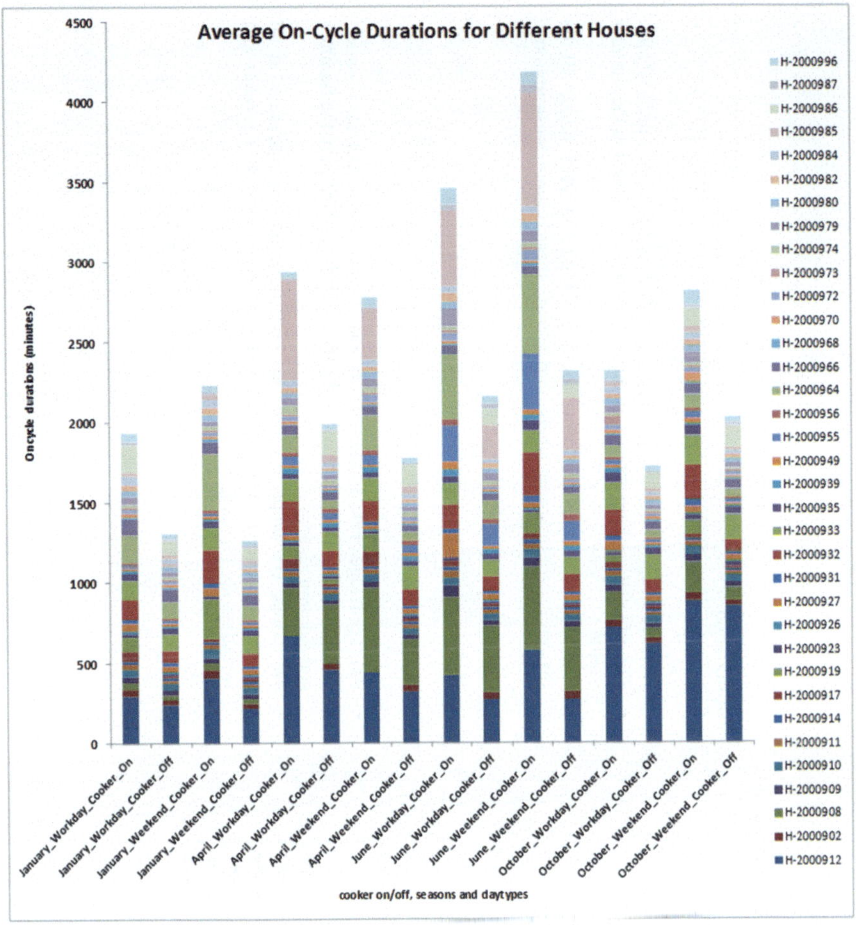

Fig. 15 Fridge consumption during cooking and non-cooking activity

inhabitants are involved in the cooking activity in most of the houses the fridge consumption cycles become longer than when there is no cooking activity.

3.4.1 Identifying Representative Behaviours

All the houses in Fig. 15 are further clustered to identify representative behaviours based on their fridge consumption based on time of season, day type and cooking activity. The houses in Table 3 are clustered based on how the cooking activity, seasons and day types (weekend/weekdays) affect the fridge consumption.

The different clusters obtained after applying k-means clustering on the data are given in Table 3. The seasons and day type is on the x-axis whereas average on-cycle

Table 3 Clusters of houses with similar energy consumption behaviour

<table>
<tr><th>Clusters</th><th>Cluster description</th></tr>
<tr><td>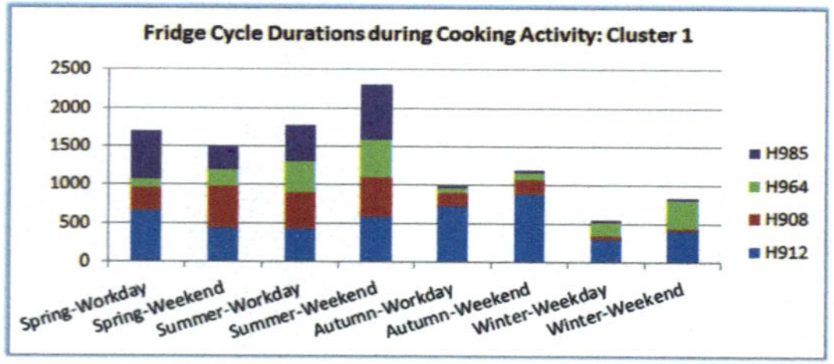
</td><td>• Average fridge cycle durations of the members of this cluster are 340 min
• The effect of cooking activity on the fridge cycles is highest in the summer season
• The fridge consumption during cooking is higher at weekends in all the seasons except in Spring where on weekdays there is more consumption</td></tr>
<tr><td>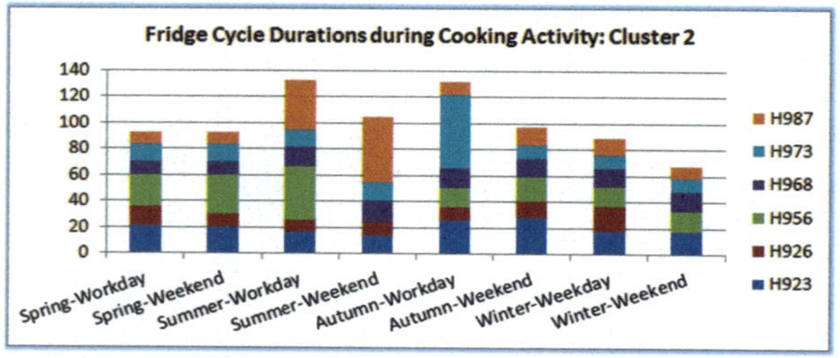
</td><td>• Average fridge cycle durations of the members of this cluster are 17 min
• The effect of cooking activity on the fridge cycles is highest in the Summer and Autumn seasons
• The fridge consumption during cooking is higher on weekdays in all the seasons except in Spring where the consumptions on weekends and weekdays are the same</td></tr>
</table>

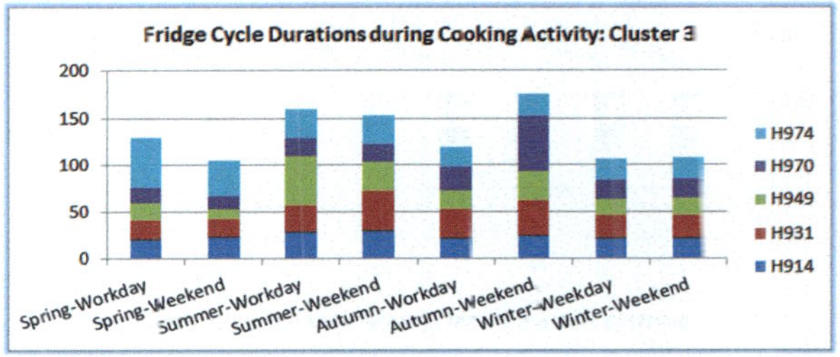	• Average fridge cycle durations of the members of this cluster are 26 min • The effect of cooking activity on the fridge cycles is highest in the Summer and Autumn season • The fridge consumption during cooking on weekends and weekdays is almost the same in Winter, whereas in Autumn there is more consumption at weekends and in Summer and Spring on weekdays
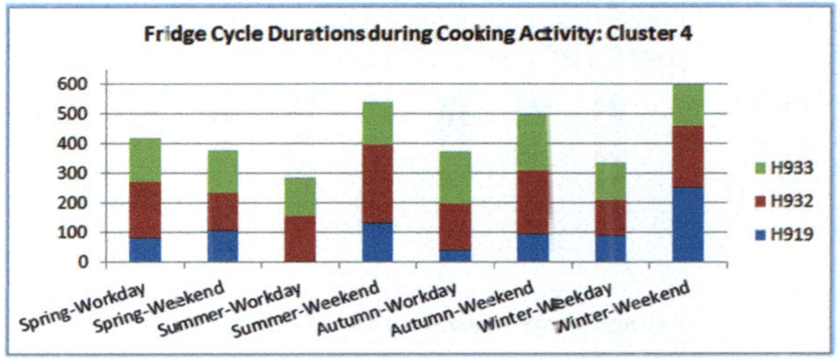	• Average fridge cycle durations of the members of this cluster are 50 min • The effect of cooking activity on the fridge cycles is highest in Summer • The fridge consumption during cooking is higher on weekdays in Spring, whereas in all other seasons it is higher at the weekend

(continued)

Table 3 (continued)

<table>
<tr>
<td>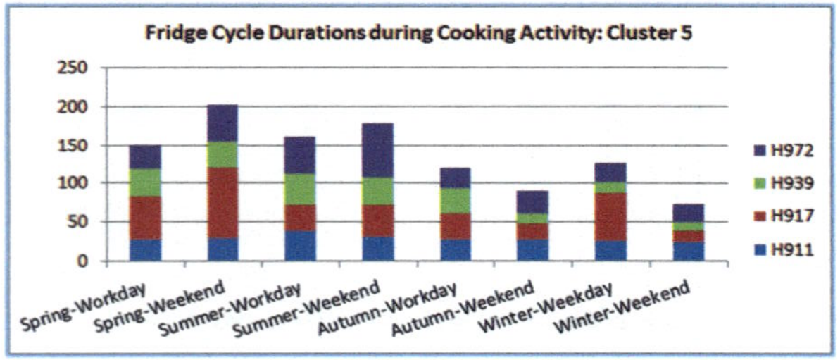
</td>
<td>
• Average fridge cycle durations of the members of this cluster are 142 min
• The effect of cooking activity on the fridge cycles is highest in Summer and Winter seasons
• The fridge consumption during cooking is higher on weekends in Spring and Summer whereas in Autumn and Winter it is high on weekdays
</td>
</tr>
<tr>
<td>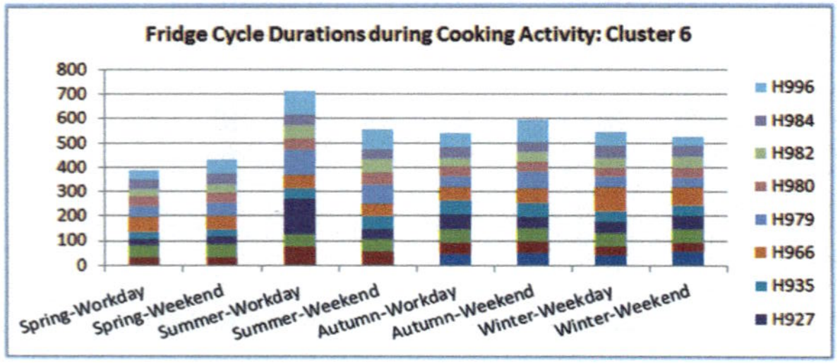
</td>
<td>
• Average fridge cycle durations of the members of this cluster are 35 min
• The effect of cooking activity on the fridge cycles is highest in Spring
• The fridge consumption during cooking is higher at weekends in Spring and Autumn whereas it is the inverse in Summer and Winter
</td>
</tr>
</table>

Season/ Daytype	Globally high consumption	High consumption on weekdyas	cluster No.	high consumption on weekends	cluster No.	no difference	cluster No.
Winter	33%	50%	2,6,5	33%	1,4	16%	
Summer	83%	50%	2,3,6	50%	1,4,5	0%	
Spring	0%	50%	1,3,4	33%	5,6	16%	2
Autumn	33%	33%	1,2	66%	1,3,4,6	0%	

Fig. 16 Similarities in clusters

duration of the fridge in minutes is on the y-axis. One month from each season is taken as representative i.e. April for Spring, June for Summer, October for Autumn and January for Winter. Each graph represents a cluster, where the consumption behaviour of the fridge is different based on the season and the day type while the occupants are involved in the cooking activity.

From the above clusters, some general consumption patterns on the population can be seen. These patterns are summarized in the similarity matrix in Fig. 16. A similarity of 50% is observed in inhabitants' consumption behaviours during Winter, Summer and Spring weekdays. The highest similarity (83%) is observed during the Summer season where globally there is more consumption as compared to the other seasons. Similarly, on Winter weekends there is 66% similarity in the behaviour of inhabitants.

3.5 *Inhabitant's Reactive, Deliberative Behaviour Modelling (Step 5)*

The inhabitants filled the information in an activity journal not only about the actions they performed on the household appliances but also the reasons that caused these actions over the period of three weeks. Hence, these results helped us to derive generic rules about how the individual and group behaviours evolve.

Figure 17 shows the model of inhabitants' behaviour that starts with perception of the environment, passes through the instinctive and cognitive phases and ends up with actions back on the environment. The outside environment includes the location and physical building models that provides the information about where the agent is. The objects, appliances, other agents, agent belongings, weather and BEMS informs the agent about what are the other things around the agent. The time provides the information about when the agent is perceiving its surroundings or taking actions. All these environmental elements are then perceived by the agent. Upon the perception, the agent will translate these elements as its beliefs, shown by the "Beliefs" part of the cycle. Beliefs represent the mental state of the inhabitant and are the first important concept in BDI architecture.

In the model in Fig. 17, however, another concept is introduced in addition to beliefs that relates to the physical state of an inhabitant. Thus the inhabitant has perception not only about the outside environment but also about its internal physical state. The question is why introduce the physical state of the inhabitant, as the agent's physical state also becomes its belief. The reason is that there are some physical phenomena that the agent could not directly perceive. For example, one can perceive the internal physical state of being thermally uncomfortable, but could not directly perceive his metabolism. Metabolism is a physical phenomenon that continues to happen in the body without notifying the person about its value. Similarly, the increase of CO2 level in the room can impact on a person's mood, but he could not directly perceive the CO2 level and identify it as a cause of his stress. That is why the physical phenomena taking place inside human body are put under Homeostasis, rather than just beliefs. The agent then has the beliefs about its homeostasis, the outside environment. Based on these beliefs the agent can have certain desires; however, due to the external environmental constraints only one of them is converted to the agent's intention. The intention then leads to the process of

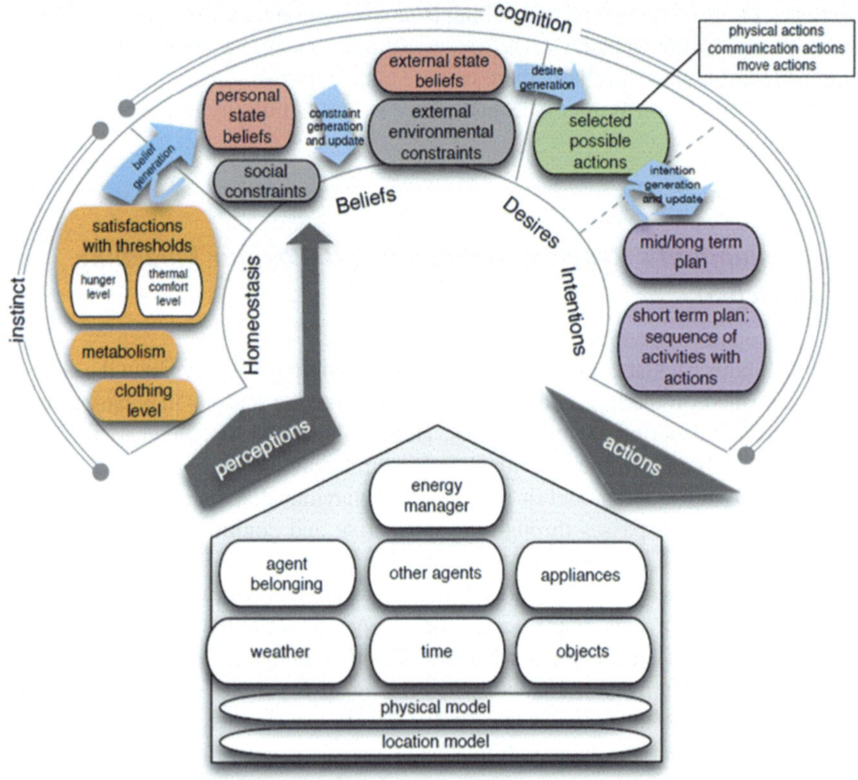

Fig. 17 H-BDI dynamic behaviour representation model

generating plans of how to fulfil the intention. Finally, the agent follows this plan to perform actions on the environment.

3.6 Implementation and Co-simulation of Behaviour Models (Step 6)

In addition to the language constructs, Brahms has models that are used in simulation [20]. The reason for selecting the Brahms simulation environment is that all the elements considered important in behaviour representation can be mapped to Brahms models, Fig. 18. The behaviour elements When, What, Why, Where, Who and How are mapped to the Timing, Object, Knowledge, Geography, Agent and Activity models, respectively.

Agent Model This model contains all the agents, the groups to which they belong, and how these groups are related to each other, resulting in a group hierarchy. Agents have common or specific beliefs, facts and attributes. Agent thoughtframes drive the thought processes and decision making capabilities of agents. Workframes define how they actually perform the activities in accordance to the decisions made.

Object Model Object model contains different objects which can make a hierarchy and inherit the attributes of the root class. They can also perceive the environment and communicate or broadcast their state to other agents and objects.

Geography Model In this model, geography is described through the concepts of area, area definitions and paths. The agents can change their location by moving from one location to another. The movement is modelled using the "move" activity.

Activity Model Figure 18d shows the activity model that is used to represent the activities performed by agents and objects. Activities take some time and may have an associated priority. Figure 18d shows different types of activities, i.e. the move, broadcast, primitive, composite and Java activities. Each activity has a set of parameters and belongs to some workframe where it is realized based on the preconditions.

Knowledge Model In this model, the agent's reasoning mechanism is represented as forward chaining production rules, called thoughtframes. Thoughtframes can be represented at group or class levels and can be inherited. Perception is modelled as conditions. These conditions are attached to workframes and are called detectable. Thus observation is dependent on what the agent is doing. Figure 19 shows a thoughtframe "tf_perceive_comfort" that receives the comfort value from the comfort calculator and changes the agent's perception of comfort at each simulation time step. The workframe that is attached to this thoughtframe is the "Watch_TV", where among the other preconditions to watch TV are that the agent's comfort value lie between 0.5 and -1. If the agents' perception of comfort lies between these values, it will continue watching TV. However, as soon as the value will be out of this range, one of the detectable in the detectables list will be triggered.

Fig. 18 Implementation of behaviour model into Brahms. (**a**) Agent model. (**b**) Object model. (**c**) Geography model. (**d**) Activity model

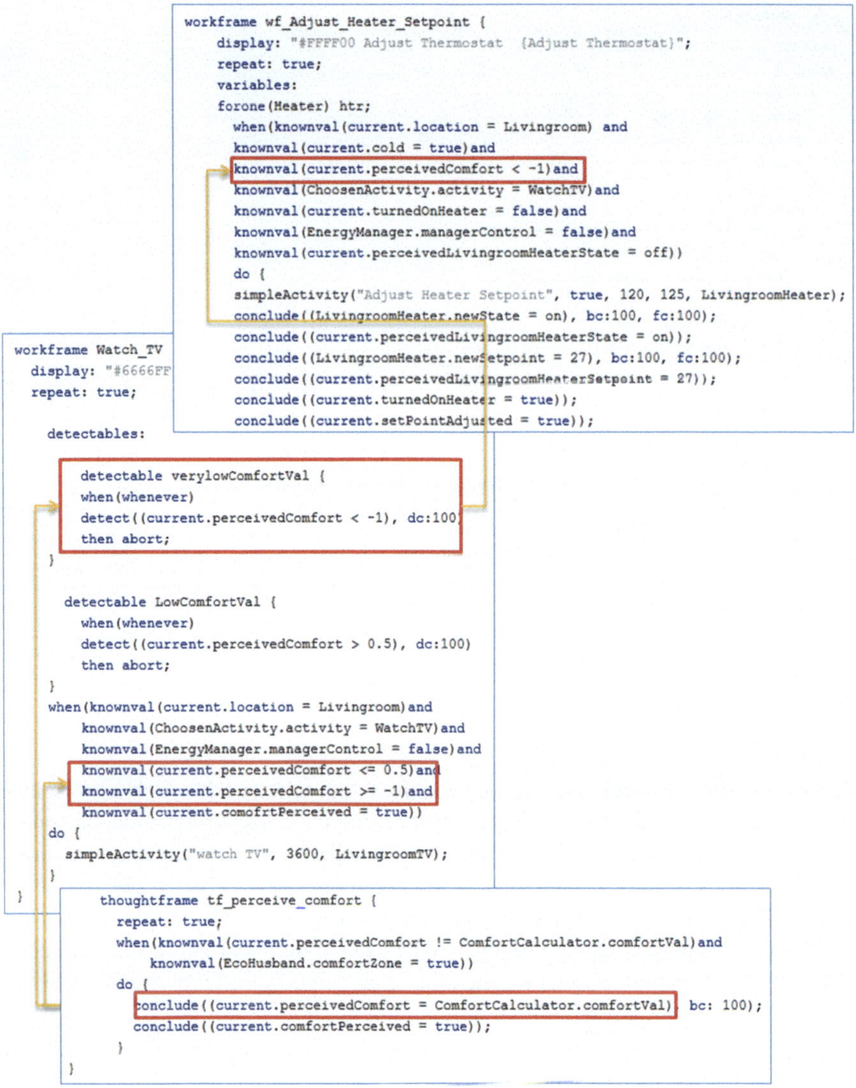

Fig. 19 Knowledge model

Timing Model This model enforces the constraints of when activities in the activity model can be performed, Fig. 20. This is represented as preconditions of situation-action rules (workframes). Activities take time (predefined duration of primitive actions) and workframes can be interrupted and resumed, making the actual length of an activity situation dependent. This model is built in the simulator by first building a clock, and then sending the clock time to every agent and object in the environment.

```
package gscop.superbat.mozart.brahms;
group SimClock {
  attributes:
    public long time;
    public long timeStep;
    public long endTime;

  activities:
    broadcast bc_announce_Time(int duration) {
      random: false;
      max_duration: duration;
      about:
        send(current.time);
      when: start;
    }

  workframes:
      workframe wf_asTimeGoesBy {
      repeat: true;
      when(knownval(current.time <= current.endTime) and knownval(current.timeStep = 60))
          do {
            bc_announce_Time(60);
            conclude((current.time = current.time + current.timeStep), bc:100, fc:100);
          }
      }
```

Fig. 20 Timing model

Communication Model This model includes the actions by which agents and objects exchange their beliefs. The communication includes telling someone something or asking a question. Conversation is modelled as an activity with communication actions. Figure 21 shows how the communication between two agents is made in order to open the window. The communication activity is realized in the workframe "DemandToOpenWIndow". When the Wife agent perceives the belief of the Husband agent, it checks for the other constraints before replying back to the Husband agent.

3.6.1 Scenario Implementation

A scenario consisting of a 2 *person* house will be considered where husband and wife are modelled as agents. It will show how the decisions taken by the agents affect the energy consumption. Figure 22a shows that the husband and wife are sitting in the living room and watching TV (perception of environment). The hunger level for the wife gradually increases with time (physical homeostasis calculation). When it reaches beyond some threshold (internal state belief generation rules), she communicates with the husband to have their meal together (generation of desire in wife and communication activity to convey the husband about desire). The husband

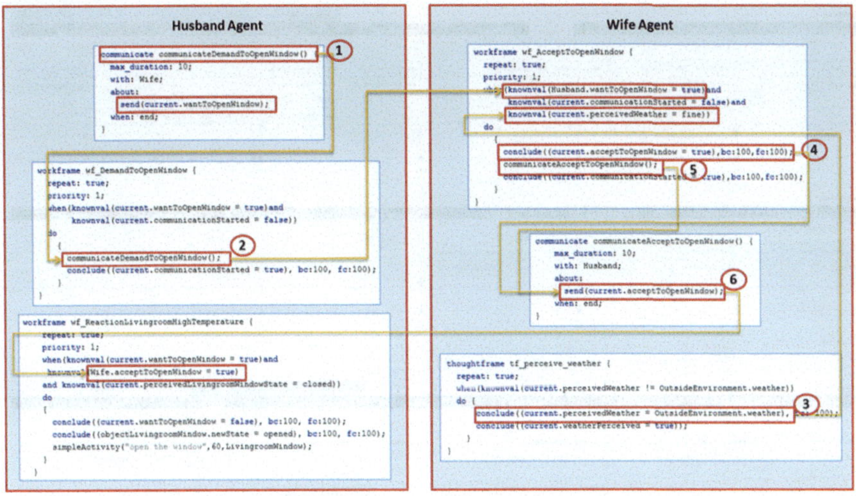

Fig. 21 Communication model

usually likes to eat at a restaurant if there is a beautiful weather outside; otherwise he prefers to eat at home (desire generation rules for husband). If the husband agrees based on perception about the weather (social behaviour as a result of external state belief), she moves to the kitchen, opens the fridge, takes the things out and prepares the dining table (plan generation to be followed in dining activity). If however, the husband does not agree to eat at home (social constraint), she puts the warm food, which she had already prepared for their meal into the fridge (action on appliance) and they go out to the restaurant.

The simulation results are presented in Fig. 22a. The output is generated randomly based on the agents' belief certainty. Belief certainty is the concept used in Brahms which assigns a probability between 0 to 100 to agents' beliefs and the facts in the environment. Beliefs and facts with varying probabilistic values influence agents' actions accordingly. For example, if for the communication between the agents, the husband agrees to eat at home, there is a higher probability that the wife will not put the warm food which she had prepared for the meal into the fridge. Also, if the husband agrees to eat at home, the duration of the activity of opening the door of the fridge and taking the things out is a random value between a minimum and maximum duration. Thus, every time the wife opens the fridge door for different durations resulting in varying behaviours of the fridge. In Fig. 22, the horizontal bar on the top represents the movements of agents to different locations. Below this is the timeline, which shows the simulation time in the agent world. The vertical bars represent the communication between agents and the broadcast activity where the agents transfer their beliefs to each other. For example in Fig. 22b, the vertical bar coming down from Wife agent to Husband agent at the moment when the Wife agent moves from the kitchen to the living room, represents the Wife agent's belief which she transfers to the husband to go to the restaurant. The bulb symbols are used to

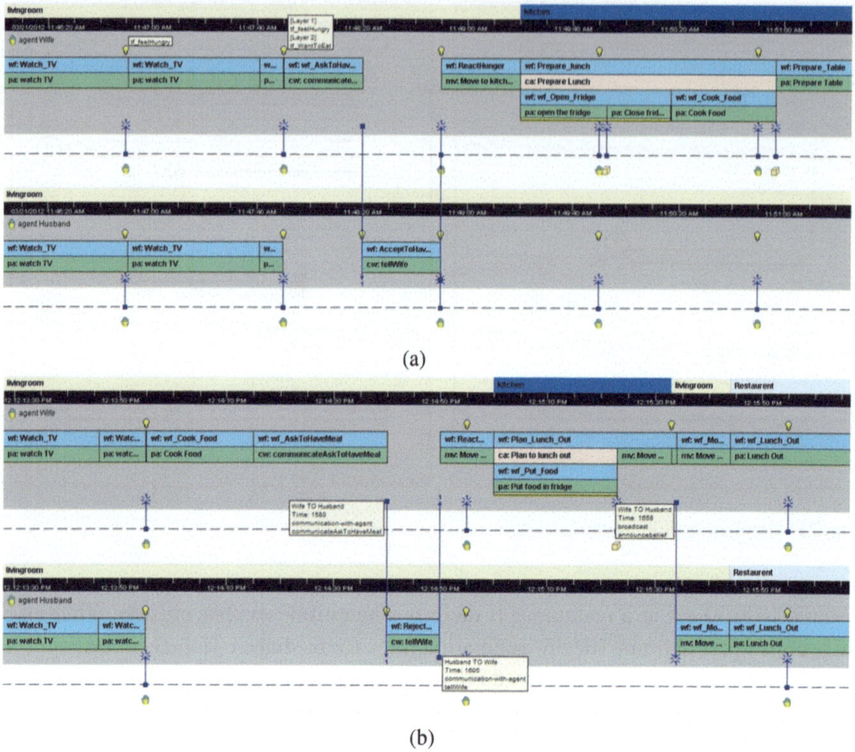

Fig. 22 Simulation results against simulated inhabitants' behaviour. (**a**) Social agreement between agents to have meal at home. (**b**) Social agreement between agents to eat out

represent the thoughtframes or beliefs of agents. Thoughtframes are changed with the passage of the simulation time and on the different perceptions of the agents from their environment.

3.7 *Co-simulate the Complex Behaviour with Physical Models (Step 6)*

As mentioned in Fig. 8, the 7th step of validation methodology is to compute the consumption distribution of an appliance from the Irise database and then compare it with the simulated consumption distribution. This could be done by modelling and simulating the behaviour of occupants from some house in the Irise database. However, in the Irise database only the consumption of appliances is available and not the activities. In this section, the inhabitants' energy consuming behaviour is extracted by analysing the appliances' consumption patterns. This is done by first pre-processing the Irise database to enrich it with some additional information. Then, the houses in Irise are clustered based on identical energy consuming behaviours. Further, the representative behaviour for some cluster is co-simulated with the selected appliance and the consumption distribution for that appliance is obtained after simulation. This simulated consumption distribution will be used in the next step where it will be compared with the actual consumption distribution for the house benchmarked for that cluster. Since, the energy consuming behaviour of inhabitants' belonging to a cluster is identical, the same simulated consumption is also compared with the actual consumption of other members of the cluster.

In the above section, the houses in the Irise database are clustered based on the generic energy consuming behaviour of inhabitants. In this section, more specific behaviour of inhabitants will be modelled and simulated. Some of the parameters and their impact on energy consumption e.g. seasons, weekday, weekend, impact of an appliance usage over another (e.g. cooking activity) are already known. However, there are still situations where high consumption is not explained by the above-mentioned parameters but some other unknown reason. In these situations, the results from field studies are used to find the reasons behind these high consumptions. Thus, the additional parameters that will be used while modelling and simulating the behaviour of inhabitants from the Irise database are the social behaviour of the family and the arrival of guests. Since there is a combination of parameters, ones that are observed from the Irise database and others from local field studies, their values need to be tuned during the simulation to see if the simulated behaviour is realistic. This simulated behaviour will generate the consumption distribution of the appliance (in this example a fridge freezer). Consumption distribution obtained from the simulated results will then be compared with the original consumption distribution of the same house to see if they follow the same trend. Further, this simulated distribution will be compared with other members of the same cluster. The proposed tuning parameters are weekend and weekday cooking probabilities, impact of weather on cooking at home, communication between agents, the involvement of factors other than normal routine e.g. the arrival of guests.

3.8 Tune Parameters of Inhabitant's Behaviour Models (Step 7)

Different combinations of values for these parameters result in different consumption simulation results, one of which is shown in Fig. 23. The values of these tuning parameters are first initialized in Brahms simulation environment. These fall between the probabilistic values of 0 and 1 and are randomly selected by the Brahms simulator during simulations. The goal is to find optimal values of these tuning parameters such that they generate the consumption distribution close to the benchmarked distribution.

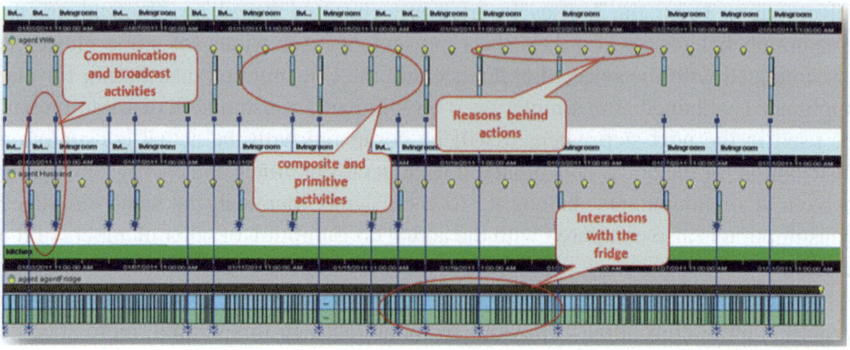

Fig. 23 Brahms scenario simulation results

Brahms has been used to implement a scenario of a husband and wife concerned with a cooking activity. The scenario is highly dynamic and random because of the probabilistic values of the tuning parameters. Figure 23 shows a snapshot of a Brahms simulation over a one-month period. The bar with the light-bulb symbols shows the thoughtframes, where the agents reason based on different perceptions (such as time, day of the week, weather) coming from the environment. The colourful vertical bars in the area just behind the thoughtframes show the workframes, where we have the activities of the agents, these may be composite activities or primitive ones. A composite activity is composed of primitive activities, e.g., the "prepare lunch" composite activity can be decomposed into the "open fridge", "close fridge" and "cook food" primitive activities. The vertical bars going from one agent's workframe area to another shows the communication between agents or the broadcasting of information (beliefs) that may in turn invoke actions in other agents.

There are many random variables in the simulation e.g. the probability of cooking on a weekday and weekend, the probability of how often the inhabitants go out to eat instead of eating at home, the probability of social agreement between inhabitants to eat at home or outside based on the weather, the probability of arrival of guests at home. Based on the combination of these probabilistic values the agents interact with the fridge more or less often, they may put hot food in the fridge, they prepare food at home or not, etc. Also, the activities performed by the agents do not always have the same duration, e.g. the cooking activity on one day may take 30 min while on another day it could take 50 min. This means that every day during the period of simulation run (1 month) not only the agents' perception about the environment and choice of actions change but also the duration of activities change as well. Thus, the probability of occurrence of some random variable along with the duration of activities needs to be averaged. Figure 24 explains the process, where, for some particular probability values for each parameter, several simulations are run (n times) and then the results are averaged. Similarly, the probability values are changed for the next runs and the results are again averaged. The process of

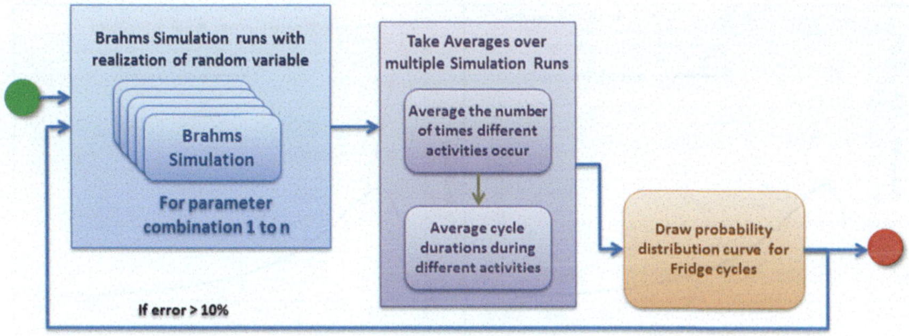

Fig. 24 How to get fridge cycle durations from simulations

changing these probabilities continues until the selected probability values match the observed behaviour of the people and the consumption trends start matching with each other. The next step is to calibrate the simulator by matching fridge freezer cycle distributions computed from the behaviour simulations with those from the Irise database. The Brahms simulation results are combined into a text file. A parser was developed to read these files and compute the energy consumption associated with the duration of fridge cycles. The initial results with the initial set of tuning parameters are presented in Fig. 25 for the reference. The probabilities for each parameter are set at the start of the simulation. For example, if the probability that the inhabitants will cook more on weekdays is set to 40%, they will cook on different days for each simulation run, but not for more or less than 40% of the time.

3.8.1 Tune Parameters of Inhabitant's Behaviour Models (Step 7)

Figure 25a shows that the weather is often not sunny during the month and inhabitants are cooking mostly on weekdays, which is not inline with the actual behaviour in cluster1, hence the actual and simulated curves do not match. In Fig. 25b there is more cooking on weekends but the social agreement's value is inconsistent with reality. Finally, in Fig. 25c the values of these parameters are tuned in accordance with the observations. In this case not only the reference distribution and the statistical curves are in compliance with each other but also the simulated behaviour is realistic.

In Fig. 25, the behaviour model is validated based on the comparison between actual and simulated energy consumption curves for the fridge. However, after clustering the energy consumption behaviour of occupants during cooking activity, the simulated energy consumption of the fridge is compared with another house (2000964) that is a member of the same cluster. Figure 25 shows the difference between the consumption distribution for house-2000964 and simulated curves for house-2000912. The difference between these distributions is bigger as compared to the benchmarked house (house-2000912). In Fig. 25, the comparison of the same

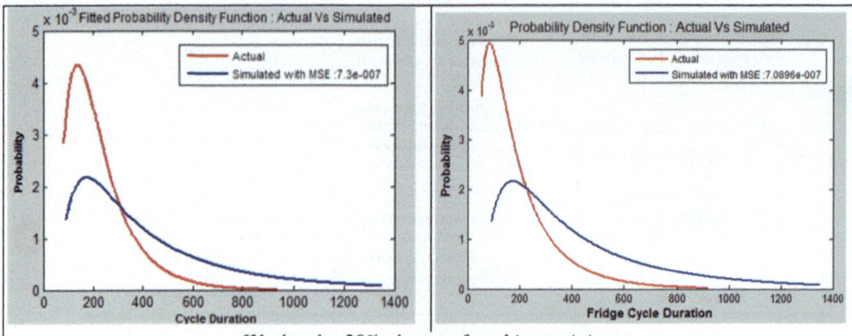

Weekend = 20% chance of cooking activity
Weekday = 80% chance of cooking activity
Weather perception = 30% that weather is sunny
Guests on weekends (more interactions with fridge) = 20%
Social agreement to eat out = 20% chance that they agree

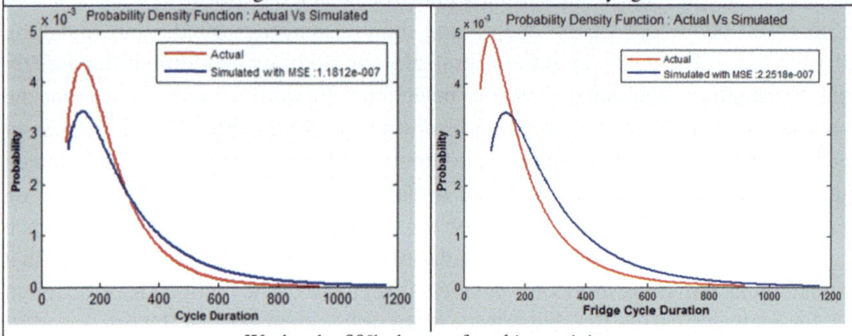

Weekend = 80% chance of cooking activity
Weekday = 20% chance of cooking activity
Weather perception = 30% that weather is sunny
Guests on weekends (more interactions with fridge) = 20%
Social agreement to eat out = 50% chance that they agree

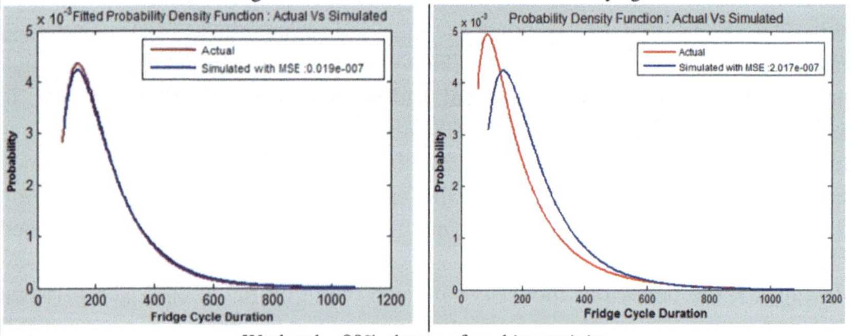

Weekend = 80% chance of cooking activity
Weekday = 20% chance of cooking activity
Weather perception = 30% that weather is sunny
Guests anytime during month (more interactions with fridge) = 20% chances
Social agreement to eat out = 100% chance that they agree

Fig. 25 Comparison with another member of the same cluster simulated distribution is made with another member of the same cluster (house-2000964) that causes the error to increase but still it is realistic and follows almost the same trend.

4 Validate the Models with Building System and BEMS

Figure 26 explains the process of how the co-simulation between different modules is performed. The inhabitants in the "Human Agent" module continuously perceive their comfort and react to the environment. The notion of comfort in the inhabitants is introduced using the Fanger's comfort model [21]. This model computes the thermal comfort conditions for inhabitants based on their clothing, activity, temperature in surroundings and some other parameters detailed in the upcoming sections. The values of these parameters are computed in separate modules i.e. the "Clothing Index Computation", "Metabolic Rate Assignment", and "Mozart Building" modules, respectively. In "Mozart Building" module the temperatures (air and radiant temperatures) are calculated by the SIMBAD thermal model and sent back to the "Thermal Comfort (PMV) Computation" module. The inhabitants based on their perceived comfort levels further control the appliances or objects in the environment through the "Control (Setpoint, Appliances and Objects)" module. This control over the environment, however, can also be taken by the "Building Energy Management System" module that maintains the thermal comfort of inhabitants. The detail about the different modules is given in the upcoming sections.

4.1 Inhabitants' Behaviour Simulation

Since the thermal model used in the simulation is of the reference house MOZART, the same house is used for developing a scenario of inhabitants' presence and their activities. The purpose of modelling the inhabitants' behaviour is to see how their choices and control of household appliances can impact the energy consumption. An important element of this behaviour is the perception of comfort, i.e. how the inhabitants' behaviour will be impacted by the feeling of comfort or discomfort and how it will lead to the choice of certain actions. The comfort is introduced in the agents through the Fanger's thermal comfort model [22].

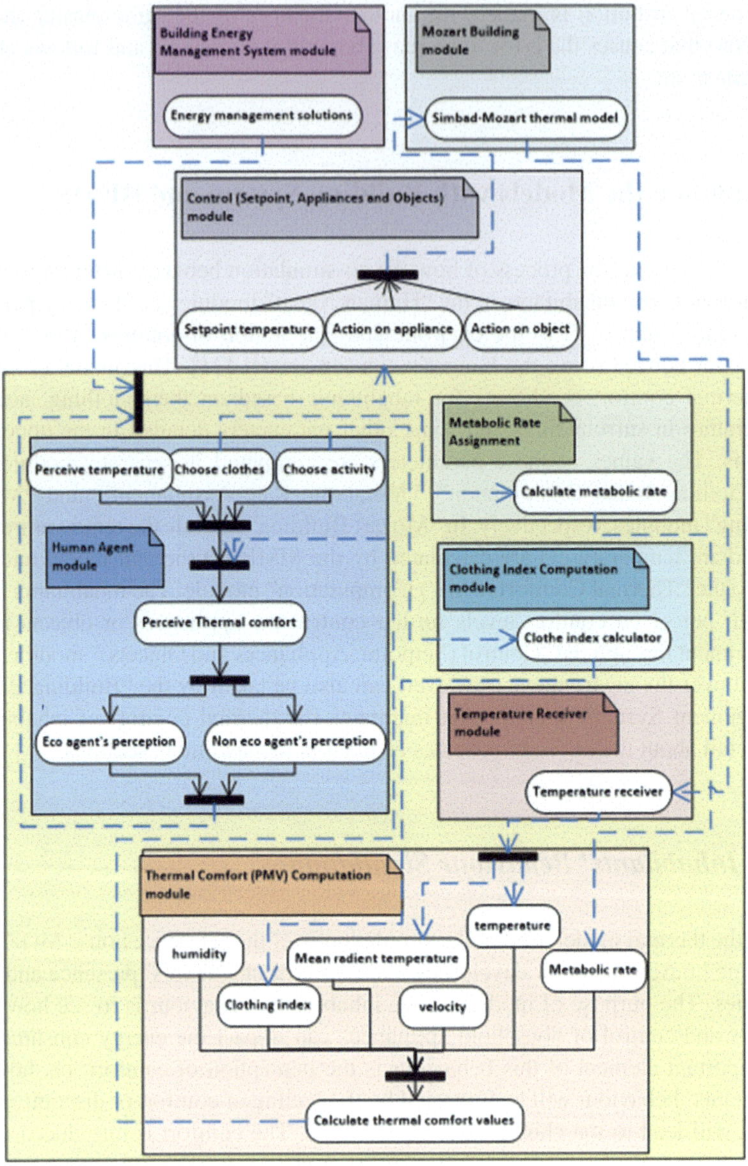

Fig. 26 Co-simulation between inhabitant's behaviour, SIMBAD and BEMS

4.2 Fanger's Thermal Comfort Model and Inhabitants' Behaviour

Occupants' comfort is given in the American Society Heating Refrigerating and Air Conditioning Engineers (ASHRAE) Standard Number 55, as "the condition of mind that expresses satisfaction with the thermal environment and is assessed by subjective evaluation". Thermal comfort is ensured by heat conduction, convection, radiation and evaporative heat loss. Thermal comfort is maintained by maintaining thermal equilibrium with the surroundings i.e. there is a balance between heat production and heat loss. Fanger describes his heat balance model as "Since the purpose of the thermoregulatory system of the body is to maintain an essentially constant internal body temperature, it can be assumed that for long exposure to a constant (moderate) thermal environment with a constant metabolic rate a heat balance will exist for the human body, i.e., the heat production will equal the heat dissipation, and there will be no significant heat storage within the body". The heat balance condition is:

$$H - E_d - E_{sw} - E_{re} - L = K = R + C$$

Where

H = the internal heat production in the human body
E_d = the heat loss by water vapour diffusion through the skin
E_{sw} = the heat loss by evaporation of sweat from the surface of the skin
E_{re} = the latent respiration heat loss
L = the dry respiration heat loss
K = the heat transfer from the skin to the outer surface of the clothed body (conduction through the clothing)
R = the heat loss by radiation from the outer surface of the clothed body
C = the heat loss by convection from the outer surface of the clothed body

Based on the heat balance equation, Fanger proposed an index in order to analyse the thermal environment. This gives the Predicted Mean Vote (PMV) of subjects according to the following psycho-physical scale (Fig. 27).
The PMV value is calculated through the following equation:

$$PMV = (0,303e - 0.036 * M + 0,028) * [(M - W) - 3,05 * 10 - 3 * 5733 - 6.99 * (M - W) - p_a - 0.42 * (M - W) - 58.15 - 1.7 * 10 - 5 * M * (5867 - p_a) - 0,0014 * M * (34 - t_a) - 3,96 * 10 - 8 * f_{cl} * (t_{cl} + 273)4 - (t_r + 273)4 - f_{cl} * h_c * (t_{cl} - t_a)]$$

M = Metabolism, W/m^2(1 met = 58.15 W/m^2)
W = External work met. Equal to zero for most metabolisms
l_{cl} = Thermal resistance of clothing, clo (1 clo = 0.155 m^{2}K/W)
f_{cl} = The ratio of the surface area of clothed body to the surface area of nude body
t_a = Air temperature, °C

t_r = Mean radiant temperature, °C
v_{ar} = Relative air velocity, m/s
p_a = Water vapour pressure, P_a
h_c = Convective heat transfer coefficient, W/m^{2}K
t_{cl} = Surface temperature of clothing, °C
Similarly, the level of discomfort called PPD (predicted percentage of dissatisfied) is calculated as
$\text{PPD} = 100\text{-}95.e^{-(0.03353.PMV4+0.2179.PMV2)}$.

Figure 27 shows how the values for different clothes are calculated. In Brahms, the agents are provided with multiple options for each piece of clothing, e.g. for the choice of shirts, pants and sweaters. The reason for making these choices for each type of clothes randomly is that the clothes impact the thermal comfort levels. Although, the choices of clothes are dependent on the season and weather, however, in order to demonstrate the impact of different clothing combinations on the calculation of PMV, the choices are made randomly by the agents during simulation. The comfort of an agent is not solely based on the temperature but a more complex model of thermal comfort i.e. Fanger's thermal comfort model. Figure 27 explains how Fanger's model is used in the co-simulation and the different input variables required for calculating the PMV value. The agents in the Brahms simulation continuously perceive their comfort. This perception of comfort is provided by the Fanger's thermal comfort model. Some of the variables i.e. the air velocity and humidity are kept constant in the simulation. The air temperature and mean radiant temperature is calculated by the SIMBAD thermal model, the metabolic rate depends upon the activities of agents and the clothing level depends upon the agents' choices of clothes. The variations in these variables impact the agents' comfort who then act on household appliances and objects to maintain

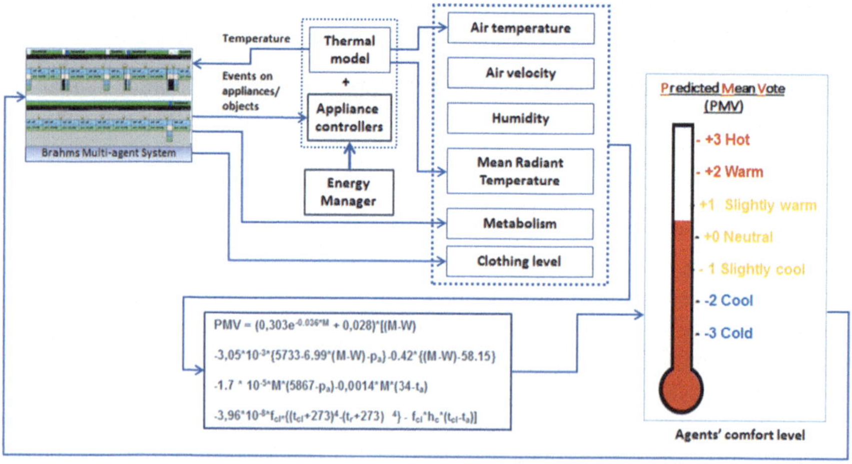

Fig. 27 Fanger's model in co-simulation

the comfort level. Figure 28 shows how the PMV value is calculated. The PMV calculator continuously perceives the input variables coming from the "temperature receiver", "clothing index computation" and "metabolic rate assignment" modules. It then uses Fanger's model in order to calculate the comfort level for each agent separately and then broadcasts it to them.

4.3 Co-Simulation Environment

The Brahms-SIMBAD-G-HomeTech[2] co-simulation environment is shown in Fig. 29. The Brahms-BEMS-Interface module provides the interconnection of SIMBAD thermal model with both the BEMS and the Brahms simulation environment. The input that goes to this module from the SIMBAD thermal model is the air temperature and mean radiant temperature. Other inputs include the electric power of appliances, the setpoint temperature and the appliance mode (on/off). The BEMS will use these variables to compute the energy plan and to control the appliances. Conversely, in Brahms these variables are perceived by the agents, who further take certain actions to control their thermal environment.

The output from this interface module either comes from the Brahms simulation environment or the BEMS. The output from Brahms simulation environment consists of occupancy data in each room in the house and the status/modes (on/off,

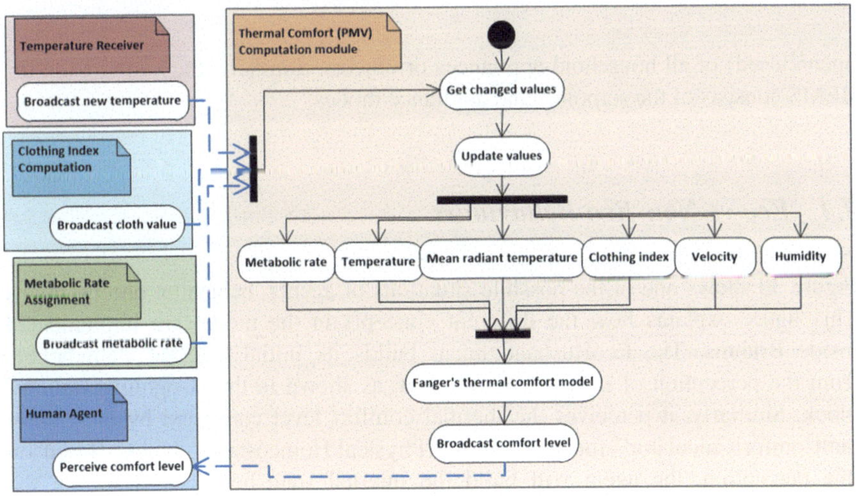

Fig. 28 Thermal comfort (PMV) computation module

[2]G-HomeTech is commercialized by Vesta System. The interconnection of BEMS with the co-simulator is established by Vesta System.

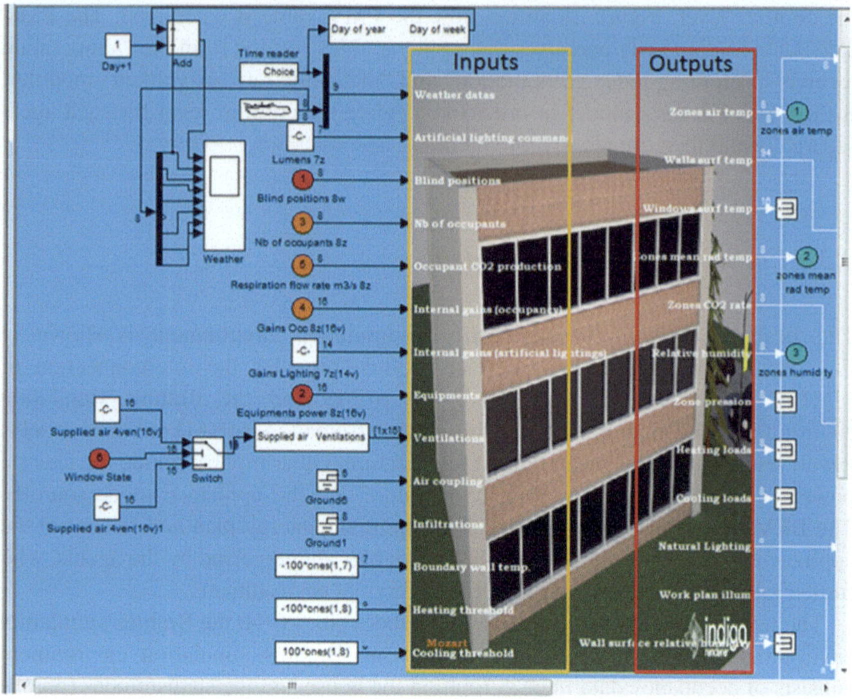

Fig. 29 SIMBAD-MOZART thermal model

open/closed) of all household appliances or objects. Similarly, the output from the BEMS consists of the setpoints and appliance modes.

4.4 *Eco vs Non-Eco Behaviours*

Figure 30 shows one of the possible situations of agents' behaviour among many. This figure explains how the different concepts in the model are implemented inside Brahms. The EcoHusband agent builds its initial external state beliefs from the perception of outside environment, as shown in the "Cognitive.Beliefs" block. Similarly, it perceives the thermal comfort level computed by the "ThermalComfortCalculator" function in the "Physical.Homeostasis" block. Based on this perception, the agent will build the internal state belief as shown in the "Cognitive.Beliefs" block. The values computed by this function lie in a range of −3 to 3 corresponding to different comfort conditions e.g. comfortable, slightly cool, etc. These comfort conditions are realized by the concept of workframes, where there are multiple workframes available at the same time. This is shown in the "Belief Generation" block that defines the agent's internal state belief generation

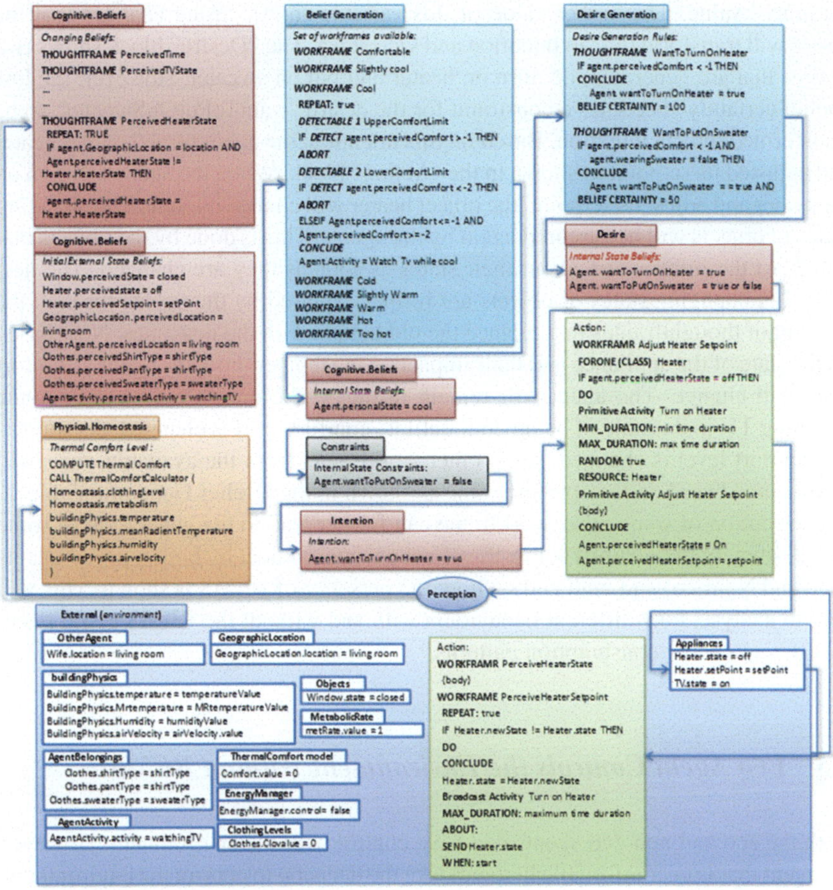

Fig. 30 A situation modelled in Brahms

rules through a set of workframes. However, depending upon the output of the "ThermalComfortCalculator" function one of them would be executed.

The agent is already in a state of watching TV as shown in the "AgentActivity" block inside "External (environment)" block which turns into its belief about its activity. If the agent is comfortable, slightly cool or slightly warm it would complete its current activity. For the other comfort conditions it could either continue the activity or abort it. An example of the "Cool" workframe is given in the "Belief Generation" block. This workframe says that if the agent's comfort level is between -1 and -2, it is cool. This will generate some desires in the agent to be comfortable. The "Desire Generation" block shows the rules that will lead to the generation of these desires. These rules are realized by the thoughtframes where based on the fact that agent's comfort level is below -1, that agent will conclude some other beliefs. These beliefs will be transformed into agent's desires based on the "belief certainty" value. The higher value of this variable shows strong chances that the desire will transform into an intention and vice versa. The "Desire" block shows two desires that are generated, i.e. turn on heater and put on sweater. However, the low "belief certainty" becomes a constraint for the desire "wantToPutOnSweater" to be transformed into an intention. Based on this intention, the agent turned on the heater and adjusted its setpoint as shown in the "Action" block. When the agent will turn on the heater and adjust its setpoint, the object heater will change its state. The changing states of objects will be captured again by the agents. This is done by the objects that broadcast the information about their states as soon as they are changed. The new beliefs of changing states of objects are further captured by the agents through the concept of thoughtframes that replace the old beliefs with the new ones. Now based on the state of the appliance and their impact on the temperature, the agent's comfort level will change. The agent will remain in the workframe "Cool" and continue watching TV in the state of being cool until its comfort level is changed. As soon as the comfort level is changed, some other workframe, from the available ones, will be executed based on the comfort value as shown in the "Belief Generation" block. The execution of some other workframe can further lead to the generation of some new desires. In the sections below the effect on environment by both types of agents (Eco and NonEco) and with and without the presence of BEMS is shown. This will help to analyse, how different behaviours with and without the BEMS could result in different energy consumption patterns.

4.5 Eco Agent Controls the Environment Without BEMS

Both the eco and non-eco agents can have control over the environment. However, the agent that is uncomfortable first will take the decision to control the environment. Figure 32c shows the PMV value of the EcoHusband agent while in the living room. At the start the PMV value is low, meaning that the agent is uncomfortable, but the agents are still watching TV comfortably. This is shown by the "watch tv comfortably" tool tip on the white coloured workframe in the EcoHusband agent's space at the start of the simulation, Fig. 31. This is due to their dynamic comfort. However, after sometime they start perceiving the real comfort value and being uncomfortable. The EcoHusband agent increases the temperature using the heater's thermostat to warm up the room. The control over the heating system is shown by yellow coloured workframes. The change in the state of heater by the EcoHusband agent is perceived by the heater, shown by the workframes in LivingroomHeater objects' space. The blue lines show the connection between the change in heater's state by the EcoHusband agent and the perception of this state by the heater. The LivingroomHeater object then broadcasts this change in its state to the other agents around, the blue lines show the signals sent to other agents. Figure 32a shows the state of the appliance as the agent acts upon it. Figure 32b shows that initially the temperature in living room was set to 18°C, it started increasing due to new thermostat settings of the heater by the EcoHusband agent.

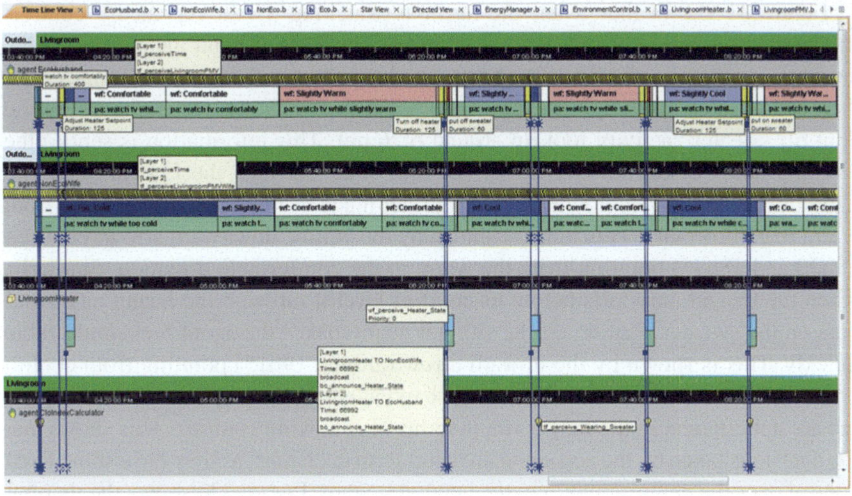

Fig. 31 Brahms simulation: perception of comfort during watching TV activity

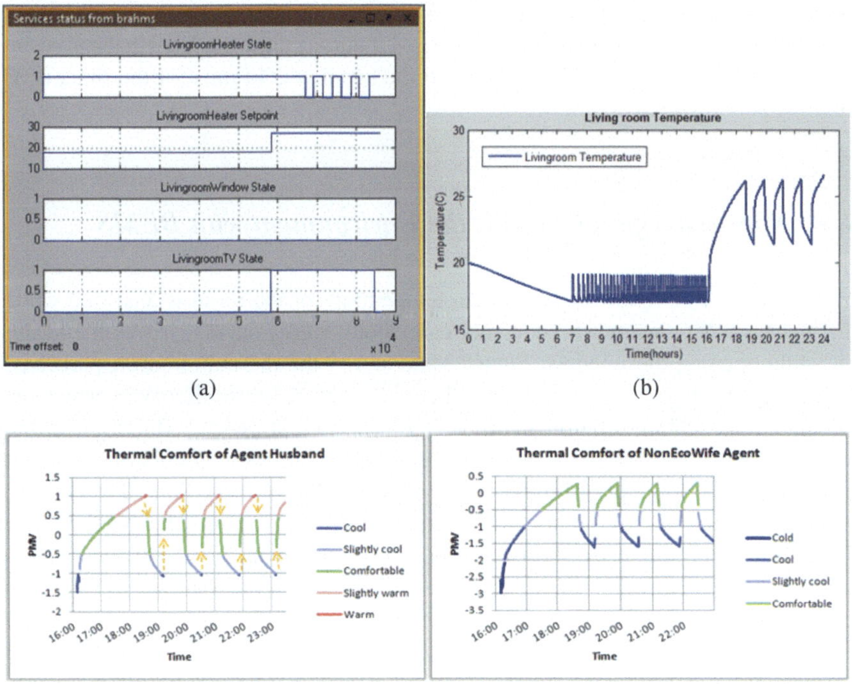

(a) (b)

(c) (d)

Fig. 32 State of the appliance/object, temperature, and PMV perceived while watching TV. (**a**) Control over appliance/object by EcoHusband. (**b**) Temperature as a result of control over environment. (**c**) PMV EcoHusband while watching TV. (**d**) PMV NonEcoWife while watching TV

Figure 31 shows that when the agent is watching TV, it repeatedly controls the heater and its clothes to achieve comfort. The EcoHusband agent puts on the sweater or takes it off which is shown by the yellow coloured workframes with "put off sweater" activity. The first time agent puts off sweater is shown by the "put off sweater" activity tool tip around 6:26 pm. This information is sent to the clothing index calculator, shown by the blue lines between the EcoHusband and CloIndexCalculator. Removing a thick sweater made its thermal comfort jump from warm to comfortable. This jump is shown by the yellow downward arrow (between 18h30 and 18h45) pointing from the warm to the comfortable direction. Similarly, when the temperature falls below its comfort level it turns on the heater again and puts on the sweater. Putting on the sweater again makes the agent feel comfortable quickly. This is shown by the upward arrow (around 19h15) pointing from cool to comfortable. The effort made by EcoHusband agent could help him to save energy, but are not efficient in the longer run in terms of achieving comfort. This shows that the decisions taken by the eco agent are short term decisions, as they have some fixed control over the environment, i.e. the heating system. Figure 32c shows the thermal comfort perceived by NonEcoWife agent during watching TV while EcoHusband agent controls the heater. At the start it feels cold (around 16h10, dark blue curve), but then after the temperature has been increased it just starts to feel cool (around 16h20, blue curve). As the heating system increasingly warms up the room it feels comfortable (between 17h30 and 18h30, green curve) until the EcoHusband agent turns off the heater again. The reason that the agent is cold most of the time, is its clothing is not warm enough.

4.6 Non-Eco Agent Controls the Environment with BEMS

In the above section, the impact on the temperature of the room is analysed while the NonEecoWife agent who does not care about energy saving, leaves the heater on while opening the window. Figure 34c,d shows the thermal comfort perceived by the agents. Figure 34c shows the thermal comfort of NonEcoWife agent. At the start it is feeling slightly cool (light blue curve at around 16h15) with the temperature set to 18°C. As the agents entered in the room, EcoHusband agent increased the setpoint temperature. This is shown by the yellow coloured workframe with "Adjust Heater Setpoint" activity in Fig. 33 that caused the NonEcoWife agent feel comfortable as shown in Fig. 34c with green curve between 16h20 and 17h30. EcoHusband agent however still remains cool (shown by the blue curve) due to its less warm clothes becoming comfortable later at around 17h30 (Fig. 34d). Figure 34b shows the temperature in the living room. As the temperature reaches above NonEcowife agent's comfort which is 24°C, it becomes slightly warm, shown by the pink curve in Fig. 34c, at around 17h30. However, as the temperature reaches 26°C, it becomes warm and then opens the window shown by the yellow coloured workframe with "Open window" activity in NonEcoWife's space. However, the BEMS would perceive that the agent has opened the window,

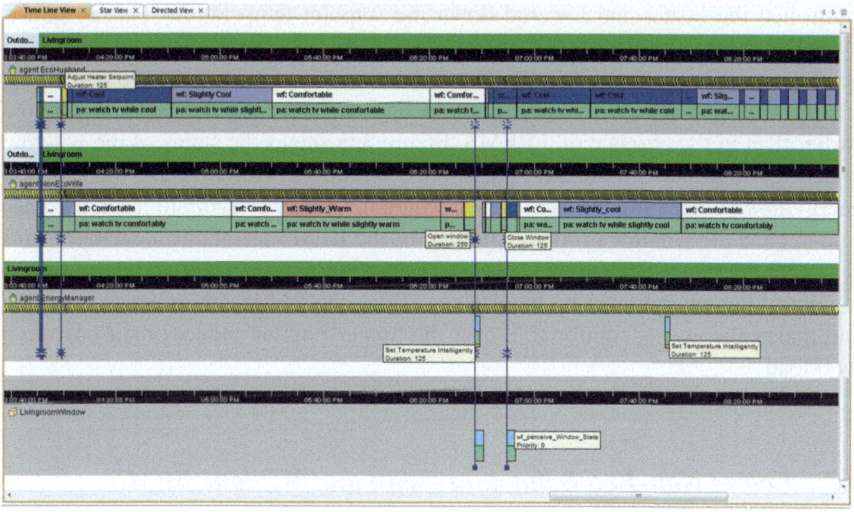

Fig. 33 Brahms simulation: NonEcoWife and BEMS controls the environment

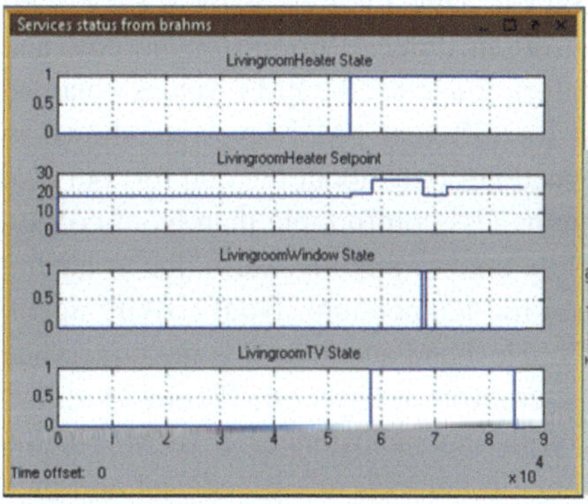

(a)

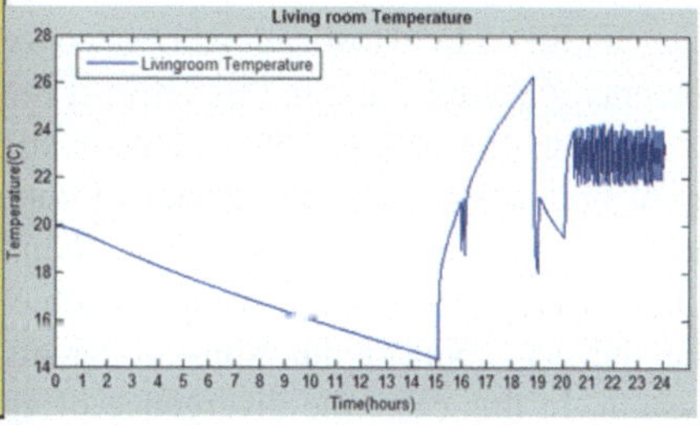

(b)

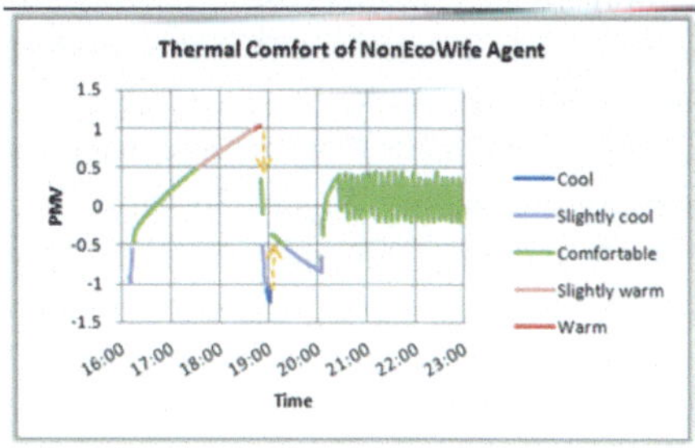

(c)

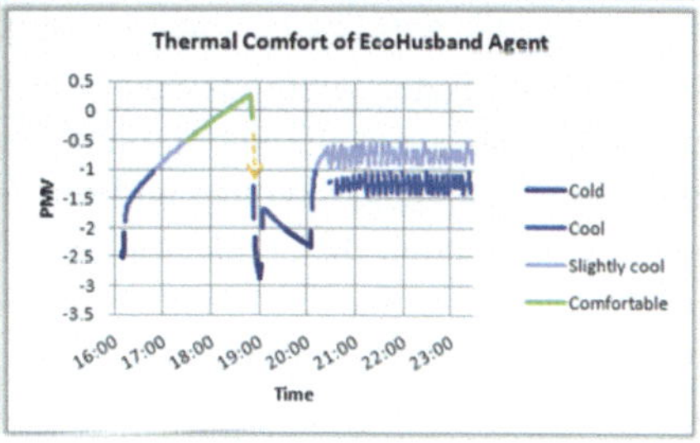

(d)

Fig. 34 State of the appliance/object, temperature, and PMV perceived while NonEcoWife controls the environment with BEMS. (**a**) Control over appliance/object by EcoHusband. (**b**) Temperature as a result of control over environment. (**c**) PMV EcoHusband while watching TV. (**d**) PMV NonEcoWife while watching TV controlled the heater and lowered down the setpoint temperature. This is shown by the workframe in EnergyManager's space with "Set Temperature Intelligently" activity. As the temperature in the living room now comes down more quickly to a level where NonEcoWife agent starts feeling slightly cool, it closes the window earlier than in the absence of a BEMS. Figure 34a shows the status of the window under "Livingroomwindow State" where the window is closed. Thus, there is less energy loss by reducing the time period where the heater is trying to reach a higher setpoint and the window is open. Afterwards, the BEMS maintains the temperature at a setpoint where the agents feel comfortable in the longer run and do not need to control the environment by themselves. Thus, the BEMS not only saves energy and makes the agents comfortable over the longer run, but reduces their cognitive workload.

4.7 Eco vs Non-Eco Behaviours with and Without BEMS

In this section, an analysis of the cost-comfort tradeoff for the situations with and without the BEMS will be given. Note that the BEMS does not take the decisions alone but the agents are also part of the control. Thus the role of BEMS becomes more challenging as it has to put more effort in order to minimize the cost and maximize the comfort. To quantify the comfort of agents, the PMV values obtained after the simulation runs are summed up for different PMV levels as shown in Fig. 35. Since EcoHusband agent is concerned not only about the comfort but also the energy savings and in this effort it remains less comfortable than NonEcoWife agent (Fig. 35a). Mostly, it remains in slightly cool or slightly warm due to having more interactions with the heater to control the temperature. NonEcoWife agent, however, remains more comfortable than EcoHusband agent, as it is not concerned about energy savings and wants to achieve comfort at any cost. Figure 34 shows the thermal comfort durations of agents with the inclusion of a BEMS in the system. In this case, the divergence of agents' comfort levels is reduced and they converge to the comfortable zone. Also, the agents remain comfortable for a longer time duration as compared to before i.e. without BEMS. In this case, EcoHusband agent's comfort is better than NonEcoWife agent. The improvement in the comfort is due to the better decisions taken by the BEMS based on the knowledge that the BEMS has about the internal and external environmental conditions, weather forecasts, inhabitant's comfort and self-learning algorithms. Figure 36 shows the power consumption of the electric heater while the environment is controlled by different agents with and without the BEMS. The highest power consumed is due to the behaviour of NonEcoWife agent since it tries to achieve comfort by opening and closing the window. This assessment of BEMS when co-simulated with building system and inhabitants shows that the BEMS is capable of not only saving the inhabitants from cognitive workload but also of providing them with better comfort and energy savings. Figure 37a shows that after 16h00 when it is in the living room and controlling the window, the heater has to put more effort to warm up the room

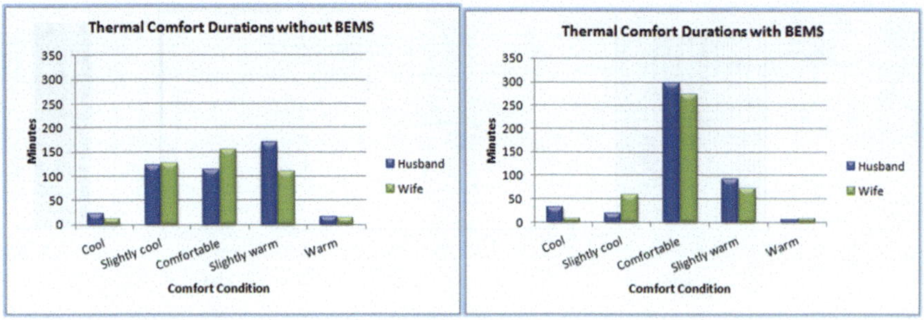

(a) (b)

Fig. 35 Comfort of agents: with and without the control of BEMS. (**a**) Agents' thermal comfort without BEMS. (**b**) Agents' thermal comfort with BEMS

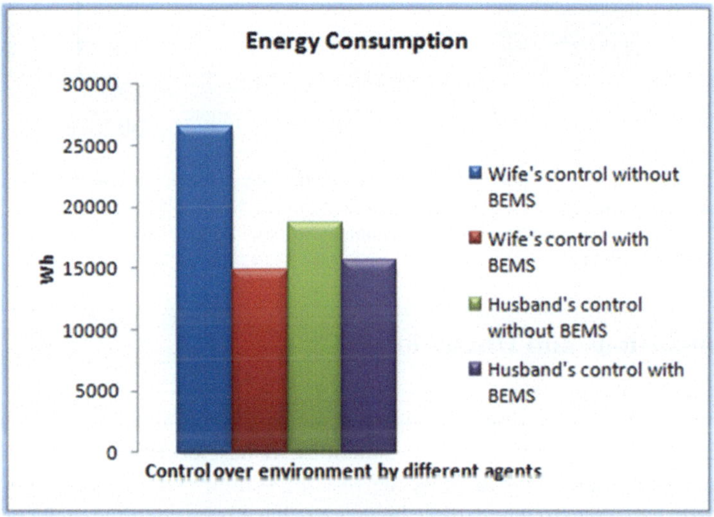

Fig. 36 Energy consumed during control over environment by different agents with/without BEMS

and the controller never stops. However, the inclusion of BEMS helps it to achieve comfort earlier by lowering the setpoint when it detects the opening of window, forcing the NonEcoWife to close the window earlier and save energy (Fig. 37b). The EcoHusband agent is however an eco-person and tries to behave the way an BEMS do, thus the energy consumption when EcoHusband is controlling the environment is much less as compared to NonEcoWife agent. However, it has to control the heating system multiple times and put extra efforts (Fig. 37c). In case of control by the BEMS, however, it helps him to control the heater and adjust the setpoint such that even if it puts on/off its extra clothing, it remains comfortable most of the time (Fig. 37b) by saving even more energy than it tries to save by its control (Fig. 37d).

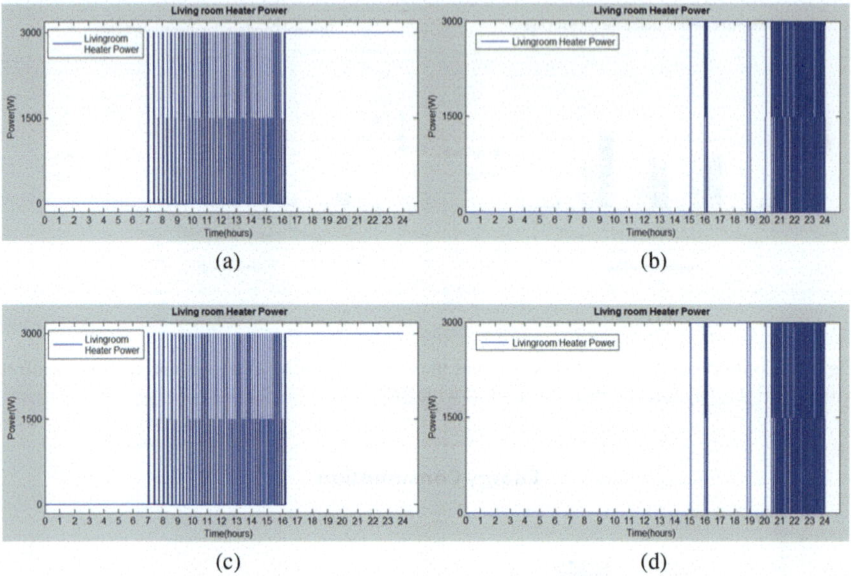

Fig. 37 Livingroom heater power while NonEcoWife and BEMS control the environment. (**a**) NonEcoWife controls environment without BEMS. (**b**) NonEcoWife controls environment with BEMS. (**c**) EcoHusband controls environment without BEMS. (**d**) EcoHusband controls environment with BEMS

5 Conclusions and Discussions

The work done in this research analyses the impact of inhabitants' behaviour on energy consumption in domestic situations. It has identified the high energy consuming activities of inhabitants; the reasons behind certain energy impacting behaviours; the extent to which these behaviours have been captured in the past; and given the motivation to improve the energy simulations with new requirements and challenges, specially, with the advancements in smart grid technology. The study has also addressed whether it is important to take into account the complex behaviours, i.e. the reactive, deliberative, social, and reasoning and cognitive elements of inhabitants' behaviour in building energy simulations and how these behaviours could be validated to ensure their representativeness. This section synthesizes the findings in order to answer the following research questions:

1. How to identify the energy impacting behaviours? The analysis of energy consumption patterns for different household appliances has revealed that these patterns are highly variable. This variability in consumption patterns is found to be linked with inhabitants' behaviour and the activities they perform in their day to day living on appliances. Hence, it is important to analyse both the consumption and behaviour patterns to identify those behaviours that are responsible for high energy consumptions. The identification of inhabitants' energy

impacting behaviours is done through data analysis. In order to perform this task, the availability of both the energy consumption data and the corresponding inhabitants' activities and behaviours data is necessary. Thus, the Irise energy consumption data is used and complemented with the inhabitants' behaviour information through field studies.

The behaviours represent not only the simple actions but a complete reasoning process on how these actions are reached. They are influenced by certain parameters that ultimately affect the energy consumption. These parameters include the environmental variables (e.g. season, weekdays, weekends etc.), specific interactions with appliances (e.g. turn on/off, put food in fridge etc.), relation between appliance usages (e.g. the impact of the cooking activity on the fridge consumption) and the reasons behind certain actions (e.g. why the cooker is used more on a particular day?). These parameters serve as important inputs to identify inhabitants' representative energy consuming behaviours from Irise database. The identified behaviours are then used in building and validating the model through the co-simulation of inhabitants' and appliances behaviours.

2. How the complex (reactive, deliberative, social and group) behaviours can be co-simulated with the thermal model of the building and physical models of appliances in residential buildings? The answer to the previous research question revealed that inhabitants' energy impacting behaviours are complex as they are based on intricate reasoning mechanisms. Thus a conceptual, BDI based model is built to capture the complete process of how the inhabitants perceive the outside environment and the internal physical homeostasis. The model describes how these perceptions convert into their beliefs, how these beliefs trigger a cognitive process of building some desires, taking into account various environmental and social constraints, how these desires turn into an intention and how based on this intention some action on the appliances, objects or building envelope are taken.

   This behaviour model is implemented in the Brahms agent based modelling and simulation environment. In this environment a complete system consisting of objects, appliances, time, inhabitants and their internal and external state beliefs is constructed. The different elements of this system interact with each other and react to change that occurs in the environment. The complexity exhibited by the inhabitants' reasoning and cognitive aspects as well as the social and group behaviours is successfully captured and simulated in Brahms. Similarly, the behaviour of an appliance or object can also be modelled to some extent inside Brahms. However, it is not easy to build the complex physical models of appliances or a thermal model of a building, etc. inside Brahms. Thus, it is better to build the physical systems outside, in an environment that is specifically built for this purpose. For example, the thermal model of the building is constructed in Matlab/Simulink, which computes the temperature in the zone and sends this information to the inhabitants in Brahms environment. The agents in Brahms then act upon the heater, air conditioner or windows, etc. inside Brahms. The information about the changing state of the appliances or objects inside Brahms goes back to the thermal model. This is used to compute the new temperature of the zone, which is then sent back to Brahms. In this way a co-simulator is built through a Java interface between the two systems. Similarly, the complex physical models of appliances can also be built this way in Matlab/Simulink e.g. a fridge freezer and co-simulated with the behaviour model in Brahms.

In addition, an energy management system is also included in the co-simulator environment. This either controls the appliances on behalf of the inhabitants or gives them advice for improving their energy consuming behaviours. In these co-simulations, the randomness and variability is introduced. Firstly, when the human agents goes through the cognitive process and acts on the building system, the variation in the state of the physical systems change their old perceptions about the environment. This will impact their cognitive process and cause them to behave differently in the new situation. Secondly, in each changing situation the agent does not necessarily behave in a single way. Rather, it could behave in multiple ways depending upon the probabilistic values for its different beliefs. These probabilities are assigned to beliefs inside the Brahms environment. Thirdly, the introduction of environmental and social constraints in the system will make the agents behave more like real humans. Fourthly, some random variables, which are difficult to model in Brahms, are also introduced through Java activities. This allows agents to make some decisions depending on the value of the random variable e.g. allowing agents to choose a combination of clothes, etc. The algorithm to compute the values for these random variables is computed in Java and sent back to the agents in Brahms. Thus a combination of all of these different elements of randomness creates interesting situations to analyse different behaviours of agents and their impact on the physical aspects of the building.

3. How can the complex behaviour models be validated to ensure its representativeness? A methodology is proposed and implemented to validate the inhabitants' behaviour model. In this methodology, the behaviour of the inhabitants in the Irise database is captured by complementing it with additional information. This information actually comprises the impact of certain parameters on inhabitants' energy consuming behaviour, e.g. seasons, weekdays and weekends, the impact of the usage of one appliance over the other, etc. Then the houses with similar behaviours are clustered to find the representative behaviours. Then the co-simulation of the inhabitants' behaviour model is done with the selected appliance. The different parameters in the model e.g. seasons, weather, weekday/weekend, social behaviour, etc. are assigned different probability values or weights to make them tuneable. This co-simulation gives the simulated energy consumption of the appliance. From the Irise database, the actual energy consumption of the appliance is also available. The appliance energy consumption distributions for both the actual and simulated situation are then compared. If the simulated behaviour is realistic, the distributions will follow the same trend. If the trends are dissimilar, the parameters are tuned such that their values come closer to the observed behaviour of that cluster and the error is significantly reduced. Similarly, the same simulated behaviour is then compared with another member of the same cluster with the same values of the tuning parameters to analyse how representative is the behaviour model of its cluster.

4. How to validate BEMS with building system and inhabitants? The BEMS controls the household appliances and objects e.g. lights and shutters, etc. and also gives advice to the inhabitants. This advice is given based on the anticipative plan that is computed based on signals coming from the grid. The anticipative plan is updated at every hour, hence the advice is given every hour. However, in order to evaluate that based on different reactions by the inhabitants, how efficiently the BEMS recomputed its strategies, whether they are feasible and whether the inhabitants are saved from cognitive workload and are provided with better comfort and energy savings, a mechanism is required. Thus, the BEMS is co-simulated with the building system and the inhabitants where the inhabitants can either directly control the appliances and objects or through building BEMS. Different stereotypes of inhabitants i.e. having Eco and non-Eco behaviours are also defined and the strategies of BEMS are assessed by putting it in different complex situations.

The work done in this research is different from the previous works in several ways. Most of the previous works focus on office buildings where human behaviour is relatively less complex as compared to home situations. In order to capture the behaviour in domestic settings the behaviour needs to be captured in much more detail than simple presence/absence profiles. Similarly, the previous works done for energy management in home situations focus on demand side predictions associated with turning on/off the electrical appliances. The work in this research is oriented more towards finding the specific usages or activities behind consumptions that impact energy consumption. These actions are the result of a complete process from perception to cognition and then to action. The introduction of inhabitants' reasoning processes towards their actions on the physical environment will give energy simulation tools more realism. By creating and putting inhabitants in different situations, it will lead them to reason differently about the situation and solve it in another way than before. Although, it is not easy to capture all different types of reasoning processes behind the different behavioural patterns, some high level categories are identified through field studies. The purpose is to analyse how the introduction of these type of reasoning processes and complex behaviours could help to bring the building energy simulations closer to reality and to reduce the gap between actual and simulated situations. In this work we have shown that complex behaviour taking into account BEMS can be managed by the proposed approach. Nevertheless, less complex behaviours, in offices for instance, can also be managed by this approach.

References

1. M. Fikru, M. Fikru, The impact of weather variation on energyconsumption in residential houses. Appl. Energy **144**, 19–30 (2015)
2. R.K. Jain, R. Gulbinas, J.E. Taylor, P.J. Culligan, Can social influence drive energy savings? detecting the impact of social influence on the energy consumption behavior of networked users exposed to normative eco-feedback. Energy Build. **66**, 119–127 (2013)

3. E. Azar, C. Menassa, A comprehensive analysis of the impact of occupancy parameters in energy simulation of office buildings. Energy Build. **55**, 841–853 (2012)
4. D. Bourgeois, C. Reinhart, I. Macdonald, Adding advanced behavioural models in whole building energy simulation: A study on the total energy impact of manual and automated lighting control. Energy Build. **38**, 814–823 (2006)
5. C. Reinhart, Lightswitch 2002: a model for manual control of electric lighting and blinds. Solar Energy **77**, 15–28 (1994)
6. S. Borgeson, G. Brager, Occupant control of windows: accounting for human behavior in building simulation. Tech. Rep., 2008
7. R. Goldstein, A. Tessier, A. Khan, Customizing the behaviour of interacting occupants using personas, in *Proceedings of 4th National conference of IBPSA*, New York, USA (2010)
8. J.A. Clarke, I.A. Macdonald, J.F. Nicol, Predicting adaptive responses-simulating occupied environments, in *Proceedings of Windsor conference on Comfort and Energy Use in Buildings-Getting them right*, Cumberland Lodge, Windsor, UK (2006)
9. N. Gilbert, *Agent-Based Modelling and Simulation in the Social and Human Sciences*, ch. Computational Social Science: Agent-Based Social Simulation, ser. ISBN 978-1-905622-01-6 (Bardwell Press, Oxford, 2007), pp. 115–134
10. P. Davidsson, M. Boman, Distributed monitoring and control of office buildings by embedded agents. Inform. Sci. **171**(4), 293–307 (2005)
11. H. Jouma, S. Ploix, S. Abras, G. De Oliveira, A MAS integrated into home automation system, for the resolution of power management problem in smart homes. Energy Procedia **6**, 786–794 (2011)
12. J. Engler, A. Kusiak, Agent-based control of thermostatic appliances, in *Green Technologies Conf., IEEE*, Grapevine, TX, USA (2010)
13. S.K. Das, D.J. Cook, A. Battacharya, E.O. Heierman, T.-Y. Lin, The role of prediction algorithm in the MavHome smart home architecture. IEEE Wireless Commun. **9**, 77–84 (2002)
14. S. Abras, S. Pesty, S. Ploix, M. Jacomino, Advantages of MAS for the resolution of a power management problem in smart homes, in *Advances in Intelligent and Soft Computing* (Springer Berlin, Heidelberg, 2010), pp. 269–278
15. C. Liao, P. Barooah, An integrated approach to occupancy modeling and estimation in commercial buildings, in *American Control Conf.* (IEEE, Baltimore, MD, USA, 2010), pp. 3130–2010
16. M.A.R. Lopes, C.H. Antunes, N. Martins, Energy behaviours as promoters of energy efficiency: A 21st century review. Renewable Sustain. Energy Rev. **16**, 4095–4104 (2012)
17. P. Hoes, J.L.M. Hensen, M.G.L.C. Loomans, B. de Vries, D. Bourgeois, User behavior in whole building simulation. Energy Build. Appl. Thermal Eng. **41**, 295–302 (2009)
18. J. Nicol, M. Humphreys, Adaptive thermal comfort and sustainable thermal standards for buildings. Energy Build. **34**(6), 563–572 (2002)
19. J. Liu, R. Yao, J. Wang, B. Li, Occupants' behavioural adaptation in workplaces with non-central heating and cooling systems. Appl. Thermal Eng. **35**, 40–54 (2012)
20. W.J. Clancey, P. Sachs, M. Sierhuis, R. van Hoof, Brahms: Simulating practice for work systems design. Energy Build. **49**, 831–865 (1998)
21. P.O. Fanger, Assessment of man's thermal comfort in practice. British J. Ind. Med. **30**, 313–324 (1973)
22. P.O. Fanger, Thermal comfort. analysis and applications. *Environmental ENgineering* (McGraw Hill, New York, 1972)
23. G. Newsham, Manual control of window blinds and electric lighting: Implications for comfort and energy consumption. *Indoor Environment* **3**, 135–144 (1994)
24. Z.M. Gill, M.J. Tierney, I.M. Pegg, N. Allan, Low-energy dwellings: the contribution of behaviours to actual performance. *Building Research and Information* **38**, 491–508 (2010)

Stochastic Prediction of Residents' Activities and Related Energy Management

Patrick Schalbart, Eric Vorger, and Bruno Peuporter

1 Introduction

1.1 Context

The building sector has a great potential to reduce energy consumptions and environmental impacts on a global level. Despite the progress in dynamic building energy simulation (DBES) models concerning deterministic phenomena [1–8], the ability to predict energy consumption is limited by the non-deterministic boundary conditions, especially those related to occupants' behaviour:

- Metabolic heat dissipation.
- Window openings.
- Action on shading devices.
- Artificial lighting.
- Electrical equipment.
- Temperature setpoints.
- Domestic hot water consumption.

This often yields higher actual energy consumptions than expected, particularly in recent high-performance buildings. The traditional use of standard scenarios shows its limitations and asks for new models. This is all the more important for architects and engineers in the design phase, when the future occupants' behaviour

P. Schalbart (✉) · B. Peuporter
CES – Centre d'efficacité energétique des systèmes, MINES ParisTech, PSL Research University, Paris, France
e-mail: patrick.schalbart@mines-paristech.fr

E. Vorger
Kocliko, Toulouse, France
e-mail: eric.vorger@kocliko.co

S. Ploix et al. (eds.), *Towards Energy Smart Homes*,
https://doi.org/10.1007/978-3-030-76477-7_17

is unknown. Typical applications include being able to predict the probability distribution of energy consumption for energy performance guarantee contracting.

Two types of models have been developed to tackle this issue. In the first type, discomfort drives actions. It needs to define discomfort and the process leading from the discomfort situation to an action. In the second type, actions derive from probability distributions generated out of observations; the physiological and psychological processes leading to an action are not described.

1.2 DBES Reliability

Evaluation of buildings energy consumption for heating and cooling through models has been developed over the last four decades. Nowadays, architects and engineers widely use DBES tools to design high energy performance buildings. Despite extensive validation campaigns of such models, significant differences have been observed between predicted consumptions and those measured during operation. Oftentimes, consumptions are larger than expected; this can be problematic for building renovation for example, because financing is associated with expected return on investment [9]. In most cases, this overconsumption is due to occupants' behaviour who have a strong influence on energy consumption [10–16]. Andersen [10] showed that the heating and domestic hot water (DHW) consumption of 290 identical dwellings varied from 1 to 20.

1.3 Various Occupants' Behaviour Modelling

Besides the traditional deterministic approach describing actions according to specified schedules and/or thresholds, agent-based and stochastic approaches have a great potential to grasp the complexity of occupants' behaviour.

1.3.1 Agent-Based Approach

Agent-based approaches represent occupants as autonomous agents interacting with each other and with their environment. The general structure of such models is shown in Fig. 1. According to MacAl and North [17], it includes:

- A population of agents with attributes and behaviour.
- A set of rules defining relations and interactions between agents.
- An environment with which agents interact.

Agents are autonomous; their states determine their behaviours. The state of all the agents characterise the state of the system. In some cases, agents can adapt their behaviour based on experience. They can have a goal, selecting actions that bring

Fig. 1 Typical agent in an agent-based model [17]

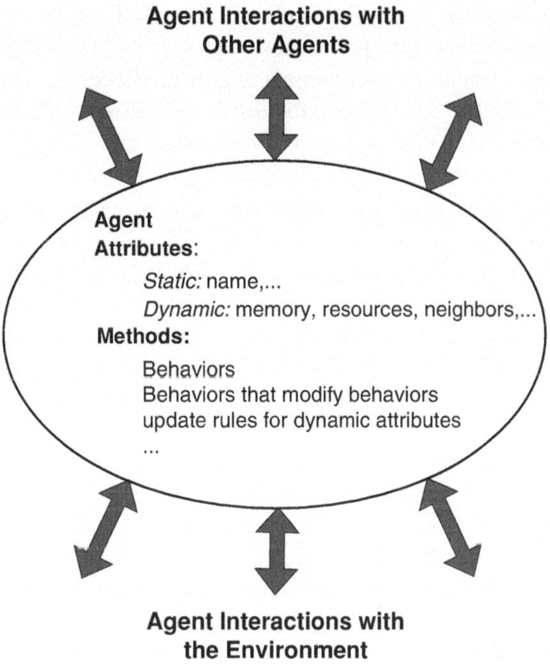

them closer to their objectives. They are connected to each other through topologies that vary from one model to another. The concept of neighbour defines how agents can interact.

Kashif et al. [18] used this type of approach to describe buildings' occupants electricity use in a household. Two types of rules are introduced: those defining thoughts and beliefs, and those defining their actions.

In a similar manner, Bonte et al. [19] created neural networks to reproduce efficient behaviour for an agent to ensure its comfort. After a training phase, agents know which actions are susceptible to increase their comfort in various possible environmental conditions.

Lee and Malkawi [20] proposed an agent-based model for office buildings based on thermal comfort, integrating perception, reflexion, action, and knowledge update. Agents can modify their activities' intensity and their clothing, can handle shading devices, turn on a ventilator, and open windows and doors. A significant number of parameters need to be specified, through assumptions and calibration.

Moujalled [21] developed an adaptive behaviour model, and linked it to a DBES tool. Windows, stores, lighting, temperature setpoint, and clothing are driven by two factors: the difference between real and expected comfort, and the expected result of actions.

Agents perceive their environment, based on their physiological and psychological characteristics, and take action based on their knowledge to achieve a goal. For instance, an agent feels discomfort due to high temperature; based on experience, he knows that opening a window is potentially interesting; he therefore wishes to open the window, asks permission to its neighbours, and if they agree, opens the window. As a result, the environment will be modified (e.g. temperature evolution evaluation through a DBES tool); the difference between the expected and actual outcome is an input to update the agent's knowledge.

This type of model is appealing because it aims at reproducing the reality of human behaviour by taking into account phenomena such as perception, memory, logic, and choice [20]. The difficulty is that the functions linking the environment to the psychological state cannot be easily defined. What is the temperature beyond which an agent is hot? When is it so unbearable that he decides to take action? The hierarchy among various aspects of comfort is hard to specify. Moreover, modelling occupants' behaviour as perfectly rational, at least from the thermal comfort point of view, is a strong assumption.

1.3.2 Stochastic Approach

Nicol et al. [22] suggested that occupants' behaviours is a non-deterministic stochastic process. For instance, they stated that there is not a precise temperature above which everybody opens their windows, but rather an increasing probability that the window is opened as the temperature increases.

The stochastic approach tries to reproduce reality without explicitly linking causes and effects. Probabilities are evaluated from observations. Thus, for an input variable, there is not one unique output but a distribution of outputs. The difficulty is to make sure that the inputs' probability distributions are built on reliable and representative data.

Stochastic behaviour models are usually established from measurement campaigns where states (e.g. if a window is open or closed) are recorded with environmental data (indoor and outdoor temperatures, solar radiation, etc.). A knowledge-based statistical approach links the explanatory variables with the variable of interest, based on the detection of significant relationships. The key elements of this type of methods are [23]:

- Being based on measurement and not on surveys, it is less susceptible to include bias.
- The relationships grasp a whole ensemble of hidden phenomena whose explicit modelling would be difficult. Thus, comfort (in all its aspects) appears as a driver in an indirect manner. The process leading from the state of the environment to an action, via perception and evaluation of the desire to take action from beliefs, is included in the function expressing the probability of action.
- Statistical models validity is based on their capacity to reproduce reality. The quality of their prediction is particularly scrutinised.

Stochastic models have a tendency to produce average behaviours, softening the variability that exists between individuals. Parys et al. [24] list three methods to reproduce diversity: the creation of statistics based on individuals [25], the calculation of standard deviations associated with the average parameters' values [26] and the division between "active" and "non-active" categories [24, 27]. The first two methods require large datasets. The third method was integrated in the model.

1.4 Chapter Outline/Objectives

This chapter presents a stochastic occupants' activities modelling approach for residential buildings, drawing from existing models from the literature for each submodel. One of the main objectives of the proposed modelling approach is to make it possible to obtain, at the output of the simulation, distributions of the building's energy consumption (heating, cooling, specific electricity), instead of single values based on deterministic scenarios. This is achieved thanks to the Monte Carlo method consisting running many simulations from randomly drawn the inputs' probability distributions. This approach is shown schematically in Fig. 2.

Section 2 introduces the 5 W approach with people's presence (Who), activities (What, When and Where) and related electrical equipment (With). Section 3 presents adaptive behaviour such as windows opening, action on shading devices or temperature setpoints. Section 4 introduces electrical equipment models to evaluate energy consumption. In Sect. 5, these models are applied to a case study in an energy performance guarantee context.

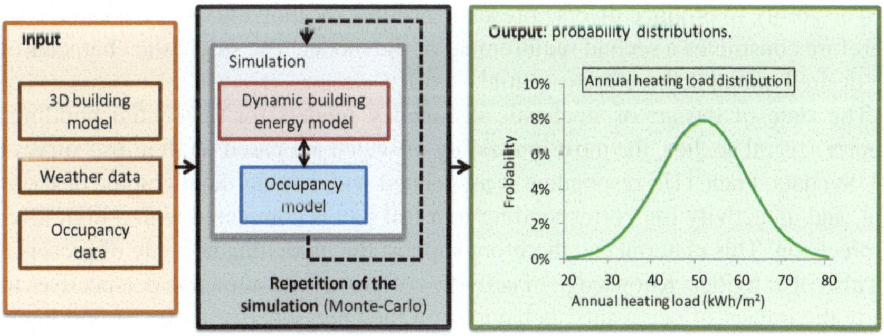

Fig. 2 General stochastic methodology

2 Modelling of Presence, Activities and Related Electrical Equipment

2.1 Introduction

The presence profiles in residential buildings are diverse and impact many aspects related to the building's energy consumption. Considering a given dwelling, it is obvious that its use is very different depending on whether it is inhabited by a retired couple, a single working person or a family with three small children. The variety of uses inevitably leads to a variety of energy consumption. A single dwelling therefore has as many potential consumption values as there are potential households.

This diversity is completely neglected in regulatory calculations since there is a single conventional scenario of presence in housing which considers that the inhabitants are absent during the day in weekdays and present the rest of the time. However, the presence profiles are infinitely more diverse and these profiles, which one could qualify as "classic", do not necessarily match the majority.

The first requirement of the model is therefore to account for the diversity of presence profiles according to the type of inhabitants. Meeting this objective by means of deterministic scenarios would require listing scenarios corresponding to all the different possibilities. This method is excluded for several reasons. First of all, establishing a typology would be extremely tedious and would need extensive data for its calibration. In fact, to create, for example, the scenario corresponding to the "retired couple" type, it would be necessary to average the presence profiles of a significant number of retired couples. Moreover, this method would neglect the differences between households of a given type. In other words, it is not enough to dissociate a working couple from a retired couple since two couples of the same category can have significantly different profiles of presence at their home.

The ability to produce diverse presence profiles for individuals of the same type therefore constitutes a second requirement of the model. The stochastic character of the desired model is therefore essential.

The state of the art of stochastic occupancy models for residential buildings reports several studies, the most interesting of which are based on time-use surveys (TUSs) data. Each TUS respondent is associated with a socio-demographic description, and an activity log corresponding to a full day is completed with a high level of precision. This material can therefore support the modelling not only of presence but also of activities. Knowledge of activities offers several attractive perspectives to study the impact of occupants' behaviour on the energy performance of buildings. They can be used as a basis to model electrical uses, and they make it possible to locate the occupants in the dwelling, provided a certain number of hypotheses which associate the activities with the rooms. They can also provide information on factors influencing windows' opening (e.g. people tend to open the kitchen window after cooking or the bathroom window after showering). Furthermore, an advantage of TUSs is linked to the size of the population samples on which they are carried out.

A better representativeness of the results is ensured compared to a study carried out on an instrumented building with its own specificities.

2.2 Presence Modelling

Wilke et al. [28] presented a simplified model. This is a zero-order (no memory) model calibrated over the 1999 French time-use survey.[1] Starting from the overall probability of presence at each time step, it makes it possible to generate presence profiles by a Bernoulli process (which directly determines the state of presence at each time step as a function of the probability of presence at this step time). As this model is independent of individual characteristics, it is denoted IIM (for Individual Independent Model).

Two other models for predicting the presence of occupants in dwellings were developed by Wilke et al. [28]. Both are based on inhomogeneous Markov chains. The first is a first-order model in which the presence state is predicted at the next time step by comparing the probability of change of state (specific to the individual, day of the week and time step) to a random number drawn according to a uniform law on [0; 1].

The second is a hybrid model, named "high-order Markov model". When a presence begins, the duration of the presence period is calculated from the probability distribution of the periods of presence (also specific to the individual, to the day of the week and to the time step). The calculation is performed by the inverse transformation method (ITM). The higher order of this second model has several advantages:

- Unlike offices for which the periods of presence and absence are highly contrasted (high probability of presence in the middle of the morning and afternoon and low at night), housing often has time slots during which the probabilities of presence have intermediate values. With a first-order model, there is a risk of fluctuations. For example, in the hypothetical case where the probability of an individual's presence is around 0.5 from 10 a.m. to noon, the first-order model risks overestimating the number of alternations between presence and absence. However, the reality is certainly closer to the statement "there is a one in two chance that the person will be absent in the morning". If this is the case, calculating the duration will allow a more realistic prediction.
- By limiting the number of presence/absence fluctuations, one can a priori better model the activities since they are necessarily interrupted each time the occupant leaves.
- The method is not more computationally expensive since the prediction of the duration is compensated by an interruption of the Markov process as long as the

[1] http://www.epsilon.insee.fr:80/jspui/handle/1/101557

presence lasts. The time steps are simply decremented and the process resumes when the occupant leaves.

The first-order Markov model meets the objectives better, in particular because the high-order model tends to homogenise the presence profiles between individuals with different characteristics.

2.2.1 Transition Probabilities

Hourly Calibration

The results of the survey are interpreted in terms of presence and absence by associating activities with one or the other situation. The transition probabilities are calibrated for hourly time steps due to the "response rounding" bias. Thus, the probability of presence at time $n + 1$ is linked to the probability of presence at time n by the relation:

$$P\,(n+1) = T_{1h}(n)P(n) \tag{1}$$

$T_{1h}(n) = \begin{pmatrix} t_{11} & t_{01} \\ t_{10} & t_{00} \end{pmatrix}_n$ is the transition matrix with an hourly resolution (n). t_{ij} is the probability of moving from state i to state j in one time step (presence: 1; absence: 0).

P is defined by:

$$P(n) = \begin{pmatrix} p(n) \\ 1 - p(n) \end{pmatrix} \tag{2}$$

$p(n)$ is the presence probability at $t = n$. Therefore, $1 - p(n)$ is the absence probability.

The transition probabilities depend on the characteristics of the individual (age, sex, day, dwelling location, age of children, income, occupation, civil status, ownership status, health…) grouped in a vector of indicator variables $x = (x_1, \ldots, x_M)$. Indicator variables are binary variables constructed from the different states that the variables can reach (a variable which can have four values is transformed into four indicator variables).

Given the high number of variables in the survey (around 30), only binary variables having a significant effect on the probability of transition are kept. For each hour, the selection is made as follows: all possibilities of dividing the sample into two subgroups are tested and the division that gives rise to the most significant difference in the proportion of transitions is kept. All possible divisions are then tested for each of the two resulting subgroups and so on until no further subgroup can be divided in the sense that the two subgroups would not be significantly different (with a 5% risk). Thus, 17 of the survey variables plus the day of the week are used to create 64 indicator variables for the presence model.

From the state in which he (or she) is (present or absent), an occupant can either remain in his (or her) state or transition into the other state. To model its probability, logistic regressions are fitted for each hour on the variables which have previously been retained for this hour. The transition probabilities obtained have the following form in which β_i are the parameters of the logit model ($M + 1$ parameters in total):

$$t_{01}(x, n) = \frac{1}{1 + \exp\left(-\left(\beta_0^{01}(x, n) + \beta_1^{01}(x, n)\, x_1 + \cdots + \beta_M^{01}(x, n)\, x_M\right)\right)}$$

$$t_{10}(x, n) = \frac{1}{1 + \exp\left(-\left(\beta_0^{10}(x, n) + \beta_1^{10}(x, n)\, x_1 + \cdots + \beta_M^{10}(x, n)\, x_M\right)\right)} \quad (3)$$

The other values of the transition matrix are obtained by the relations:

$$t_{11} + t_{10} = t_{01} + t_{00} = 1 \quad (4)$$

Derivation for Smaller Time Steps

For more detailed scenarios, smaller time steps are required. In the following, a 10-min time step is selected. The Markov chain matrices are obtained thanks to the Chapman-Kolmogorov equation according to which at a given time t:

$$T_{10\ \min}(t) = T_{1h}(t)^{1/6} \quad (5)$$

The values are attributed to the time step at the middle of the hour. The values for the other time steps are determined by linear interpolation.

When calculating the sixth roots of the matrices $T_{1h}(n)$, in the rare occasion (0.5% of the cases) that the second eigenvalue which is $1 - (t_{10h}(n) + t_{01h}(n))$ is negative (the first being 1), a complex sixth root is obtained. In this case, it is set to 0. For the positive eigenvalues, the positive values of the sixth roots are selected.

At this point, the 24-h transition probabilities are determined with a 10-min resolution depending on the characteristics of the individual and the day of the week. Figure 3 gives the average, over the entire survey population, of the values of t_{01} and t_{10} with a resolution of 10 min.

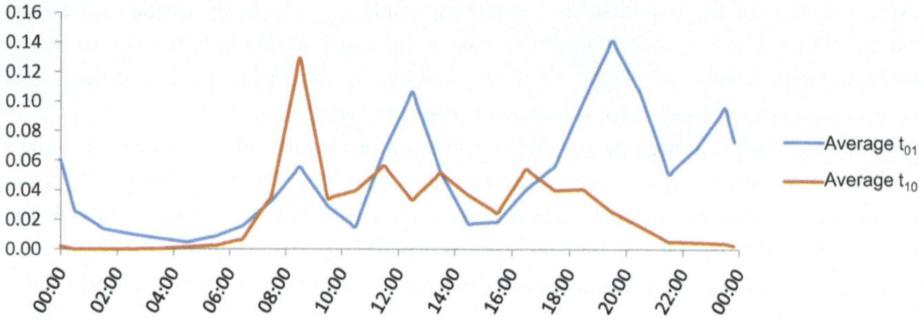

Fig. 3 Average over the entire survey population of the values of t_{01} and t_{10} with a resolution of 10 min

2.2.2 Presence Duration

Disaggregation According to Individual Characteristics and Approximation by Weibull Laws

Due to the same biases that affected the transition probabilities, the distributions of the duration of the presence periods are calculated for hourly time steps. The individual characteristics and the day of the week are taken into account using a recursive method which has similarities to the method used to select the variables influencing the transition probabilities. From the survey, for a given hour, the number of arrivals during that hour is recorded as well as each of the associated presence duration. All the possible divisions into two subgroups, relative to the set of binary variables describing the individual and the day, are tested by comparing the average attendance times of each of the two subgroups (*Z*-tests). The division resulting in the greatest difference between the mean durations of the subgroups is kept, and the operation is repeated within each subgroup until no further distinction can be made with a risk of error smaller than 5%. Thus, the number of final subgroups varies from hour to hour. For example, it is 91 between 12 a.m. and 1 a.m.; it is 1 between 5 a.m. and 6 a.m. (no significant division is detected).

For each final subgroup, the distribution of the durations of presence is modelled by a Weibull law. Considering a subgroup of the hour h ($h \in [\![1;24]\!]$) associated with the pair of Weibull parameters (λ, k). For an individual and a day corresponding to this subgroup, the probability density function PDF of the durations of presence starting during the hour h (i.e. at the time step t_s with $t_s \in [\![(h-1)\times 6;(h-1)\times 6+5]\!]$) is given by:

$$f_{t_s}(t) = \frac{k}{\lambda}\left(\frac{t}{\lambda}\right)^{k-1} \exp\left(-\left(\frac{t}{\lambda}\right)^{k}\right) \quad (6)$$

The shape and scale parameters, respectively, λ and k, are determined by the maximum likelihood method. This means that their values are set in such a way as to maximise the probability of reproducing, by a Weibull law, the observed distribution.

Dealing with the Midnight Discontinuity

Due to the format of the survey, all periods of attendance which have not ended earlier are interrupted in the evening at midnight. There is every reason to believe that they should have continued beyond that; they are qualified as "right-truncated".

In order to account for this discontinuity, the distribution of durations is corrected based on the assumption that the durations interrupted at midnight can be replaced by the sum of the duration remaining until midnight and a new duration which would begin at midnight the next day. For set of characteristics x,denoting $\rho_{t_s}(x)$ the percentage of durations starting at t_s that were truncated, $f_{t_s,u}(x,t)$ the PDF of the non-truncated durations starting at t_s, and $f_1(x,t)$ the PDF of the durations starting during the first hour of the day, the corrected PDF is given by:

$$f_{t_s}(x,t) = \left(1 - \rho_{t_s}(x)\right) f_{t_s,u}(x,t) + \rho_{t_s}(x) f_1\left(x, t - (24\text{h} - t_s)\right) \quad (7)$$

The characteristics are structured by dissociating the criteria linked to the individual and those linked to the day in order to be able to calculate weekly presence profiles. Therefore, the PDF $f_1(x,t)$ of the next day can be used (e.g. to a duration which starts on Friday evening and which is censored, is added a duration calculated from the distribution of durations from Saturday at 0:00 and not from Friday at midnight).

Binary Trees

The distributions of the duration of presence depend on the time of day, the day of the week and the characteristics of each individual. They are characterised by their parameters k and λ (Weibull laws). A classification of the different distributions in the form of a binary tree for each hour of the day follows logically from the procedure of decomposition into successive subgroups described above. As a reminder, the most significant division into two subgroups (with respect to a binary variable) is sought for each hour. The operation is repeated within each subgroup. For example, the most significant division separates Sunday from the other days of the week and then within the subgroup which includes the "other days of the week" the most significant division splits the population between retirees and non-retirees, etc.

In the binary tree structure used, the nodes are associated with:

- A possible value of a binary variable.
- A "left child" and a "right child" which each represents one of the two values of the same binary variable (which can have child nodes, and so on).

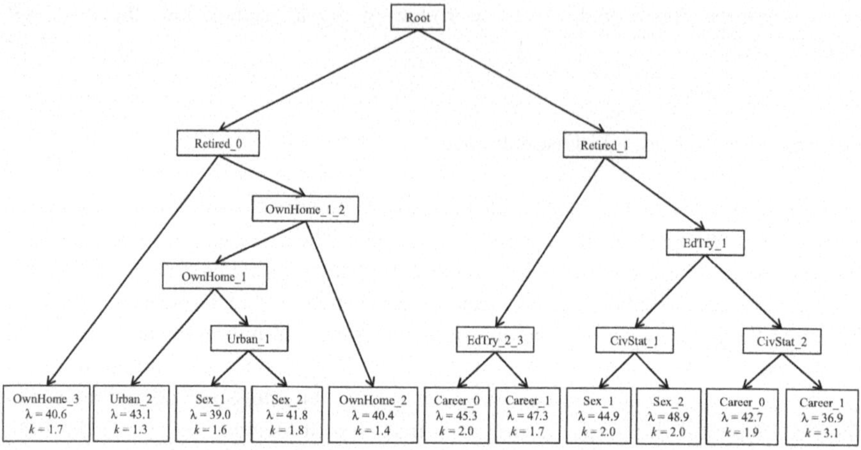

Fig. 4 Presence durations binary tree at noon

The last nodes, the only ones that have no children, are called "leaves" and correspond to the final subgroups of the differentiation process. They are associated with the parameters of the Weibull law of the distributions of the durations of presence corresponding to this subgroup.

Figure 4 shows the binary tree of the durations of presence which start at noon. The names of the labels correspond to the variables that characterise the individuals as presented. The tree given as an example reads like this: if an occupant arrives in his dwelling at noon, the time he will stay in his home depends first of all on whether or not he is retired. If he is (Retired_1), the next main distinction is his level of education. If he does not have a high school diploma or equivalent (EdTry_1), the following distinction is made according to his civil situation (living with a partner or not). If he lives as a couple (CivStat_1), the most representative distinction separates men and women. Within the subgroups thus formed ("retired, no diploma, couple, man" and "retired, no diploma, couple, woman"), no distinction is significant, they can no longer be subdivided. If the occupant is a man (Sex_1), the period of presence which begins is determined by the ITM on a Weibull law with parameters $\lambda = 44.9$ and $k = 2.0$.

Some distinctions are not necessarily the most intuitive, but one must not lose sight of the existence of multi-collinearities between the characteristics of an individual. In the example, the level of education of a retired person is certainly correlated with his standard of living, his state of health, his leisure activities, etc. Thus, the fact that this variable is the most significant to divide a group means that it "contains" the most information, this information going beyond the variable itself if it were considered alone.

Computer Implementation

The model was developed in Delphi environment. It comprises 24 trees for the durations of presence and 20 × 24 trees for the durations of the 20 activities. Each of them is built at the start of the simulation (common step to all occupants) from the corresponding text file, according to an ascending grouping method: the algorithm creates a new node when it identifies two nodes which constitute the left and right children of the same father until reaching the initial node.

2.2.3 First-Order Model Selection

High-order model tends to perform better than first-order model regarding the durations of presence (even if the Weibull laws present limits, for example, to take into account the bimodal character of the distribution of the durations at certain times). However, the first-order model is better at predicting the probabilities of presence over all the time steps at certain hours, especially at night. The zero-order model is inferior to the two Markov models in terms of both duration and profile prediction.

Figure 5 compares the mean profile from the survey and the mean profiles from the simulation with the Markov models (first-order FOMP and high-order HOMP). For each individual in the survey, 1 day was simulated (i.e. 15,441 days), considering its characteristics and the day of the week. Although the statistical quality indicators show similar performances, it appears that the first-order model (FOMP) generates presence profiles that are closer on average to the results of the survey. In particular, the night presence is better represented. There are two main reasons for this:

- Respondents begin to fill in activity diaries at midnight, generally with the "sleep" activity; this results in an underestimation of arrivals in the early hours of the day.
- The representation of durations by Weibull's laws in the HOMP model leads to predictions of durations that are too short for the presences which start at the first hour of the day.

Presence is underestimated in the morning by the first-order model while it is overestimated in the afternoon by the high-order model. In general, it seems that an underestimation of presence is less of a problem than an overestimation of the same magnitude. Indeed, the individuals of the same household being simulated completely independently, the model probably naturally tends to overestimate the total duration during which the dwelling—inhabited by several people—is occupied, due to the overlap of the periods of presence.

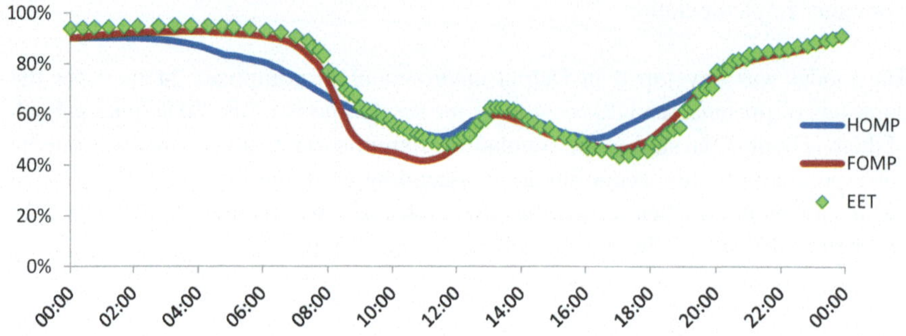

Fig. 5 Comparison of the average presence profiles for all TUS individuals, generated by the HOMP (blue) and FOMP (red) models and measured by the survey (green)

Note

With regard to the modelling of activities, it is necessary to proceed according to a principle equivalent to that of the high-order presence model, i.e. calculating the duration of a period when it begins. Indeed, the choice no longer concerns only two states (present/absent) but 20 states, and certain choices are unlikely. A model which would work according to a Markov principle of order one, choosing an activity at each time step, would tend to quickly interrupt the unlikely activities and to homogenise the series. By calculating the duration of actions, the scarcity of an activity at a certain period is reflected only by its probability of occurrence and not by its duration. In addition, the inability of Weibull's laws to represent bimodal distributions is less problematic in the case of activities. Indeed, for the presence model, an arrival at 1 p.m., for example, could be associated with two main modes, the first corresponding to "going home for lunch break" followed by a departure at the beginning of the afternoon (i.e. a mode centred on a duration of around 1 h) and a second mode corresponding to "going home for the night" (i.e. centred over a duration of around 18 h). There is no equivalent case with regard to activities.

2.2.4 Presence Model Results

In Fig. 6, the presence profile of the French 2012 standard (RT2012) calculation method is compared to the presence profile generated by the model (FOMP) for the 15,441 individuals in the survey. Each individual is simulated for a full week. It appears that the regulatory scenario greatly underestimates the presence rate during the day for the weekdays (which in reality hardly ever drops below the 40% mark) and just as greatly overestimates it on weekends. Moreover, the model results indicate that weekend days (on average over the 15,441 individuals) are not that different from weekdays. The presence rate is generally lower in the evening and at night while it is higher in the morning and afternoon, but the occupants do not remain confined to their homes. The presence rate never reaches 100% but tops out

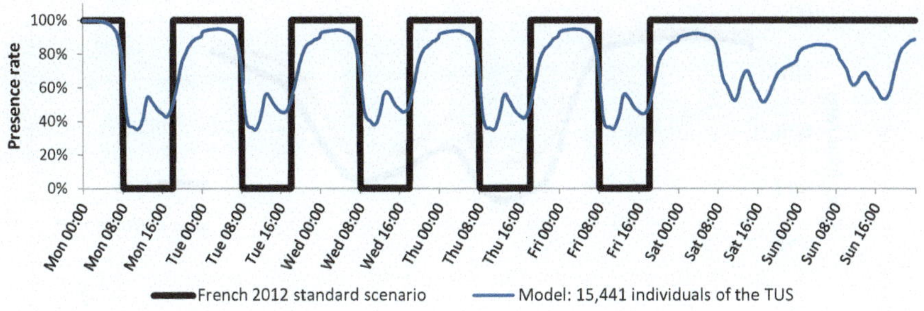

Fig. 6 Weekly presence profiles generated for the 15,441 individuals in the survey by the model (in red), and according to the French 2012 standard scenario (in black)

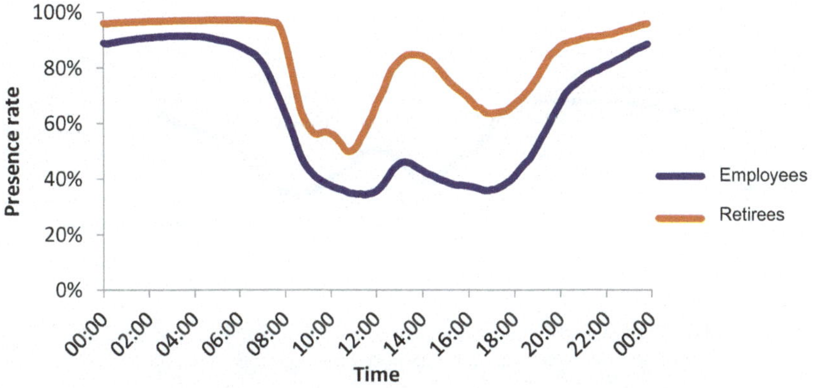

Fig. 7 Weekly presence profiles generated by the model for the employees (in purple), and the retired (in orange)

at around 90% during the night. This is in part due to the fact that some people do not sleep at home or work at night.[2]

The following figures (Figs. 7, 8, 9, and 10) present the average daily profiles generated by the model for different subcategories of the survey: employees–retirees (Fig. 7), men–women (Fig. 8), weekdays–week-ends (Fig. 9), urban–rural area (Fig. 10).

[2]In France, 15% of employees (i.e. 3.5 million people) work at night (between midnight and 5 a.m. Source: http://travail-emploi.gouv.fr/IMG/pdf/2014-062.pdf

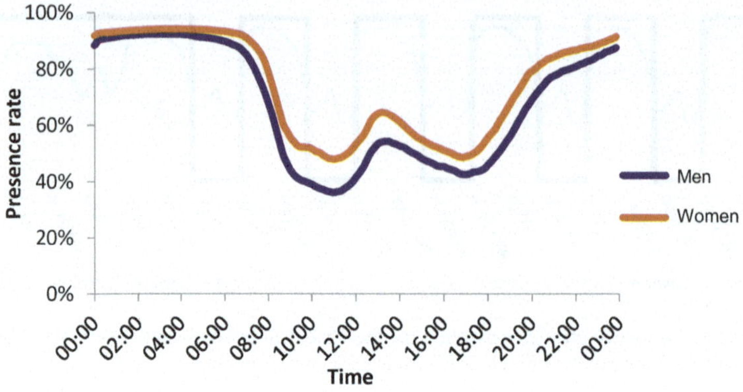

Fig. 8 Weekly presence profiles generated by the model for the men (in purple), and the women (in orange)

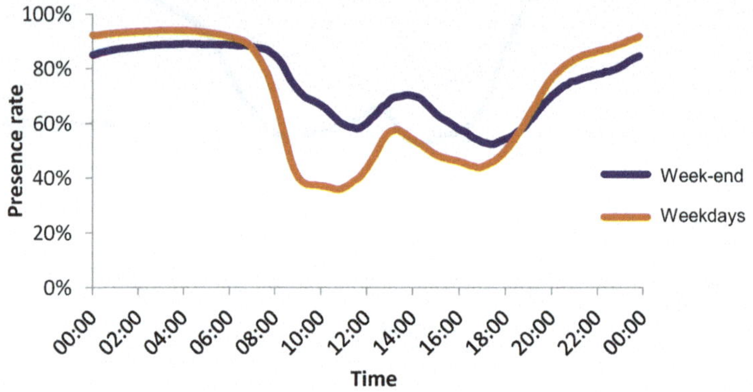

Fig. 9 Weekly presence profiles generated by the model for the week-ends (in purple), and the weekdays (in orange)

2.3 *Residential Activities Modelling*

The activity prediction model developed by Wilke et al. [28] works as follows. For each occupant, the activities are modelled after his presence profile is established. When a period of presence begins, an activity also begins, and a duration is assigned to it. If the activity ends before the occupant leaves, a second activity starts. As soon as the occupant is absent, the activity in progress is interrupted. The process is resumed when the occupant returns home. The selection of the activity is carried out by a multinomial logit model which integrates the day of the week and the characteristics of the individual. The duration of the activity is calculated by the ITM from the distribution of the durations of this activity at the time considered, which also depends on the day of the week and the characteristics of the individual.

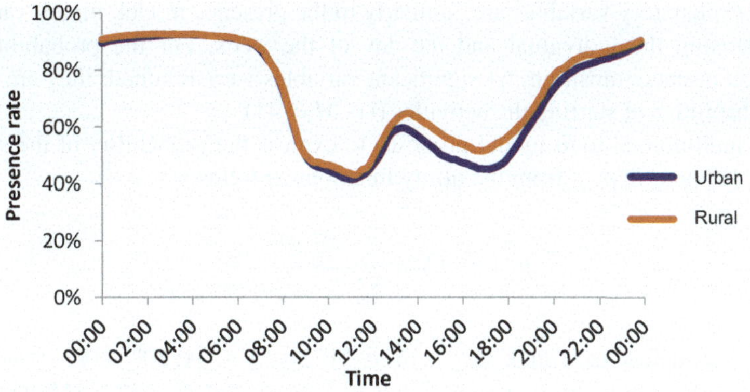

Fig. 10 Weekly presence profiles generated by the model for the urban areas (in purple), and the rural areas (in orange)

Note

Several methods used in this part are inspired by research in the field of economics/econometrics [29] but especially by research in the field of transport and mobility. Discrete choice models have been the subject of important developments and are commonly used to predict, for example, modal choices or journeys as a function of time and of household characteristics. In particular, Ben-Akiva and Lerman [30], Ben-Akiva and Bierlaire [31] and Bierlaire et al. [32] studied the theory on discrete choice models and applied it to the field of transport. Zhang et al. [33] used the same type of techniques based on TUS data.

2.3.1 Activities' Probabilities

Multinomial Logit Model

The expressions of the transition probabilities t_{01} and t_{10} as a function of the explanatory variables (individual characteristics and day of the week) were obtained by means of logistic regressions. They are in fact binary variables: the occupant changes state or remains in his current state. For activities, this involves evaluating the probability one activity starts out of the 20 possible ones. The corresponding model is a multinomial discrete choice model. The influence of the explanatory variables is captured by a random utility model (RUM). In order to take into account the dependence of the choice probabilities of activities on the time of day, 24 random utility models have been developed. For each hour h, the utility functions $V_{h,j}$ of the activities j ($j \in [\![1; 20]\!]$) are expressed by linear functions of the explanatory variables x_i, $i \in [\![1; M]\!]$:

$$\mathrm{V_{h,j}}\,(\mathrm{x}) = \alpha_{\mathrm{h,j}} + \beta_{\mathrm{h,j,1}}\mathrm{x}_1 + \beta_{\mathrm{h,j,2}}\mathrm{x}_2 + \cdots + \beta_{\mathrm{h,j,M}}\mathrm{x}_\mathrm{M} \quad (8)$$

The explanatory variables are, similarly to the presence model, binary variables characterising the individual and the day of the week. For the probabilities of presence/absence transition, 64 significant variables were retained; they are 41 for the probabilities of starting the activities (i.e. $M = 41$).

The multinomial logit model proposes to express the probability of the start of activity j at hour h, $p_{s,j}$, from the utility functions as follows:

$$p_{s,j}(x,h) = \frac{e^{V_{h,j}(x)}}{\sum_{j'=1}^{20} e^{V_{h,j'}(x)}} \tag{9}$$

The parameters $\alpha_{h,j}$ and $\beta_{h,j,i}$, ($h \in [\![1;24]\!]$, $j \in [\![1;20]\!]$, $i \in [\![1;41]\!]$) are calculated using the free software Biogeme[3] developed by EPFL (Michel [34]) for the estimation of discrete choice models. The parameters are set to maximise the likelihood. The existence of several optimal solutions requires that certain parameters be fixed.

Eliminating Non-influential Parameters and Model Variants

24 models linking 20 activities to a hundred variables represent a total of several tens of thousands of parameters. Somehow, a large number of parameters can be eliminated. First of all, all the estimated β coefficients whose significance is insufficient are discarded.

The significance of each coefficient is evaluated using a Student test. This test verifies whether the null hypothesis (H0), namely that the regression coefficient is equal to zero, can be rejected and this with a desired degree of confidence (generally 90, 95 or 99%). The test is based on the calculation of the following statistic: $t = \frac{\hat{\beta}}{S_{\hat{\beta}}}$ (ratio of the estimated value of the parameter by the estimated standard deviation of this parameter). If the null hypothesis is true (and under certain conditions, notably the normal distribution of the residuals of the regression), the test statistic follows a Student's law with n-k-1 degrees of freedom (where n is the number of observations and k the number of explanatory variables). The probability of observing a value greater than t under the hypothesis H0 (called p-value) is therefore known and can be compared with a chosen threshold α (e.g. 5%) corresponding to a risk of error. If the p-value is less than this threshold, then the H0 hypothesis can be discarded with the desired degree of confidence (e.g. 95%) and the regression coefficient $\hat{\beta}$ is kept in the model. In the opposite case (p-value$> \alpha$), even if the test does not strictly allow to validate the hypothesis H0, the coefficient $\hat{\beta}$ is considered to be equal to zero.

An additional step is proposed by Wilke et al. [28] in order to retroactively eliminate some of the parameters selected in the previous step. Let $k \in \mathbb{N}$ be

[3] http://biogeme.epfl.ch/

the number of parameters in the model, the k models with $k-1$ parameters are compared with the initial model with k parameters by a likelihood[4] ratio test. If there is at least one $k-1$ parameters model which cannot be considered to be significantly different from the initial model, then the $k-1$ parameter models with the lowest likelihood ratio replaces the initial model and the operation is repeated with the $k-1$ models with $k-2$ parameters. This process was carried out with two values of the threshold α (5 and 10%), leading to the creation of two new models.

The three models presented above are noted S5, S10 and S100 (S: Starting, followed by the threshold value). Not applying the process is equivalent to its application with a threshold of 100%.

Two additional models of the probabilities of starting activities are proposed, bringing the total to five. The fourth model integrates an additional effect: the influence of the previous activity j on the starting activity j' (e.g. the probability of washing the dishes is higher if the occupant finished his meal). This can be achieved by adding, for all combinations (j,j'), additional terms $\beta_{j',j}x_j$ to the utility functions, x_j being the binary variable indicating that the previous activity was j. In order not to avoid unnecessary burden, these terms are included only if, at the time considered, the probability that activity j' follows activity j is significantly different (Z-test) from the probability that j' starts independently of previous activity. Furthermore, if the number of occurrences of the sequence $j \rightarrow j'$ is less than five over the entire survey, the terms $\beta_{j',j}$ are not added. The integration of the influence of the previous activity constitutes somehow a Markov property. Consequently, this model is denoted SM (Starting Markov).

The size of the survey is sometimes insufficient to correctly capture the influence of variables on certain entry probabilities. This problem was treated by merging at each time step the activities, qualified as "small", having started less than 50 times at this time step in the survey. An activity $\tilde{j}$ which encompasses these activities is created at each time step. The activities that it groups together do not have their own starting probability. A starting probability is assigned to $\tilde{j}$ for this hour. It is calculated from the non-disaggregated survey results. If the stochastic process starts activity $\tilde{j}$, activity j included in $\tilde{j}$ is chosen relative to the proportion of j in $\tilde{j}$.

The evaluation of the models which justified our choice is carried out on the complete models which include, in addition to the starting probabilities, the duration distributions. The modelling of the latter is the subject of the next paragraph.

Note

Another model corresponding to a calibration on aggregated data was also built. It will be noted SG (Starting Generic) when comparing the performance of the models. The probability of starting activity j is calculated directly for each hour on the entire

[4]The likelihood ratio, in case of equality of models, follows a χ^2 law to a degree of freedom. The probability of obtaining the calculated ratio (p-value) can therefore be compared to the risk of error α that was set.

survey population. It is defined as the ratio of the number of starts of activity j to the total number of activities that started at the same time.

2.3.2 Activities' Duration

The modelling of activities' durations is carried out in a very similar way to that of the periods of presence (§ 2.2), each of the 20 activities being associated with 24 distributions of durations approximated by Weibull laws (6).

Distributions of Non-individualised Durations

A specific treatment is necessary when there are few occurrences of certain activities at certain times (mainly at night). In this case, the distributions are not tailored according to the characteristics of the individuals. Assuming that the distribution of durations does not change abruptly from 1 h to another, the applied distribution can be calculated by grouping together several adjacent time steps for more significance. This is carried out after making sure that the distributions calculated independently of the two time steps are not significantly different (application of the Z-test to the parameters λ and k of the two laws).

Out of the $20 \times 24 = 480$ distributions of activities' durations, this concerned 108 out of which 94 correspond to hours between midnight and 8 a.m. For the others, the number of activities' starts on the time step was sufficient to study the influence of individual characteristics on the distribution of the durations.

Disaggregation of Distributions According to Individual Characteristics

When possible, the model attempts to capture the influence of the day of the week and individual characteristics on the activity duration distributions. The same methodology as the one used for the modelling of the presence durations is applied (§ 2.2.2). The start of activities plays a role equivalent to the arrivals. The binary trees structure of the data is identical. There were 24 binary trees containing the parameters of the Weibull laws of the presence durations; there are 480 binary trees for the activities' durations of which 108 are composed of a single leaf.

In order to assess the interest of the disaggregation of durations, a model composed simply of non-individualised durations was included in the evaluation. It is noted DG for "Duration PDFs modelled Generically". The model comprising the disaggregated durations is denoted DI for "Duration Individual-specific".

2.3.3 Evaluation of the Models and Selection

The five variants of starting probabilities modelling combined with the two variants of duration modelling constitute a total of ten "complete" models. Their respective performances were evaluated by comparing their predictions to the observations of the survey. To do this, Wilke [35] suggested reasoning on two indicators.

The first characterises the average relative deviation between forecasts and observations for all activities over the total aggregate population. It is characterised by the parameter D (Deviation):

$$D = \frac{1}{\overline{p}} \frac{1}{N.t_{\text{end}}} \sum_{j=1}^{N} \sum_{t=1}^{t_{\text{end}}} \left| p_{j,\text{sim}}(t) - p_{j,\text{obs}}(t) \right| \tag{10}$$

with N the number of activities (equal to 20); $t_{\text{end}} = 24$ because the simulated and observed probabilities are averaged hour by hour to avoid the rounding bias of the survey (§ 2.2.1); $\overline{p} = \frac{1}{N}$, the average probability of doing one of the N activities; $p_{j,\text{sim}}(t)$ and $p_{j,\text{obs}}(t)$, respectively, the simulated and observed probabilities of doing activity j at time step t. The best model with regard to this criterion is the one with the smallest D value.

The second indicator (noted A for Activity) is the percentage of time steps for which the predicted activity is exact. It consists in comparing the series of activities generated and observed at the level of the individual. There is a single indicator for the entire population: the number of time steps for which the activity is correctly planned over the number of time steps of presence at home. The ratio is therefore 1 if the predictions are perfect and 0 if the predictions and observations are discordant over all the time steps. Due to its length, the "sleep" activity was excluded from the calculation in a variant of the indicator denoted A_{ns} ("A no sleep"). The model that best predicts the series of activities is the one with the highest A and A_{ns} values.

Two validation processes were carried out with these indicators. Firstly, a cross validation was performed. Each of the ten versions of the model was calibrated on a random sample made up of one tenth of the individuals in the survey, then its predictions were evaluated against the observations on the remaining nine tenths. Secondly, the models were calibrated and tested on all the individuals in the survey. The indicators were also calculated for subsets of the population, for example, distinguishing between individuals according to whether or not they have a job.

From the results of these multiple evaluations, detailed in Table 1, the SMDI (Starting Markov, Durations Individualised) model was selected. In this model, the calculation of the starting probabilities takes into account the previous activity, and the durations' distributions depend on individual characteristics. This model obtained on the whole the best results for the various evaluations. In fact, this is the most refined model so these results are quite consistent. The better precision of this model justifies its higher level of complexity in terms of implementation time and calculation time.

Table 1 Comparison of the performance of different activity prediction models, source (Urs [35])

Model	Whole population			Cross-validation			Sub-population	
	D	A	A_{ns}	D	A	A_{ns}	$D(C_w)$	$D(C_{nw})$
SGDG	10.10	39.11	14.83	12.72 ± 0.34	39.09 ± 0.15	14.81 ± 0.07	16.17	14.28
S5DG	10.22	40.35	16.48	12.58 ± 0.36	40.33 ± 0.20	16.43 ± 0.09	11.22	11.14
S10DG	10.19	40.44	16.54	12.57 ± 0.35	40.38 ± 0.16	16.48 ± 0.08	11.20	11.04
S100DG	10.29	40.71	16.87	12.59 ± 0.47	40.62 ± 0.19	16.75 ± 0.11	10.54	11.83
SGDI	9.01	40.18	14.92	11.48 ± 0.46	40.86 ± 0.18	14.89 ± 0.07	15.03	13.52
S5DI	8.44	41.47	16.66	11.22 ± 0.41	42.07 ± 0.21	16.54 ± 0.08	9.09	10.08
S10DI	8.39	41.55	16.73	11.18 ± 0.40	42.12 ± 0.20	16.61 ± 0.08	9.12	9.99
S100DI	8.41	41.80	17.03	11.14 ± 0.44	42.32 ± 0.22	16.86 ± 0.08	8.77	9.91
SMDG	9.06	41.48	16.79	12.42 ± 0.38	40.42 ± 0.16	16.54 ± 0.11	9.73	10.71
SMDI	8.15	42.04	17.19	10.79 ± 0.44	41.89 ± 0.29	16.91 ± 0.10	9.54	8.83

SG and SM stand for "Starting Generic" and "Starting Markov"; the number (5, 10 or 100) corresponds to the risk of error in the procedure eliminating non-significant parameters; DG and DI stand for "Duration–Generic" and "Duration–Individualised"

2.4 *Presence and Activities Simulation Results*

2.4.1 Algorithm

The algorithm generates randomly a weekly presence and activities profile for all occupants of a dwelling (it can be part of a larger building). With a resolution of 10 min, each occupant is attributed a state of presence/absence and, in case of presence, an activity. The same week is then reproduced throughout the year. Several reasons justify this decision:

- The month of the year, recorded in the TUS was not found to have a significant influence on the presence and activities profiles.
- The calculation is only performed once instead of 52.
- Intuitively, the timetables must be relatively similar from 1 week to another (same days and hours of work, repetition of leisure activities on the same days at the same hours, etc.).

Note
Due to the non-continuity of the survey at midnight, it was necessary to stop activities every evening at midnight. Otherwise, the activities in progress logically continued beyond that time and the "sleep" activity was greatly underestimated. Indeed, the survey bias created an over-concentration of the start of sleep at the first time step of the day. When this time step was exceeded, the probability of starting a sleep phase during the following time steps was low and the "sleep" activity was clearly under-represented.

The complete algorithm is described in Fig. 11 for the FOMP variant which is the one that will be used. The dark blue box (in the top) corresponds to the initialisation

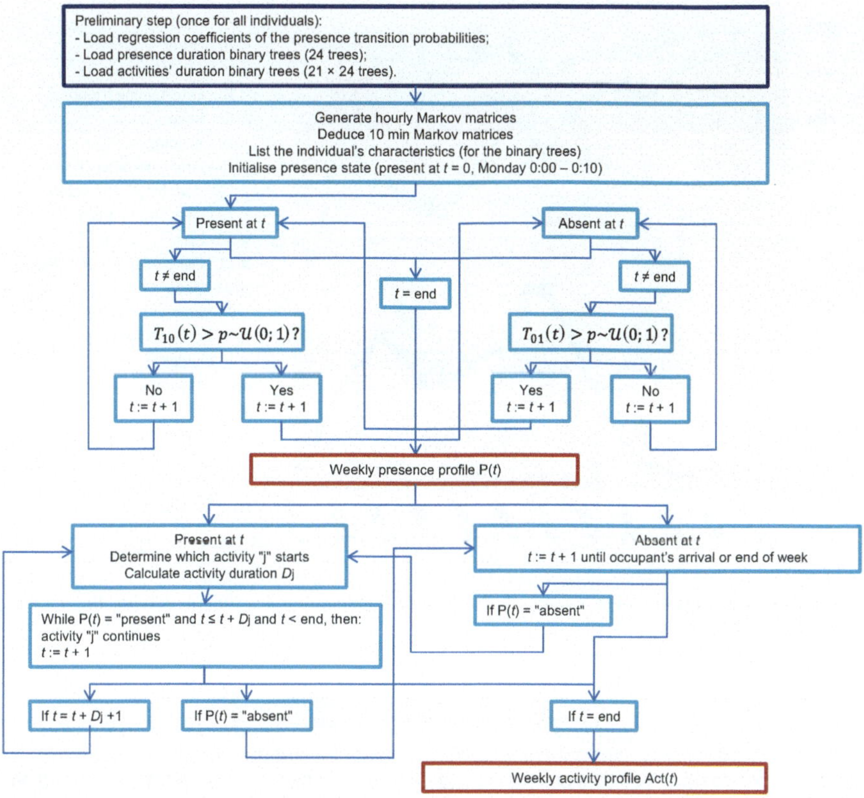

Fig. 11 Algorithmic diagram of the presence and activities model in dwellings with the FOMP variant of the presence model

step (carried out once for all occupants) and the sky blue boxes correspond to the steps reproduced for each occupant. The algorithm first generates the presence scenario $P(t)$, $t \in [\![0, 1007]\!]$, and then the activity scenario $Act(t)$, $t \in [\![0, 1007]\!]$.

2.4.2 Simulation Results

Figure 12 shows the daily activities profile (when the occupants are present in their home) averaged for the 15,441 individuals in the survey. The profile corresponds well to the TUS. It is also in agreement with common sense. The most important "sleep" activity is largely at night and decreases rapidly in the morning. It is also found at the beginning of the afternoon in a much more attenuated way. Meals, as well as the "kitchen-dishwashing" activity, are grouped around noon and 8 p.m. The "television" activity is mainly represented in the evening and therefore undergoes interruptions that are doubtless inconsistent at midnight (due to the continuity problem). The "dressing-shower-toilet" activity takes place rather at the start and

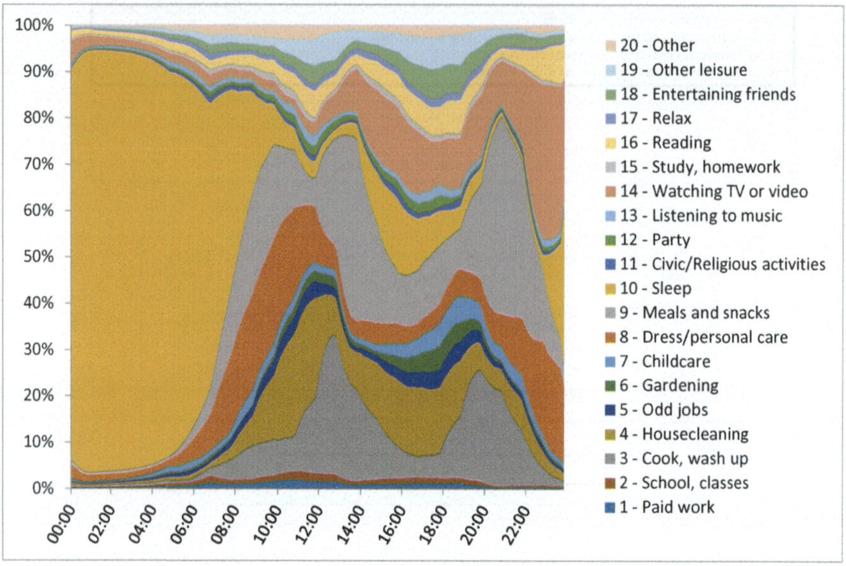

Fig. 12 Average profile over 1 day of the activities generated by the model for the 15,441 individuals in the survey

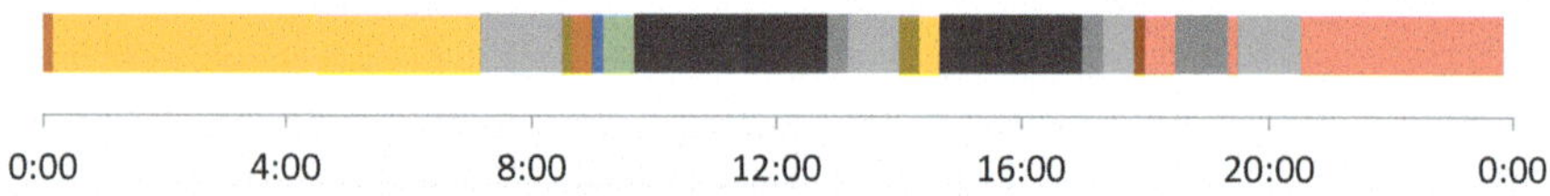

Fig. 13 Example of an individual's daily activities scenario; the colour legend is the same as Fig. 12

end of the day. The other activities are either more marginal or distributed more evenly throughout the day.

A daily scenario generated by the model is presented in Fig. 13. It is a Monday, the individual is a 51-year-old woman living in a couple without children, owner of a home in an urban area, whose household is in the third quartile of the population in terms of income, not having a diploma, employed full time, working between 21 and 40 h per week, not suffering from disability, having no person to help on a daily basis, with a level of health ranging between good and very good, having a computer at home and not owning a car.

Figure 14 shows the average of the weekly profiles. It contains both information on absence and activities during periods of presence. The survey bias generating a discontinuity at midnight is clearly visible. This discontinuity, present in the survey records, is logically reproduced by the model. Figures 15 and 16 compare the presence and activity profiles averaged over a day of two subgroups of the survey, retirees (3791 individuals) and full-time employees (6359 individuals).

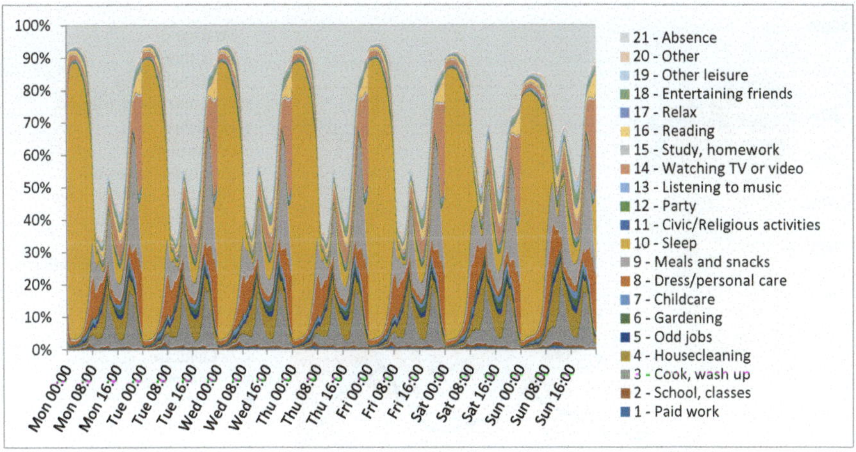

Fig. 14 Simulated weekly profiles of presence and activities of the 15,441 individuals from the 1999 TUS

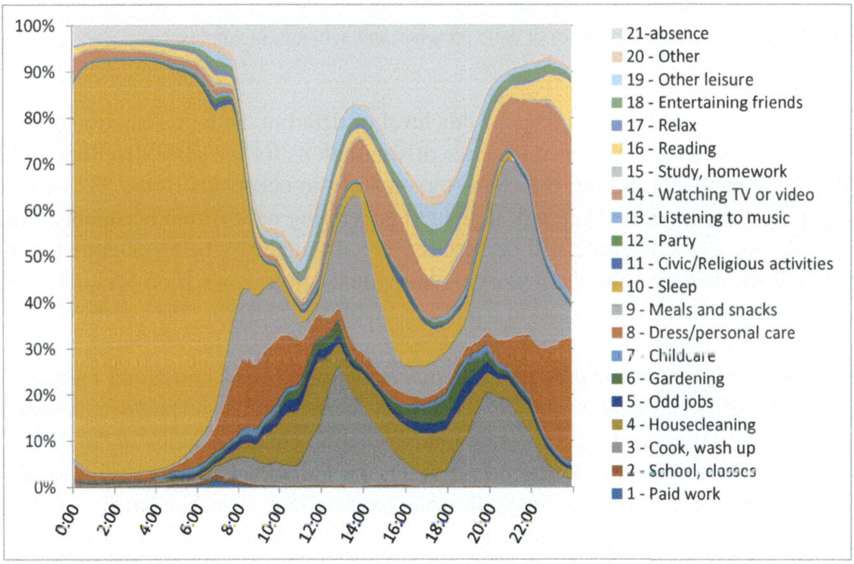

Fig. 15 Average simulated profiles of daily presence and activities of retirees

2.5 Creation of a Household

2.5.1 Introduction

The implemented model makes it possible to take into account at a very fine level the influence of socio-demographic parameters of individuals on their timetables.

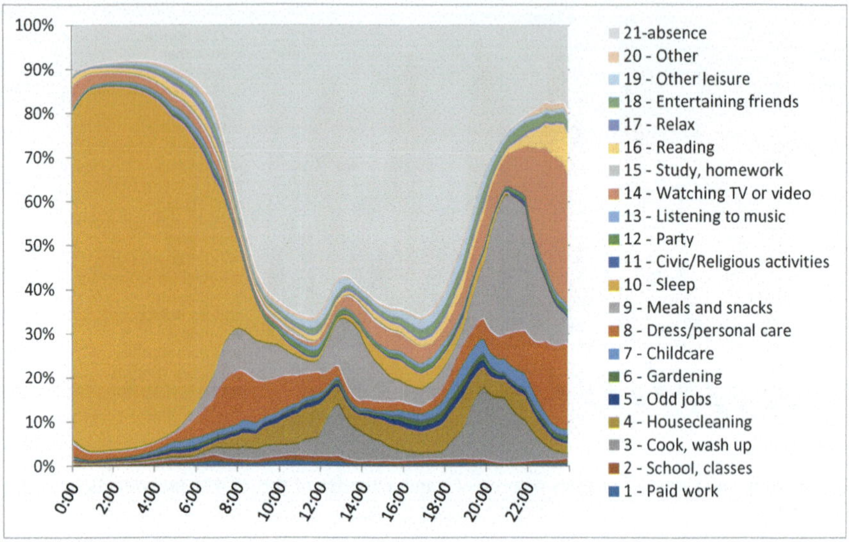

Fig. 16 Average simulated profiles of daily presence and activities of employees

Our objective is to take advantage of this level of detail in the representation of the diversity of occupants to carry out series of simulations for the same dwelling with different inhabitants. So far, the characteristics of the occupants of the TUS were used to test the models. They will now help generating new virtual occupants. For example, the same dwelling will be successively simulated by being occupied by a large family, then by a couple of working individuals, then by a retired person living alone, etc. The next step is to build a model capable of answering the question: who can live in the dwelling?

As it stands, the occupancy model requires the number of household members and their individual characteristics. This step is not feasible in practice in most cases, because users of DBES tools do not have sufficient information in the design phase. The rare models that take into account the individual characteristics of the occupants propose simple archetypal arrangements (e.g. "working man"), based on purely intuitive characteristics. The variety of situations is only very partially reproduced; moreover, the detailed composition of the household is left to the user. However, a large amount of data is necessary to accurately represent occupants.

The probability of a dwelling being occupied by one type of household or another depends strongly on the its characteristics. A house of 150 m^2 will more likely be inhabited by a large household than a studio apartment of 20 m^2. With the same living area, a dwelling will on average be more densely populated in urban areas than in rural areas. A large apartment in an urban area is unlikely to be occupied by an individual living alone with a very low income, etc. In order to carry out successive simulations by varying occupancy, it is necessary to generate types of households according to realistic probabilistic characteristics.

The presence and activities model described above requires, for all the members of the households, a number of socio-demographic characteristics. These are strongly intercorrelated. Age, marital status, age of children, employment status, income, retired or student status . . . , obviously cannot be determined independently. It is therefore necessary to ensure the good coherence between these characteristics when creating an individual.

The characteristics of the different members of a household must be realistic at the individual level but also at the household level. Some are common to all its members, such as household income, possession of a vehicle or the age of the youngest child. Others are specific to the dwelling and therefore characterise the household, such as location (urban/rural) or ownership status (tenant or owner). For the rest, the characteristics of the members of the same household must be consistent with each other (e.g. if an individual in the household is reported as living in a couple or with a child under 5 years old, the spouse or child in question must be created in the household).

These three objectives of consistency (dwelling, individual characteristics, household characteristics) were met by using statistics from INSEE[5] (*Institut national de la statistique et des études économiques*) from the 2010 Population Census and associated surveys, such as the 2010 Housing Survey and the 2010 Household Survey. For certain variables that are strongly correlated and for which no usable statistics were found, the population sample of the 1999 TUS survey was used.

2.5.2 Household Model Description

The model developed by Vorger [23] was designed with minimal required inputs while leaving the possibility of integrating as much information as possible (optional inputs). The thermal zones of the building model are grouped to form housing zones, office zones or other types of zones. For each housing zone, an object is created that will contain the information about the zone. Housing zones are grouped into dwellings.

Each "housing zone" inherits parameters from the thermal zone (DBES model) such as the floor area. It is also associated with:

- A type (e.g. "bathroom", "kitchen", "kitchen" + "living room" + "bedroom", etc.) which will make it possible to locate the occupants and place the household electrical equipment.
- A "household" which aims at establishing the link between a space and the inhabitants who will be attached to it. The areas associated with this object are considered to be fully included in the dwelling. A dwelling cannot encompass several households. On the other hand, a dwelling can be made up of several

[5]INSEE is the French "National Institute of Statistics and Economic Studies": https://www.insee.fr/

housing zones. This is even recommended to benefit from the precision made possible by the location of occupants and equipment. In order for the capabilities of the model to be exploited to the maximum, a relatively fine distinction between thermal zones (e.g. one room per zone) is desirable. However, the model is flexible and can deal with a dwelling with a unique zone comprising all the functions.

Information related to the dwelling need to be filled:

- The "type of dwelling", namely house, apartment or other (hostel, separate room, hotel room, etc.). This characteristic is not a parameter of the presence and activities model, but is required during the household generation procedure.
- The "location", urban–suburban or rural–semi-rural.
- The "number of rooms in the dwelling". By default, it is the number of housing zones ("living room", "kitchen" or "bedroom") attached to the dwelling.

These three parameters (type of dwelling, location and number of rooms) must be entered by the user. All subsequent parameters, on the other hand, are defined by default as "unknown" and the user is asked to specify them only if he/she has the correct information. The values of the unknown parameters are determined randomly by the household model from their PDFs. For each of the parameters listed below, the value "unknown" is proposed. It is not indicated here to lighten this presentation.

The "ownership status" parameter, which defines whether occupants are owners, tenants or free occupants of the dwelling, can sometimes be a characteristic of the dwelling (this is the case, for example, for social housing where the inhabitants are tenants). But more generally, there is reason to believe that DBES user probably has access to this information. In any case, filling in this parameter is not required.

The following parameters characterise the household and are therefore common to all its members:

- Type of household (single individual, single couple, couple + others or other type of household).
- Age of the youngest child in the household (0–4, 5–12, 13–17, 18+).
- Monthly household income (<700 €, 700–1900 €, 1900–4000 €, >4000 €).
- Possession of a motorised vehicle (none, 1, 2+).
- Possession of a computer (yes, no).

The last set of parameters that the user can enter is specific to each member of the household. It is first proposed to indicate their value for the reference occupant of the household and then, possibly, for other inhabitants. These parameters are:

- Age (between 0 and 120).
- Gender (man, woman).
- Marital status (adult between 18 and 39 years old without a minor co-resident, adult with a co-resident under 5 years old, adult with a co-resident between 5 to 17 years old and no co-resident under 5 years old, adult over 40 years old

without a minor co-resident, minor living with parents/guardians, minor living under a different or unknown arrangement).

- Single parent (yes, no).
- Civil status (not in a couple, lives with his spouse/partner).
- Employment status (full-time, part-time, employed - status unknown, not working).
- Retired (yes, no).
- Student (yes, no).
- Number of hours worked per week (1–20, 21–40,> 40).
- Level of studies (no secondary studies, secondary studies completed at *baccalauréat*[6] level, university licence degree or equivalent, higher university degree).
- General state of health (from poor to fair, from good to very good).
- Takes care of/helps (daily) a person with health problems (yes, no).
- Invalidity (yes, no).

Note

- The information entered is kept in the form of a text file in the working folder. If the Monte-Carlo method is applied, they do not have to be filled in at each iteration.
- The procedure to generate the occupants of each dwelling is stochastic. Their characteristics are drawn randomly from calibrated probability distributions. The inhabitants created are therefore different for each simulation, unless the user fills in exactly all the characteristics of the inhabitants.

The generation of occupants of a household is carried out in several steps. The concept of reference household member,[7] commonly used in economic and social statistics, plays a central role. From the attributes of the dwelling, the number of household members are determined, then the characteristics of the reference occupant, then the characteristics of his co-residents.

The procedure is described below if the user only entered the mandatory information, namely the type of accommodation and its location. The aim is to present how the different characteristics are determined as clearly as possible. In

[6]French secondary education diploma.

[7]In the population census, the reference household member is determined automatically using a rule which only takes into account the three oldest persons of the household (ranked in descending order) and considers their potential relationships:

- If there is only one person in the household, this person is the reference person.
- If the household comprises two people: if they are of different sex and identified as forming a couple, the man is the reference person; otherwise the reference person is the oldest active person, or if neither of the two is active, the oldest person.
- If the household comprises three or more people: if a couple made up of a man and a woman is identified, the man of the couple is the reference person; otherwise the reference person is the oldest active person, or if none of the three considered persons is active, the oldest person.

Source: http://www.insee.fr/fr/methodes/default.asp?page=definitions/pers-ref-menage-exp prin rrp.htm practice, the procedure is flexible and allows the user to enter any characteristic. This requires dealing with clearly contradictory or inconsistent data (e.g. a retiree cannot be full-time employed). This also requires to take into account the consequences of filling in a variable on the other variables for a given occupant and for the members of the same household (e.g. if the user specifies that an occupant is in a couple and works, then the type of household, the age of this occupant and, by extension, that of the other members must be determined accordingly). This requirement which consists in allowing the user to fill in as much information as possible generates a series of particular cases that are automatically managed (not detailed here).

2.5.3 Number of Household Members

The most relevant data on this subject come from the 2010 French Housing Survey.[8]It lists the main residences by type of dwelling, number of rooms and household size, for each municipality. Access to the disaggregated results via the INSEE site makes it possible to divide the municipalities according to whether they have more or less than 20,000 inhabitants. Any other value could be used, but this corresponds to the border between urban and rural locations in the 1999 TUS (according to which the presence and activities model is calibrated). The objective of this division is to take into account the difference in terms of occupancy density between dwellings located in urban areas and others. The data format allows a more detailed breakdown which would isolate, for example, big cities. However, it was not considered necessary to further increase the precision of the model at this point, especially since the "house/apartment" distinction is already strongly correlated with urban or rural location.

As an example, the distributions of the number occupants are shown for five-room houses and two-room apartments (Figs. 17 and 18), for which the proportion of "rural" is, respectively, 82% (out of a total of approximately 4,500,000 dwellings) and 29% (out of a total of approximately 2,900,000 dwellings). It should be noted that the separation assigns the "rural" and "urban" categories, respectively, to 16 and 11 million dwellings.

The number of rooms is between 1 and "6 and more" as is the number of people (when the number of members, determined by the ITM, is "6 or more", it is set randomly to 6, 7 or 8 with equal probabilities), and there are three types of accommodation. By adding the distinction between urban site and rural site, the number of values integrated into the model amounts to $6 \times 6 \times 3 \times 2 = 216$. They are stored in a text file which contains all the statistical data useful for the creation procedure of a virtual household.

[8]Main residences by type of dwelling, number of rooms and household size: https://www.insee.fr/fr/statistiques/2051951?sommaire=2403791&q=r%C3%A9sidences+principales+en+2010

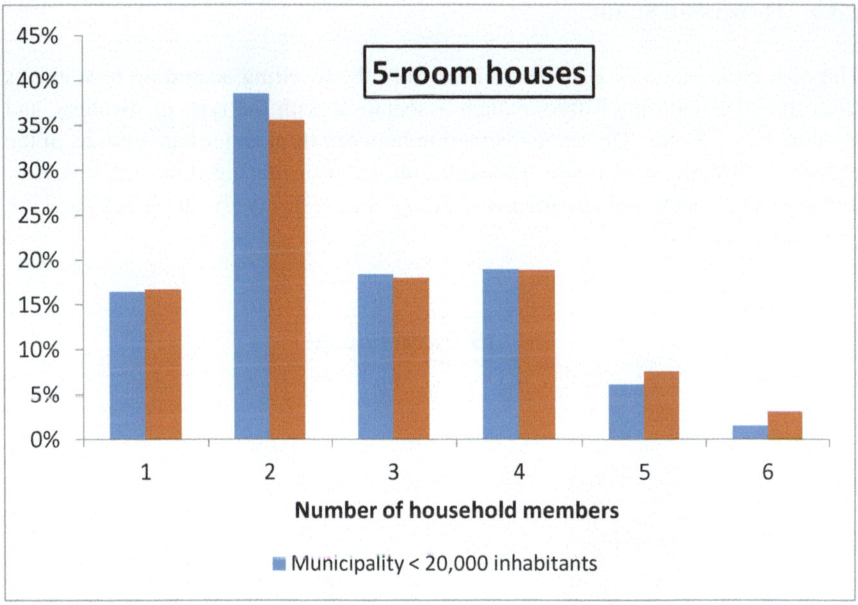

Fig. 17 Distribution of the number of household members living in 5-room houses

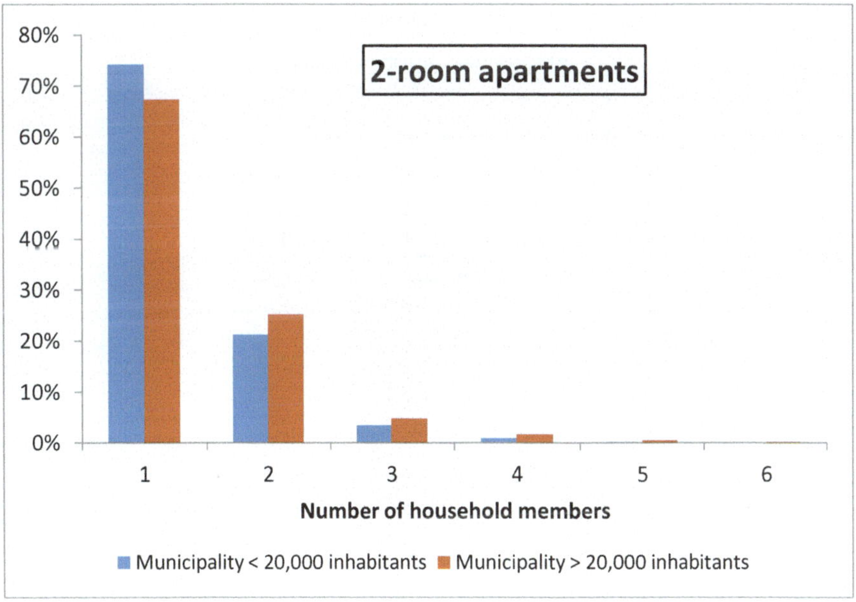

Fig. 18 Distribution of the number of household members living in 2-room apartments

2.5.4 Ownership Status

The ownership status is also determined from the dwelling according to statistics from the 2010 Housing Survey which associate it with the type of dwelling and the number of rooms. The same distinction between rural and urban areas as in the previous subsection was made. The distributions of ownership status are given for houses with 5 rooms and apartments with 2 rooms, respectively, in Figs. 19 and 20.

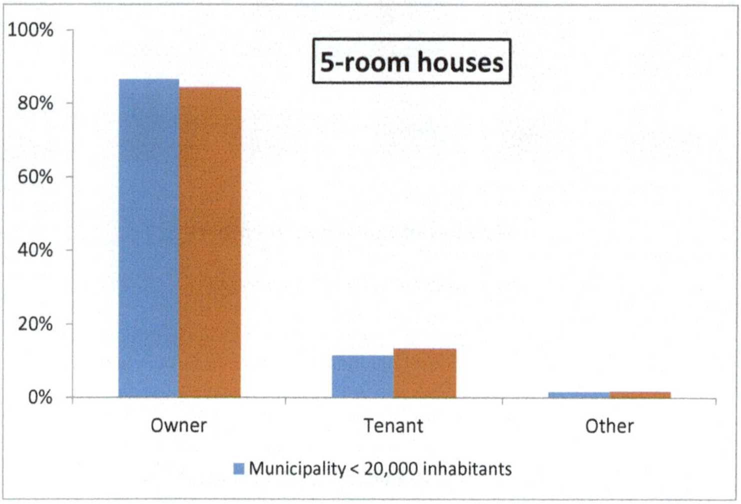

Fig. 19 Distribution of the ownership status for 5-room houses

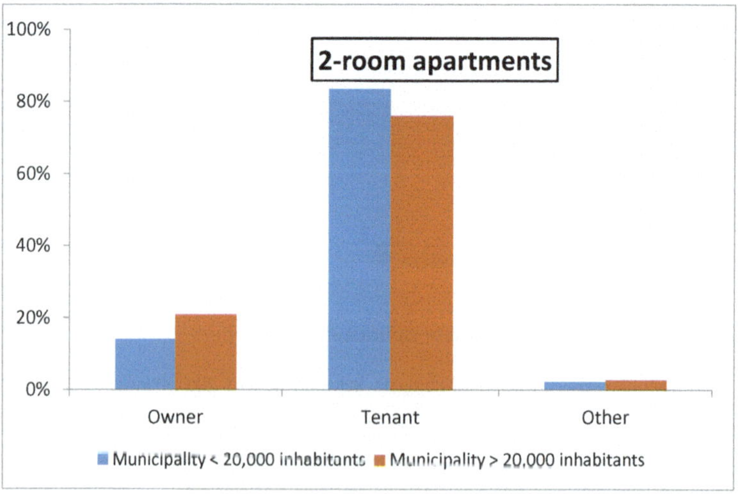

Fig. 20 Distribution of the ownership status for 2-room apartments

Note
The difference between municipalities with more and less than 20,000 inhabitants is not very large for houses, but it is not negligible in the case of apartments or dwellings in the "other" category.

2.5.5 Type of Household and Reference Household Member's Characteristics

The type of household (single individual, single couple, couple + others or other type of household) is determined according to the number of members. Several characteristics of the reference person which depend on the type of household are filled: the age of the youngest child, marital status, single parent status, gender, age and marital status. Many possible cases exist and the reader is invited to consult Vorger [23] for the details. An example for a single individual is given hereafter.

If there is only one member, the household is of the "single individual" type. In this case, the household characteristic indicating the age group of the youngest child is set to "no child". The reference person is informed as not being in a couple and not being a single parent. Its gender is randomly determined from population census statistics which indicate a proportion of 42% men and 58% women among people living alone in their accommodation. The distribution of people living alone by age groups for each of the sexes (Fig. 21) makes it possible to obtain the age group; the age is drawn randomly (according to a uniform law) from the interval. Depending on the age, the family situation is set to "adult aged between 18 and 39 years without a minor co-resident", "adult over 40 years without a minor co-resident" or "minor living with another or unknown arrangement".

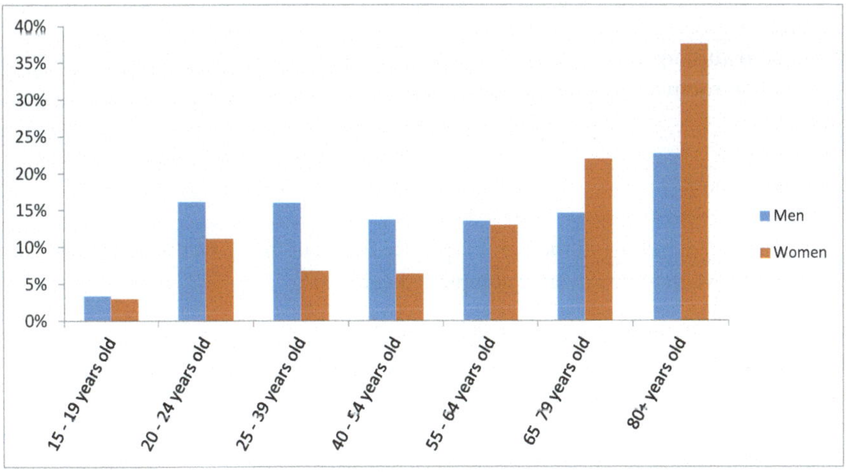

Fig. 21 Age distribution of people living alone

2.5.6 Remaining Household Member's Characteristics

At this point, the following variables, describing the household or its reference person, are known:

- Type of household.
- Age of the youngest child.
- Age.
- Gender.
- Family status.

The remaining household member's characteristics (employment status, income, education, health, vehicle, computer . . .) are determined in a similar manner [23].

2.6 Activities' Location

2.6.1 Association of Occupants and Zones

All the dwelling's zones are associated with one or more of the following types[9]:

- Living room/main room.
- Master bedroom.
- Kitchen.
- Main bathroom.
- second bedroom
- third bedroom
- Other bedroom*.
- Other living room*.
- Office*.
- second bathroom
- Other bathroom*.
- Cellar*.
- Corridor*.
- Laundry room*.
- Other*.
- T1 to T5[10] (including all the types typically found in these dwellings to speed up data input when zone covers a complete dwelling).

From this information, each member of the household is allocated five zones in which he will live: "kitchen", "bedroom", "living room", "bathroom" and "office" according to the following rules, fixed a priori.

[9]The asterisk "*" indicates zones that can appear multiple times in a dwelling.

[10]In France, T*N* refers to an apartment with *N* main rooms (living room, bedroom).

The reference person and his (or her) possible spouse use the main bedroom, bathroom and living room. The other occupants are distributed over the remaining rooms as long as they are not all assigned. If the number of rooms is insufficient, several people (especially children) occupy the same rooms. If there is no more available room for adults, an "other living area" type area or the main living area may be designated as their bedroom. There is at most one kitchen per dwelling which is therefore the kitchen for all members of the household.

Members other than the reference person (and his/her spouse) are assigned to the second, third... bathrooms (if they exist). When there is no unused bathroom, occupants are randomly assigned to those already in use.

If there are several living rooms, the occupants other than the reference occupant (and his/her spouse) are randomly assigned to one of them (including the living room/main room).

The available offices are allocated primarily to the reference person and then to their spouse. When there is no more office available, an occupant's office area is his bedroom, unless the household has only one or two members in which case the occupants' office area is the main room.

2.6.2 Association of Occupants' Activities and Zones

From the activities scenario, it is possible to locate an occupant inside the accommodation thanks to rules associating activities with his five zones. These rules are grouped together in Table 2. When the total of the probabilities for an activity does not reach 100%, it means that it sometimes take place outside the dwelling.

When an activity takes place in a zone, there is a probability that it is associated with an electrical equipment (§ 3). This results in electricity consumption and internal heat input. Moreover, metabolic heat is taken into account with default values (Table 3).

2.7 Discussion on Presence and Activities Modelling

Several limitations of the presence and activities model can be noted. First of all, the scenarios are generated independently for each individual whereas it would be more realistic to take into account interactions between household members. It is for instance likely that they will eat their meals together. The algorithm could be completed in this sense to force the occupants of a household to synchronise with the reference person. By integrating interaction rules between occupants, the model would take an "agent-oriented" tint. One can imagine, for example, that all the occupants present when the reference occupant begins to eat, interrupt their current activity to join him. It would then be necessary to count the occurrences of the "meal" activity during a day to prevent an individual from taking several

Table 2 Zones activities probabilities

Activity	Living room	Kitchen	Bedroom	Office	Bathroom
1-paid work	–	–	–	100%	–
2-school, classes	30%	–	–	70%	–
3-cook, wash up	–	100%	–	–	–
4-housecleaning	25%	25%	15%	10%	25%
5-odd jobs	25%	–	–	25%	–
6-gardening	25%	–	–	–	–
7-childcare	100%	–	–	–	–
8-dress/personal care	–	–	10%	–	90%
9-meals and snacks	25%	75%	–	–	–
10-sleep	–	–	100%	–	–
11-civic/religious activities	75%	–	25%	–	–
12-party	100%	–	–	–	–
13-listening to music	50%	–	50%	–	–
14-watching TV or video	75%	–	25%	–	–
15-study, homework[a]	–	–	–	100%	–
16-reading	25%	–	75%	–	–
17-relax	25%	–	75%	–	–
18-entertaining friends	100%	–	–	–	–
19-other leisure	50%	–	50%	–	–
20-other	25%	25%	25%	–	25%

[a]Few children have a dedicated office room. Therefore, their homework activity takes place in their bedroom or in the living room

Table 3 Metabolic heat input values

	Sleep	Wake
Age >10 years old	80 W	100 W
Age ≤10 years old	50 W	70 W

consecutive meals. Such complexity was not added to the model. More attention is paid to this synchronisation problem when modelling electrical equipment (§ 3).

The number of predicted activities is high and partly superfluous for DBES. It was kept intact because the calibration of a model comprising a reduced number of activities according to the method employed by Wilke et al. [28] would have been cumbersome. The detailed level of activities remains useful to position the occupants in the zones and to model the use of equipment (including electrical appliances).

The presence and activities model takes into account a large number of socio-demographic characteristics. Due to its stochastic nature, it generates different scenarios for individuals with identical characteristics. It would be interesting to evaluate the share of diversity respectively linked to characteristics and stochastic nature.

The data used for the construction of the model has flaws which have been addressed several times. Thus, the model reproduces certain TUS biases such as the discontinuity at midnight. More data, collected with particular care with regard to the sources of errors, would significantly improve the quality of the model.

Holidays and sick leave were not included in the model and are among the potential improvements.

The allocation of rooms in a dwelling to its occupants and then their location based on their current activity are based solely on pragmatic assumptions. The results of surveys could provide more information on the way in which the members of a household are distributed among the available rooms, or on the rooms in which they prefer to have their meals, watch television, etc.

If the model is applied to buildings comprising a large number of dwellings, imposing at least one thermal zone per dwelling leads to an increased simulation time. However, with an identical number of thermal zones, the increase in computation time is quite negligible. In a collective housing comprising 16 dwellings, the creation of the inhabitants and the generation of their activity scenarios over a year (52 identical weeks) took less than a second on a PC with a capacity of 8.00 GB of RAM and with an Intel (R) Core (TM) i7-3520M processor with a frequency of 2.90 GHz. Markov processes are inexpensive given the speed of the pseudo-random number generation function.

3 Electrical Equipment Modelling

Specific electricity is electricity used for services that cannot be provided by any other source of energy. The specific electricity consumption of a dwelling therefore includes the consumption of electrical appliances for cooking, cooling, washing, leisure and office automation. Heating, air conditioning and domestic hot water (DHW) do not fall under specific electricity. From the point of view of building energy, the use of electrical devices results, in addition to the electricity consumption, in a release of heat due to the Joule effect. This induces a heating load reduction in winter, and discomfort increase in summer (or cooling load increase if air conditioning is used).

Specific electricity consumption varies greatly from one household to another. Households own and use more or less electrical appliances depending on the number and age of their members, their income, their activities, etc. Behavioural diversity is currently neglected in standard calculations, which consider simple deterministic scenarios inducing an error in the prediction of the comfort and the heating and cooling loads.

3.1 General Principles

Electrical equipment are modelled explicitly. Compared to a direct conversion "activity → electricity consumption" as proposed by Tanimoto et al. [36] or Widén et al. [37] in the first version of their model, explicit modelling of devices offers several advantages:

- It is possible to account for households' variability in terms of equipment.
- It is possible to generate variability between power inputs of devices of the same type (not all televisions will have the same operating and standby power).
- It is possible to generate variability for the same device (e.g. washing machines have several operating options).
- It is possible to specify the exact characteristics if the information is available.
- The model is scalable and can be updated easily. This point appears essential due to the evolution of specific electricity consumption, which is very rapid with regard to the lifespan of the buildings. The performance of existing devices can be changed as technology evolves, and new devices can be integrated without questioning the structure of the model.
- In connection with its evolutionary nature, the model can be used in a prospective approach focusing on specific electricity consumption, but also the thermal behaviour of buildings (in particular regarding summer comfort). It makes it possible to simulate a proliferation of devices, improvements in performance, technological breakthroughs, or even behavioural changes linked to new constraints on electricity costs or to awareness campaigns (in which case the integration of socio-demographic factors is also of interest).

The model consists in randomly populating the dwelling with electrical devices (based on statistics), defining probabilities of associating an activity with an equipment, specifying the operating characteristics (including sleep mode), and calculating the electrical load [38] and internal heat input. These steps are detailed in Vorger [23].

3.2 Simulation Results

3.2.1 Single Dwelling Electricity Load

Load curves generated for three random dwellings during the first week of the year are shown in Figs. 22, 23, and 24. For dwelling N°1 (Fig. 22), repetitive consumption of lighting and audiovisual can be observed in the evening as well as two marked peaks related to kitchen occurring on Tuesday noon and Sunday noon. The laundry cycles appear in pairs on Wednesday and Saturday indicating that they would be washing machine cycles followed by tumble dryer cycles. Dwelling N°2 consumes more energy on average (Fig. 23). There is a significant proportion of audiovisual standby consumption. Dwelling N°3 stand out by daily use of cooking appliances at noon (Fig. 24).

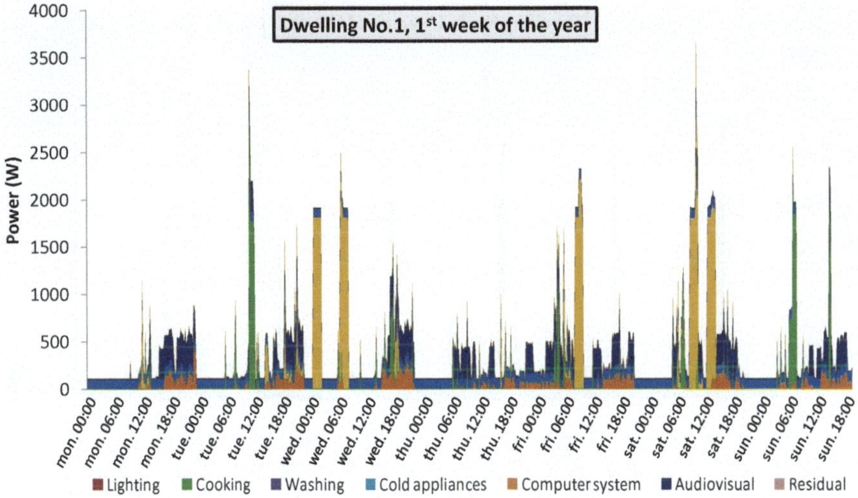

Fig. 22 Detailed load curve during the first week of the year for dwelling No. 1

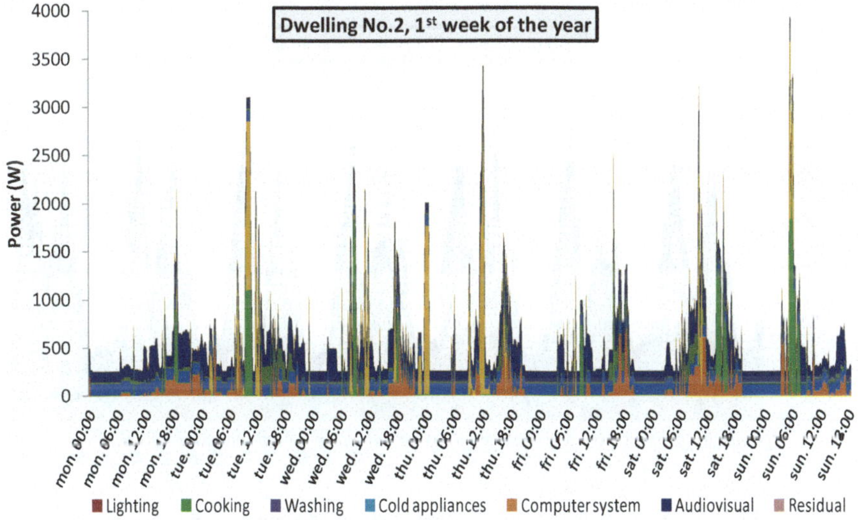

Fig. 23 Detailed load curve during the first week of the year for dwelling No. 2

3.2.2 Aggregated Electricity Load

In Fig. 25, the power load of 100 random dwellings is aggregated. The peaks that appear in the individual curves are no longer visible. The maximal electricity load for dwelling n°2 was 3937 W; the average aggregated maximal load is 690 W. Limitations of the lighting model are visible, in particular an excessively sudden

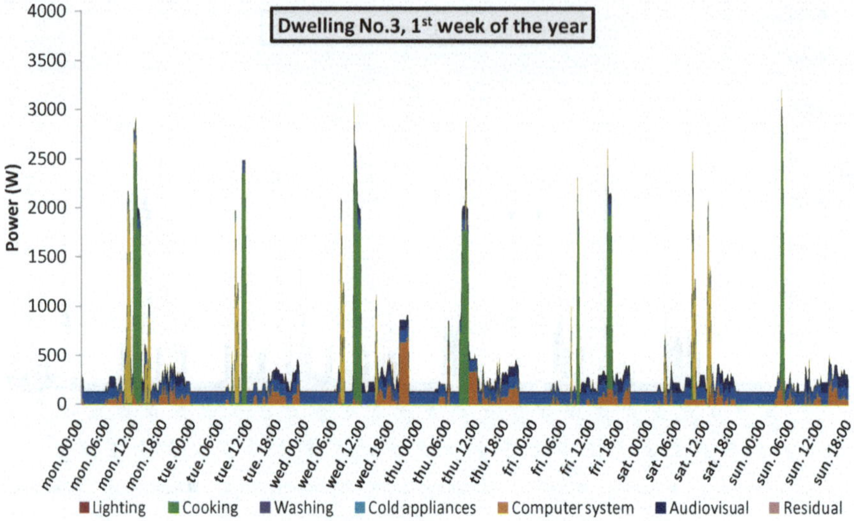

Fig. 24 Detailed load curve during the first week of the year for dwelling No. 3

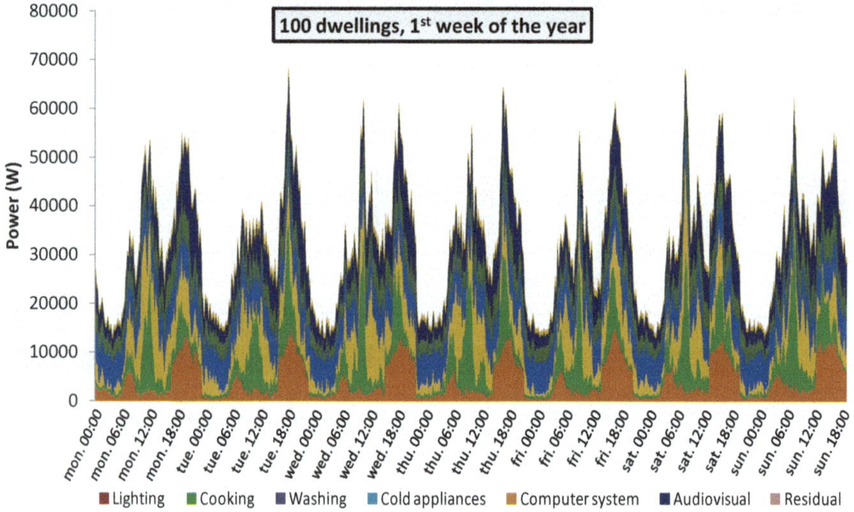

Fig. 25 Detailed load curve during the first week of the year, aggregated for 100 random dwellings

variation in lighting consumption at the end of the afternoon and at midnight. Lighting consumption at the end of the day could be smoothed by introducing a gradual switch-on probability. The sharp drop at midnight is an artefact of the activity model (§ 2.4).

Figures 26 and 27 display, respectively, the average (100 dwellings) "summer" (May 15th to October 14th) and "winter" (October 15th to May 14th) electricity

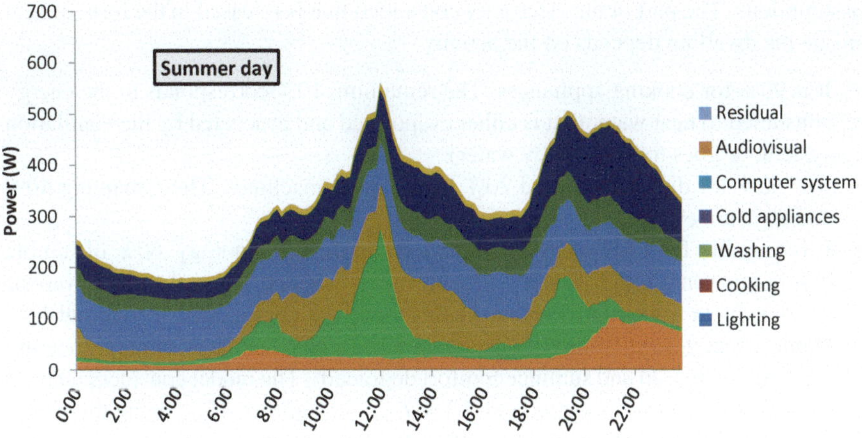

Fig. 26 Average daily summer load curve (100 dwellings)

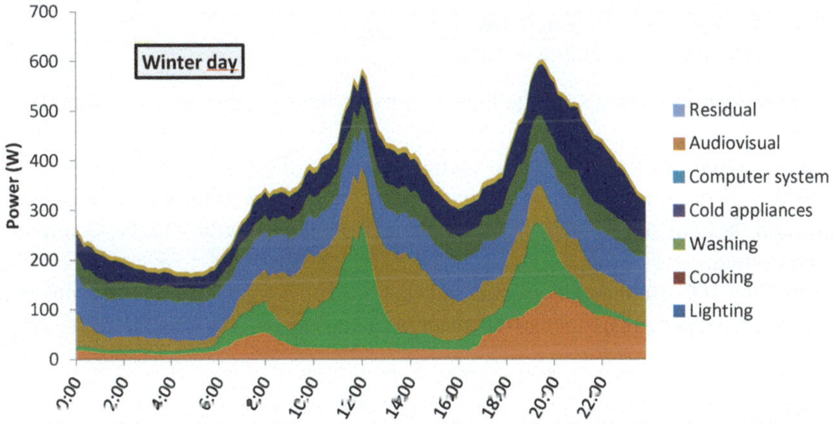

Fig. 27 Average daily winter load curve (100 dwellings)

load. The daily consumption is 8.06 kWh and 8.55 kWh, respectively, in summer and winter. The consumption for cooling (refrigerator) is higher in summer, but the washing, cooking and lighting consumptions are lower. The peaks are around noon and 7:30 p.m. in both cases. In summer, the highest peak is at noon while in winter it is at 7:30 p.m.

3.2.3 Internal Heat Input

The electrical equipment model is used to generate scenarios of internal heat inputs for the DBES. Electricity consumption is converted into heat input through a few assumptions. The part of the electricity consumed that is released in the form of heat inside the dwelling depends on the activity:

- It is 90% for cooking appliances. The remaining 10% corresponds to the energy consumed to heat water that is either evaporated and evacuated by the ventilation system, or goes to waste (grey water).
- It is 60% for dishwashers and 20% for washing machines. The remaining 40% and 80% are evacuated with grey water.
- It is variable for tumble dryers depending on their technology. 60% of tumble dryers are vented, that is, they take in indoor air and reject the water vapour outdoor or in a ventilated room. In this case, the contribution to internal heat (Joule effect + ventilation) is negative in winter and positive in summer (heating loads are increased and summer comfort degraded). The model considers internal heat of -0.3 × P_{fonc} in winter and 0.3 × P_{fonc} in summer, P_{fonc} being the power load of the dryer. The remaining 40% are thermodynamic clothes dryers that condense water vapour instead of rejecting it. The (sensible) heat balance for the room is therefore limited to the Joule effect; 100% of the electricity consumed is released in the form of heat in the room.
- It is 100% for all remaining equipment.

As a reminder, the devices can be located outside the heated space, for example, in garages.

The internal heat input scenario corresponding to the first week of the year for dwelling n°1 is shown in Fig. 28 alongside the electric load. If the dwelling consists in several thermal zones, the heat inputs are distributed according to the location of the equipment.

From the 100 simulated dwellings, it is possible to suggest average internal heat input scenarios per square meter (the average surface area of the simulated dwellings

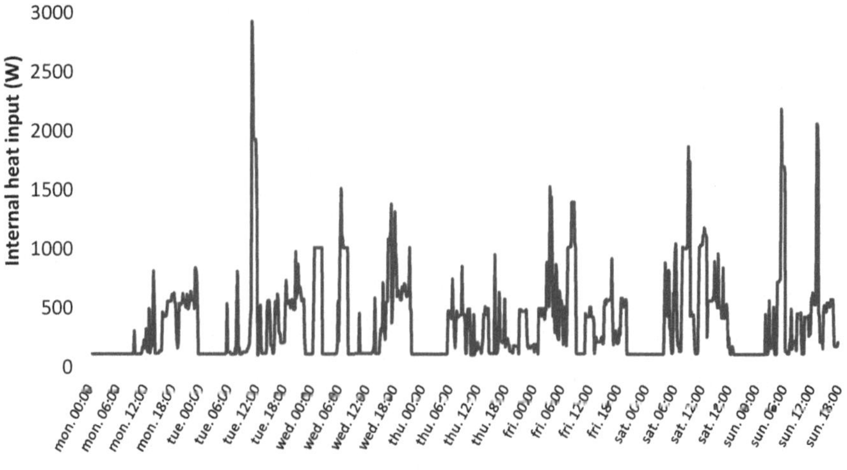

Fig. 28 Internal heat input curve—dwelling No. 1

is 93 m^2). A distinction between "summer" and "winter" and between weekdays and weekends is proposed (Figs. 29 and 30).

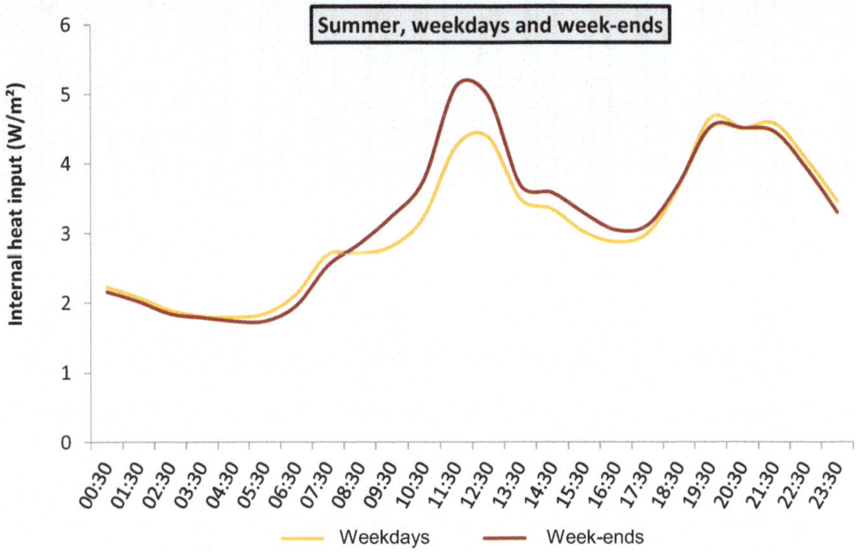

Fig. 29 Daily average internal heat input scenarios for summer

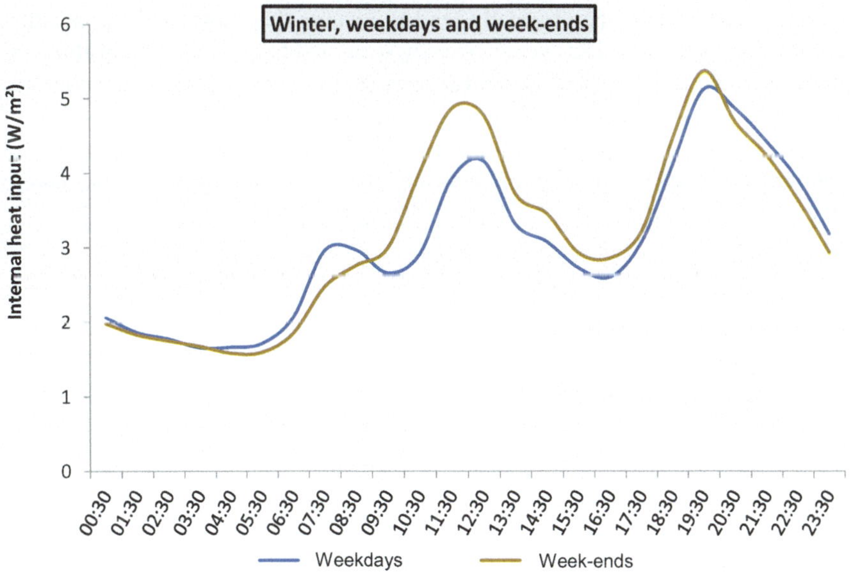

Fig. 30 Daily average internal heat input scenarios for winter

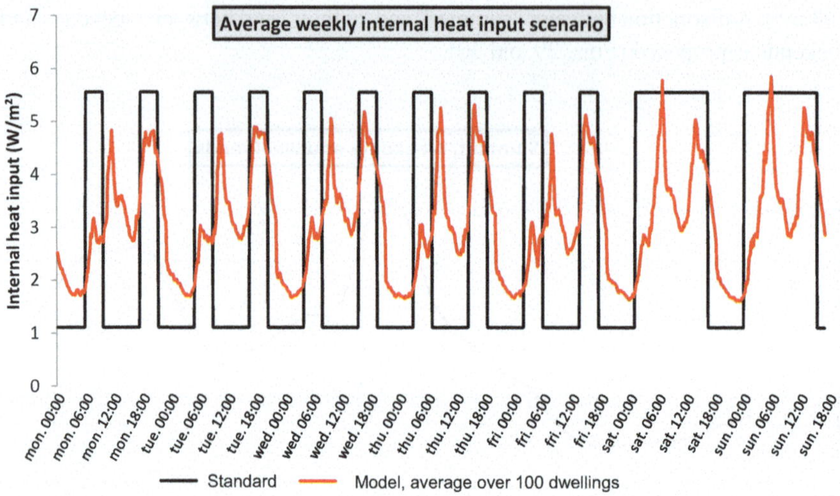

Fig. 31 Weekly scenario of internal heat input, modelled yearly average for 100 dwellings (red) and 2012 French standard (black)

Differences are relatively small between weekdays and weekends; these two types of day are not fundamentally different on average across the population. The introduction of simplified differences between weekdays and weekends in conventional scenarios results from a subjectivity bias. Since their authors have "classic" work weeks, they logically tend to assume that most of the population is in the same situation. The average weekly scenario obtained for 100 dwellings is compared to the 2012 French standard scenario in Fig. 31. Similarly to the presence rates, the standard scenario underestimates the internal heat input in the middle of weekdays and at night, and overestimates them during the weekends. Internal heat input are also overestimated in the mornings (between 6 a.m. and 10 a.m.) and in the evening (between 6 p.m. and 10 p.m.) during the week.

4 Adaptive Behaviour

4.1 Windows Opening

The opening of windows by occupants is one of the main factors influencing the ventilation of buildings. According to studies conducted in Japan and Denmark respectively, Iwashita and Akasaka [66] and Kvisgaard and Collet [67] attributed 87% and 63% of air renewal in dwellings, respectively, to occupants' behaviour. The influence on the thermal state of the building depends on the difference between indoor and outdoor temperatures. The management of openings is therefore a major topic in the literature on occupants' behaviour, especially regarding its impact on the energy consumption of buildings.

Since Dick and Thomas [39], it has been established that actions on windows are correlated with outdoor temperature and marginally with wind speed. This has been subsequently corroborated by several studies.

Warren and Parkins [40] in particular, from measurements in five offices in Great Britain, evaluated the contribution of outdoor temperature, solar radiation and wind speed at 76%, 8% and 4%, respectively, in the explanation of the variance of the opening state of windows (considering two states: open or closed). They performed linear regressions to establish a relationship between the percentage of open windows and outdoor temperature, differentiating between small and large opening times. Their measurements revealed that actions on small openings were much less correlated with temperature than those on large ones, which seemed to indicate that the former were driven by air quality and the latter by thermal comfort.

Fritsch et al. [41] constructed the first predictive model of the state of windows opening on the basis of measurements made at LESO-PB[11] (Solar Energy and Building Physics Laboratory). A Markov process defined the transition probabilities between six possible opening angles over four temperature ranges. Air quality was not taken into account in this model. The indoor temperature was the only parameter. According to the authors, this model is only valid for outdoor temperatures below 18 °C.

At the end of the 1990s, several measurement campaigns were conducted in Europe and Pakistan as part of research on the adaptive approach to thermal comfort [22, 42, 43]. The data made it possible to establish a first predictive model of window condition as a function of indoor or outdoor temperatures based on logistic regressions. As a first step, Nicol [44] recommends using the outdoor temperature as a parameter, since this is an input to simulation software. Subsequently, Nicol et al. [22] demonstrated that indoor temperature is a more relevant parameter. Even if the outdoor temperature appears to be more strongly correlated with the actions on the windows than the indoor temperature, a model calibrated on the outdoor temperature alone will be unreliable when it is used on another building (since the indoor temperature dynamics are different even for the same outdoor climate).

The original model of Nicol [44] was improved and implemented in the DBES ESP-r software as "Humphreys Adaptive Algorithm". This calculates the comfort temperature as a function of the sliding average outdoor temperature. If the difference between the operative temperature and the comfort temperature is greater than 2 °C, the probability of action is calculated by a logit model depending on the outdoor and indoor (operative) temperatures. The probability is then compared to a random number to determine whether or not the action takes place. In order to avoid possible oscillations (succession of openings-closings), a hysteresis effect of 2 °C on the indoor temperature and of 5 °C on the outdoor temperature was introduced.

[11] https://www.epfl.ch/labs/leso/

Herkel et al. [45], on the basis of measurements carried out in 21 offices over 13 months in Freiburg (Germany), proposed a stochastic model of management of windows by occupants which uses Markov chains with probabilities of transitions as a function of outdoor temperature. The model considers three distinct phases to account for the observation that the frequencies of actions are significantly higher at the time of the arrival or departure of occupants. This phenomenon has also been noted by Haldi [46], Mahdavi and Pröglhöf [47] and Yun and Steemers [48].

Page [49] developed a model in which the occupants open windows if the pollutant concentration exceeds a critical threshold—following Fanger [50], or if the atmosphere no longer corresponds to the conditions of comfort thermal defined according to the rational approach.

Based on measurements taken in summer (3 months) in six offices in two separate buildings, Yun and Steemers [48] proposed a stochastic model based on Markov chains to predict windows opening. Indoor temperature is preferred over outdoor temperature as the main explanatory variable, in accordance with the principles of adaptive comfort. The analysis showed that outdoor temperature is not significant (however, the measurements only covered the summer period). Windows are assumed to be systematically closed when occupants leave. During the arrival periods, the opening probabilities are calculated by a logit model as a function of indoor temperature, while the probabilities of actions during the intermediate periods follow a linear function of indoor temperature. An extension was made to take into account the possibility of night ventilation. Differences in behaviour between different individuals were highlighted; therefore, the model defines three categories of occupants: active, average and passive.

Haldi and Robinson [51] undertook a large measurement campaign, in 14 offices of the LESO-PB building during 7 years. They developed and compared several discrete models (with 10 min resolution). The comparisons demonstrated the superiority of Markov models over Bernoulli models, that is, the prediction of transitions over the prediction of positions. In the best-performing Markov model, the probabilities of actions are determined by logit models (one for each type of period of presence). This Markov model is completed by a continuous process. When a window is open, the duration during which it will remain in this state is calculated. In case of departure, the closing is decided or not according to a probability. The performance of the model was evaluated through various tests on the LESO-PB measurement samples but also by cross-validations from measurements taken on other buildings. Haldi et al. [52] showed that the model calibrated on the measurements at LESO-PB allowed a good prediction of the actions recorded in an office building in Austria and vice versa. Schweiker et al. [53] extended the process to three apartments located in Switzerland and one student residence located in Japan. Validation tests between Swiss homes and offices revealed an acceptable robustness of the models. Japanese data was not correctly predicted, even by the model calibrated on Japanese data, indicating that its formulation and/or the selection of explanatory variables were unsuitable in this context. This can be explained by significant variations in terms of climate (humidity in particular) and habits (e.g. use of air conditioning). According to Schweiker et al. [53], the calibrated model of Haldi and Robinson [51] can be used for office buildings or bedrooms and living rooms of dwellings in a climatic and cultural context close to Switzerland and Austria.

Andersen et al. [54] developed a window opening model dedicated specifically to housing. Indoor and outdoor environmental conditions were measured in ten apartments and five houses for 8 months in Denmark. The dwellings were divided into four groups according to whether they were inhabited by tenants or owners and ventilated mechanically or naturally. By logistic regressions, the opening and closing probabilities were evaluated for the four groups by distinguishing bedrooms and living rooms as well as periods of the day ("morning", "day", "evening" and "night") and, optionally, season. The resolution of both models and measurements was 10 min. Depending on the groups, certain variables appeared to be dependent on the dwelling and were therefore removed from the models (e.g. indoor and outdoor temperatures were removed from the window opening model, and indoor temperature for the group "owner households, mechanical ventilation"). Overall, the most influential variable on the probability of opening is the CO_2 concentration in the room, while the probability of closing depends mainly on outdoor temperature.

Only the models of Andersen et al. [54] and Haldi and Robinson [51] are candidates to model windows opening in residential buildings. Both were only calibrated and evaluated for bedrooms and living areas.

The model of Andersen et al. [54] has not been validated. The quality of its predictions is unknown, nor its validity under conditions different from the Danish climate. Moreover, it is not possible to diversify individual behaviours, for example, by creating "active" and "passive" categories with different properties.

According to Schweiker et al. [53], the model of Haldi and Robinson [51] developed for offices is sufficiently robust to be transposed to homes under certain conditions. The predictions of the actions in three apartments in Switzerland are in relatively good agreement with the observations.

Following the state of the art, the model of Haldi and Robinson [51] was selected. It is based on the largest sample of measurements and was submitted to several validation procedures. Important limitations remain: IAQ is not taken into account and the potential for generalisation needs to be confirmed. The model is described in Vorger [23]. It calculates air flowrates that are input to DBES models.

4.2 *Temperature Setpoint*

4.2.1 State of the Art

In France, the value of 19 °C has been in the 1974 standard following the oil crisis. According to Brisepierre [55], this normative approach is based on the following erroneous assumptions:

- Residents have the means to precisely adjust the temperature in their homes during heating periods.
- Thermal needs are uniform in all rooms regardless of the socio-demographic profile of the inhabitants, their way of living.

Huebner et al. [56] confirmed that the first hypothesis is not realistic. Their study is based on the measurement of the temperatures of the main rooms of 248 various dwellings in the United Kingdom for 1 year at a time step of 45 min. With an algorithm that analysed the evolution of indoor temperature, they found that the (desired) setpoint temperature was 20.6 °C on average, while the average indoor temperature when the heating is on was only 19.5 °C. The presence of additional heating equipment, as well as the testimonies of residents, attests to the inability to always reach the desired temperature [55].

Numerous data from measurements or surveys indicate that the second hypothesis is also false. Wei et al. [57] reviewed the literature on the factors that influence heating behaviours. Table 4 lists the most influential factors and specifies whether they relate to occupants' behaviour or the ability to reach the setpoint (by the heating system).

From a behavioural modelling perspective, the difference between the setpoint and the actual temperature is not essential. The goal is to predict the temperature that occupants want to reach at a given time. Depending on the characteristics of the heating system, the available power may be insufficient, in which case the setpoint is not reached. The inability of the system to meet residents' expectations can influence behaviour. One can easily imagine residents not reducing the setpoint in case of absence knowing that it will be difficult to get back to it, or even overheating and storing heat when they have the possibility in anticipation of future cold wave. These aspects are not integrated in the proposed model: the inhabitants define a setpoint independently of the system; then, the system succeeds or fails to meet the demand.

4.2.2 Temperature Setpoint Model Principles

Parys et al. [58] proposed a model consisting in assigning a setpoint to each dwelling, by drawing randomly in a distribution resulting from measurements. In this section, it is completed by integrating spatial and dynamic variations:

- In a first step, a main comfort temperature T_{base} corresponding to the desired temperature is set randomly for each home from a distribution resulting from measurements.
- In a second step, T_{base} is modified according to the characteristics of the household, according to Table 4.
- A comfort temperature for each zone is deduced depending on the type of room. The setpoint reduction is drawn randomly from a uniform range, for example, the temperature in a bedroom is between 0 and 2 °C lower than the main comfort temperature. The thermal zone temperature setpoint is calculated by a weighted average of the rooms it contains.

Table 4 Factors influencing the temperature setpoint or the ability to reach it

Factor	Description of the influence
Type of room	The main rooms have higher setpoints and have longer heating periods
Type of control	Centralised thermostats are generally counterproductive due to their complexity, which causes residents to use them in "constant temperature" mode Thermostatic valves are sometimes used for the sake of financial or energy savings. Insufficient technical knowledge hinders their proper use. It is often assumed that the taps control the power and therefore the setpoint will be reached more quickly by the highest setting (whereas they only cut off the water circulation once the setpoint is reached)
Age of occupants	The elderly who are generally less active and more often at home prefer higher temperatures. Families with young children also have a tendency to heat their homes more, for health reasons
Household size	The temperature in dwellings with large households tends to be warmer than average. However, the cause-and-effect relationship between household size and setpoint is blurred due to the fact that internal and metabolic heat are greater for large households. Setpoint reductions are less frequent, probably because the periods of occupancy overlap
Ownership status	Tenants set higher setpoints
Time	The setpoints change with time depending on the presence and activities of the occupants. They are lower during the night (whereas the conventional scenarios consider a reduction during the day) Contrary to conventional scenarios, weekdays and weekends do not appear to be fundamentally different
Gender	Women seem to prefer higher temperatures. Men, often more interested in technical aspects, use thermostats more frequently
Climate	Paradoxically, several studies reveal higher winter temperatures in colder climates
Type of dwelling	Apartments are generally warmer in winter than houses. Several reasons can explain this phenomenon: Collective dwellings are warmer because they are more compact, while houses have higher surface to volume ratios. In multi-family buildings, the hottest dwellings cause the temperature of the entire building to rise due to the flow of heat passing through the uninsulated interior walls. The correlation between the type of housing and the ownership status is another explanatory factor for the observed differences Residents lower the setpoint more frequently in apartments than in houses. We can assume that, since the setpoint can be more easily reached by the heating system, residents are less reluctant to lower the temperature when they are away from home
Age of dwelling, insulation	The age of the building is strongly correlated with its level of insulation. The temperatures are lower in the less insulated buildings, probably no by choice of the occupants but rather because of the difficulty of the systems to reach the setpoint

- A probability of reducing the temperature during periods of absence or sleep is assigned to each household.
- A final random parameter is set, indicating whether the unused rooms (bedrooms, bathrooms, or even living rooms for spacious dwellings) are heated or not, in which case a reduction is considered. This parameter aims at accounting for the partitioning strategy [59] observed among residents who only heat the inhabited part of their home.
- A heating scenario is generated for the dwelling: when at least one inhabitant is present and not asleep, all zones are heated to their comfort temperature (except unused zones if the parameter indicates that they are not heated). If a period of sleep or absence begins and if the probability of going into reduced mode is greater than a random number between 0 and 1, the reduced temperature is considered in all the zones until the return/awakening of an inhabitant.
- The heating scenario is constructed during a pre-process, and then used as input to DBES tools.

The same comfort temperatures and probabilities of temperature reduction are assigned to all members of one household. This is a reasonable approximation according to Lomas and Kane [60] who found that a household can be seen as a single organism with its shared perception of what is or is not comfortable. Fabi et al. [61] also observed variable temperature preferences from one household to another but homogeneous setpoints within the same household.

Note

- Individual heating and collective heating are not differentiated. In the case of district heating, this corresponds to assuming an optimal situation in which the occupants can control the temperature according to their wishes thanks to the thermostatic valves.
- The influence of the type of heating element on behaviour is not integrated. The use of underfloor heating, for example, should be dealt separately.
- The case of electric heating is special because, with the exception of programmable radiators, the setpoint is not explicitly defined. It is therefore likely that the equivalent setpoint is more variable with electric heating, just as it is likely that occupants change it more frequently. This aspect is not modelled but could easily be added, by incorporating a modification of the setpoint upon the arrival of residents, for example.

4.2.3 Temperature Setpoint Data

The data used in this part come from a census of several measurement campaigns carried out by Enertech[12] (Table 5). The sample includes collective (in a large majority) and individual buildings, new and renovated buildings which are all highly

[12] https://www.enertech.fr/

Table 5 List of dwellings in which the temperature setpoint was measured

	Building	City (France)	Number of monitored dwellings	Temperature setpoint (°C)
New buildings	Damidot	Villeurbanne (69)	10	21.3
	Le Concerto	Grenoble (38)	6	21.5
	Le Henri IV	Grenoble (38)	3	19.8
	Le Connestable	Grenoble (38)	12	20.4
	Le Carré d'Or	Grenoble (38)	4	20.2
	Jardins de Bonne + Pallium dauphinois	Grenoble (38)	7	21.5
	Le Vendôme	Grenoble (38)	4	21.3
	Patio Lumière	Grenoble (38)	4	21.1
	Zac du Fort	Bron (69)	5	21.9
	Residence Le claret	Revel (38)	3	22.7
	Les Santolines	Ancône (26)	4	21.2
	MINERGIE	Epagny (74)	11	20.7
	Le Pérenne	Epagny (74)	5	22.2
	Ambroise Croisat	Venissieux (69)	6	22.7
Renovated buildings: Quartier Franklin		Mulhouse (68)	3	21.9
			3	20.3
			5	18.5

insulated. The indoor temperature was monitored in the living rooms and sometimes in bedrooms. The setpoint (during the heating period) is assumed to be equal to the average temperature during the coldest 2 months of the year. For each building, the value indicated is an average of the monitored dwellings of the building.

The overall average temperature setpoint is 21.1 °C. It is consistent with the mean deduced by Huebner et al. [56] for 248 homes in Great Britain (20.6 °C, with a standard deviation of 2.5 °C). It is also in agreement with the results of a survey of passive houses which showed that the desired temperatures were between 17 and 25 °C with an average of 20.5 °C [62]. The standard deviation cannot be estimated from the aggregated results. For a design stage, a value of 2 °C is set by default.

For each dwelling, a temperature T_{base} is drawn randomly according to a normal PDF with mean 21.1 °C and standard deviation 2 °C. In order to avoid inconsistent values, the law is truncated between 17.5 °C and 25 °C.

4.2.4 Thermal Zones Temperature Setpoint

The influence of the households' socio-demographic characteristics is reflected in a modification of T_{base}. The values of the coefficients X_V associated with influential variables V are determined randomly according to laws based on assumptions inspired by the qualitative remarks from the literature (Table 6). A new temperature

Table 6 Coefficients quantifying the influence of socio-demographic characteristics on the temperature setpoint

Coefficient	PDF
$X_{\hat{a}ge}$	If at least one occupant is older above 70 years old or under 12 years old, then $X_{age} \sim \mathcal{U}(0; 1)$ Else $X_{\hat{a}ge} \sim \mathcal{U}(-1; 0)$
X_{gender}	If there is a single female occupant, $X_{sex} \sim \mathcal{U}(0; 1)$ If there is a single male occupant, $X_{sex} \sim \mathcal{U}(-1; 0)$ Else $X_{sex} = 0$
X_{fees}	If the occupants are tenants who do not pay for heating (50% of tenants by default), then $X_{fees} \sim \mathcal{U}(0; 2)$. Else $X_{fees} \sim \mathcal{U}(-0.6; 2)$.

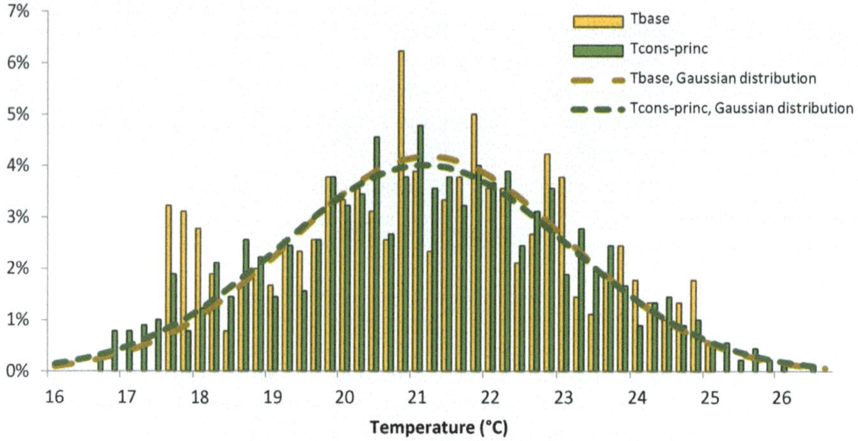

Fig. 32 Distribution of $\boldsymbol{T}_{\mathbf{base}}$ and $\boldsymbol{T}_{\mathbf{cons-princ}}$ for 1000 simulations

called the main setpoint temperature, $T_{cons-princ}$, is obtained by adding the coefficients (positive or negative) to T_{base}.

$$\mathrm{T_{cons-princ}} = \mathrm{T_{base}} + \sum_{\mathrm{V}} \mathrm{X_V} \tag{11}$$

Figure 32 presents the distributions of T_{base} and $T_{cons-princ}$ obtained from 1000 simulations. Even if T_{base} is bounded between 17.5 °C and 25 °C, $T_{cons-princ}$ can cross these bounds. The theoretical minimal and maximal values for $T_{cons-princ}$ are, respectively, 15.9 and 29 °C; however, the observed values in this experiment were between 16.5 and 26.8 °C. The normal distributions with same means and standard deviations are also displayed.

From $T_{cons-princ}$, a setpoint temperature is calculated for each zone in the dwelling. The setpoint in the bedrooms is supposed to be lower. The temperature difference between a room "i" and the living room, ΔT_i, is drawn randomly (one value per dwelling) according to a uniform law between 0 and 3 °C. The kitchens, bathrooms and hallways are supposed to have the same setpoint as the living room. Sometimes rooms are not used: bedrooms in particular (for example, a single person in a house with two bedrooms) but also bathrooms or living rooms. Some rooms are systematically unused (from the point of view of the temperature setpoint model) such as cellars. The algorithm considers by default a probability of 75% that unused rooms temperature setpoint will be: $T_{\text{cons - princ}} - \Delta T_i$. ΔT_i is drawn randomly (one value per dwelling) in a uniform law between 1 and 4 °C. The setpoint for a zone is finally obtained by averaging the temperatures of its rooms. Since the surface areas associated with the rooms are not known, the living rooms are weighted by a factor of 3 (estimated average ratio of their surface area to the surface areas of other rooms). Therefore, for a zone comprising a living room associated with a 21 °C setpoint and two bedrooms associated with 19 °C setpoints, the global setpoint will be (21 × 3 + 19 × 2) /5 = 20.2 °C. The general setpoint formula is:

$$T_{\text{cons-zone}} = \frac{\sum_{i=1}^{\text{Nb of rooms}} \alpha_i \left(T_{\text{cons-princ}} - \Delta T_i\right)}{\sum_{i=1}^{\text{Nb of rooms}} \alpha_i} \tag{12}$$

$\alpha_i = 3$ if room i is a living room, otherwise $\alpha_i = 1$. $\Delta T_i \sim \mathcal{U}(0; 3)$ if room i is a bedroom; $\Delta T_i \sim \mathcal{U}(1; 4)$ if room i is unused.

Note

The difference between apartments and houses is not directly integrated by a coefficient modifying the main setpoint. In a context of efficient buildings, the inhabitants of houses are expected to have the same requirements as those of apartments; on the contrary, in less insulated houses, the inhabitants know that they cannot maintain temperatures at high levels). On the other hand, houses have on average more unoccupied rooms than apartments, and therefore the model will generate lower zone setpoints in houses than in apartments.

4.2.5 Temperature Setpoint Management

The model considers that residents are likely to lower the setpoint when they are away or when they go to bed. As soon as they are present and awake, the setpoint corresponds to their desired level of comfort. When a potential reduction period begins, the probability of a decrease is compared to a random number between 0 and 1. The probabilities, P_{reduced}, are set randomly to reflect the diversity of households in this area. Thus, some people will always lower the setpoint when they are away, others will sometimes do it, and others will never. The probabilities intervals as well as the reduction values corresponding to the periods of absence and of sleep are given in Table 7.

Setpoint reduction probabilities do not depend on their amplitude. It is likely that in reality these parameters are correlated. Frugal occupants, for example, will focus

Table 7 Possible setpoint reduction periods and corresponding values

Period	Setpoint reduction probability	PDF
Sleep	$P_{\text{reduced-sleep}} \sim \mathcal{U}(0; 0,5)$	$\Delta T_{\text{sleep}} \sim \mathcal{U}(0,5; 1,5)$
Long absence (>2 days)	$P_{\text{reduced-long abs}} \sim \mathcal{U}(0; 1)$	$\Delta T_{\text{long abs}} \sim \mathcal{U}(3; 5)$
Short absence (>4 h)	$P_{\text{reduced-short abs}} \sim \mathcal{U}\left(0; P_{\text{reduced-long abs}}\right)$	$\Delta T_{\text{short abs}} \sim \mathcal{U}(1,5; 2,5)$

Table 8 Locations and types of dwellings probabilities

Location	House	Apartment	Other
Urban area	10.8%	29.4%	0.5%
Rural area	46.1%	12.8%	0.4%

Table 9 Results of the setpoint model for 1000 random dwellings

Parameter	Average value (1000 dwellings)
Number of used rooms	1.2
Number of setpoint reductions—long absence	2.10^{-4} day^{-1}
Number of setpoint reductions—short absence	0.05 day^{-1}
Number of setpoint reductions—sleep	0.18 day^{-1}
Duration of setpoint reductions—long absence	27 min/day
Duration of setpoint reductions—short absence	0.6 min/day
Number of setpoint reductions—sleep	92 min/day

their attention on both setpoint and control. If the data confirms this, the model could be modified accordingly.

4.2.6 Temperature Setpoint Results

The model was tested on 1000 random dwellings generated from French statistics Table 8. The distributions of the number of rooms for each type of dwelling are known from the population census. From this information, households are created automatically following the stochastic process described in § 2.5. The results of the model for 1000 random dwellings (and therefore 1000 households) are shown in Table 9.

The average number of unused rooms is not negligible. The proportion of households not heating unused rooms is set by default at 75%. The number of setpoint reductions to long absence (greater than 2 days) is not significant since these are not frequent. The integration of holidays in the presence model for housing constitutes a research perspective. The number of setpoint reductions corresponding to short absence or periods of sleep are more frequent and constitute significant periods of time.

5 Application

5.1 Implementation

The model was implemented in a DBES tool (Pleiades[13]) according to the algorithm described in Fig. 33. It can be used for various objectives including robust optimisation [63], uncertainty propagation [64], and energy performance contracting [65].

5.2 Case Study

The case study is one of the experimental houses built at INES (National Institute of Solar Energy) in Chambéry, France. It is a two-storey 90 m^2 house with concrete walls reaching the "Passive house" performance. PV panels were installed on the roof to potentially reach the zero-energy level. A detailed description is available in Munaretto [5]. For the simulation, it was divided in 11 thermal zones (Fig. 34).

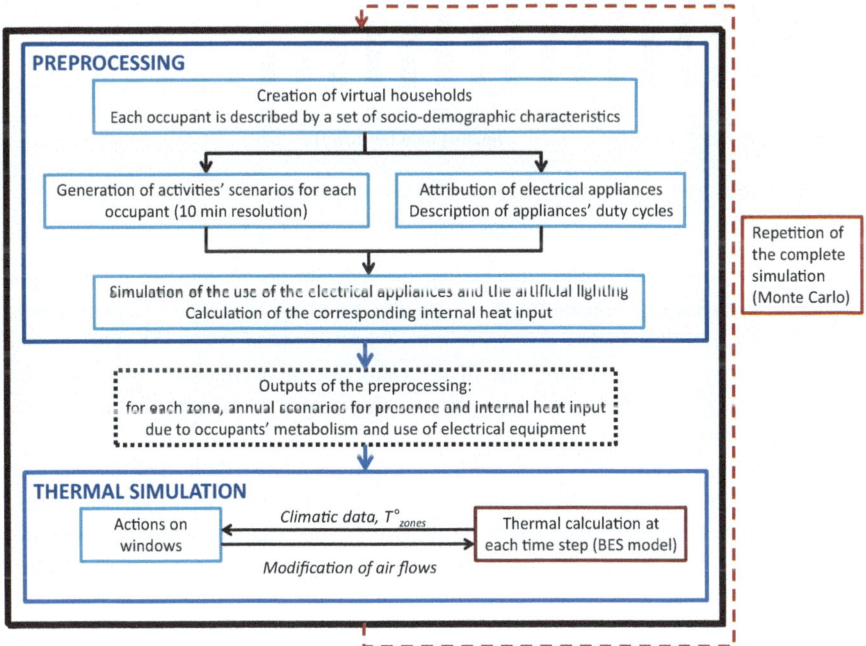

Fig. 33 Algorithm of the integrated behavioural model coupled to the DBES tool

[13]https://www.izuba.fr/logiciels/outils-logiciels/std-comfie/

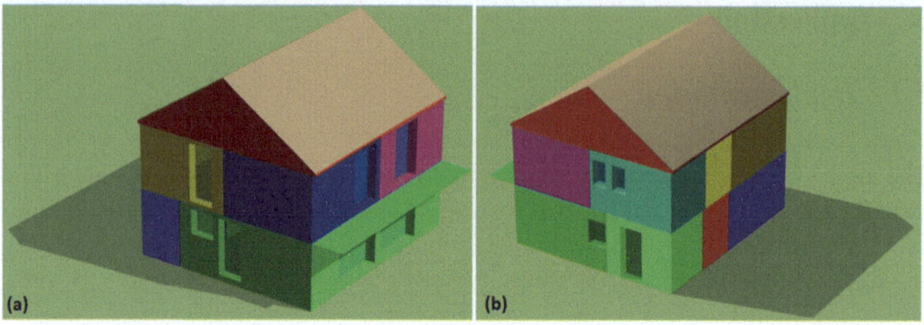

Fig. 34 3D model of the house, southwest corner (**a**) and northeast corner (**b**) (Pleiades software)

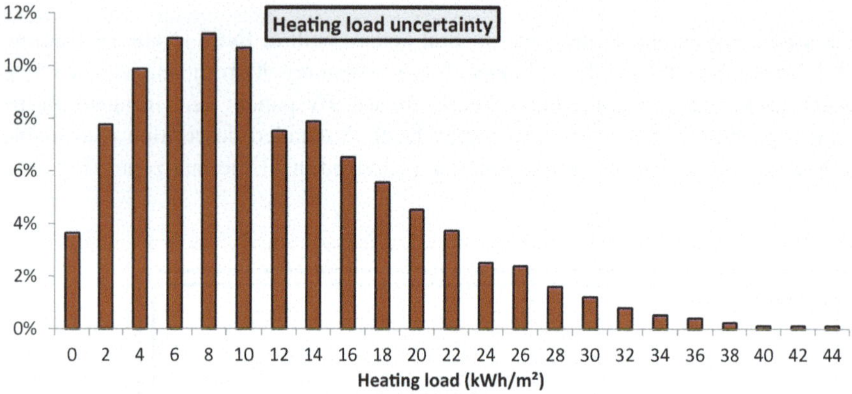

Fig. 35 Heating load distribution of the house due to occupancy

The models developed in this chapter can be used in an uncertainty propagation procedure. Assuming constant values for the building envelope characteristics, occupancy variability yields a distribution of heating load (Fig. 35). The annual primary energy balance is symbolic of the influence of occupants' behaviour: the objective of a positive balance (energy production larger than energy consumption) is reached in 70% of cases (Fig. 36).

The level of comfort also varies with occupancy. The distribution of the summer comfort indicator over the 3000 simulations is shown in Fig. 37. In 40% of cases, there is no discomfort due to high temperature. The average deviation from the comfort zone is less than 1 °C in 40% of cases, but for some households it reaches 5.5 °C, which means that on average, when the inhabitants were present, the temperature was 5.5 °C above the comfort zone.

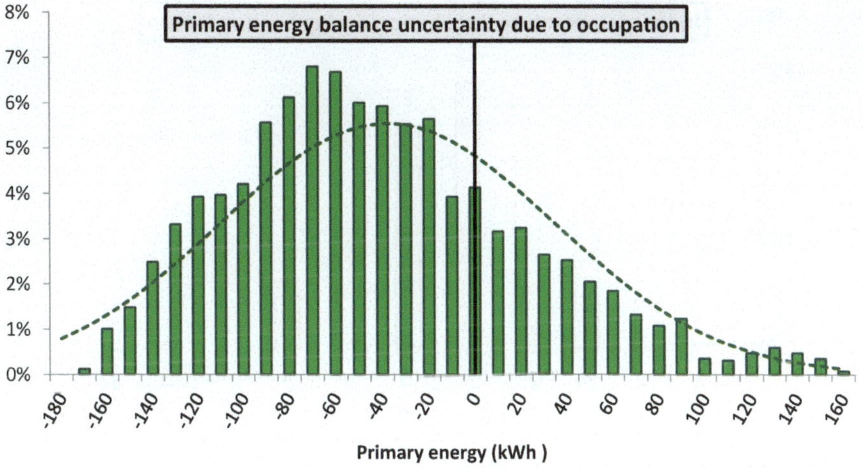

Fig. 36 Primary energy balance distribution due to occupancy

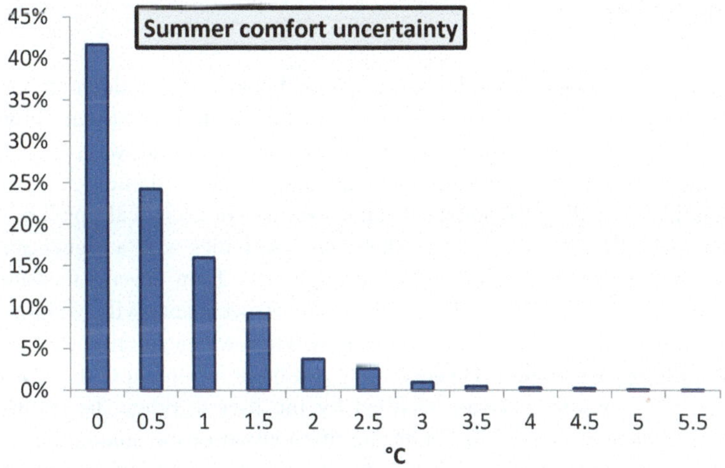

Fig. 37 Comfort indicator distribution due to occupancy

The distribution of the average indoor temperatures for the period between May 15th and October 15th is shown in Fig. 38. Globally, the average is 26 °C. Nevertheless, in 9% of cases, the average temperature is larger than 28 °C. The issue of summer comfort in highly insulated buildings calls for particular vigilance. Night ventilation would improve summer comfort in this concrete house.

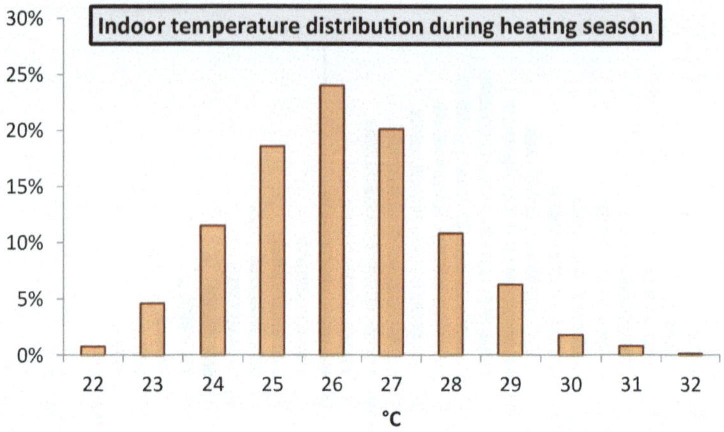

Fig. 38 Indoor temperature distribution due to occupancy (heating season)

6 Conclusion

A comprehensive stochastic model of occupants' behaviour in residential buildings is proposed. It integrates an original model for the creation of virtual individuals described by a set of socio-demographic parameters. This allows a high degree of refinement in the generation of schedules and in the attribution of equipment to households according to statistical data. The use of electrical appliances and lighting is modelled on the basis of inhabitants' activities with a higher accuracy than existing models in the literature, through data from several measurement campaigns. A reference model for interactions of occupants with windows was adapted. The whole model is integrated to a DBES tool with no more necessary input than the building description. However, any available information on inhabitants' characteristics or equipment can be filled by the user to refine the results. The Monte Carlo method is used to obtain the distribution of the simulation outputs. Applications of this type of model include among others uncertainty propagation, energy performance guarantee, and robust optimisation.

References

1. D.P. Bloomfield, Overview of validation methods for energy and environmental software. ASHRAE Transactions **105**, 685–693 (1999)
2. S.Ø. Jensen, Validation of building energy simulation programs: a methodology. Energ. Buildings **22**(2), 133–144 (1995). https://doi.org/10.1016/0378-7788(94)00910-C
3. R. Judkoff, J. Neymark, International energy agency building energy simulation test (BESTEST) and diagnostic method, in *NREL/TP-472-6231*, (National Renewable Energy Lab, Golden, CO, 1995). https://doi.org/10.2172/90674

4. R. Judkoff, J. Neymark, Twenty years on!: Updating the Iea Bestest building thermal fabric test cases for Ashrae standard 140, in *Proceedings of BS 2013: 13th Conference of the International Building Performance Simulation Association*, (2013), pp. 63–70
5. F. Munaretto, Étude de l'influence de l'inertie Thermique Sur Les Performances Énergétiques Des Bâtiments. Thèse de doctorat, École nationale supérieure des mines de Paris (2014). http://pastel.archives-ouvertes.fr/pastel-01068784
6. M. Robillart, Étude de Stratégies de Gestion En Temps Réel Pour Des Bâtiments Énergétiquement Performants. Thèse de doctorat, École nationale supérieure des mines de Paris (2015)
7. C. Spitz, 'Analyse de La Fiabilité Des Outils de Simulation et Des Incertitudes de Métrologie Appliquée à l'efficacité Énergétique Des Bâtiments'. Thèse de doctorat, Université de Grenoble (2012). http://tel.archives-ouvertes.fr/docs/00/76/85/06/PDF/These_-_Clara_-_Spitz_-_version_-_definitive.pdf
8. T. Recht, Étude de l'écoconception de Maisons à Énergie Positive. Thèse de doctorat, Paris: École nationale supérieure des mines de Paris (2016). https://www.theses.fr/2016PSLEM024
9. M. Rivallain, Étude de l'aide à La Décision Par Optimisation Multicritère Des Programmes de Réhabilitation Énergétique Séquentielle Des Bâtiments Existants. Thèse de doctorat, Université Paris-Est (2013). http://pastel.archives-ouvertes.fr/pastel-00861172
10. R. Andersen, The influence of occupants' behaviour on energy consumption investigated in 290 identical dwellings and in 35 apartments, in *10th International Conference on Healthy Buildings*, vol. 3, (2012), pp. 2279–2280
11. G. Branco, B. Lachal, P. Gallinelli, W. Weber, Predicted versus observed heat consumption of a low energy multifamily complex in Switzerland based on long-term experimental data. Energ. Buildings **36**(6), 543–555 (2004). https://doi.org/10.1016/j.enbuild.2004.01.028
12. A.F. Emery, C.J. Kippenhan, A long term study of residential home heating consumption and the effect of occupant behavior on homes in the Pacific northwest constructed according to improved thermal standards. Energy **31**(5), 677–693 (2006). https://doi.org/10.1016/j.energy.2005.04.006
13. R. Haas, H. Auer, P. Biermayr, The impact of consumer behavior on residential energy demand for space heating. Energ. Buildings **27**(2), 195–205 (1998). https://doi.org/10.1016/s0378-7788(97)00034-0
14. L.K. Norford, R.H. Socolow, E.S. Hsieh, G.V. Spadaro, Two-to-one discrepancy between measured and predicted performance of a "low-energy" office building: Insights from a reconciliation based on the DOE-2 model. Energ. Buildings **21**(2), 121–131 (1994). https://doi.org/10.1016/0378-7788(94)90005-1
15. C. Seligman, J.M. Darley, L.J. Becker, Behavioral approaches to residential energy conservation. Energ. Buildings **1**(3), 325–337 (1978). https://doi.org/10.1016/0378-7788(78)90012-9
16. R.C. Sonderegger, Movers and stayers: The resident's contribution to variation across houses in energy consumption for space heating. Energ. Buildings **1**(3), 313–324 (1978). https://doi.org/10.1016/0378-7788(78)90011-7
17. C.M. MacAl, M.J. North, Tutorial on agent-based modelling and simulation. J. Simul. **4**(3), 151–162 (2010). https://doi.org/10.1057/jos.2010.3
18. A. Kashif, X.H.B. Le, J. Dugdale, S. Ploix, Agent based framework to simulate inhabitants' behaviour in domestic settings for energy management, in *ICAART 2011 - Proceedings of the 3rd International Conference on Agents and Artificial Intelligence*, vol. 2, (2011), pp. 190–199
19. M. Bonte, A. Perles, B. Lartigue, F. Thellier, An occupant behavior model based on artificial intelligence for energy building simulation, in *Proceedings of BS 2013: 13th Conference of the International Building Performance Simulation Association*, (2013), pp. 1467–1474
20. Y.S. Lee, A.M. Malkawi, Simulating multiple occupant behaviors in buildings: An agent-based modeling approach. Energ. Buildings **69**, 407–416 (2014). https://doi.org/10.1016/j.enbuild.2013.11.020
21. B. Moujalled, Modélisation Dynamique Du Confort Thermique Dans Les Bâtiments Naturellement Ventilés (2007). http://www.theses.fr/2007ISAL0005/document

22. J.F. Nicol, M.A. Humphreys, B. Olesen, A stochastic approach to thermal comfort - occupant behavior and energy use in buildings. ASHRAE Transactions **110**(PART II), 554–568 (2004)
23. E. Vorger, Étude de l'influence Du Comportement Des Occupants Sur La Performance Énergétique Des Bâtiments. Thèse de doctorat, École nationale supérieure des mines de Paris (2014)
24. W. Parys, D. Saelens, H. Hens, Coupling of dynamic building simulation with stochastic modelling of occupant behaviour in offices - a review-based integrated methodology. J. Build. Perform. Simul. **4**(4), 339–358 (2011). https://doi.org/10.1080/19401493.2010.524711
25. F. Haldi, D. Robinson, Adaptive actions on shading devices in response to local visual stimuli. J. Build. Perform. Simul. **3**(2), 135–153 (2010). https://doi.org/10.1080/19401490903580759
26. F. Haldi, A probabilistic model to predict building occupants' diversity towards their interactions with the building envelope, in *Proceedings of BS 2013: 13th Conference of the International Building Performance Simulation Association*, (2013), pp. 1475–1482
27. C.F. Reinhart, K. Voss, Monitoring manual control of electric lighting and blinds. Light. Res. Technol. **35**(3), 243–258 (2003). https://doi.org/10.1191/1365782803li064oa
28. U. Wilke, F. Haldi, J.-L. Scartezzini, D. Robinson, A bottom-up stochastic model to predict building occupants' time-dependent activities. Build. Environ. **60**, 254–264 (2013). https://doi.org/10.1016/j.buildenv.2012.10.021
29. C.F. Manski, S.R. Lerman, The estimation of choice probabilities from choice based samples. Econometrica **45**(8), 1977–1988 (1977). https://doi.org/10.2307/1914121
30. M. Ben-Akiva, S.R. Lerman, *Discrete Choice Analysis: Theory and Application to Travel Demand* (MIT Press, Cambridge, MA, 1985) https://mitpress.mit.edu/books/discrete-choice-analysis
31. M. Ben-Akiva, M. Bierlaire, Discrete choice models with applications to departure time and route choice, in *Handbook of Transportation Science*, ed. by R. W. Hall, (Springer, Boston, MA, 2003), pp. 7–37. https://doi.org/10.1007/0-306-48058-1_2
32. M. Bierlaire, D. Bolduc, D. McFadden, The estimation of generalized extreme value models from choice-based samples. Transp. Res. B Methodol. **42**(4), 381–394 (2008). https://doi.org/10.1016/j.trb.2007.09.003
33. J. Zhang, H.J.P. Timmermans, A. Borgers, A model of household task allocation and time use. Transp. Res. B Methodol. **39**(1), 81–95 (2005). https://doi.org/10.1016/j.trb.2004.03.001
34. M. Bierlaire, *BIOGEME: A Free Package for the Estimation of Discrete Choice Models* (2003). http://transp-or.epfl.ch/documents/proceedings/Bier03.pdf
35. U. Wilke, Probabilistic bottom-up modelling of occupancy and activities to predict electricity demand in residential buildings. EPFL (2013). https://doi.org/10.5075/epfl-thesis-5673
36. J. Tanimoto, A. Hagishima, H. Sagara, Validation of probabilistic methodology for generating actual inhabitants' behavior schedules for accurate prediction of maximum energy requirements. Energ. Buildings **40**(3), 316–322 (2008). https://doi.org/10.1016/j.enbuild.2007.02.032
37. J. Widén, M. Lundh, I. Vassileva, E. Dahlquist, K. Ellegård, E. Wäckelgård, Constructing load profiles for household electricity and hot water from time-use data-modelling approach and validation. Energ. Buildings **41**(7), 753–768 (2009). https://doi.org/10.1016/j.enbuild.2009.02.013
38. E. Vorger, P. Schalbart, B. Peuportier, Integration of a comprehensive stochastic model of occupancy in building simulation to study how inhabitants influence energy performance, in *Proceedings PLEA 2014*, Ahmedabad (2014), p. 8
39. J.B. Dick, D.A. Thomas, Ventilation research in occupied houses. J. Inst. Heat. Vent. Eng. **19**, 279–305 (1951)
40. P.R. Warren, L.M. Parkins, Window-opening behavior in office buildings. ASHRAE Transactions **90**, 1056–1076 (1984)
41. R. Fritsch, A. Kohler, M. Nygård-Ferguson, J.-L. Scartezzini, A stochastic model of user behaviour regarding ventilation. Build. Environ. **25**(2), 173–181 (1990). https://doi.org/10.1016/0360-1323(90)90030-U
42. K.J. McCartney, J. Fergus Nicol, Developing an adaptive control algorithm for Europe. Energ. Buildings **34**(6), 623–635 (2002). https://doi.org/10.1016/S0378-7788(02)00013-0

43. I.A. Raja, J.F. Nicol, K.J. McCartney, M.A. Humphreys, Thermal comfort: Use of controls in naturally ventilated buildings. Energ. Buildings **33**(3), 235–244 (2001). https://doi.org/10.1016/S0378-7788(00)00087-6
44. J.F. Nicol, Characterising occupant behaviour in buildings: towards a stochastic model of occupant use of windows, lights, blinds, heaters and fans, in *Seventh International IBPSA Conference*, de Janeiro (2001)
45. S. Herkel, U. Knapp, J. Pfafferott, Towards a model of user behaviour regarding the manual control of windows in office buildings. Build. Environ. **43**(4), 588–600 (2008). https://doi.org/10.1016/j.buildenv.2006.06.031
46. F. Haldi, Towards a unified model of occupants' behaviour and comfort for building energy simulation. EPFL (2010). https://doi.org/10.5075/epfl-thesis-4587
47. A. Mahdavi, C. Pröglhöf, Toward empirically-based models of People's presence and actions in buildings, in *IBPSA 2009 - International Building Performance Simulation Association 2009*, (2009), pp. 537–544
48. G.Y. Yun, K. Steemers, Time-dependent occupant behaviour models of window control in summer. Build. Environ. **43**(9), 1471–1482 (2008). https://doi.org/10.1016/j.buildenv.2007.08.001
49. J. Page, Simulating occupant presence and behaviour in buildings. EPFL (2007). https://doi.org/10.5075/epfl-thesis-3900
50. P.O. Fanger, Introduction of the Olf and the Decipol units to quantify air pollution perceived by humans indoors and outdoors. Energ. Buildings **12**(1), 1–6 (1988). https://doi.org/10.1016/0378-7788(88)90051-5
51. F. Haldi, D. Robinson, Interactions with window openings by office occupants. Build. Environ. **44**(12), 2378–2395 (2009). https://doi.org/10.1016/j.buildenv.2009.03.025
52. F. Haldi, D. Robinson, C. Pröglhöf, A. Mahdavi, A partial double blind evaluation of a comprehensive window opening model, in *BauSim 2010*, (2010)
53. M. Schweiker, F. Haldi, M. Shukuya, D. Robinson, Verification of stochastic models of window opening behaviour for residential buildings. J. Build. Perform. Simul. **5**(1), 55–74 (2012). https://doi.org/10.1080/19401493.2011.567422
54. R. Andersen, V. Fabi, J. Toftum, S.P. Corgnati, B.W. Olesen, Window opening behaviour modelled from measurements in Danish dwellings. Build. Environ. **69**, 101–113 (2013). https://doi.org/10.1016/j.buildenv.2013.07.005
55. G. Brisepierre, Les Conditions Sociales et Organisationnelles Du Changement Des Pratiques de Consommation d'énergie Dans l'habitat Collectif (2011)
56. G.M. Huebner, M. McMichael, D. Shipworth, M. Shipworth, M. Durand-Daubin, A. Summerfield, Heating patterns in English homes: Comparing results from a National Survey against common model assumptions. Build. Environ. **70**, 298–305 (2013). https://doi.org/10.1016/j.buildenv.2013.08.028
57. S. Wei, R. Jones, P. De Wilde, Driving factors for occupant-controlled space heating in residential buildings. Energ. Buildings **70**, 36–44 (2014). https://doi.org/10.1016/j.enbuild.2013.11.001
58. W. Parys, B. Souyri, M. Woloszyn, Agent-based Behavioural models for residential buildings in dynamic building simulation: State-of-the-art and integrated model assembly. IBPSA-FR **2014**, 1–8 (2014)
59. G. Brisepierre, Les Ménages Français Choisissent-Ils Réellement Leur Température de Chauffage?: La Norme Des 19°C En Question, in *Sociologie de l'énergie : Gouvernance et Pratiques Sociales*, Sociologie, ed. by C. Beslay, M.-C. Zélem, (CNRS Éditions, Paris, 2015), pp. 273–281. http://books.openedition.org/editionscnrs/25998
60. K.J. Lomas, T. Kane, Summertime temperatures and thermal comfort in UK homes. Build. Res. Inf. **41**(3), 259–280 (2013). https://doi.org/10.1080/09613218.2013.757886
61. V. Fabi, R.V. Andersen, S.P. Corgnati, Influence of Occupant's heating set-point preferences on indoor environmental quality and heating demand in residential buildings. HVAC & R Res. **19**(5), 635–645 (2013). https://doi.org/10.1080/10789669.2013.789372
62. J. Schnieders, A. Hermelink, CEPHEUS results: Measurements and occupants' satisfaction provide evidence for passive houses being an option for sustainable building. Energy Policy **34**(2 SPEC), 151–171 (2006). https://doi.org/10.1016/j.enpol.2004.08.049

63. T. Recht, P. Schalbart, B. Peuportier, Ecodesign of a "plus-energy" house using stochastic occupancy model, life-cycle assessment and multi-objective optimisation, in *Building Simulation & Optimisation*, vol. 8, (Newcastle, Great North Museum, 2016)
64. M.-L. Pannier, Etude de La Quantification Des Incertitudes En Analyse de Cycle de Vie Des Bâtiments'. Thèse de doctorat, MINES ParisTech PSL (2017)
65. S. Ligier, M. Robillart, P. Schalbart, B. Peuportier. Energy performance contracting methodology based upon simulation and measurement, in *Building Simulation 2017*, San Francisco (2017), p. 10
66. G. Iwashita, H. Akasaka, The Effects of Human Behavior on Natural Ventilation Rate and Indoor Air Environment in Summer - A Field Study in Southern Japan. Energy and Buildings **25**(3), 195–205 (1997)
67. B. Kvisgaard, P. Collet, The User's Influence on Air Change. In STP1067-EB Air Change Rate and Airtightness in Buildings, ed. M. Sherman. West Conshohocken, PA: ASTM International. (1990), pp. 67–76. https://doi.org/10.1520/STP17205S

Index

S. Ploix et al. (eds.), *Towards Energy Smart Homes*,
https://doi.org/10.1007/978-3-030-76477-7

S